BARRON'S
Regents Exams and Answers

Earth Science— Physical Setting

Revised Edition

EDWARD J. DENECKE, JR., M.A.
Formerly, Science Department
William H. Carr J.H.S. 194Q
Whitestone, N.Y.

© Copyright 2021, 2020, 2019, 2018, 2017, 2016, 2015, 2014, 2013, 2012, 2011, 2010, 2009, 2008, 2007, 2006, 2005, 2004, 2003, 2002, 2001, 2000, 1999, 1997 by Kaplan North America, LLC, dba Barron's Educational Series

All rights reserved.

No part of this book may be reproduced or distributed in any form or by any means without the written permission of the copyright owner.

Published by Kaplan North America, LLC, dba Barron's Educational Series
1515 West Cypress Creek Road
Fort Lauderdale, Florida 33309
www.barronseduc.com

ISBN: 978-1-5062-6465-3

10 9 8 7 6 5 4

Kaplan North America, LLC, dba Barron's Educational Series print books are available at special quantity discounts to use for sales promotions, employee premiums, or educational purposes. For more information or to purchase books, please call the Simon & Schuster special sales department at 866-506-1949.

Contents

Preface v

How to Use This Book 1

    Format of the Physical Setting/Earth Science Regents Examination 1
    Contents of This Book 4

Test-Taking Tips 6

    General Helpful Tips 6
    Specific Helpful Tips for the Multiple-Choice Questions 10
    Helpful Tips for Questions on Parts B and C 20

Topic Outline and Question Index 27

    How to Use the Topic Outline for the Physical Setting/
        Earth Science Examination 27
    Topic Outline and Question Index—Physical Setting/
        Earth Science Examination 28

Earth Science Reference Tables and Charts 39

Glossary of Earth Science Terms 58

Regents Examinations, Answers, and Self-Analysis Charts **69**

August 2017 71
June 2018 160
August 2018 252
June 2019 345
August 2019 436

Preface

This book is designed to strengthen your understanding and mastery of the material in the New York State Physical Setting/Earth Science Core Curriculum, a comprehensive course of study in earth science on the secondary level. This book has been specifically written to assist you in preparing for the Regents examination covering this course. It contains the following special features that will help you in studying for this exam:

(1) *Complete sets of questions from previous Physical Setting/Earth Science Regents examinations are provided.* Attempting to answer these questions will make you familiar with the topics tested on the examination and the degree of difficulty to which you are expected to master each topic. It will also increase your confidence as you become familiar with the nature and language of the various question types that will appear on the examination.

(2) *Correct answers to all Regents questions are fully explained.* Explanations of correct answers are detailed, yet simply and clearly stated. Each explanation has been written to help you grasp the reasoning used to arrive at the correct answer rather than just stating the correct choice. Careful study of the step-by-step explanations of how the correct answer is arrived at for each question will improve your mastery of the subject. Even if you have answered the question correctly, you should read the explanation to check your reasoning and gain insight into the topic. This insight may be valuable when answering a more difficult question on the same topic.

(3) *Wrong choices are explained.* If you've ever answered a test question incorrectly and not understood *why* your answer was wrong, you will find this feature helpful. Understanding why a choice is incorrect will help you to clear up misconceptions you may have and to see errors in your reasoning.

(4) *Reference Tables for Physical Setting/Earth Science.* These tables are provided to all students for use while taking the Physical Setting/Earth Science Regents examination. Some questions *require* the use of these tables, and that is

usually indicated in the question. When answering other questions, you may find information in the tables *useful*. These tables are included in this book so that you may become *thoroughly familiar* with all of the information contained in them.

(5) *Topic Outline and Self-Analysis Charts.* The New York State Physical Setting/Earth Science Core Curriculum has been cross-referenced to questions from past Physical Setting/Earth Science Regents examinations. This will help you to identify your areas of weakness and to locate questions to review particular earth science concepts. Refer to "How to Use the Topic Outline" and Self-Analysis Charts for more detailed instructions.

(6) *Test-Taking Tips.* These helpful tips and strategies will help you to increase your grade on the Physical Setting/Earth Science Regents examination.

The author would like to acknowledge his editor, Linda Turner, for her invaluable assistance in writing and producing this book; his wife, Gerry, for her infinite patience; and his children, Merry, Abby, and Ben, for being a constant source of joy.

How to Use This Book

FORMAT OF THE PHYSICAL SETTING/EARTH SCIENCE REGENTS EXAMINATION

The examination consists of a performance portion containing three tasks to measure laboratory skills and a written portion containing multiple-choice, constructed-response, and extended constructed-response questions. You will be required to answer ALL of the questions on the Physical Setting/Earth Science Regents examination.

THE LABORATORY PERFORMANCE TEST

Since laboratory experiences are an essential part of a science course, a portion of the Physical Setting/Earth Science Regents examination is devoted to assessing laboratory skills. Tasks have been identified from laboratory experiments that you will have performed during the school year. These tasks, which represent skills that you are expected to have mastered, change only slightly, if at all, from year to year.

The performance portion of the examination is administered separately from the written portion, normally two weeks earlier. Arrangements for administering the performance portion are made at each school in accordance with guidelines set by the New York State Education Department.

The scoring for each task is based upon accuracy. Values within a certain range are granted the full point value allotted to each task.

Additional information regarding the performance portion, including an indication of the three tasks to be completed, will be provided by your teacher when this portion of the examination is given. The following is an outline of these tasks. The time allowed for completing the tasks at each station is 9 minutes.

Note: The following description represents the information that the State Education Department has stated may be shared with students before taking the performance portion of the examination. You should be familiar with the skills being assessed because you have used them in laboratory activities throughout the year. However, you will not be allowed to practice the entire test or any of the individual stations before this performance component is administered.

Station 1 . . *Mineral and Rock Identification*

The student determines the properties of a mineral and identifies that mineral using a flowchart. Then the student classifies two different rock samples and states a reason for each classification based on observed characteristics.

Station 2 . . *Locating an Epicenter*

The student determines the location of an earthquake epicenter using various types of data that were recorded at three seismic stations.

Station 3 . . *Constructing and Analyzing an Asteroid's Elliptical Orbit*

The student constructs a model of an asteroid's elliptical orbit and compares the eccentricity of the orbit with that of a given planet.

THE WRITTEN TEST

The written portion of the Physical Setting/Earth Science Regents examination represents 85 points of the total score and has three parts: A, B, and C. You should be prepared to answer questions in multiple-choice, constructed-response, and extended constructed-response formats. Questions will be content- and skills-based and may require you to graph data, complete a data table, label or draw diagrams, design experiments, make calculations, or write short or extended responses. In addition, you may be required to hypothesize, to interpret, analyze, or evaluate data, or to apply your scientific knowledge and skills to real-world situations. Some of the questions will require use of the 2011 Edition of the *Reference Tables for Physical Setting/Earth Science*.

> You will be required to answer ALL of the questions in the Physical Setting/Earth Science Regents examination!

Format of the Physical Setting/Earth Science Regents Exam

The following is a brief description of the types of questions comprising each part of the examination:

Part A—Multiple Choice A multiple-choice question offers several answers from which you choose the one that best answers the question or completes the statement. Part A of the exam focuses on earth science content from Standard 4 (see Topic Outline) and represents 30–40 percent of the examination. Many practice questions of this type from previous Regents examinations are included at the end of each chapter.

Part B—Multiple Choice (B-1) and Constructed Response (B-2) In a constructed-response question there is no list of choices from which to choose an answer; rather you are required to provide the answer. Constructed-response questions test skills ranging from constructing graphs or topographic maps to formulating hypotheses, evaluating experimental designs, and drawing conclusions based upon data. In Part B, which represents 25–35 percent of the examination, you are asked to demonstrate skills identified in Standards 1, 2, 6, and 7 in the context of earth science. Practice questions of this type are also included at the end of each chapter.

Part C—Extended Constructed Response These are constructed-response questions that require more time (15–20 minutes per item) and effort on your part to answer. Questions in Part C require you to apply your earth science knowledge and skills to real-world problems and applications. You may be asked to write short essays, design controlled experiments, predict outcomes, or analyze the risks and benefits of various solutions to a problem. Part C represents 15–25 percent of the examination. Again, practice questions are provided at the end of most chapters.

You should review constructed-response and extended constructed-response questions from previous examinations to familiarize yourself with these types of questions. Parts B and C are recent additions, and each examination may contain questions unlike those in previous exams. Keep these points in mind:

- Write complete sentences whenever you are asked to answer in your own words.
- Show all work, and use correct units in mathematical problems.
- Be able to interpret map scales and legends, to construct isolines, and to draw profiles.

Although the questions on each written test are unique, they are similar to questions on past Regents examinations. By studying questions that have been asked in the past, therefore, you can strengthen your skills and knowledge in preparation for the examination you will take.

CONTENTS OF THIS BOOK

THE TOPIC OUTLINE AND THE REFERENCE TABLES FOR PHYSICAL SETTING/EARTH SCIENCE

The first part of this book consists of two special features, the Topic Outline and the Earth Science Reference Tables. For detailed instructions on how to use the *Topic Outline*, see page 28. The *Reference Tables for Physical Setting/Earth Science* in this book are the same ones that are provided to students for use while taking the Regents examination. You should be familiar with these tables at the time of the exam, since you will have used them during the school year. Some questions require the use of these tables, and this fact is usually indicated in the question. For other questions you may find information in the tables helpful. It is therefore a good idea to become generally familiar with what is contained in the tables. For the Physical Setting/Earth Science exam, use the tables beginning on page 41.

REGENTS EXAMINATIONS AND ANSWERS

In the second, much longer part of the book there are five Physical Setting/Earth Science Regents examinations with Answer Keys. Every correct answer is fully explained. Also, to help clear up misconceptions that may lead students astray, in many cases explanations for wrong answers are provided as well.

THE SELF-ANALYSIS CHART

At the end of the *Answers* section for each examination, a *Self-Analysis Chart* is provided to aid you in pinpointing your areas of weakness. The upper portion of the chart, labeled "Standards 1, 2, 6, and 7: Skills and Applications" lists the numbers of questions requiring the use of specific skills, or the ability to apply knowledge. These skills and abilities are described more completely in the *Major Understandings* of Standards 1, 2, 6, and 7 listed in the Topic Outline. The lower portion of the chart, labeled "Standard 4: The Physical Setting/Earth Science" lists the numbers of questions requiring knowledge of specific earth science concepts and is subdivided into sections labeled *"Astronomy,"* *"Geology,"* and *"Meteorology."* These concepts are described more completely in the *Major Understandings* of Standard 4 listed in the Topic Outline.

Many question numbers may appear in both portions of the chart because many questions involve using a skill, such as deductive reasoning, together with your knowledge of specific earth science concepts to answer the question. Therefore, you should consider these two sections of the chart independently, using the upper section of the Self-Analysis Chart to pinpoint areas of weakness in your skills, and the lower section of the chart to pinpoint areas of weakness in your knowledge of earth science concepts.

Test-Taking Tips

The following pages contain several tips to help you achieve a good grade on the Physical Setting/Earth Science Regents exam. They are divided into GENERAL HELPFUL TIPS and SPECIFIC HELPFUL TIPS.

GENERAL HELPFUL TIPS

TIP 1

Be Confident and Prepared

SUGGESTIONS

- Review previous tests.
- Use a clock or watch, and take previous exams at home under examination conditions (i.e., don't have the radio or television on).
- Get a review book. (The preferred book is Barron's *Let's Review: Earth Science*.)
- Talk over the answers to questions on these tests with someone else, such as another student in your class or someone at home.
- Finish all your homework assignments.
- Look over classroom exams that your teacher gave during the term.
- Take class notes carefully.
- Practice good study habits.
- Know that there are answers for every question.
- Be aware that the people who made up the Regents exam want you to pass.
- Remember that thousands of students over the last few years have taken and passed an Earth Science Regents. You can pass too!
- On the night prior to the exam day, lay out all the things you will need, such as clothing, pens, and admission cards.

General Helpful Tips

- Go to bed early; eat wisely.
- Bring at least two pens to the exam room.
- Bring your favorite good luck charm/jewelry to the exam.
- Once you are in the exam room, arrange things, get comfortable, be relaxed, attend to personal needs (the bathroom).
- Keep your eyes on your own paper; do not let them wander over to anyone else's paper.
- Be polite in making any reasonable requests of the exam room proctor, such as changing your seat or having window shades raised or lowered.

TIP 2
Read Test Instructions and Questions Carefully

SUGGESTIONS

- Be familiar with the test directions ahead of time.
- Decide upon the task(s) that you have to complete.
- Know how the test will be graded.
- Know which question or questions are worth the most points.
- Give only the information that is requested.
- Where a choice of questions exists, read all of them and answer *only* the number requested.
- Underline important words and phrases.
- Ask for assistance from the exam room proctor if you do not understand the directions.

TIP 3

Budget Your Test Time in a Balanced Manner

SUGGESTIONS

- Bring a watch or clock to the test.
- Know how much time is allowed.
- Arrive on time; leave your home earlier than usual.
- Prepare a time schedule and try to stick to it. Remember that Regents exams are longer than classroom tests, so you will need to pace yourself accordingly.
- Answer the easier questions first.
- Devote more time to the harder questions and to those worth more credit.
- Don't get "hung up" on a question that is proving to be very difficult; go on to another question and return later to the difficult one.
- Ask the exam room proctor for permission to go to the lavatory, if necessary, or if only to "take a break" from sitting in the room.
- Plan to stay in the room for the entire three hours. If you finish early, read over your work—there may be some things that you omitted or that you may wish to add. You also may wish to refine your grammar, spelling, and penmanship.

TIP 4

Be "Kind" to the Exam Grader/Evaluator

SUGGESTIONS

- Assume that you are the teacher grading/evaluating your test paper.
- Answer questions in an orderly sequence.
- Write legibly.
- Answer Part C questions with complete sentences.
- Proofread your answers prior to submitting your exam paper.

TIP 5

Use Your Reasoning Skills

SUGGESTIONS

- Answer *all* questions.
- Relate (connect) the question to anything that you studied, wrote in your notebook, or heard your teacher say in class.
- Relate (connect) the question to any film you saw in class, any project you did, or to anything you may have learned from newspapers, magazines, or television.
- Decide whether your answers would be approved by your teacher.
- Look over the entire test to see whether one part of it can help you answer another part.
- Be cautious when changing an answer. Try to remember why you selected the first answer to be sure that the new answer is better.

TIP 6

Don't Be Afraid to Guess

SUGGESTIONS

- In general, go with your first answer choice.
- Eliminate obvious incorrect choices.
- If you are still unsure of an answer, make an educated guess.
- There is no penalty for guessing; therefore, answer ALL questions. An omitted answer gets no credit.

Let's now review the six GENERAL HELPFUL TIPS for short-answer questions:

SUMMARY OF TIPS

1. Be confident and prepared.
2. Read test instructions and questions carefully.
3. Budget your test time in a balanced manner.
4. Be "kind" to the exam grader/evaluator.
5. Use your reasoning skills.
6. Don't be afraid to guess.

SPECIFIC HELPFUL TIPS FOR THE MULTIPLE-CHOICE QUESTIONS

TIP 1

Read the Question and Try to Answer It Before Looking at the Possible Choices

Thinking out the answer to a question before looking at the possible choices stimulates your memory. Key facts, concepts, and relationships come to mind. If the answer you anticipated is among the possible choices, it is likely to be the correct one. But **keep reading** through all the choices! By thinking of the answer first, you are less likely to be fooled by a wrong answer. But if you don't read all of the choices, you might miss a more correct or more complete choice farther down the list.

Example

Earth's hydrosphere is best described as the

(1) solid outer layer of Earth
(2) liquid outer layer of Earth
(3) magma layer located below Earth's stiffer mantle
(4) gaseous layer extending several hundred kilometers from Earth into space

As you read the question, you will probably think of the terms *lithosphere, hydrosphere,* and *atmosphere.* You may recall that "hydro" refers to water and remember that the hydrosphere is all the liquid water resting on the lithosphere. Therefore, you anticipate an answer referring to Earth's layer of liquid water. With this in mind, you recognize that the correct answer is choice 2, because choice 1 refers to a solid layer, choice 4 refers to a gaseous layer, and choice 3 refers to a layer of magma, which is liquid rock—not liquid water.

TIP 2

Always Read *All* of the Choices Before Choosing an Answer

Read the question and all of the choices carefully. Make sure you are reading exactly what has been written and not just skimming for words that you are anticipating. Continue reading even if you find your anticipated answer because there may be a more complete or more correct answer farther down the list.

Specific Helpful Tips for the Multiple-Choice Questions   11

Example

The length of an Earth year is based on Earth's

(1) rotation of 15°/hr
(2) revolution of 15°/hr
(3) rotation of approximately 1°/day
(4) revolution of approximately 1°/day

When reading this question you might anticipate that Earth's year is based on its revolution around the Sun. If you rushed through the answers and chose the first one to mention revolution—choice 2—you would miss this question. If you carefully read *all* of the choices, you realize that both choice 2 and choice 4 refer to revolution, and that choice 4 also correctly states the rate at which Earth revolves.

TIP 3

Answer the Easy Questions First; Do Not Spend Too Much Time on Any One Question

In this part of the test, all of the questions have the same value. You get as much credit for an easy question as a hard one. Don't waste a lot of time on questions you can't answer quickly. Answer all of the easy questions first, then go back and figure out the hard ones in the time left over. Easy questions usually contain short sentences and answers with few words. The question can usually be answered quickly from the information presented.

Example

An air mass classified as mT usually forms over which type of Earth surface?

(1) cool land
(2) cool water
(3) warm land
(4) warm water

The correct answer, choice 4, is obvious if you know that mT stands for maritime tropical. Since tropical means warm and maritime refers to the sea, such an air mass usually forms over warm water. But if you don't know what mT means, and don't realize you can find its meaning in the *Reference Tables for Physical Setting/Earth Science*, don't waste a lot of time on the question.

TIP 4

Look for and Underline Key Words or Phrases in the Question That Tell You What the Questioner Wants You to Do

Read every word of the question. Make sure you understand what is being asked before you move to the possible answers. Sometimes there are key words or phrases that will help you pick the correct answer. Some examples of key words are *all, none, some, most, never, always, least, greatest, less, more, maximum, minimum, highest, lowest, average, most nearly,* and *best*. Pay careful attention to these words. If you overlook a key word, you may miss a question to which you really know the answer.

Example

A student read in a newspaper that the maximum length of the daylight period for the year in Syracuse, New York, had just been reached. What was the date of the newspaper?

(1) March 22
(2) June 22
(3) September 22
(4) December 22

The key phrase in the question is "the maximum length of the daylight period for the year" and the question asks for the "date" on which this was reported in a newspaper. The questioner wants you to choose that *date* on which the *daylight period* is at a *maximum*. Since the longest period of daylight in New York State occurs at the summer solstice, June 21, the newspaper probably reported it the next day, June 22—choice 2.

TIP 5

Use Partial Knowledge to Eliminate Wrong Choices

When you have been through all the questions once, go back and find questions about which you have some knowledge. Read all of the possible choices and use your partial knowledge to eliminate one or two of the answers. Cross off incorrect choices to narrow possible answers. Then choose between the remaining answers. If you can eliminate two wrong answers, your chance of guessing the right answer is greater.

Example

In which list are celestial features correctly shown in order of increasing size?

(1) galaxy → solar system → universe → planet
(2) solar system → galaxy → planet → universe
(3) planet → solar system → galaxy → universe
(4) universe → galaxy → solar system → planet

Let's suppose you *know* that the universe is the biggest feature because it contains all the others. You *think* a planet might be the smallest of the features, but you are not absolutely sure. You have no idea of the sizes of the other features. You can use this partial knowledge to greatly increase your chances of guessing correctly. Because the question asks which list shows the features in *order of increasing size*, the list should end with the largest feature—the universe. Therefore, you can eliminate choices 1 and 4. Of the remaining two choices, you would then guess the answer to be choice 3, since you think a planet might be the smallest feature. And choice 3 is the correct answer.

TIP 6

Look for Clues Among the Choices in the Question, and Use Information from One Part of the Test to Help You with Others

By reading questions and choices carefully, you may often find words or phrases that provide clues to an answer. This tip is important because, if you are alert, you may find links and connections between various questions on the exam. One question may give you a clue (or even an answer) to another question.

Example

The diagram below represents part of Earth's latitude-longitude system.

What is the latitude and longitude of point **L**?

(1) 5° E 30° N
(2) 5° W 30° S
(3) 5° N 30° E
(4) 5° S 30° W

The first clue is that all of the possible answers give 5° and 30° as the coordinates of point *L*. If you look at the axes, you see that 5° is the coordinate of point *L* on the vertical axis to the right. Next you should realize that the vertical axis is a North-South axis, and you can eliminate choices 1 and 2 because they give the 5° coordinates as East or West. If the vertical axis is a North-South axis, coordinates above the 0° mark (the Equator) are N; those below it are S. Since point *L* is 5° below the 0° mark, it is an S coordinate, as shown in only one of the remaining choices—choice 4.

Specific Helpful Tips for the Multiple-Choice Questions 15

TIP 7
Know What Is in the *Reference Tables for Physical Setting/ Earth Science,* and Carefully Examine Any Diagrams, Graphs, or Tables Given with a Question

Many questions expect you to apply what these graphs, tables, and diagrams contain. If you can't answer a question with the information given, there is a good chance that you will find the information you need in the *Reference Tables for Physical Setting/Earth Science.*

Example 1
You need information given in the *Reference Tables for Physical Setting/ Earth Science* to answer the question.

The air outside a classroom has a dry-bulb temperature of 10°C and a wet-bulb temperature of 4°C. What is the relative humidity of the air?

(1) 1%
(2) 14%
(3) 33%
(4) 54%

In this question, you must know that there is a chart in the *Reference Tables for Physical Setting/Earth Science* that can be used to determine relative humidity given wet- and dry-bulb temperatures. If you know how to use this chart, you look for the intersection of the dry-bulb temperature and the difference between the wet- and dry-bulb temperatures, that is, the intersection of the row for 10°C and the column for 6°C, where you find the value for the % Relative Humidity—33—choice 3.

Example 2

You need to apply information in a chart, graph, or diagram given with the question.

The table below shows the noontime data for air pressure and air temperature at a location over a period of 1 week.

Date	Nov. 9	Nov. 10	Nov. 11	Nov. 12	Nov. 13	Nov. 14	Nov. 15
Air Temperature (°C)	1	6	0	−2	−4	5	10
Air pressure (mb)	1,024	998	1,015	1,021	1,030	1,013	?

Based on the data provided, which air pressure would most likely occur at noon on November 15?

(1) 987 mb
(2) 1,015 mb
(3) 1,017 mb
(4) 1,022 mb

In this question, you need to apply the information in the chart. By carefully examining the data in the table, you see that as air temperature increases, air pressure decreases. Note that as temperature goes from 1°C to 5°C to 6°C, the air pressure goes from 1,024 mb to 1,013 mb to 998 mb. Therefore, if the temperature increases to 10°C, the air pressure should drop even lower than 998 mb. Now read the choices and note that only one is lower than 998 mb, choice 1—987 mb.

TIP 8

Do Not Choose an Answer That Contains an Element of Truth but That Is Incorrect as It Relates to the Question

It is very important to remember that just because a statement is true, or contains an element of truth, does not mean that it is the correct answer to the question. Statements that are true, or partially true, make attractive distractors (wrong answers). Several answers may contain an element of truth, but you must decide which one is related most directly to the question.

Specific Helpful Tips for the Multiple-Choice Questions 17

Example

Major ocean and air currents appear to curve to the right in the Northern Hemisphere because

(1) Earth has seasons
(2) Earth's axis is tilted
(3) Earth rotates on its axis
(4) Earth revolves around the Sun

Each of the answer choices is a true statement. However, the question focuses on the curving of major ocean and air currents, that is, the Coriolis Effect. If you understand the Coriolis Effect, you know that its primary cause is Earth's rotation on its axis—choice 3.

TIP 9

Beware of Broad, Sweeping Generalizations in the Answer Choices—They Are Generally Incorrect

Pay careful attention to the use of words such as *always*, *never*, *only*, *must*, and so on. Such words usually make general statements improbable. This tip is particularly helpful when two or more choices may be correct. Often, the broadest, most inclusive choice is correct because it may include the other choice(s).

Example

According to the Rock Cycle in Earth's Crust diagram in the *Reference Tables for Physical Setting/Earth Science*, which type(s) of rock can be the source of deposited sediments?

(1) igneous and metamorphic rocks, only
(2) metamorphic and sedimentary rocks, only
(3) sedimentary rocks, only
(4) igneous, metamorphic, and sedimentary rocks

In this question, the best answer is choice 4—it is the least restrictive, includes all the other choices, and gives the broadest possibilities.

TIP 10
Don't Be Afraid to Guess If You're Not Sure of the Correct Answer, but Do Not Keep Changing Your Answer

If you are not sure of an answer, don't be afraid to guess. You have nothing to lose if you guess wrong. A wrong answer and no answer are both worth zero points. However, the correct answer is right there on the page, and if you guess correctly, you can increase your point score. If you can eliminate one or more wrong answers before guessing, your chance of choosing the correct answer jumps from one in four to one in three, or one in two. But don't think that you can pass a Regents exam by guessing; the odds are that if you guessed on every question, you would get one in four correct—and score a 25 on the exam! There is no substitute for careful, diligent preparation for the exam.

If you do have to guess, make an educated guess. An educated guess is made by considering the options and focusing on what you *do* know. Use what you *do* know to eliminate improbable answers. Then read the question and answer as a complete sentence and ask yourself if it is true, or sounds true. Often, when reading the choices to yourself, one of the answers will "sound right."

Once you have made your choice, be slow to change your answer without a good reason, because research shows that a person's first choices are usually correct. Most people who change their answers change from a correct one to a wrong one. Choose the answer that seems best to you and move on to the next question. Change your answer only if you are absolutely sure you made a mistake (for example, if another question on the test reminds you of the correct answer).

Example

In Earth's geologic past there were long periods that were much warmer than the present climate. What is the primary evidence that these long warm periods existed?

(1) United States National Weather Service records
(2) polar magnetic directions preserved in the rock record
(3) radioactive decay rates
(4) plant and animal fossils

Suppose you had no idea how to answer this question. You could use what you *do* know to make an educated guess. You do know that Earth's geologic past covers billions of years. You do know that the United States didn't exist until the late 1700s. So you can eliminate choice 1 because it is unlikely that USNWS

Specific Helpful Tips for the Multiple-Choice Questions 19

records go back far enough to tell you about long warm periods in Earth's geologic past. You also know that compasses work in both winter and summer, so you have the idea that Earth's magnetic field doesn't really change with temperature. So now you eliminate choice 2. You remember reading something about radioactive decay and heat, and you know that the plants and animals that live in places with warm climates are very different from those that live in cold climates. At least, though, you've narrowed it down to two choices. Because you are more sure of the connection between climate, plants, and animals, you select choice 4 as your answer—and get the question right!

SUMMARY OF TIPS

1. Read the question and try to answer it before looking at the possible choices.
2. Always read *all* of the choices before choosing an answer.
3. Answer the easy questions first; do not spend too much time on any one question.
4. Look for and underline key words or phrases in the question that tell you what the questioner wants you to do.
5. Use partial knowledge to eliminate wrong choices.
6. Look for clues among the choices in the question, and use information from one part of the test to help you with others.
7. Know what is in the *Reference Tables for Physical Setting/ Earth Science*, and carefully examine any diagrams, graphs, or tables given with a question.
8. Do not choose an answer that contains an element of truth but that is incorrect as it relates to the question.
9. Beware of broad, sweeping generalizations in the answer choices—they are generally incorrect.
10. Don't be afraid to guess if you're not sure of the correct answer, but do not keep changing your answer.

HELPFUL TIPS FOR QUESTIONS ON PARTS B AND C

Part B includes both multiple-choice and constructed-response questions and represents 25–35 percent of the examination. Part C consists of extended constructed-response questions and represents 15–25 percent of the examination. The precise number of questions in each part may vary from year to year. Some questions may require the use of the *Reference Tables for Physical Setting/Earth Science*.

TIP 1
Know What a Constructed-Response Question Is

"Constructed-response" means that you must provide, in writing, your own response to a question, rather than choosing your answer from a list of possible choices. There are two types of constructed-response questions: open-response and free-response. For an open-response question, there is only one correct answer, while a free-response question has many alternative answers.

Example of an open-response question:

Base your answers to questions 80 and 81 on the diagram below and on your knowledge of earth science. The diagram represents the apparent path of the Sun on the dates indicated for an observer in New York State. The diagram also shows the angle of Polaris above the horizon.

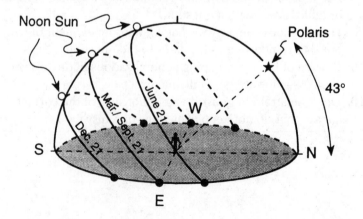

Helpful Tips for Questions on Parts B and C   21

80 State the latitude of the location represented by the diagram to the *nearest degree*. Include the latitude direction in your answer. [2]

This is an open-response question; it has only one possible answer: 43°N.

Example of a free-response question.

81 The climate of most locations near the Equator is warm and moist.

(a) Explain why the climate is warm. [1]

_____
_____

(b) Explain why the climate is moist. [1]

_____
_____

Although an acceptable response must be scientifically correct, there are many alternative ways of expressing your answer to this question. Therefore, it is a free-response question.

Acceptable answers to (a) might include statements such as these: *"Locations near the Equator receive more direct sunlight"* and *"The intensity of sunlight is greater near the Equator."* Both responses, although different, would receive full credit.

Tip 2
Know What an Extended Constructed-Response Question Is

Extended constructed-response questions are constructed-response questions that require more time and effort on your part to answer. These questions often have multiple parts and higher point values and may require you to graph data, complete a data table, label or draw diagrams, design experiments, make calculations, or write short or extended responses. In addition, you may be required to hypothesize, interpret, analyze, or evaluate data, or apply your scientific knowledge and skills to real-world situations.

Tip 3

Understand Why These Kinds of Questions Are on the Examination

Constructed-response and extended constructed-response questions are designed to:

- Assess your *understanding* of earth science concepts rather than straight recall of factual information.
- Require you to demonstrate that you can "pull all the pieces together," that is, integrate data, to solve a problem.
- Provide insight into how you think by requiring you to defend your answer in writing or to show all of your work.
- Assess knowledge and skills that cannot easily be assessed using a multiple-choice format.

Tip 4

Know What Will Be Expected of You for These Parts of the Examination

Parts B-2 and C of the examination will require you to demonstrate in a variety of ways that you have certain knowledge, skills, and abilities. You may be asked to construct a graph or draw a diagram. You may be asked to solve a mathematical problem, showing all of your work. You may be asked to provide explanations for physical phenomena or to write an essay on a topic in earth science.

Some of the skills and abilities that may be assessed in Parts B-2 and C of the examination are:

- Apply the concepts of eccentricity, rate, gradient, standard error of measurement, and density in context.
- Determine the relationships among velocity, slope, sediment size, channel shape, and volume of a stream.
- Understand the relationships among the planets' distances from the Sun, gravitational forces, periods of revolution, and speeds of revolution.
- In a field, use isolines to determine a source of pollution.
- Show how observations of celestial motions supports the early idea that stars moved around a stationary Earth (the geocentric model), but further investigation led scientists to understand that most of these motions are a result of Earth's revolution around the Sun (the heliocentric model).

Helpful Tips for Questions on Parts B and C

- Design an experiment to test sediment properties and the rate of deposition of particles.
- Graph the changing length of a shadow based on the motion of the Sun.
- Analyze patterns of topography and drainage, and design solutions to deal effectively with runoff.
- Analyze weather maps to predict future weather events.
- Critique printed or electronic materials that exemplify miscommunication and/or misconceptions of current commonly accepted scientific knowledge.
- Discuss how early warning systems can protect society and the environment from natural disasters such as hurricanes, tornadoes, earthquakes, tsunamis, floods, and volcanoes.
- Analyze a depositional-erosional system of a stream.
- Draw a simple contour map of a model landform.
- Construct and interpret a profile based on an isoline map.
- Use flowcharts to identify rocks and minerals.
- Develop a scale model to represent planet size and/or distance from the Sun.
- Develop a scale model of units of geologic time.
- Use topographical maps to determine distances and elevations.
- Analyze the interrelationship between gravity and inertia and its effects on the orbits of planets or satellites.
- Graph and interpret the nature of cyclic changes such as sunspots, tides, and atmospheric carbon dioxide concentrations.
- Determine, based on current data of plate movements, past and future positions of land masses.
- Using given weather data, identify the interface between air masses, such as cold fronts, warm fronts, and stationary fronts.
- Compare and contrast the effects of human activities as they relate to quality of life on Earth systems (global warming, land use, preservation of natural resources, pollution).
- Analyze the issues related to local energy needs and develop a viable energy-generation plan for a community.
- Evaluate the political, economic, and environmental impacts of global distribution and use of mineral resources and fossil fuels.
- Consider the environmental and social implications of various solutions to a specific environmental Earth resources problem.
- Using a topographic map, determine the safest and most efficient route for rescue purposes.

A complete list of skills and abilities is described in greater detail in Standards 1, 2, 6, and 7 of the Topic Outline and Index.

Tip 5
Work to Improve Your Score on These Types of Questions

1. **Determine what the question is asking you to do. Underline key words in the question.**

    Most questions of this type have an action word that identifies what you must do to earn points.

    Below, the action words in a question you looked at earlier are bold and underlined. Other key words are underlined.

    State the latitude of the location represented by the diagram to the nearest degree. **Include** the latitude direction in your answer. [2]

2. **Check to see what points are being awarded for.**

    The points are listed in brackets after each question. If more than one point is being awarded, you can usually make a good guess about what each point will be given for if you read the question again carefully. In the example above, you have probably guessed that one point is given for stating the latitude to the nearest degree, and one point is given for including the correct direction. The answer is 43° North. If you wrote just 43°, you would get only one point. For full credit you must state both 43° and North (or N).

3. **Always include units in your answers, and check to make sure that the form of your answer matches the form asked for in the question.**

    Consider this example:

Air Temperature	21°C
Barometric Pressure	993.1 mb
Wind Direction	From the east
Windspeed	25 knots

State the barometric pressure in its proper form, as used on a station model.

What you are asked to do is state the barometric pressure. However, if you write 993.1 mb, your answer will be marked incorrect because the form asked for is "as used on a station model." On a station model, the decimal point is dropped and the last three digits only are listed. Your answer should be given as 931 because that is what would appear on a station model.

Helpful Tips for Questions on Parts B and C

4. **When asked to respond in the form of a statement or essay, use complete sentences and organize your answer carefully.**

    A complete sentence begins with a capital letter, has a subject and a verb, and ends with a period. Very often, a point is awarded solely for answering in a complete sentence—even if the answer is wrong! Therefore, read over your statements or essays carefully.

5. **Examine the answer sheet for Part B-2 or C for clues to how the question is to be answered.**

    The answer sheet for Part B-2 or C will often give you a format for answering the question. For example, the answer sheet for the question below provides several blank lines, so you are probably being asked to write a short essay.

82 Using one or more complete sentences, state one reason that ultraviolet rays are dangerous.   [2]

_____

_____

_____

If the answer sheet has a blank graph, you are probably being asked to construct a graph. Be sure to include all elements asked for in the question. For example, on the graph below you must not only plot points but also plot them with specific shapes, label axes, and choose an appropriate scale for each axis.

83 Plot the data for rock sample *A* for the 20 minutes of the investigation. Surround each point with a small circle and connect the points.   [1]

   Example:

84. Plot the data for rock sample *B* for the 20 minutes of the investigation. Surround each point with a small triangle and connect the points.   [1]

   Example:

85. Plot the data for rock sample *C* for the 20 minutes of the investigation. Surround each point with a small square and connect the points.   [1]

   Example:

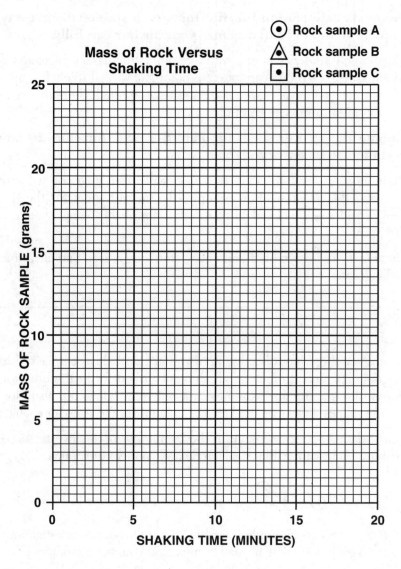

6. Review questions from Parts B-2 and C of the sample exams.

Use the Answers Explained sections of this book to review how questions of these types are answered and how point values are awarded. Focus on the kinds of questions being asked rather than on specific questions.

Topic Outline and Question Index

HOW TO USE THE TOPIC OUTLINE FOR THE PHYSICAL SETTING/EARTH SCIENCE EXAMINATION

The Topic Outline follows the organization of the Physical Setting/Earth Science Core Curriculum, which was written to accompany the New York State Mathematics, Science, and Technology Learning Standards, which are organized in the following way:

STANDARDS—statements of what you are expected to know and be able to do

1. **KEY IDEAS**—broad, unifying, general statements of the most important ideas relating to the Standard
2. **PERFORMANCE INDICATORS**—statements describing how you should be able to demonstrate that you understand the Key Ideas
3. **MAJOR UNDERSTANDINGS**—specific statements of underlying concepts needed to do the things described in the Performance Indicators

The Topic Outline is divided into the five Standards (Standards 1, 2, 4, 6, and 7) that will be addressed on the Physical Setting/Earth Science Regents examination. Each Standard is subdivided into Key Ideas. Each Key Idea is further subdivided into Performance Indicators and their associated Major Understandings, so that you may locate questions for individual concepts. Suppose, for example, you wish to check your mastery of concepts related to tectonic plate boundaries. If you look in the Topic Outline under Standard 4, Key Idea 2, Performance Indicator 2.1, Major Understanding 2.1*l*, you will find *"The lithosphere consists of separate plates that ride on the more fluid asthenosphere and move slowly in relationship to one another, creating convergent, divergent, and transform plate boundaries."* By following a parallel line across the page, you can locate appropriate questions from past examinations. On the June 2012 exam, for example, question 85 tests concepts related to plate boundaries. In a similar manner, you can locate questions on this topic from other past examinations.

Topic Outline and Question Index

When you have identified questions from the Topic Outline, turn to the appropriate examination in the book for the first keyed question. Try to answer it, and then check your response, reviewing explanations where desirable. Continue in this manner for other keyed questions.

You will notice that for some headings in the Topic Outline there is more than one test item on a particular examination. This usually occurs because more than one concept is involved. To find a question on a particular concept, you may find it necessary to check the keyed questions on several examinations.

TOPIC OUTLINE AND QUESTION INDEX—PHYSICAL SETTING/EARTH SCIENCE EXAMINATION

STANDARD 1—Analysis, Inquiry, and Design: Students will use mathematical analysis, scientific inquiry, and engineering design, as appropriate, to pose questions, seek answers, and develop solutions.

	EXAMINATION QUESTIONS				
	August 2017	June 2018	August 2018	June 2019	August 2019
MATHEMATICAL ANALYSIS	60, 69, 70	5, 49, 60, 64, 69, 74, 80	30, 63, 74, 76	11, 53, 60, 83	46, 53, 57, 66, 67, 72, 84
Key Idea 1: Abstraction and symbolic representation					
Key Idea 2: Deductive and inductive reasoning	28, 49, 51, 63, 70, 76	13, 16, 25, 27, 36, 43, 44, 65	20, 43, 45, 47, 66, 77, 78, 83	26, 36, 37, 51	14, 42, 43, 66, 68, 79, 80
Key Idea 3: Critical thinking skills	28, 67, 68	48, 60, 66	62	66, 81	
SCIENTIFIC INQUIRY	2, 10, 12, 26, 36, 39, 47, 48, 50, 55, 56, 65, 67, 71, 74, 79, 81	4, 22, 30, 38, 42, 46, 54, 56, 57, 63, 71, 76, 77, 84	5, 8, 33, 34, 46, 58–60, 66, 68, 79, 84, 85	9, 32, 43, 49, 57, 58, 61–65, 67, 71, 73, 74, 76, 77, 79	8, 10, 12, 21 39, 61, 62, 71, 72, 73, 76, 77, 81
Key Idea 1: The central purpose of scientific inquiry is to develop explanations of natural phenomena in a continuing, creative process.					
Key Idea 2: Beyond the use of reasoning and consensus, scientific inquiry involves the testing of proposed explanations involving the use of conventional techniques and procedures and usually requiring considerable ingenuity.			59	71	2
Key Idea 3: The observations made while testing proposed explanations, when analyzed using conventional and invented methods, provide new insights into phenomena.	6, 8–10, 13, 15, 17–19, 24, 25, 29, 32–35, 37, 38, 40, 50, 55, 57, 60, 62–66, 73, 77, 80, 82–85	2, 3, 5, 6, 8, 10, 13, 15, 16, 21, 24, 25, 27–29, 31, 35, 39, 41, 45, 47, 49, 51–54, 59, 61, 62, 64, 65, 67, 68, 70, 71, 78–80, 85	2, 3, 7, 10, 11, 15, 17, 18, 19, 20, 21, 23, 24, 25, 26, 28, 30, 35, 37, 39, 42, 44, 50, 55, 56, 57, 61, 65, 72, 75, 76, 80, 81, 82	4, 5, 7, 8, 10, 11, 13, 14, 19, 21, 25–27, 30, 33–35, 40, 45–47, 50, 52, 53, 55, 56, 59, 63, 66–68, 72, 75, 77, 79, 80, 83, 85	1, 3, 4, 5, 14, 18, 20, 21, 22, 23, 24, 25, 28, 33, 34, 35, 36, 40, 41, 42,43, 44, 47, 48,49, 50, 52, 54, 55, 58, 59, 60, 64, 65, 66, 69, 70, 71, 74, 78
ENGINEERING DESIGN					
Key Idea 1: Engineering design is an iterative process involving modeling and optimization (finding the best solution within given constraints); this process is used to develop technological solutions to problems within given constraints.					

Topic Outline for the Physical Setting/Earth Science Examination 29

	August 2017	June 2018	August 2018	June 2019	August 2019
STANDARD 2—Students will access, generate, process, and transfer information, using appropriate technologies.					
INFORMATION SYSTEMS					
Key Idea 1: Information technology is used to retrieve, process, and communicate information as a tool to enhance learning.		68			
Key Idea 2: Knowledge of the impacts and limitations of information systems is essential to its effective and ethical use.					
Key Idea 3: Information technology can have positive and negative impacts on society, depending upon how it is used.	72	7			
STANDARD 6—Interconnectedness: Common Themes: Students will understand the relationships and common themes that connect mathematics, science, and technology and apply the themes to these and other areas of learning.					
INTERCONNECTEDNESS COMMON THEMES **Key Idea 1: Systems Thinking** Through systems thinking, people can recognize the commonalities that exist among all systems and how parts of a system interrelate and combine to perform specific functions.	7, 30, 45, 78	20, 32, 52, 54, 56, 83, 84	8, 41, 67, 68, 85	15, 29, 70	10, 30, 38
Key Idea 2: Models Models are simplified representations of objects, structures, or systems used in analysis, explanation, interpretation, or design.	6, 9, 17, 20, 23, 26, 27, 30–32, 34–36, 39, 41–46, 52–55, 57–59, 61, 64, 66, 67, 69, 73, 75, 76, 81	1, 8–12, 18–20, 22–24, 30, 32–37, 40, 41, 44, 45, 47–50, 55, 58, 59, 61, 66, 67, 70, 71, 73–75, 80–84	1, 4, 13, 14, 21, 24, 27–29, 31-38, 40, 42, 43, 45, 47–49, 51–54, 59–63, 65, 69–74, 80-82	1–4, 7, 12, 14–16, 19, 21–23, 27–29, 32, 38–46, 50, 52, 55, 56, 62–64, 66–69, 73–78, 81–85	7, 9, 15, 17, 20, 25, 26, 27, 28, 29, 30, 36, 37, 39, 44, 45, 46, 47, 48, 49, 50, 51, 52, 54, 55, 56, 60, 61, 62, 63, 64, 67, 69, 70, 75, 80, 82, 83, 84, 85
Key Idea 3: Magnitude and Scale The grouping of magnitudes of size, time, frequency, and pressures or other units of measurement into a series of relative order provides a useful way to deal with the immense range and the changes in scale that affect the behavior and design of systems.	1, 76	48, 50, 75	64	59, 84	
Key Idea 4: Equilibrium and Stability Equilibrium is a state of stability due either to a lack of change (static equilibrium) or a balance between opposing forces (dynamic equilibrium).				80	
Key Idea 5: Patterns of Change Identifying patterns of change is necessary for making predictions about future behavior and conditions.	7, 16, 30, 40, 43, 45, 52, 53, 62, 79	5, 9, 12, 19, 23, 35, 37, 41–43, 55, 57, 68, 84	6, 41, 45, 47, 53, 59, 78	19, 36, 37, 51, 54, 62, 71, 75, 77, 79, 80	2, 5, 23, 36, 51, 61, 71, 83
Key Idea 6: Optimization In order to arrive at the best solution that meets criteria within constraints, it is often necessary to make trade-offs.		14, 33			
STANDARD 7—Interdisciplinary Problem Solving: Students will apply the knowledge and thinking skills of mathematics, science, and technology to address real-life problems and make informed decisions.					
INTERDISCIPLINARY PROBLEM SOLVING **Key Idea 1: Connections** The knowledge and skills of mathematics, science, and technology are used together to make informed decisions and solve problems, especially those relating to issues of science/technology/society, consumer decision making, design, and inquiry into phenomena.			12		

Topic Outline and Question Index

	EXAMINATION QUESTIONS				
	August 2017	June 2018	August 2018	June 2019	August 2019
Key Idea 2: Strategies Solving interdisciplinary problems involves a variety of skills and strategies, including effective work habits; gathering and processing information; generating and analyzing ideas; realizing ideas; making connections among the common themes of mathematics, science, and technology; and presenting results.	72	7		48	
STANDARD 4—Students will understand and apply scientific concepts, principles, and theories pertaining to the physical setting and living environment and recognize the historical development of ideas in science.					
Physical Setting Key Idea 1: **EARTH AND CELESTIAL PHENOMENA CAN BE DESCRIBED BY PRINCIPLES OF RELATIVE MOTION AND PERSPECTIVE.**					
Intro 1: People have observed the stars for thousands of years, using them to find direction, note the passage of time, and to express their values and traditions. As our technology has progressed, so has understanding of celestial objects and events.					
Intro 2: Theories of the universe have developed over many centuries. Although to a casual observer celestial bodies appeared to orbit a stationary Earth, scientific discoveries led us to the understanding that Earth is one planet that orbits the Sun, a typical star in a vast and ancient universe. We now infer an origin and an age and evolution of the universe, as we speculate about its future.					
Intro 3: As we look at Earth, we find clues to its origin and how it has changed through nearly five billion years, as well as the evolution of life on Earth.					
Physical Setting Performance Indicator 1.1: Explain complex phenomena, such as tides, variations in day length, solar insolation, apparent motion of the planets, and annual traverse of the constellations.					
MAJOR UNDERSTANDINGS 1.1a Most objects in the solar system are in regular and predictable motion. • These motions explain such phenomena as the day, the year, seasons, phases of the moon, eclipses, and tides. • Gravity influences the motions of celestial objects. The force of gravity between two objects in the universe depends on their masses and the distance between them. *See Standard 6, Key Idea 4*	6, 40, 51–53, 66	1, 9, 81	39, 40, 41	3, 38–40, 60, 61, 64	2, 82, 83, 84, 85
1.1b Nine planets move around the Sun in nearly circular orbits. • The orbit of each planet is an ellipse with the Sun located at one of the foci. • Earth is orbited by one Moon and many artificial satellites. *See Standard 1: Mathematical Analysis, Key Idea 2, and Standard 6, Key Ideas 3 and 4*			67, 68	59	3
1.1c Earth's coordinate system of latitude and longitude, with the Equator and Prime Meridian as reference lines, is based upon Earth's rotation and our observation of the Sun and stars.	69	39			37, 38, 39

Topic Outline for the Physical Setting/Earth Science Examination

	EXAMINATION QUESTIONS				
	August 2017	June 2018	August 2018	June 2019	August 2019
1.1d Earth rotates on an imaginary axis at a rate of 15 degrees per hour. To people on Earth, this turning of the planet makes it seem as though the Sun, the Moon, and the stars are moving around Earth once a day. Rotation provides a basis for our system of local time; meridians of longitude are the basis for time zones. *See Standard 1: Scientific Inquiry, Key Idea 3*	39, 41, 68	56	1, 29, 30	1, 62	7, 36
1.1e The Foucault pendulum and the Coriolis effect provide evidence of Earth's rotation.	3	4	6	9, 65	45, 46
1.1f Earth's changing position with regard to the Sun and the Moon has noticeable effects. • Earth revolves around the Sun with its rotational axis tilted at 23.5 degrees to a line perpendicular to the plane of its orbit, with the North Pole aligned with Polaris. • During Earth's one-year period of revolution, the tilt of its axis results in changes in the angle of incidence of the Sun's rays at a given latitude; these changes cause variation in the heating of the surface. This produces seasonal variation in weather.	26, 67	5, 11, 57	5, 51, 52, 53, 54	41, 63	9
1.1g Seasonal changes in the apparent positions of constellations provide evidence of Earth's revolution.	5	12	4	6	5
1.1h The Sun's apparent path through the sky varies with latitude and season.		55			79, 81
1.1i Approximately 70 percent of Earth's surface is covered by a relatively thin layer of water, which responds to the gravitational attraction of the Moon and the Sun with a daily cycle of high and low tides.	4	36–38		36, 37	
Physical Setting Performance Indicator 1.2: **Describe current theories about the origin of the universe and solar system.**					
1.2a The universe is vast and estimated to be over ten billion years old. The current theory is that the universe was created from an explosion called the big bang. Evidence for this theory includes: • cosmic background radiation; • a redshift (the Doppler effect) in the light from very distant galaxies.	2	3, 10		2, 5	6, 75, 76, 77
1.2b Stars form when gravity causes clouds of molecules to contract until nuclear fusion of light elements into heavier ones occurs. Fusion releases great amounts of energy over millions of years. • The stars differ from each other in size, temperature, and age. • Our Sun is a medium-sized star within a spiral galaxy of stars known as the Milky Way. Our galaxy contains billions of stars, and the universe contains billions of such galaxies.	1, 54–57	40–42, 79	45, 46, 47		78
1.2c Our solar system formed about five billion years ago from a giant cloud of gas and debris. Gravity caused Earth and the other planets to become layered according to density differences in their materials. • The characteristics of the planets of the solar system are affected by each planet's location in relationship to the Sun. • The terrestrial planets are small, rocky, and dense. The Jovian planets are large and gaseous, and have low density.	36, 37	2, 80, 82	2, 3, 66	66–68	4

Topic Outline and Question Index

		EXAMINATION QUESTIONS			
		August 2017	June 2018	August 2018	June 2019	August 2019
1.2d	Asteroids, comets, and meteors are components of our solar system. • Impact events have been correlated with mass extinction and global climatic change. • Impact craters can be identified in Earth's crust.				58	1, 8
1.2e	Earth's early atmosphere formed as a result of the out-gassing of water vapor, carbon dioxide, nitrogen, and lesser amounts of other gases from its interior.					
1.2f	Earth's oceans formed as a result of precipitation over millions of years. The presence of an early ocean is indicated by sedimentary rocks of marine origin, dating back about four billion years.					
1.2g	Earth has continuously been recycling water since the outgassing of water early in its history. This constant recirculation of water at and near Earth's surface is described by the hydrologic (water) cycle. • Water is returned from the atmosphere to Earth's surface by precipitation. Water returns to the atmosphere by evaporation or transpiration from plants. A portion of the precipitation becomes runoff over the land or infiltrates into the ground to become stored in the soil or groundwater below the water table. Soil capillarity influences these processes. • The amount of precipitation that seeps into the ground or runs off is influenced by climate, slope of the land, soil, rock type, vegetation, land use, and degree of saturation. • Porosity, permeability, and water retention affect runoff and infiltration.	7, 42, 43	20, 63, 64	8, 9	11, 12	10, 11, 12
1.2h	The evolution of life caused dramatic changes in the composition of Earth's atmosphere. Free oxygen did not form in the atmosphere until oxygen-producing organisms evolved.					21
1.2i	The pattern of evolution of life-forms on Earth is at least partially preserved in the rock record. • Fossil evidence indicates that a wide variety of life-forms has existed in the past and that most of these forms have become extinct. • Human existence has been very brief compared to the expanse of geologic time.	29	16, 69, 72	17	4, 21	22
1.2j	Geologic history can be reconstructed by observing sequences of rock types and fossils to correlate bedrock at various locations. • The characteristics of rocks indicate the processes by which they formed and the environments in which 69 these processes took place. • Fossils preserved in rocks provide information about past environmental conditions. • Geologists have divided Earth history into time units based upon the fossil record. • Age relationships among bodies of rocks can be determined using principles of original horizontality, superposition, inclusions, cross-cutting relationships, contact metamorphism, and unconformities. The presence of volcanic ash layers, index fossils, and meteoritic debris can provide additional information. • The regular rate of nuclear decay (half-life time period) of radioactive isotopes allows geologists to determine the absolute age of materials found in some rocks.	31, 47–50, 74, 75, 77	17, 19, 23, 43, 44, 70, 71, 85	18, 19, 21, 55, 56, 59, 74, 75, 76	13, 22, 32, 47, 51–53, 78	23, 26, 27, 57, 58, 59, 60, 61, 62, 74

Topic Outline for the Physical Setting/Earth Science Examination 33

	EXAMINATION QUESTIONS				
	August 2017	June 2018	August 2018	June 2019	August 2019
Physical Setting Key Idea 2: **MANY OF THE PHENOMENA THAT WE OBSERVE ON EARTH INVOLVE INTERACTIONS AMONG COMPONENTS OF AIR, WATER, AND LAND.**					
Intro 1: Earth may be considered a huge machine driven by two engines, one internal and one external. These heat engines convert heat energy into mechanical energy.					
Intro 2: Earth's external heat engine is powered primarily by solar energy and influenced by gravity. Nearly all the energy for circulating the atmosphere and oceans is supplied by the Sun. As insolation strikes the atmosphere, a small percentage is directly absorbed, especially by gases such as ozone, carbon dioxide, and water vapor. Clouds and Earth's surface reflect some energy back to space, and Earth's surface absorbs some energy. Energy is transferred between Earth's surface and the atmosphere by radiation, conduction, evaporation, and convection. Temperature variations within the atmosphere cause differences in density that cause atmospheric circulation, which is affected by Earth's rotation. The interaction of these processes results in the complex atmospheric occurrence known as weather.					
Intro 3: Average temperatures on Earth are the result of the total amount of insolation absorbed by Earth's surface and its atmosphere and the amount of long-wave energy radiated back into space. However, throughout geologic time, ice ages occurred in the middle latitudes. In addition, average temperatures may have been significantly warmer at times in the geologic past. This suggests that Earth had climate changes that were most likely associated with long periods of imbalances of its heat budget.					
Intro 4: Earth's internal heat engine is powered by heat from the decay of radioactive materials and residual heat from Earth's formation. Differences in density resulting from heat flow within Earth's interior caused the changes explained by the theory of plate tectonics: movement of the lithospheric plates, earthquakes, volcanoes, and the deformation and metamorphism of rocks during the formation of young mountains.					
Intro 5: Precipitation resulting from the external heat engine's weather systems supplies moisture to Earth's surface that contributes to the weathering of rocks. Running water erodes mountains that were originally uplifted by Earth's internal heat engine and transports sediments to other locations, where they are deposited and may undergo the processes that transform them into sedimentary rocks.					
Intro 6: Global climate is determined by the interaction of solar energy with Earth's surface and atmosphere. This energy transfer is influenced by dynamic processes such as cloud cover and Earth rotation, and the positions of mountain ranges and oceans.					
Physical Setting Performance Indicator 2.1: Use the concepts of density and heat energy to explain observations of weather patterns, seasonal changes, and the movements of Earth's plates.					
MAJOR UNDERSTANDINGS					
2.1a Earth systems have internal and external sources of energy, both of which create heat.					
2.1b The transfer of heat energy within the atmosphere, the the hydrosphere, and Earth's interior results in the formation of regions of different densities. These density differences result in motion.		45, 46	8		

Topic Outline and Question Index

		EXAMINATION QUESTIONS				
		August 2017	June 2018	August 2018	June 2019	August 2019
2.1c	Weather patterns become evident when weather variables are observed, measured, and recorded. These variables include air temperature, air pressure, moisture (relative humidity and dewpoint), precipitation (rain, snow, hail, sleet, etc.), wind speed and direction, and cloud cover.		6	10	10	43
2.1d	Weather variables are measured using instruments such as thermometers, barometers, psychrometers, precipitation gauges, anemometers, and wind vanes.	28, 34, 71		73	57	42, 44
2.1e	Weather variables are interrelated. For example: • Temperature and humidity affect air pressure and probability of precipitation. • Air pressure gradient controls wind velocity.	27, 70	66	36	43	13
2.1f	Air temperature, dewpoint, cloud formation, and precipitation are affected by the expansion and contraction of air due to vertical atmospheric movement.					71
2.1g	Weather variables can be represented in a variety of formats including radar and satellite images, weather maps (including station models, isobars, and fronts), atmospheric cross sections, and computer models. *See Standard 2, Information Systems, Key Idea 1*	8, 9	47	37, 38, 72	14, 55, 56	69
2.1h	Atmospheric moisture, temperature, and pressure distributions; jet streams; wind; air masses and frontal boundaries; and the movement of cyclonic systems and associated tornadoes, thunderstorms, and hurricanes occur in observable patterns. Loss of property, personal injury, and loss of life can be reduced by effective emergency preparedness. *Also see Standard 6, Key Idea 5, and Standard 7, Key Idea 2*	10, 11, 72	7, 8, 67, 68	31, 32, 69, 70, 71	42, 44, 54	14, 15, 70
2.1i	Seasonal changes can be explained using concepts of density and heat energy. These changes include the shifting of global temperature zones, the shifting of planetary wind and ocean current patterns, the occurrence of monsoons, hurricanes, flooding, and severe weather.			34		16
2.1j	Properties of Earth's internal structure (crust, mantle, inner core, and outer core) can be inferred from the analysis of the behavior of seismic waves, including velocity and refraction. • Analysis of seismic waves allows the determination of the location of earthquake epicenters, and the measurement of earthquake magnitude; this analysis leads to the inference that Earth's interior is composed of layers that differ in composition and states of matter.	38, 62, 63, 65	13, 25	20, 42, 43, 44	26	24, 25, 50
2.1k	The outward transfer of Earth's internal heat drives convective circulation in the mantle that moves the lithospheric plates comprising Earth's surface.					

Topic Outline for the Physical Setting/Earth Science Examination

		EXAMINATION QUESTIONS			
		August 2017	June 2018	August 2018	June 2019	August 2019
2.1l	The lithosphere consists of separate plates that ride on the more fluid asthenosphere and move slowly in relationship to one another, creating convergent, divergent, and transform plate boundaries. These motions indicate Earth is a dynamic geologic system. • These plate boundaries are the sites of most earthquakes, volcanoes, and young mountain ranges. • Compared to continental crust, ocean crust is thinner and denser. New ocean crust continues to form at mid-ocean ridges. • Earthquakes and volcanoes present geologic hazards to humans. Loss of property, personal injury, and loss of life can be reduced by effective emergency preparedness. See Standard 7, Key Idea 2	18, 35	24, 59, 60	80, 81	25, 27, 45, 46, 48	48, 49
2.1m	Many processes of the rock cycle are consequences of plate dynamics. These include the production of magma (and subsequent igneous rock formation and contact metamorphism) at both subduction and rifting regions, regional metamorphism within subduction zones, and the creation of major depositional basins through downwarping of the crust.			15		
2.1n	Many of Earth's surface features, such as mid-ocean ridges/rifts, trenches/subduction zones/island arcs, mountain ranges (folded, faulted, and volcanic), hot spots, and the magnetic and age patterns in surface bedrock, are a consequence of forces associated with plate motion and interaction.	17, 19	61, 62	82	7	51, 52, 53
2.1o	Plate motions have resulted in global changes in geography, climate, and the patterns of organic evolution. See Standard 6, Key Idea 5	32			19	
2.1p	Landforms are the result of the interaction of tectonic forces and the processes of weathering, erosion, and deposition.					
2.1q	Topographic maps represent landforms through the use of contour lines that are isolines connecting points of equal elevation. Gradients and profiles can be determined from changes in elevation over a given distance. See Standard 6, Key Ideas 2 and 3, and Standard 7, Key Idea 2	58–61	48–50, 73–75	24, 62, 63, 64	81–84	66, 67, 68
2.1r	Climate variations, structure, and characteristics of bedrock influence the development of landscape features including mountains, plateaus, plains, valleys, ridges, escarpments, and stream drainage patterns.	20, 21	26, 32	33, 65	23, 24	
2.1s	Weathering is the physical and chemical breakdown of rocks at or near Earth's surface. Soils are the result of weathering and biological activity over long periods of time.		30	58	31	31
2.1t	Natural agents of erosion, generally driven by gravity, remove, transport, and deposit weathered rock particles. Each agent of erosion produces distinctive changes in the material that it transports and creates characteristic surface features and landscapes. In certain erosional situations, loss of property, personal injury, and loss of life can be reduced by effective emergency preparedness. See Standard 7, Key Idea 2			60		

Topic Outline and Question Index

		EXAMINATION QUESTIONS				
		August 2017	June 2018	August 2018	June 2019	August 2019
2.1u	The natural agents of erosion include:					
	• **Streams (running water):** Gradient, discharge, and channel shape influence a stream's velocity and the erosion and deposition of sediments. Sediments transported by streams tend to become rounded as a result of abrasion. Stream features include V-shaped valleys, deltas, flood plains, and meanders. A watershed is the area drained by a stream and its tributaries. *See Standard 1: Engineering Design, Key Idea 1; Standard 1: Mathematical Analysis, Key Idea 2; and Standard 6, Key Idea 1*	44, 45, 79	27	22, 48, 49, 50	69–72	29, 30, 55, 56, 63, 64, 65
	• **Glaciers (moving ice):** Glacial erosional processes include the formation of U-shaped valleys, parallel scratches, and grooves in bedrock. Glacial features include moraines, drumlins, kettle lakes, finger lakes, and outwash plains.	76	83, 84	83, 84	28, 29	32
	• **Wave Action:** Erosion and deposition cause changes in shoreline features, including beaches, sandbars, and barrier islands. Wave action rounds sediments as a result of abrasion. Waves approaching a shoreline move sand parallel to the shore within the zone of breaking waves.		22, 34	27		
	Other • **Wind:** Erosion of sediments by wind is most common in arid climates and along shorelines. Wind-generated features include dunes and sand-blasted bedrock. • **Mass Movement:** Earth materials move downslope under the influence of gravity.	22, 23			49	
2.1v	Patterns of deposition result from a loss of energy within the transporting system and are influenced by the size, shape, and density of the transported particles. Sediment deposits may be sorted or unsorted. *See Standard 1: Scientific Inquiry, Key Idea 2*	46, 78	65	85		
2.1w	Sediments of inorganic and organic origin often accumulate in depositional environments. Sedimentary rocks form when sediments are compacted and/or cemented after burial or as the result of chemical precipitation from seawater.		28			
Physical Setting Performance Indicator 2.2: **Explain how incoming solar radiation, ocean currents, and land masses affect weather and climate.**						
2.2a	Insolation (solar radiation) heats Earth's surface and atmosphere unequally due to variations in: • the intensity caused by differences in atmospheric transparency and angle of incidence, which vary with time of day, latitude, and season; • characteristics of the materials absorbing the energy, such as color, texture, transparency, state of matter, and specific heat; • duration, which varies with seasons and latitude.	13–16	58, 76–78	7, 14, 16, 77, 78, 79	17, 74	15, 19, 80
2.2b	The transfer of heat energy within the atmosphere, the hydrosphere, and Earth's surface occurs as the result of radiation, convection, and conduction. • Heating of Earth's surface and atmosphere by the Sun drives convection within the atmosphere and oceans, producing winds and ocean currents.			11		18, 20, 47

Topic Outline for the Physical Setting/Earth Science Examination 37

	EXAMINATION QUESTIONS				
	August 2017	June 2018	August 2018	June 2019	August 2019
2.2c A location's climate is influenced by latitude, proximity to large bodies of water, ocean currents, prevailing winds, vegetative cover, elevation, and mountain ranges.	12, 80, 81	15, 18, 21, 33	13, 15	16, 73, 75, 76	17, 72, 73
2.2d Temperature and precipitation patterns are altered by: • natural events, such as El Niño and volcanic eruptions; • human influences, including deforestation, urbanization, and the production of greenhouse gases such as carbon dioxide and methane.	30	14	12	18, 20	

Physical Setting Key Idea 3:
MATTER IS MADE UP OF PARTICLES WHOSE PROPERTIES DETERMINE THE OBSERVABLE CHARACTERISTICS OF MATTER AND ITS REACTIVITY.

Intro 1: Observation and classification have helped us understand the great variety and complexity of Earth materials. Minerals are the naturally occurring inorganic solid elements, compounds, and mixtures from which rocks are made. We classify minerals on the basis of their chemical composition and observable properties. Rocks are generally classified by their origin (igneous, metamorphic, and sedimentary), texture, and mineral content.					
Intro 2: Rocks and minerals help us understand Earth's historical development and its dynamics. They are important to us because of their availability and properties. The use and distribution of mineral resources and fossil fuels have important economic and environmental impacts. As limited resources, they must be used wisely.					

Physical Setting Performance Indicator 3.1: Explain the properties of materials in terms of the arrangement and properties of the atoms that compose them.

3.1a Minerals have physical properties determined by their chemical composition and crystal structure. • Minerals can be identified by well-defined physical and chemical properties, such as cleavage, fracture, color, density, hardness, streak, luster, crystal shape, and reaction with acid. • Chemical composition and physical properties determine how minerals are used by humans. *See Standard 6, Key Idea 2*	24	51, 52	57	34, 35	34, 35
3.1b Minerals are formed inorganically by the process of crystallization as a result of specific environmental conditions. These include: • cooling and solidification of magma; • precipitation from water caused by such processes as evaporation, chemical reactions, and temperature changes; • rearrangement of atoms in existing minerals subjected to conditions of high temperature and pressure.		35		79, 80	
3.1c Rocks are usually composed of one or more minerals. • Rocks are classified by their origin, mineral content, and texture. • Conditions that existed when a rock formed can be inferred from the rock's mineral content and texture. • The properties of rocks determine how they are used and also influence land usage by humans.	25, 33, 64, 73, 82–85	29, 31, 53, 54	23, 25, 26, 28, 35, 61	30, 33, 50, 77, 85	33, 40, 41, 54

Topic Outline and Question Index

	EXAMINATION QUESTIONS				
	August 2017	June 2018	August 2018	June 2019	August 2019
Reference Tables for Physical Settings/Earth Science 2011 Edition (Revised)	8–11, 13, 15, 18, 19, 24, 25, 29, 32–35, 37, 38, 48, 50, 55, 57, 60, 62–65, 73, 77, 80, 82–85	2, 3, 5, 6, 8, 10, 13, 15, 16, 21, 24, 25, 27–29, 31, 35, 39, 41, 45, 47, 49, 51–54, 59, 61, 62, 64, 65, 67, 68, 70, 71, 78, 80, 85	2, 3, 7, 10, 11, 15, 17–21, 23–26, 28, 30, 34, 35, 37, 42, 44, 50, 55–57, 61, 65, 72, 75, 76, 80–82	4, 5, 7, 8, 10, 11, 13, 14, 19, 21, 25–27, 30, 33–35, 40, 45–47, 50, 52, 53, 55, 56, 59, 63, 66, 67, 68, 72, 75–77, 79, 80, 83, 85	1, 3, 4, 14, 18, 20, 21, 22, 23, 24, 25, 28, 33, 34, 35, 36, 40, 41, 42, 43, 44, 47, 48, 49, 50, 52, 54, 55, 58, 59, 60, 64, 65, 66, 69, 70, 71, 74, 78

Earth Science Reference Tables and Charts

Reference Tables for Physical Setting/EARTH SCIENCE

Radioactive Decay Data

RADIOACTIVE ISOTOPE	DISINTEGRATION	HALF-LIFE (years)
Carbon-14	$^{14}C \rightarrow {}^{14}N$	5.7×10^3
Potassium-40	$^{40}K \nearrow {}^{40}Ar \searrow {}^{40}Ca$	1.3×10^9
Uranium-238	$^{238}U \rightarrow {}^{206}Pb$	4.5×10^9
Rubidium-87	$^{87}Rb \rightarrow {}^{87}Sr$	4.9×10^{10}

Equations

$$\text{Eccentricity} = \frac{\text{distance between foci}}{\text{length of major axis}}$$

$$\text{Gradient} = \frac{\text{change in field value}}{\text{distance}}$$

$$\text{Rate of change} = \frac{\text{change in value}}{\text{time}}$$

$$\text{Density} = \frac{\text{mass}}{\text{volume}}$$

Specific Heats of Common Materials

MATERIAL	SPECIFIC HEAT (Joules/gram•°C)
Liquid water	4.18
Solid water (ice)	2.11
Water vapor	2.00
Dry air	1.01
Basalt	0.84
Granite	0.79
Iron	0.45
Copper	0.38
Lead	0.13

Properties of Water

Heat energy gained during melting	334 J/g
Heat energy released during freezing	334 J/g
Heat energy gained during vaporization	2260 J/g
Heat energy released during condensation	2260 J/g
Density at 3.98°C	1.0 g/mL

Average Chemical Composition of Earth's Crust, Hydrosphere, and Troposphere

ELEMENT (symbol)	CRUST Percent by mass	CRUST Percent by volume	HYDROSPHERE Percent by volume	TROPOSPHERE Percent by volume
Oxygen (O)	46.10	94.04	33.0	21.0
Silicon (Si)	28.20	0.88		
Aluminum (Al)	8.23	0.48		
Iron (Fe)	5.63	0.49		
Calcium (Ca)	4.15	1.18		
Sodium (Na)	2.36	1.11		
Magnesium (Mg)	2.33	0.33		
Potassium (K)	2.09	1.42		
Nitrogen (N)				78.0
Hydrogen (H)			66.0	
Other	0.91	0.07	1.0	1.0

2011 EDITION

This edition of the Earth Science Reference Tables should be used in the classroom beginning in the 2011–12 school year. The first examination for which these tables will be used is the January 2012 Regents Examination in Physical Setting/Earth Science.

Eurypterus remipes

New York State Fossil

Generalized Bedrock Geology of New York State

modified from
GEOLOGICAL SURVEY
NEW YORK STATE MUSEUM
1989

GEOLOGIC PERIODS AND ERAS IN NEW YORK

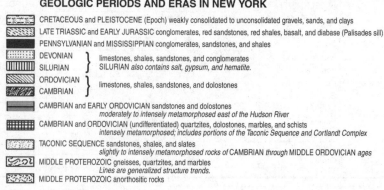

CRETACEOUS and PLEISTOCENE (Epoch) weakly consolidated to unconsolidated gravels, sands, and clays
LATE TRIASSIC and EARLY JURASSIC conglomerates, red sandstones, red shales, basalt, and diabase (Palisades sill)
PENNSYLVANIAN and MISSISSIPPIAN conglomerates, sandstones, and shales
DEVONIAN } limestones, shales, sandstones, and conglomerates
SILURIAN } SILURIAN also contains salt, gypsum, and hematite.
ORDOVICIAN } limestones, shales, sandstones, and dolostones
CAMBRIAN }
CAMBRIAN and EARLY ORDOVICIAN sandstones and dolostones
    moderately to intensely metamorphosed east of the Hudson River
CAMBRIAN and ORDOVICIAN (undifferentiated) quartzites, dolostones, marbles, and schists
    intensely metamorphosed; includes portions of the Taconic Sequence and Cortlandt Complex
TACONIC SEQUENCE sandstones, shales, and slates
    slightly to intensely metamorphosed rocks of CAMBRIAN through MIDDLE ORDOVICIAN ages
MIDDLE PROTEROZOIC gneisses, quartzites, and marbles
    Lines are generalized structure trends.
MIDDLE PROTEROZOIC anorthositic rocks

Earth Science Reference Tables and Charts 43

Tectonic Plates

Earth Science Reference Tables and Charts

Rock Cycle in Earth's Crust

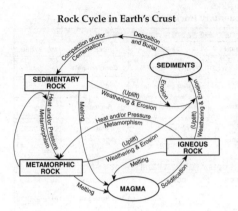

Relationship of Transported Particle Size to Water Velocity

This generalized graph shows the water velocity needed to maintain, but not start, movement. Variations occur due to differences in particle density and shape.

Scheme for Igneous Rock Identification

Scheme for Sedimentary Rock Identification

INORGANIC LAND-DERIVED SEDIMENTARY ROCKS

TEXTURE	GRAIN SIZE	COMPOSITION	COMMENTS	ROCK NAME	MAP SYMBOL
Clastic (fragmental)	Pebbles, cobbles, and/or boulders embedded in sand, silt, and/or clay	Mostly quartz, feldspar, and clay minerals; may contain fragments of other rocks and minerals	Rounded fragments	Conglomerate	
			Angular fragments	Breccia	
	Sand (0.006 to 0.2 cm)		Fine to coarse	Sandstone	
	Silt (0.0004 to 0.006 cm)		Very fine grain	Siltstone	
	Clay (less than 0.0004 cm)		Compact; may split easily	Shale	

CHEMICALLY AND/OR ORGANICALLY FORMED SEDIMENTARY ROCKS

TEXTURE	GRAIN SIZE	COMPOSITION	COMMENTS	ROCK NAME	MAP SYMBOL
Crystalline	Fine to coarse crystals	Halite	Crystals from chemical precipitates and evaporites	Rock salt	
		Gypsum		Rock gypsum	
		Dolomite		Dolostone	
Crystalline or bioclastic	Microscopic to very coarse	Calcite	Precipitates of biologic origin or cemented shell fragments	Limestone	
Bioclastic		Carbon	Compacted plant remains	Bituminous coal	

Scheme for Metamorphic Rock Identification

TEXTURE		GRAIN SIZE	COMPOSITION	TYPE OF METAMORPHISM	COMMENTS	ROCK NAME	MAP SYMBOL
FOLIATED	MINERAL ALIGNMENT	Fine	MICA, QUARTZ, FELDSPAR, AMPHIBOLE, GARNET, PYROXENE	Regional (Heat and pressure increases)	Low-grade metamorphism of shale	Slate	
		Fine to medium			Foliation surfaces shiny from microscopic mica crystals	Phyllite	
					Platy mica crystals visible from metamorphism of clay or feldspars	Schist	
	BAND-ING	Medium to coarse			High-grade metamorphism; mineral types segregated into bands	Gneiss	
NONFOLIATED		Fine	Carbon	Regional	Metamorphism of bituminous coal	Anthracite coal	
		Fine	Various minerals	Contact (heat)	Various rocks changed by heat from nearby magma/lava	Hornfels	
		Fine to coarse	Quartz	Regional or contact	Metamorphism of quartz sandstone	Quartzite	
			Calcite and/or dolomite		Metamorphism of limestone or dolostone	Marble	
		Coarse	Various minerals		Pebbles may be distorted or stretched	Metaconglomerate	

Earth Science Reference Tables and Charts

Inferred Properties of Earth's Interior

GEOLOGIC HISTORY

Earth Science Reference Tables and Charts

OF NEW YORK STATE

Earth Science Reference Tables and Charts

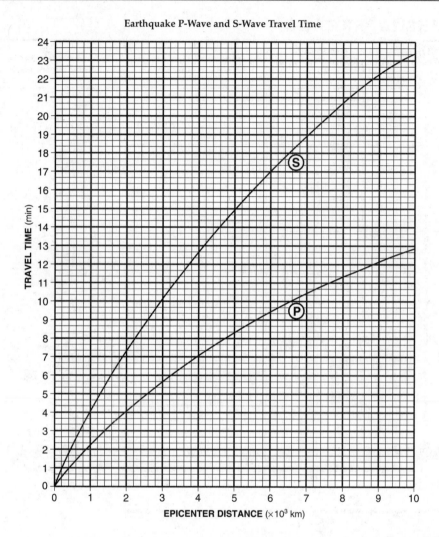

Earthquake P-Wave and S-Wave Travel Time

Dewpoint (°C)

Dry-Bulb Temperature (°C)	Difference Between Wet-Bulb and Dry-Bulb Temperatures (C°)															
	0	1	2	3	4	5	6	7	8	9	10	11	12	13	14	15
-20	-20	-33														
-18	-18	-28														
-16	-16	-24														
-14	-14	-21	-36													
-12	-12	-18	-28													
-10	-10	-14	-22													
-8	-8	-12	-18	-29												
-6	-6	-10	-14	-22												
-4	-4	-7	-12	-17	-29											
-2	-2	-5	-8	-13	-20											
0	0	-3	-6	-9	-15	-24										
2	2	-1	-3	-6	-11	-17										
4	4	1	-1	-4	-7	-11	-19									
6	6	4	1	-1	-4	-7	-13	-21								
8	8	6	3	1	-2	-5	-9	-14								
10	10	8	6	4	1	-2	-5	-9	-14	-28						
12	12	10	8	6	4	1	-2	-5	-9	-16						
14	14	12	11	9	6	4	1	-2	-5	-10	-17					
16	16	14	13	11	9	7	4	1	-1	-6	-10	-17				
18	18	16	15	13	11	9	7	4	2	-2	-5	-10	-19			
20	20	19	17	15	14	12	10	7	4	2	-2	-5	-10	-19		
22	22	21	19	17	16	14	12	10	8	5	3	-1	-5	-10	-19	
24	24	23	21	20	18	16	14	12	10	8	6	2	-1	-5	-10	-18
26	26	25	23	22	20	18	17	15	13	11	9	6	3	0	-4	-9
28	28	27	25	24	22	21	19	17	16	14	11	9	7	4	1	-3
30	30	29	27	26	24	23	21	19	18	16	14	12	10	8	5	1

Relative Humidity (%)

Dry-Bulb Temperature (°C)	Difference Between Wet-Bulb and Dry-Bulb Temperatures (C°)															
	0	1	2	3	4	5	6	7	8	9	10	11	12	13	14	15
-20	100	28														
-18	100	40														
-16	100	48														
-14	100	55	11													
-12	100	61	23													
-10	100	66	33													
-8	100	71	41	13												
-6	100	73	48	20												
-4	100	77	54	32	11											
-2	100	79	58	37	20	1										
0	100	81	63	45	28	11										
2	100	83	67	51	36	20	6									
4	100	85	70	56	42	27	14									
6	100	86	72	59	46	35	22	10								
8	100	87	74	62	51	39	28	17	6							
10	100	88	76	65	54	43	33	24	13	4						
12	100	88	78	67	57	48	38	28	19	10	2					
14	100	89	79	69	60	50	41	33	25	16	8	1				
16	100	90	80	71	62	54	45	37	29	21	14	7	1			
18	100	91	81	72	64	56	48	40	33	26	19	12	6			
20	100	91	82	74	66	58	51	44	36	30	23	17	11	5		
22	100	92	83	75	68	60	53	46	40	33	27	21	15	10	4	
24	100	92	84	76	69	62	55	49	42	36	30	25	20	14	9	4
26	100	92	85	77	70	64	57	51	45	39	34	28	23	18	13	9
28	100	93	86	78	71	65	59	53	47	42	36	31	26	21	17	12
30	100	93	86	79	72	66	61	55	49	44	39	34	29	25	20	16

Earth Science Reference Tables and Charts

Selected Properties of Earth's Atmosphere

Planetary Wind and Moisture Belts in the Troposphere

The drawing on the right shows the locations of the belts near the time of an equinox. The locations shift somewhat with the changing latitude of the Sun's vertical ray. In the Northern Hemisphere, the belts shift northward in the summer and southward in the winter.

(Not drawn to scale)

Electromagnetic Spectrum

(Not drawn to scale)

Characteristics of Stars

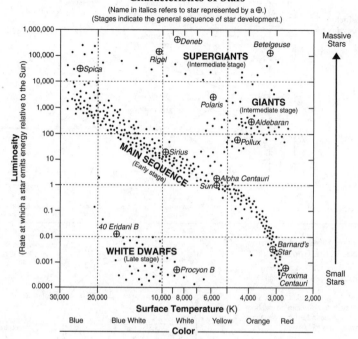

(Name in italics refers to star represented by a ⊕.)
(Stages indicate the general sequence of star development.)

Solar System Data

Celestial Object	Mean Distance from Sun (million km)	Period of Revolution (d=days) (y=years)	Period of Rotation at Equator	Eccentricity of Orbit	Equatorial Diameter (km)	Mass (Earth = 1)	Density (g/cm^3)
SUN	—	—	27 d	—	1,392,000	333,000.00	1.4
MERCURY	57.9	88 d	59 d	0.206	4,879	0.06	5.4
VENUS	108.2	224.7 d	243 d	0.007	12,104	0.82	5.2
EARTH	149.6	365.26 d	23 h 56 min 4 s	0.017	12,756	1.00	5.5
MARS	227.9	687 d	24 h 37 min 23 s	0.093	6,794	0.11	3.9
JUPITER	778.4	11.9 y	9 h 50 min 30 s	0.048	142,984	317.83	1.3
SATURN	1,426.7	29.5 y	10 h 14 min	0.054	120,536	95.16	0.7
URANUS	2,871.0	84.0 y	17 h 14 min	0.047	51,118	14.54	1.3
NEPTUNE	4,498.3	164.8 y	16 h	0.009	49,528	17.15	1.8
EARTH'S MOON	149.6 (0.386 from Earth)	27.3 d	27.3 d	0.055	3,476	0.01	3.3

Properties of Common Minerals

LUSTER	HARD-NESS	CLEAVAGE	FRACTURE	COMMON COLORS	DISTINGUISHING CHARACTERISTICS	USE(S)	COMPOSITION*	MINERAL NAME
Metallic luster	1–2	✓		silver to gray	black streak, greasy feel	pencil lead, lubricants	C	Graphite
Metallic luster	2.5	✓		metallic silver	gray-black streak, cubic cleavage, density = 7.6 g/cm³	ore of lead, batteries	PbS	Galena
Metallic luster	5.5–6.5		✓	black to silver	black streak, magnetic	ore of iron, steel	Fe_3O_4	Magnetite
Metallic luster	6.5		✓	brassy yellow	green-black streak, (fool's gold)	ore of sulfur	FeS_2	Pyrite
Either	5.5–6.5 or 1		✓	metallic silver or earthy red	red-brown streak	ore of iron, jewelry	Fe_2O_3	Hematite
Nonmetallic luster	1	✓		white to green	greasy feel	ceramics, paper	$Mg_3Si_4O_{10}(OH)_2$	Talc
Nonmetallic luster	2		✓	yellow to amber	white-yellow streak	sulfuric acid	S	Sulfur
Nonmetallic luster	2	✓		white to pink or gray	easily scratched by fingernail	plaster of paris, drywall	$CaSO_4 \cdot 2H_2O$	Selenite gypsum
Nonmetallic luster	2–2.5	✓		colorless to yellow	flexible in thin sheets	paint, roofing	$KAl_3Si_3O_{10}(OH)_2$	Muscovite mica
Nonmetallic luster	2.5	✓		colorless to white	cubic cleavage, salty taste	food additive, melts ice	$NaCl$	Halite
Nonmetallic luster	2.5–3	✓		black to dark brown	flexible in thin sheets	construction materials	$K(Mg,Fe)_3$ $AlSi_3O_{10}(OH)_2$	Biotite mica
Nonmetallic luster	3	✓		colorless or variable	bubbles with acid, rhombohedral cleavage	cement, lime	$CaCO_3$	Calcite
Nonmetallic luster	3.5	✓		colorless or variable	bubbles with acid when powdered	building stones	$CaMg(CO_3)_2$	Dolomite
Nonmetallic luster	4	✓		colorless or variable	cleaves in 4 directions	hydrofluoric acid	CaF_2	Fluorite
Nonmetallic luster	5–6	✓		black to dark green	cleaves in 2 directions at 90°	mineral collections, jewelry	$(Ca,Na)(Mg,Fe,Al)$ $(Si,Al)_2O_6$	Pyroxene (commonly augite)
Nonmetallic luster	5.5	✓		black to dark green	cleaves at 56° and 124°	mineral collections, jewelry	$CaNa(Mg,Fe)_4 (Al,Fe,Ti)_3$ $Si_6O_{22}(O,OH)_2$	Amphibole (commonly hornblende)
Nonmetallic luster	6	✓		white to pink	cleaves in 2 directions at 90°	ceramics, glass	$KAlSi_3O_8$	Potassium feldspar (commonly orthoclase)
Nonmetallic luster	6	✓		white to gray	cleaves in 2 directions, striations visible	ceramics, glass	$(Na,Ca)AlSi_3O_8$	Plagioclase feldspar
Nonmetallic luster	6.5		✓	green to gray or brown	commonly light green and granular	furnace bricks, jewelry	$(Fe,Mg)_2SiO_4$	Olivine
Nonmetallic luster	7		✓	colorless or variable	glassy luster, may form hexagonal crystals	glass, jewelry, electronics	SiO_2	Quartz
Nonmetallic luster	6.5–7.5		✓	dark red to green	often seen as red glassy grains in NYS metamorphic rocks	jewelry (NYS gem), abrasives	$Fe_3Al_2Si_3O_{12}$	Garnet

*Chemical symbols:
Al = aluminum   Cl = chlorine   H = hydrogen   Na = sodium   S = sulfur
C = carbon   F = fluorine   K = potassium   O = oxygen   Si = silicon
Ca = calcium   Fe = iron   Mg = magnesium   Pb = lead   Ti = titanium

✓ = dominant form of breakage

Glossary of Earth Science Terms

abrasion the wearing away of rock material that results when one rock particle strikes another. In stream abrasion, particles carried by the stream may strike other particles, causing pieces to break off.

adiabatic change the change in temperature of a gas caused by expansion or compression. When a gas expands, for example, it cools and its temperature drops.

air mass a large area in the atmosphere in which the temperature and moisture conditions are similar.

altitude the angle between the line of sight to a star and the horizon; also, the elevation above sea level.

angle of insolation the angle at which the Sun's rays strike the Earth's surface.

anticline a series of folded rock layers that bends upward near the center. *See also* **syncline**.

apparent diameter the size that an object appears to an observer. When objects are close by, they appear larger than when they are farther away.

asthenosphere a region of the upper mantle between 100 and 350 kilometers in depth that behaves like a fluid.

atmosphere the envelope of air that encircles Earth. It has been divided into zones based largely on temperature differences.

axis an imaginary line around which an object rotates. Earth's axis extends from the North Pole to the South Pole.

barometer an instrument for measuring atmospheric pressure.

barometric pressure the amount of force exerted by the air per square inch or centimeter at a particular location.

bedrock the rock layer nearest Earth's surface, lying directly below any soil layers.

Big Bang theory the idea that the universe started out with all of its matter in a small volume and then expanded outward in all directions.

bioclastic a sedimentary rock consisting of fragmental or broken remains of organisms, such as limestone composed of shell fragments.
boiling point the temperature at which a liquid changes to a vapor.

capillarity (capillary action) the rising of water against gravity, as when water rises above the water table in soil because of the attraction between the water and the soil.
cementation the process by which sediments are bonded together by material dissolved in water when the water evaporates.
chemical weathering weathering that occurs because of the chemical reaction between material dissolved in water and local rock material. *See also* **physical weathering**.
cleavage the splitting of a mineral in distinctive directions caused by the arrangement of the atoms in a mineral. For example, the mineral mica cleaves in layers because the atoms in the crystal structure of mica are arranged in layers.
climate the average weather conditions, in terms of temperature and moisture, of an area over a long period of time.
clouds masses of water droplets suspended in the air. They form when air is cooled and moisture condenses.
cold front the boundary between two air masses of different temperatures, the point where the colder air mass moves under and pushes up the warmer air mass. *See also* **warm front**.
compaction the process that results when buried sediments are subjected to pressure, which packs them together. Together with cementation, this process causes sediments to be converted to rock.
condensation the process whereby, when moisture in the atmosphere is cooled, it changes from a vapor to a liquid.
conduction the method of heat transfer in solids in which faster moving molecules strike other molecules, causing them to speed up.
contact metamorphism changes in rock that result from the extreme heat produced by contact with magma or lava.
continental air mass an air mass that forms over land and therefore is relatively dry.
continental drift a hypothetical slow movement of the continents, which forms the basis of the theory that the present continents were once part of one large landmass that broke up. Since that time, the continents have been drifting apart.
continental glacier a large sheet of ice that covered much of a continent during a period of Earth's past geologic history.
contour interval the differences in elevation between contour lines on a contour or topographic map.

contour line a line on a topographic map that connects points having the same elevation.

contour map *See* **topographic map**.

convection the method of energy transfer in fluids in which the fluid expands when it is heated and rises. When the fluid cools, it contracts and sinks.

convection cell the circular pattern of movement in a fluid caused by the rising and sinking of the fluid due to differences in density caused by differences in temperature.

convergent boundary places where edges of adjacent plates are colliding.

core the innermost zone of Earth's interior. A solid inner core is surrounded by a molten outer core.

coriolis effect an apparent force, due to the rotation of the Earth, that deflects winds toward the right in the Northern Hemisphere and toward the left in the Southern Hemisphere.

correlation the matching of rock layers at different locations, based on composition, thickness, and in some cases fossil content, for the purpose of establishing that they represent the same rock layer.

cosmic background radiation a remnant of radiation left over from the original Big Bang that fills the universe.

crust the outermost solid layer of Earth, extending across the continents and under the oceans. It is thicker under the continents than under the oceans.

density the mass per unit volume of a substance.

deposition the process by which material carried by running water, glaciers, or wind settles out when the velocity of the carrier slows down.

dew moisture that condenses at Earth's surface when moist air touches a cool area.

dewpoint temperature the temperature at which the air becomes saturated and excess moisture begins to condense.

dike a rock layer that forms when molten rock material flows through breaks in rock layers and then cools and hardens.

discharge the total volume of water flowing in a river or stream per unit of time.

divergent boundary places where edges of adjacent plates are spreading apart.

Doppler effect the shift in wavelength of a spectrum line from its normal position due to relative motion between the source and the observer.

duration of insolation the number of hours that the Sun's rays strike Earth's surface over a 24-hour period; the number of hours of daylight.

earthquake the large-scale and rapid motion of rock layers that occurs when pressure is released.

Glossary of Earth Science Terms

electromagnetic spectrum the range of wavelengths of energy released by the Sun. A small range of wavelengths represents visible light. Other bands within the spectrum include ultraviolet radiation, cosmic rays, infrared radiation, and radio waves.

elevation the height above sea level.

ellipse a curve that has two centers, or foci. The sum of the distance between either focus and any point on the ellipse and the distance between the other focus and that point is a constant. The shape of Earth's orbit around the Sun is an ellipse with the Sun at one focus.

epicenter the location on Earth's surface directly above the point of origin or focus of an earthquake.

equator an imaginary circle around Earth lying halfway between the poles and dividing the Earth's surface into the Northern and Southern hemispheres.

equinox the two times a year when the Sun is directly overhead at the equator at noon. At these times, about March 21 and September 23, the number of hours of day and of night is the same.

erosion the process by which weathered rock material is carried away by agents such as running water, ice, wind, and gravity.

evaporation the process by which water is converted from a liquid to a vapor.

evapotranspiration the combined processes of evaporation and transpiration, representing the method by which water vapor returns to the atmosphere.

extinct referring to a plant or animal species that lived in the past but is no longer found alive on Earth.

fault movement within rock layers where pressure has caused the layers to break. The layers on one side of the break move up, while the opposite layers move down.

felsic referring to igneous rocks composed of minerals with a high content of aluminum. These rocks tend to be lighter in color than others.

focus the point of origin within Earth's crust or mantle of an earthquake.

foliated a metamorphic rock texture characterized by thin, leaflike layers caused by the flattening of mineral grains under heat and pressure.

formation (rock) a sequence of rock layers that cover a large area and were formed over a period of time.

fossil the preserved remains or traces of an animal or plant that lived in the past.

fossil record groups of fossils found in a rock layer that are used to interpret how and when the rock layer was formed and what the environment was like at that time.

Foucault pendulum a freely swinging pendulum whose path appears to change direction relative to Earth's surface in a predictable manner due to Earth's rotation.

Glossary of Earth Science Terms

fracture a term used to describe the irregular way in which some minerals break.
freezing point the temperature at which a liquid changes to a solid.
front the boundary between two different air masses. *See also* **cold front; warm front**.
frost moisture that condenses directly from a vapor to a solid when moist air touches a cold surface. When the air temperature is below freezing, frost forms instead of dew.

galaxy a system consisting of hundreds of billions of stars.
gradient the slope of the land or of a river or stream.
greenhouse effect the process by which longer wavelength radiation emitted from Earth's surface is absorbed by carbon dioxide in the atmosphere, causing a rise in air temperature.

half-life the amount of time it takes for half the mass of a radioactive substance to decay. For example, the half-life of carbon-14 is 5.6×10^3 years.
high-pressure center a location on a weather map where winds blow outward, in a clockwise direction, away from the center. This occurs because the air pressure is lower at the center than over the surrounding area. *See also* **low-pressure center**.
humidity the amount of moisture in the air.
hydrosphere the outer zone of Earth, which consists of the oceans and seas.
hypothesis the attempt to explain a scientific phenomenon on the basis of observations and other relevant information.

igneous referring to rocks that form when molten rock material cools and solidifies.
impermeable referring to a layer of material through which water cannot pass. Most soil layers are permeable, while most rock layers are not.
index fossil a fossil that is found over widespread areas and that formed from an organism that existed for a relatively short geologic period of time.
inference an interpretation or explanation of a natural phenomenon based on observations.
infiltration the process by which water at Earth's surface penetrates and filters down through porous soil and rock layers.
insolation the radiation reaching Earth's surface from the Sun. This term is a contraction of "incoming solar radiation."
intrusion the forcible entry of solidified molten rock material between rock layers or into breaks within a rock layer.
isobar a line on a weather map that connects points having the same air or barometric pressure.

Glossary of Earth Science Terms

land breeze a wind blowing from over the land to over a lake or ocean. It occurs when the water temperature is warmer than the land temperature. *See also* **sea breeze**.

landscape region a grouping of landscapes with similar relief, stream patterns, and soil associations.

latent heat of fusion the amount of energy required to convert one gram of ice at 0°C to water at 0°C.

latent heat of vaporization the amount of energy required to convert one gram of water at 100°C to vapor at 100°C.

latitude imaginary circles around Earth parallel to the equator and between the North and South poles.

lava molten rock material that reaches Earth's surface through volcanoes.

light-year the distance light travels in one year; roughly 9.5 trillion kilometers, or 6 trillion miles.

lithosphere the solid outer portion of Earth.

longitude imaginary circles around Earth that pass through the North and South poles.

low-pressure center a location on a weather map where winds blow inward, in a counterclockwise direction, into a center. This occurs because the air pressure is lower at the center than in the surrounding area. *See also* **high-pressure center**.

luminosity the total amount of energy radiated by a star in 1 second.

lunar eclipse the phenomenon that occurs when Earth passes between the Moon and the Sun, causing light from the Sun to be blocked from reaching the surface of the Moon. *See also* **solar eclipse**.

luster the type of shine exhibited by a mineral, based on the way its surface reflects light. Examples are metallic, glassy, and dull lusters.

mafic referring to igneous rocks composed of minerals with a high content of magnesium and iron. They tend to be darker in color than other rocks.

magma molten rock material beneath the Earth's surface.

magnetic north (pole) the point near the geographic North Pole toward which the needle on a compass points.

mantle the zone within Earth that lies between the crust and the outer core.

maritime air mass an air mass that forms over water and therefore has a relatively high moisture content.

mass the measure of the amount of matter contained in a sample.

meander a curve in a river or stream.

meridian a line of longitude measured in degrees. The prime meridian (0°) passes through Greenwich, England.

metamorphic referring to rocks that form when existing rocks are subjected to enough heat and pressure to cause partial melting of the minerals present.

meter the standard unit of length in the metric system; 1 meter = 39.37 inches.

mid-ocean ridge a chain of undersea mountains running through the center of an ocean.

millibar a unit of air pressure in the atmosphere, commonly used on weather maps.

mineral a naturally occurring substance that is always made of the same elements in a fixed proportion.

modified Mercalli scale a system that measures the strength of an earthquake based upon perception of motion by human observers and damage to structures built by humans.

Moho the boundary between the dense rock of the mantle and the less dense rock of the crust. It was named for the seismologist Andrija Mohorovicic, who recognized its existence after analyzing seismic wave behavior inside Earth.

moraine a large deposit formed when a glacier melts and leaves behind the material it is carrying.

mountain any part of Earth's crust that projects at least 300 m (1000 ft) above the surrounding land, has a limited summit area (as opposed to a plateau), steep sides, and considerable bare-rock surface.

observation a description of what is perceived by the senses. It can often be made more accurate by using instruments.

occluded front a weather front that forms when a cold front moves in behind a warm front, lifting the warm front off the ground.

orbit the path followed by one object as it revolves around another; for example, the orbit of Earth around the Sun.

outgassing the release of gases and water vapor from molten rocks, leading to the formation of Earth's atmosphere and oceans.

permeability the ability of water to penetrate through a material. The permeability of a soil depends upon the size of the soil grains and the amount of space between the grains.

physical weathering the breakdown of rocks due to physical changes, such as changes in temperature or the freezing and thawing of water, that fills cracks in the rock. *See also* **chemical weathering**.

plain a flat area at low elevation.

plateau a large, flat region elevated more than 150–300 m (500–1000 ft) above the surrounding land, or above sea level.

polar air mass an air mass that forms over the polar regions and therefore contains relatively cold air.

pollution a condition of the air, water, or land in which there is a surplus of materials present that may be harmful to living things.
porosity the percentage of open space in a soil sample.
precipitation all forms of moisture that reach the Earth's surface from the atmosphere, including rain, snow, hail, and sleet.
prevailing westerlies the wind belt stretching across the United States in which the general direction of wind movement is from southwest to northeast.
***P*-wave (primary wave)** the compression type of wave emitted by an earthquake that travels through both liquids and solids. *See also* **S-wave**.

radiation the method of energy transfer by which energy from the Sun reaches the Earth.
radioactive dating the use of the radioactive isotope to determine the age of a fossil or rock layer. This technique is possible because each radioactive isotope decays at a unique and fixed rate.
radioactive decay the process by which a radioactive element breaks down to emit particles and radiation that form a new element.
radioactive isotope an unstable form of an element that breaks down by radioactive decay. For example, carbon-14 is a radioactive isotope of carbon that breaks down by emitting electrons to form nitrogen-14.
rain gauge an instrument that collects atmospheric precipitation so that the amount of rainfall can be measured.
regional metamorphism changes in rock over an extensive area that occur due to the pressure and high temperatures associated with either deep burial or movements of Earth's crust.
relative humidity the percentage of moisture present in the air as compared to the maximum amount of moisture the air can hold at the prevailing temperature.
residual soil soil that is formed by the weathering of local rock material.
Richter scale a scale with a range of 1 to 10 that indicates the magnitude of an earthquake. For each successive number, the magnitude increases by a factor of 10.
rock naturally occurring materials that are composed of one or more minerals.
rock cycle the process by which each of the three rock types (igneous, metamorphic, and sedimentary) can be converted into each other type.
runoff excess precipitation reaching Earth's surface that flows into rivers and streams because the surface soil is saturated.

salinity a measure of the amount of salt dissolved in water.
saturation temperature *See* **dewpoint temperature**.

Glossary of Earth Science Terms

sea breeze a wind blowing from over a lake or ocean to over land. It occurs when the water temperature is cooler than the land temperature. *See also* **land breeze**.

sea floor spreading the concept that portions of the ocean floor are moving away from a central ridge because new material moving upward at the ridge is pushing the old material outward.

sedimentary referring to rocks that form by the compaction and cementation of sediments.

seismograph an instrument used to measure the disturbances caused by an earthquake.

sill a rock layer that forms when molten rock material forces its way between existing rock layers and then cools and solidifies.

sling psychrometer an instrument used to determine relative humidity.

solar eclipse the phenomenon that occurs when the Moon passes between Earth and the Sun, causing light from the Sun to be blocked from reaching the surface of Earth. *See also* **lunar eclipse**.

solar system the Sun and the various objects that orbit it, including the planets and their moons, comets, and asteroids.

solstice one of the two times a year, about June 22 and December 22, when the Sun is directly overhead at 23½° north and south latitude. The solstices represent the longest and shortest days of the year.

specific heat the relative amount of energy needed to raise the temperature of one gram of a material by one degree Celsius.

spectroscope an instrument used to break down the visible portion of the electromagnetic spectrum into individual wavelengths or colors.

spectrum the band of colors formed when a beam of white light is passed through a prism; each wavelength of light in the beam is refracted at a slightly different angle, causing the light to be arrayed in order of its constituent wavelengths.

stationary front a front that forms when the opposing warm and cold air masses are of equal energy.

station model a pattern of symbols on a weather map that is used to describe local weather conditions such as temperature, pressure, humidity, wind speed and direction, and cloud cover.

stratosphere the layer of the atmosphere directly above the troposphere.

streak a more accurate method of identifying the color of a mineral by rubbing it against a plate, causing powder to form.

stream load the total amount of material carried by a stream, including material that is carried in solution or suspension or is pushed along the bottom.

subduction the sliding of a denser ocean plate beneath a less dense continental plate resulting in the melting of the ocean plate as it plunges into the hot mantle.

subsoil the layer of soil just below the topsoil. It does not contain the organic matter found in the topsoil and has not been as extensively weathered.

S-wave (shear wave) the transverse type of wave emitted by an earthquake that travels through solids but is absorbed by liquids. *See also* **P-wave**.

syncline a series of folded rock layers that dips downward near the center. *See also* **anticline**.

terminal moraine the material deposited by the meltwater of a glacier at the point of its farthest advance.

texture the grain size of a rock. A course texture represents large grains; a fine texture, small grains.

theory an explanation for scientific observations. A theory is formed from a hypothesis when there is substantial evidence to support it.

till the material found in a glacial moraine and usually representing a wide range of particle sizes.

time zone a region in which the same time is used throughout, instead of the local time at each place within it.

topographic (contour) map a map of an area with contour lines to show the elevations at all locations, as well as other geologic features such as rivers, streams, and mountain peaks.

topsoil the uppermost layer of soil. It contains the most highly weathered rock fragments, as well as organic remains.

transform boundary places where edges of plates are sliding laterally past each other.

transpiration the process by which plants release moisture to the atmosphere.

transported soils soils formed from weathered rock material that have been carried from other locations and deposited.

trench a large valley on the ocean floor.

tropical air mass an air mass that has formed over the tropics and therefore contains relatively warm air.

Tropic of Cancer the line of latitude at 23½° North, which represents the farthest north that the noon Sun can be overhead.

Tropic of Capricorn the line of latitude at 23½° South, which represents the farthest south that the noon Sun can be overhead.

troposphere the layer of the atmosphere closest to Earth's surface. All weather phenomena occur within this zone.

unconformity a gap in a sequence of rock layers, resulting either from the removal of layers by erosion or failure of layers to form for long periods.

uniformitarianism the principle that the geologic processes acting today are the same as those that occurred in the past.

valley glacier a glacier that forms in a valley between mountain peaks.

vesicular a rock texture characterized by cavities formed by gas bubbles escaping from a lava as it cools and solidifies.

volcano a mountain formed when lava erupts from beneath the surface and then cools and solidifies.

warm front the boundary between two air masses of different temperatures; the point where the warmer air mass moves over the colder air mass and pushes it backward. *See also* **cold front**.

water cycle the cyclic process during which water leaves Earth's surface by evaporation and transpiration to enter the atmosphere. It returns to the surface as precipitation.

watershed the area drained by a stream and its tributaries.

water table the upper boundary of saturated rock or soil beneath Earth's surface.

weathering the processes in nature by which rock materials are broken down into smaller pieces to form sand, soil, and so on. *See also* **chemical weathering; physical weathering**.

wind belts zones around Earth in which the winds blow in the same general direction.

Regents Examinations, Answers, and Self-Analysis Charts

Examination August 2017
Physical Setting/Earth Science

PART A
Answer all questions in this part.

Directions (1–35): For *each* statement or question, choose the word or expression that, of those given, best completes the statement or answers the question. Some questions may require the use of the *2011 Edition Reference Tables for Physical Setting/Earth Science*. Record your answers in the space provided.

1 In which sequence are the celestial objects correctly listed in order from the smallest mass to the largest mass?

(1) Saturn, solar system, Milky Way, universe
(2) Saturn, universe, Milky Way, solar system
(3) Milky Way, Saturn, solar system, universe
(4) Milky Way, universe, solar system, Saturn

1 _____

2 The red shift of light from distant galaxies provides evidence that these galaxies are

(1) decreasing in size
(2) increasing in size
(3) decreasing in distance from Earth
(4) increasing in distance from Earth

2 _____

3 The best evidence of Earth's rotation is provided by the

(1) Foucault pendulum and global warming
(2) Foucault pendulum and Coriolis effect
(3) Moon phases and global warming
(4) Moon phases and Coriolis effect

3 _____

4 Which sphere of Earth covers approximately 70% of Earth's surface?

(1) atmosphere
(2) lithosphere
(3) hydrosphere
(4) asthenosphere

5 Some of the constellations that are visible to New York State observers at midnight in December are different from the constellations that are visible at midnight in June because

(1) constellations rotate on an axis
(2) constellations revolve around Earth
(3) Earth rotates on its axis
(4) Earth revolves around the Sun

6 The diagram below represents Earth and the Moon as viewed from above the North Pole. Points A, B, C, and D are locations on Earth's surface.

Moon

A D
Earth
B C

(Not drawn to scale)

According to the diagram, where will high ocean tides and low ocean tides most likely be located?

(1) high tides at A and B; low tides at C and D
(2) high tides at B and D; low tides at A and C
(3) high tides at A and C; low tides at B and D
(4) high tides at C and D; low tides at A and B

7 Urbanization affects the amount of vegetation and runoff in an area by

(1) decreasing vegetation and decreasing runoff
(2) decreasing vegetation and increasing runoff
(3) increasing vegetation and decreasing runoff
(4) increasing vegetation and increasing runoff

8 A severe thunderstorm warning was issued on a warm summer afternoon. Which present weather symbol represents the dangerous solid form of precipitation that is commonly associated with some of these severe thunderstorms?

9 The station model below represents the weather conditions for a location in New York State.

The barometric trend for the past three hours at this location indicates a steady increase of

(1) 0.2 mb   (3) 0.002 mb
(2) 2.0 mb   (4) 0.02 mb

10 Which New York State location is most often affected by lake-effect snow storms caused by winds blowing over Lake Ontario?

(1) Jamestown   (3) Oswego
(2) Plattsburgh (4) Riverhead

11 Which type of air mass would most likely form over the Pacific Ocean north of the Aleutian Trench?

(1) mP   (3) cP
(2) mT   (4) cT

12 Mount Kilimanjaro is located in eastern Africa at 3° S. Which climate factor best explains the presence of permanent snow on its peak?

(1) latitude   (3) prevailing winds
(2) elevation   (4) ocean currents

13 In which portion of the electromagnetic spectrum is the maximum intensity of Earth's outgoing radiation?
   (1) visible light
   (2) gamma rays
   (3) infrared
   (4) ultraviolet   13 _____

14 A solar water heater contains fluid-filled tubing that absorbs sunlight energy on its outside surface. Which tubing exterior will best absorb insolation?
   (1) dark-colored and rough
   (2) dark-colored and smooth
   (3) light-colored and rough
   (4) light-colored and smooth   14 _____

15 Equal masses of granite, iron, copper, and lead are placed in sunlight. Based on specific heat, which material will warm up the fastest?
   (1) granite
   (2) iron
   (3) copper
   (4) lead   15 _____

16 During explosive volcanic eruptions, large amounts of ash entering Earth's atmosphere often rise to an altitude of 20 kilometers. What is the most likely effect of this ash cloud?
   (1) a decrease in the insolation reaching Earth's surface
   (2) a decrease in the thickness of Earth's stratosphere layer
   (3) an increase in the insolation reaching Earth's surface
   (4) an increase in the thickness of Earth's stratosphere layer   16 _____

17 The map below shows some tectonic plate boundaries near South America and Africa. Letters *A*, *B*, *C*, and *D* represent locations on Earth's surface.

Which location most likely has the youngest bedrock?

(1) *A* (3) *C*
(2) *B* (4) *D*   17 ____

18 Compared to the average density and composition of oceanic crust, continental crust is

(1) less dense and more felsic
(2) less dense and more mafic
(3) more dense and more felsic
(4) more dense and more mafic   18 ____

19 The Hawaiian Islands were formed as a result of

(1) lava flowing over Earth's surface where two tectonic plates move apart
(2) an oceanic plate moving over a mantle hot spot
(3) two oceanic plates colliding to form an island arc
(4) tectonic plates sliding past each other   19 ____

20 The block diagram below represents the underlying bedrock structure of a landscape region.

Which diagram represents the most likely stream drainage pattern on the surface of this landscape?

20 ____

21 Which characteristics identify mountain landscape regions?
 (1) steep slopes with deformed bedrock
 (2) steep slopes with horizontal bedrock
 (3) gentle slopes with deformed bedrock
 (4) gentle slopes with horizontal bedrock   21 ____

22 Which agent of erosion causes the sandblasting of bedrock?
 (1) glaciers   (3) running water
 (2) wind   (4) wave action   22 ____

23 The photograph below shows wire netting installed over a steep rock outcrop.

This wire netting has been installed to prevent loss of property or life resulting from

(1) crosscutting and downwarping
(2) folding and faulting
(3) weathering and erosion
(4) high winds and flooding   23 _____

24 Which two minerals are commercial sources of iron?

(1) galena and graphite
(2) muscovite mica and biotite mica
(3) garnet and fluorite
(4) hematite and magnetite   24 _____

25 Which mineral can be found in the rocks phyllite, sandstone, and granite?

(1) quartz   (3) gypsum
(2) pyroxene   (4) calcite   25 _____

26 Which diagram best represents Earth's axis position relative to Earth's orbital plane?

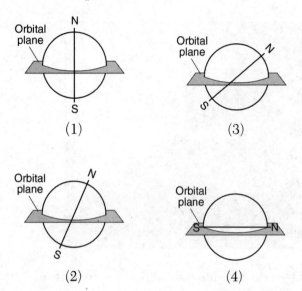

27 In which cross section do the arrows represent the most probable direction of air movement over land and water surfaces at an ocean coast on a hot, sunny, summer afternoon?

28 The table below shows the average diameter and falling velocity of five types of precipitation. The data for moderate rain have been left blank.

Type of Precipitation	Average Diameter (millimeters)	Falling Velocity (meters/second)
drizzle	0.96	4.1
light rain	1.24	4.8
moderate rain		
heavy rain	2.05	6.7
excessive rain	2.40	7.3

What are the probable values for the average diameter and falling velocity of the raindrops found in moderate rain?

(1) average diameter = 1.20 mm
    falling velocity = 4.6 m/sec
(2) average diameter = 1.20 mm
    falling velocity = 5.7 m/sec
(3) average diameter = 1.60 mm
    falling velocity = 4.6 m/sec
(4) average diameter = 1.60 mm
    falling velocity = 5.7 m/sec

29 Which New York State index fossil is classified as a coral?

(1)   (2)   (3)   (4)

30 The map below shows the weak trade winds and strong equatorial countercurrent in the Pacific Ocean during El Niño conditions. This causes warm surface ocean water to migrate eastward, lowering the atmospheric pressure above this warm water.

El Niño Conditions

What are the most likely changes to atmospheric temperature and precipitation along the west coast of South America during El Niño conditions?

(1) lower temperatures and lower amounts of precipitation
(2) lower temperatures and higher amounts of precipitation
(3) higher temperatures and lower amounts of precipitation
(4) higher temperatures and higher amounts of precipitation

31 The cross sections below represent two bedrock outcrops 15 kilometers apart. Numbers 1 through 9 indicate rock layers. Some layers contain index fossils. No overturning of rock layers has occurred.

Which layers most likely formed during the same geologic time period?

(1) 1 and 8
(2) 2 and 9
(3) 3 and 7
(4) 4 and 5

31 _____

32 Labeled lines on the map below show the inferred location of Earth's equator during the middle of several geologic periods.

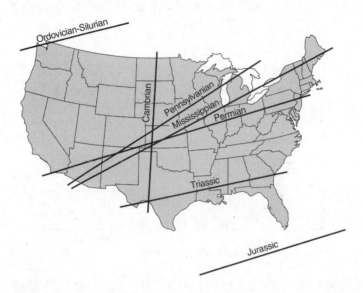

Approximately how many million years ago (mya) was the region around current-day Watertown, New York, located the nearest to the equator?

(1) 270 mya (3) 340 mya
(2) 300 mya (4) 450 mya

33 Which table correctly matches rock textures with a sedimentary rock that exhibits each texture?

Rock Texture	Sedimentary Rock
clastic	sandstone
crystalline	breccia
bioclastic	rock gypsum

(1)

Rock Texture	Sedimentary Rock
clastic	shale
crystalline	dolostone
bioclastic	sandstone

(3)

Rock Texture	Sedimentary Rock
clastic	dolostone
crystalline	limestone
bioclastic	siltstone

(2)

Rock Texture	Sedimentary Rock
clastic	conglomerate
crystalline	rock salt
bioclastic	limestone

(4)

34 The data table below shows the dry-bulb and wet-bulb temperatures measured with a psychrometer on four different days at the same location.

Temperatures Measured with a Psychrometer

Day	1	2	3	4
Dry-bulb temperature (°C)	0	5	10	15
Wet-bulb temperature (°C)	−5	0	5	10

According to the data shown in the table, which day had the highest relative humidity?

(1) 1
(2) 2
(3) 3
(4) 4

34 _____

35 The cross section below represents a lithospheric plate boundary.

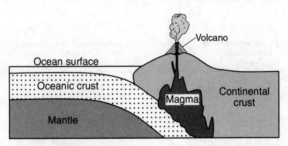

(Not drawn to scale)

In which diagram do the arrows show the relative directions of plate movement at this type of plate boundary?

35 _____

PART B–1
Answer all questions in this part.

Directions (36–50): For *each* statement or question, choose the word or expression that, of those given, best completes the statement or answers the question. Some questions may require the use of the *2011 Edition Reference Tables for Physical Setting/Earth Science*. Record your answers in the space provided.

Base your answers to questions 36 through 38 on the information and diagram below and on your knowledge of Earth science. The diagram represents a simplified model of the early formation of Earth's interior.

Early in its formation, Earth was a molten mass of evenly mixed composition. During the next few million years, the heavier and more dense elements sank to the center, while lighter and less dense elements rose toward the surface. This is called chemical fractionation.

(Not drawn to scale)

36 Chemical fractionation is most likely caused by
  (1) solidification
  (2) gravity
  (3) magnetic force
  (4) chemical weathering   36 ____

37 Approximately how many years ago did Earth and other planets in our solar system begin the process of chemical fractionation?

(1) 8.2 billion years ago
(2) 13.8 billion years ago
(3) 542 million years ago
(4) 4600 million years ago   37 _____

38 Which pair of elements sank to Earth's center during chemical fractionation?

(1) aluminum and silicon
(2) carbon and sulfur
(3) iron and nickel
(4) oxygen and potassium   38 _____

Base your answers to questions 39 through 41 on the diagram below and on your knowledge of Earth science. The diagram represents the apparent path of the Sun through the sky as viewed by an observer in the Northern Hemisphere. Points A, B, C, and D represent four positions of the Sun.

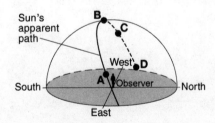

39 This apparent path of the Sun through the sky is caused by

(1) Earth's revolution around the Sun
(2) Earth's rotation on its axis
(3) the Sun's revolution around Earth
(4) the Sun's rotation on its axis   39 _____

40 The observer has the longest shadow when the Sun is at position

(1) A   (3) C
(2) B   (4) D   40 _____

41 What is the approximate time of day when the Sun is at position *C*?

(1) 6 a.m.
(2) 9 a.m.
(3) 3 p.m.
(4) 6 p.m.

41 ____

Base your answers to questions 42 and 43 on the diagram below and on your knowledge of Earth science. The diagram represents three tubes, *A*, *B*, and *C*, each containing an equal volume of uniform-sized spherical beads. The bottom of each tube is covered with a wire screen. *XY* is a reference line.

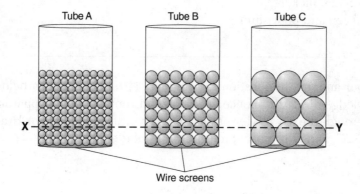

42 Which bar graph would best represent the rate of water infiltration through tubes *A*, *B*, and *C*?

42 _____

43 These tubes are placed in water up to the level of line *XY* to demonstrate capillarity. After one hour, the height of the water above line *XY* will be

(1) highest in tube *A*
(2) highest in tube *B*
(3) highest in tube *C*
(4) the same height in all three tubes

43 _____

Base your answers to questions 44 through 46 on the block diagram below and on your knowledge of Earth science. The block diagram represents a river drainage system. A portion of the river, seen in box A, has been enlarged. Points X and Y are on opposite sides of the river. Letter B indicates the location where the river enters the ocean.

44 The area of land drained by this river and its tributaries is best described as the river's

(1) topography
(2) watershed
(3) water table
(4) floodplain

44 _____

45 Which cross section best represents the profile of the bottom of the river between points X and Y at location A?

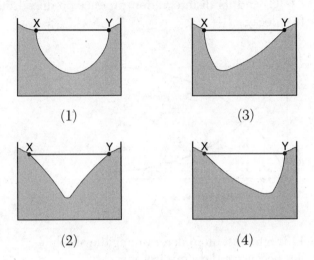

46 Which cross section represents the most likely pattern of sediments deposited where the river enters the ocean at location B?

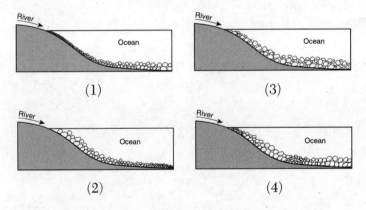

45 ____

46 ____

Base your answers to questions 47 through 50 on the graph below and on your knowledge of Earth science. The graph shows the percentages of the radioactive isotope carbon-14 (^{14}C) and its disintegration product produced during four half-lives of radioactive decay.

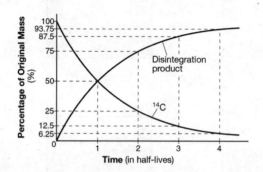

47 Radioactive carbon-14 is often useful in determining the absolute age of geologic samples because radioactive isotopes

(1) decay at a regular rate
(2) become less stable during decay
(3) remain unchanged over time
(4) stabilize after four half-lives

47 _____

48 Which disintegration product is represented on the graph?
(1) ^{206}Pb
(2) ^{87}Sr
(3) ^{40}Ar
(4) ^{14}N

48 _____

49 How many half-lives have passed if a sample contains 25% of its original carbon-14?

(1) 1 half-life
(2) 2 half-lives
(3) 3 half-lives
(4) 4 half-lives

49 _____

50 The age of which index fossil could be determined by using carbon-14?

50 _____

PART B–2
Answer all questions in this part.

Directions (51–65): Record your answers in the spaces provided. Some questions may require the use of the *2011 Edition Reference Tables for Physical Setting/Earth Science*.

Base your answers to questions 51 through 53 on the graph below and on your knowledge of Earth science. The graph shows the percentage of the lighted portion of the Moon that is visible to an observer in New York State through eight consecutive Moon phases.

51 The phases of the Moon are said to be waxing when the lighted portion of the Moon gradually increases over time. Identify the numbered phase of the Moon when waxing begins and the numbered phase when waxing ends. [1]

Waxing begins: Phase _____

Waxing ends: Phase _____

52 On the diagram *below*, place an **X** on the Moon's orbit to represent the Moon's position at phase 5. [1]

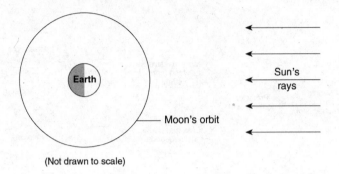

(Not drawn to scale)

53 The diagram below represents the appearance of the Moon at phase 3 as seen by an observer in New York State.

In the circle below, shade the part of the Moon that appears dark to an observer in New York State when the Moon is at phase 7. [1]

Base your answers to questions 54 through 57 on the flowchart below and on your knowledge of Earth science. The flowchart represents possible pathways in the evolution of stars.

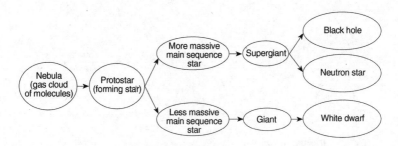

54 Based on this flowchart, identify the characteristic of a main sequence star that determines whether the star becomes a giant or a supergiant. [1]

_____

55 State the name of *one* star labeled on the *Characteristics of Stars* graph in the *Earth Science Reference Tables* that may become either a black hole or neutron star. [1]

_____

56 Identify the nuclear process that occurs when lighter elements in a star combine to form heavier elements, producing the star's radiant energy. [1]

_____

57 Circle the relative surface temperature and relative luminosity of the main sequence star *Sirius* compared with the surface temperature and luminosity of the Sun. [1]

Relative surface temperature of *Sirius* (circle one):

lower   **higher**   **the same**

Relative luminosity of *Sirius* (circle one):

less   **greater**   **the same**

Base your answers to questions 58 through 61 on the topographic map below and on your knowledge of Earth science. Points *A*, *B*, *C*, and *D* represent locations on Earth's surface. Elevations are measured in meters.

58 On the map *above*, draw the 200-meter contour line in the southern portion of the map. Extend the contour line to the edges of the map. [1]

59 On the grid *below*, construct a topographic profile along line *AB* by plotting the elevation of *each* contour line that crosses line *AB*. The elevations of points *A* and *B* have been plotted on the grid. Connect *all ten* plots with a line from *A* to *B* to complete the profile. [1]

60 Calculate the gradient between points *C* and *D*. Label your answer with the correct units. [1]

_____

61 Identify the compass direction toward which Kim Brook flows. Describe the evidence shown on the map that indicates the water flows downhill in that compass direction. [1]

Compass direction: _____

Evidence: _____

_____

Base your answers to questions 62 through 65 on the information below and on your knowledge of Earth science.

Adirondack Earthquake

On October 7, 1983, a magnitude 5.3 earthquake occurred in New York State's Adirondack region. The earthquake's epicenter was at Blue Mountain Lake, which is located approximately 32 miles (50 kilometers) southwest of Mt. Marcy.

62 Circle the New York State location (Old Forge *or* New York City) that recorded the greater amount of time between the arrival of the first *P*-wave and the arrival of the first *S*-wave from the Blue Mountain Lake earthquake. Explain why this location had the greater difference between the *P*-wave and *S*-wave arrival times. [1]

Circle one: **Old Forge**   **New York City**

Explanation: _____

_____

63 Determine how long it took the first *P*-wave to travel from Blue Mountain Lake to a seismic station 1200 kilometers away. [1]

_____ **min** _____ **s**

64 Identify *one* type of metamorphic surface bedrock where this earthquake epicenter was located. [1]

_____

65 Perth, Australia, is located almost directly on the opposite side of Earth from the epicenter of this earthquake. A seismograph in Perth received *P*-waves but *not* *S*-waves from this earthquake. Identify the interior layer of Earth and the characteristic of this layer that prevented the *S*-waves from arriving at Perth. [1]

Earth's interior layer: _____
Characteristic of this layer: _____

PART C
Answer all questions in this part.

Directions (66–85): Record your answers in the spaces provided. Some questions may require the use of the *2011 Edition Reference Tables for Physical Setting/Earth Science*.

Base your answers to questions 66 through 68 on the diagram below and on your knowledge of Earth science. The diagram represents Earth on the first day of a season. The equator, several lines of longitude, and the North and South Poles have been labeled. Letters *A* through *D* represent locations on Earth's surface.

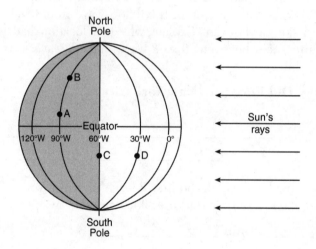

66 Identify *one* possible date that is represented by the position of Earth in this diagram. [1]

_____

67 State whether the relative altitude of *Polaris* at location A is lower or higher than at location B. Explain why this difference is observed. [1]

Relative altitude of *Polaris* at location A: _____

Explanation: _____

68 State the solar time at location D if the solar time at location C is 6:00 a.m. Indicate a.m. or p.m. in your answer. [1]

_____

Base your answers to questions 69 through 72 on the data table below, on the map below, and on your knowledge of Earth science. The data table shows latitude and longitude locations of the center of Hurricane Odile recorded at the same time each day from September 12 to September 18, 2014. The data table also shows the hurricane's barometric pressure in millibars (mb) and wind speed in knots (kt). The location of La Paz, Mexico, is indicated on the map.

Hurricane Odile

Date	Location		Barometric Pressure (mb)	Wind Speed (kt)
	Latitude (° N)	Longitude (° W)		
September 12	15	105	993	50
September 13	16	106	983	65
September 14	19	107	918	120
September 15	23	110	941	110
September 16	27	113	987	55
September 17	30	114	995	40
September 18	31	112	1003	25

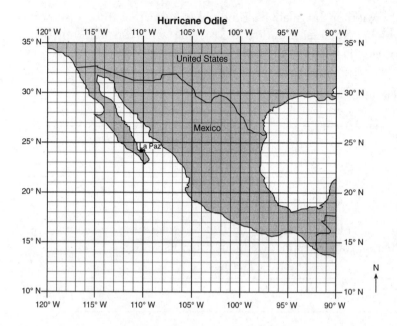

69 On the map *above*, plot the seven locations of Hurricane Odile indicated by the latitudes and longitudes shown in the data table. Connect *all seven* plots with a line. [1]

70 Using the set of axes *below*, draw a line to represent the general relationship between barometric pressure and wind speed associated with Hurricane Odile. [1]

71 Identify *one* weather instrument that was used to measure the wind speed of Hurricane Odile. [1]

_____

72 Describe *two* actions that a person living in La Paz, Mexico, could take to prepare for an approaching hurricane. [1]

Action 1: _____

Action 2: _____

_____

Base your answers to questions 73 through 75 on the cross section below and on your knowledge of Earth science. On the cross section, letters A through F represent rock units. Line XX' indicates an unconformity and line YY' indicates a fault. No overturning of rock layers has occurred.

73 Identify the name of the rock formed in the zone of contact metamorphism between rock units *A* and *F*. [1]

_____

74 List *two* processes that produced unconformity *XX'*. [1]

Process 1: _____

Process 2: _____

75 List the letters of rock units *D*, *E*, *F*, and fault *YY'* in the correct sequence from oldest to youngest. [1]

_____ → _____ → _____ → _____

Oldest ─────────────────────→ Youngest

Base your answers to questions 76 through 79 on the passage and cross section below, and on your knowledge of Earth science. The cross section represents the elevation and depth of each Finger Lake in New York State. The gray shading in each lake represents the sediment fill at the bottom of each lake.

Formation of the Finger Lakes

The Finger Lakes originated as a series of south-flowing rivers that existed in what is now central New York State. Around 1.7 to 1.8 million years ago, a continental glacier advanced southward, beginning an ice age that consisted of many advances and retreats of glaciers. The Finger Lakes were carved by several of these advances, which gouged the valleys of the rivers into deep troughs, some of which extend below present-day sea level. As the glaciers advanced, they pushed a great amount of soil and rock ahead of them. During the time when the ice sheets were melting and retreating, glacial moraines were left behind. These deposits dammed the glacial valleys at their southern ends, and the valleys then flooded to form the Finger Lakes.

76 Identify *one* Finger Lake that has a maximum water depth of approximately 175 feet. [1]

_____Lake

77 Identify the geologic time period and epoch when these continental glaciers formed the Finger Lakes of New York State. [1]

_____ **Period**

_____ **Epoch**

78 Describe *one* way that the arrangement of sediments in the glacial moraines is different from the arrangement of sediments deposited in the earlier river channels. [1]

_____

_____

79 Describe the cross-sectional shape of the original river valleys before they were gouged by the advancing glaciers. [1]

_____

_____

Base your answers to questions 80 and 81 on the map below and on your knowledge of Earth science. The map shows the locations of the Coast Range and Cascade Range in the Pacific Northwest of the United States and two cities in this region.

80 Identify the name of the cool surface ocean current that influences the climate of this region. [1]

_____ **Current**

81 Explain why the difference between average winter and summer temperatures is smaller in Long Beach than in Richland. [1]

_____

_____

_____

Base your answers to questions 82 through 85 on the diagram below and on your knowledge of Earth science. The diagram represents a portion of the scheme for igneous rock identification. Line *AB* represents the percentages of the minerals in igneous rock 1 and line *CD* represents the percentages of the minerals in igneous rock 2.

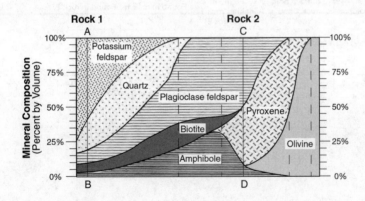

82 Rock 1 is a glassy, vesicular igneous rock. Identify the name of rock 1. [1]

_____

83 Compared to the color and density of igneous rock 1, describe how the relative color and density of igneous rock 2 are different. [1]

Relative color of rock 2: _____

Relative density of rock 2: _____

84 Explain why andesite and diorite can both have the same percentage of mineral composition by volume, and yet be two different igneous rocks. [1]

_____

_____

85 The table *below* shows the composition of three minerals commonly found in igneous rocks. Complete the table by placing one **X** in each row to indicate if that mineral is found in both rock 1 and rock 2, neither rock 1 nor rock 2, rock 1 only, or rock 2 only. [1]

Mineral Composition	Found in both rock 1 and rock 2	Found in neither rock 1 nor rock 2	Found in rock 1, only	Found in rock 2, only
$(Na,Ca)AlSi_3O_8$				
$KAlSi_3O_8$				
$(Fe,Mg)_2SiO_4$				

Answers August 2017
Physical Setting/Earth Science

Answer Key

PART A

1. 1	8. 3	15. 4	22. 2	29. 1
2. 4	9. 1	16. 1	23. 3	30. 4
3. 2	10. 3	17. 3	24. 4	31. 3
4. 3	11. 1	18. 1	25. 1	32. 3
5. 4	12. 2	19. 2	26. 2	33. 4
6. 2	13. 3	20. 4	27. 4	34. 4
7. 2	14. 1	21. 1	28. 4	35. 2

PART B–1

36. 2	39. 2	42. 3	45. 4	48. 4
37. 4	40. 4	43. 1	46. 2	49. 2
38. 3	41. 3	44. 2	47. 1	50. 3

PART B–2 and **PART C**. *See* **Answers Explained**.

Answers Explained

PART A

1. **1** A solar system consists of a single star, and all of the objects (such as planets) that orbit it. All of these objects have mass. Thus, a solar system contains more mass than a planet, such as Saturn. The Milky Way is a galaxy, which is a group of hundreds of billions of stars, many of which have solar systems. Thus, the Milky Way contains more mass than a solar system. The universe is everything that we can perceive and includes all of the stars, gas clouds, galaxies, and other objects that we can detect by the radiation they emit. The universe contains billions of galaxies. Thus, the universe has more mass than a galaxy. Thus, the correct sequence of celestial objects listed in order from smallest mass to largest mass is Saturn, solar system, Milky Way, universe.

2. **4** Find the chart labeled Electromagnetic Spectrum in the *Reference Tables for Physical Setting/Earth Science*. Note that visible light at the red end of the spectrum has a longer wavelength than light at the blue end of the spectrum. If a source of electromagnetic waves is moving *away* from an observer at the same time as it is emitting light of a particular wavelength, fewer wave crests will reach the eye of the observer each second. The eye will interpret this as meaning that the light has a longer wavelength than it actually has. In other words, the light will appear shifted toward the red end of the spectrum. Therefore, the fact that the light from most galaxies shows a red shift is evidence that most galaxies are moving away from Earth. If all galaxies are moving away from Earth, these galaxies are increasing in distance from Earth.

WRONG CHOICES EXPLAINED:
(1) and (2) Galaxies of many different sizes show a red shift because a red shift is the result of the motion of a light source relative to an observer, not the size of the source.

(3) If a galaxy's distance from Earth is decreasing, the galaxy is moving toward Earth. If a galaxy is moving toward an observer on Earth at the same time as it is emitting light of a particular wavelength, more wave crests will reach the eye of the observer each second. The eye will interpret this as meaning that the light has a shorter wavelength than it actually has. In other words, the light will appear shifted toward the blue end of the spectrum. Thus, if distant galaxies are decreasing in distance from Earth, their light would appear blue shifted, not red shifted.

3. **2** In 1851, the French physicist Jean Foucault suspended a heavy iron ball on a long steel wire from the top of the dome of the Pantheon in Paris. As this pendulum swung back and forth, it appeared to change direction slowly, passing over different lines on the floor until eventually coming full circle to its original position after 24 hours. Since Foucault knew that the path of a freely swinging pendulum would not change on its own, he concluded that the apparent shift in the direction of swing of the pendulum was due to the floor (Earth's surface) rotating beneath the pendulum. Thus, the apparent change in direction of swing of a Foucault pendulum provides evidence of Earth's rotation.

Planetary winds tend to blow in a straight line from regions of high pressure toward regions of low pressure. As these winds are blowing, Earth is turning on its axis. This rotation causes the winds to appear to be turning toward the right in the Northern Hemisphere and toward the left in the Southern Hemisphere. This phenomenon is called the Coriolis effect. Since there would be no Coriolis effect if Earth did not rotate, the Coriolis effect is also considered evidence of Earth's rotation.

Thus, the best evidence of Earth's rotation is provided by the Foucault pendulum and Coriolis effect.

WRONG CHOICES EXPLAINED:

(1) The Foucault pendulum does provide evidence of Earth's rotation. However, global warming is an increase in average global temperatures, not evidence that Earth is rotating.

(3) and (4) Moon phases refer to changes in the illuminated portion of the Moon visible to an observer on Earth. Moon phases occur because the Moon revolves into different positions relative to Earth and the Sun. The Moon would revolve around Earth even if Earth did not rotate. Therefore, Moon phases are not evidence of Earth's rotation.

4. **3** Oceans cover more than 70% of Earth's surface. Earth's hydrosphere consists of all the liquid water that rests on the lithosphere. Thus, the sphere of Earth that covers approximately 70% of Earth's surface is the hydrosphere.

WRONG CHOICES EXPLAINED:

(1) Earth's atmosphere is a thin shell of gases, water, dust, and other particles bound to Earth by gravity. Earth's atmosphere is in contact with all of the lithosphere and hydrosphere that comprise Earth's surface. Thus, the atmosphere covers 100% of Earth's surface.

(2) The lithosphere is Earth's solid, outer layer of rock. Earth's crust corresponds to the upper portion of the lithosphere. Thus, the lithosphere comprises 100% of Earth's solid surface.

(4) Find the Inferred Properties of Earth's Interior chart in the *Reference Tables for Physical Setting/Earth Science*. In the cross section, locate the asthenosphere. Note that the asthenosphere is located beneath Earth's surface. Thus, the asthenosphere covers 0% of Earth's surface.

5. **4** As Earth revolves around the Sun, the side of Earth facing the Sun experiences day and the side facing away from the Sun experiences night. Midnight occurs when a location on Earth is directly opposite the Sun in the center of the night side of Earth. At different times of the year, Earth is at different points along its orbit and the night side of Earth faces different portions of the universe. Since stars are visible only at night, only constellations in the portion of the universe facing Earth's night side are visible from Earth. In December, Earth's night side faces one portion of the universe and its constellations. Six months later, in June, Earth has revolved to a position in its orbit directly opposite the Sun from its December position. The night side of Earth faces a different portion of the universe and its constellations. Therefore, the constellations visible to New York State observers at midnight in December are different than those visible at midnight in June because Earth revolves around the Sun.

WRONG CHOICES EXPLAINED:
(1) A constellation is an *imaginary* pattern formed by a group of stars in an area of the sky as seen by an observer on Earth. Therefore, constellations do not have an axis about which they rotate.

(2) Constellations consist of stars. Earth's small mass does not exert enough gravitational force to hold larger, more massive objects such as stars in orbit.

(3) As Earth spins on its axis, the boundary between daylight and darkness does not spin along with it; Earth's night side is always located directly opposite the Sun. Only Earth's motion in relation to the Sun (revolution) affects the direction in which the night side faces and the constellations visible to an observer, not Earth's rotation on its axis.

(Not drawn to scale)

6. **2** The gravitational attraction of the Moon creates tidal bulges in Earth's oceans on the side of Earth directly facing the Moon and on the side of Earth directly facing away from the Moon. As Earth rotates on its axis, the positions of the tidal bulges remain fixed in line with the Moon. Locations experience high tide when they align with the bulges and experience low tide when between bulges. According to the diagram, location *D* is on the side of Earth directly facing the Moon and location *B* is on the side of Earth directly facing away from the Moon. Therefore, tidal bulges and their associated high tides will likely be located at *B* and *D*. Locations *A* and *C* are between the bulges and will experience low tides. Thus, high tides will be located at *B* and *D*; low tides at *A* and *C*.

7. **2** Urbanization is the process by which towns and cities are formed and become larger as more and more people begin living and working in central areas. As towns and cities grow, more and more of the land surface is paved over or covered by buildings and other structures. Vegetation is not able to grow on land covered by pavement and buildings. Therefore, urbanization results in a decrease in vegetation. Pavement and the roofs of buildings are impermeable. Therefore, rainwater that falls onto pavement and buildings runs off rather than infiltrates. Thus, urbanization decreases vegetation and increases runoff.

8. **3** A dangerous form of solid precipitation associated with severe thunderstorms is hail. In the upper portion of the troposphere, temperatures below −40°C are the norm year-round. For this reason, thunderstorms whose cloud tops extend into the upper troposphere may produce hail. Hailstones form when water droplets carried into the upper troposphere freeze to form tiny balls of ice (hailstones) and then fall. Again and again, hailstones are hurled up by updrafts and then fall through layers of air that alternate above and below freezing. Each cycle adds a layer to the hailstone. The more violent the updrafts, the larger and heavier the hailstones can become before falling. Large hailstones are dangerous because they can cause great damage when they strike crops, buildings, and vehicles. In some cases, hailstones have resulted in loss of life. Find the Key to Weather Map

Symbols in the *Reference Tables for Physical Setting/Earth Science*. Locate the section labeled "Present Weather." Note the symbol representing hail, △, which corresponds to the symbol shown in choice (3).

9. **1** Find the Key to Weather Map Symbols in the *Reference Tables for Physical Setting/Earth Science,* and locate the section labeled "Station Model Explanation." Note that barometric trend over the past three hours is shown to the center right of a station model. According to the station model explanation, a "+" indicates an increase in barometric pressure. The two-digit value indicates the number of millibars that the barometric pressure has changed but does not include a decimal point. The decimal should be placed between the first and second digits. Finally, the "Station Model Explanation" indicates that an upward-sloping line after the millibar value means a steady rise. The value representing the barometric trend for the past three hours at the location in New York State shown on the station model in the question is "+02/." Therefore, the barometric trend for the past three hours at this location indicates a steady increase of 0.2 mb.

10. **3** Winter lake-effect snow storms occur when cold winds move across large stretches of warm lake water. Evaporation of the warm lake water adds water vapor to the air. The water vapor is picked up by the cold wind, freezes, and is deposited as snow onto the cold land of the leeward shores of the lake. Find the Generalized Bedrock Geology of New York State map in the *Reference Tables for Physical Setting/Earth Science.* Locate Lake Ontario, and note that it is located at about 43°–44° N latitude. Find the Planetary Wind and Moisture Belts in the Troposphere diagram in the *Reference Tables for Physical Setting/Earth Science.* Note that at the latitudes corresponding to Lake Ontario, the planetary winds generally blow from west to east. Thus, the cold winds are moving from west to east over the lake. So the eastern shores of Lake Ontario are the leeward shores that would experience lake-effect snow storms. On the Generalized Bedrock Geology of New York State map in the *Reference Tables for Physical Setting/ Earth Science,* locate the eastern shores of Lake Ontario. Note that of the cities listed as answer choices, only Oswego is located on the eastern shores of Lake Ontario. Thus, the New York State location most often affected by lake-effect snow storms caused by winds blowing over Lake Ontario is Oswego.

11. **1** The characteristics of an air mass are the result of the geographical region over which it formed, or its *source region*. Air resting on or moving very slowly over a region tends to take on the characteristics of that region. In general, air masses that form near the poles are cold and are called polar air masses; air masses that form near the equator are warm and are called tropical air masses. Air masses that form over water are moist and are called maritime air masses; air masses that form over land are dry and are called continental air masses.

The type of air mass that would most likely form over the Pacific Ocean is a maritime air mass. Find the Tectonic Plates map in the *Reference Tables for Physical Setting/Earth Science*. Locate the Aleutian Trench. Trace horizontally right to the latitude scale along the vertical edge of the map. Note that the Aleutian Trench is located at about 60° N latitude (in the Arctic Circle). Thus, the type of air mass that would form north of the Aleutian Trench is a polar air mass. Therefore, an air mass that formed over the Pacific Ocean north of the Aleutian Trench would most likely be a maritime polar air mass. Find the Key to Weather Map Symbols in the *Reference Tables for Physical Setting/Earth Science*. In the section labeled "Air Masses," note that the symbol for a maritime polar air mass is mP, as shown in choice (1).

12. **2** Temperature is a major climate factor. Mountains are, by definition, regions of high elevation. Find the Selected Properties of Earth's Atmosphere chart in the *Reference Tables for Physical Setting/Earth Science*. Locate the graph labeled "Temperature (°C)." Note that temperature decreases rapidly with increasing altitude (elevation) in the troposphere. Elevation is a climate factor because there is a decrease in temperature with an increase in elevation. The fact that temperature decreases about 1°C for every 100-meter rise in elevation explains why high mountains may have tropical vegetation at their bases but permanent ice and snow at their peaks. Thus, the climate factor that best explains the presence of permanent snow on the peak of Mount Kilimanjaro is elevation (altitude above sea level).

WRONG CHOICES EXPLAINED:

(1) It is given that the latitude of Mount Kilimanjaro is 3° S, which is near the equator. Near the equator, insolation strikes Earth's surface almost vertically. Therefore, throughout the year, the insolation reaching the tropics is more concentrated (i.e., more intense) than that reaching midlatitude or polar regions. The more intense the insolation, the more Earth's surface is warmed by that insolation. More warming would likely melt snow, not allow the snow to remain permanently. Therefore, it is unlikely that the latitude of Mount Kilimanjaro is the climate factor that best explains the permanent snow on the mountain's peak.

(3) Find the Planetary Winds and Moisture Belts in the Troposphere diagram in the *Reference Tables for Physical Setting/Earth Science*. Note that 3° S latitude lies in the planetary wind belt labeled "S.E. Winds." Find the Surface Ocean Currents map in the *Reference Tables for Physical Setting/Earth Science*. In the key, note the symbols for warm currents and cool currents. Note that the surface ocean currents off the coast of east Africa are all warm ocean currents. Thus, the prevailing winds that reach Mount Kilimanjaro would come from over warm water. Air moving over warm water becomes warm and moist. Warm, moist winds would not explain the cold temperatures necessary to support the permanent snow on the peak of Mount Kilimanjaro.

(4) Find the Surface Ocean Currents map in the *Reference Tables for Physical Setting/Earth Science*. In the key, note the symbols for warm currents and cool currents. Note that the surface ocean currents off the coast of east Africa are all warm ocean currents. Warm ocean currents would not explain the cold temperatures necessary to support permanent snow on the peak of Mount Kilimanjaro.

13. **3** The wavelength of radiation emitted by a source depends on the temperature of the source. The cooler the source, the longer the wavelength of the radiation it emits. At Earth's average surface temperature, the radiation emitted would be in the range of long-wavelength infrared radiation. Thus, the maximum intensity of Earth's outgoing radiation is in the infrared portion of the electromagnetic spectrum.

14. **1** A major component of insolation is visible light. Color is a fairly good indicator of whether the substance absorbs more light than it reflects. Light-colored substances tend to reflect light. Dark-colored substances tend to absorb light. A smooth surface is more likely to reflect light than is a rough surface. The irregularities on a rough surface can cause some of the light to hit the surface several times before leaving. Each time the light strikes the irregular surface, a little more of the light's energy is absorbed. Therefore, a dark-colored and rough surface would most likely absorb the greatest amount of visible light. Thus, a dark-colored and rough tubing exterior will best absorb insolation.

15. **4** Find the Specific Heats of Common Materials chart in the *Reference Tables for Physical Setting/Earth Science*. Note that specific heat is measured as $J/g \cdot °C$. This means that the numerical value of each material's specific heat is the number of joules of heat energy that must be added to 1 gram of the substance to increase that material's temperature by 1°C. For example, if 0.79 J/g of heat energy are added to 1 gram of granite, the temperature of the granite will increase

by 1°C. However, if 0.79 J of heat energy is added to 1 gram of lead, the temperature of the lead will increase by about 6°C (0.79 J/g ÷ 0.13 J/g · °C = 6.07°C). Thus, the smaller the specific heat value, the smaller the amount of heat energy needed to raise the temperature of the material by 1°C and the faster the material will warm up. It is given that equal masses of the materials were placed in sunlight. It is reasonable to infer that all received the same amount of energy from the sunlight per unit of time. Therefore, the material that will warm up the fastest is the one with the smallest specific heat value. Of the choices given, the material with the smallest specific heat is lead.

16. **1** Volcanic ash consists of tiny, solid particles of igneous rock. These solid particles block sunlight, thereby decreasing the amount of insolation that reaches Earth's surface. Thus, the most likely effect of this ash cloud is a decrease in insolation reaching Earth's surface.

WRONG CHOICES EXPLAINED:
(2) and (4) Find the Selected Properties of Earth's Atmosphere chart in the *Reference Tables for Physical Setting/Earth Science*. Note that Earth's atmospheric layers correspond to temperature zones within the atmosphere. In each zone, the temperature profile is different—with temperature either increasing or decreasing with altitude. The tiny particles of igneous rock in a volcanic ash cloud cool rapidly. By the time they reach the stratosphere, they would have cooled to the surrounding temperature and spread out. Therefore, a volcanic ash cloud would not change the temperature profile upon which the thickness of the stratosphere is based. Thus, the ash cloud would not change the thickness of the stratosphere.

(3) Solid particles of volcanic ash block sunlight. The more sunlight that is blocked by particles, the less that reaches Earth's surface. Thus, the most likely effect of this ash cloud is to decrease, not increase the insolation reaching Earth's surface.

17. **3** Find the Tectonic Plates map in the *Reference Tables for Physical Setting/Earth Science*. Locate the region corresponding to the map in the question. Note that letter *A* is located on the Pacific Ocean floor of the Nazca Plate. Locations *B* and *C* are located on ocean floor on either side of the Mid-Atlantic Ridge. Location *D* is on the continent of Africa. The youngest crustal bedrock is formed when molten rock emerges from volcanoes and in the rift zones of the mid-ocean ridges and then hardens into bedrock. As the tectonic plates diverge along a mid-ocean ridge, the bedrock fractures and new molten rock emerges, pushing the bedrock aside. As this process is repeated, new bedrock is formed, and the bedrock on either side of the mid-ocean ridges is pushed sideways. Thus, the youngest, most recently formed bedrock is found nearest a mid-ocean ridge, and the oldest bedrock is found on either side farthest from the mid-ocean ridge. Therefore, the location that most likely has the youngest bedrock will be nearest a mid-ocean ridge. Location *C* is nearest a mid-ocean ridge and will most likely have the youngest bedrock.

WRONG CHOICES EXPLAINED:

(1) Find the Tectonic Plates map in the *Reference Tables for Physical Setting/Earth Science*. Note that the tectonic plate boundary along the west coast of South America near location *A* corresponds to a convergent plate boundary (subduction zone). Note that location *A* is near the edge of the subducting plate, which is being forced beneath the overriding South American Plate. Therefore, the bedrock near location *A* is already existing bedrock, not newly formed, young bedrock.

(2) Recall that the youngest, most recently formed bedrock is found nearest a mid-ocean ridge and the oldest bedrock is found on either side farthest from the ridge. Note that *B* is farther from the Mid-Atlantic Ridge than *C*. Thus, the bedrock at *B* is older than the bedrock at *C* and, therefore, is not the youngest bedrock.

(4) Find the Geologic History of New York State chart in the *Reference Tables for Physical Setting/Earth Science*. Locate the column labeled "Inferred Positions of Earth's Landmasses." Note that the African landmass existed for

hundreds of millions of years prior to the initial opening of the Atlantic Ocean. Thus, it is unlikely that the bedrock at D on the African landmass is composed of the youngest crustal bedrock.

18. **1** Find the Inferred Properties of Earth's Interior chart in the *Reference Tables for Physical Setting/Earth Science*. Locate the "Density (g/cm^3)" scale along the right side of the cross section. Note that granitic continental crust has a density of 2.7 and that basaltic oceanic crust has a density of 3.0. Thus, continental crust is granitic while oceanic crust is basaltic. Note, too, that continental crust is less dense than oceanic crust. Find the Scheme for Igneous Rock Identification in the *Reference Tables for Physical Setting/Earth Science*. Locate granite and basalt in the upper portion of the chart. Trace the columns corresponding to granite and basalt downward to the arrow labeled "Composition" in the center portion of the chart labeled "Characteristics." Note that granite is near the end of the arrow labeled "Felsic (rich in Si, Al)" and that basalt is near the end of the arrow labeled "Mafic (rich in Fe, Mg)." Thus, granitic continental crust is more felsic than basaltic ocean crust. Therefore, compared to the average density and composition of oceanic crust, continental crust is less dense and more felsic.

19. **2** Find the Tectonic Plates map in the *Reference Tables for Physical Setting/Earth Science*. Note the symbol labeled "Hawaii Hot Spot" over the islands of Hawaii. Locate this symbol in the key along the bottom of the Tectonic Plates map, and note that it corresponds to a mantle hot spot. A hot spot is a long-lasting zone of rising hot magma beneath moving plates. Large plumes of magma rise from the mantle at these hot spots and work their way upward through the plate above. The magma rises because it is more buoyant than the surrounding rock, wedges apart cracks in the plate, and melts through to erupt, forming a volcano. When the volcano rises above the level of the ocean, it forms an island. As the plate continues to move, the volcano that formed over the hot spot is carried away from the hot spot in the direction in which the plate is moving. At the same time, the plate motion carries a new section of the plate over the hot spot and a new volcano forms. As this process continues, a series of volcanic islands forms with the youngest located nearest the hot spot and the oldest farthest from the hot spot. Thus, the Hawaiian Islands were formed as a result of an oceanic plate moving over a mantle hot spot.

WRONG CHOICES EXPLAINED:
(1) A plate boundary where two tectonic plates move apart is called a divergent plate boundary. Divergent boundaries occur along plate edges. Find the Tectonic Plates map in the *Reference Tables for Physical Setting/Earth Science*.

Note that the Hawaiian Islands are located in the center of the Pacific Plate far from any divergent boundaries along the edges of the plate. Thus, it is unlikely that the Hawaiian Islands formed as a result of lava flowing over Earth's surface where two tectonic plates move apart.

(3) A collision between two oceanic plates would take place at the edges of the two plates. Find the Tectonic Plates map in the *Reference Tables for Physical Setting/Earth Science*. Note that the Hawaiian Islands are located in the center of the Pacific Plate far from the edges of the plate. Therefore, it is unlikely that the Hawaiian Islands formed as a result of two oceanic plates colliding to form an island arc.

(4) Tectonic plates slide past each other along their edges. Find the Tectonic Plates map in the *Reference Tables for Physical Setting/Earth Science*. Note that the Hawaiian Islands are located in the center of the Pacific Plate far from the edges of the plate. Therefore, it is unlikely that the Hawaiian Islands formed as a result of tectonic plates sliding past each other.

20. **4** Water flows from higher elevations to lower elevations. Therefore, water flows downhill in channels along either side of the tops of the long, parallel ridges shown in the block diagram into the valleys between the ridges, forming larger streams flowing down the centers of the valleys parallel to the ridges. Thus, the most likely stream drainage pattern on the surface of this landscape is shown in diagram (4).

21. **1** Mountain landscape regions consist of mountains. By definition, mountains are parts of Earth's crust that project at least 300 meters above the surrounding land and have great relief (steep slopes), a restricted summit, and considerable bare-rock surface. Most mountains are formed by crustal motions that fold or fault rock and have an underlying bedrock structure composed of folded and faulted rock layers. Thus, the characteristics that identify mountain landscape regions are steep slopes with deformed bedrock.

22. **2** Sandblasting refers to using a stream of sand projected by compressed air or steam to clean, polish, or decorate the surface of something by abrasion. In nature, windblown sand directed against bedrock can have a similar effect. Smooth, polished surfaces on bedrock in dry, sandy areas are most likely the result of abrasion by wind-borne sand. Thus, the agent of erosion that causes sandblasting of bedrock is wind.

23. **3** On steep slopes, the dominant type of erosion that occurs is mass wasting. In erosion by mass wasting, fragments of rock broken loose by weathering move downslope under the direct influence of gravity. As you can see in the photograph, the road is at the bottom of the steep slope. Any rocks on the steep slope that are broken loose by weathering would be carried downhill to the roadway. If rocks fell on or in the path of cars traveling on the roadway, they could cause serious damage to the cars and possibly loss of life. Accidents resulting from cars swerving to avoid the rocks could also cause loss of property or life. Thus, the wire netting was installed to prevent rocks that were broken loose by weathering from falling onto the roadway. Therefore, the wire netting has been installed to prevent loss of property or life resulting from weathering and erosion.

WRONG CHOICES EXPLAINED:
(1) Crosscutting refers to a geologic feature that cuts across another geologic feature. It does not involve rocks on a steep slope that would be held back from falling by wire netting. Downwarping refers to a bending of Earth's crust that results in a bowl-like depression, not rocks on a steep slope that would be held back from falling by wire netting.

(2) Folding and faulting result from enormous forces acting on Earth's crust. These forces are too enormous for wire netting to constrain the crust from folding or faulting.

(4) The wire netting would not prevent high winds or flooding because wind and water would be able to easily pass through the large openings in the wire netting.

24. **4** A naturally occurring solid material from which a metal or valuable mineral can be profitably extracted is called an ore. Find the Properties of Common Minerals table in the *Reference Tables for Physical Setting/Earth Science*. In the column labeled "Use(s)," locate the phrase "ore of iron." Note that "ore of iron" occurs in two rows. Trace each of these rows to the right to the column labeled "Mineral Name." Note that two minerals that are ores of iron are magnetite and hematite. Thus, two minerals that are commercial sources of iron are hematite and magnetite.

25. **1** Find the Scheme for Metamorphic Rock Identification in the *Reference Tables for Physical Setting/Earth Science*. In the column labeled "Rock Name," locate phyllite. From phyllite, trace left horizontally to the column labeled "Composition." Note that phyllite is composed of quartz, feldspar, amphibole, garnet, and pyroxene. Now find the Scheme for Sedimentary Rock Identification in the *Reference Tables for Physical Setting/Earth Science*. In the column labeled "Rock Name," locate sandstone. From sandstone, trace left horizontally to the column labeled "Composition." Note that sandstone is "mostly quartz, feldspar, and clay minerals." Finally, find the Scheme for Igneous Rock Identification in the *Reference Tables for Physical Setting/Earth Science*. Locate granite in the upper part of the chart. Trace the column containing granite vertically down to the section labeled "Mineral Composition (relative by volume)." Note that granite is composed of the minerals potassium feldspar, quartz, plagioclase feldspar, biotite, and amphibole. Thus of the choices, the mineral that can be found in the rocks phyllite, sandstone, and granite is quartz.

26. **2** Earth's axis of rotation is tilted at an angle of 23.5° from a perpendicular to Earth's orbital plane. This position relative to Earth's orbital plane is best represented in diagram (2).

WRONG CHOICES EXPLAINED:

(1) In this diagram, Earth's axis is perpendicular to Earth's orbital plane instead of being tilted 23.5° from a perpendicular to the orbital plane.

(3) In this diagram, Earth's axis is tilted 23.5° from the orbital plane itself, not 23.5° from a perpendicular to the orbital plane.

(4) In this diagram, Earth's axis of rotation is aligned with Earth's orbital plane instead of being tilted 23.5° from a perpendicular to the orbital plane.

27. **4** On a hot, sunny, summer afternoon, sunlight causes land surfaces to increase in temperature more than water surfaces because water has a higher specific heat than land. The cooler air over the water exerts more pressure than the warmer air over the land. Therefore, air pressure over the water will be high, and air pressure over the land will be low. As a result, the air will move from the region of higher air pressure over the water toward the region of lower air pressure over the land, as shown in cross section (4).

Type of Precipitation	Average Diameter (millimeters)	Falling Velocity (meters/second)
drizzle	0.96	4.1
light rain	1.24	4.8
moderate rain		
heavy rain	2.05	6.7
excessive rain	2.40	7.3

28. **4** By interpolation, it can be inferred that the average diameter and falling velocity of moderate rain will be between the average diameters and falling velocities of light rain and heavy rain. According to the table, the average diameter of light rain is 1.24 millimeters and the average diameter of heavy rain is 2.05 millimeters. Therefore, the average diameter of moderate rain will be between 1.24 and 2.05 millimeters. According to the table, the falling velocity of light rain is 4.8 meters/second and the falling rate of heavy rain is 6.7 meters/second. Therefore, the falling velocity of moderate rain will be between 4.8 and 6.7 meters/second. Only choice (4) has values that fall into these ranges: average diameter = 1.60 mm, falling velocity = 5.7 m/s.

29. **1** Find the Geologic History of New York State chart in the *Reference Tables for Physical Setting/Earth Science*. Among the diagrams of index fossils along the bottom of the chart, locate the four index fossils shown in the answer choices. Note the circled letter associated with each of the four index fossils. Note the index fossil corresponding to each choice: 1—Ⓣ, 2—Ⓟ, 3—Ⓘ, and 4—Ⓧ. In the section of the chart labeled "Time Distribution of Fossils (including important fossils of New York)," locate the vertical gray bar labeled "Corals." Note that

the circled letters representing index fossils on this bar are Ⓣ, Ⓤ, and Ⓥ. Thus, the index fossils at the bottom of the chart corresponding to the circled letters Ⓣ, Ⓤ, and Ⓥ are corals. Therefore, of the choices given, the New York State index fossil that is classified as a coral is Ⓣ, which corresponds to the index fossil shown in choice (1).

30. **4** Note the region in the map labeled "Warm water" adjacent to the west coast of South America. The warm water along the west coast of South America warms the air that comes into contact with it. Therefore, atmospheric temperature along the west coast of South America increases. The higher the temperature, the higher the evaporation rate. So the moisture level of the air over the warm water increases. The higher temperature and moisture levels result in air with a low density. As this low-density air rises and cooler, denser air rushes in to replace it, numerous convection cells and thunderstorms form. Therefore, there is an increase in precipitation. Thus, the most likely changes to atmospheric temperature and precipitation along the west coast of South America during El Niño conditions are higher temperatures and higher amounts of precipitation.

WRONG CHOICES EXPLAINED:
(1) and (2) The presence of warm water along the west coast of South America during El Niño conditions would raise temperatures, not lower temperatures.

(3) The presence of warm water along the west coast of South America warms the air and results in higher temperatures. Higher temperatures cause an increase in evaporation rates, which adds more moisture to the air. The more moisture the air contains, the greater the chance of precipitation. Therefore, during El Niño conditions, the amounts of precipitation would be higher, not lower.

Outcrop 1 Outcrop 2
(Not drawn to scale)

31. **3** Note that in the outcrops, all of the rock units are flat and horizontal. The presence of fossils indicates that these are sedimentary rocks. The principle of superposition states that the bottom layer of a sedimentary series is the oldest, unless it was overturned or had older rock thrust over it, because the bottom layer was deposited first. Similarly, in a sequence of rock layers, a rock layer is older than those above it and is younger than those below it. Therefore, in outcrop 1, the layers from oldest to youngest are 4, 3, 2, and 1. In outcrop 2, the layers from oldest to youngest are 9, 8, 7, 6, and 5.

Rock layers can sometimes be correlated on the basis of distinct similarities in physical characteristics such as composition, color, thickness, and fossil remains. Thus, it is reasonable to correlate the rock layers in these two outcrops by their composition and fossil content.

Find the Scheme for Sedimentary Rock Identification in the *Reference Tables for Physical Setting/Earth Science*. In the column labeled "Map Symbol," locate the map symbols representing the rock layers in the outcrops. From each map symbol, trace left to the column labeled "Rock Name." Note the name of the rock corresponding to each map symbol present in the outcrops: rock layers 1, 4, 5, and 8—shale; rock layers 2, 6, and 9—sandstone; rock layers 3 and 7—limestone.

Now find the Geologic History of New York State chart in the *Reference Tables for Physical Setting/Earth Science*. Among the diagrams of index fossils along the bottom of the chart, locate the five index fossils shown in rock layers 1, 3, 4, 6, and 8. Note the circled letter associated with each of the five index fossils: layer 1—Ⓡ, layer 3—Ⓥ, layer 4—Ⓤ, layer 6—Ⓩ, and layer 8—Ⓨ. Locate the column labeled "Time Distribution of Fossils (including important fossils of

New York)." Note that at the top of the column it states: "The center of each lettered circle indicates the approximate time of existence of a specific index fossil (e.g., Fossil Ⓐ lived at the end of the Early Cambrian)." Locate the lettered circles Ⓡ, Ⓥ, Ⓤ, Ⓨ, and Ⓩ. From each lettered circle, trace horizontally to the left to the column labeled "Period." Note that fossil Ⓡ lived during the late Devonian, fossil Ⓥ during the early Devonian, fossil Ⓤ during the early Silurian, fossil Ⓨ during the early Silurian, and fossil Ⓩ during the middle Devonian. Since layers 4 and 8 both formed during the early Silurian, they are the same age. Therefore, the shale rock layer 4 in outcrop 1 can be correlated with the shale rock layer 8 in outcrop 2. Using the principle of superposition, we can now correlate the ages from youngest to oldest of the rock layers of the two outcrops as follows: shale layers 1 and 5; sandstone layers 2 and 6; limestone layers 3 and 7; shale layers 4 and 8; sandstone layer 9. Thus, of the choices given, the two layers that most likely formed during the same geologic time period are 3 and 7.

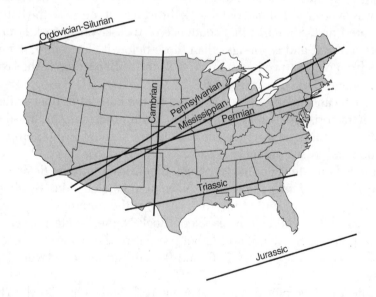

32. **3** Find the Generalized Bedrock Geology of New York State map in the *Reference Tables for Physical Setting/Earth Science*. Locate Watertown, New York, and note its relative position in New York State. On the map in the question, locate the position that corresponds to Watertown, New York. Note that the line showing the inferred position of Earth's equator during the middle of several geologic periods closest to this position is labeled "Mississippian." Find the Geologic History of New York State chart in the *Reference Tables for Physical Setting/*

Earth Science. In the column labeled "Period," locate "Mississippian." From Mississippian, trace right to the time scale labeled "Million years ago," and note that the Mississippian extended from 318 to 359 million years ago. Of the choices, only 340 mya falls within this range. Thus, the region around current-day Watertown, New York, was located nearest to the equator 340 million years ago.

33. **4** Find the Scheme for Sedimentary Rock Identification in the *Reference Tables for Physical Setting/Earth Science*. Find the column labeled "Rock Name," and locate the names of the sedimentary rocks listed in the tables. From each rock name, trace left to the column labeled "Texture," and note the texture of that rock. For example, in choice (1), sandstone has a clastic texture, breccia has a clastic texture, and rock gypsum has a crystalline texture. Thus, the table in choice (1) does not correctly match rock textures with a sedimentary rock that exhibits each texture. In this way, check the names and textures of the rocks in each of the tables. Note that only the table in choice (4) correctly matches rock textures with a sedimentary rock that exhibits each texture.

Temperatures Measured with a Psychrometer

Day	1	2	3	4
Dry-bulb temperature (°C)	0	5	10	15
Wet-bulb temperature (°C)	−5	0	5	10

34. **4** According to the data table, the difference between the wet-bulb and the dry-bulb temperatures on each of the four days was 5°C. Find the Relative Humidity (%) chart in the *Reference Tables for Physical Setting/Earth Science*. Note the vertical scale labeled "Dry-Bulb Temperature (°C)" along the left side of the chart. Now locate the column labeled "5" along the horizontal scale at the top of the chart labeled "Difference Between Wet-Bulb and Dry-Bulb Temperatures (°C)." Trace this column downward. Note that as the dry-bulb temperature increases, the values in this column also increase. Thus, the higher the dry-bulb temperature when the difference between the wet-bulb and dry-bulb temperature is 5°C, the higher the relative humidity. Thus, the day that had the highest relative humidity was day 4.

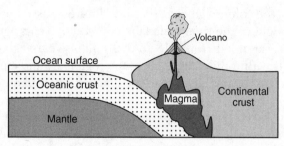

(Not drawn to scale)

35. **2** According to the cross section, the oceanic crust is plunging beneath the continental crust. This is consistent with a subduction zone. Recall that in a subduction zone, denser ocean crust collides with and is pushed beneath less dense continental crust. During this collision, the edge of the continental crust crumples and is uplifted as the denser ocean crust plunges beneath it. As the plunging plate is forced into the mantle and melts, molten plate material rises beneath the overlying crust, causing further uplift. If the molten material reaches the surface through cracks, volcanoes form. Find the Tectonic Plates map in the *Reference Tables for Physical Setting/Earth Science*. In the key along the bottom of the map, locate "Convergent plate boundary (subduction zone)." At a convergent plate boundary, the plates are moving toward one another and colliding. Therefore, the arrows showing the relative directions of plate movement should be pointed toward one another as in diagram (2).

PART B–1

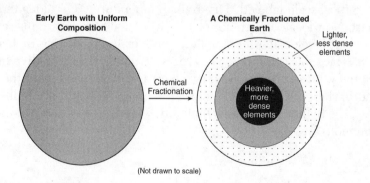

(Not drawn to scale)

36. **2** The force of gravitational attraction between any two objects is directly proportional to their masses and indirectly proportional to the square of the distance between them. Density is the ratio of mass to volume. Denser elements

have more mass per unit volume than less dense elements. Chemical fractionation occurred while Earth was molten, that is, while Earth was a vast globe of hot liquid. Now imagine that in this vast, hot liquid, there is a volume of denser elements and an equal volume of less dense elements that are the same distance from Earth's center. The volume of denser elements will have more mass. Therefore, there will be a greater force of gravitational attraction between Earth and the volume of denser elements than between Earth and the volume of less dense elements. The denser elements will be drawn more forcefully toward Earth's center and sink in the liquid Earth, displacing the less dense elements upward. Thus, chemical fractionation is most likely caused by gravity.

WRONG CHOICES EXPLAINED:

(1) Solidification is a change from the liquid phase to the solid phase. Chemical fractionation occurred while Earth was in the liquid (molten) phase, before solidification occurred. Therefore, chemical fractionation was not caused by solidification.

(3) A magnetic force would act on only magnetic elements. Most elements are not magnetic. Additionally, the shape of Earth's magnetic field is very different from the layering shown by the chemically fractionated Earth. Therefore, it is unlikely that chemical fractionation was caused by magnetic force.

(4) Chemical weathering is the breaking down of rock by changing its chemical composition. Rocks are solids. Chemical fractionation occurred while Earth was in the liquid (molten) phase. Therefore, chemical fractionation was not caused by chemical weathering.

37. **4** The question states that chemical fractionation occurred early in Earth's formation. Find the Geologic History of New York State chart in the *Reference Tables for Physical Setting/Earth Science*. In the column labeled "Era," locate the statement "Estimated time of origin of Earth and the solar system." Trace left to the time scale labeled "Million years ago," and note that this occurred 4600 million years ago. Thus, Earth and the other planets in our solar system began the process of chemical fractionation approximately 4600 million years ago.

38. **3** According to the diagram labeled "A Chemically Fractionated Earth," the heavier, denser elements are located nearest Earth's center. Find the Inferred Properties of Earth's Interior chart in the *Reference Tables for Physical Setting/Earth Science*. In the cross section at the top of the chart, note that the layer nearest Earth's center is labeled "Inner core (iron & nickel)." Thus, the pair of elements that sank to Earth's center during chemical fractionation was iron and nickel.

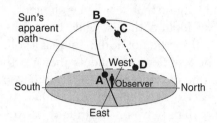

39. **2** The Sun's apparent path through the sky from sunrise to sunset is an arc, or part of a circle, like all other celestial objects. The Sun appears to move across the sky from east to west at a constant rate of 15° per hour, or one complete circle every day (24 hours/day × 15°/hour = 360°/day) due to Earth's rotation on its axis.

WRONG CHOICES EXPLAINED:
(1) Earth revolves around the Sun once every 365.26 days, or about 1° per day. If the apparent path of the Sun was caused by Earth's revolution around the Sun, the Sun would appear to move at a rate of 1° per day. However, the Sun appears to move across the sky from east to west at a constant rate of 15° per hour. Therefore, the apparent path of the Sun is not caused by Earth's revolution around the Sun.

(3) The Sun does not revolve around Earth. Therefore, this motion could not cause the apparent rising and setting of the Sun as viewed from Earth.

(4) In order for the Sun to change position in the sky, it must change position relative to an observer on Earth. The Sun's rotation on its axis does not change its position relative to an observer on Earth. The Sun could remain stationary to an observer on Earth and still rotate. Therefore, the apparent path of the Sun through the sky cannot be caused by the Sun's rotation on its axis.

40. **4** As the angle at which the Sun's rays strike Earth's surface decreases, the length of the shadow cast by an object increases. Thus, an observer's shadow is longest when the altitude of the Sun is lowest, that is, when the Sun is closest to the horizon. According to the diagram, the Sun is closest to the horizon at position *D*. Thus, the observer has the longest shadow when the Sun is at position *D*.

41. **3** The Sun appears to move across the sky from east to west at a constant rate of 15° per hour. Thus, sunrise occurs along the eastern horizon, and sunset occurs along the western horizon. Therefore, the Sun moves across the sky from position *A*, to *B*, to *C*, and then to *D*. Solar noon occurs when the Sun reaches its highest point in the sky for the day. At this moment, the Sun crosses an observer's meridian, a semicircle connecting the North and South Poles that passes through the observer's zenith. Therefore, position *B* represents solar noon. Since the Sun

arrives at position C after position B, position C is in the afternoon. Thus, the approximate time of day when the Sun is at position C is 3 p.m.

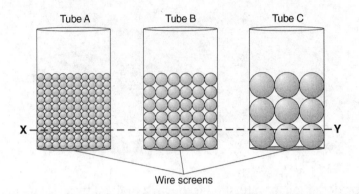

42. **3** The infiltration rate of a substance depends on two factors: the size of its pores (the openings between the particles) and the degree to which the pores are interconnected. Small pores constrict the flow of water, slowing the rate of infiltration. If water cannot get from one pore to another, it cannot flow through a substance. Since all four tubes contain uniform-sized, spherical beads, all four have interconnecting pores. Thus, the size of the pores determines the infiltration rate. In general, the larger the sediment particles, the larger the pore spaces are between them. Therefore, as particle diameter increases, the rate of infiltration increases. This relationship is best shown by graph (3).

43. **1** Capillarity is the tendency of water to rise in narrow openings in solid particles. The water rises because its attraction to the surrounding particles pulls the water upward and surface tension drags the water surface between the materials upward along with it. The narrower the opening is, the greater the force of attraction is between the water and the surrounding materials compared to the weight of the water column. So the narrower the opening between particles, the higher the water rises. Smaller particles fit closer together than larger ones. Therefore, the smaller the particles, the narrower the openings between particles and the higher water rises due to capillarity. Tube A contains the smallest beads. Thus, after one hour, the height of the water above line XY will be highest in tube A.

44. **2** By definition, a watershed is the area drained by a stream and its tributaries. A river is a stream. The question states that the block diagram represents a river drainage system. Therefore, the area of land drained by this river and its tributaries is best described as the river's watershed.

WRONG CHOICES EXPLAINED:
(1) Topography refers to the shape of Earth's surface, not the area drained by a river.
(3) The water table is the upper surface of underground pore spaces that are filled with groundwater, not the area drained by a river.
(4) A floodplain is a low plain adjacent to a stream that is likely to be flooded if a stream overflows, not the area drained by a river.

45. **4** Stream erosion is greatest where the velocity of the water is greatest. Along a straight section of a stream's channel, velocity is greatest in the center of the stream farthest from the friction of the banks or bed. When the stream curves, or meanders, velocity is greatest along the outside of the curve and lowest along the inside of the curve. In the diagram, point Y is located at the outside of a curve in the stream where the velocity of the water and erosion are the greatest. Point X is located at the inside of a curve in the stream where the velocity of the water and erosion are the least. Therefore, the streambed will be more deeply eroded near point Y than it is near point X—a shape best represented by the cross section in choice (4).

46. **2** As the water flowing in the river enters the standing water of the ocean, the velocity of the river decreases. Find the Relationship of Transported Particle Size to Water Velocity graph in the *Reference Tables for Physical Setting/Earth*

Science. Note that as stream velocity decreases, the particle size that can be transported also decreases. Therefore, as distance from where the river enters the ocean increases, stream velocity decreases and the sediment size deposited decreases. Thus, the arrangement of the sediments deposited where the river enters the ocean is largest particles deposited in the shallow waters near the mouth of the river and smallest particles deposited farther out in the ocean in deeper water. This pattern of sediments deposited where the river enters the ocean at location *B* is best represented by the cross section in choice (2).

47. **1** Find the Radioactive Decay Data table in the *Reference Tables for Physical Setting/Earth Science*. Note that carbon-14 is a radioactive isotope. Radioactive isotopes break apart, or decay, at a steady, constant rate that is not affected by outside factors such as changes in temperature, pressure, or chemical state. Therefore, the decay process may be used as a clock to determine the actual age of geologic samples. Radioactive isotopes start to decay when the geologic sample forms. The decay of a radioactive isotope occurs at a statistically predictable rate known as a half-life. The half-life is the time required for one-half of the unstable radioisotope to change into a stable decay product. By comparing the relative amounts of radioactive isotope and stable decay product in a rock, the number of half-lives that elapsed may be determined and may be used as a clock to determine the actual ages of these geologic samples. Therefore, radioactive carbon-14 is often useful in determining the absolute age of geologic samples because radioactive isotopes decay at a regular rate.

WRONG CHOICES EXPLAINED:

(2) When a radioactive isotope decays, its atoms break apart and form more stable atoms of a different element. Therefore, radioactive isotopes become more stable during decay, not less stable.

(3) The most common isotopes of the elements that comprise rocks are stable, meaning their atoms do not change. However, some isotopes are unstable.

In a process called radioactive decay, these atoms break apart. Such isotopes are called radioactive isotopes. Thus, radioactive isotopes are, by definition, unstable and do not remain unchanged over time.

(4) Radioactive isotopes continue to decay at a constant rate until all of the radioactive atoms have decayed. According to the graph, after four half-lives, 6.25% of the original mass of radioactive carbon-14 still remains. Therefore, radioactive carbon-14 does not stabilize after four half-lives because some still remains undecayed.

48. **4** Find the Radioactive Decay Data table in the *Reference Tables for Physical Setting/Earth Science*. In the column labeled "Radioactive Isotope," locate carbon-14. Trace right to the column labeled "Disintegration," and note that ^{14}C (carbon-14) decays into ^{14}N (nitrogen-14). Therefore, the disintegration product represented in the graph is ^{14}N.

49. **2** On the vertical axis of the graph labeled "Percentage of Original Mass (%)," locate "25." Trace horizontally right until you intersect the line labeled "14C." From this intersection, trace vertically down to the horizontal axis of the graph labeled "Time (in half-lives)." Note the value "2." Thus, if a sample contains 25% of its original carbon-14, 2 half-lives have passed.

50. **3** Find the Radioactive Decay Data table in the *Reference Tables for Physical Setting/Earth Science*. In the column labeled "Radioactive Isotope," locate carbon-14. Trace right to the column labeled "Half-life (years)," and note that carbon-14 has a half-life of 5.7×10^3 (5700) years. After about 10 half-lives (~50,000 years), there is too little carbon-14 left in a fossil to accurately date it. Therefore, carbon-14 can be used to date only relatively recent organic remains. Now find the Geologic History of New York State chart in the *Reference Tables for Physical Setting/Earth Science*. Among the diagrams of index fossils along the bottom of the chart, locate the four index fossils shown as choices. Note the circled letter associated with each of the four index fossils: 1—Ⓐ, 2—Ⓛ, 3—Ⓞ, and 4—Ⓠ. Locate the column labeled "Time Distribution of Fossils (including important fossils of New York)." Note that at the top of this column it states: "The center of each lettered circle indicates the approximate time of existence of a specific index fossil (e.g., Fossil Ⓐ lived at the end of the Early Cambrian)." Locate the lettered circles Ⓐ, Ⓛ, Ⓞ, and Ⓠ. From each lettered circle, trace horizontally to the left to the time scale labeled "Million years ago." Note how many million years ago each index fossil existed: Ⓐ—488–542; Ⓛ—200–251; Ⓞ—0–1.8; and Ⓠ—359–416. Of the choices, index fossil Ⓞ existed most recently. The drawing of index fossil Ⓞ is labeled "Mastodont." Mastodonts existed until

the end of the last ice age about 10,000 years ago, well within the 50,000-year range during which carbon-14 can be used to date a fossil. Thus, the only index fossil whose age could be determined by using carbon-14 is the mastodont shown in choice (3).

PART B–2

51. Locate the vertical axis of the graph labeled "Percentage of Lighted Moon Visible (%)." Note that the percentage increases from the bottom of the graph to the top. Therefore, the phases of the Moon begin waxing when the line on the graph begins to ascend. The waxing ends when the line has reached the maximum percentage of lighted Moon visible and begins to descend. Locate the point on the graph when the line begins to ascend and note that it corresponds to "1" on the horizontal axis labeled "Numbered Phases of Moon." Locate the point at which the line reaches its maximum and begins to descend. Trace vertically downward to the horizontal axis labeled "Numbered Phases of Moon," and note that it corresponds to the numbered phase "5." Thus, the numbered phase when waxing begins is 1, and the numbered phase when waxing ends is 5. The phase of the Moon at which 0% of the lighted Moon is visible is called the New Moon phase. The phase of the Moon at which 100% of the lighted portion of the Moon is visible is called the Full Moon Phase.

One credit is allowed for **waxing begins at Phase 1 *or* New Moon and waxing ends at Phase 5 *or* Full Moon.**

52. The full Moon phase occurs when the Moon is on the opposite side of Earth from the Sun. Thus, the **X** should be placed on the orbit, as shown below.

One credit is allowed if **the center of the X is within or touches the clear banded region shown below**.

Note: Allow credit if a symbol other than **X** is used.

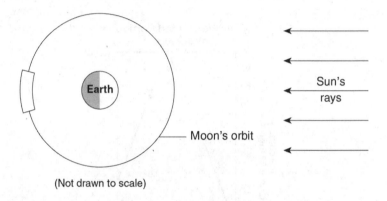

(Not drawn to scale)

53. According to the graph, when the Moon is at phase 1, 0% of the lighted portion of the Moon is visible. Thus, phase 1 corresponds to the New Moon phase. When the Moon is at phase 5, 100% of the lighted portion of the Moon is visible. Thus, phase 5 corresponds to the Full Moon phase. Therefore, phase 3 occurs midway between the New Moon and Full Moon phases and corresponds to the First Quarter phase. Similarly, phase 7 occurs midway between the Full Moon and New Moon phases and corresponds to the Last Quarter phase. Note that at the First Quarter phase, the 50% of the lighted portion of the Moon visible to an observer in New York State is to the observer's right and the left side of the Moon is shaded. When the Moon is at the Last Quarter phase, it is on the side of Earth directly opposite where it was at the First Quarter phase. Thus, an observer would see the lighted portion of the Moon on the opposite side of the Moon, or to the observer's left. Therefore, the right side of the circle should be shaded.

One credit is allowed for **shading half of the circle on the right side. The edge of the shading should be within or touching the clear rectangle shown below**.

Examples of 1-credit responses:

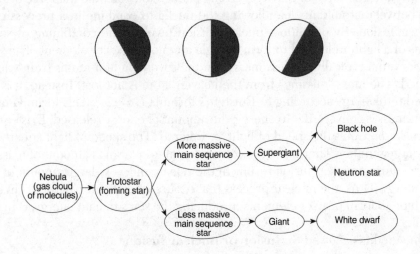

54. According to the flowchart, supergiants form from more massive main sequence stars and giants form from less massive main sequence stars. Thus, the characteristic of a main sequence star that determines whether the star becomes a giant or a supergiant is mass. Mass is a measure of the amount of matter in a star. Typically, the more mass a main sequence star contains, the larger the star.
   One credit is allowed. Acceptable responses include but are not limited to:

- **mass**
- **amount of matter in a star**
- **how massive the star is**
- **size**

Note: Credit is *not* allowed for "luminosity" because even though luminosity is related to the mass of a star, this is not how scientists classify the life cycle of stars based on this flowchart.

55. According to the flowchart, black holes and neutron stars form from supergiants. Find the Characteristics of Stars chart in the *Reference Tables for Physical Setting/Earth Science*. Locate the row labeled "Supergiants (Intermediate stage)" at the top of the chart. Note the names of stars in this row: *Rigel, Deneb, Betelgeuse, Spica,* and *Polaris.*
   One credit is allowed for **Rigel, Deneb, Betelgeuse, Spica, or Polaris.**

56. Most of the radiant energy given off by stars is the result of nuclear fusion. Within stars, the force of gravity is strong enough to overcome the force of repulsion between atomic nuclei, allowing the nuclei to combine in a process called nuclear fusion. By definition, nuclear fusion involves the combining of several atoms of a lighter element to form a single atom of a heavier element. The single heavier atom typically has less mass than the several lighter atoms from which it formed. The mass "missing" from the heavier atom is not lost. Instead, it is converted into energy according to Einstein's formula $E = mc^2$. This formula means that if mass is converted into energy, the amount of energy released, E, is equal to the mass, m, times the speed of light, c, squared. The speed of light squared is a very large number. Therefore, the conversion of even a small amount of mass into energy during nuclear fusion results in the release of a very large amount of radiant energy. Thus, the nuclear process that occurs when lighter elements in a star combine to form heavier elements, producing the star's radiant energy, is nuclear fusion.

One credit is allowed for **fusion or nuclear fusion**.

57. Find the Characteristics of Stars chart in the *Reference Tables for Physical Setting/Earth Science*. Locate "*Sun.*" Trace horizontally left to the vertical scale along the left-hand side of the chart labeled "Luminosity (Rate at which a star emits energy relative to the Sun)." Note that the Sun has a luminosity of 1. From the Sun, trace vertically downward to the horizontal scale along the bottom of the chart labeled "Surface Temperature (K)." Note that the Sun has a surface temperature of about 5,700K. Repeat this procedure for *Sirius*. Note that *Sirius* has a luminosity of about 25 and a surface temperature of nearly 10,000K. Thus, *Sirius* has a higher relative surface temperature and a greater relative luminosity than the Sun.

One credit is allowed if **both higher relative surface temperature and greater relative luminosity are circled**.

58. Contour lines are isolines that connect points of equal elevation on a topographic map. To draw the 200-meter isoline, connect all of the points labeled "200" with a smooth curve. Not every point that has an elevation of 200 meters has been labeled. However, between any locations, the elevation changes steadily from one location to the other. Begin with the point labeled "200" near the lower right corner of the map. Make sure you extend this line to the right edge of the map, as shown below. Note the two nearby points labeled "175" and "250." Somewhere between these two points, the elevation is 200 meters. Therefore extend the 200-meter contour line westward between these two points. Farther west are two points labeled "240" and "180." Extend the 200-meter contour line

between these two points to the point labeled "200." From this point, extend the 200-meter contour line westward and then north between the two points labeled "210" and "180" to the point labeled "200" near the west edge of the map and extend the line to the edge of the map, as shown below.

One credit is allowed for **correctly drawing the 200-meter contour line extended to the edges of the map**.

Note: If additional contour lines are drawn, all must be drawn correctly to receive credit.

Example of a 1-credit response:

59. To construct a topographic profile along line *AB*, proceed as follows. Place the straightedge of a piece of scrap paper along the solid line connecting point *A* to point *B*. Mark the edge of the paper at points *A* and *B* and wherever the paper intersects a contour line. Wherever the paper intersects a contour line, label the mark with the elevation of that contour line, as shown below.

Now place this scrap paper along the lower edge of the grid provided in your answer booklet so that points *A* and *B* on the paper align with points *A* and *B* on the grid. Then, at each point where a contour line crosses the edge of the paper, draw a dot on the grid at the appropriate elevation.

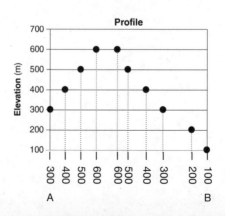

Finally, connect all of the dots in a smooth curve to form the finished profile, as shown below. Since the contour interval is 100 meters and no additional contour line exists within the 600-meter contour, the elevation in that area is greater than 600 meters but less than 700 meters. Therefore, the high point of the profile must be greater than 600 meters but less than 700 meters.

One credit is allowed if **the centers of *all eight* plots are within or touch the clear rectangles shown below and *all ten* plots are correctly connected with a line that passes within or touches the rectangles. The line must show the highest elevation above 600 m but below 700 m.**

Note: Credit is allowed if the line does not pass through the plots but is still within or touches the rectangles.

60. Find the Equations section of the *Reference Tables for Physical Setting/Earth Science*, and note the equation for gradient:

$$\text{Gradient} = \frac{\text{change in field value}}{\text{change in distance}}$$

The map shown is a topographic map. The field value on a topographic map is elevation. Note that point C is located directly on the 600-meter contour line and that point D is located directly on the 300-meter contour line. Thus, the change in field value (elevation) from C to D is 200 meters.

On a piece of scrap paper, mark off the distance along the straight line between points C and D. Compare the distance marked off on the scrap paper to the scale printed beneath the map to determine the distance between points C and D—about 4 kilometers. Substitute these values in the equation, as shown below:

$$\text{Gradient} = \frac{600 \text{ meters} - 300 \text{ meters}}{4 \text{ kilometers}}$$

Solve the equation, as shown below:

$$\text{Gradient} = \frac{300 \text{ m}}{4 \text{ km}} = 75 \frac{\text{m}}{\text{km}}$$

Thus, the approximate gradient along the straight line between points C and D is 75 meters/kilometer.

One credit is allowed for **any value from 73 to 77 or –73 to –77 with the correct units**.

Acceptable responses include but are not limited to:

- **m/km**
- **meters/kilometers**

61. Note that Kim Brook begins near the 300-meter contour line and ends in the ocean. The elevation of the ocean is sea level, or 0 meters. Water flows from higher to lower elevations. So Kim Brook flows from its source at an elevation near 300 meters to the ocean at an elevation of 0 meters. Thus, Kim Brook flows toward the ocean or to the northeast.

Another way to determine the direction in which Kim Brook flows is to note the shape of the contour lines where they cross Kim Brook. The bed of a stream slopes downhill and is lower than its banks. A person standing on a streambed is at a lower elevation than a person standing on the stream's bank. A person standing on the streambed has to walk uphill along the streambed, that is, upstream, to reach the same elevation as a person on the bank. On a topographic map, contour lines connect points of equal elevation. Therefore, the contour line has to bend upstream of the stream's bank to intercept the same elevation at the center of the streambed and then downstream on the opposite side to reach the same elevation on the opposite bank. Therefore, contour lines bend upstream when they cross a streambed, forming a distinctive V-shaped curve with the apex pointing upstream and the sides opening toward the downstream direction. Note that where the contour lines cross Kim Brook, they bend with the apex of each bend pointing southwest and the sides opening toward the northeast. Therefore, Kim Brook is flowing toward the northeast.

One credit is allowed if *both* responses are correct. Acceptable responses include but are not limited to:

Compass direction:

- **NE**
- **east northeast**
- **NNE**
- **from SW to NE**

Evidence:

- **Contour lines bend upstream when they cross Kim Brook.**
- **Contour lines make a V shape at Kim Brook and the V points uphill.**
- **The brook flows out of the Vs.**
- **Kim Brook flows from a higher to a lower elevation.**
- **Elevations decrease toward the northeast.**
- **Kim Brook flows to sea level/the ocean.**

Note: Do *not* accept the response "from the SW" only, because this does *not* indicate the direction *toward* which Kim Brook flows. Do *not* accept the response "water flows downhill" because this is given in the question.

62. When an earthquake occurs, both types of seismic waves, P-waves and S-waves, start moving outward from the focus at the same time. However, since they travel at different speeds, these two different types of seismic waves do not arrive at a seismic station at the same time. The faster P-waves arrive first, followed by the slower S-waves some time later. The farther a seismic station is located from the epicenter, the greater the difference between the arrival times of the P-waves and the S-waves recorded on a seismogram. Find the Generalized Bedrock Geology of New York State map in the *Reference Tables for Physical Setting/Earth Science*. Locate the map scale in the lower right-hand corner of the map. On the straightedge of a piece of scrap paper, mark off a distance equivalent to 50 kilometers. Now locate Mt. Marcy. Using the scrap paper, mark a point 50 kilometers to the southwest of Mt. Marcy. This point represents the location of the epicenter of the Blue Mountain Lake earthquake. Next locate Old Forge and New York City. Note that Old Forge is much closer to the epicenter than New York City. Therefore, the difference between the arrival time of the first P-wave and the arrival of the first S-wave is greater in New York City than in Old Forge because New York City is farther from the epicenter than Old Forge.

One credit is allowed for **both circling New York City and providing an acceptable explanation.**

Acceptable responses include but are not limited to:

- **New York City is farther from the epicenter, so there is a greater difference between the arrival of the first P-wave and the first S-wave.**
- **As distance to the epicenter increases, the difference in arrival times increases.**
- **Old Forge is closer to the epicenter near Blue Mountain Lake (*or* Mt. Marcy), so the arrival times are closer together.**

Note: Credit is allowed if New York City is not circled but is correctly used in the explanation. All responses must correctly refer to the earthquake epicenter or earthquake origin to receive credit.

63. It is given that the seismic station is 1200 kilometers from the epicenter. Find the Earthquake P-Wave and S-Wave Travel Time graph in the *Reference Tables for Physical Setting/Earth Science*. Locate the point corresponding to 1200 kilometers along the horizontal axis at the bottom of the graph labeled "Epicenter Distance ($\times 10^3$ km)." From 1200 km, trace vertically until you intersect the bold line labeled "*P*." From this intersection, trace horizontally to the left to the vertical axis labeled "Travel Time (min)." Read the *P*-wave travel time—about 2 minutes 35 seconds. Thus, the time needed for a *P*-wave to travel the 1200 km from the epicenter to the seismic station is 2 min 35 s.

One credit is allowed for **any value from 2 min 30 s to 2 min 40 s**.

64. Find the Generalized Bedrock Geology of New York State map in the *Reference Tables for Physical Setting/Earth Science*. Then find the point representing the Blue Mountain Lake earthquake epicenter you located in question 62. Note the map symbol for the surface bedrock at that point. Depending on how precisely you determined the direction to the southwest of Mt. Marcy, you might have located your point on one of two different map symbols. Now locate the two map symbols closest to the epicenter in the key labeled "Geologic Periods and Eras in New York." Note that one is labeled "MIDDLE PROTEROZOIC gneisses, quartzites, and marbles" and the other is labeled "MIDDLE PROTEROZOIC anorthositic rocks." Thus, the type of metamorphic surface bedrock where this earthquake was located is gneiss, quartzite, marble, or anorthositic rocks.

One credit is allowed for an acceptable response. Acceptable responses include but are not limited to:

- **gneiss**
- **quartzite**
- **marble**
- **anorthositic rocks**

65. In order to reach Perth, Australia, on the opposite side of Earth, the seismic waves from this earthquake would have to pass through Earth's center and all of Earth's interior layers. Find the Inferred Properties of Earth's Interior chart in the *Reference Tables for Physical Setting/Earth Science*. In the upper section of the diagram, find the layer labeled "Outer Core (Iron & Nickel)." Next trace the dashed lines marking the upper and lower boundaries of the outer core downward to the graph of temperature versus depth at the bottom of the chart. Note that in the sec-

tion of the graph corresponding to the outer core, the interior temperature (solid line) is higher than the melting temperature (dotted line), indicating that the outer core is liquid. S-waves are transverse waves that twist rock back and forth, deforming the rock's shape in a direction perpendicular to that of wave travel. S-waves can be transmitted only through solids. S-waves cannot travel through liquids and gases because when liquids and gases are deformed, they do not return to their original shape. Therefore, S-waves from this earthquake that travel toward Earth's center are absorbed by the liquid outer core rather than being transmitted.

One credit is allowed for **both outer core and an acceptable characteristic**.

Acceptable responses of characteristics of this layer include but are not limited to:

- **liquid/fluid**
- **molten rock**
- **melted iron and nickel**
- **temperature greater than melting point**

Note: Do *not* accept the response "core" alone because it does not indicate the liquid part of the core.

PART C

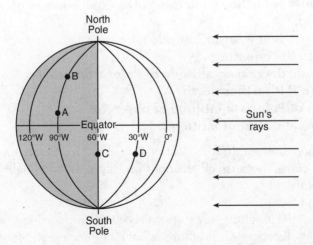

66. When sunlight strikes the spherical Earth, exactly one-half of Earth's sphere is illuminated and experiences daylight and exactly one-half is in darkness. The boundary between daylight and darkness forms a great circle that bisects Earth's sphere. On an equinox, the Sun's direct rays are at the equator. On that

day, the boundary between daylight and darkness passes through both the North and South Poles as it bisects Earth's spherical surface. Thus, the date represented by the position of Earth in this diagram is an equinox.

One credit is allowed. Acceptable responses include but are not limited to:

- **March 19, 20, 21, or 22**
- **September 21, 22, 23, or 24**
- **Spring/vernal equinox**
- **Autumn/fall equinox**
- **Equinox**
- **3/20**

67. Recall that the latitude of the North Pole is 90° and the latitude of the equator is 0°. Thus, location A, which is nearer the equator, has a lower latitude than location B, which is nearer the North Pole. Simple geometry shows that to an observer in the Northern Hemisphere, the altitude of *Polaris* is the same as the observer's latitude. (Refer to Figure 4.5 in *Let's Review: Earth Science, The Physical Setting* for a fuller explanation.) Therefore, the relative altitude of *Polaris* at location A is lower than at location B because the latitude of location A is lower than the latitude of location B.

Allow 1 credit for **both lower and an acceptable explanation**.

Acceptable responses include, but are not limited to:

- **Location A is farther from the North Pole, where *Polaris* is directly overhead.**
- **Location A is at a lower latitude.**
- **Closer to the equator**
- **As latitude decreases, altitude of *Polaris* decreases.**
- **Location B is farther north.**
- **Altitude of *Polaris* = latitude of observer**
- **They are at different latitudes.**

Note: Do *not* allow credit for "location B is closer to *Polaris*" or "location A is farther from *Polaris*" because all locations on Earth are essentially the same distance from *Polaris*.

68. Note that the longitude of location C is 60° W and the longitude of location D is 30° W. Thus, locations C and D are 30° of longitude apart. Since Earth makes one 360° rotation every 24 hours, in one hour Earth rotates 360°/24, or 15°. Thus, every 15° difference in longitude corresponds to a 1-hour difference in solar time. Since points C and D are 30° of longitude apart, the solar time difference between points C and D is 30/15, or 2 hours. Note that location D is to the east of location C.

ANSWERS August 2017

Since Earth rotates from west to east, for every 15° longitude you travel east of a location, the solar time is 1 hour later. Therefore, when the solar time at location C is 6:00 a.m., the solar time at location D is two hours later or 8:00 a.m.

One credit is allowed for **a response that indicates a time value of 8 a.m**. Acceptable responses include but are not limited to:

- **8:00 a.m.**
- **8 a.m.**
- **8 o'clock in the morning**
- **0800**

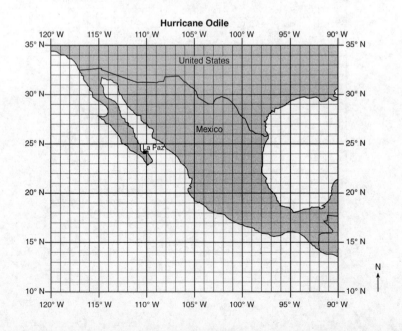

Hurricane Odile

Date	Location		Barometric Pressure (mb)	Wind Speed (kt)
	Latitude (° N)	Longitude (° W)		
September 12	15	105	993	50
September 13	16	106	983	65
September 14	19	107	918	120
September 15	23	110	941	110
September 16	27	113	987	55
September 17	30	114	995	40
September 18	31	112	1003	25

Hurricane Odile

69. To plot the seven locations of Hurricane Odile indicated by the latitudes and longitudes shown in the data table, proceed as follows. From September 12, trace right to the column labeled "Latitude (°N)." Note the value "15." Continue tracing right to the column labeled "Longitude (°W)." Note the value "105." Thus on September 12, the coordinates of Hurricane Odile were 15° N, 105° W. On the map provided, locate 15° N on the vertical axis along the left side of the map. Now locate 105° W along the horizontal axis at the bottom of the map. Trace horizontally right from 15° N and vertically upward from 105° W until the lines intersect. Plot this point. Repeat this procedure for each of the dates listed in the data table. When completed, connect the plots with a solid line, as shown below.

One credit is allowed if **the centers of *all seven* plots are within or touch the circles shown and are correctly connected with a line that passes within or touches each circle**.

Note: Credit is allowed if the line does *not* pass through the student's plots but is still within or touches the circles.

Example of a 1-credit response:

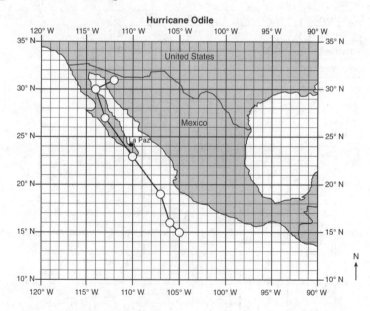

70. Note that the data table is ordered according to date. This makes it easy to track the motion of the hurricane over time. However, in order to see the relationship between barometric pressure and wind speed more easily, these data should be ordered by barometric pressure, as shown below:

Barometric Pressure (mb)	Wind Speed (kt)
918	120
941	110
983	65
987	55
993	50
995	40
1003	25

In this format, the relationship is readily apparent. As barometric pressure increases, wind speed decreases—an inverse relationship. Thus, the line on the graph should be drawn with a negative slope, that is, sloping downward from top left to bottom right.

One credit is allowed for **a line that shows a negative slope**.

Examples of 1-credit responses:

71. One of the most common instruments used to measure wind speed is an anemometer. A simple anemometer has cups mounted on a shaft that cause the shaft to spin when wind blows against them. The faster the wind blows, the faster the cups spin the shaft and the higher the number that registers on the anemometer's scale. Small handheld anemometers, called wind gauges, contain small fan blades that spin in the wind. There are also a variety of other instruments that can be used to measure wind speed. Some handheld digital wind speed meters contain a metal wire that is heated by an electrical current. The wire is cooled by air flow, and the meter instantly calculates wind speed. Wind speed can also be

measured using a pitot tube. A pitot tube is a narrow tube containing a fluid that is closed at one end. The open end of the tube faces toward the wind, which exerts a pressure on the fluid that compresses the fluid. The higher the wind speed, the more the liquid is compressed. Doppler radar can also be used to determine wind speed. For Doppler radar, atmospheric objects moving inbound (toward the radar) produce a positive shift in frequency of the radar signal. Objects moving away from the radar (outbound) produce a negative shift in frequency. This change in frequency can then be analyzed to determine wind speed.

One credit is allowed. Acceptable responses include but are not limited to:

- **anemometer**
- **wind speed meter**
- **wind gauge**
- **Doppler radar/radar**
- **pitot tube**

72. According to the National Weather Service, the major hazards associated with hurricanes are storm surge and storm tide, heavy rains, inland flooding, high winds, rip currents, and tornadoes. Actions that people who live in hurricane-prone areas should take in order to prepare for future hurricanes include finding out if you live in an evacuation area and learning your evacuation routes/shelter locations; assessing your risks and knowing your home's vulnerabilities to wind, storm surge, or flooding; having on hand the materials needed to secure your home; finding out what types of emergencies might occur, learning how you should respond, and obtaining emergency equipment; and obtaining water and nonperishable food.

One credit is allowed for *two* correct actions. Acceptable responses include but are not limited to:

- **Move to an emergency shelter, shelter, or bunker.**
- **Evacuate/learn evacuation routes.**
- **Stock up on supplies of food, water, and medicine.**
- **Assemble an emergency kit.**
- **Cover windows/board up windows.**
- **Secure outdoor furniture.**
- **Get a generator/make sure generator is working.**
- **Ensure vehicle has full tank of gas.**
- **Charge or stock up on batteries for electronic equipment.**
- **Stack sandbags where needed.**

73. Note the map symbol for rock unit *F*. According to the key, the map symbol for rock unit *F* corresponds to basalt, an igneous rock. Note the map symbol for rock unit *A*. Find the Scheme for Sedimentary Rock Identification in the *Reference Tables for Physical Setting/Earth Science*. Locate the map symbol for rock unit *A*, trace left to the column labeled "Rock Name," and note that rock unit *A* is sandstone. From sandstone, trace left to the column labeled "Composition." Note that sandstone is composed mostly of quartz. Thus, the rock formed in the zone of contact metamorphism between rock units *A* and *F* is contact metamorphosed quartz sandstone. Find the Scheme for Metamorphic Rock Identification in the *Reference Tables for Physical Setting/Earth Science*. In the column labeled "Comments," locate the statement "Metamorphism of quartz sandstone." Trace this row left to the column labeled "Type of Metamorphism," and note the statement "Regional or contact." Now trace this row right to the column labeled "Rock Name." Note that the rock formed by the contact metamorphism of quartz sandstone is quartzite. Additionally, in the column labeled "Comments," locate the statement "Various rocks changed by heat from nearby magma/lava." Trace this row left to the column labeled "Type of Metamorphism," and note the statement "Contact (heat)." Now trace this row right to the column labeled "Rock Name." Note that the rock formed by the contact metamorphism of various rocks by heat from nearby magma or lava is hornfels.

One credit is allowed for **quartzite *or* hornfels**.

74. Layers of rock are generally deposited in an unbroken sequence. However, if forces within Earth cause rocks to be uplifted, deposition ceases. Weathering and erosion may wear away layers of rock before the land surface is low enough for another layer to be deposited. The uplift may cause rock layers to tilt or fold. The result is an unconformity, which is a break or gap in the sequence of a series of rock layers. Thus, the rocks above an unconformity are quite a bit younger than those below it. There are several types of unconformities. *Angular unconformities* form when rock layers are tilted or folded before being eroded. When new layers are deposited, they form horizontally. The layers below the unconformity are at an angle to those above it. *Disconformities* are irregular erosional surfaces between parallel layers of rock. Disconformities occur when deposition stops and layers are eroded but no tilting or folding occurs. These surfaces are not easy to discern and are often found when fossils of very different ages are discovered in adjacent layers. *Nonconformities* are places where sedimentary layers lie on top of igneous or metamorphic rocks and are not metamorphosed in any way. According to the cross section, the rock units above the unconformity are flat, horizontal layers and the rock units below the unconformity are tilted and folded. There is also a fault along line YY' and the layers to the right of the fault moved upward relative to the layers to the left of the fault. Thus, unconformity XX' is most likely an angular unconformity that resulted when tilted and folded rock layers were uplifted and then worn down by weathering and erosion. This was followed by subsidence and submersion during which time rock layers A, B, and C were deposited. Then the land was again uplifted, exposing these rocks at the surface.

One credit is allowed for *two* correct responses. Acceptable responses include but are not limited to:

- **uplift/emergence**
- **weathering**
- **erosion**
- **subsidence/submergence**
- **deposition**
- **burial**

75. Note that fault YY' cuts across rock unit D and that rock unit D has been displaced along the fault. In order to be displaced, a rock must already have existed when the fault occurred. Therefore, fault YY' is younger than rock unit D. Note, however, that rock unit E cuts across fault YY' and has not been displaced. Therefore, rock layer E had not yet formed when the fault occurred. So rock layer E is younger than fault YY'. Note the map symbol for rock unit E. Locate this map symbol in the key. Note that it corresponds to diorite, which is an igneous rock. Note, too, the symbol for contact metamorphism along the boundary between

rock units D and E. Thus, rock unit E is an igneous intrusion that cuts across and contact metamorphosed rock unit D. An igneous intrusion is younger than any rock layer or structure that it cuts across. In order to be contact metamorphosed, a rock layer must have already existed when the igneous intrusion occurred. Therefore, rock unit E is younger than the rock unit D. Now note the map symbol for rock unit F. According to the key, the map symbol for rock unit F corresponds to basalt, which is an igneous rock. Note that rock unit F cuts across rock unit E and that rock unit E has been contact metamorphosed. Therefore, rock unit F is younger than rock unit E. Thus, from oldest to youngest, the rock units and fault YY' are D, YY', E, and F.

One credit is allowed for $D \rightarrow \textbf{\textit{fault YY'}} \rightarrow E \rightarrow F$.

Note: Credit is allowed for fault, *or* YY' alone, *or* Y alone in place of fault YY'. Credit is allowed if the correct rock names are substituted for D (gneiss), E (diorite), and F (basalt).

76. In the key along the right side of the cross section, note how water and sediment in the Finger Lakes are indicated. The maximum water depth of a Finger Lake is the vertical distance between the top of the sediment at the bottom of a lake and the surface of that lake in the cross section (the region corresponding to "Water" in the key). Find the vertical scale labeled "Elevation (ft)" along the left side of the cross section. Note that the scale is marked off in subdivisions of 100. Using the straightedge of a piece of scrap paper, mark off a distance equivalent to 175 ft, or about 1¾ subdivisions. Holding the piece of scrap paper vertically, compare the marked off distance of 175 ft on the scrap paper with the distance between the top of the sediment at the bottom of each lake and the water surface of the lake. Note that Keuka Lake and Owasco Lake have a maximum water depth of about 175 ft.

One credit is allowed for **Owasco Lake *or* Keuka Lake**.

77. According to the reading passage, "Around 1.7 to 1.8 million years ago, a continental glacier advanced southward, beginning an ice age that consisted of many advances and retreats of glaciers. The Finger Lakes were carved by some of these advances." Thus, the continental glaciers formed the Finger Lakes around 1.7 to 1.8 million years ago. Find the Geologic History of New York State chart in the *Reference Tables for Physical Setting/Earth Science*. On the vertical time scale labeled "Million years ago" to the right of the column labeled "Epoch," locate the row corresponding to between 0.01 and 1.8 million years ago. This range includes 1.7 to 1.8 million years ago. Trace this row left to the column labeled "Epoch," and note that this range corresponds to the Pleistocene epoch. Now trace this row left to the column labeled "Period," and note that this range corresponds to the Quaternary period. Thus, these continental glaciers formed the Finger Lakes of New York State during the Pleistocene epoch of the Quaternary Period.

One credit is allowed for **both the Quaternary period and Pleistocene epoch**.

78. The sediments deposited in the earlier river channels were deposited in water. When rock particles are deposited in water, they settle at different rates and tend to become sorted in layers. Glacial moraine is material transported by a glacier and then deposited. Glaciers carry sediments of all sizes on their surface and frozen in the ice. These sediments are deposited directly from the glacier as it melts, and the sediments simply fall to the ground. Since they do not settle through a medium such as water, the sediments do not become sorted or layered. Glaciers also push and drag sediments along as they move. Here again, the sediments do not settle through a medium but, instead, get jumbled as they get pushed and dragged around by the glacier. Thus, sediments directly deposited by a glacier are an unsorted, unlayered mixture of particle sizes. Thus, sediments deposited in glacial moraines are unsorted and unlayered, while sediments deposited in earlier river channels are sorted and layered.

One credit is allowed. Acceptable responses include but are not limited to:

- **Glacial moraine deposits are unsorted/mixed.**
- **Glacial deposits in a moraine are unlayered.**
- **Unsorted**
- **River deposits are sorted.**

79. In most streams, water continues to flow even when there is no rainfall or runoff from the surrounding slopes. Therefore, the stream erodes downward into the underlying bedrock even when the sides of the valley are not being eroded by runoff. As a result, the stream valley becomes deeper faster than it becomes wider, forming a narrow valley bottom with sloping sides, or a V shape. Thus, the

cross-sectional shape of the original river valleys before they were gouged by the advancing glacier was most likely a V shape, with a narrow valley bottom with sloping sides.

One credit is allowed. Acceptable responses include but are not limited to:

- **V-shaped valleys**
- **narrow valley bottom with gently sloped sides**

80. Find the Surface Ocean Currents map in the *Reference Tables for Physical Setting/Earth Science*. Locate the region corresponding to the Pacific Northwest Coast of the United States. Note that the California Current flows southward along the coast in this region. According to the key, the California Current is a cool current. Therefore, the name of the cool surface ocean current that influences the climate of this region is the California Current.

One credit is allowed for **California Current**.

81. According to the map, Long Beach and Richland are located at about the same latitude. Therefore, both cities receive the same intensity of insolation. However, Long Beach is located on the coast next to the Pacific Ocean while Richland is an inland location. Due to the high specific heat of water, land surfaces increase in temperature more than water surfaces when insolation strikes. Land surfaces also cool more rapidly than water surfaces. Furthermore, moist air has a higher specific heat than dry air, so dry air heats and cools more rapidly than moist air. Therefore, large bodies of water tend to moderate the temperatures of coastal regions of nearby landmasses by warming them in winter and cooling them in summer. Thus Long Beach, which is a coastal city, experiences a smaller range of temperatures throughout the year because the nearby Pacific Ocean moderates its temperatures.

One credit is allowed for an acceptable response. Acceptable responses include but are not limited to:

- **The Pacific Ocean moderates the climate of Long Beach.**
- **Water has a higher specific heat, and temperatures change more slowly.**
- **Long Beach is near a large body of water.**
- **Richland is farther inland.**
- **Long Beach has a marine climate, whereas Richland's is more continental.**
- **It is located closer to a large body of water.**

82. The question states that rock 1 is a glassy, vesicular igneous rock. Find the Scheme for Igneous Rock Identification in the *Reference Tables for Physical Setting/Earth Science*. On the graph labeled "Mineral Composition (relative by volume)," locate the position corresponding to the line representing rock 1 on the diagram in the question. Trace vertically upward from this position to the section of the Scheme for Igneous Rock Identification labeled "Igneous Rocks." In the column labeled "Texture," locate the row corresponding to both a "Glassy" and "Vesicular (gas pockets)" texture. Trace this row to the left to where it intersects with the position corresponding to rock 1. Note the name "Pumice." Thus, the name of rock 1 is pumice.

One credit is allowed for **pumice**.

83. Find the Scheme for Igneous Rock Identification in the *Reference Tables for Physical Setting/Earth Science*. On the graph labeled "Mineral Composition (relative by volume)," locate the position corresponding to the line representing rock 1 on the diagram in the question. Trace upward to the section of the Scheme for Igneous Rock Identification labeled "Characteristics." Note that an igneous rock with the composition of rock 1 intersects the arrow labeled "Color" near the end of the range labeled "Lighter" and intersects the arrow labeled "Density" near the end of the range labeled "Lower."

On the graph labeled "Mineral Composition (relative by volume)," locate the position corresponding to the line representing rock 2 on the diagram in the question. Trace upward to the section of the Scheme for Igneous Rock Identification labeled "Characteristics." Note that an igneous rock with the composition of rock 2 intersects the arrow labeled "Color" near the end of the range labeled "Darker" and intersects the arrow labeled "Density" near the end of the range labeled "Higher." Thus, compared to the color and density of igneous rock 1, igneous rock 2 is darker in color and has a higher density.

Another way to compare color is to compare the colors of the major minerals that comprise rocks 1 and 2. According to the diagram, rock 1 is composed mainly of potassium feldspar, quartz, and plagioclase feldspar. Rock 2 is composed mainly of plagioclase feldspar, pyroxene, and amphibole. Find the Properties of Common Minerals table in the *Reference Tables for Physical Setting/Earth Science*. In the column labeled "Mineral Name," locate potassium feldspar, quartz, plagioclase feldspar, pyroxene, and amphibole. From each of these minerals, trace left to the column labeled "Common Colors." Note the common colors of these minerals: potassium feldspar—white to pink, quartz—colorless or variable, plagioclase feldspar—white to gray, pyroxene—black to dark green, and amphibole—black to dark green. Therefore, rock 1 is composed mainly of light-colored minerals, and rock 2 is composed mainly of dark-colored minerals, particularly black to dark green pyroxene.

One credit is allowed if *both* responses are acceptable. Acceptable responses include but are not limited to:

Relative color of rock 2:

- **Igneous rock 2 is darker/dark.**
- **Blacker/black**
- **Greener/green**
- **Rock 1 is lighter than rock 2.**

Relative density of rock 2:

- **Igneous rock 2 has a higher density.**
- **Greater**
- **More dense/denser**
- **Rock 1 is less dense than rock 2.**

Note: The response "more mafic" is *not* accepted for either relative color or relative density of rock 2 as this refers to composition.

84. Find the Scheme for Igneous Rock Identification in the *Reference Tables for Physical Setting/Earth Science*. In the upper section of the scheme labeled "Igneous Rocks," locate andesite and diorite. Trace these two rows left to the column labeled "Environment of Formation." Note that the row corresponding to andesite is in the range labeled "Extrusive" and that the row corresponding to diorite is in the range labeled "Intrusive." Thus, andesite is an extrusive (volcanic) rock formed by the cooling of lava, and diorite is an intrusive (plutonic) rock formed by the cooling of magma. Now trace these two rows right to the column labeled "Texture." Note that the row corresponding to andesite is in the range labeled "Fine" and that the row corresponding to diorite is in the range labeled "Coarse." Therefore, andesite and diorite can both have the same percentage of mineral composition by volume and yet be two different igneous rocks because they formed in different environments and have different textures.

One credit is allowed. Acceptable responses include but are not limited to:

- **The rocks may have different environments of formation.**
- **Diorite is plutonic, and andesite is volcanic.**
- **One forms from cooling magma, and one forms from cooling lava.**
- **They cooled at different rates.**
- **They have different grain sizes or textures.**
- **One is extrusive, and one is intrusive.**
- **They formed differently.**

85. Find the Properties of Common Minerals table in the *Reference Tables for Physical Setting/Earth Science*. In the column labeled "Composition," locate "$(Na,Ca)AlSi_3O_8$." From this chemical formula, trace right to the column labeled "Mineral Name." Note that this composition corresponds to plagioclase feldspar. Repeat this procedure for the other two chemical formulas listed in the table. Note that "$KAlSi_3O_8$" corresponds to potassium feldspar and that "$(Fe,Mg)_2SiO_4$" corresponds to olivine.

According to the diagram in the question, rock 1 is composed of potassium feldspar, quartz, plagioclase feldspar, biotite, and amphibole. Rock 2 is composed of plagioclase feldspar, pyroxene, and amphibole. Therefore, plagioclase feldspar [$(Na,Ca)AlSi_3O_8$] is found in both rock 1 and rock 2, potassium feldspar [$KAlSi_3O_8$] is found in rock 1 only, and olivine [$(Fe,Mg)_2SiO_4$] is found in neither rock 1 nor rock 2. This information should be recorded on the answer table, as shown below:

One credit is allowed if *all three* **X**s are only placed in the correct boxes, as shown below.

Note: Allow credit if a symbol other than an **X** is used.

Mineral Composition	Found in both rock 1 and rock 2	Found in neither rock 1 nor rock 2	Found in rock 1, only	Found in rock 2, only
$(Na,Ca)AlSi_3O_8$	X			
$KAlSi_3O_8$			X	
$(Fe,Mg)_2SiO_4$		X		

SELF-ANALYSIS CHART August 2017

Topic	Question Numbers (Total)	Wrong Answers (x)	Grade
Standards 1, 2, 6, and 7: Skills and Application			
Skills			
Standard 1 Analysis, Inquiry, and Design	2, 6, 8–10, 12, 13, 15, 17–19, 24–26, 28, 29, 32–40, 47–51, 55–57, 60, 62–71, 73, 74, 76, 77, 79–85		$\frac{100(55-x)}{55} = \%$
Standard 2 Information Systems	72		$\frac{100(1-x)}{1} = \%$
Standard 6 Interconnectedness, Common Themes	1, 6, 7, 9, 16, 17, 20, 23, 26, 27, 30–32, 34–36, 39–46, 52–55, 57–59, 61, 62, 64, 66, 67, 69, 73, 75, 76, 78, 79, 81		$\frac{100(43-x)}{43} = \%$
Standard 7 Interdisciplinary Problem Solving	72		$\frac{100(1-x)}{1} = \%$
Standard 4: The Physical Setting/Earth Science			
Astronomy			
The Solar System (MU 1.1a, b; 1.2d)	6, 40, 51–53, 66		$\frac{100(6-x)}{6} = \%$
Earth Motions and Their Effects (MU 1.1c, d, e, f, g, h, i)	3–5, 26, 39, 41, 67–69		$\frac{100(9-x)}{9} = \%$
Stellar Astronomy (MU 1.2b)	1, 54–57		$\frac{100(5-x)}{5} = \%$
Origin of Earth's Atmosphere, Hydrosphere, and Lithosphere (MU 1.2e, f, h)			
Theories of the Origin of the Universe and Solar System (MU 1.2a, c)	2, 36, 37		$\frac{100(3-x)}{3} = \%$

SELF-ANALYSIS CHART August 2017

Topic	Question Numbers (Total)	Wrong Answers (x)	Grade
Meteorology			
Energy Sources for Earth Systems (MU 2.1a, b)			
Weather (MU 2.1c, d, e, f, g, h)	8–11, 27, 28, 34, 70–72		$\frac{100(10-x)}{10} = \%$
Insolation and Seasonal Changes (MU 2.1i; 2.2a, b)	13–16		$\frac{100(4-x)}{4} = \%$
The Water Cycle and Climates (MU 1.2g; 2.2c, d)	7, 12, 30, 42, 43, 80, 81		$\frac{100(7-x)}{7} = \%$
Geology			
Minerals and Rocks (MU 3.1a, b, c)	24, 25, 33, 64, 73, 82–85		$\frac{100(9-x)}{9} = \%$
Weathering, Erosion, and Deposition (MU 2.1s, t, u, v, w)	22, 23, 44–46, 76, 78, 79		$\frac{100(8-x)}{8} = \%$
Plate Tectonics and Earth's Interior (MU 2.1j, k, l, m, n, o)	17–19, 32, 35, 38, 62, 63, 65		$\frac{100(9-x)}{9} = \%$
Geologic History (MU 1.2i, j)	29, 31, 47–50, 74, 75, 77		$\frac{100(9-x)}{9} = \%$
Topographic Maps and Landscapes (MU 2.1p, q, r)	20, 21, 58–61		$\frac{100(6-x)}{6} = \%$
ESRT			
2011 Edition Reference Tables for Physical Setting/ Earth Science	8–11, 13, 15, 18, 19, 24, 25, 29, 32–35, 37, 38, 48, 50, 55, 57, 60, 62–65, 73, 77, 80, 82–85		$\frac{100(33-x)}{33} = \%$

To further pinpoint your weak areas, use the Topic Outline in the front of the book.
MU = Major Understanding (see Topic Outline)

Examination June 2018
Physical Setting/Earth Science

PART A

Answer all questions in this part.

Directions (1–35): For *each* statement or question, choose the word or expression that, of those given, best completes the statement or answers the question. Some questions may require the use of the *2011 Edition Reference Tables for Physical Setting/Earth Science*. Record your answers in the space provided.

1 The photographs below show two types of solar eclipses. Letters *A* and *B* represent two celestial objects.

Total Solar Eclipse

Partial Solar Eclipse

Which two celestial objects are represented by letters *A* and *B*?

(1) *A*-Moon; *B*-Sun
(2) *A*-Moon; *B*-Earth
(3) *A*-Sun; *B*-Moon
(4) *A*-Sun; *B*-Earth

1 _____

2 Compared to the terrestrial planets, the Jovian planets
   (1) are less massive
   (2) are more dense
   (3) have greater orbital velocities
   (4) have shorter periods of rotation

3 Which event occurred more than 10 billion years ago?
   (1) Big Bang
   (2) origin of life on Earth
   (3) Pangaea begins to break up
   (4) origin of Earth and its Moon

4 In 1851, French physicist Léon Foucault used a swinging pendulum to demonstrate that Earth
   (1) is rotating
   (2) is revolving
   (3) has a curved surface
   (4) has a gravitational pull

5 Approximately how many degrees does Earth travel in its orbit in one month?
   (1) 1°
   (2) 15°
   (3) 30°
   (4) 360°

6 What is the relative humidity when the drybulb temperature is 16°C and the wet-bulb temperature is 10°C?
   (1) 6%
   (2) 14%
   (3) 33%
   (4) 45%

7 Boarding up windows would be one emergency action most likely taken to prepare for which natural disaster?

(1) earthquake
(2) hurricane
(3) flood
(4) tsunami

8 Which diagram best represents the general position and direction of flow of the polar front jet stream in the Northern Hemisphere during the winter months?

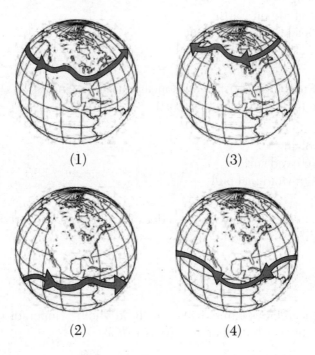

9 The diagram below represents four positions of the Moon, labeled A through D, as it orbits Earth.

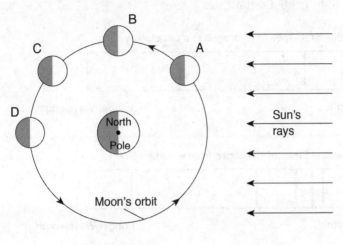

Which diagram best represents the sequence of Moon phases, as seen by an observer in New York State, when the Moon travels from position A to position D in its orbit around Earth?

9 _____

10 The diagrams below represent spectral lines of hydrogen gas observed in a laboratory and the spectral lines of hydrogen gas observed in the light from a distant star.

Spectral Lines of Hydrogen in a Laboratory

Shorter Wavelength   Longer Wavelength

Spectral Lines of Hydrogen from a Distant Star

Shorter Wavelength   Longer Wavelength

Compared to the spectral lines observed in the laboratory, the spectral lines observed in the light from the distant star have shifted toward the

(1) red end of the spectrum, indicating the star's movement toward Earth
(2) red end of the spectrum, indicating the star's movement away from Earth
(3) blue end of the spectrum, indicating the star's movement toward Earth
(4) blue end of the spectrum, indicating the star's movement away from Earth

10 _____

11 The diagram below represents a cross-sectional view of the plane of Earth's orbit around the Sun. A line drawn perpendicular to the plane of Earth's orbit is shown on the diagram.

How many degrees is Earth's rotational axis tilted with respect to the perpendicular line shown in the diagram?

(1) 15°
(2) 23.5°
(3) 90°
(4) 180°

11 _____

12 The larger white dots in the diagrams below represent stars in the constellations Scorpius and Orion. Information indicating when these constellations are visible from New York State is provided below the diagrams.

Scorpius
Visible in the New York State nighttime sky during July; not visible at all in January

Orion
Visible in the New York State nighttime sky during January; not visible at all in July

Which statement best explains why these two constellations are visible in the night sky in the months identified?

(1) Earth spins on its axis at a constant rate during a 24-hour period.
(2) Earth spins on its axis at a variable rate during the year.
(3) The nighttime side of Earth is facing different parts of our galaxy as Earth orbits the Sun.
(4) The nighttime side of Earth is facing different parts of our galaxy as the stars orbit Earth.

13 Which table correctly shows the interior temperature, melting point, and state (phase) of matter of the materials located 4000 kilometers below Earth's surface?

Interior Temperature (°C)	Melting Point (°C)	State of Matter
5700	5400	solid

(1)

Interior Temperature (°C)	Melting Point (°C)	State of Matter
5700	5400	liquid

(2)

Interior Temperature (°C)	Melting Point (°C)	State of Matter
5400	5700	solid

(3)

Interior Temperature (°C)	Melting Point (°C)	State of Matter
5400	5700	liquid

(4)

14 Which gas is a greenhouse gas that has increased in Earth's atmosphere partly as a result of deforestation over the last 100 years?

(1) ozone
(2) oxygen
(3) nitrogen
(4) carbon dioxide

15 Which ocean current brings warm water to the southeastern tip of Africa?

(1) Brazil Current
(2) Agulhas Current
(3) Guinea Current
(4) Benguela Current

16 Which pie graph is shaded to best represent the approximate percentage of time that humans have existed during Earth's entire history?

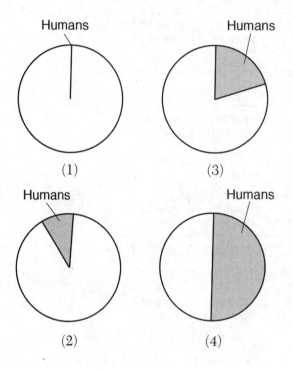

16 _____

17 Volcanic ash can be used as a time marker to correlate rock layers because the ash

(1) is deposited rapidly over a large area
(2) represents a buried erosional surface
(3) forms intrusive igneous rock
(4) cuts across rock layers

17 _____

18 The cross section below represents a mountain range. Points A and B represent locations on Earth's surface.

Compared to the climate of location A, the climate of location B is most likely

(1) cooler and wetter
(2) cooler and drier
(3) warmer and wetter
(4) warmer and drier   18 _____

19 The photograph below shows conglomerate composed of pebbles cemented together with calcite.

Compared to the ages of the calcite cement and the conglomerate, the relative age of the pebbles is

(1) younger than both the calcite cement and the conglomerate
(2) younger than the calcite cement, but the same age as the conglomerate
(3) older than both the calcite cement and the conglomerate
(4) older than the calcite cement, but the same age as the conglomerate   19 _____

20 The cross section below represents some parts of Earth's water cycle. Letters A, B, C, and D represent processes that occur during the cycle.

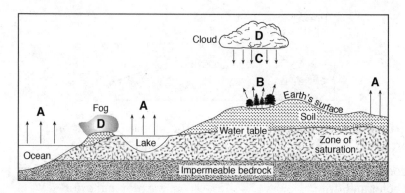

Which table correctly matches each letter with the process it represents?

Letter	Process
A	Condensation
B	Transpiration
C	Precipitation
D	Evaporation

(1)

Letter	Process
A	Condensation
B	Evaporation
C	Precipitation
D	Transpiration

(3)

Letter	Process
A	Evaporation
B	Precipitation
C	Transpiration
D	Condensation

(2)

Letter	Process
A	Evaporation
B	Transpiration
C	Precipitation
D	Condensation

(4)

20 _____

21 Which table best shows the relationship between latitude and general climate conditions on Earth?

Latitude	Climate Conditions
90°N	Arid
60°N	Arid
30°N	Humid
0°	Humid
30°S	Humid
60°S	Arid
90°S	Arid

(1)

Latitude	Climate Conditions
90°N	Humid
60°N	Arid
30°N	Humid
0°	Humid
30°S	Humid
60°S	Arid
90°S	Humid

(3)

Latitude	Climate Conditions
90°N	Arid
60°N	Humid
30°N	Arid
0°	Humid
30°S	Arid
60°S	Humid
90°S	Arid

(2)

Latitude	Climate Conditions
90°N	Humid
60°N	Arid
30°N	Humid
0°	Arid
30°S	Humid
60°S	Arid
90°S	Humid

(4)

21 _____

22 The photograph below shows different-sized rounded sediment.

Which table shows the most likely process and agent of erosion responsible for this rounded sediment?

Process	Agent of Erosion
sandblasting	running water

(1)

Process	Agent of Erosion
abrasion	wave action

(2)

Process	Agent of Erosion
land slide	mass movement

(3)

Process	Agent of Erosion
deposition	wind

(4)

23 The photograph below shows an outcrop with two basaltic intrusions, labeled A and B, in a rock unit, labeled C.

What is the relative age of these three rock units from oldest to youngest?

(1) $B \to A \to C$
(2) $B \to C \to A$
(3) $C \to A \to B$
(4) $C \to B \to A$

23 _____

24 The world map below shows Earth's major tectonic plate boundaries. Letters *A* through *D* represent four surface locations.

Which location is on a major rift valley?
(1) *A*
(2) *B*
(3) *C*
(4) *D*

24 _____

25 The first *P*-wave of an earthquake took 11 minutes to travel to a seismic station from the epicenter of the earthquake. What is the seismic station's distance to the epicenter of the earthquake and how long did it take for the first *S*-wave to travel that distance?

(1) Distance to epicenter: 3350 km
   S-wave travel time: 4 min 50 sec
(2) Distance to epicenter: 3350 km
   S-wave travel time: 6 min 10 sec
(3) Distance to epicenter: 7600 km
   S-wave travel time: 9 min
(4) Distance to epicenter: 7600 km
   S-wave travel time: 20 min

25 _____

26 The Catskills are commonly called mountains, but are actually part of the Allegheny Plateau. The Catskills are classified as a plateau because of their

(1) low elevation
(2) bedrock structure
(3) bedrock age
(4) high degree of metamorphism

27 The minimum stream velocity necessary to transport a sediment particle that is 0.1 centimeter in diameter is closest to

(1) 0.1 cm/s
(2) 0.002 cm/s
(3) 5.5 cm/s
(4) 10.0 cm/s

28 Which rock is classified as an evaporite?

(1) clastic shale
(2) foliated phyllite
(3) nonfoliated marble
(4) crystalline rock salt

29 Which pair of elements makes up most of Earth's crust by volume?

(1) nitrogen and potassium
(2) oxygen and silicon
(3) hydrogen and oxygen
(4) potassium and oxygen

30 The cross section below represents zones of soil labeled *A*, *B*, and *C*. Letter *D* represents underlying bedrock.

Which letter identifies the zone having the most organic and weathered material?

(1) *A* (3) *C*
(2) *B* (4) *D*   30 _____

31 Which type of surface bedrock is most commonly found in the Utica, New York area?

(1) sedimentary, with limestone, shale, sandstone, and doloston
(2) sedimentary, with limestone, shale, sandstone, and conglomerate
(3) metamorphic, with quartzite, dolostone, marble, and schist
(4) metamorphic, with gneiss, quartzite, marble, and slate   31 _____

32 The diagram below represents a geologic landscape.

Which type of stream drainage pattern formed on this landscape?

32 _____

33 The north polar view maps below show the average area covered by Arctic Sea ice in September of 1980, 2000, and 2011.

The maps best support the inference that Earth's climate is

(1) cooling, because the average area covered by Arctic Sea ice is decreasing
(2) cooling, because the average area covered by Arctic Sea ice is increasing
(3) warming, because the average area covered by Arctic Sea ice is decreasing
(4) warming, because the average area covered by Arctic Sea ice is increasing

33 _____

34 The photographs below show two depositional features labeled A and B.

Which terms correctly identify depositional features A and B?

(1) A-delta; B-barrier island
(2) A-sand bar; B-island arc
(3) A-barrier island; B-delta
(4) A-island arc; B-sand bar

34 _____

35 Diagrams *A* and *B* represent magnified views of the arrangement of mineral crystals in a rock before and after being subjected to geologic processes.

Which geologic processes are most likely responsible for the banding and alignment of mineral crystals represented in diagram *B*?

(1) melting and solidification
(2) heating and increasing pressure
(3) compaction and cementation
(4) weathering and erosion

PART B–1
Answer all questions in this part.

Directions (36–50): For *each* statement or question, choose the word or expression that, of those given, best completes the statement or answers the question. Some questions may require the use of the *2011 Edition Reference Tables for Physical Setting/Earth Science*. Record your answers in the space provided.

Base your answers to questions 36 through 39 on the graph below and on your knowledge of Earth science. The graph shows the observed water levels, in feet (ft), for a tide gauge located at Montauk, New York, on the easternmost end of Long Island, from January 24, 2008 to noon on January 25, 2008.

36 What was the height of the water above average low tide level at noon on January 24?

(1) 1.2 ft
(2) 1.6 ft
(3) 2.2 ft
(4) 2.6 ft

36 _____

37 These changing water levels at Montauk can best be described as
   (1) cyclic and predictable
   (2) cyclic and not predictablee
   (3) noncyclic and predictable
   (4) noncyclic and not predictable   37 _____

38 What causes the water-level variation pattern shown by the graph?
   (1) changes in wind velocity produced by coastal storms
   (2) changes in magnetic orientation of the North American Plate
   (3) Earth's revolution and the distance from the equator
   (4) Earth's rotation and the gravitational pull of the Moon   38 _____

39 What is the approximate latitude and longitude of the tide gauge?
   (1) 40°30′ N 72°00′ W
   (2) 40°30′ N 74°00′ W
   (3) 41°00′ N 72°00′ W
   (4) 41°00′ N 74°00′ W   39 _____

Base your answers to questions 40 through 42 on the diagram below and on your knowledge of Earth science. The diagram represents some of the inferred stages in the life cycle of stars according to their original mass.

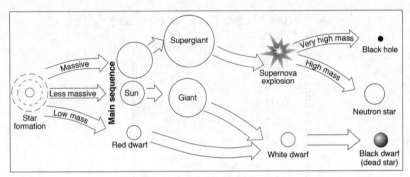

(Not drawn to scale)

40 The final stage in the life cycle of the most massive stars is a

(1) black hole
(2) black dwarf
(3) supergiant
(4) white dwarf   40 _____

41 Which star may once have been similar to our Sun in mass and luminosity?

(1) *Deneb*
(2) *Spica*
(3) *Procyon B*
(4) *Proxima Centauri*   41 _____

42 Energy is produced in the cores of main sequence stars when

(1) lighter elements undergo fusion into heavier elements
(2) heavier elements undergo fusion into lighter elements
(3) cosmic background radiation is absorbed
(4) cosmic background radiation is released   42 _____

Base your answers to questions 43 and 44 on the graph below and on your knowledge of Earth science. The graph shows the number of radioactive Isotope X atoms present as a sample of the isotope undergoes radioactive decay.

43 Based on the graph, the half-life of this radioactive isotope is
(1) 6 h
(2) 9 h
(3) 3 h
(4) 12 h

43 _____

44 Based on the graph, what is the approximate number of radioactive atoms of Isotope X that are present when 8 hours of decay has occured?
(1) 90
(2) 115
(3) 155
(4) 200

44 _____

Base your answers to questions 45 through 47 on the diagram below and on your knowledge of Earth science. The arrows in the diagram show air movement in a thunderstorm cloud. Point *A* represents a location in the atmosphere.

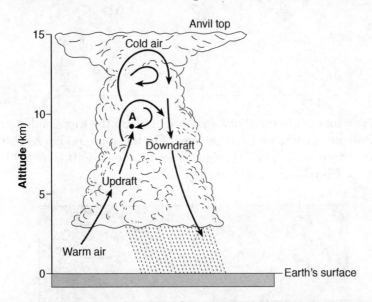

45 In which temperature zone of the atmosphere is point *A* located?

(1) thermosphere (3) stratosphere
(2) mesosphere   (4) troposphere   45 _____

46 The updrafts and downdrafts represented within this cloud are primarily caused by differences in

(1) altitude above sea level
(2) air density
(3) relative humidity
(4) specific heat   46 _____

47 Which weather symbol would be placed on a station model to represent this weather event?

(1) (2) (3) (4)

47 _____

Base your answers to questions 48 through 50 on the topographic map below and on your knowledge of Earth science. On the map, points *A*, *B*, *C*, and *D* represent surface locations. The dashed line between points *C* and *D* represents a hiking trail. Elevations are in feet (ft).

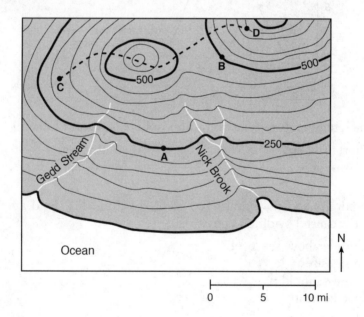

48 What is the contour interval on this map?
 (1) 25 ft
 (2) 50 ft
 (3) 150 ft
 (4) 250 ft

48 _____

49 The gradient between location A and location B is approximately

(1) 0.04 ft/mi
(2) 25 ft/mi
(3) 40 ft/mi
(4) 50 ft/mi

50 A person walks along the trail from location C to location D. The person will be walking

(1) downhill then uphill, only
(2) downhill, then uphill, then downhill again
(3) uphill then downhill, only
(4) uphill, then downhill, then uphill again

PART B–2
Answer all questions in this part.

Directions (51–65): Record your answers in the spaces provided. Some questions may require the use of the *2011 Edition Reference Tables for Physical Setting/Earth Science.*

Base your answers to questions 51 through 54 on the passage below and on your knowledge of Earth science.

The Mica Family

The familiar term "mica" is not the name of a specific mineral, but rather the name for a family of more than 30 minerals that share the same properties. All members of the mica family have high melting points and are similar in density, luster, hardness, streak, type of breakage, and crystal shape. As a result, telling the micas apart can be difficult. However, some common members of the family can be identified by color. For example, biotite is black to dark brown while muscovite can be light shades of several colors, or even colorless. When less common members of the mica family have any of these colors, or have similar colors, chemical tests are needed to tell them apart.

51 Identify the *two* chemical elements present in biotite mica that are *not* present in muscovite mica. [1]

_____ and _____

52 Identify the luster, hardness, and dominant form of breakage for members of the mica family. [1]

Luster: _____

Hardness: _____

Dominant form of breakage: _____

53 State the name of the igneous rock in which crystals of biotite mica are larger than 10 millimeters in diameter. [1]

_____

54 Large crystals of mica, sometimes weighing several hundred tons, have been found in igneous rock in Canada. Identify the environment of formation and the relative rate of cooling of the magma that formed the igneous rock containing these large crystals. [1]

Environment of formation: _____

Relative rate of cooling: _____

Base your answers to questions 55 through 58 on the diagram below and on your knowledge of Earth science. The diagram represents the Sun's apparent daily path for the first day of three seasons at 43° North latitude. The solid lines represent daytime paths as seen by an observer at this latitude. The dashed lines represent the nighttime paths that can *not* be seen by the observer.

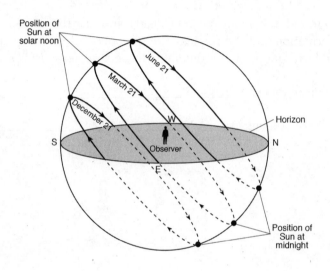

55 On the diagram *above*, draw an **X** to represent the solar noon position of the Sun as seen by the observer on April 21. [1]

56 Identify the rate of the Sun's apparent movement, in degrees per hour, along its path on December 21. [1]

_____ °/h

57 Identify the compass direction toward which the observer's shadow would point at solar noon on March 21. [1]

58 List the three dates shown on the diagram from the least number of nighttime hours to the greatest number of nighttime hours. [1]

Base your answers to questions 59 through 62 on the information below, on the map below, and on your knowledge of Earth science. The map shows a portion of the tectonic plates map from the *2011 Edition Reference Tables for Physical Setting/Earth Science*. Letters A and B represent locations on the ocean floor.

The area between North America and South America is a tectonically active region of Earth. This region contains all of the types of tectonic plate boundaries, and it has frequent earthquake and volcanic activity. The tectonic plates on either side of the East Pacific Ridge move at an average rate of 7.5 cm/year.

59 On the map *above*, draw *one* arrow in each of the two boxes to show the relative motion of the Caribbean Plate and the North American Plate. [1]

60 On the set of axes *below*, draw a line to represent the relative age of the ocean floor bedrock from location *A* to location *B*. [1]

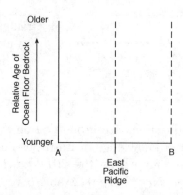

61 Identify the name of the hot spot shown on the map, and identify the name of the tectonic plate under which the center of this hot spot is located. [1]

_____ **Hot Spot**

_____ **Plate**

62 Identify the type of mafic igneous bedrock that is most likely to make up the oceanic crust at location *A*, and state the average density of this oceanic crust. [1]

Type of bedrock: _____

Density: _____ **g/cm³**

Base your answers to questions 63 through 65 on the diagram and data table below and on your knowledge of Earth science. The diagram represents laboratory materials used for an investigation of the effects of particle diameter on permeability and porosity (percentage of pore space). Four separate plastic tubes were filled to the same level with different particles.

(Not drawn to scale)

Particle Type	Particle Diameter (cm)	Time for Water to Infiltrate (s)	Porosity (%)
Sand	0.1	7	42.0
Clay	0.0003	322	40.0
Mixture	from 0.0003 to 0.8	15	34.0
Plastic beads	0.4	4	44.0

63 Explain why the particle sizes fit together more closely in the mixture, resulting in the lowest porosity of all these particle types. [1]

_____

_____

64 The height of the column of sand is 28 centimeters. Calculate the rate of infiltration, in centimeters per second, for the water that flowed through the column of sand. [1]

_____ cm/s

65 Based on the particle diameter of the plastic beads, identify the type of sediment represented by these beads. [1]

_____

PART C
Answer all questions in this part.

Directions (66–85): Record your answers in the spaces provided. Some questions may require the use of the *2011 Edition Reference Tables for Physical Setting/Earth Science*.

Base your answers to questions 66 through 68 on the weather map below and on your knowledge of Earth science. The map shows the location of a wintertime low-pressure system over Lake Ontario with two fronts extending into New York State. Isobar values are recorded in millibars. Partial weather station data are shown for several locations.

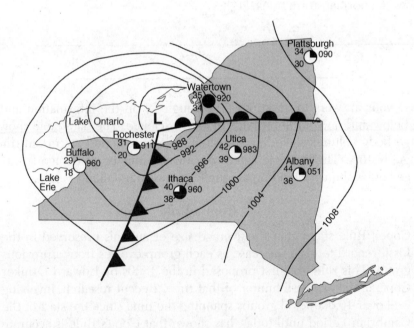

66 Describe the evidence shown on the map that indicates that the highest wind speeds occurred near Watertown, New York. [1]

_____

_____

67 Complete the table *below* by recording the weather data shown on the station model for Albany, New York. [1]

Albany, New York

Weather Variable	Weather Data
Dewpoint	°F
Cloud cover	%
Actual barometric pressure	mb

68 State the compass direction toward which the center of this low-pressure system moved over the next two days if the low followed a normal storm track. [1]

_____

_____

Base your answers to questions 69 through 72 on the information and data table below and on your knowledge of Earth science. The data table shows the average body volume, including the shell, of a brachiopod at certain times in geologic history. The geologic ages are shown in million years ago (mya). The average body volumes including the shell are shown in milliliters (mL).

Cope's Rule

Cope's Rule states that the average size of animals preserved in the fossil record tends to increase as each group evolves from a previous group. This rule was first proposed in the 1800s by Edward Drinker Cope, a famous fossil hunter of that time. Recent research, involving well over 10,000 fossil groups spanning the time since the start of the Cambrian Period until today, has shown that Cope's Rule is accurate for most animal groups. Brachiopod data support Cope's Rule.

Brachiopod Data Table

Geologic Age (mya)	Average Body Volume Including the Shell (mL)
480	0.1
460	0.2
430	0.6
410	1.0
380	1.1

69 On the grid *below*, plot the average brachiopod body volume for each of the geologic ages listed in the data table. Connect *all five* plots with a line. [1]

70 Identify, by name, *two* geologic periods when the brachiopods represented in the data table were living. [1]

_____ **Period**

and _____ **Period**

71 State the names of the *two* brachiopod index fossils found in New York State bedrock. [1]

_____ and _____

72 The earliest horses appeared in the Eocene epoch and were about the size of a large dog of today. Explain how the evolution of horses supports Cope's Rule. [1]

_____

_____

_____

EXAMINATION June 2018 199

Base your answers to questions 73 through 75 on the snowfall map below and on your knowledge of Earth science. The snowfall map shows some average yearly snowfall values, measured in inches, recorded for a portion of New York State. Some average yearly snowfall isolines have been drawn. Line *XY* is a reference line on the map. The cities of Watertown and Oswego are shown on the map.

73 On the map *above*, draw the 240-inch average yearly snowfall isoline. [1]

74 On the grid *below*, construct a profile of the average annual snowfall along line XY by plotting the value of each isoline that crosses line XY. Connect *all six* plots with a line to complete the profile. [1]

75 The diagram *below* represents an observer standing next to the side of a building. Using the scale shown, draw an **X** on the side of the building to represent the height of the greatest amount of average yearly snowfall that is indicated on the map. [1]

Base your answers to questions 76 through 78 on the diagram below and on your knowledge of Earth science. The diagram represents a cutaway view of a flat-plate solar collector used to heat water at a New York State location.

76 Identify the energy transfer process by which light travels through space from the Sun to the solar collector. [1]

_____ ++

77 Explain why the flow tubes and collector plate inside the solar collector are black in color. [1]

_____

_____

_____

78 The glass cover on this solar collector allows visible light to enter the collector. Identify the type of electromagnetic energy emitted by the flow tubes and collector plate that is trapped inside the collector by the glass cover. Also, circle the relative wavelength of this trapped electromagnetic energy compared to wavelengths of visible light. [1]

Emitted electromagnetic energy:

_____

Relative wavelength (circle one):

        shorter   longer   the same

Base your answers to questions 79 through 82 on the passage and diagram below, on the data table on the next page, and on your knowledge of Earth science. The diagram compares the inner planets of our solar system to the planetary system surrounding the star *Kepler-62*, which is located in our galaxy. The data table shows some data for the planets in the *Kepler-62* system.

Kepler-62 Planetary System

Five planets orbit a seven-billion-year-old star, *Kepler-62*, which has a surface temperature of approximately 4900 Kelvin. Two of these planets are located within the habitable zone, which is the region around a star where life may exist due to the possible presence of water in the liquid phase. The shaded areas in the orbital diagrams below indicate the habitable zone of each system.

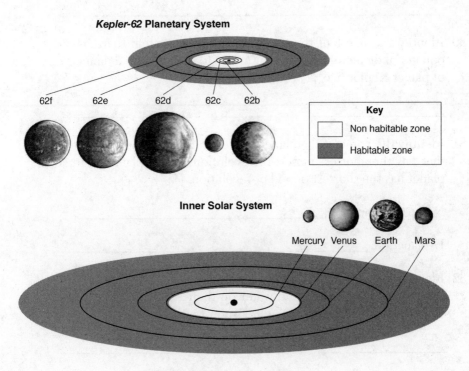

(Planets and orbits are drawn to scale)

Data Table

Name of Planet	Distance from *Kepler-62* (million kilometers)	Equatorial Diameter (compared to Earth's diameter)
62b	8.23	1.31
62c	13.76	0.54
62d	17.95	1.95
62e	63.88	1.6
62f	107.41	1.4

79 Identify the name of the galaxy where the *Kepler-62* planetary system is located. [1]

_____

80 Identify the name of the planet in our solar system that has an equatorial diameter most similar in size to the equatorial diameter of planet Kepler-62c. [1]

_____

81 Identify the name of the planet in the *Kepler-62* planetary system that has the shortest period of revolution, and explain why this planet has the shortest period of revolution. [1]

Name of planet: _____

Explanation: _____

82 Identify the names of the *two* planets in the *Kepler-62* planetary system that may have liquid water on their surfaces, and explain why these planets may have liquid water on their surfaces. [1]

_____ and _____

Explanation: _____

_____

Base your answers to questions 83 through 85 on the block diagram below and on your knowledge of Earth science. The diagram represents glacial features formed by a continental glacier and its melt water.

83 Describe the arrangement of the sediments found within the terminal moraine, which marks the farthest advance of the glacier. [1]

_____

_____

84 The cross sections below, labeled A, B, C, and D, represent four stages in the development of a kettle lake. The stages are *not* shown in the correct order.

Stages in Kettle Lake Formation

On the lines below, place the letters in the correct order to indicate the sequence of development of a kettle lake from earliest stage to latest stage. [1]

_____ → _____ → _____ → _____

Earliest stage ⟶ Latest stage

85 Terminal moraines found on Long Island were deposited during the advance and retreat of glacial ice during the last ice age. Identify, by name, the geologic epoch during which these moraines were deposited. [1]

_____ **epoch**

Answers June 2018
Physical Setting/Earth Science

Answer Key

PART A

1. 3	8. 1	15. 2	22. 2	29. 4
2. 4	9. 4	16. 1	23. 3	30. 1
3. 1	10. 2	17. 1	24. 1	31. 1
4. 1	11. 2	18. 4	25. 4	32. 1
5. 3	12. 3	19. 3	26. 2	33. 3
6. 4	13. 2	20. 4	27. 3	34. 1
7. 2	14. 4	21. 2	28. 4	35. 2

PART B–1

36. 2	39. 3	42. 1	45. 4	48. 2
37. 1	40. 1	43. 3	46. 2	49. 2
38. 4	41. 3	44. 3	47. 3	50. 4

PART B–2 and **PART C**. *See* **Answers Explained**.

Answers Explained

PART A

1. 3 A solar eclipse occurs when the Moon passes directly between Earth and the Sun, casting a shadow on Earth. An observer on Earth would see the Moon blocking the Sun. The Moon does not produce visible light of its own. The Moon is visible only due to light from the Sun that is reflected from the Moon's surface. When the Moon passes between Earth and the Sun, the side of the Moon facing the Sun is illuminated and the side of the Moon facing Earth is in darkness. Thus, it is reasonable to infer that the dark object B with the circular surface is the Moon and the bright object A in the partial solar eclipse photograph is the Sun.

The Sun is surrounded by a layer of hot gases called its atmosphere. The outermost part of the Sun's atmosphere is called the corona. The corona is usually hidden by the bright light of the Sun's surface. During a total solar eclipse, though, the Moon blocks the light of the Sun's surface. So the corona can be seen as a glowing aura of light around the dark surface of the Moon. Therefore, object A in the total solar eclipse photograph is the corona, which is a part of the Sun, and object B is the Moon. Thus, in both photographs, A is the Sun and B is the Moon.

2. 4 The planets can be divided by mass and density into the terrestrial (Earth-like) and the Jovian (Jupiter-like) planets. The terrestrial planets are small, dense, and rocky. They are composed mostly of metal silicates and iron. The terrestrial planets include the four innermost planets: Mercury, Venus, Earth, and Mars. The Jovian planets are large, low in density, and gaseous. They are composed mainly of hydrogen and helium. The Jovian planets include the outer planets: Jupiter, Saturn, Uranus, and Neptune.

Find the Solar System Data table in the *Reference Tables for Physical Setting/Earth Science*. Locate the column labeled "Period of Rotation at Equator." Note that the periods of rotation for the terrestrial planets range from 23 h 56 min 4 s to 243 d. Note that the periods of rotation of the Jovian planets range from 9 h 50 min 30 s to 17 h 14 min. Thus, compared to the terrestrial planets, the Jovian planets have shorter periods of rotation.

WRONG CHOICES EXPLAINED:

(1) Find the Solar System Data table in the *Reference Tables for Physical Setting/Earth Science*. Locate the column labeled "Mass (Earth = 1)." Note that the terrestrial planets range in mass from 0.06 to 1.00 Earth masses. Note that the Jovian planets range from 14.54 to 317.83 Earth masses. Thus, compared to terrestrial planets, Jovian planets are more massive, not less massive.

(2) Find the Solar System Data table in the *Reference Tables for Physical Setting/Earth Science*. Locate the column labeled "Density (g/cm^3)." Note that the densities of the terrestrial planets range from 3.9 g/cm^3 to 5.5 g/cm^3 while the densities of the Jovian planets range from 0.7 g/cm^3 to 1.8 g/cm^3. Thus, compared to terrestrial planets, Jovian planets are less dense, not more dense.

(3) Increasing the distance between a star and a planet orbiting the star decreases the gravitational attraction between the two objects. This, in turn, causes the orbital velocity of the planet to decrease. Find the Solar System Data table in the *Reference Tables for Physical Setting/Earth Science*. In the column labeled "Mean Distance from Sun (million km)," note that from Mercury to Neptune, mean distance from the Sun increases. Thus, Jovian planets are farther from the Sun than terrestrial planets. Therefore, compared to terrestrial planets, the Jovian planets have lesser, not greater, orbital velocities.

3. **1** According to the Big Bang theory, the universe started out with all of its matter and energy in a small volume and then expanded outward in all directions, a motion very much like an explosion. The explosion associated with the Big Bang theory and the formation of the universe occurred *before* the formation of Earth and the solar system. In other words, the universe is older than either Earth or the solar system. Find the Geologic History of New York State chart in the *Reference Tables for Physical Setting/Earth Science*. Locate the column labeled "Eon." Note that the entry "Estimated time of origin of Earth and solar system" corresponds to 4600 million years ago (4.6 billion years ago) on the time scale. Thus, the explosion associated with the Big Bang theory took place more than 4.6 billion years ago. Based on the current expansion rate of the universe, scientists extrapolate back to when the universe was all concentrated in a small volume and infer that the Big Bang occurred approximately 13.8 billion years ago. Thus, of the choices given, the event that occurred more than 10 billion years ago is the Big Bang.

WRONG CHOICES EXPLAINED:

(2) Earth had to exist before life on Earth could originate. Therefore, the origin of life occurred after the estimated time of origin of Earth and the solar system 4.6 billion years ago. Thus, the origin of life on Earth occurred less than 10 billion years ago, not more than 10 billion years ago.

(3) Find the Geologic History of New York State chart in the *Reference Tables for Physical Setting/Earth Science*. Locate the column labeled "Important Geologic Events in New York." Note the entry "Pangaea begins to break up." Trace horizontally to the left to the time scale labeled "Million years ago," and note that Pangaea began to break up about 200 million years ago. Thus, Pangaea began to break up less than, not more than, 10 billion years ago.

(4) The explosion associated with the Big Bang theory and the formation of the universe occurred before the formation of Earth and the solar system 4.6 billion years ago. Thus, the origin of Earth and its Moon occurred less than, not more than, 10 billion years ago.

4. **1** In 1851, French physicist Jean Foucault suspended a heavy iron ball on a long steel wire from the top of the dome of the Pantheon in Paris. As this pendulum swung back and forth, it appeared to change direction slowly, passing over different lines on the floor until eventually coming full circle to its original position after 24 hours. Since Foucault knew that the path of a freely swinging pendulum would not change on its own, he concluded that the apparent shift in the direction of swing of the pendulum was due to the floor (Earth's surface) rotating beneath the pendulum. Thus, the apparent change in direction of swing of a Foucault pendulum provides evidence that Earth is rotating.

WRONG CHOICES EXPLAINED:
(2) The apparent direction of swing of a Foucault pendulum changes cyclically every 24 hours. Earth takes 365 days to revolve around the Sun. If the cyclic change in a Foucault pendulum was due to Earth revolving, the path would change over a period of 365 days, not over a period of 24 hours.
(3) The shape of the surface beneath a freely swinging pendulum, whether that surface is flat or spherical, has no effect on the pendulum's direction of swing because there is no contact between the pendulum and the surface. Thus, a Foucault pendulum provides no evidence that Earth has a curved surface.
(4) The swing of any pendulum is the result of Earth's gravity acting on the suspended weight. However, Earth's gravity would act on a pendulum's suspended weight causing it to swing even if Earth was not rotating. Foucault used the cyclic change in direction of a swinging pendulum to demonstrate Earth's rotation, not that Earth had a gravitational pull.

5. **3** Earth revolves around the Sun once every 365.26 days, or about 1° per day. A month contains approximately 30 days. Therefore, in one month Earth travels approximately 30° in its orbit.

6. **4** It is given that the dry-bulb temperature is 16°C and the wet-bulb temperature is 10°C. Thus, the difference between the wet-bulb and dry-bulb temperatures is 6°C. Find the Relative Humidity (%) table in the *Reference Tables for Physical Setting/Earth Science*. Locate the column headed "6" in the "Difference Between Wet-Bulb and Dry-Bulb Temperatures (°C)" scale along the top of the chart. Find the "Dry-Bulb Temperature (°C)" scale along the left side of the chart. Locate the row labeled "16," trace to the right until it

intersects the column headed "6," and note the value of 45. Thus, when the dry-bulb temperature is 16°C and the wet-bulb temperature is 10°C, the relative humidity is 45%.

7. **2** People board up windows to protect the windows from being broken by very high winds and windblown debris. If windows break and high winds can penetrate a house, these winds can put pressure onto the roof, causing it to lift off the building. Of the natural disasters listed as choices, only a hurricane produces very high winds. Thus, boarding up windows is a common emergency action taken to prepare for a hurricane.

WRONG CHOICES EXPLAINED:
(1) The motions of an earthquake shake a building and can cause the building to break and, in turn, can cause windows to shatter. However, boarding up windows would not prevent the building from shaking and thus would not prevent the windows from breaking.

(3) Windows are typically placed above floor level. Rising water entering through openings at floor level would cause extensive damage before it reached high enough to penetrate a window. Thus, boarding up windows would not prevent a home from being damaged by flooding and is not an emergency action taken to prepare for a flood.

(4) Tsunamis are waves caused by undersea movements of Earth's crust. Tsunamis may be only a few meters high, but they have very long wavelengths and travel much more rapidly than ordinary ocean waves. When tsunamis reach shallow water, they are slowed by friction with the ocean bottom and begin to pile up as large waves. In bays and narrow channels, their high speed and long wavelength may be funneled into huge breaking waves more than 40 m high. The force exerted by such a huge, fast-moving mass of water can do extensive damage and can cause sudden, widespread coastal flooding. Most of the deaths caused by tsunamis are drownings. Boarding up windows would do little to prevent flooding and drowning and is therefore not an emergency action taken to prepare for a tsunami.

8. **1** When two different air masses meet, the boundary between them is called a front. Find the Planetary Wind and Moisture Belts in the Troposphere diagram in the *Reference Tables for Physical Setting/Earth Science*. Locate the polar front jet stream indicated by the symbol ⊗ at the boundary between the convection cell located between the North Pole and 60° N latitude and the convection cell located between 30° N and 60° N latitude. The polar front marks the boundary along which cold air masses of the polar convection cell meet warm tropical air masses carried north by the convection cell between 30° N and 60° N.

The sharp temperature gradient at this boundary results in high winds. Combined with the effects of Earth's rotation, these high winds form jet streams. The polar jet stream that tracks through the United States moves from west to east. Cold fronts pushing against the polar front from the north and warm air masses pushing against the polar front from the south cause bends to form in the polar front and its jet stream. In the winter, extremely cold air masses pushing against the boundary from the north can cause this boundary to dip to the south of its usual position at 60° N.

Now, find the Tectonic Plates map in the *Reference Tables for Physical Setting/Earth Science*. Locate 60° N along the right-hand edge of the map, and trace horizontally to the left across North America. Note that the Great Lakes are located south of latitude 60° N. Note that in diagram (1), the polar front jet stream dips below the Great Lakes and south of 60° N latitude. Thus, diagram (1) best represents the general position and flow of the polar front jet stream in the Northern Hemisphere during the winter months.

WRONG CHOICES EXPLAINED:

(2) Diagram (2) shows the polar front jet stream located near the equator. Although on rare occasions the polar front jet stream has dipped as far south as northern Florida, the polar front jet stream has never been observed to reach as far south as the equator. Thus, the general position of the polar front jet stream during the winter months is not near the equator.

(3) The polar jet stream that tracks through the United States moves from west to east. The arrows on diagram (3) show the direction of flow of the polar front jet stream to be from east to west, not from west to east.

(4) Diagram (4) shows the polar front jet stream near the equator with a direction of flow from east to west. As explained above for choices (2) and (3), this does not represent the general position and direction of flow of the polar front jet stream.

9. **4** An observer on Earth can see only that portion of the Moon that is illuminated by the Sun's rays and only from the side of Earth that is in darkness. In the diagram, the part of the Moon that is *not* shaded represents the illuminated portion of the Moon. According to the diagram, when the Moon is in position D, its entire illuminated portion faces Earth. An observer in New York State would see the full moon phase. When the Moon is in position C, most, but not all, of the illuminated portion would be visible to the observer's right. The observer would see the new gibbous moon phase. When the Moon is in position B, half of the illuminated portion would be visible to the observer's right. The observer would see the first quarter moon phase. When the Moon is in position A, only a thin crescent of the illuminated portion of the Moon's surface would be visible to the observer's right. The observer

would see the new crescent moon phase. Thus diagram (4), which shows the full moon at position D and the illuminated portions at positions A, B, and C to the observer's right, best represents the sequence of Moon phases as seen by an observer in New York State.

WRONG CHOICES EXPLAINED:

(1) and (2) As explained above, an observer in New York State would see the illuminated portions of the Moon to the observer's right. Both of these choices show the illuminated portions of the Moon to the observer's left.

(3) This choice shows the full moon at position A, not at position D.

10. **2** According to the diagrams, compared to the spectral lines of hydrogen observed in a laboratory, the spectral lines of hydrogen observed in the light from a distant star are shifted toward the end of the spectrum labeled "longer wavelength." Find the Electromagnetic Spectrum chart in the *Reference Tables for Physical Setting/Earth Science*. Note that visible light at the red end of the spectrum has a longer wavelength than light at the blue end of the spectrum. Thus, a shift toward the longer wavelength end of the spectrum is a shift toward the red end of the spectrum. If a source of electromagnetic waves is moving *away* from an observer at the same time as the source is emitting light of a particular wavelength, fewer wave crests will reach the eye of the observer each second. The eye will interpret this as meaning that the light has a longer wavelength than it actually has. In other words, the light will appear shifted toward the red end of the spectrum. The fact that the spectral lines of hydrogen from a distant star show a red shift indicates that the distant star is moving away from Earth. Therefore, compared to the spectral lines observed in the laboratory, the spectral lines observed in the light from the distant star have shifted toward the red end of the spectrum, indicating the star's movement away from Earth.

WRONG CHOICES EXPLAINED:

(1) If a source of electromagnetic waves is moving toward an observer at the same time as it is emitting light of a particular wavelength, more wave crests will reach the eye of the observer each second. The eye will interpret this as meaning that the light has a shorter wavelength than it actually has. In other words, the light will appear shifted toward the blue end of the spectrum. The diagrams in this question show the opposite.

(3) and (4) According to the diagrams, compared to the spectral lines of hydrogen observed in a laboratory, the spectral lines of hydrogen observed in the light from a distant star are shifted toward the red end of the spectrum, not toward the blue end of the spectrum.

11. **2** Earth's rotational axis is tilted at an angle of 23.5° from a perpendicular to Earth's orbital plane.

12. **3** A constellation is an imaginary pattern formed by a group of stars in an area of the sky as seen by an observer on Earth. As Earth revolves around the Sun, the side of Earth facing the Sun experiences day and the side of Earth facing away from the Sun experiences night. Stars are visible only at night. Therefore, the stars (and constellations) seen by an observer on Earth vary as Earth revolves around the Sun and as Earth's night side faces different portions of the universe. During winter nights in January, an observer in New York State sees constellations in the region of the universe facing Earth's night side. Six months later during summer nights in July, Earth has revolved to a position in its orbit directly opposite the Sun from its winter position. Earth's night side now faces the region of the universe directly opposite the region it faced in winter. So different constellations are visible. Therefore, the best explanation for why these two constellations are visible in the night sky in the months identified is that the nighttime side of Earth is facing different parts of our galaxy as Earth orbits the Sun.

WRONG CHOICES EXPLAINED:

(1) Earth's daily rotation causes a daily apparent motion of constellations, not a seasonal change in constellations visible from Earth.

(2) The rate at which Earth spins on its axis affects the daily apparent motion of constellations, not the seasonal change in constellations visible from Earth.

(4) The idea that the stars in constellations revolve around Earth is part of the geocentric theory, which has been proven to be incorrect. Furthermore, Earth is too small and too far from the stars in constellations to exert enough gravitational attraction to hold the stars in orbit around Earth and cause them to revolve around Earth.

13. **2** Find the Inferred Properties of Earth's Interior chart in the *Reference Tables for Physical Setting/Earth Science*. At the bottom of the chart, find the horizontal scale labeled "Depth (km)" and locate the value "4000." Trace the vertical line corresponding to 4000 km upward until it intersects the dashed line labeled "Melting Point." From this intersection, trace horizontally left to the scale labeled "Temperature (°C)." Note that the melting point corresponds to "5400." Now, trace the vertical 4000 km depth line upward until it intersects the solid line labeled "Interior Temperature." From this intersection, trace horizontally left to the scale labeled "Temperature (°C)." Note that the interior temperature corresponds to "5700." Thus, the temperature of the materials at a depth of 4000 km is

5700°C and their melting point is 5400°C. Since the temperature of the materials exceeds their melting point, they are in the liquid state (phase) of matter. Thus, the table that correctly shows the interior temperature, melting point, and state (phase) of matter is shown in choice (2).

14. **4** Deforestation is the removal of a forest or a stand of trees to clear the land for other uses such as farms, ranches, and construction projects. Trees absorb carbon dioxide (CO_2) and emit oxygen (O_2) as they carry out photosynthesis. Photosynthesis ties up CO_2 in the form of sugars, starches, and cellulose. Therefore, deforestation decreases the amount of carbon dioxide removed from the atmosphere by trees, causing a net increase in carbon dioxide in the atmosphere. Thus, the greenhouse gas that has increased in Earth's atmosphere partly as a result of deforestation over the last 100 years is carbon dioxide.

15. **2** Find the Surface Ocean Currents map in the *Reference Tables for Physical Setting/Earth Science*. Locate Africa. Note that the continent of Africa is split, with the western half on the right side of the map and the eastern half on the left side of the map. Locate the southeastern tip of Africa on the left side of the map. Note that the ocean current along the southeastern tip of southern Africa is the Agulhas Current, which flows to the south. According to the key, the Agulhas Current is a warm current. Thus, the ocean current that brings warm water to the southeastern coast of Africa is the Agulhas Current.

16. **1** Find the Geologic History of New York State chart in the *Reference Tables for Physical Setting/Earth Science*. In the column labeled "Life on Earth," find the entry "Humans, mastodonts, mammoths." Trace left to the scale labeled "Million years ago," and note that humans appeared 1.8 million years ago. In the column labeled "Era," find the statement "Estimated time of origin of Earth and the solar system." Trace left to the scale labeled "Million years ago," and note that the origin of Earth occurred 4600 million years ago. Thus, the approximate percentage of geologic time that humans have existed on Earth since its origin is 0.039% (1.8/4600 × 100 = 0.039%), or less than 1%. Thus, the pie graph that is shaded to best represent the approximate percentage of time that humans have existed during Earth's entire history is shown in choice (1).

17. **1** Volcanic ash is ejected high into the atmosphere during a volcanic eruption. The ash is then carried by global winds, quickly spreads out, and finally settles to the ground, forming a volcanic ash layer over a wide area. Once the volcanic eruption is over, the source of volcanic ash is cut off and deposition quickly tapers off. Thus, a volcanic ash layer can be used as a time marker for correlating rock layers because the ash is deposited rapidly over a large area.

WRONG CHOICES EXPLAINED:

(2) A buried erosional surface represents a gap in the sequence of rock layers due to uplift and erosion, causing some of the rock layers to be removed. A layer of volcanic ash is the result of deposition, not erosion.

(3) Intrusive igneous rock forms when molten magma cools beneath Earth's surface. Volcanic ash consists of tiny particles of rock that form from lava ejected from a volcano and is a type of extrusive igneous rock. Thus, volcanic ash forms an extrusive igneous rock, not an intrusive igneous rock.

(4) Volcanic ash is deposited from the air and forms a layer on top of Earth's exposed surface, not across rock layers.

18. **4** According to the cross section, locations A and B are at the same elevation on different sides of the mountain. The arrow labeled "Prevailing wind" indicates that location A is on the windward side of the mountain and that location B is on the leeward side of the mountain. As moist air is carried upward and over the windward side of the mountain by the prevailing wind, it expands and cools. When the air temperature reaches the dewpoint, moisture begins to condense and clouds form, causing precipitation at location A. As the air descends on the leeward side of the mountain, it is compressed and the air temperature increases. Thus, location B has a warmer climate than location A. The increase in air temperature causes a decrease in relative humidity, resulting in less precipitation at location B. Thus, location B has a drier climate than location A. Therefore, compared to the climate at location A, the climate at location B is warmer and drier.

19. **3** Find the Scheme for Sedimentary Rock Identification in the *Reference Tables for Physical Setting/Earth Science*. In the "Rock Name" column, locate "conglomerate." Trace left to the "Texture" column, and note that conglomerate has a "clastic (fragmental)" texture. Clastic sedimentary rocks are composed of fragments of rock, or sediments. Thus, sediments are rocks that already existed and that were broken into fragments by weathering and erosion. Sediments become a clastic sedimentary rock through the process of lithification. Lithification begins when the sediments are buried and become compacted. Then certain minerals begin to recrystallize in the fluids trapped in the pores among the rock fragments that make up the sediments, cementing the fragments together. Thus, the rock particles of the sediments had to exist already in order to be compacted and cemented together to form the clastic sedimentary rock. Therefore, compared to the age of the calcite cement and the conglomerate, the relative age of the pebbles is older than both the calcite cement and the conglomerate.

20. **4** All processes of the water cycle involve the movement of water. The arrows at letter *A* extend from the surface of the ocean, lake, and Earth's surface to the atmosphere. Water on land surfaces and in oceans and lakes exists mainly as liquid water. Water in the atmosphere exists mainly as water vapor. Therefore, the process at letter *A* involves liquid water entering the atmosphere as water vapor. The process by which water changes from a liquid to a vapor is called evaporation, or vaporization. Therefore, water cycle process *A* is evaporation.

The arrows at letter *B* extend from trees to the atmosphere. Trees are plants. Water in plants exists mainly as liquid water. Water in the atmosphere exists mainly as water vapor. Therefore, letter *B* represents a process by which liquid water enters the atmosphere as water vapor from plants. Plants remove liquid water from the ground and release it into the atmosphere in the form of water vapor by a process called transpiration. In transpiration, liquid water absorbed by plant roots from the soil increases the water pressure inside the lower parts of the plant. Simultaneously, evaporation of water through plant stems or leaf stomata decreases the water pressure in the upper parts of a plant. This difference in pressure causes the liquid water to move upward from the roots and toward the leaves. When the liquid water reaches the leaves, it evaporates. Then more water is drawn upward from the roots, and the process of transpiration continues. Thus, water cycle process *B* best represents the process of transpiration because it shows water moving into the atmosphere from plants. The arrows at letter *C* extend from a cloud toward Earth's surface. Clouds are composed of tiny droplets of liquid water. Therefore, it is a process by which the liquid water droplets in the cloud fall to the ground. The process by which water in the form of rain, snow, hail, or sleet falls to the ground is called precipitation. Therefore, water cycle process *C* is precipitation. The process labeled with the letter *D* is shown taking place in a cloud and in fog. Clouds and fog are composed of tiny droplets of liquid water that form when water vapor in air cools to its dewpoint and changes from the gas phase to the liquid phase. The change from a gas to a liquid is called condensation. Therefore, water cycle process *D* is condensation. Thus, the table in choice (4) correctly matches each letter with the process it represents: *A*—evaporation, *B*—transpiration, *C*—precipitation, and *D*—condensation.

21. **2** Arid climates are dry, and humid climates are wet. Find the Planetary Wind and Moisture Belts in the Troposphere diagram in the *Reference Tables for Physical Setting/Earth Science*. Note the zones labeled "wet" and "dry." Note that dry (arid) zones exist at 90° N and 90° S (the poles) and at 30° N and 30° S. Note that wet (humid) zones exist at 0° (the equator) and at 60° N and 60° S. Thus, table (2) best shows the relationship between latitude and general climate conditions on Earth.

22. **2** Wave action is turbulent. The turbulence of waves crashing against the shore roils up sediments. As sediments are thrown about by wave action, they bounce off of and rub against one another and bounce, roll, and scrape against the bottom. These collisions cause smaller pieces to break off of sharp corners and edges on the surface of the sediments, particularly at corners that protrude—a process called abrasion. As a result, the sediments become rounded. Thus, table (2) shows the most likely process (abrasion) and agent of erosion (wave action) responsible for the rounded sediments shown in the photograph.

WRONG CHOICES EXPLAINED:
(1) Sandblasting is the process by which sand carried by the wind collides with and wears away solid objects. When the sand particles collide with the solid object, the impact causes small pieces of the object to break off. This can cause the object being hit with particles to become smooth or worn. The agent of erosion responsible for sandblasting is wind, not running water.

(3) Mass movements involve the downhill movement of sediments under the direct influence of gravity. When pieces of bedrock are deposited by a landslide, the particles move downhill rapidly through air. Air resistance is so small that particles of all sizes move through air at virtually the same speed. Thus, the particles do not become sorted by size as they fall. During the landslide, turbulence causes the pieces of bedrock material to collide and break apart into angular fragments. Thus, the sediments deposited by a landslide are best described as angular and unsorted. Thus, a mass movement such as a landslide is unlikely to be responsible for this rounded sediment.

(4) By definition, deposition is the process by which transported sediment is dropped in a new place. Deposition by wind would cover the sediments shown in the photograph and protect them from abrasion, not cause them to become rounded.

23. **3** Note that basaltic intrusion B cuts across both rock unit C and basaltic intrusion A. In order for basaltic intrusion B to cut across both of these rocks, A and C already had to exist when B formed. Therefore, basaltic intrusion B is the youngest rock. Note that basaltic intrusion A cuts across rock unit C. Therefore, rock unit C already had to exist when basaltic intrusion A formed. Therefore, rock unit C is older than basaltic intrusion A. Thus, the relative age of these three rock units from oldest to youngest is $C \rightarrow A \rightarrow B$.

24. **1** A rift valley is an elongated, low-lying region that forms where Earth's tectonic plates move apart, or rift, allowing the land to drop down between parallel faults. These valleys also form when oceanic plates are moving apart and forming a divergent boundary. Find the Tectonic Plates map in the *Reference Tables*

for *Physical Setting/Earth Science*. On the map, locate the positions corresponding to surface locations *A*, *B*, *C*, and *D*. Location *A* corresponds to the East African Rift, and the arrows indicate that the plates are diverging in this location, or moving apart. Location *B* corresponds to a convergent plate boundary. Location *C* corresponds to the Hawaii Hot Spot. Location *D* corresponds to a transform plate boundary. Thus, the location that is on a major rift valley is *A*.

WRONG CHOICES EXPLAINED:

(2) Rift valleys form at divergent boundaries. According to the Tectonic Plates map in the *Reference Tables for Physical Setting/Earth Science*, location *B* corresponds to a convergent boundary.

(3) Rift valleys form at divergent boundaries. Location *C* is in the center of the Pacific Plate, far from any plate boundaries. According to the Tectonic Plates map in the *Reference Tables for Physical Setting/Earth Science*, location *C* corresponds to the Hawaii Hot Spot, which is associated with the formation of volcanic mountains, not rift valleys.

(4) Rift valleys form at divergent boundaries. According to the Tectonic Plates map in the *Reference Tables for Physical Setting/Earth Science*, location *D* is along the northern boundary between the Scotia Plate with the South American Plate. Location *D* corresponds to a transform boundary, not a divergent boundary.

25. **4** It is given that the *P*-wave travel time was 11 minutes. Find the Earthquake P-Wave and S-Wave Travel Time graph in the *Reference Tables for Physical Setting/Earth Science*. Locate 11 minutes on the vertical axis labeled "Travel Time (min)." From this value, trace horizontally to the right until you intersect the bold line labeled "P." From the intersection, trace vertically downward to the horizontal axis labeled "Epicenter Distance ($\times 10^3$ km)" at the bottom of the graph and note the value 7.6×10^3, or 7600 km. Thus, the distance to the epicenter is 7600 km. To determine the *S*-wave travel time, trace vertically upward from 7600 km until you intersect the bold curve marked "S," and then trace horizontally to the left to the vertical scale labeled "Travel Time (min)." Note that this corresponds to 20 min. Thus, the distance to the epicenter was 7600 km, and it took the first *S*-wave 20 minutes to travel that distance.

26. **2** It is given that the Catskills are part of the Alleghany Plateau. They are commonly called mountains because of their high elevation and hilly surface. Mountains are typically formed by folding or faulting due to tectonic motions or by volcanism. In contrast, plateaus generally have an underlying structure of horizontal layers of rock. Find the Generalized Bedrock Geology of New York State

map in the *Reference Tables for Physical Setting/Earth Science.* Locate the region corresponding to the Catskills, and note the map symbol for the surface bedrock. Locate this map symbol in the key labeled "Geologic Periods and Eras in New York," trace right, and note the statement "limestones, shales, sandstones, and conglomerates." Find the Scheme for Sedimentary Rock Identification in the *Reference Tables for Physical Setting/Earth Science.* Locate the column labeled "Rock Name," and note that limestone, shale, sandstone, and conglomerate are all sedimentary rocks. Streams and other agents of erosion can cut deep valleys into the surface of a plateau, and the walls of these valleys can be eroded into steep hills. This is most likely how the Catskills formed. Thus, although they look like small mountains, the Catskills are classified as a plateau because of their bedrock structure.

WRONG CHOICES EXPLAINED:
(1) Plateaus are located at high elevations, not at low elevations.
(3) A region is classified as a plateau because of its elevation, relief, and underlying bedrock structure. Bedrock age alone does not determine how a landscape region was formed, its structure, or the processes that shape it. For example, bedrock of many different ages may have been formed by volcanic activity, metamorphism, or deposition in water, giving rise to very different types of landscape regions.
(4) As explained above, the Catskills consist of limestones, shales, sandstones, and conglomerates—all sedimentary rocks. If the Catskills had undergone a high degree of metamorphism, these sedimentary rocks would have been changed into metamorphic rocks. That is not the case.

27. **3** Find the Relationship of Transported Particle Size to Water Velocity graph in the *Reference Tables for Physical Setting/Earth Science*. Locate 0.1 cm on the vertical axis labeled "Particle Diameter (cm)." Trace horizontally to the right until you intersect the bold curve. At this point, trace down to the horizontal axis labeled "Stream Velocity (cm/s)." Read the minimum stream velocity that can maintain the transportation of a particle 0.1 cm in diameter. That minimum stream velocity is 5.5 cm/s.

28. **4** Evaporites are sediments chemically precipitated due to evaporation of an aqueous solution (such as seawater). Find the Scheme for Sedimentary Rock Identification in the *Reference Tables for Physical Setting/Earth Science.* In the "Comments" column, note the statement "crystals from chemical precipitates and evaporites." Trace this section to the right to the column labeled "Rock Name." Note that rock salt, rock gypsum, and dolostone can form from evaporites. Thus, the rock that is classified as an evaporite is crystalline rock salt.

29. **4** Find the Average Chemical Composition of Earth's Crust, Hydrosphere, and Troposphere chart in the *Reference Tables for Physical Setting/Earth Science*. Locate the column labeled "Percent by volume" in the section labeled "Crust." In this column, find the row corresponding to the greatest percent by volume (94.04). Trace this row left to the column labeled "Element (symbol)." Note that the element that makes up the greatest percent by volume of Earth's crust is oxygen. Now locate the row corresponding to the second greatest percent by volume (1.42). Trace this row left to the column labeled "Element." Note that the element that makes up the second greatest percent by volume of Earth's crust is potassium. Thus, the pair of elements that makes up most of Earth's crust by volume is potassium and oxygen.

30. **1** According to the diagram, soil zone A is located just beneath the surface and contains plant roots. The surface of zone A shows grass, plants, and trees. Leaves and other plant materials that die fall to the ground and begin to decompose. Rain washes these organic materials into the pore spaces in zone A. Thus, zone A contains both fresh and partially decomposed organic matter. Note that the number of larger particles decreases as you move upward from the underlying bedrock in zone D to zone A. This is because the particles near the surface have been exposed to the weathering process for a longer period of time and have been broken down into smaller and smaller pieces by weathering. Thus, the zone having the most organic and weathered material is zone A.

31. **1** Find the Generalized Bedrock Geology of New York State map in the *Reference Tables for Physical Setting/Earth Science*. Locate Utica, New York. Note the rock symbol for the bedrock in that area—lines slanting downward to the right. Now find this symbol in the "Geologic Periods and Eras in New York" key at the lower left of the map. Note that the symbol corresponds to Ordovician-age rocks consisting of limestones, shales, sandstones, and dolostones. Find the Scheme for Sedimentary Rock Identification in the *Reference Tables for Physical Setting/Earth Science*. Note that limestone, shale, sandstone, and dolostone are all sedimentary rocks. Thus, the type of surface bedrock most commonly found in the Utica, New York, area is sedimentary, with limestone, shale, sandstone, and dolostone.

32. **1** Water flows from higher elevations to lower elevations. Therefore, water will flow to either side of the tops of the circular ridges shown in the block diagram and then into the valleys between the ridges, forming a stream drainage pattern, as shown in map (1).

WRONG CHOICES EXPLAINED:

(2) On a flat plain, the slope is gentle and streams flow downhill slowly. Streams typically form a dendritic stream drainage pattern. In a dendritic stream drainage pattern, numerous small streams join together to form larger streams in a treelike pattern, as shown in map (2). The geologic landscape shown in the diagram in the question shows a series of circular ridges, not a flat plain.

(3) The trellis stream pattern in map (3) forms in tilted or folded rock layers of unequal resistance to erosion. Major streams run through long, parallel valleys following belts of weak rock between parallel ridges of stronger rock. Tributary streams enter major streams at right angles. The geologic landscape shown in the diagram in the question shows a series of circular ridges, not a series of long, parallel ridges.

(4) The geologic landscape shown in the diagram in the question shows a series of circular ridges with the highest point in the center. Map (4) shows a stream pattern in which the streams are flowing toward the center from larger streams into smaller streams. In order for this to occur, water would have to flow uphill, which it does not.

33. **3** Note the symbol for ice in the key. Find the area corresponding to ice in all three maps. Note that the maps indicate that the average area covered by sea ice is decreasing. In order for the amount of ice to decrease, some of the ice must have melted. Ice must absorb heat in order to melt. Thus, the maps best support the inference that Earth's climate is warming because the average area covered by sea ice is decreasing.

34. **1** Photograph A shows a large, fan-shaped deposit where a stream flows into a large body of water. When a stream flows into a standing body of water, the stream slows down and most of the stream's sediment is deposited. The resulting landform is a large, flat, fan-shaped pile of sediment at the mouth of the stream called a delta. Thus, depositional feature A is a delta.

Photograph B shows an elongated island running parallel to the coastline. Note the inlets at the northeast and southwest ends of the island. Barrier bars are long, narrow, offshore deposits of sand that run parallel to the coastline. Barrier bars form when waves deposit enough sand onto a sandbar so that the sandbar is above water. When tidal inlets break a barrier bar into elongated islands, the islands are called barrier islands. Thus, depositional feature B is a barrier island. Therefore, depositional features A and B are correctly identified as A—delta; B—barrier island.

35. **2** In diagram A, the mineral crystals of the rock are randomly oriented. In diagram B, the crystals occur in a pattern of parallel layers of flat, elongated crystals separated into light- and dark-colored bands. The type of mineral crystal arrangement shown in diagram B is called a foliated texture; it shows mineral alignment and banding. Foliation occurs when mineral grains recrystallize, or are flattened, under heat and pressure. Grains that may have been randomly oriented form parallel layers or become aligned when deformed under high heat and pressure. Banding occurs when minerals of different densities recrystallize under heat and pressure. As they recrystallize, minerals of different densities separate into layers like a mixture of oil and water. Since light-colored minerals tend to be less dense than dark-colored ones, the layers that form are alternating light and dark. The changes that occur in solid rock due to heat, pressure, and chemical activity are called metamorphism. The rock in diagram A has been changed by recrystallizing under heat and pressure into the metamorphic rock shown in diagram B. Thus, the geologic processes most likely responsible for the banding and alignment of mineral crystals represented in diagram B are heating and increased pressure.

WRONG CHOICES EXPLAINED:
(1) Melting and solidification would lead to the formation of a new igneous rock, not a change in an existing rock.
(3) Compaction and cementation are processes that transform loose particles of sediment into solid sedimentary rocks, not processes that change an existing rock into a metamorphic rock.
(4) Weathering and erosion are processes that break down rock into smaller fragments called sediments and transport them to a new location, not processes that change an existing rock into a metamorphic rock.

PART B–1

36. **2** Locate the point corresponding to 12:00 p.m. (noon) on January 24 on the horizontal axis labeled "Time" at the bottom of the graph. From this point, trace vertically upward until you intersect the bold curved line. From this intersection, trace horizontally to the left until you intersect the vertical axis labeled "Height Above Average Low Tide Level (ft)." Note the value of 1.6. Thus, the height of the water above average low tide level at noon on January 24 was 1.6 ft.

37. **1** Note that the line representing water levels at Montauk rises and falls in a repeating pattern. In other words, the changes in water level are cyclic. Note, too, that all of these changes repeat in a specific period of time. Therefore, they are predictable. For example, three low tides are shown on the graph. The first

low tide occurs at 3:00 a.m. on January 24, the second at 3:30 p.m. on January 24, and the third at 4:00 a.m. on January 25. Each of the low tides are 12 hours 30 minutes apart. Thus, the changing water levels at Montauk can best be described as cyclic and predictable.

38. **4** Both the Sun and Moon exert a force of gravity on Earth that causes bulges in the oceans. As Earth rotates on its axis, the positions of the bulges remain fixed in line with the Moon and Sun. The water levels at a particular location on Earth rise and fall as that location rotates into and out of these bulges. The repeated pattern of Earth's rotation and the orbital motions of the Moon and Earth cause ocean water levels to change in a cyclic pattern called tides. Thus, the water-level variation pattern shown by the graph is caused by Earth's rotation and the gravitational pull of the Moon.

WRONG CHOICES EXPLAINED:
(1) The changes in water levels at Montauk, New York, are cyclic and predictable. The changes in wind velocity produced by coastal storms are neither cyclic nor predictable. Therefore, it is unlikely that a noncyclic, nonpredictable event would produce a cyclic and predictable pattern.

(2) Water is not a magnetic substance. So changes in the magnetic orientation of the North American Plate would not affect the water or change its depth.

(3) The distance between Montauk and the equator is constant, not cyclic. Therefore, this distance is unlikely to produce a cyclic change in water levels. Earth's revolution is cyclic, but each cycle is one year. A yearly cyclic change is unlikely to result in a 12 hour 30 minute cyclic change.

39. **3** It is given that the tide gauge is located at Montauk, New York, at the easternmost end of Long Island. Find the Generalized Bedrock Geology of New York State map in the *Reference Tables for Physical Setting/Earth Science*. Locate the easternmost end of Long Island, and note that it is very close to the coordinates 41°, 72°. Note the compass rose in the lower right-hand corner of the map. Note that the latitude values increase from 41° to 45° as you move northward on the map in New York State. Therefore, these values represent north latitude. Note that the longitude values increase from 72° to 79° as you move westward on the map in New York State. Therefore, these values represent west longitude. Thus, the approximate latitude of the tide gauge is 41°00′ N 72°00′ W.

40. **1** The question refers to the "most massive stars." In the diagram, begin at "Star formation." Follow the arrow labeled "Massive" to the main sequence, and note that supergiant stars are formed. Supergiants then undergo supernova explosions.

The most massive stars have a very high mass. Follow the arrow labeled "Very high mass," and note that black holes are the result. Thus, the final stage in the life cycle of the most massive stars is a black hole.

41. **3** On the diagram, locate the star labeled "Sun." Note that the next stage in the Sun's life cycle is a giant star. After becoming a giant, the next stage is a white dwarf, which then ends its life cycle as a black dwarf. Thus, giants, white dwarfs, and black dwarfs may once have been similar to our Sun in mass and luminosity. Find the Characteristics of Stars chart in the *Reference Tables for Physical Setting/Earth Science*. Locate the four stars listed as choices. Note that *Deneb* is a supergiant, *Spica* is a main sequence star, *Procyon B* is a white dwarf, and *Proxima Centauri* is a main sequence star. Thus, *Procyon B*, which is the white dwarf, may once have been similar to our Sun in mass and luminosity.

42. **1** In the cores of stars, gravity is strong enough to overcome the force of repulsion between atomic nuclei, allowing the nuclei to combine in a process called nuclear fusion. By definition, nuclear fusion involves the combining of several atoms of a lighter element to form a single atom of a heavier element. The single heavier atom typically has less mass than the lighter atoms from which it formed. The mass "missing" from the heavier atom is not lost. Instead, it is converted into energy according to Einstein's formula $E = mc^2$. This formula states that if mass is converted into energy, the amount of energy released, E, is equal to the mass, m, times the speed of light, c, squared. The speed of light squared is a very large number. Therefore, the conversion of even a small amount of mass into energy during nuclear fusion results in the release of a very large amount of energy. Thus, energy is produced in the cores of main sequence stars when lighter elements combine and form heavier elements during the process of nuclear fusion.

WRONG CHOICES EXPLAINED:
(2) By definition, fusion is the process or result of joining two or more things together to form a single thing. Nuclear fusion involves the combining of several atoms of a lighter element to form a single atom of a heavier element, not combining heavier elements to form lighter elements.

(3) Cosmic background radiation is radiation coming from all directions in the universe. It is believed to be radiation left over from the Big Bang that formed the universe. In the mid-1960s, this radiation was detected by Arno Penzias and Robert Wilson at Bell Laboratories in New Jersey. It was found to fill the universe uniformly in every direction and had a temperature of about 3 kelvins (K). Fusion requires temperatures of about 100 million kelvins. Thus, absorbing cosmic radiation would not increase the temperature of a star's core enough to support fusion.

(4) Releasing energy would decrease the energy in the star, not produce energy in the star's core.

43. **3** Radioactive isotopes break apart, or decay, at a steady, constant rate that is not affected by outside factors such as changes in temperature, pressure, or chemical state. Therefore, the decay process may be used as a clock to determine the actual age of the rocks in which the isotopes are found. Radioactive isotopes start to decay when a rock forms. The decay of a radioactive isotope occurs at a statistically predictable rate known as its half-life. Half-life is the time required for one-half of the unstable radioisotope to change into a stable decay product. According to the graph, at time = 0 there were 1000 atoms of radioactive Isotope X in the sample. After 1 half-life, half of those atoms would have decayed into a stable daughter product and only half of the original 1000 atoms (500 atoms) would still be radioactive. Locate "500" on the vertical axis labeled "Number of Radioactive Atoms Present," and trace right until you intersect the bold curve. From this intersection, trace vertically downward to the horizontal axis labeled "Time (h)." Note that the value is 3. Thus, based on the graph, the half-life of this radioactive isotope is 3 h.

44. **3** On the horizontal axis labeled "Time (h)," locate the point corresponding to "8" between the values 6 and 9. From this point, trace vertically upward until you intersect the bold curve. From this point, trace horizontally to the left until you intersect the vertical axis labeled "Number of Radioactive Atoms Present." Note that this point corresponds to approximately 155 atoms. Thus, the approximate number of radioactive atoms of Isotope X present when 8 hours of decay has occurred is 155.

45. **4** From point A on the diagram, trace horizontally until you intersect the vertical axis labeled "Altitude (km)." Note that point A is located at an altitude of about 9 km. Find the Selected Properties of Earth's Atmosphere chart in the *Reference Tables for Physical Setting/Earth Science*. On the vertical axis labeled "Altitude," locate the point corresponding to 9 km. Note that km are shown on the left side of the axis and miles on the right side. So be sure to use the values on the left side of the axis to locate 9 km. From 9 km, trace horizontally to the right to the portion of the chart labeled "Temperature Zones." Note that the temperature zone at an altitude of 9 km is called the troposphere. Thus, point A is located in the troposphere.

46. **2** Air is a fluid. Heated fluids are less dense than cooler fluids. When air is warmed, it becomes less dense than the surrounding cooler air and rises upward. Similarly, when air is cooled, it becomes denser than the surrounding warmer air

and sinks downward. Note that the arrow labeled "Updraft" starts in warm air. Note that the arrow labeled "Downdraft" starts in cold air. The updraft is therefore warm air that is rising, and the downdraft is cold air that is sinking. Thus, the updrafts and downdrafts are primarily caused by differences in air density.

WRONG CHOICES EXPLAINED:
(1) Note that the updrafts and downdrafts are occurring at the same altitude above sea level. Therefore, it is unlikely that altitude causes the updrafts and downdrafts.

(3) Relative humidity is a measure of the concentration of water vapor in the atmosphere. When air temperature reaches the dewpoint, the relative humidity is 100%, and water vapor in the air begins to condense and form clouds. Thus, the relative humidity in a cloud is 100%. Note that both the updrafts and the downdrafts are occurring in the cloud, where relative humidity is 100%. Therefore, it is unlikely that relative humidity causes the updrafts and downdrafts.

(4) Specific heat is a measure of the heat energy needed to raise the temperature of 1 gram of a substance by 1 degree Celsius. It is a characteristic of matter, not a force that causes motion.

47. **3** It is given that the diagram shows a thunderstorm cloud. Thus, this weather event is a thunderstorm. Find the Key to Weather Map Symbols in the *Reference Tables for Physical Setting/Earth Science*. Find the section labeled "Present Weather," and locate the symbol for a thunderstorm. Note that this symbol corresponds to the symbol in choice (3).

48. **2** There are 5 contour intervals between the bold 250-foot index contour line and the bold 500-foot index contour line. Thus, the contour interval on this map is 50 feet ($250 \div 5 = 50$).

49. **2** Find the Equations section of the *Reference Tables for Physical Setting/Earth Science* and note the equation for gradient:

$$\text{Gradient} = \frac{\text{change in field value}}{\text{change in distance}}$$

The map shown is a topographic map. The field value on a topographic map is elevation. Note that point A is located directly on the 250-foot contour line and that point B is located directly on the 500-foot contour line. Thus, the change in field value (elevation) from A to B is 250 feet.

On a piece of scrap paper, mark off the distance along the straight line between points A and B. Compare the distance marked off on the scrap paper to the scale printed beneath the map to determine the distance between points A and B. The distance is about 10 miles. Substitute these values in the equation and solve:

$$\text{Gradient} = \frac{500 \text{ ft} - 250 \text{ ft}}{10 \text{ mi}}$$

$$\text{Gradient} = \frac{250 \text{ ft}}{10 \text{ mi}} = 25 \text{ ft/mi}$$

Thus, the gradient between location A and location B is approximately 25 ft/mi.

50. **4** Note that the bold index contour to the west of location C is labeled "250" and the bold index contour east-northeast of location C is labeled "500." Thus, the elevation increases between these two index contours. From location B to location D, the elevation continues to increase. As explained earlier, the contour interval on this map is 50 feet. Thus, the elevation of location C is between 300 and 350 feet.

Note that the trail leads through the center of a series of concentric contour lines within the oval-shaped 500-foot index contour. On a topographic map, concentric contour lines typically represent a hill. Thus, the elevation increases within these contours. The innermost concentric contour has an elevation of 600 feet. Therefore, the top of the hill lies within this 600-foot contour at an elevation of more than 600 feet but less than 650 feet. Thus, a person walking along the trail from location C to the center of the 600-foot contour would be walking uphill. From the top of the hill, the trail continues back to the 500-foot contour. So a person would now be walking downhill. Note that the trail then crosses the 500-foot contour and continues across the area between it and the next 500-foot contour line to the east. The elevation between the two 500-foot contours is lower than 500 feet but greater than 450 feet. Thus, a person walking along the trail would walk downhill to about the midpoint in this area and then begin walking uphill again to cross the next 500-foot contour. Note that location D lies on the fourth contour line from the 500-foot contour, so the elevation of location D is 700 feet. Therefore, a person walking from the 500-foot contour line to location D would be walking uphill. When taken together, a person walking along the trail from location C to location D would walk uphill, then downhill, then uphill again.

PART B–2

51. Find the Properties of Common Minerals chart in the *Reference Tables for Physical Setting/Earth Science*. In the column labeled "Mineral Name," locate "Muscovite mica" and locate "Biotite mica." Trace left to the column labeled "Composition*." Note that the chemical formula for muscovite mica is $KAl_3Si_3O_{10}(OH)_2$ and the chemical formula for biotite mica is $K(Mg,Fe)_3AlSi_3O_{10}(OH)_2$. Refer to the key labeled "Chemical symbols" at the bottom of the chart. Note that both muscovite mica and biotite mica contain potassium (K), aluminum (Al), silicon (Si), oxygen (O), and hydrogen (H). However, biotite mica also contains iron (Fe) and/or magnesium (Mg), while muscovite mica does not. Thus, two chemical elements present in biotite mica that are not present in muscovite mica are magnesium (Mg) and iron (Fe).

One credit is allowed for **magnesium (Mg) and iron (Fe)**.

52. The passage states that all members of the mica family are similar in density, luster, hardness, streak, type of breakage, and crystal shape. Thus, identifying these characteristics for two members of this family will provide information about all members of this family. Find the Properties of Common Minerals chart in the *Reference Tables for Physical Setting/Earth Science*. In the column labeled "Mineral Name," locate "Muscovite mica" and locate "Biotite mica." In the row for each of these minerals, trace left to the column labeled "Cleavage/Fracture." Note that both micas display cleavage. Recall that cleavage is the tendency of a mineral to split along definite crystalline planes, yielding smooth surfaces. Note in the column labeled "Distinguishing Characteristics" that both micas are "flexible in thin sheets." Thus, both muscovite mica and biotite mica cleave into thin sheets.

Now trace left to the column labeled "Hardness." Note that muscovite mica has a hardness of 2–2.5 and that biotite mica has a hardness of 2.5–3. Next, trace left to the column labeled "Luster" and note that both micas have a nonmetallic luster. Nonmetallic lusters can be described by a variety of terms, such as vitreous/glassy or pearly/silky. However, these terms are somewhat subjective.

One credit is allowed for a correct luster, hardness, and dominant form of breakage. Acceptable responses include but are not limited to:

Luster:
- **nonmetallic**
- **vitreous/glassy**
- **pearly**
- **silky**

Hardness:

- **any value or range from 2.0 to 3.0**

Dominant form of breakage:

- **cleavage/basal cleavage**
- **breaks into thin sheets**
- **one direction of cleavage**

53. Find the Scheme for Igneous Rock Identification in the *Reference Tables for Physical Setting/Earth Science*. In the column labeled "Crystal Size," locate the row corresponding to "10 mm or larger." Trace this row to the left, and note that the only igneous rock listed in this row is pegmatite. From the column containing pegmatite, trace vertically down to the section labeled "Mineral Composition (relative by volume)." Note that one of the sections representing various minerals that the pegmatite intersects is the black section labeled "Biotite (black)." Thus, pegmatite is an igneous rock in which crystals of biotite mica are larger than 10 millimeters in diameter.

One credit is allowed for **pegmatite**.

54. Large crystals of mica that weigh several hundred tons are clearly larger than 10 millimeters in diameter. Thus, as explained in answer 53, the only igneous rock with mica crystals in this size range is pegmatite. Find the Scheme for Igneous Rock Identification in the *Reference Tables for Physical Setting/Earth Science*. Locate "Pegmatite," and trace this row left to the column labeled "Environment of Formation." Note that the row corresponding to pegmatite is in the range labeled "Intrusive (Plutonic)" igneous rock. Thus, pegmatite is an intrusive igneous rock. Intrusive igneous rocks form by the cooling and solidification of molten magma deep underground where the magma is insulated by surrounding rock. Therefore, the magma cools slowly and takes a long time to solidify to form intrusive rock. Very slow cooling over a long period of time allows very large crystals to form before the rock is completely solid. Thus, an igneous rock with very large crystals formed deep underground in an intrusive environment and its relative rate of cooling was slower than that of other igneous rocks.

One credit is allowed if *both* responses are correct. Acceptable responses include but are not limited to:

Environment of formation:

- **intrusive**
- **plutonic**
- **deep underground**

Relative rate of cooling:

- **slower/slowly**
- **took a longer time/long time**

Note: Credit is *not* allowed for "long" alone, as this does not accurately describe a rate of cooling but, instead, describes a duration.

55. Note that the Sun's position at solar noon increases in altitude from December 21 to March 21 to June 21. From the diagram, it can be inferred that the position of the Sun at solar noon on April 21 will lie between its position on March 21 and its position on June 21 because April occurs between March and June. From March 21 to June 21 is three months. From March 21 to April 21 is one month, or about one-third the time between March 21 and June 21. Therefore, the position of the Sun at solar noon on April 21 will be about one-third of the distance between its position on March 21 and its position on June 21. Therefore, the **X** marking the Sun's position at noon on April 21 should be drawn, as shown below.

One credit is allowed if **the center of the X is within or touches the boxed area shown below**.

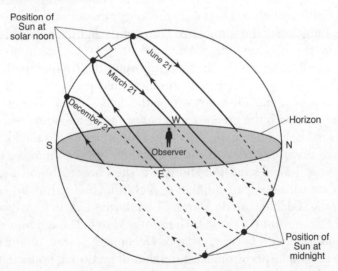

[**NOTE:** It is recommended that an overlay of the same scale as the student answer booklet be used to ensure reliability in rating.]

56. Due to Earth's rotation, all celestial objects, including the Sun, appear to move across the sky from east to west along a path that is an arc, or part of a circle. Earth completes one complete 360° rotation every day (including December 21). Thus, Earth rotates from west to east at a constant rate of

15°/hour (360° day ÷ 24 hr/day = 15°/hr). As Earth rotates on its axis from west to east at 15° per hour, the Sun appears to move through the sky from east to west at 15° per hour.

One credit is allowed for **15°/h**.

57. To determine the direction in which the observer's shadow would point at solar noon on March 21, do the following. Draw a straight line connecting north and south on the horizon. Note that this line passes beneath the observer's feet. Now draw a straight line connecting the position of the Sun at solar noon on March 21 and the tip of the observer's head. Extend that line to the line beneath the observer's feet. This intersection represents the tip of the observer's shadow. Now draw a bold horizontal line connecting the observer's feet to the tip of the observer's shadow; this represents the observer's shadow. Note that the observer's shadow lies directly on the north-south line with the tip of the shadow to the north. Thus, the observer's shadow at solar noon on March 21 points north.

One credit is allowed for **North/N**.

58. The Sun appears to move across the sky at a constant rate of 15° of arc per hour due to Earth's rotation. Therefore, the longer the arc of the Sun's path above the horizon, the greater the number of hours of daylight. Conversely, the longer the arc below the horizon, the greater the number of nighttime hours. It is given that the solid lines on the diagram represent the daytime paths of the Sun and that the dashed lines represent the nighttime paths. Thus, the date with the least number of nighttime hours will be the date with the shortest dashed line path, and the date with the greatest number of nighttime hours will be the date with the longest dashed line path. Note that December 21 has the longest dashed line path and therefore represents the date with the greatest number of nighttime hours. Note that June 21 has the shortest dashed line path and therefore represents the date with the least number of nighttime hours. Finally, note that the length of the dashed line path on March 21 is intermediate in length between that of December 21 and June 21 paths. Therefore, March 21 has a number of nighttime hours intermediate between that of December 21 and June 21. Thus, the dates on the diagram listed from least number of nighttime hours to the greatest number of nighttime hours is June 21 → March 21 → December 21.

One credit is allowed for **a correct sequence, as shown below**.

June 21	March 21	December 21
Least number of nighttime hours	⟶	Greatest number of nighttime hours

59. It is given that the map in your answer booklet shows a portion of the Tectonic Plates map in the *Reference Tables for Physical Setting/Earth Science*. Find the Tectonic Plates map in the *Reference Tables for Physical Setting/Earth Science*. On the map, locate the plate boundary corresponding to the one shown on the map in your answer booklet. According to the key at the bottom of the map, a bold arrow represents relative motion at a plate boundary. Note that the Tectonic Plates map in the *Reference Tables for Physical Setting/Earth Science* shows the North American Plate moving to the WSW at this plate boundary and the Caribbean Plate moving to the ENE. Thus, an arrow pointing WSW should be drawn in the box on the North American Plate, and an arrow pointing ENE should be drawn in the box on the Caribbean Plate, as shown below.

One credit is allowed for **the correct direction of both arrows, as shown below**.

60. Find the Tectonic Plates map in the *Reference Tables for Physical Setting/Earth Science*. Locate the plate boundary corresponding to the one labeled "East Pacific Ridge" on the map in your answer booklet. Note that this corresponds to the boundary between the Pacific Plate and the Cocos Plate. Thus, location *A* is on the Pacific Plate, and location *B* is on the Cocos Plate. According to the key at the bottom of the map, the boundary between these two plates is a divergent boundary. Along divergent boundaries, adjacent plates move apart and open rifts through which magma can rise to the surface and solidify to form new ocean floor. As the plates continue to move apart, this process is repeated, constantly creating new ocean floor and pushing the older ocean floor on either side of the ridge away from the ridge. Thus, the age of the ocean floor rock increases with distance away from the East Pacific Ridge. Therefore, the line representing the relative age of the ocean floor bedrock from location *A* to location *B* should indicate the oldest bedrock ages at locations *A* and *B* and the youngest at the East Pacific Ridge, as shown below.

One credit is allowed for **a V-shaped or U-shaped line showing the oldest bedrock ages above A and B and the youngest bedrock age at the East Pacific Ridge**.
Examples of 1-credit responses:

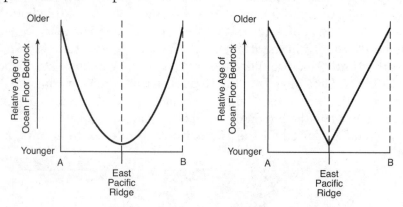

Note: Credit is allowed even if the oldest bedrock ages above A and B are not the same relative age.

61. Find the Tectonic Plates map in the *Reference Tables for Physical Setting/ Earth Science*. In the key, note the symbol for a hot spot. On the Tectonic Plates map, locate the hot spot corresponding to the one on the map in your answer booklet. It is the Galapagos Hot Spot. On the Tectonic Plates map, locate the plate boundary near the Galapagos Hot Spot. Note that this plate boundary is between the Cocos Plate to the north and the Nazca Plate to the south. Locate the center of the symbol representing the Galapagos Hot Spot, and note that it lies south of the boundary on the Nazca Plate. Thus, the hot spot on the map is the Galapagos Hot Spot, and the center of this hot spot is located on the Nazca Plate.

One credit is allowed for **Galapagos Hot Spot and Nazca Plate**.

62. The bedrock at location A formed when magma rose to Earth's surface beneath the ocean through cracks at the plate boundary and then solidified to form new ocean floor. Thus, the ocean floor at location A is considered an extrusive igneous rock, because it cooled at Earth's surface. Find the Scheme for Igneous Rock Identification in the *Reference Tables for Physical Setting/Earth Science*. Locate the column labeled "Environment of Formation." In that column, locate the rows labeled "Extrusive (Volcanic)." It is given that bedrock at location A is mafic igneous rock. Locate the arrow labeled "Composition" in the "Characteristics" section in the middle of the chart. Note that the arrow extends from "Felsic (rich in Si, Al)" on the left to "Mafic (right in Fe, Mg)" on the right. Trace the

columns on the mafic side of the scale upward until they intersect with the rows corresponding to an extrusive environment of formation. Thus, the extrusive igneous rocks that are mafic in composition are basaltic glass, scoria, vesicular basalt, basalt, and diabase. Collectively, these are considered to be basaltic rocks. Thus, the bedrock at location A is either basalt and/or other basaltic rocks. Find the Inferred Properties of Earth's Interior chart in the *Reference Tables for Physical Setting/Earth Science*. Locate the "Density (g/cm^3)" scale along the right side of the cross section. Note that oceanic crust is listed as basaltic and has a density of 3.0 g/cm^3. Thus, the mafic igneous bedrock that is most likely to make up the oceanic crust at location A is basalt or basaltic igneous rocks, and the average density of oceanic crust is 3.0 g/cm^3.

One credit is allowed for **basalt/basaltic bedrock and a density of 3.0 g/cm^3** *or* **3 g/cm^3**.

63. Porosity is a measure of the amount of empty space among the particles in a rock or sediment sample. It is often expressed as the percentage of the total volume of a rock or a sediment sample that is open pore space. It is given that the mixture consists of pebbles, sand, silt, and clay. Find the Relationship of Transported Particle Size to Water Velocity graph in the *Reference Tables for Physical Setting/Earth Science*. Note the axis labeled "Particle Diameter (cm)" along the left side of the graph, and note the names of sediments corresponding to size ranges along the right side of the graph. From largest to smallest, the particles in the mixture are pebbles, sand, silt, and clay. In a mixture of different particle sizes, the smaller particles can fit into and fill the open pore spaces among the larger particles. This decreases the total amount of open pore space, or porosity, of the mixture. For example, sand, silt, and clay particles are small enough to fit into and fill the open pore spaces among pebbles. Silt and clay can fill the empty pore spaces among grains of sand. Clay can fill the empty pore spaces among silt. Thus, the particle sizes fit more closely in the mixture. This results in the lowest porosity because smaller particles fill the open pore spaces among larger particles, decreasing the total amount of open pore space.

One credit is allowed. Acceptable responses include but are not limited to:

- **The smaller particles filled the pore spaces among the larger particles and decreased the total amount of open pore space.**
- **Pore spaces among the pebbles were filled by sand, silt, or clay.**
- **Small particles take up spaces between larger particles.**

Note: Credit is *not* allowed for "it has less pores" or "less pore space" because that is restating the question.

Note: Credit is *not* allowed for "particles fit together more closely" or "particles are more closely packed" alone because this is stated in the question.

64. Infiltration is the downward movement of liquid water into and through interconnected openings among soil particles due to gravity. It is given that the rate of infiltration is measured in centimeters per second. Thus, the infiltration rate is the distance water moves downward through a material in a given amount of time.

Find the Equations chart in the *Reference Tables for Physical Setting/Earth Science*. Locate the equation for rate of change:

$$\text{Rate of change} = \frac{\text{change in value}}{\text{time}}$$

The change in value is the distance from the top of the column of sand to the bottom. It is given that this distance is 28 centimeters. On the table in the question, find "Sand" in the column labeled "Particle Type." Trace this row right to the column labeled "Time for Water to Infiltrate (s)," and note the value of 7 seconds. Thus, the time the water takes to infiltrate the 28 centimeters from the top to the bottom of the column is 7 seconds. Substitute these values in the rate of change equation and solve:

$$\text{Rate of change} = \frac{28 \text{ cm}}{7 \text{ s}} = 4 \frac{\text{cm}}{\text{s}}$$

One credit is allowed for **4 cm/s**.

65. On the table in the question, find "Plastic beads" in the column labeled "Particle Type." Trace this row right to the column labeled "Particle Diameter (cm)," and note the value of 0.4. Find the Relationship of Transported Particle Size to Water Velocity graph in the *Reference Tables for Physical Setting/Earth Science*. On the axis labeled "Particle Diameter (cm)" along the left side of the graph, locate 0.4 cm. Note that 0.4 lies between 0.1 and 1.0. Note that on this segment of the axis, each subdivision represents 0.1 cm, so 0.4 is the third subdivision above 0.1. From 0.4 cm, trace horizontally to the right side of the graph. Note that 0.4 cm falls into the range of particle sizes called "Pebbles." Thus, based on particle diameter, the plastic beads represent pebbles.

One credit is allowed for **pebbles**.

PART C

66. Surface wind speed is directly related to the air pressure gradient. The closer together the isobars on the map, the greater the pressure gradient and, therefore, the greater the wind speed. On the given map, the isobars are closest together near Watertown, New York. Therefore, the highest wind speed was most

likely recorded near Watertown, New York. Thus, the evidence shown on the map that indicates that the highest wind speeds occurred near Watertown, New York, is that the isobars there are closely spaced, showing a steep pressure gradient.

One credit is allowed. Acceptable responses include but are not limited to:

- **The isobars are closest together.**
- **The steepest pressure gradient occurs at Watertown.**
- **The closer the isolines/lines are, the faster the wind speed.**

67. Find the Key to Weather Map Symbols in the *Reference Tables for Physical Setting/Earth Science*, and locate the section labeled "Station Model Explanation." Note the positions in which the values for dewpoint, amount of cloud cover, and barometric pressure are placed on a station model. Note that dewpoint (°F) is shown to the lower left of the station model. Thus, the dewpoint in Albany, New York, is 36°F. Note that the amount of cloud cover is represented by how much of the circle at the center of the station model is shaded. One-quarter of the circle at the center of the station model is shaded. Thus, the cloud cover at Albany, New York, is 25%. Finally, note that barometric pressure is shown at the upper right of the station model and that the three digits listed represent the last three digits of the air pressure in millibars. Since air pressure rarely rises above 1050.0 mb or drops below 950.0 mb, the general rule is to place a 9 in front of the three digits on the station model if they are greater than 500 and place a 10 in front of the three digits if they are less than 500. Then place a decimal point between the last two digits. According to the station model, the barometric pressure is represented by the digits 051. Since the three digits are less than 500, place a 10 in front of the digits and a decimal point between the last two digits. Therefore, the barometric pressure at the weather station is 1005.1 mb. Record these values in the table provided in your answer booklet, as shown below.

One credit is allowed if *all three* weather variables are correct, as shown in the table below.

Albany, New York

Weather Variable	Weather Data
Dewpoint	36°F
Cloud cover	25%
Actual barometric pressure	**1005.1 mb**

68. Find the Generalized Bedrock Geology of New York State map in the *Reference Tables for Physical Setting/Earth Science*. Note that New York State lies between 40° N latitude and 45° N latitude.

Find the Planetary Wind and Moisture Belts in the Troposphere diagram in the *Reference Tables for Physical Setting/Earth Science*. Note that New York State is located in the belt of southwest planetary winds that lie between 30° N latitude and 60° N latitude. As these winds travel northward, they curve more and more to the east. Thus, if this low-pressure system follows a normal storm track over the next two days, it will move in a northeasterly direction and curve toward the east.

One credit is allowed. Acceptable responses include but are not limited to:

- **East/E**
- **East northeast/ENE**
- **Northeast/NE**
- **North northeast/NNE**

69. To plot the average body volume for each of the geologic ages listed in the Brachiopod Data Table, proceed as follows. In the column labeled "Geologic Age (mya)," locate 480. Trace this row right to the column labeled "Average Body Volume Including the Shell (mL)" and note the value of 0.1. Thus, 480 mya, the average body volume of brachiopods was 0.1 mL. On the graph provided in your answer booklet, locate the point corresponding to 0.1 mL on the vertical axis labeled "Average Brachiopod Body Volume (mL)." Now locate the point corresponding to 480 mya along the horizontal axis labeled "Geologic Age (mya)." Trace horizontally right from 0.1 and vertically upward from 480 until the lines intersect. Plot this point. Repeat this procedure for each of the geologic ages listed in the data table. When completed, connect the plots with a solid line, as shown below.

One credit is allowed if **the centers of *all five* plots are within or touch the circles shown and the plots are correctly connected with a line that passes within or touches the circles.**

Note: Credit is allowed if the student-drawn line does *not* pass through the student plots but is still within or touches the circles. It is recommended that an overlay of the same scale as the student answer booklet be used to ensure reliability in rating.

70. The brachiopods represented on the Brachiopod Data Table existed between 380 and 480 mya. Find the Geologic History of New York State chart in the *Reference Tables for Physical Setting/Earth Science*. Locate the time scale labeled "Million years ago" along the right edge of the column labeled "Epoch." On this time scale, locate the positions corresponding to 380 and 480 million years ago. From each of these times, trace left to the column labeled "Period." Note that 380 mya corresponds to the Devonian period and 480 mya corresponds to the Ordovician period. Note that this time range encompasses the Ordovician, Silurian, and Devonian periods. Record the names of any two of these three periods in your answer booklet.

One credit is allowed for **Ordovician period, Silurian period, and/or Devonian period**.

Note: Credit is *not* allowed for Early, Middle, or Late Ordovician, for Early or Late Silurian, or for Early, Middle or Late Devonian because these are epochs, *not* periods.

71. Find the Geologic History of New York State chart in the *Reference Tables for Physical Setting/Earth Science*. Locate the column headed "Time Distribution of Fossils (including important fossils of New York State)." Note the bold black lines labeled with names and circled letters. The names are of types of fossil organisms found in New York State bedrock. The circled letters are keyed to the illustrations of specific important fossils printed along the bottom of the chart. Along the black line labeled "Brachiopods," note the circled letters "Y" and "Z." Locate drawings "Y" and "Z" at the bottom of the chart. Note that brachiopod "Y" is named *Eospirifer* and that brachiopod "Z" is named *Mucrospirifer*. Thus, two brachiopod index fossils found in New York State bedrock are *Eospirifer* and *Mucrospirifer*.

One credit is allowed for ***Eospirifer* and *Mucrospirifer***.

72. According to the reading passage, "Cope's Rule states that the average size of animals preserved in the fossil record tends to increase as each group evolves from a previous group." It is given that the earliest horses appeared in the Eocene epoch and were about the size of a large dog today. Find the Geologic History of New York State chart in the *Reference Tables for Physical Setting/Earth Science*. Locate Eocene in the column labeled "Epoch." Note along the right edge of the

column that the Eocene extended from about 33.9 to 55.8 million years ago. It is logical to infer that since the Eocene epoch, horses evolved into the horses of today. The average horse today is a much larger animal than the earliest horses were in the Eocene when they were only the size of a large dog today. Therefore, the evolution of horses supports Cope's Rule because the fossil record shows that horses have become larger over geologic time.

One credit is allowed. Acceptable responses include but are not limited to:

- **The fossil record shows that horses have become larger over geologic time.**
- **Large dogs are much smaller than horses of today.**
- **The average horse of today is a large animal, so horses must have become larger since Eocene time.**

73. An isoline connects points of equal field value, and the field values shown on this map are average yearly snowfall measured in inches. The 240-inch average yearly snowfall isoline connects all points that experienced 240 inches of average yearly snowfall. To draw the 240-inch isoline, do the following. Note that only one of the points specified on this map received exactly 240 inches of snowfall. However, that does not mean that there are no other places on the map with that value. Note that there are values greater than 240 adjacent to values less than 240. For example, note the point labeled "235" just south and slightly east of a point labeled "270." It is logical to infer that somewhere between the locations that recorded 235 and 270 inches of snowfall is a location that recorded 240 inches of snowfall. Therefore, the 240-inch isoline should pass between these two points. Similarly, as the 240-inch isoline moves west from that point, it should pass between the points labeled "270" and "236." Further west, it should pass between the point labeled "248" and the 220-inch isoline. Then the isoline should swing northeast to between the points labeled "225" and "241" and then north between the points labeled "241" and "238." Continue the isoline north between the point labeled "257" and the 220-inch isoline. From there, the isoline should turn south and run between the 220-inch isoline and the points labeled "257," "270," and "248." The isoline should end up back at the point labeled "240."

One credit is allowed for **correctly drawing the 240-inch isoline. The isoline must pass through or touch the dot labeled 240**.

Note: If additional lines are drawn, all must be drawn correctly to receive credit.

Example of a 1-credit response:

74. To construct a profile of the average annual snowfall along line XY, proceed as follows. Place the straight edge of a piece of scrap paper along the solid line connecting point X to point Y. Mark the edge of the paper at points X and Y and wherever the paper intersects an average yearly snowfall isoline. Note that the interval between isolines is 20 inches. Wherever the paper intersects an isoline, label the mark with the value of the average yearly snowfall of that isoline, as shown below. (You may want to label the isolines on the Oswego side of line XY. If your paper is not on the Oswego side, it will cover the labels.)

Average Yearly Snowfall Map

Now, place the scrap paper along the lower edge of the grid provided on your answer sheet so that points X and Y on the paper align with points X and Y on the grid, as shown below.

At each point where an isoline crossed the edge of the paper, draw a dot on the grid at the appropriate number of inches of average yearly snowfall directly above it. Finally, connect all of the dots in a smooth curve to form the finished profile, as shown below. Note that the top of the curve should extend above 200 inches. Since no additional isoline exists within the 200-inch isoline, and the interval between isolines is 20 inches, the average yearly snowfall in that area is more than 200 inches but less than 220 inches. Therefore, the highest point of the profile must be greater than 200 inches but less than 220 inches.

One credit is allowed if **the centers of *all six* plots are within or touch the rectangles shown and are correctly connected with a line that passes within or touches each rectangle. The line must extend above the 200-inch line but remain below the 220-inch line**.

Note: Credit is allowed if the student-drawn line does *not* pass through the student plots but is still within or touches the rectangles. It is recommended that an overlay of the same scale as the student answer booklet be used to ensure reliability in rating.

75. Note that the greatest amount of average annual snowfall indicated on the map is 270 inches. Locate the scale beneath the diagram. Note that the scale is marked in increments of 30 inches and labeled every 60 inches. Note, too, that the scale only extends to 180. On the straight edge of a piece of scrap paper, mark off a distance equivalent to 180 inches. Then move the 180 mark so that it lines up with 0 on the scale. Mark off an additional 90 inches, and label this as 270 (180 + 90 = 270.) Now place the scrap paper next to the wall. Mark off the distance equivalent to 270 inches above the ground on the wall, as shown below. Mark the point on the wall that is 270 inches above the ground with an **X**.

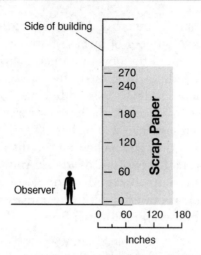

One credit is allowed if the center of **the X is within or touches the edge of the box shown**.

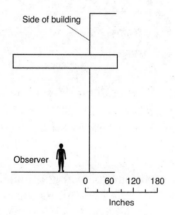

Note: Credit is allowed if a symbol other than an **X** is used. It is recommended that an overlay of the same scale as the student answer booklet be used to ensure reliability in rating.

76. The solar collector is located on Earth. Energy from the Sun must travel through a vacuum to reach Earth. Energy cannot be transmitted through a vacuum by either conduction (molecular collisions) or convection (movements of fluids due to density differences) because both of these methods require a medium. Therefore, energy is transferred from the Sun to Earth mainly in the form of radiation consisting of electromagnetic waves, many of which are light waves. Thus, the energy transfer process by which light travels through space from the Sun to the solar collector is radiation.

One credit is allowed for **radiation**.

77. It is given that the solar collector is used to heat water. Sunlight consists of electromagnetic waves of many wavelengths, but peaks occur in the range of visible light. Clear glass is transparent to visible light but blocks most other wavelengths. Thus, most of the sunlight that passes through the clear glass is visible light. When the light that passes through the clear glass strikes the collector plate and flow tubes, some of the light is absorbed and some of it is reflected.

The energy that is absorbed by the collector plate and flow tubes causes their molecules to move faster, increasing their temperature. The faster-moving molecules of the hot collector plate and flow tubes then transfer some of their energy to molecules of the air by molecular collisions (conduction). The collector plate and flow tubes also reradiate energy in the form of infrared (heat) waves, which is trapped by the clear glass cover. These infrared rays are then absorbed by the molecules of the air inside the collector, causing that air to heat up even more. The heat energy in the air is then transferred first to the flow tubes and then to the water inside the flow tubes by conduction, causing the water to heat up.

Substances of all colors absorb light energy. In general, though, dark colors absorb more light energy than light colors. Dark colors are dark because so much of the light energy that strikes them is absorbed and so little is reflected back to our eye. Dark colors are also better radiators than light colors. The more energy that is radiated by the collector plate and flow tubes, the more the air in the collector heats up. Black is such a dark color because it is a particularly good absorber (and radiator). Therefore, by painting the collector plate and the flow tubes black, the amount of energy absorbed and reradiated by these objects is increased, which, in turn, increases the temperature of the air in the collector. The hotter the air is in the collector, the greater the amount of energy that is transferred to the flow tubes and then to the water inside them. The more energy that is transferred to the water, the more the water heats up.

Therefore, the collector plate and flow tubes are painted black because the color black is a good absorber and is a good radiator of electromagnetic energy such as sunlight.

One credit is allowed. Acceptable responses include but are not limited to:

- **Black is a good absorber of electromagnetic energy/sunlight/insolation.**
- **Black is a good absorber and a good radiator.**
- **Dark colors take in radiation better than light colors.**
- **Black absorbs more energy.**

Note: Credit is *not* allowed for "black absorbs energy" alone because all colors absorb energy. Black is just a better absorber of that energy.

78. Temperature is a measure of the speed at which the molecules in a substance are moving. The faster the molecules in the substance are moving, the higher is the temperature of that substance. The speed at which molecules are moving (temperature) also determines the wavelength of the electromagnetic energy emitted. The faster the molecules of a substance are moving, the shorter the wavelength of the electromagnetic waves it emits. Conversely, the slower the molecules in a substance are moving, the lower its temperature and the longer the wavelength of the electromagnetic waves it emits. Most substances begin to radiate electromagnetic waves in the red visible range when they get "red hot." The light energy that is absorbed by the collector plate and flow tubes raises their temperature. However, that light energy is never high enough to cause the plate and tubes to get red hot and to emit visible light. Thus, the molecules of the collector plate and flow tubes are moving slower than molecules that would emit visible light. So the plate and tubes emit electromagnetic waves that have a longer wavelength than visible light.

Find the Electromagnetic Spectrum chart on the *Reference Tables for Physical Setting/Earth Science*. Locate the segment labeled "Visible light." Note the arrow labeled "Increasing wavelength" on the right on the spectrum. Note that the electromagnetic waves to the right of visible light are labeled "Infrared." Thus, the type of electromagnetic energy emitted by the flow tubes and collector plate is infrared, or IR. Infrared waves cannot pass through glass and are trapped inside the collector by the glass cover. As explained above, infrared waves have a longer wavelength than visible light waves. Therefore, "longer" should be circled in your answer booklet.

One credit is allowed for **infrared/IR and circling longer**.

79. It is given that the star *Kepler-62* is located in our galaxy. Our solar system is located in the Milky Way galaxy. Therefore, *Kepler-62* and its planetary system are located in the Milky Way galaxy.

One credit is allowed for **Milky Way**.

80. In the Data Table in the question, find the column labeled "Name of Planet" and locate "62c." Trace this row right to the column labeled "Equatorial Diameter (compared to Earth's diameter)." Note the value of 0.54. Thus, the equatorial diameter of Kepler-62c is 0.54 Earth diameters. Find the Solar System Data chart in the *Reference Tables for Physical Setting/Earth Science*. In the column labeled "Celestial Object," locate Earth. Trace this row right to the column labeled "Equatorial Diameter (km)," and note the value of "12,756." Thus, the equatorial diameter of Earth is 12,756 km. Since the equatorial diameter of Kepler-62c is 0.54 Earth diameters, multiply 12,756 km by 0.54 and solve to find

the equatorial diameter of Kepler-62c: 12,756 km × 0.54 = 6888.24 km. Thus, Kepler-62c has an equatorial diameter of 6888.24 km. On the Solar System Data chart, locate the column labeled "Equatorial Diameter (km)." Note that the value nearest to 6888.24 is 6794. From this value, trace left to the column labeled "Celestial Object." Note that this value is the equatorial diameter of the planet Mars. Thus, the planet in our solar system that has an equatorial diameter most similar in size to the equatorial diameter of planet Kepler-62c is Mars.

One credit is allowed for **Mars**.

81. The smaller the distance between a planet and the star it orbits, the greater the force of gravitational attraction between the planet and the star. The greater the force of gravitational attraction between a planet and a star, the faster the planet has to move in order to remain in orbit and not be drawn into the star. In addition, the closer a planet is to the star it orbits, the smaller the planet's orbit is and the less distance the planet has to travel to complete one revolution. The combination of a small orbit and high speed means that the planet closest to the star has the shortest period of revolution.

In the Data Table in the question, locate the column labeled "Distance from *Kepler-62* (million kilometers)." Note that the smallest value (closest to *Kepler-62*) is 8.23. From this value, trace left to the column labeled "Name of Planet." Note that the planet located closest to the star *Kepler-62* is Kepler-62b. Thus, planet Kepler-62b has the shortest period of revolution. Kepler-62b has the shortest period of revolution because it is the closest planet to its star. Kepler-62b also orbits at a faster speed and in a smaller orbit than planets farther from the star, resulting in a shorter period of revolution.

One credit is allowed for Kepler-62b *or* 62b and for an acceptable explanation. Acceptable explanations include but are not limited to:

- **Kepler-62b is closest to *Kepler-62*.**
- **Kepler-62b; the closer a planet is to a star, the shorter its period of revolution.**
- **62b; closer planets orbit faster.**
- **62b has the shortest orbital path.**
- **Kepler-62b is closest to its sun.**

82. The reading passage states, "The habitable zone, which is the region around a star where life may exist due to the possible presence of water in the liquid phase." Thus, the two planets that may have liquid water on their surfaces are located within the habitable zone around the star *Kepler-62*. According to the key to the right of the diagram of the *Kepler-62* Planetary System, the habitable zone in the diagram is shaded. According to the diagram, the two planets that

orbit *Kepler-62* in the habitable zone and therefore may have liquid water on their surfaces are Kepler-62e and Kepler-62f.

In order to have liquid water on its surface, a planet must have a surface temperature at which water can exist in the liquid phase. The surface temperature of a planet depends on its distance from the star it orbits. The intensity of radiation a planet receives from its star decreases with distance from the star. If a planet is too close to the star, the radiation is so intense that the planet's surface reaches temperatures above those at which liquid water can exist. If a planet is too far from a star, the radiation is so weak in intensity that the planet's surface does not reach a temperature high enough for liquid water to exist. Thus, Kepler-62e and Kepler-62f may have liquid water on their surface because they are in the habitable zone, the zone in which planets are at a distance from the star that allows them to have surface temperatures at which liquid water can exist.

One credit is allowed for Kepler-62e and Kepler-62f and for an acceptable explanation. Acceptable explanations include but are not limited to:

- **Kepler-62e and Kepler-62f are located in the habitable zone.**
- **Kepler-62e and Kepler-62f are located at a distance from *Kepler-62* to allow water to remain in the liquid phase.**
- **Kepler-62e and Kepler-62f; they have the correct temperature.**
- **Kepler-62e and Kepler-62f; these planets are at the correct distance from the star.**

Note: Credit is *not* allowed for "located farthest from the star" alone because this does not refer to the habitable zone.

83. The terminal moraine is a ridgelike accumulation of debris marking the farthest point that a glacier has advanced. A glacier stops advancing when its leading edge melts at a faster rate than new ice is formed at the glacier's source. At the very end of a melting glacier, all of the rocks, soil, and sediments that were picked up and frozen into, or fell on top of, the glacier and that were pushed to the front of the glacier are released as the ice around them melts. Air resistance is so small that particles of all sizes fall through air at virtually the same speed. Therefore, sediments deposited by a melting glacier do not become sorted by size as they fall and do not form layers due to different settling rates. Particles of mixed shapes and sizes, ranging from boulders to clay, drop to the ground in an unorganized jumble. The result is a ridgelike pile of debris running along the very end of the glacier that is unsorted and unlayered—a terminal moraine. Thus, the arrangement of the sediments found within the terminal moraine is an unsorted and unlayered mixture of sediment sizes.

One credit is allowed. Acceptable responses include but are not limited to:
- **The sediments are unsorted.**
- **The sediments are unlayered.**
- **Mixed sediment sizes ranging from boulders to clay**
- **Unorganized arrangement**

84. Kettles are bowl-shaped depressions formed when buried blocks of glacial ice melt. If the depressions fill with water, they are called kettle lakes. Kettle lakes form as follows. First, a block of glacial ice falls off the face of the glacier and comes to rest on the ground, as shown in diagram C. As the glacier melts, sediments that were trapped in or on the glacier fall to the ground and are deposited around the block of glacial ice, as shown in diagram B. Eventually, the block of glacial ice is buried beneath sediments and begins to melt, as shown in diagram D. Once the ice has melted, it no longer supports the surrounding sediments and they collapse. This forms a circular depression called a kettle, which can fill with water to form a kettle lake, as shown in diagram A. Thus, the correct order in which the letters should be placed to indicate the sequence of development of a kettle lake from earliest stage to latest stage is $C \rightarrow B \rightarrow D \rightarrow A$.

One credit is allowed for $C \rightarrow B \rightarrow D \rightarrow A$.

C	B	D	A
Earliest Stage		$\longrightarrow$	Latest stage

85. Find the Geologic History of New York State chart in the *Reference Tables for Physical Setting/Earth Science*. In the column labeled "Important Geologic Events in New York," locate the statement "Advance and retreat of last continental ice." Thus, this advance and retreat occurred during the last ice age. Trace this row to the left to the column labeled "Epoch." Note that the advance and retreat of the last continental ice occurred during the Pleistocene epoch. Thus, the terminal moraines found on Long Island were deposited during the Pleistocene epoch.

One credit is allowed for **Pleistocene epoch**.

SELF-ANALYSIS CHART June 2018

Topic	Question Numbers (Total)	Wrong Answers (x)	Grade
Standards 1, 2, 6, and 7: Skills and Application			
Skills			
Standard 1 Analysis, Inquiry, and Design	2–6, 8, 10, 13, 15, 16, 21, 22, 24, 25, 27–31, 35, 36, 38, 39, 41–49, 51–57, 59–71, 74, 76–80, 84, 85		$\frac{100(60-x)}{60} = \%$
Standard 2 Information Systems	7, 68		$\frac{100(2-x)}{2} = \%$
Standard 6 Interconnectedness, Common Themes	1, 5, 8–12, 14, 18–20, 22–24, 30, 32–37, 40–45, 47–50, 52, 54–59, 61, 66–68, 70, 71, 73–75, 80–84		$\frac{100(52-x)}{52} = \%$
Standard 7 Interdisciplinary Problem Solving	7		$\frac{100(1-x)}{1} = \%$
Standard 4: The Physical Setting/Earth Science			
Astronomy			
The Solar System (MU 1.1a, b; 1.2d)	1, 9, 81		$\frac{100(3-x)}{3} = \%$
Earth Motions and Their Effects (MU 1.1c, d, e, f, g, h, i)	4, 5, 11, 12, 36–39, 55–57		$\frac{100(11-x)}{11} = \%$
Stellar Astronomy (MU 1.2b)	40–42, 79		$\frac{100(4-x)}{4} = \%$
Origin of Earth's Atmosphere, Hydrosphere, and Lithosphere (MU 1.2e, f, h)			
Theories of the Origin of the Universe and Solar System (MU 1.2a, c)	2, 3, 10, 80, 82		$\frac{100(5-x)}{5} = \%$

SELF-ANALYSIS CHART June 2018

Topic	Question Numbers (Total)	Wrong Answers (x)	Grade
Meteorology			
Energy Sources for Earth Systems (MU 2.1a, b)	45, 46		$\dfrac{100(2-x)}{2}=\%$
Weather (MU 2.1c, d, e, f, g, h)	6–8, 47, 66–68		$\dfrac{100(7-x)}{7}=\%$
Insolation and Seasonal Changes (MU 2.1i; 2.2a, b)	58, 76–78		$\dfrac{100(4-x)}{4}=\%$
The Water Cycle and Climates (MU 1.2g; 2.2c, d)	14, 15, 18, 20, 21, 33, 63, 64		$\dfrac{100(8-x)}{8}=\%$
Geology			
Minerals and Rocks (MU 3.1a, b, c)	29, 31, 35, 51–54		$\dfrac{100(7-x)}{7}=\%$
Weathering, Erosion, and Deposition (MU 2.1s, t, u, v, w)	22, 27, 28, 30, 34, 65, 83, 84		$\dfrac{100(8-x)}{8}=\%$
Plate Tectonics and Earth's Interior (MU 2.1j, k, l, m, n, o)	13, 24, 25, 59–62		$\dfrac{100(7-x)}{7}=\%$
Geologic History (MU 1.2i, j)	16, 17, 19, 23, 43, 44, 69–72, 85		$\dfrac{100(11-x)}{11}=\%$
Topographic Maps and Landscapes (MU 2.1p, q, r)	26, 32, 48–50, 73–75		$\dfrac{100(8-x)}{8}=\%$
ESRT			
2011 Edition Reference Tables for Physical Setting/Earth Science	2, 3, 5, 6, 8, 10, 13, 15, 16, 21, 24, 25, 27–29, 31, 35, 39, 41, 45, 47, 49, 51–54, 59, 61, 62, 64, 65, 67, 68, 70, 71, 78, 80, 85		$\dfrac{100(38-x)}{38}=\%$

To further pinpoint your weak areas, use the Topic Outline in the front of the book.
MU = Major Understanding (see Topic Outline)

Examination August 2018
Physical Setting/Earth Science

PART A
Answer all questions in this part.

Directions (1–35): For *each* statement or question, choose the word or expression that, of those given, best completes the statement or answers the question. Some questions may require the use of the *2011 Edition Reference Tables for Physical Setting/Earth Science*. Record your answers in the space provided.

1 The diagram below represents a sundial positioned in New York State. During daylight, the shadow cast by the gnomon (pointer) moves across the disc, with the tip of the shadow pointing to the time of day.

This motion of the gnomon's shadow on the sundial is mainly due to

(1) Earth's rotation
(2) Earth's revolution
(3) the Sun's rotation
(4) the Sun's revolution   1 ____

2 The formation of the planet Uranus is estimated to have occurred approximately

(1) 100,000 million years ago
(2) 2.0 billion years ago
(3) 4.6 billion years ago
(4) 13.7 billion years ago

2 _____

3 Compared to the Jovian planets in our solar system, the terrestrial planets have

(1) less mass and are less dense
(2) less mass and are more dense
(3) more mass and are less dense
(4) more mass and are more dense

3 _____

4 The diagram below represents the constellation Leo that can be seen by an observer in New York State at midnight during March.

Leo is *not* visible to this observer at midnight during September because

(1) Leo has rotated on its axis
(2) Leo has revolved in its orbit around the Sun
(3) Earth has rotated on its axis
(4) Earth has revolved in its orbit around the Sun

4 _____

5 An observer in New York City measured the angle of insolation at solar noon each day. During which month did this observer see the noontime angle of insolation increase each day?
   (1) April
   (2) July
   (3) September
   (4) December

6 The Coriolis effect occurs as a result of Earth's
   (1) rotation
   (2) revolution
   (3) tilted axis
   (4) magnetic field

7 During the process of condensation, water vapor
   (1) releases 334 J/g of heat energy
   (2) releases 2260 J/g of heat energy
   (3) gains 334 J/g of heat energy
   (4) gains 2260 J/g of heat energy

8 Infiltration is generally greater than runoff where the land has a
   (1) gentle slope and permeable soil
   (2) gentle slope and impermeable bedrock
   (3) steep slope and permeable soil
   (4) steep slope and impermeable bedrock

9 Nearly 90% of the water vapor that enters Earth's atmosphere comes from the evaporation of Earth's surface waters. Most of the remaining 10% is water vapor that enters the atmosphere through
   (1) precipitation from clouds
   (2) transpiration from plants
   (3) condensation within the troposphere
   (4) melting of polar ice caps

10 What is the relative humidity if the dry-bulb temperature is 26°C and the wet-bulb temperature is 18°C?
   (1) 13%
   (2) 33%
   (3) 45%
   (4) 51%

11 Most of the long-wave energy radiated from Earth and lost to space on a cloudless night is

(1) ultraviolet
(2) infrared
(3) visible light
(4) gamma rays

11 _____

12 In addition to carbon dioxide, two other major greenhouse gases in Earth's atmosphere are

(1) oxygen and nitrogen
(2) oxygen and methane
(3) water vapor and nitrogen
(4) water vapor and methane

12 _____

13 The arrows in the diagram below represent the movement of air over a mountain.

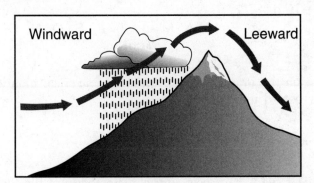

Clouds are forming on the windward side of this mountain because the air is

(1) expanding and cooling to the dewpoint
(2) expanding and warming to the dewpoint
(3) compressing and cooling to the dewpoint
(4) compressing and warming to the dewpoint

13 _____

14 The graph below shows the average monthly amount of insolation received throughout a year at four locations (A, B, C, and D) on Earth.

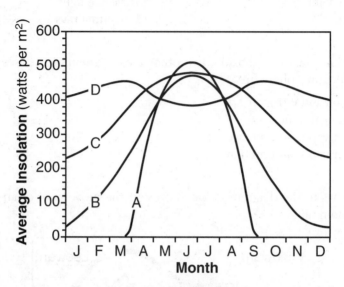

Which line on the graph best represents the average monthly insolation received at the equator?

(1) A
(2) B
(3) C
(4) D

14 ____

15 Earth's polar regions have cold, dry climates because the Sun's rays are at a

(1) low angle, and upper atmospheric air is sinking
(2) low angle, and upper atmospheric air is rising
(3) high angle, and lower atmospheric air is sinking
(4) high angle, and lower atmospheric air is rising

15 ____

16 Atmospheric transparency will increase when

(1) volcanic eruptions occur
(2) fog is produced
(3) insolation is reflected by clouds
(4) precipitation removes dust particles from the air

16 ____

17 The existence of which group of organisms spans the shortest geologic time?
   (1) birds
   (2) humans
   (3) dinosaurs
   (4) placoderm fish

18 Which New York State geologic event occurred most recently?
   (1) Taconian orogeny
   (2) Grenville orogeny
   (3) formation of the Catskill delta
   (4) dome-like uplift of the Adirondack region

19 The only dinosaur fossils found in New York State are footprints found on 210-million-year-old bedrock. In which New York State landscape region were these dinosaur fossils found?
   (1) Tug Hill Plateau
   (2) Newark Lowlands
   (3) Allegheny Plateau
   (4) Adirondack Mountains

20 The first *P*-wave of an earthquake travels 5600 kilometers from the epicenter and arrives at a seismic station at 10:05 a.m. At what time did this earthquake occur?
   (1) 9:49 a.m.
   (2) 9:56 a.m.
   (3) 10:02 a.m.
   (4) 10:14 a.m.

21 The photograph below shows a New York State index fossil.

What is the best classification of this fossil, and during which geologic time period did the organism that produced this fossil exist?

(1) Classification: Coral
    Geologic time period: Permian
(2) Classification: Coral
    Geologic time period: Ordovician
(3) Classification: Trilobite
    Geologic time period: Permian
(4) Classification: Trilobite
    Geologic time period: Ordovician   21 ____

22 As a quartz pebble is transported by a stream, the pebble will become more rounded as a result of

(1) dissolving as water is running over the rock
(2) abrasion by colliding with other rocks
(3) deposition in well-sorted layers
(4) resistance to weathering and erosion   22 ____

23 Which three minerals are most likely used in the construction of a house?

(1) graphite, pyrite, and halite
(2) garnet, galena, and sulfur
(3) talc, amphibole, and fluorite
(4) selenite gypsum, dolomite, and muscovite mica

24 The topographic map below shows a portion of a volcanic island in the Pacific Ocean. Elevations are shown in feet. Letters A and B represent locations on Earth's surface. Locations A and B are 2.5 miles apart.

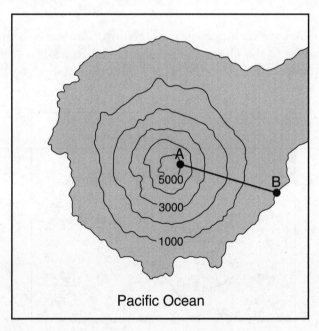

What is the approximate gradient from point A to point B on the island?

(1) 1000 ft/mi
(2) 1250 ft/mi
(3) 2000 ft/mi
(4) 2500 ft/mi

25 Which two rocks usually consist of only one mineral, but may contain additional minerals?

(1) hornfels and diorite
(2) quartzite and dunite
(3) rock salt and basalt
(4) gabbro and bituminous coal   25 ____

26 Which rock has never melted, but was produced by great heat and pressure, which distorted and rearranged its minerals?

(1) siltstone   (3) pegmatite
(2) breccia   (4) metaconglomerate   26 ____

27 The map below shows a portion of the North Carolina coastline, including some of the Outer Banks. The Outer Banks is a string of narrow barrier islands consisting of well-sorted sand along the Atlantic Ocean coast.

Which agent of erosion is primarily responsible for the formation of these barrier islands?

(1) wave action   (3) streams
(2) landslides   (4) glacial ice   27 ____

28 The photograph below shows a magnified view of a portion of a rock that can float if placed in water.

Which terms best describe this rock?
(1) non-crystalline and vesicular
(2) coarse and non-vesicular
(3) clastic and fragmental
(4) foliated and banded

28 _____

29 The map below shows three locations, labeled A, X, and B, on Earth's surface.

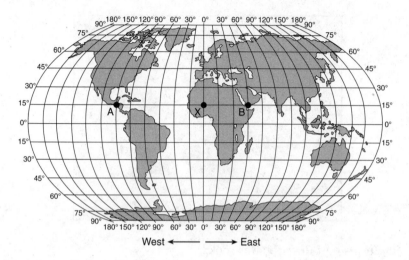

Which table correctly indicates the solar times at locations A and B when it is 12 noon at location X?

Location	Solar Time
A	6 a.m.
B	9 a.m.

(1)

Location	Solar Time
A	6 p.m.
B	9 a.m.

(3)

Location	Solar Time
A	6 a.m.
B	3 p.m.

(2)

Location	Solar Time
A	6 p.m.
B	3 p.m.

(4)

29 _____

30 Which equation is used to determine the approximate rate of Earth's revolution?

(1) Approximate rate of Earth's revolution = $\dfrac{365°}{360 \text{ days}}$

(2) Approximate rate of Earth's revolution = $\dfrac{360°}{24 \text{ hours}}$

(3) Approximate rate of Earth's revolution = $\dfrac{360°}{365 \text{ days}}$

(4) Approximate rate of Earth's revolution = $\dfrac{24°}{360 \text{ hours}}$

31 Which cross section best represents a cold front?

32 The isolines on the map below show snowfall totals from a lake-effect storm that affected a portion of New York State.

The surface winds that produced this storm came from which direction?

(1) northwest
(2) northeast
(3) southeast
(4) southwest

32 _____

33 The photograph below shows Mount Rainier, a volcano in the state of Washington.

Which map best shows the complete stream drainage pattern for this mountain?

(1)

(2)

(3)

(4)

33 _____

34 Which table best represents the characteristics of the continental crust and the oceanic crust?

Type of Crust	Density (g/cm³)	Composition	Relative Thickness
Continental	3.0	basaltic	thicker
Oceanic	2.7	granitic	thinner

(1)

Type of Crust	Density (g/cm³)	Composition	Relative Thickness
Continental	3.0	granitic	thicker
Oceanic	2.7	basaltic	thinner

(2)

Type of Crust	Density (g/cm³)	Composition	Relative Thickness
Continental	2.7	granitic	thinner
Oceanic	3.0	basaltic	thicker

(3)

Type of Crust	Density (g/cm³)	Composition	Relative Thickness
Continental	2.7	granitic	thicker
Oceanic	3.0	basaltic	thinner

(4)

34 _____

35 The map below shows the locations of some oil and gas fields in New York State.

Which type of bedrock contains these oil and gas deposits?

(1) extrusive igneous rock
(2) intrusive igneous rock
(3) metamorphic rock
(4) sedimentary rock

35 _____

PART B–1
Answer all questions in this part.

Directions (36–50): For *each* statement or question, choose the word or expression that, of those given, best completes the statement or answers the question. Some questions may require the use of the 2011 *Edition Reference Tables for Physical Setting/Earth Science*. Record your answers in the space provided.

Base your answers to questions 36 through 38 on the map below and on your knowledge of Earth science. The map shows typical weather systems over North America. Letters X, Y, and Z represent locations on the map. The isobars on the map are measured in millibars (mb).

36 Which map information indicates that the wind velocity is greater at location Z than at location X?

  (1) Location Z is closer to the ocean.
  (2) The isobars are closer together at Z.
  (3) The latitude of location X is greater.
  (4) Location X is closer to the front.   36 _____

37 Which type of front extends northwest from location Y?

  (1) warm front   (3) occluded front
  (2) cold front   (4) stationary front   37 _____

38 Which map best shows the locations for the centers of high pressure (**H**) and low pressure (**L**)?

38 ____

Base your answers to questions 39 through 41 on the passage below and on your knowledge of Earth Science.

Supermoon Eclipse

On September 27, 2015, a rare total lunar eclipse of a supermoon occurred. A supermoon occurs when the entire lighted half of the Moon faces Earth (full Moon phase) and the Moon is at its closest point to Earth in its orbit. At this time, the Moon will appear 14% larger and 30% brighter than normal. Supermoon events are rare, but a total lunar eclipse during a supermoon is even more rare. There have been only six total supermoon lunar eclipses since 1900. The next one will not happen until 2033.

39 Supermoon total lunar eclipses are celestial events that

(1) are random occurrences
(2) are predictable
(3) will never happen again after 2033
(4) will happen every full Moon

39 ____

40 The diagram below represents the Moon in four positions, A through D, in its orbit around Earth.

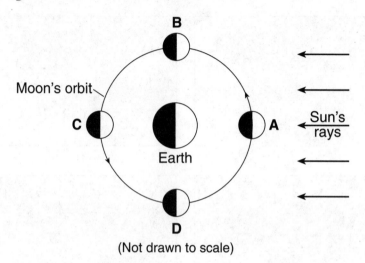

(Not drawn to scale)

At which position in its orbit was the Moon located during the 2015 supermoon total lunar eclipse?

(1) A (3) C
(2) B (4) D   40 ____

41 The time it took for the Moon to go from this supermoon to the next full moon phase was

(1) 15 days   (3) 29.5 days
(2) 27.3 days   (4) 365 days   41 ____

Base your answers to questions 42 through 44 on the diagram and graph below and on your knowledge of Earth science. The diagram represents a portion of Earth's interior. Letters A, B, and C represent interior layers. The graph shows the velocity of P-waves and S-waves at various depths in Earth's interior.

42 Which layers of Earth's interior are represented by letters A and B?
  (1) A is the crust and B is the rigid mantle.
  (2) A is the lithosphere and B is the asthenosphere.
  (3) A is the asthenosphere and B is the crust.
  (4) A is the rigid mantle and B is the lithosphere.   42 ____

43 What is the approximate velocity in kilometers/second of the P-waves at a depth of 1000 kilometers?
  (1) 6.2 km/s   (3) 11.3 km/s
  (2) 7.2 km/s   (4) 13.8 km/s   43 ____

44 Some locations within layer C have an inferred density of
(1) 3.4 g/cm³
(2) 5.6 g/cm³
(3) 11.5 g/cm³
(4) 12.9 g/cm³

44 _____

Base your answers to questions 45 through 47 on the graph below and on your knowledge of Earth science. The graph shows the changes in a single star's luminosity and relative temperature from its formation (point 1) to its late stage (point 4) relative to the Sun.

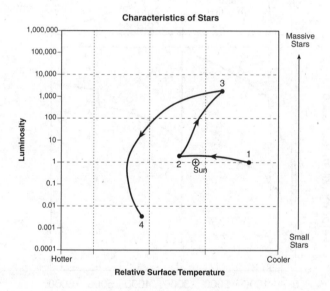

45 Which is a possible surface temperature of this star at point 2?
(1) 3000 K
(2) 5000 K
(3) 7000 K
(4) 10,000 K

45 _____

46 Between points 1 and 3, this star is visible to observers on Earth because it emits light energy. This energy is released by the process of nuclear fusion when
(1) dust collides with the star
(2) dust is broken apart by radiation
(3) lighter elements combine to form heavier elements
(4) heavier elements are broken down to form lighter elements

46 _____

47 Which table correctly classifies this star at points 3 and 4?

Point	Classification
3	Giant
4	White Dwarf

(1)

Point	Classification
3	Supergiant
4	Main Sequence

(3)

Point	Classification
3	White Dwarf
4	Supergiant

(2)

Point	Classification
3	Giant
4	Main Sequence

(4)

47 _____

Base your answers to questions 48 through 50 on the photograph below and on your knowledge of Earth science. The photograph shows a meandering stream in a wooded area. Points A and B represent locations on the streambanks. Letter X represents a flat area near the stream.

48 The streambank at location B is steeper than the streambank at location A because the water near location B is moving

(1) slower than the water near location A, causing more erosion
(2) slower than the water near location A, causing more deposition
(3) faster than the water near location A, causing more erosion
(4) faster than the water near location A, causing more deposition

48 _____

49 The area labeled letter X represents a portion of a
  (1) delta   (3) finger lake
  (2) sand dune   (4) floodplain   49 ____

50 Most of the particles deposited where the stream velocity *decreases* from 50 centimeters per second to 5 centimeters per second are
  (1) small cobbles and large pebbles
  (2) small pebbles and large sand
  (3) small sand and large silt
  (4) small silt and large clay   50 ____

PART B–2
Answer all questions in this part.

Directions (51–65): Record your answers in the spaces provided. Some questions may require the use of the *2011 Edition Reference Tables for Physical Setting/Earth Science.*

Base your answers to questions 51 through 54 on the diagram below and on your knowledge of Earth science. In the diagram, letters *A, B, C,* and *D* represent Earth's location on the first day of the four seasons as it orbits the Sun. Aphelion (Earth's farthest distance from the Sun) and perihelion (Earth's closest distance to the Sun) are labeled to show the approximate positions where they occur in Earth's orbit. The dashed lines represent Earth's axis, and the North Pole is labeled N.

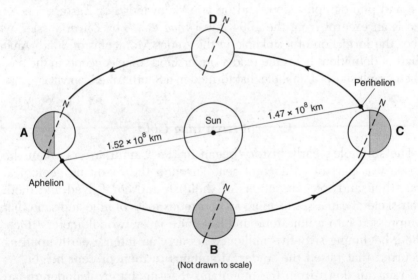

(Not drawn to scale)

51 State the number of degrees that Earth's axis is tilted from a line perpendicular to the plane of its orbit at each lettered location. [1]

_____°

52 Identify the season in New York State during which Earth is at perihelion. [1]

_____

53 State the number of hours of daylight that an observer in New York State will experience when Earth is at position *D*. [1]

_____ h

54 Identify the name of the star that is aligned with Earth's axis above the North Pole. [1]

_____

Base your answers to questions 55 through 58 on the passage, chart of definitions, and photograph below, and on your knowledge of Earth science. The passage is an excerpt from the 1994 novel *Inca Gold*, by Clive Cussler, which describes the formation of a sinkhole in the Andes Mountains of South America. The chart of definitions helps the reader understand some concepts in the passage. The photograph shows a sinkhole that formed in a South American village.

Excerpt from *Inca Gold*

...The sinkhole's early history began in the Cambrian era when the region was part of an ancient sea. Through the following geological eras, thousands of generations of shellfish and coral lived and died, their skeletal carcasses forming an enormous mass of lime and sand that compressed into a limestone and dolomite layer two kilometers thick. Then, beginning sixty-five million years ago, an intense earth uplifting occurred that raised the Andes Mountains to their present height. As the rain ran down from the mountains it formed a great underground water table that slowly began dissolving the limestone. Where it collected and pooled, the water ate upward until the land surface collapsed and created the sinkhole....

Chart of Definitions

Sinkhole	A natural depression or hole in the ground surface caused by some form of collapse of the bedrock beneath. Most are caused when slightly acidic groundwater chemically breaks down the carbonate rocks or the carbonate cement holding the rock particles together. The removal of the carbonates by groundwater gradually forms a hollow space or cavern under the surface layer. As the roof of the cavern weakens, it sometimes collapses, forming a sinkhole.
Lime	A general term for material containing the element calcium that combined with oxygen.

Photograph of a Sinkhole

55 In the first line of the passage, referring to Cambrian as an "era" is scientifically incorrect. State the unit of geologic time that should be substituted for the word "era." [1]

56 Identify *one* group of animals that became extinct at the time the passage states that intense uplifting of the Andes Mountains began. [1]

57 Using chemical symbols, state the chemical composition of the mineral found in limestone. [1]

_____

58 Describe the chemical weathering that contributes to the formation of sinkholes. [1]

_____

_____

Base your answers to questions 59 through 61 on the geologic cross section below and on your knowledge of Earth science. The cross section represents rock units, labeled A through K, that have *not* been overturned. Two unconformities and a volcanic ash layer are indicated.

59 List the letters E, H, and K to indicate the correct order of rock unit formation, from oldest to youngest, that formed this portion of Earth's crust. [1]

Letters: _____, _____, _____

Oldest ⸺⸺⸺⸺⸺⸺⸺⸺⸺⸺⟶ Youngest

60 Identify *two* processes that most likely caused the formation of both unconformities. [1]

(1) _____

(2) _____

61 Identify *one* metamorphic rock that most likely formed within rock unit G at the boundary of rock unit K. [1]

Base your answers to questions 62 through 65 on the field map below and on your knowledge of Earth science. The map shows the depth of Cuba Lake, located in New York State at latitude 42°14′ N, longitude 78°18′ W. Isoline values indicate water depth, in feet. Points A and B represent locations on the shoreline of Cuba Lake. Points W, X, Y, and Z represent locations on the bottom of the lake. The 30-foot isoline has been partially drawn.

Cuba Lake

62 On the map above, complete the 30-foot water depth isoline from point W to point X. [1]

63 On the grid below, construct a profile of the bottom of Cuba Lake from point A to point B. Plot each point where an isoline showing depth is crossed by line AB. Connect the plots with a line, starting at A and ending at B, to complete the profile. [1]

64 State the compass direction and distance in feet (ft) from point Y to point Z. [1]

Direction: _____

Distance: _____ ft

65 Identify the New York State landscape region where Cuba Lake is located. [1]

PART C

Answer all questions in this part.

Directions (66–85): Record your answers in the spaces provided. Some questions may require the use of the *2011 Edition Reference Tables for Physical Setting/Earth Science*.

Base your answers to questions 66 through 68 on the table below and on your knowledge of Earth science. The data table lists some information about the dwarf planet Pluto, which revolves around our Sun, and the five known moons that orbit Pluto.

Data for Dwarf Planet Pluto and Its Five Moons

Object Name	Classification	Period of Revolution in Earth Days (d)	Eccentricity of Orbit	Diameter (km)
Pluto	Dwarf Planet	90,511.4 (247.8 years)	0.2488	2370
Charon	Moon	6.4	0.0022	1208
Styx	Moon	20.2	0.0058	10 to 15 *
Nix	Moon	24.9	0.0020	40
Kerberos	Moon	32.2	0.0033	13 to 34 *
Hydra	Moon	38.2	0.0059	33 to 43 *

* There is a range in diameters for these moons due to their irregular shapes.

66 Identify the name of Pluto's moon that most likely has an orbit farthest from Pluto. Explain how the data indicate that this moon's orbit has the greatest distance from Pluto. [1]

Moon of Pluto: _____

Explanation: _____

_____

67 Describe the shape of the orbit of Pluto and the orbits of its five moons. [1]

_____

68 Explain why Pluto and its five moons are considered to be part of our solar system. [1]

_____

_____

_____

Base your answers to questions 69 through 73 on the passage and map below and on your knowledge of Earth science. The map shows a satellite image of a nor'easter that influenced the weather of the northeastern United States. The white areas represent clouds associated with this storm system. The locations of North Carolina and Albany, New York, are labeled on the map. The storm's low-pressure center is represented by letter L. Letters cP and mT represent two air masses.

Nor'easters

A nor'easter is a large, low-pressure storm system that moves along the east coast of the United States. The wind over the land blows generally from the northeast as the center of the low passes by a location, hence the name nor'easter. Due to the circulation of winds around the center of the low-pressure system, large amounts of precipitation occur as moist air is carried from the ocean to the land. These storms usually intensify off of the North Carolina coast as they track toward the northeast.

69 Describe *two* characteristics of the circulation pattern of the surface winds around the center of the low-pressure area represented on the map. [1]

Characteristic 1: _____

Characteristic 2: _____

70 Circle the terms that best describe the relative moisture and relative temperature characteristics of the mT air mass compared to the cP air mass shown on the map. [1]

Relative moisture of mT air mass (circle one):   **more humid**   **less humid**   **the same**

Relative temperature of mT air mass (circle one):   **cooler**   **warmer**   **the same**

71 The map below shows some of the principal storm tracks across the United States and the names of these storm tracks.

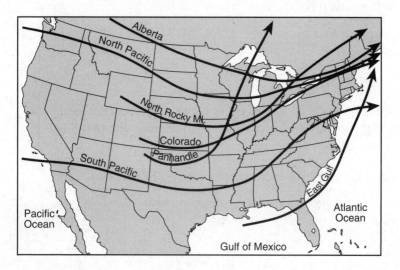

Identify the name of the storm track that this nor'easter most closely followed. [1]

72 The table below shows weather conditions recorded in Albany, New York, at the time that the satellite image was taken.

Weather Conditions

Dewpoint	22°F
Barometric Pressure	988.0 mb
Cloud Cover	100%
Present Weather	Snow

On the station model below, use the correct symbols and proper format to indicate the four conditions in the table. [1]

73 Identify *one* weather instrument that was most likely used to determine the dewpoint at Albany, New York. [1]

_____

Base your answers to questions 74 through 76 on the data table below and on your knowledge of Earth science. The data table shows the mass of a sample of radioactive carbon-14 remaining after each half-life.

Data Table

Number of Half-lives	Mass of Carbon-14 (g)
0	64
1	32
2	16
3	8
4	4
5	2
6	1

74 On the grid below, construct a line graph by plotting the data for the mass of carbon-14 in the sample for *each* half-life shown on the data table. Connect the plots with a line. [1]

75 Identify the stable disintegration product of the radioactive isotope carbon-14. [1]

_____

76 Determine the number of grams of carbon-14 remaining in this sample at 17,100 years. [1]

_____ g

Base your answers to questions 77 through 79 on the data tables below and on your knowledge of Earth science. Data table 1 shows the average maximum and average minimum water temperatures at different depths in Lake Michigan. Data table 2 shows the average maximum and average minimum land temperatures at different depths in soil at St. Paul, Minnesota (MN).

Table 1

Water in Lake Michigan, 45° N		
Water Depth Range (m)	Average Temperature (°C)	
	Maximum (Occurs in Summer)	Minimum (Occurs in Winter)
0–10	24.0	4.0
10–20	19.5	4.0
20–30	12.5	4.0
30–40	7.5	4.0
40–60	5.5	4.0
60–110	4.5	4.0
110–150	4.0	4.0

Table 2

Land in St. Paul, MN, 45° N		
Soil Depth Range (m)	Average Temperature (°C)	
	Maximum (Occurs in Summer)	Minimum (Occurs in Winter)
0–0.1	25.5	–6.0
0.1–1	23.0	–1.5
1–2	19.5	2.0
2–3	16.0	4.5
3–4	13.5	7.0
4–5	12.5	8.5
5–6	11.0	10.0

77 Identify the water depth range and the soil depth range, in meters, that have the same average maximum temperature of 12.5°C. [1]

Water depth range: _____ m

Soil depth range: _____ m

78 Describe the general relationship between depth and average temperature for both water and soil in the summer. [1]

_____

_____

79 Explain why the winter water surface is warmer than the winter land surface. [1]

_____

_____

Base your answers to questions 80 through 82 on the map below and on your knowledge of Earth science. The map shows an enlargement of a portion of the *Tectonic Plates* map from the *Physical Setting/Earth Science Reference Tables*. Arrows showing plate motion have been omitted. Points Y and Z represent locations on plate boundaries.

80 Identify the type of tectonic plate boundary found at location Z. [1]

_____

81 State the names of the tectonic plates on *each* side of the Bouvet Hot Spot. [1]

_____ **Plate** and _____ **Plate**

82 The cross section below represents a portion of Earth's interior beneath point Y. On this cross section, draw an arrowhead on *each* of the *four* bold lines, to represent the direction of the convection currents in the asthenosphere. [1]

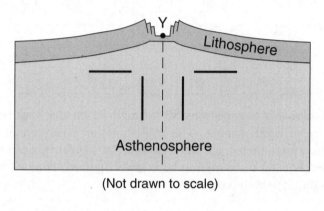

(Not drawn to scale)

Base your answers to questions 83 through 85 on the passage below and on your knowledge of Earth science.

Glacier Movement

Glaciers are thick sheets of ice in motion. Mountain glaciers tend to move down the slopes of mountains from higher elevations to lower elevations, while continental ice sheets move over large areas of continents. The bottom of a glacier is under great pressure due to the weight of the thick sheet of ice. This pressure causes the bottom of the glacier to partially melt, allowing the glacier to move. As the glacier thickens, more pressure is created and the glacier moves faster. Different parts of a glacier can move at different rates, depending on the amount of pressure and friction between the glacier and the underlying bedrock.

83 Describe the relationship between the thickness of a glacier and its rate of movement downhill. [1]

84 Describe the most likely shape of the cross section of a valley formed as a result of erosion by a mountain glacier. [1]

85 Compared to sediments deposited by meltwater from a glacier, describe the difference in the arrangement of the sediment deposited directly by a glacier. [1]

Answers August 2018
Physical Setting/Earth Science

Answer Key

PART A

1. 1	8. 1	15. 1	22. 2	29. 2
2. 3	9. 2	16. 4	23. 4	30. 3
3. 2	10. 3	17. 2	24. 3	31. 3
4. 4	11. 2	18. 4	25. 2	32. 1
5. 1	12. 4	19. 2	26. 4	33. 3
6. 1	13. 1	20. 2	27. 1	34. 4
7. 2	14. 4	21. 4	28. 1	35. 4

PART B–1

36. 2	39. 2	42. 2	45. 3	48. 3
37. 4	40. 3	43. 3	46. 3	49. 4
38. 1	41. 3	44. 4	47. 1	50. 2

In **PART B–2** and **PART C**, you are required to show how you arrived at your answers. For sample methods of solutions, see the *Answers Explained* section.

Answers Explained

PART A

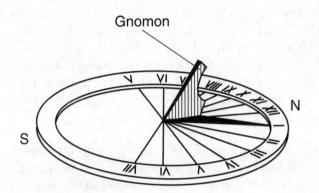

1. **1** The shadow cast by an object will always point in the direction opposite the position of the Sun in relation to the object. Therefore, a line drawn from the tip of the shadow to the tip of the object casting the shadow will point toward the Sun. It is given that the shadow cast by the gnomon moves across the disc pointing to the time of day. In order for the shadow to move, the position of the Sun in relation to the gnomon must change. Thus, when the shadow cast by the gnomon points to XI (A.M.), the Sun is located to the east of south. When the shadow cast by the gnomon points to I (P.M.), the Sun is located to the west of south. Thus, the Sun appears to be moving from east to west across the sky. The Sun's apparent path through the sky from sunrise to sunset is an arc, like all other celestial objects, and is the result of Earth's rotation. As Earth (and any object on Earth's surface) rotates on its axis from west to east, the Sun appears to move from east to west. Thus, the motion of the gnomon's shadow on the sundial is mainly due to Earth's rotation.

2. **3** The planet Uranus, like Earth, is part of the solar system. Find the Geologic History of New York State chart in the *Reference Tables for Physical Setting/Earth Science* and locate the time scale along the left edge of the column labeled "Eon." Locate 4600 million years ago (4.6 billion) on the time scale, trace right to the column labeled "Era," and note the statement "Estimated time of origin of Earth and our solar system." Thus, the formation of Earth and our solar system (which includes the planet Uranus) occurred approximately 4.6 billion years ago.

3. **2** The planets can be divided by mass and density into the terrestrial (Earth-like) and Jovian (Jupiter-like) planets. The terrestrial planets include the four innermost planets: Mercury, Venus, Earth, and Mars. The Jovian planets include the four outermost planets: Jupiter, Saturn, Neptune, and Uranus. Find the Solar System Data chart in the *Reference Tables for Physical Setting/Earth Science*. Locate the column labeled "Mass (Earth=1)" and note that the four terrestrial planets range in mass from 0.06 to 1, while the four Jovian planets range in mass from 14.54 to 317.83. Thus, compared to the Jovian planets, terrestrial planets have *less* mass. Now, locate the column labeled "Density (g/cm^3)" and note that the four terrestrial planets range in density from 3.9 to 5.5, while the four Jovian planets range in density from 0.7 to 1.8. Thus, compared to the Jovian planets, terrestrial planets are denser. Therefore, compared to the Jovian planets, the terrestrial planets have less mass and are more dense.

4. **4** As Earth revolves around the Sun, the side of Earth facing the Sun experiences day and the side facing away experiences night. Since the stars are only visible at night, the portion of the universe whose stars are visible to an observer on Earth varies cyclically as Earth revolves around the Sun. It is given that the constellation Leo is visible to an observer in New York State at midnight during March. Therefore, Earth's night side is facing Leo in March. In September, six months later, Earth would have revolved halfway through its orbit and would be on the opposite side of the Sun. Thus, its nighttime side would be facing away from Leo, and Leo would not be visible to an observer at night. See the diagram below. Therefore, the constellation Leo is not visible to this observer at midnight during September because of the revolution of Earth around the Sun.

At various times of the year, different constellations are visible at night because Earth's nighttime side faces different parts of the universe.

5. **1** In New York State, the angle of insolation at solar noon is greatest on June 21 and is least on December 21. From June 21 through December 21 (which includes July, September, and December), the angle of insolation at solar noon decreases every day. From December 21 through June 21 (which includes April), the angle of insolation at solar noon increases every day. Thus, an observer in New York City would see the noontime angle of insolation increase each day during April.

6. **1** Planetary winds tend to blow in a straight line from regions of high pressure toward regions of low pressure. As these winds are blowing, Earth is turning on its axis. This rotation causes the winds to appear to be turning toward the right in the Northern Hemisphere and toward the left in the Southern Hemisphere. This phenomenon is called the Coriolis effect. Thus, the Coriolis effect occurs as a result of Earth's rotation.

7. **2** Condensation is the phase change from gas to liquid. Find the Properties of Water chart in the *Reference Tables for Physical Setting/Earth Science*. Note the entry "Heat energy released during condensation . . . 2260 J/g." Thus, during the process of condensation, when water changes phase from water vapor to liquid water, the water vapor releases 2260 J/g of heat energy.

8. **1** Infiltration is the downward movement of liquid water into and through interconnected openings among soil particles due to gravity. Runoff is precipitation that does not evaporate or infiltrate, but, instead, runs downhill along Earth's surface.

A substance through which water can flow is said to be permeable; a substance through which water cannot flow is impermeable. Thus, the land surface must be permeable in order for infiltration to occur.

Water runs downhill more slowly on gentle slopes than on steep slopes, so on gentle slopes there is more time for the water to infiltrate before it runs off. Therefore, in an area with gentle slopes, more water will infiltrate and less will run off. Thus, infiltration is generally greater than runoff where the land has a gentle slope and permeable soil.

9. **2** The process that comprises the remaining 10% of water that enters the atmosphere is one that releases water vapor into the atmosphere. Plants cover a significant portion of Earth's land surface. Plants remove liquid water from the ground and release it into the atmosphere in the form of water vapor by a process called transpiration. In transpiration, liquid water absorbed by plant roots from the soil increases the water pressure inside the lower parts of the plant. Simultaneously, evaporation of water through plant stems or leaf stomata decreases the water pressure in the upper parts of a plant. This difference in pressure causes the liquid water to move upward from the roots and toward the leaves. When the liquid water reaches the leaves, it evaporates. Then more water is drawn upward from the roots, and the process of transpiration continues. Thus, most of the remaining 10% of water vapor enters the atmosphere through transpiration from plants.

WRONG CHOICES EXPLAINED:

(1) Precipitation is the process by which water in the form of rain, snow, hail, or sleet falls to the ground. Thus, water vapor does not enter the atmosphere through precipitation.

(2) Condensation is the process by which water vapor changes phase to form liquid water. Thus, water vapor does not enter the atmosphere due to condensation; water vapor leaves the atmosphere due to condensation.

(3) Melting is the process by which solid water changes phase to form liquid water. Thus, melting does not add water vapor to the atmosphere.

10. **3** It is given that the dry-bulb temperature is 26°C and the wet-bulb temperature is 18°C. Thus, the difference between the wet-bulb and dry-bulb temperatures is 8°C. Find the Relative Humidity (%) chart in the *Reference Tables for Physical Setting/Earth Science*. Locate the column headed "8" in the "Difference Between Wet-Bulb and Dry-Bulb Temperatures (°C)" scale along the top of the chart. Find the "Dry-Bulb Temperature (°C)" scale along the left

side of the chart. Locate the row labeled "26," trace to the right until it intersects the column headed "8," and note the value of 45. Thus, when the dry-bulb temperature is 26°C and the wet-bulb temperature is 18°C, the relative humidity is 45%.

11. **2** Every object not at absolute zero radiates energy. The type of energy radiated depends on the temperature of the object. The Sun, which is at a high temperature, radiates energy of relatively short wavelengths, such as ultraviolet and visible light rays. Earth's surface, which is at a much lower temperature, radiates energy at longer wavelengths, mainly in the infrared range. Thus, most of the long-wave energy radiated and lost to space on a cloudless night is infrared.

12. **4** When sunlight strikes Earth's surface, some of it is reflected back toward space, and some of it is absorbed and then re-radiated as infrared radiation (or heat). Over time, the amount of energy received from the Sun should be about the same as the amount of energy radiated back into space, leaving the temperature of the Earth's surface roughly constant. However, many chemical compounds found in the Earth's atmosphere act as "greenhouse gases." These gases allow sunlight to enter the atmosphere and strike Earth's surface, but absorb the infrared radiation re-radiated by Earth's surface, trapping the heat in the atmosphere resulting in global warming. Some of these "greenhouse gases" occur in nature (water vapor, carbon dioxide, methane, ozone, and nitrous oxide), while others are exclusively human-made (like the chlorofluorocarbons used in aerosol sprays). Thus, in addition to carbon dioxide, only choice 4—water vapor and methane—contains two other major greenhouse gases found in Earth's atmosphere.

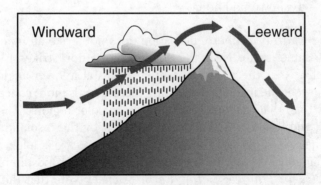

13. **1** According to the arrows in the diagram, air is moving upward on the windward side of the mountain. As air is carried upward to higher elevations, it expands and cools. When the air temperature cools to the dewpoint, moisture in the air begins to condense and clouds form. Thus, clouds are forming on the windward side of the mountain because the air is expanding and cooling to the dewpoint.

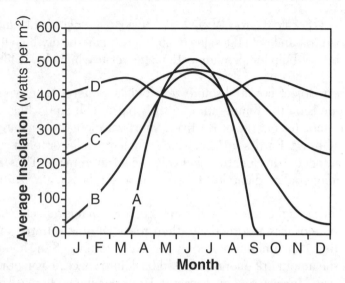

14. **4** The average monthly insolation received at the equator will be greatest when the sunlight striking Earth's surface at the equator is most direct. At the equinoxes, Earth's axis of rotation tilts neither toward nor away from the Sun, and the Sun's direct rays strike Earth's surface at the equator. Thus, the average monthly insolation received at the equator will peak at the equinoxes (March and September). At other times of the year, Earth is tilted either toward or away from the Sun and sunlight strikes Earth's surface at less than 90°, and the average monthly insolation received at the equator will be lower. Thus, the line on the graph representing the equator should show two peaks—one in March and one in September—as shown on line D.

15. **1** Since Earth is a sphere, insolation does not strike all points on Earth's surface at the same angle. Near the equator, insolation strikes Earth's surface almost vertically. Near the poles, insolation strikes at a lower angle. Therefore, throughout the year the insolation reaching the tropics is more concentrated (i.e., more intense) than that reaching mid-latitude or polar regions. The less intense the insolation, the less Earth's surface is warmed by the insolation. Therefore, polar regions have cold climates because the Sun's rays are at a low angle.

Find the Planetary Winds and Moisture Belts in the Troposphere chart in the *Reference Tables for Physical Setting/Earth Science.* Note that the arrows indicating air movements show that, near the poles, air in the upper troposphere is sinking. As the air sinks, it is compressed and warms (although it is still very cold). This increases the difference between the air temperature and the dew-point temperature, resulting in lower relative humidity. Thus, Earth's polar regions also have dry climates. Therefore, Earth's polar regions have cold, dry climates because the Sun's rays are at a low angle, and upper atmosphere air is sinking.

16. **4** Air is mainly composed of transparent gases. However, air also contains varying amounts of substances that are opaque and block light, or that reflect or scatter light. The more of these substances the air contains, the lower the atmospheric transparency. Common substances in air that block, reflect, or scatter light include water droplets, ice crystals, salt particles, or particles of dust, ash, or smoke. Removing any of these substances from the air will increase atmospheric transparency. One common process that removes dust and other particles from the air is precipitation. Precipitation is condensed moisture (such as rain, snow, hail, or sleet) in the atmosphere that falls to the ground. Precipitation removes particles of dust from the air in a number of ways. For example, dust particles are often the condensation nuclei on which water condenses. When the condensed moisture falls to the ground, these condensation nuclei are carried to the ground. Dust particles can also come into contact with the condensed moisture as they fall through the air, adhere to its surface, and be carried to the ground. Thus, atmospheric transparency will increase when precipitation removes dust particles from the air.

WRONG CHOICES EXPLAINED:
(1) Volcanic eruptions spew gases, ash, and dust into the atmosphere, which are then carried around the globe by winds in the upper atmosphere. Volcanic ash and dust are opaque and will block, reflect, or scatter incoming solar radiation. Thus, a volcanic eruption would decrease atmospheric transparency.

(2) Fog is composed of water droplets. Water droplets can reflect and scatter light. Therefore, fog decreases atmospheric transparency.

(3) When insolation is reflected by clouds, less passes through the cloud and reaches Earth's surface. Insolation consists of electromagnetic waves, and electromagnetic waves are totally transparent to other electromagnetic waves. Thus, atmospheric transparency is not affected by the quantity of insolation passing through the air. Therefore, if insolation is reflected by clouds, atmospheric transparency would neither increase nor decrease.

17. **2** Find the Geologic History of New York State chart in the *Reference Tables for Physical Setting/Earth Science*. In the column labeled "Life on Earth," locate "humans." From humans, trace left to the time scale labeled "Million years ago." Note that humans appeared less than 1.8 million years ago. Humans still exist today. Therefore, humans have existed for less than 1.8 million years.

Now find the column labeled "Time Distribution of Fossils (including important fossils of New York)." Note the vertical gray bars labeled with the names of different groups of organisms. The vertical gray bars indicate the range of time during which each group of organisms existed on Earth. The bottom of the gray bar indicates when the group first appeared on Earth, and the top of the bar indicates when the group became extinct or, if it extends to the top of the column, that the group still exists. Locate the gray bar labeled "Birds." From the bottom of this gray bar, trace left to the time scale labeled "Million years ago." Note that birds first appeared a little more than 146 million years ago. From the top of this gray bar, trace left to the time scale labeled "Million years ago," and note that birds still exist. Therefore, birds have existed for more than 146 million years. Now, locate the gray bar labeled "Dinosaurs." From the bottom of this gray bar, trace left to the time scale labeled "Million years ago." Note that dinosaurs first appeared between 200 and 251, or about 225 million years ago. From the top of this gray bar, trace left to the time scale labeled "Million years ago," and note that dinosaurs became extinct 65.5 million years ago. Thus, dinosaurs existed for about 159.5 million years (225 − 65.5 = 159.5). Finally, locate the gray bar labeled "Placoderm fish." From the bottom of this gray bar, trace left to the time scale labeled "Million years ago." Note that placoderm fish first appeared between 416 and 444, or about 430 million years ago. From the top of this gray bar, trace left to the time scale labeled "Million years ago," and note that placoderm fish became extinct 359 million years ago. Thus, placoderm fish existed for about 71 million years (430 − 359 = 71). Therefore, the group whose existence spans the shortest geologic time is humans.

18. **4** Find the Geologic History of New York State chart in the *Reference Tables for Physical Setting/Earth Science*. Locate the time scale labeled "Million years ago" to the left of the column labeled "Life on Earth." Note that, on the scale, age increases from top to bottom. Now, locate the column labeled "Important Geologic Events in New York." Note that, according to the time scale, the most recent events are nearest the top of the column. Now, locate "Taconian Orogeny," "Grenville Orogeny," "formation of Catskill delta," and "dome-like uplift of the Adirondack region." Note that, of these events, the dome-like uplift of the Adirondack region is nearest the top of the chart. Therefore, the New York State geologic event that occurred most recently is the dome-like uplift of the Adirondack region.

19. **2** It is given that the dinosaur footprints were found in 210-million-year-old bedrock. Find the Geologic History of New York State chart in the *Reference Tables for Physical Setting/Earth Science*. Locate the time scale labeled

"Million years ago" to the left of the column labeled "Life on Earth." On the time scale, locate the point corresponding to 210 million years ago. From this point, trace left to the column labeled "Period," and note that 210 million years ago corresponds to the Late Triassic period. Next, find the Generalized Bedrock Geology of New York State map in the *Reference Tables for Physical Setting/Earth Science*. In the key labeled "Geologic Periods and Eras in New York," locate the map symbol corresponding to Late Triassic and Early Jurassic conglomerates, red sandstones, red shales, basalt, and diabase (Palisades sill). On the map, locate the region shaded with this map symbol (along the border between New York and New Jersey). Now, find the Generalized Landscape Regions of New York State map in the *Reference Tables for Physical Setting/Earth Science*. Note that the region of Late Triassic rocks in New York State corresponds to the Newark Lowlands. Thus, the New York State landscape region in which these fossils were found is the Newark Lowlands.

20. **2** It is given that the first P-wave traveled 5600 kilometers from the epicenter of the earthquake to the seismic station. Find the Earthquake P-wave and S-wave Travel Time graph in the *Reference Tables for Physical Setting/Earth Science*. Locate 5600 kilometers (5.6×10^3 km) along the horizontal Epicenter Distance axis at the bottom of the graph. From this point, trace vertically until you intersect the bold line labeled "P." From this intersection, trace horizontally to the left to the Travel Time axis and read the P-wave travel time—9 minutes. If the first P-wave takes 9 minutes to travel the 5600 kilometers from the epicenter to the seismic station and arrives at 10:05 A.M., it must have left the epicenter 9 minutes before 10:05 A.M., or 9:56 A.M. Thus, the earthquake occurred at 9:56 A.M.

21. **4** It is given that the photograph shows a New York State index fossil. Find the Geologic History of New York State chart in the *Reference Tables for Physical Setting/Earth Science*. Locate the diagrams of index fossils along the bottom of the chart. Note that the New York State index fossil that most closely resembles the one in the photograph is B—*Cryptolithus*. In the column labeled "Time Distribution of Fossils," locate the circled letter B and note that it lies on the gray bar labeled "Trilobites." Thus, the index fossil in the photograph is a trilobite. Note, too, the explanation at the top of the column: "The center of each lettered circle indicates the approximate time of existence of a specific index fossil." From the center of the circled letter B representing *Cryptolithus*, trace left to the column labeled "Period," and note that *Cryptolithus* lived during the Ordovician period. Thus, the best classification of this fossil and geologic time period during which it existed is Classification: Trilobite; Geologic time period: Ordovician.

22. **2** As sediment particles, such as quartz pebbles, are carried along by the moving water in a stream, they bounce off of and rub against one another and bounce, roll, and scrape against the bottom of the streambed. These collisions cause smaller pieces to break off of sharp corners and edges on the surface of the sediments, particularly at corners that protrude—a process called abrasion. As a result, the sediments eroded by the water in this stream become more rounded. Thus, as a quartz pebble is transported by a stream, the pebble will become more rounded as a result of abrasion by colliding with other rocks.

WRONG CHOICES EXPLAINED:
(1) The mineral quartz is almost insoluble in water. Therefore, a quartz pebble would not become rounded by water running over it.

(3) A quartz pebble is deposited when the stream stops transporting it, not as it is transported. If it is deposited in layers, the surrounding sediments shield it from further abrasion. Therefore, it would not become rounded as a result of deposition in layers.

(4) If a quartz pebble is resistant to weathering and erosion, it is less susceptible to abrasion and likely to become less rounded as it is transported by a stream.

23. **4** Find the Properties of Common Minerals table in the *Reference Tables for Physical Setting/Earth Science*. In the column labeled "Uses," note uses that relate to the construction of a house, such as "plaster of paris, drywall," "paint, roofing," and "building stones." From each of these uses, trace right to the column labeled "Mineral Name," and note that they correspond to selenite gypsum, muscovite mica, and dolomite, as shown in choice 4.

WRONG CHOICES EXPLAINED:
(1), (2) and (3) Note that there are other minerals with uses relating to the construction of a house, such as "cement, lime," "ceramics, glass," and "construction materials," but none of the minerals in answer choices 1–3 has a construction-related use for all three of the minerals listed in the choice.

24. **3** Find the Equations section of the *Reference Tables for Physical Setting/Earth Science* and note the equation for gradient:

$$\text{gradient} = \frac{\text{change in field value}}{\text{change in distance}}$$

The map shown is a topographic map. The field value on a topographic map is elevation. Note that point A is located directly on the 5000-foot contour line and that point B is located directly on the coastline of the island, which corresponds to sea level, or elevation 0 feet. Thus, the change in field value (elevation) from A to B is 5000 feet.

It is given that the distance between locations A and B is 2.5 miles. Substitute these values in the equation and solve:

$$\text{gradient} = \frac{500 \text{ ft}}{2.5 \text{ mi}} = 2000 \text{ ft/mi}$$

Thus, the gradient between location A and location B is approximately 2000 ft/mi.

25. **2** Find the Scheme for Metamorphic Rock Identification in the *Reference Tables for Physical Setting/Earth Science*. In the column labeled "Rock Name," locate quartzite. From quartzite, trace left to the column labeled "Composition," and note that quartzite is composed of quartz. Thus, quartzite usually contains only one mineral. Now, trace right to the column labeled "Comments," and note that quartzite forms by the metamorphism of quartz sandstone. Quartz sandstone is composed of grains of sand that are cemented together. Not all grains of sand are pure quartz, and not all cement is pure quartz. Therefore, although quartzite is usually composed of only quartz, it may contain other minerals.

Find the Scheme for Igneous Rock Identification in the *Reference Tables for Physical Setting/Earth Science*. In the upper portion of the scheme labeled "Igneous Rocks," locate dunite. Trace the column for dunite down to the graph labeled "Mineral Composition." Note that dunite is composed of olivine. Note, too, that the upper left edge of the column touches the region labeled "pyroxene." Thus, dunite may contain traces of pyroxene. Therefore, two rocks which usually consist of only one mineral, but may contain additional minerals, are quartzite and dunite.

WRONG CHOICES EXPLAINED:

(1) Find the Scheme for Metamorphic Rock Identification in the *Reference Tables for Physical Setting/Earth Science*. In the column labeled "Rock Name," locate hornfels. Trace the row for hornfels to the left to the column labeled "Composition," and note that hornfels is composed of various minerals. Thus, hornfels does not usually consist of only one mineral.

(3) Find the Scheme for Igneous Rock Identification in the *Reference Tables for Physical Setting/Earth Science*. In the upper portion of the scheme labeled "Igneous Rocks," locate basalt. Trace the column for basalt down to the graph labeled "Mineral Composition." Note that basalt is composed of plagioclase feldspar, pyroxene, biotite, amphibole, and olivine. Thus, basalt does not usually consist of only one mineral.

(4) Find the Scheme for Igneous Rock Identification in the *Reference Tables for Physical Setting/Earth Science*. In the upper portion of the scheme labeled "Igneous Rocks," locate gabbro. Trace the column for gabbro down to the graph labeled "Mineral Composition." Note that gabbro is composed of plagioclase feldspar, pyroxene, biotite, amphibole, and olivine. Thus, gabbro does not usually consist of only one mineral.

26. **4** By definition, metamorphism refers to the changes that occur in solid rock due to great heat, great pressure, and the chemical activity that sometimes results from the combination of these two factors. Metamorphism occurs while rocks are still in the solid state and does not involve molten rock in the liquid state. Thus, a rock that has never melted, but was produced by great heat and pressure that distorted and rearranged its minerals, is a metamorphic rock. Find the Scheme for Metamorphic Rock Identification in the *Reference Tables for Physical Setting/Earth Science*. In the column labeled "Rock Name," note that only answer choice 4—metaconglomerate—is a metamorphic rock.

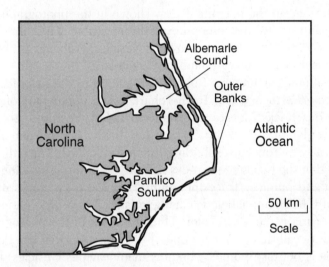

27. **1** Barrier islands form when tidal inlets break a barrier bar into elongated islands. Barrier bars are formed when waves deposit enough sand on a sandbar so that it is above water. Thus, the agent of erosion that is primarily responsible for the formation of these barrier islands is wave action.

WRONG CHOICES EXPLAINED:
(2) During mass movements such as landslides, sediment is pulled downhill by gravity and is deposited at the base of a slope when it stops moving. Since the barrier islands are located offshore, not at the base of a slope, it is unlikely that they formed as a result of landslides.

(3) When a stream flows into a standing body of water, such as the Atlantic Ocean, it stops moving, and most of the stream's sediment is deposited. The resulting landform is a large, flat, fan-shaped pile of sediment at the mouth of the stream called a delta. These barrier islands are elongated and run parallel to the coast; therefore, it is unlikely that they were formed by streams.

(4) Glaciers do not currently exist in North Carolina. At the end of the last ice age, vast amounts of water were locked up in polar ice caps, and sea level was much lower than it is today. As a result, at the time of the last ice age, the shoreline of what is now North Carolina was as much as 40 miles east of the current coastline at the edge of the North American continental shelf. In the thousands of years since the last ice age, rising sea level and the action of waves and currents would have eroded and drowned any deposits left by glaciers even if they had existed along the North Carolina coast of the Atlantic Ocean. Therefore, it is unlikely that these barrier islands were formed by glacial ice.

28. **1** Note that, in the magnified view shown in the photograph, the rock is light in color and no distinct mineral crystals are visible. Therefore, this rock is non-crystalline. Note, too, the many cavities (vesicles) visible in the surface of the rock. Thus, this rock is non-crystalline and vesicular.

Find the Scheme for Igneous Rock Identification in the *Reference Tables for Physical Setting/Earth Science*. In the chart in the upper part of the scheme labeled "Igneous Rocks," locate the column labeled "Texture." Locate the row labeled "Vesicular (gas pockets)." Trace left to the column labeled "Crystal Size," and locate the row corresponding to non-crystalline. Trace this row to the left and note that the two igneous rocks that are both non-crystalline and vesicular are scoria and pumice. Trace the column for these two rocks downward to the section of the scheme labeled "Characteristics." Note that pumice is lighter in color and scoria is darker in color. Thus, the rock in the photograph is most likely pumice. Recall that pumice is one of the few rocks that can float if placed in water. This is because pumice contains many cavities (vesicles) that formed when molten rock solidified as gases were still bubbling out of it. The presence of these vesicles decreases the overall density of the pumice sample enough that its density may be less than that of water and it will float.

Thus, the rock in the photograph that can float is most likely pumice and is best described by the terms non-crystalline and vesicular.

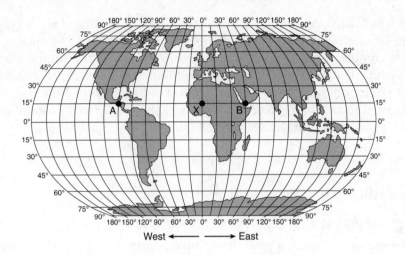

29. **2** By convention, maps are oriented with north at the top. Lines of longitude run north–south (vertically) and lines of latitude run east–west (horizontally). On the given map, longitude values are read along the top and bottom of the map; latitude values are read along the left and right side of the map. Note that location X is on the Prime Meridian (0° longitude). Earth rotates eastward at a rate of 15° per hour. Therefore, for every 15° longitude you travel west of the Prime Meridian, the Sun rises one hour later and solar time is one hour earlier than at the Prime Meridian. Conversely, for every 15° longitude you travel east of the Prime Meridian, the Sun rises one hour earlier and solar time is one hour later than at the Prime Meridian. It is given that it is 12 noon at location X. Note that location B is 45° of longitude east of the Prime Meridian. Therefore, when it is 12 noon at location X, the solar time at location B will be 3 hours later (45°/15°/hr = 3 hr), or 3 P.M. Note that location A is 90° of longitude west of the Prime Meridian. Therefore, when it is 12 noon at location X, the solar time at location B will be 6 hours earlier (45°/15°/hr = 3 hr), or 6 A.M. This corresponds to the table shown in choice 2.

30. **3** Find the Equations chart in the *Reference Tables for Physical Setting/Earth Science*. Locate the equation for rate of change:

$$\text{Rate of change} = \frac{\text{change in value}}{\text{time}}$$

$$\text{Rate of revolution} = \frac{\text{change in Earth's position relative to the Sun during one revolution}}{\text{time it takes Earth to complete one revolution}}$$

Changes in Earth's position relative to the Sun are measured in degrees. In completing one revolution, Earth changes position by 360°. The time it takes Earth to complete one revolution is called its period of revolution. Find the Solar System Data chart in the *Reference Tables for Physical Setting/Earth Science* and locate the column labeled "Period of Revolution." Note that Earth's period of revolution is 365.26 days. Substitute these values in the rate of change equation to solve for Earth's rate of revolution:

$$\text{Rate of revolution} = \frac{360°}{365 \text{ days}}$$

This equation is best represented by the equation in choice 3.

31. **3** Cold fronts form along the leading edge of a cooler air mass advancing against a warmer air mass. A cool, dry air mass is denser than the warm, moist air ahead of it. So, the cool, dry air mass bulges into the warm, moist air, pushing against and under it like a wedge. The curved slope of the front is rather steep, especially if it is moving fast. Thus, a cold front slopes downward from the cool air mass toward the warm air mass in a steep-sided, curved bulge as shown in choice 3.

WRONG CHOICES EXPLAINED:

(1) and (2) Cold fronts form along the leading edge of a cooler air mass advancing against a warmer air mass. Both of these cross sections show the opposite—warm, moist air advancing against cool, dry air.

(4) In this cross section, cool, dry air is advancing against warm, moist air. However, the boundary shown slopes upward away from the cool, dry air, indicating that the denser, cool, dry air moves up and over the less dense, warm, moist air. This will not happen because denser materials tend to sink beneath less dense materials, not rise above them.

32. **1** Winter lake-effect snow storms occur when cold winds move across large stretches of warm lake water. Evaporation of the warm lake water adds water vapor to the air. The water vapor is picked up by the cold wind, freezes, and is deposited as snow on the cold land of the leeward shores of the lake. Find the Generalized Bedrock Geology of New York State map in the *Reference Tables for Physical Setting/Earth Science*. Locate Lake Ontario and note that it is located at about 43–44° N latitude. Find the Planetary Winds and Moisture Belts in the Troposphere diagram in the *Reference Tables for Physical Setting/Earth Science*. Note that, at the latitudes corresponding to Lake Ontario, the planetary winds generally blow from west to east. According to the map, the area of snowfall from the lake-effect storm is on the southeast shoreline of the lake. Note, too, that the regions encompassed by the isolines are elongated along an axis running northwest to southeast. Thus, the cold winds are moving from northwest to southeast over the lake, and the southeastern shores of Lake Ontario are the leeward shores that would be affected by lake-effect snow from this storm. Therefore, the surface winds that produced this storm came from the northwest.

33. **3** According to the photograph, Mount Rainier is a roughly cone-shaped volcano. Water flows from higher elevations to lower elevations. A cone-shaped mountain is highest at the peak and is lower in all directions outward from the peak. Thus, precipitation running off the surface of a hill will flow outward in all directions from the central, higher areas, forming a radial stream pattern as shown in diagram (3).

WRONG CHOICES EXPLAINED:

(1) The rectangular stream drainage pattern shown in map (1) forms in regions where there are faults or joints that break the surface bedrock into rectangular blocks. The main streams and their tributaries in such regions follow these faults and joints, displaying right-angle bends at the corners of the rectangular blocks and perpendicular sections of approximately the same length as they flow along their edges. The photograph in the question shows a cone-shaped volcanic mountain, not a landscape in which bedrock has broken into rectangular blocks by faulting.

(2) The trellis stream pattern in map (2) forms in tilted or folded rock layers of unequal resistance to erosion. Major streams run through long, parallel valleys following belts of weak rock between parallel ridges of stronger rock. Tributary streams enter major streams at right angles. The photograph in the question shows a cone-shaped volcanic mountain, not a series of long, parallel ridges.

(4) On a flat plain, the slope is gentle and streams will flow downhill slowly and typically form a dendritic stream drainage pattern. In a dendritic stream drainage pattern, there are numerous small streams that join together to form

larger streams in a tree-like pattern as shown in map (4). The photograph in the question shows a cone-shaped volcanic mountain with steep slopes, not a gently sloped, flat plain.

34. **4** Find the Inferred Properties of Earth's Interior chart in the *Reference Tables for Physical Setting/Earth Science*. Locate the "Density (g/cm^3)" scale along the right side of the cross section. Note the following densities: granitic continental crust—2.7 and basaltic oceanic crust—3.0.

Find the Scheme for Igneous Rock Identification in the *Reference Tables for Physical Setting/Earth Science*. Locate granite and basalt in the upper portion of the chart. Trace downward to the center section of the scheme labeled "Characteristics." Note that granite is lower in density and basalt is higher in density. A lower-density substance occupies more volume per given mass than a higher-density substance. Thus, compared to a given mass of oceanic crust, an equivalent mass of continental crust will occupy more volume. Therefore, continental crust tends to be thicker than an equal area of oceanic crust of the same mass. Thus, the characteristics of continental crust and oceanic crust are best represented by the table in choice (4.)

Source: New York State Museum, State Geological Survey

35. **4** According to the map key, the oil fields are found primarily in sandstone and carbonated [sic] rocks (rocks composed of carbonate minerals such as calcite and dolomite), and gas fields are found primarily in sandstone, limestone, and shale. Find the Scheme for Sedimentary Rock Identification in the *Reference*

Tables for Physical Setting/Earth Science. Locate the column labeled "Rock Name," and note that sandstone, limestone, dolostone, and shale are all sedimentary rocks. Carbonate rocks are composed of carbonate minerals. Find the Properties of Common Minerals table in the *Reference Tables for Physical Setting/Earth Science.* In the column labeled "Composition," locate chemical formulas that indicate that the mineral is a carbonate, that is, formulas ending in CO_3. Note the two formulas $CaCO_3$ and $CaMg(CO_3)_2$. Trace left to the column labeled "Mineral Name," and note that these formulas correspond to the minerals Calcite and Dolomite, respectively. Thus, the sedimentary rocks limestone and dolostone are carbonate rocks. Thus, the type of bedrock that contains these oil and gas deposits is sedimentary rock.

PART B–1

36. **2** Surface wind velocity is directly related to the air pressure gradient. The closer together the isobars on the map, the greater the pressure gradient and, therefore, the greater the wind velocity. On the given map, the isobars are closer together at location Z than they are at location X. Therefore, the wind velocity is greater at location Z than at location X because the isobars are closer together at Z.

37. **4** Find the Weather Map Symbols chart in the *Reference Tables for Physical Setting/Earth Science* and locate the key for Front Symbols. Note that the symbol shown in the weather map on the front that extends northwest from location Y corresponds to a stationary front.

38. **1** Isobars are lines connecting points of equal air pressure. Isobars are labeled with the air pressure in millibars of all points on that line. Low-pressure centers are surrounded by air with higher pressures. The lowest air pressure shown on the map is the circular isobar labeled "1008" just north of Florida. Note that this region is surrounded by higher-pressure air. Thus, the region within the 1008-mb isobar should be labeled as a low-pressure center (**L**.) High-pressure centers are surrounded by air with lower pressures. The highest air pressure shown on the map is the semi-circular isobar labeled "1032" in northwestern Canada. Note that this region is surrounded by lower-pressure air. Thus, the region within the 1032-mb isobar should be labeled as a high-pressure center (**H**). There is another region of high pressure surrounded by lower pressures just east of the Great Lakes centered on the circular 1028-mb isobar. Thus, the region within this circular 1028-mb isobar should also be labeled as a high-pressure center (**H**). Therefore, the map that best shows the locations for the centers of high pressure and low pressure is map (1).

39. **2** Lunar eclipses are predictable because the Moon's revolution around Earth and Earth's revolution around the Sun are cyclic; that is, they occur in a constantly repeating pattern with reference to time and space. The pattern of repetition of these cyclic motions allows predictions to be made about future positions of the Sun, Earth, and Moon relative to one another. Thus, scientists can predict the occurrence of lunar eclipses. Furthermore, the reading passage states that "The next one will not happen until 2033," indicating that supermoon eclipses can be predicted. Thus, supermoon total eclipses are celestial events that are predictable.

WRONG CHOICES EXPLAINED:
(1) By definition, random occurrences are unpredictable. Supermoon total eclipses are the result of the predictable cyclic motions of Earth and its Moon in relation to one another and the Sun. Therefore, supermoon total lunar eclipses are predictable, not random. Furthermore, the reading passage states that the next supermoon total lunar eclipse will occur in 2033. The year 2033 is in the future; therefore, this statement is a prediction—a prediction that could not be made if supermoon total lunar eclipses were random occurrences.

(3) The occurrence of supermoon total lunar eclipses is the result of the Moon's cyclic revolution around Earth and Earth's cyclic revolution around the Sun. As long as these cyclic motions continue, supermoon total eclipses will continue to occur. There is no evidence that the motions of Earth and the Moon will cease after 2033. Therefore, there is no evidence that supermoon total eclipses will never happen again after 2033.

(4) The reading passage states that "supermoon events are rare, but a total lunar eclipse during a supermoon is even rarer." A full Moon is not rare; it occurs every 29.5 days. Therefore, supermoon total lunar eclipses do not happen every full Moon.

40. **3** The reading passage states that "A supermoon occurs when the entire lighted half of the Moon faces Earth (full Moon phase) and the Moon is at its closes point to Earth in its orbit." According to the diagram, when the Moon is in position C, its entire lighted half faces Earth and a full Moon phase occurs. Thus, during the 2015 supermoon total eclipse, the Moon was located at position C in its orbit.

41. **3** Lunar phases are cyclic and predictable. Find the Solar System Data chart in the *Reference Tables for Physical Setting/Earth Science*. In the column labeled "Celestial Object," locate Earth's Moon. Trace right to the column labeled "Period of Revolution." Note that the Moon takes 27.3 days to orbit Earth once. Although the Moon makes one revolution in 27.32 days, the Moon takes 29.5 days to go through a complete cycle of phases from the full Moon position represented in the diagram to the full Moon the following month. Why the extra two days? Let's start at full Moon, when the Sun, Earth, and Moon align when the Moon is on the side of Earth directly opposite the Sun. At the same time that the Moon revolves around Earth, Earth is revolving around the Sun at a rate of about 1° per day. Thus, by the time the Moon has completed one revolution, Earth's position relative to the Sun has changed. In 27.3 days, Earth has moved about 27.3° in its orbit. The Moon moves at about 13° per day. At that rate, it takes the Moon a bit more than two more days to catch up to still moving Earth and align again with Earth and the Sun in a full Moon phase. Thus, the Moon completes a cycle of phases in 29.5 days, not 27.3 days. Therefore, the time it took for the Moon to go from this supermoon to the next full Moon phase was 29.5 days.

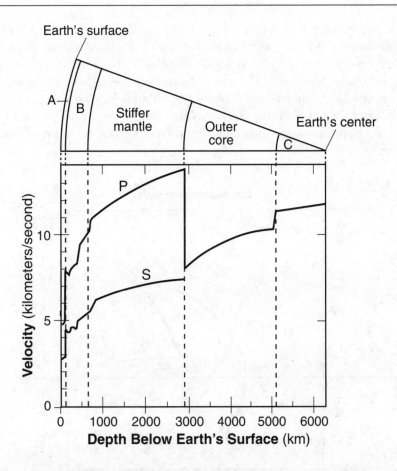

42. **2** Find the Inferred Properties of Earth's Interior chart in the *Reference Tables for Physical Setting/Earth Science*. Locate the cross section of Earth's interior at the top of the chart. Note that the region corresponding to layer *B* is labeled "Asthenosphere." Note, too, that the region corresponding to layer *A* is labeled "Lithosphere." Thus, the layers of Earth's interior represented by letters *A* and *B* are as follows: *A* is the lithosphere and *B* is the asthenosphere.

43. **3** On the horizontal axis of the graph labeled "Depth Below Earth's Surface (km)," locate 1000 kilometers. From this point, trace vertically upward until you intersect the bold curve labeled "P." At this intersection, trace horizontally left to the vertical axis labeled "Velocity (kilometers/second)," and note the value at this point—about 11.3. Thus, the approximate velocity in kilometers per second of the P-waves at a depth of 1000 kilometers is 11.3 km/s.

44. **4** According to the diagram, layer C is located between Earth's center and the outer core. Find the Inferred Properties of Earth's Interior chart in the *Reference Tables for Physical Setting/Earth Science*. Locate the cross section of Earth's interior at the top of the chart. Note that the region corresponding to layer C is labeled "Inner Core." Locate the column labeled "Density (g/cm³)" along the right side of the cross section. Note that the density of the inner core is 12.7–13.0. Therefore, some locations within layer C have an inferred density of 12.9 g/cm³.

45. **3** Find the Characteristics of Stars chart in the *Reference Tables for Physical Setting/Earth Science*. On this chart, locate the position corresponding to the star at point 2 in the question. From this position, trace downward until you intersect the horizontal axis labeled "Surface Temperature (K)" and read the value 7000. Thus, a possible surface temperature of the star in the question at point 2 is 7000 K.

46. **3** In the cores of stars, gravity is strong enough to overcome the force of repulsion between atomic nuclei, allowing the nuclei to combine in a process called nuclear fusion. By definition, nuclear fusion involves the combining of several atoms of a lighter element to form a single atom of a heavier element. The single heavier atom typically has less mass than the lighter atoms from which it formed. The mass "missing" from the heavier atom is not lost. Instead, it is converted into energy according to Einstein's formula $E = mc^2$. This formula states that if mass is converted into energy, the amount of energy released, E, is equal to the mass, m, times the speed of light, c, squared. The speed of light squared is a very large number. Therefore, the conversion of even a small amount of mass into energy during nuclear fusion results in the release of a very large amount of energy. Thus, between points 1 and 3, energy is released by the process of nuclear fusion when lighter elements combine to form heavier elements.

47. **1** Find the Characteristics of Stars chart in the *Reference Tables for Physical Setting/Earth Science*. Locate the position corresponding to the star at point 3 in the chart in the question. Note that point 3 on the chart in the question corresponds to a region labeled "Giants" on the Characteristics of Stars chart. Now, locate the position on the Characteristics of Stars chart corresponding to point 4 on the chart in the question. Note that point 4 on the chart in the question corresponds to a region labeled "White Dwarfs" on the Characteristics of Stars chart. Thus, at point 3 in the life cycle, the star is a Giant, and at point 4 in its life cycle, it is a White Dwarf, as shown in table (1).

48. **3** Note that, in the photograph, the area containing A and B cuts across a section of the stream where the stream channel curves in a meander. When a stream curves, or meanders, water velocity is higher along the outside of the curve and lower along the inside of the curve because the water is forced to cover

a greater distance along the outside of the curve. Since stream erosion is greatest where the velocity of the water is highest, the outside of the curve of a meandering channel experiences more erosion than the inside of the curve. In the diagram, point B is located at the outside of a curve in the stream, where the velocity of the water is highest and erosion is the greatest. Therefore, the streambank at location B is steeper than the streambank at location A because the water near location B is moving faster than the water near location A, causing more erosion.

49. **4** Location X is a flat area adjacent to a stream. When a stream is floods, it carries much more sediment than normal because of its increased speed and volume and often overflows its banks, and the sediment-laden water spreads out over the surrounding land. Outside of the channel, the water is shallower and slower moving, so the flood water begins to deposit its sediment load. Over time, these deposits form a broad, flat area adjacent to the stream called a floodplain. Therefore, the area labeled letter X represents a portion of a floodplain.

50. **2** Find the Relationship of Transported Particle Size to Water Velocity graph in the *Reference Tables for Physical Setting/Earth Science*. Note the dashed lines labeled with the size ranges corresponding to clay, silt, sand, pebbles, cobbles, and boulders. It is given that the stream is initially flowing at 50 cm/sec and is slowing to 5 cm/sec. Find the 50 cm/sec value on the axis labeled "Stream Velocity," trace upward to the bold curve, and then trace right to the axis labeled with particle sizes. Note that, at 50 cm/sec, a stream can transport particles as large as pebbles. Note that, at this point, the pebbles are near the lower end of the pebble size range. Find the 5 cm/sec value on the axis labeled "Stream Velocity," trace upward to the bold curve, and then trace right to the axis labeled with particle sizes. Note that, at 5 cm/sec, a stream can transport particles as large as sand. Also note that, at this point, the sand particles are near the upper end of the sand size range. Therefore, most of the particles deposited where the stream velocity decreases from 50 centimeters per second to 5 centimeters per second are small pebbles and large sand.

PART B–2

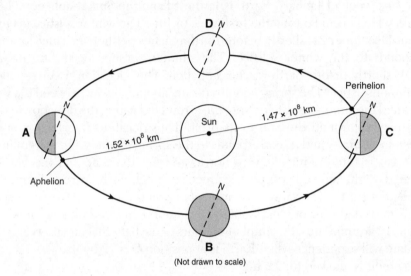

(Not drawn to scale)

51. Earth's spin axis is tilted 23.5° from a line perpendicular to the plane of its orbit. As Earth revolves, its spin axis always points in the same direction in space; that is, the direction of its axis at any given time is parallel to its direction at any other time. Thus, the number of degrees that Earth's axis is tilted from a line perpendicular to the plane of its orbit at each lettered location is 23.5°.

One credit is allowed for any value from **23.4° to 23.5°** or its mathematical equivalent.

Note: Credit is not allowed for a number with a direction, such as 23.5° N, because N indicates a latitude, not an angle.

52. According to the diagram, when Earth is at perihelion, it is closest to location C, which represents the first day of one of the four seasons. Note that, at location C, Earth's Northern Hemisphere is tilted away from the Sun. Therefore, the Sun's rays will strike Earth's surface less directly in the Northern Hemisphere and will be less intense, resulting in the lower temperatures associated with winter. The shading in the diagram shows that locations surrounding the North Pole are in darkness and will experience 24 hours of darkness as Earth rotates. Thus, location C corresponds to the winter solstice, December 21. Thus, when Earth is at perihelion, the season in New York State is winter.

One credit is allowed for **winter**.

53. Earth is closest to the Sun at perihelion (1.47×10^8 km), which occurs during the first week of January. Earth is farthest from the Sun at aphelion (1.52×10^8 km), which occurs during the first week of July. The winter solstice occurs on or about December 21, shortly before Earth reaches perihelion. Thus, location C corresponds to the winter solstice. The summer solstice occurs on or about June 21, shortly before Earth reaches aphelion. Thus, location A corresponds to the summer solstice. The spring equinox occurs about halfway between the winter solstice and the summer solstice. Note the arrows indicating the direction in which Earth revolves around the Sun in its orbit. Thus, location D, which is halfway between locations C and A, corresponds to the spring equinox. On the equinoxes, Earth's axis is tilted neither toward nor away from the Sun, and the boundary between daylight and darkness passes directly through the North and South Poles, bisecting the Earth. Therefore, as Earth rotates on these dates, every location on Earth is in daylight for one-half rotation (12 hours) and in darkness for one-half rotation (12 hours). Thus, the number of hours of daylight that an observer in New York State will experience when Earth is at position D is 12 hours.

One credit is allowed for **12 h.**

54. When you observe the sky, you find that every celestial object (such as stars) changes position over time or is in motion. If you keep track of this motion, you find that, with very few exceptions, all celestial objects appear to move across the sky from east to west along a path that is an arc, or part of a circle. If you measure the rate at which celestial objects are moving, they appear to follow a circular path at a constant rate of 15° per hour, or one complete circle every day (24 h/day × 15°/h = 360°/day). This motion is called "apparent daily motion." In the Northern Hemisphere, all the circles formed by completing the arcs along which celestial objects move are centered very near the star *Polaris*. This is because *Polaris* is located directly above the North Pole and is roughly aligned with Earth's axis of rotation. Because of its position, *Polaris* is often referred to as the North Star. Thus, the name of the star that is aligned with Earth's axis above the North Pole is *Polaris*.

One credit is allowed for ***Polaris* or North Star.**

55. Find the Geologic History of New York State chart in the *Reference Tables for Physical Setting/Earth Science*. In the column labeled "Era," note that the geologic eras are the Cenozoic, Mesozoic, and Paleozoic. Now, in the column labeled "Period," locate Cambrian. Thus, the Cambrian is not a geologic Era; it is a geologic Period. Therefore, the unit of geologic time that should be substituted for the word "Era" is "Period."

One credit is allowed for **Period.**

56. The passage states that "Then, beginning sixty-five million years ago, an intense earth uplifting occurred that raised the Andes Mountains to their present height." Find the Geologic History of New York State chart in the *Reference Tables for Physical Setting/Earth Science*. On the time scale labeled "Million years ago," locate 65. Trace right to the column labeled "Life on Earth," and note the statement "Mass extinction of dinosaurs, ammonoids, and many land plants." Dinosaurs were animals. In the column labeled "Time Distribution of Fossils," locate the gray bar labeled "ammonoids," and note the circled letter *G*. On the key at the bottom of the chart, locate the drawing of index fossil "*G—Manticoceras*." Note that this drawing shows a shell similar in shape to modern nautiloids and snails, so it is reasonable to infer that ammonoids were animals, not plants.

One credit is allowed for an acceptable answer. Acceptable answers include, but are not limited to:

- **dinosaurs**
- **ammonoids**

Note: Credit is *not* allowed for land plants alone because the question asks for animals.

57. Find the Scheme for Sedimentary Rock Identification table in the *Reference Tables for Physical Setting/Earth Science*. In the column labeled "Rock Name," locate limestone. Trace left to the column labeled "Composition," and note that limestone is composed of calcite. The passage also states that the limestone formed from an "enormous mass of lime and sand that compressed into a limestone and dolomite layer . . . ." According to the Chart of Definitions, lime is "a general term for material containing the element calcium that combined with oxygen." Therefore, the limestone may be mixed with dolomite.

Find the Properties of Common Minerals table in the *Reference Tables for Physical Setting/Earth Science*. In the column labeled "Mineral Name," locate calcite and dolomite. Trace left to the column labeled "Composition," and note that the chemical composition of calcite and dolomite are $CaCO_3$ and the $CaMg(CO_3)_2$, respectively.

One credit is allowed for an acceptable answer. Acceptable answers include, but are not limited to:

- **$CaCO_3$**
- **Ca, C, and O (All three elements must be present for credit.)**
- **$CaMg(CO_3)_2$**

58. It is given in the passage that "As rain ran down from the mountains it formed a great underground water table that slowly began dissolving the limestone." As explained earlier, this limestone may also contain the mineral dolomite. Find the Scheme for Sedimentary Rock Identification in the *Reference Tables for Physical Setting/Earth Science*. In the column labeled "Rock Name," locate limestone. From limestone, trace left to the column labeled "Composition." Note that limestone is composed of calcite. Find the Properties of Common Minerals chart in the *Reference Tables for Physical Setting/Earth Science*. In the column labeled "Mineral Name," locate both calcite and dolomite. From each of these minerals, trace left to the column labeled "Distinguishing Characteristics," and note the statement "bubbles with acid." This indicates that both calcite and dolomite chemically react with acid. Acid can completely dissolve the minerals calcite and dolomite. Carbon dioxide in the atmosphere reacts with water to form carbonic acid. Carbonic acid changes calcite and dolomite into new substances that are soluble in water; that is, they dissolve. Therefore, carbonic acid in rainwater and groundwater that seeps through limestone and dolomite bedrock may almost completely dissolve away the bedrock, leaving large holes or caverns. If the roof of the cavern collapses, a depression called a sinkhole is formed in the ground above it. Thus, the chemical weathering that contributes to the formation of sinkholes is best described as acid in water chemically reacting with the calcite and/or dolomite in the rocks. (**Note:** Calcite and dolomite are also called carbonate minerals because they contain "CO_3," which is called the carbonate anion.)

One credit is allowed for an acceptable answer. Acceptable answers include, but are not limited to:

- **The acid in water chemically reacts with the carbonates in the rocks.**
- **Limestone is chemically altered and changed into new materials.**
- **Slightly acidic groundwater chemically breaks down calcite and/or dolomite.**
- **Water flowing underground dissolves the limestone.**

59. According to the key, rock unit K is an igneous intrusion. Note that rock unit K cuts across rock unit H and that the boundary between rock units K and H is marked with a symbol representing contact metamorphism. An igneous intrusion is younger than any rock layer or structure that it cuts across. In order to be contact metamorphosed, a rock layer must have already existed when the igneous intrusion occurred. Therefore, rock unit K is younger than the rock unit H. However, rock unit E has not undergone contact metamorphism along its boundary with rock unit K. Therefore, rock unit E did not yet exist when the igneous intrusion of rock unit K occurred. Therefore, rock unit E is younger than rock unit K. Thus, the correct order of rock unit formation, from oldest to youngest, is H, K, E.

One credit is allowed for the sequence shown below:

Letters: ___H___, ___K___, ___E___

Oldest ──────────────→ Youngest

60. Layers of rock are generally deposited in an unbroken sequence. However, if forces within Earth cause rocks to be uplifted, deposition ceases. Weathering and erosion may wear away layers of rock before the land surface is low enough for another layer to be deposited. The uplift may cause rock layers to tilt or fold. The result is an unconformity, a break or gap in the sequence of a series of rock layers. Thus, the rocks above an unconformity are quite a bit younger than those below it. Note that the layers beneath the upper unconformity are tilted relative to the unconformity. Note that the layers beneath the lower unconformity are also tilted relative to the unconformity. Thus, it is reasonable to infer that the lowest rock units were uplifted until they emerged from their depositional environment.

Then, their surface was weathered and eroded to form the lower unconformity. After that, the rock units subsided or were worn down until they were low enough for deposition to begin once again. Then, several layers were deposited atop the unconformity, each burying the layers below it. Finally, the whole process occurred again, forming the upper unconformity and uppermost layers. Thus, the formation of the unconformities involved many processes, including uplift, emergence, tilting, weathering, erosion, subsidence, deposition, and burial. Any two of these processes should be listed as your answer.

One credit is allowed for two correct responses. Acceptable responses include, but are not limited to:

- **uplift/emergence/tilting/folding**
- **erosion**
- **submergence/subsidence**
- **weathering**
- **deposition**
- **burial**

61. Contact metamorphism describes changes in rock that result from the extreme heat produced by contact with magma or lava. When magma and lava cool, they form igneous rock. Thus, a zone of contact metamorphism would most likely be found where one of two adjoining bedrock types is igneous. The rock that will undergo contact metamorphism is the one that came in contact with the igneous rock when it was still molten magma or lava. According to the cross section and key, rock unit K is igneous rock, and rock unit G has undergone contact metamorphism. Find the Scheme for Sedimentary Rock Identification in the *Reference Tables for Physical Setting/Earth Science*. In the column labeled "Map Symbol," locate the symbol corresponding to rock unit G. Note that rock unit G corresponds to limestone. Now, find the Scheme for Metamorphic Rock Identification in the *Reference Tables for Physical Setting/Earth Science*. In the column labeled "Comments," note that metamorphism of limestone or dolostone forms marble, and metamorphism of various rocks by contact with magma or lava forms hornfels. Therefore, the metamorphic rock that most likely formed within rock unit G at the boundary of rock unit K is most likely marble or hornfels.

One credit is allowed for **marble** *or* **hornfels**.

Cuba Lake

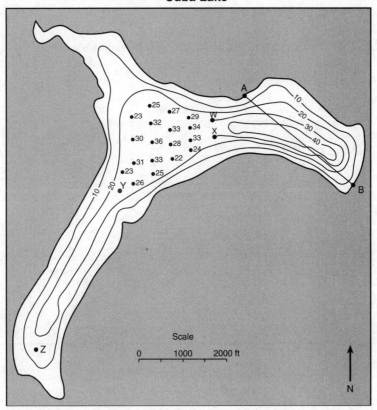

62. An isoline connects points of equal field value, and the field values shown on this map are water depth measured in feet. The 30-foot isoline connects all points that are at a depth of 30 feet. To draw the 30-foot isoline, do the following. Begin at point W and note that there are no points marked 30 near it. However, that does not mean that there are no other places on the map with that value. Note that there are values greater than 30 adjacent to values less than 30. For example, note the points labeled "34" and "29" just west of point W. It is logical to infer that somewhere between the locations at 34 feet deep and 29 feet deep is a location that is 30 feet deep. Therefore, the 30-foot isoline should pass between these two points. In this way, continue drawing the line west between the points labeled "27" and "33," and then between "25" and "32." Then, bend south between "32" and "23," and connect to the point labeled "30." From there, continue southward between "23" and "31," then turn east and draw the isoline between "35" and "25," then between "28" and "33," and finally between "24" and "33," extending the isoline to point X. A correctly drawn 30-foot isoline is shown below.

One credit is allowed if the 30-foot isoline is drawn correctly. If additional isolines are drawn, all must be correct to receive credit.

Note: The isoline must touch points X, W, and the point for 30.

Example of a 1-credit response:

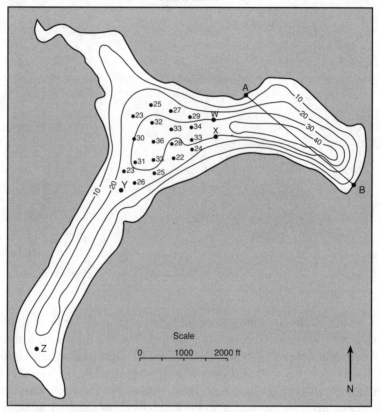

63. To construct a profile of the bottom of Cuba Lake along line AB, proceed as follows. Place the straight edge of a piece of scrap paper along the solid line connecting point A to point B, and mark the edge of the paper at points A and B and wherever the paper intersects a depth isoline. Wherever the paper intersects a depth isoline, label the mark with the depth of the isoline as shown below. Note that points A and B are on the shoreline of Cuba Lake. Therefore, the water depth at points A and B is "0."

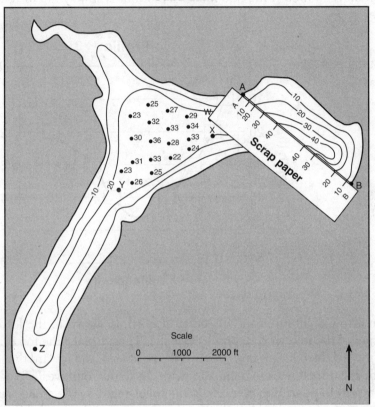

Cuba Lake

Then, place this paper along the *lower* edge of the grid provided in your answer booklet so that points A and B on the scrap paper align with points A and B on the grid. Note that the water depths at points A and B have already been plotted. At each of the other points where a depth isoline crossed the edge of the scrap paper, draw a plot on the grid at the appropriate depth. Finally, connect all of the plots in a smooth curve to form the finished profile as shown below.

One credit is allowed if the centers of *all eight* plots are within or touch the rectangles shown below and are correctly connected with a line from A to B that passes within or touches each rectangle. The low point of the line should extend below 40 feet but remain above 50 feet.

Note: Credit is allowed if the line misses a plot but is still within or touches the rectangle.

64. An arrow pointing in the direction of north is shown in the lower right-hand corner of the map in your answer booklet. Thus, toward the top of the map is north, toward the bottom south, toward the right east, and toward the left west. Therefore, the direction from point Y to point Z is to the southwest. To determine the distance from point Y to point Z, do the following. Use the straight edge of a piece of scrap paper to mark off the distance between points Y and Z. Compare the distance marked off on the scrap paper to the scale printed beneath the map to determine the distance between points Y and Z. The distance is about 4000 feet.

One credit is allowed if both responses for compass direction and distance are correct.

Compass Direction:
- Southwest
- SW
- SSW

Distance:
- Any value from 3800 to 4200 feet

65. It is given that the coordinates of Cuba Lake are 42°14¢ N, 78°18¢ W. Find the Generalized Bedrock Geology of New York State map in the *Reference Tables for Physical Setting/Earth Science*. Locate the approximate position of 42°14¢ N along the right side of the map, and the approximate position of 78°18¢ W along the bottom of the map. Trace horizontally left from 42°14¢ N and vertically upward from 78°18¢ W until the two intersect. Note that Cuba Lake is located between Jamestown and the Genesee River. Now, find the Generalized Landscape Regions of New York State map in the *Reference Tables for Physical Setting/Earth Science*. Locate the point corresponding to the location of Cuba Lake, and note that it lies within the landscape region labeled "Allegheny Plateau." Note that the Allegheny Plateau in New York State is part of a major geographic province that encompasses several states—the Appalachian Plateau (Uplands).

One credit is allowed for **Allegheny Plateau** *or* **Appalachian Plateau** *or* **Appalachian Uplands**.

PART C

Data for Dwarf Planet Pluto and its Five Moons

Object Name	Classification	Period of Revolution in Earth Days (d)	Eccentricity of Orbit	Diameter (km)
Pluto	Dwarf Planet	90,511.4 (247.8 years)	0.2488	2370
Charon	Moon	6.4	0.0022	1208
Styx	Moon	20.2	0.0058	10 to 15*
Nix	Moon	24.9	0.0020	40
Kerberos	Moon	32.2	0.0033	13 to 34*
Hydra	Moon	38.2	0.0059	33 to 43*

*There is a range in diameters for these moons due to their irregular shapes

66. Find the Solar System Data chart in the *Reference Tables for Physical Setting/Earth Science* and locate the columns labeled "Mean Distance from Sun" and "Period of Revolution." Note that, as distance from the Sun increases, the period of revolution increases. This is generally true of all primaries and their satellites, such as planets and their moons. Thus, the moon of Pluto that has the longest period of revolution is likely the greatest distance from Pluto. According to the data table, Hydra has the longest period of revolution. Therefore, the name of Pluto's moon that most likely has an orbit farthest from Pluto is Hydra, because Hydra has the longest period of revolution.

One credit is allowed for both the moon Hydra and a correct explanation. Acceptable responses include, but are not limited to:

- **Hydra has the longest period of revolution.**
- **As a moon's distance from Pluto increases, the time to make one revolution also increases.**
- **Hydra travels the greatest distance in its orbit because it has the longest period of revolution.**

67. An ellipse is an oval shape obtained by cutting a circular cone with a plane. An ellipse has a major axis and a minor axis, which are lines connecting the two points farthest apart and the two points closest together on the ellipse. It also contains two special points along the major axis, each called a focus (plural is foci), instead of one center. The distance from one focus to any point on the ellipse and back to the other focus is always the same. The closer together the two foci, the more circular the ellipse. The farther apart the foci, the flatter the ellipse. The flatness of an ellipse is called its *eccentricity*. Eccentricity is expressed as the ratio between the distance between the foci and the length of the major axis. A perfect circle would have an eccentricity of 0; a straight line has an eccentricity of 1. According to the data table, neither Pluto nor Pluto's moons have an eccentricity of 0 or 1. Therefore, Pluto and its moons have elliptical, eccentric orbits. Pluto's moons have very small eccentricities of orbit, indicating that their orbits, though elliptical, are nearly circular. Pluto, however, has a much more eccentric orbit— 0.2488 compared to 0.0059 or less for its moons. Thus, Pluto's orbit is much more elliptical (less circular) than that of its moons.

One credit is allowed. Acceptable responses include, but are not limited to:

- **elliptical**
- **The orbits are eccentric.**
- **The orbits are nearly, but not perfectly, circular.**
- **oval**
- **Pluto's orbit is elliptical, and the moons of Pluto have a more nearly circular orbit.**

68. Our solar system includes the Sun and all of the objects that orbit it, such as planets, their moons, dwarf planets, asteroids, meteoroids, and numerous comets. All of these objects orbit the Sun because the gravitational attraction of the Sun influences their motion. It is given that Pluto revolves around our Sun and its five moons revolve around Pluto. As Pluto's moons orbit Pluto, they are carried along with Pluto as Pluto revolves around the Sun, so Pluto's moons also revolve around the Sun. Therefore, Pluto and its five moons are considered part of our solar system because they revolve around the Sun.

One credit is allowed. Acceptable responses include, but are not limited to:

- **Pluto and its five moons revolve around the Sun.**
- **All go around/orbit the Sun.**
- **The gravitational attraction of the Sun influences the motion of Pluto and its moons.**
- **Pluto revolves around the Sun, and its moons revolve around Pluto.**
- **Pluto orbits the Sun.**

69. The air near the center of a low-pressure system exerts less pressure than the surrounding air because it is less dense than the surrounding air. Therefore, the denser surrounding air moves inward toward the center of the low-pressure system. As the air moves inward toward the low-pressure center, the Coriolis effect causes the moving air to be deflected to the right in the Northern Hemisphere. The combination of these two motions results in winds that blow inward in a counterclockwise spiral. Thus, two characteristics of the circulation pattern of the surface winds around the center of the low-pressure area represented on the map is that it is counterclockwise and inward toward the center.

One credit is allowed for an acceptable response. Acceptable responses include, but are not limited to:

- **counterclockwise and in toward the center**
- **counterclockwise and spiraling toward the center**
- **inward and counterclockwise**
- **counterclockwise and converging**

70. Find the Key to Weather Map Symbols in the *Reference Tables for Physical Setting/Earth Science* and locate the section labeled "Air Masses." Note that cP is the symbol for a continental polar air mass, and mT is the symbol for a maritime tropical air mass. Air masses take on the characteristics of the surface over which they form. Continental air masses form over land and are dry; maritime air masses form over water and are moist. Polar air masses form near the poles and are cold; tropical air masses form near the equator and are warm. Thus, continental polar air masses are cold and dry; maritime tropical air masses are warm and moist. Thus, compared to the cP air mass shown on the map, the mT air mass is warmer and more humid.

One credit is allowed for circling *both* **more humid** and **warmer**.

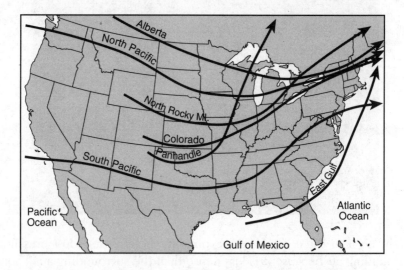

71. According to the passage, "A nor'easter is a large, low-pressure storm system that moves along the east coast of the United States." Note that, of the principal storm tracks shown on the map, the one that moves along the east coast of the United States is labeled "East Gulf."

One credit is allowed for **East Gulf**.

Weather Conditions

Dewpoint	22°F
Barometric Pressure	988.0 mb
Cloud Cover	100%
Present Weather	Snow

72. Find the Key to Weather Map Symbols in the *Reference Tables for Physical Setting/Earth Science*, and locate the section labeled "Station Model Explanation." Note the positions in which the values for temperature, dewpoint, barometric pressure, cloud cover, and present weather are placed on a station model. Note the format in which each of these weather factors is shown on the station model. For example, dewpoint and barometric pressure are shown as values without units, cloud cover is shown by the percent of the inner circle filled in, and present weather is shown as a symbol. Note that the value for barometric pressure is recorded as the last three digits with no decimal point. Therefore, on the station model in your answer booklet, the 22°F dewpoint should be recorded as "22" to the lower left. The 988.0 mb barometric pressure should be recorded as "880" to the upper right. The 100% cloud cover should be represented by completely filling in the central circle. Now, find the section labeled "Present Weather," and note that the symbol representing snow is an asterisk. Thus, the present weather of snow should be shown as an asterisk to the center left. A completed station model is shown on the diagram below.

One credit is allowed if all four weather conditions are correctly indicated, using the proper format.

A 1-credit response is shown below.

73. Find the Dewpoint Temperatures chart in the *Reference Tables for Physical Setting/Earth Science*. Note that, in order to determine the dewpoint using this chart, the dry-bulb temperature and wet-bulb temperature must be known. The instrument that records both the wet-bulb and dry-bulb temperature is called a psychrometer. A psychrometer consists of two thermometers, one that is dry and one that is kept moist by covering it with a wick moistened with distilled water. Evaporation of water from the wick lowers the temperature (at temperatures above freezing). The temperatures on these two thermometers coincide when the air is fully saturated with water vapor. The drier the air, the greater the difference in their temperatures.

The dewpoint is the temperature to which the air must be cooled to become fully saturated. The dewpoint is related to humidity; the higher the dewpoint, the greater the humidity. Thus, the dewpoint can be determined from the wet-bulb and dry-bulb temperatures using an appropriate chart.

Another instrument used to determine dewpoint is a hygrometer. Modern hygrometers measure the water vapor levels in air by how moisture changes characteristics such as electrical resistance or thermal conductivity. Some older hygrometers use a human hair under tension. Human hair retains moisture, changing length with humidity. Thus, one weather instrument that was most likely used to determine the dewpoint at Albany was a psychrometer *or* a hygrometer.

One credit is allowed for an acceptable response. Acceptable responses include, but are not limited to:

- **psychrometer/sling psychrometer**
- **dry-bulb thermometer/thermometer**
- **wet-bulb**
- **hygrometer**

Data Table

Number of Half-lives	Mass of Carbon-14 (g)
0	64
1	32
2	16
3	8
4	4
5	2
6	1

74. To construct a graph that shows the radioactive decay of carbon-14, use the data in the table to plot the points for the Mass of Carbon-14 (g) after each half-life as follows. For example, the table states that at 0 half-lives, 64 g of carbon-14 remains. To plot these coordinates, locate "0" on the horizontal scale labeled "Number of Half-Lives," trace upward along the "0" line until you intersect the line corresponding to "64" on the vertical scale labeled "Mass of Carbon-14 (g)," and mark a plot at the intersection of the two lines. Repeat this process for each of the remaining coordinates: (1, 32), (2, 16), (3, 8), (4, 4), (5, 2) and (6, 1). Then, connect the plots with a line as shown in the graph below.

One credit is allowed if the centers of *all seven* of the student's plots are within or touch the circles shown, and *all seven* plots are correctly connected with a line that passes within or touches each circle.

75. When a radioactive isotope decays, its atoms break apart (disintegrate) and form more stable atoms of a different element. Find the Radioactive Decay Data chart in the *Reference Tables for Physical Setting/Earth Science*. In the column labeled "Radioactive Isotope," locate Carbon-14. Trace right to the column labeled "Disintegration," and note that the product of the disintegration of carbon-14 (^{14}C) is nitrogen-14 (^{14}N.)

One credit is allowed for an acceptable response. Acceptable responses include, but are not limited to:

- ^{14}N
- **nitrogen-14**
- **N-14**

Note: Credit is *not* allowed for nitrogen alone because nitrogen has more than one isotope, and the stable disintegration product is needed.

76. Find the Radioactive Decay Data chart in the *Reference Tables for Physical Setting/Earth Science*. In the column labeled "Radioactive Isotope," locate Carbon-14. Trace right to the column labeled "Half-Life (years)," and note that carbon-14 has a half-life of 5.7×10^3 (5,700) years. Thus, in 17,100 years, three half-lives would have occurred. According to the data table, after 3 half-lives, 8 grams of carbon-14 would remain in this sample.

One credit is allowed for **8 g**.

Table 1

Water in Lake Michigan, 45° N		
Water Depth Range (m)	Average Temperature (°C)	
	Maximum (Occurs in Summer)	Minimum (Occurs in Winter)
0–10	24.0	4.0
10–20	19.5	4.0
20–30	12.5	4.0
30–40	7.5	4.0
40–60	5.5	4.0
60–110	4.5	4.0
110–150	4.0	4.0

Table 2

Land in St. Paul, MN, 45° N		
Soil Depth Range (m)	Average Temperature (°C)	
	Maximum (Occurs in Summer)	Minimum (Occurs in Winter)
0–0.1	25.5	–6.0
0.1–1	23.0	–1.5
1–2	19.5	2.0
2–3	16.0	4.5
3–4	13.5	7.0
4–5	12.5	8.5
5–6	11.0	10.0

77. On Table 1, in the column labeled "Average Temperature (°C)," locate the sub-column labeled "Maximum (occurs in summer)." In this sub-column, locate "12.5." Trace left to the column labeled "Water Depth (m)," and note that the maximum temperature of 12.5°C occurs at a water depth of 20–30 m. On Table 2, in the column labeled "Average Temperature (°C)," locate the sub-column labeled

"Maximum (occurs in summer)." In this sub-column, locate "12.5." Trace left to the column labeled "Soil Depth Range (m)," and note that the maximum temperature of 12.5°C occurs at a soil depth of 4–5 m. Thus, the water depth range and soil depth range, in meters, that have the same average maximum temperature of 12.5°C are water depth range of 20–30 m and soil depth range of 4–5 m.

One credit is allowed for a **water depth range of 20–30 m** and a **soil depth range of 4–5 m**.

78. On Table 1, the average temperature in the summer is recorded in the sub-column labeled "Maximum (occurs in summer)." Note that in summer, as water depth range increases, the average water temperature decreases. On Table 2, the average temperature in the summer is also recorded in the sub-column labeled "Maximum (occurs in summer)." Note that in summer, as soil depth range increases, the average temperature also decreases. Thus, the general relationship between depth and average temperature for both water and soil in the summer is an inverse relationship; that is, as depth increases, temperature decreases.

One credit is allowed for an acceptable response. Acceptable responses include, but are not limited to:

- **As depth increases, temperature decreases.**
- **It gets colder, the greater the depth.**
- **inverse relationship/negative relationship**
- **The shallower the depth, the greater the average temperature for soil and water.**

79. According to the data tables, Lake Michigan and St. Paul, MN are at the same latitude. Therefore, both places receive roughly the same intensity of insolation. However, land surfaces have a different composition from water surfaces. Find the Specific Heats of Common Materials chart in the *Reference Tables for Physical Setting/Earth Science*. Note that liquid water has a specific heat of 4.18 Joules/gram • C°. Note, too, that basalt and granite (common materials from which land is derived) have specific heats of 0.84 and 0.79 Joules/gram • C°, respectively. This means that it takes more energy to change the temperature of water than it takes to change the temperature of land. For example, if 1 gram of water loses 4.18 Joules of heat, the temperature of the water will decrease by 1°C (-4.18 J/g ÷ 4.18 J/g • °C = -1°C). However, if 1 gram of basalt loses 4.18 Joules of heat, the temperature of the basalt will decrease by about 5°C (-4.18 J/g ÷ 0.84 J/g • °C = -4.97°C). Thus, when they lose the same amount of heat, a land surface will become cooler than a water surface. Therefore, the winter water surface is warmer than the winter land surface because they are composed of different substances, which have different specific heat values. Water has a higher specific heat than land and changes temperature more slowly. Therefore, water

will not cool off as quickly as land in the winter, and the winter water surface will be warmer than the winter land surface.

One credit is allowed for an acceptable response. Acceptable responses include, but are not limited to:

- **Water has a higher specific heat.**
- **Water changes temperature more slowly.**
- **Land has a lower specific heat.**
- **It takes more energy to change the temperature of water than land.**
- **Land cools off more quickly than water.**

80. Note that location Z is along the northern boundary of the Scotia Plate. Find the Tectonic Plates map in the *Reference Tables for Physical Setting/Earth Science*. Locate the Scotia Plate, and note the symbol along its northern border—a solid black line with arrows pointing in different directions north and south of the boundary. According to the key at the bottom of the map, this represents a Transform plate boundary (transform fault). Transform faults are a special category of strike-slip faults, a type of fault in which the two sides of the fault slide past one another laterally instead of vertically.

One credit is allowed for an acceptable response. Acceptable responses include, but are not limited to:

- **transform boundary**
- **transform fault**
- **strike-slip fault**

81. Note that the tectonic plate to the south of the Bouvet Hot Spot extends across the entire lower edge of the map. Note, too, that the tectonic plate to the north of the Bouvet Hot Spot encompasses the southern tip of Africa. Find the Tectonic Plates map in the *Reference Tables for Physical Setting/Earth Science*. Locate the Bouvet Hot Spot near the lower right-hand corner. Note that the tectonic plate stretching across the bottom of the map that is south of the Bouvet Hot Spot is the Antarctic Plate. Note that the tectonic plate north of the Bouvet Hot Spot that encompasses the southern tip of Africa is the African Plate. Thus, the tectonic plates on each side of the Bouvet Hot Spot are the Antarctic Plate and the African Plate.

One credit is allowed for **African Plate and Antarctic Plate**.

82. Note that point Y is located directly on the Mid-Atlantic Ridge. Find the Inferred Properties of Earth's Interior chart in the *Reference Tables for Physical Setting/Earth Science*. In the cross section, locate the region labeled "Mid-Atlantic Ridge." Note that beneath the Mid-Atlantic Ridge the arrows showing motion in the asthenosphere indicate rising currents that diverge beneath the lithosphere.

Thus, the arrowheads should be drawn on each of the four bold lines to indicate rising currents and a divergent movement in the asthenosphere as shown in the example below.

One credit is allowed for *four* arrowheads/arrows that indicate rising currents and a divergent movement in the asthenosphere.

Example of a 1-credit response:

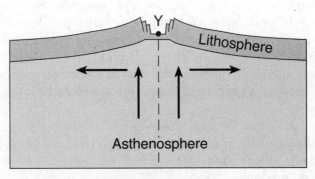

(Not drawn to scale)

83. According to the passage, "As the glacier thickens, more pressure is created and the glacier moves faster." Thus, the relationship between the thickness of a glacier and its rate of downhill motion is a direct relationship—as the thickness of a glacier increases, the rate of movement also increases.

One credit is allowed for an acceptable response. Acceptable responses include, but are not limited to:

- **As the thickness of a glacier increases, the rate of movement increases.**
- **Thicker glaciers move faster.**
- **direct relationship**

84. Glaciers move down the slopes of mountains from higher elevations to lower elevations along the easiest paths. Those paths are usually stream valleys that were eroded before the glacier formed. When a glacier moves down a stream valley, the glacier alters the valley's shape. The ice does not meander back and forth as readily as water. So, the ice often breaks through obstacles, forming a straighter path. The ice also cuts back the walls of the valley, making them steeper. The result is a wider, straighter, steeper-sided, U-shaped valley. Thus, the most likely shape of the cross section of a valley formed as a result of erosion by a mountain glacier is U-shaped.

One credit is allowed for an acceptable response. Acceptable responses include, but are not limited to:

- **U-shaped**
- **wide valley with steep sides**

85. When a glacier melts, numerous streams of meltwater spread out over the land and deposit glacial sediments in a wide sheet. Sediments deposited by running water are typically sorted. Horizontal sorting occurs when a stream gradually slows and deposits smaller and smaller sediment particles that settle through water. Thus, sediments deposited by meltwater are typically sorted and layered.

When a glacier melts, sediments carried in and on the ice fall directly to the ground. Air resistance is so small that particles of all sizes fall through air at virtually the same speed. Therefore, sediments deposited directly by a melting glacier do not become sorted by size and do not form layers due to different settling rates.

Thus, compared to sediments deposited by meltwater from a glacier, sediments deposited directly by a glacier are unsorted and not layered.

One credit is allowed for an acceptable response. Acceptable responses include, but are not limited to:

- **unsorted**
- **mixed piles**
- **not layered**

SELF-ANALYSIS CHART August 2018

Topic	Question Numbers (Total)	Wrong Answers (x)	Grade
Standards 1, 2, 6, and 7: Skills and Application			
Skills Standard 1 Analysis, Inquiry, and Design	2, 3, 5, 7, 8, 10, 11, 15, 17–21, 23–26, 28, 30, 33, 34, 35, 37, 39, 42–47, 50, 55-63, 65, 66, 68, 72, 74–85		$\frac{100(56-x)}{56} = \%$
Standard 2 Information Systems			
Standard 6 Interconnectedness, Common Themes	1, 4, 6, 8, 13, 14, 21, 24, 27–29, 31–38, 40–43, 45, 47–49, 51–54, 59–65, 67–74, 78, 80–82, 85		$\frac{100(51-x)}{51} = \%$
Standard 7 Interdisciplinary Problem Solving	12		$\frac{100(1-x)}{1} = \%$
Standard 4: The Physical Setting/Earth Science			
Astronomy The Solar System (MU 1.1a, b; 1.2d)	39, 40, 41, 67, 68		$\frac{100(5-x)}{5} = \%$
Earth Motions and Their Effects (MU 1.1c, d, e, f, g, h, i)	1, 4, 5, 6, 29, 30, 51, 52, 53, 54		$\frac{100(10-x)}{10} = \%$
Stellar Astronomy (MU 1.2b)	45, 46, 47		$\frac{100(3-x)}{3} = \%$
Origin of Earth's Atmosphere, Hydrosphere and Lithosphere, (MU 1.2e, f, h)			
Theories of the Origin of the Universe and Solar System (MU 1.2a,c)	2, 3, 66		$\frac{100(3-x)}{3} = \%$

SELF-ANALYSIS CHART August 2018

Topic	Question Numbers (Total)	Wrong Answers (x)	Grade
Meteorology Energy Sources for Earth Systems (MU 2.1a, b)			
Weather (MU 2.1c, d, e, f, g, h)	10, 31, 32, 36-38, 69–73		$\frac{100(11-x)}{11} = \%$
Insolation and Seasonal Changes (MU 2.1i; 2.2a,b)	7, 11, 14, 16, 34, 77–79		$\frac{100(8-x)}{8} = \%$
The Water Cycle and Climates (MU 1.2g; 2.2c,d)	8, 9, 12, 13, 15		$\frac{100(5-x)}{5} = \%$
Geology Minerals and Rocks (MU 3.1a, b, c)	23, 25, 26, 28, 35, 57, 61		$\frac{100(7-x)}{7} = \%$
Weathering, Erosion, and Deposition (MU 2.1s, t, u, v, w)	22, 27, 48, 49, 50, 58, 60, 83–85		$\frac{100(10-x)}{10} = \%$
Plate Tectonics and Earth's Interior (MU 2.1j, k, l, m, n, o)	20, 42–44, 80-82		$\frac{100(7-x)}{7} = \%$
Geologic History (MU 1.2 i, j)	17–19, 21, 55, 56, 59, 74–76		$\frac{100(10-x)}{10} = \%$
Topographic Maps and Landscapes (MU 2.1p, q, r)	24, 33, 62, 63, 64, 65		$\frac{100(6-x)}{6} = \%$
ESRT *2011 Edition Reference Tables for Physical Setting/Earth Science*	2, 3, 7, 10, 11, 15, 17–21, 23–26, 28, 30, 34, 35, 37, 42, 44, 50, 55–57, 61, 65, 72, 75, 76, 80–82		$\frac{100(34-x)}{34} = \%$

To further pinpoint your weak areas, use the Topic Outline in the front of the book. MU = Major Understanding (see Topic Outline)

Examination
June 2019
Physical Setting/Earth Science

PART A
Answer all questions in this part.

Directions (1–35): For *each* statement or question, choose the word or expression that, of those given, best completes the statement or answers the question. Some questions may require the use of the *2011 Edition Reference Tables for Physical Setting/Earth Science*. Record your answers in the space provided.

1 The map below shows four major time zones of the United States. The locations of Boston and San Diego are shown.

What is the time in Boston when it is 11 A.M. in San Diego?

(1) 8 a.m. (3) 3 p.m.
(2) 2 p.m. (4) noon

1 _____

2 The diagram below represents the spectral lines from the light emitted from a mixture of two gaseous elements in a laboratory on Earth.

If the same two elements were detected in a distant star that was moving away from Earth, how would the spectral lines appear?

(1) The entire set of spectral lines would shift toward the red end.
(2) The entire set of spectral lines would shift toward the blue end.
(3) The spectral lines of the shorter wavelengths would move closer together.
(4) The spectral lines of the longer wavelengths would move closer together.

2 _____

3 The diagram below shows Earth in orbit around the Sun, and the Moon in orbit around Earth. M_1 and M_2 indicate positions of the Moon in its orbit where eclipses might be seen from Earth.

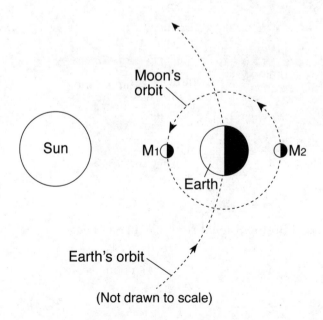

Which table correctly matches each type of eclipse with the orbital position of the Moon and the cause of each eclipse?

Type of Eclipse	Moon's Position	Cause of Eclipse
Solar	M_1	Earth's shadow falls on Moon
Lunar	M_2	Moon's shadow falls on Earth

(1)

Type of Eclipse	Moon's Position	Cause of Eclipse
Solar	M_1	Moon's shadow falls on Earth
Lunar	M_2	Earth's shadow falls on Moon

(2)

Type of Eclipse	Moon's Position	Cause of Eclipse
Solar	M_2	Earth's shadow falls on Moon
Lunar	M_1	Moon's shadow falls on Earth

(3)

Type of Eclipse	Moon's Position	Cause of Eclipse
Solar	M_2	Moon's shadow falls on Earth
Lunar	M_1	Earth's shadow falls on Moon

(4)

4 The timeline below represents the entire geologic history of Earth. The lettered dots on the timeline represent events in Earth's history.

Which lettered dot best indicates the geologic time when humans first appeared on Earth?

(1) A (3) C
(2) B (4) D

5 Compared to our solar system, the universe is

(1) younger, smaller, and contains fewer stars
(2) younger, larger, and contains more stars
(3) older, smaller, and contains fewer stars
(4) older, larger, and contains more stars

6 Which motion allows an observer on Earth to view different constellations throughout the year?

(1) Earth orbiting the Sun
(2) constellations orbiting Earth
(3) Earth orbiting the constellations
(4) constellations orbiting the Sun

7 What is the approximate location of the Tasman Hot Spot in the Pacific Ocean?

(1) 36° N 160° W (3) 160° N 36° W
(2) 36° S 160° E (4) 160° S 36° E

8 Most ozone is found in a region of Earth's atmosphere between 10 and 20 miles above Earth's surface. This temperature zone of the atmosphere is known as the

(1) thermosphere (3) stratosphere
(2) mesosphere (4) troposphere

9 The Coriolis effect, which causes the curving of planetary winds, is a direct result of the

(1) distance between Earth and the Sun
(2) inclination of Earth's axis
(3) orbiting of Earth around the Sun
(4) spinning of Earth on its axis   9 _____

10 What is the dewpoint if the dry-bulb temperature is 18°C and the relative humidity is 64%?

(1) 14°C   (3) 9°C
(2) 11°C   (4) 4°C   10 _____

11 When one gram of liquid water at its boiling point is changed into water vapor

(1) 334 J/g is gained from the surrounding environment
(2) 334 J/g is released into the surrounding environment
(3) 2260 J/g is gained from the surrounding environment
(4) 2260 J/g is released into the surrounding environment   11 _____

12 The cross section below represents some processes of the water cycle. Arrows represent the infiltration of water. The dashed line labeled X represents the uppermost level of Earth material that is saturated by groundwater.

What is indicated by the dashed line labeled *X*?

(1) watershed
(2) water table
(3) impermeable bedrock
(4) impermeable soil   12 _____

13 Which index fossil in sedimentary surface bedrock most likely indicates that a marine environment once existed in the region where the sediments were deposited?

(1) Mastodont
(2) Condor
(3) Eospirifer
(4) Coelophysis   13 _____

14 The three diagrams below represent three frontal boundaries with the surface locations of the fronts labeled *A*, *B*, and *C*. Arrows indicate direction of air movement.

Which table correctly matches each letter with the type of frontal boundary it represents?

Letter	Type of Frontal Boundary
A	Cold Front
B	Warm Front
C	Occluded Front

(1)

Letter	Type of Frontal Boundary
A	Warm Front
B	Cold Front
C	Occluded Front

(2)

Letter	Type of Frontal Boundary
A	Cold Front
B	Warm Front
C	Stationary Front

(3)

Letter	Type of Frontal Boundary
A	Warm Front
B	Cold Front
C	Stationary Front

(4)

15 The cross section below indicates the geologic ages of the bedrock beneath the state of Michigan. These rocks formed from sediments deposited in an ancient depositional basin. This region is called the Michigan Basin. Glacial deposits cover most of the surface.

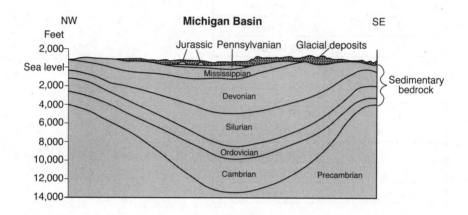

Which process most likely caused the formation of the Michigan Basin?

(1) uplift
(2) faulting
(3) metamorphism
(4) downwarping   15 _____

16 The cross section below represents prevailing winds, shown by arrows, moving over a coastal mountain range. Letters *A* through *D* represent locations on Earth's surface.

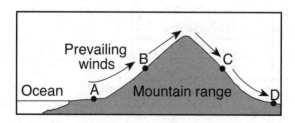

Which location would most likely have cloud cover and precipitation?

(1) *A*
(2) *B*
(3) *C*
(4) *D*   16 _____

17 Which Earth surface would most likely absorb the greatest amount of insolation on a sunny day if all of these surfaces have equal areas?

(1) blacktop parking lot
(2) white sand beach
(3) surface of a calm lake
(4) snow covered mountain slope   17 ____

18 Most of the infrared radiation given off by Earth's surface is absorbed in Earth's atmosphere by greenhouse gases such as water vapor, carbon dioxide, and

(1) hydrogen   (3) oxygen
(2) nitrogen   (4) methane   18 ____

19 The X on the map below indicates the region where the state of Washington is located on the present-day North American continent.

During which geologic period was the region of Washington State closest to the equator?

(1) Cretaceous   (3) Mississippian
(2) Triassic   (4) Ordovician   19 ____

20 A climate event that occurs when surface water in the eastern equatorial area of the Pacific Ocean becomes warmer than normal and may cause a warm, dry winter in New York State is

(1) an air mass formation
(2) the Doppler effect
(3) an El Niño
(4) a monsoon   20 ____

21 In the New York State rock record, there is *no* bedrock from the Permian Period. Which two other entire geologic periods have no sediment or bedrock in the New York State rock record?

(1) Cretaceous and Quaternary Periods
(2) Paleogene and Neogene Periods
(3) Triassic and Jurassic Periods
(4) Mississippian and Pennsylvanian Periods   21 _____

22 The cross sections below represent two widely separated bedrock outcrops, 1 and 2. Letters *A*, *B*, *C*, and *D* identify some rock layers. Line *XY* represents a fault. The rock layers have *not* been overturned.

Which lettered rock layer is the youngest?

(1) *A*   (3) *C*
(2) *B*   (4) *D*   22 _____

23 The block diagram below represents the surface features in a landscape region.

Which diagram best represents the general stream drainage pattern of this entire region?

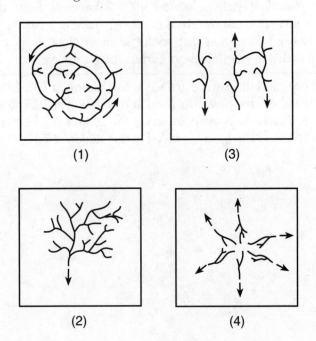

23 ____

24 Compared to a landscape that develops in a cool, dry climate, a landscape that develops in a warm, rainy climate will most likely weather and erode

(1) slower, so the landforms are more angular
(2) slower, so the landforms are more rounded
(3) faster, so the landforms are more angular
(4) faster, so the landforms are more rounded   24 _____

25 Earth's crustal bedrock at the Mid-Atlantic Ridge is composed mostly of

(1) basalt, with a density of 2.7 g/cm^3
(2) basalt, with a density of 3.0 g/cm^3
(3) granite, with a density of 2.7 g/cm^3
(4) granite, with a density of 3.0 g/cm^3 25 _____

26 A seismic recording station located 4000 kilometers from the epicenter of an earthquake received the first *P*-wave at 7:10:00 P.M. (h:min:s). Other information that could be determined from this recording was that the earthquake occurred at approximately

1 7:03:00 P.M., and the S-wave arrived at this station at 7:12:40 P.M.
2 7:03:00 P.M., and the S-wave arrived at this station at 7:15:40 P.M.
3 7:17:00 P.M., and the S-wave arrived at this station at 7:12:40 P.M.
4 7:17:00 P.M., and the S-wave arrived at this station at 7:15:40 P.M.   26 _____

27 Which graph best represents the percentage by mass of elements of Earth's crust?

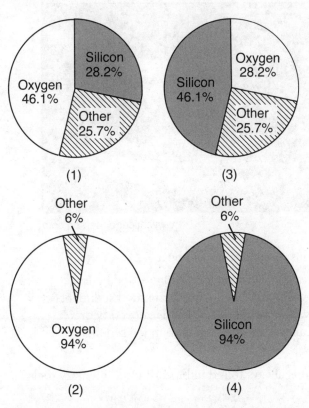

28 The aerial photograph below shows an elongated hill near the Finger Lakes formed by glacial deposition in New York State.

This elongated hill is best identified as a

(1) kettle
(2) moraine
(3) sand dune
(4) drumlin

29 The photograph below shows a valley glacier altering the landscape.

www.alicehenderson.com

The result of this glacial action will be a

(1) U-shaped valley with scratched and grooved bedrock
(2) U-shaped valley with well-sorted, rounded sediment
(3) V-shaped valley with scratched and grooved bedrock
(4) V-shaped valley with well-sorted, rounded sediment   29 _____

30 Which element is always found in bioclastic sedimentary rocks?
   (1) iron   (3) carbon
   (2) sodium   (4) sulfur   30 _____

31 What are two major factors that control the development of soil in a given location?
   (1) vegetative cover and slope
   (2) tectonic activity and elevation
   (3) erosion and transpiration
   (4) bedrock composition and climate   31 _____

32 The cross section below represents rock units within Earth's crust. The geologic ages of two of the layers are shown.

Which sequence of geologic events occurred after the formation of the Cambrian limestone and before the formation of the Silurian limestone?

(1) uplift → weathering → erosion → subsidence
(2) uplift → subsidence → erosion → weathering
(3) subsidence → weathering → erosion → uplift
(4) subsidence → erosion → uplift → weathering

33 Which two rocks contain the mineral quartz?

(1) gabbro and schist
(2) dunite and sandstone
(3) granite and gneiss
(4) pumice and scoria

34 Which chemical formula represents the composition of a mineral that usually exhibits fracture, *not* cleavage?

(1) $Mg_3Si_4O_{10}(OH)_2$
(2) $NaCl$
(3) $CaCO_3$
(4) $(Fe,Mg)_2SiO_4$

35 Which mineral is commonly used as the "lead" in pencils?

(1) pyrite
(2) graphite
(3) galena
(4) fluorite

PART B–1
Answer all questions in this part.

Directions (36–50): For *each* statement or question, choose the word or expression that, of those given, best completes the statement or answers the question. Some questions may require the use of the *2011 Edition Reference Tables for Physical Setting/Earth Science*. Record your answers in the space provided.

Base your answers to questions 36 and 37 on the graph below and on your knowledge of Earth science. The graph shows changing ocean water levels, over a 3-day period, at a shoreline location at Kings Point, New York on Long Island.

36 Based on the graph, the first low tide on December 26 occurred at approximately

(1) 6 A.M. (3) 6 P.M.
(2) 11 A.M. (4) 11 P.M.   36 ____

37 These Long Island tides show a pattern that is

(1) cyclic and predictable
(2) cyclic and unpredictable
(3) noncyclic and predictable
(4) noncyclic and unpredictable   37 ____

Base your answers to questions 38 through 41 on the diagram below and on your knowledge of Earth science. The diagram represents Earth orbiting the Sun. Four positions of Earth in its orbit are labeled A, B, C, and D. Letter N represents the North Pole. Distances are indicated for aphelion (Earth's farthest position from the Sun around July 4) and perihelion (Earth's closest position to the Sun around January 3). Arrows indicate directions of movement.

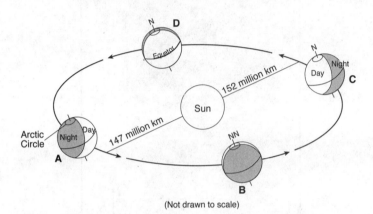

(Not drawn to scale)

38 During which season in the Northern Hemisphere is Earth at aphelion?
(1) winter
(2) spring
(3) summer
(4) fall

38 _____

39 Between which pair of lettered positions is the Sun's vertical ray moving from the equator southward to the Tropic of Capricorn?
(1) A and B
(2) B and C
(3) C and D
(4) D and A

39 _____

40 Approximately how many times does Earth rotate as it moves in its orbit from position A back to position A?
(1) 1 time
(2) 15 times
(3) 24 times
(4) 365 times

40 _____

41 What is the tilt of Earth's rotational axis relative to a line perpendicular to the plane of Earth's orbit?
(1) 15°
(2) 23.5°
(3) 66.5°
(4) 90°

41 _____

Base your answers to questions 42 through 44 on the map below and on your knowledge of Earth science. The map shows a composite of Doppler radar images. Darker shadings indicate the precipitation pattern of a large storm system over the eastern United States.

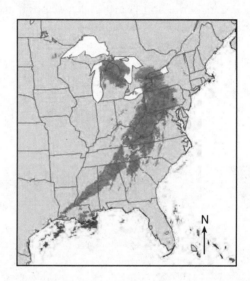

42 The surface wind circulation pattern around the center of this storm system is

(1) inward and clockwise
(2) inward and counterclockwise
(3) outward and clockwise
(4) outward and counterclockwise

43 The best evidence on a weather map to indicate high-speed winds near the center of this storm system would most likely be

(1) cloud cover of 100%
(2) type of precipitation
(3) temperature and dewpoint values
(4) isobars drawn close together

44 As this storm system follows a normal storm track, it will most likely move toward the

(1) southeast
(2) southwest
(3) northeast
(4) northwest

Base your answers to questions 45 through 48 on the passage and map below and on your knowledge of Earth science. The map shows the location of the epicenter (✱) of a major earthquake that occurred about 1700 years ago. Point A represents a location on a tectonic plate boundary. Plates X and Y represent major tectonic plates. The island of Crete; the Anatolian Plate, which is a minor tectonic plate; and the Hellenic Trench have been labeled. Arrows indicate the relative directions of plate motion.

Crete Earthquake

Scientists have located the geological fault, off the coast of Crete in the Mediterranean Sea, that likely shifted, causing a huge earthquake in the year 365 that devastated life and property on Crete. The southwestern coastal region of Crete was uplifted, as evidenced by remains of corals and other sea life now found on land 10 meters above sea level. Scientists measured the age of these corals to verify when this event occurred. This earthquake caused a tsunami that devastated the southern and eastern coasts of the Mediterranean Sea. It is estimated that earthquakes along the fault, associated with the Hellenic Trench, may occur about every 800 years.

45 Which type of plate boundary is represented at point *A*?
  (1) divergent   (3) transform
  (2) convergent   (4) complex   45 _____

46 What are the names of the major tectonic plates *X* and *Y*?
  (1) *X* = Eurasian Plate; *Y* = African Plate
  (2) *X* = Eurasian Plate; *Y* = Arabian Plate
  (3) *X* = Indian-Australian Plate; *Y* = African Plate
  (4) *X* = Indian-Australian Plate; *Y* = Arabian Plate   46 _____

47 Which two New York State index fossils are most closely related to the corals that were radioactively dated in this study?
  (1) *Eucalyptocrinus* and *Ctenocrinus*
  (2) *Elliptocephala* and *Phacops*
  (3) *Maclurites* and *Platyceras*
  (4) *Lichenaria* and *Pleurodictyum*   47 _____

48 Which activity could best prepare residents along the Mediterranean coast to reduce the loss of human life during a future tsunami?
  (1) Board up windows.
  (2) Remove heavy objects from the walls in homes.
  (3) Plan evacuation routes to higher ground.
  (4) Build reinforced basements.   48 _____

Base your answers to questions 49 and 50 on the photograph below and on your knowledge of Earth science. The photograph shows a sandstone erosional feature that formed near the Grand Canyon, in southwestern United States.

www.nationalgeographic.com

49 Which erosional agent is most likely sandblasting this rock formation?

(1) wind
(2) waves
(3) running water
(4) moving ice

49 _____

50 What is the range of grain sizes that are most commonly found in rock making up this feature?

(1) 0.0004 cm–0.006 cm
(2) 0.006 cm–0.2 cm
(3) 0.2 cm–6.4 cm
(4) 6.4 cm–25.6 cm

50 _____

PART B–2
Answer all questions in this part.

Directions (51–65): Record your answers in the spaces provided. Some questions may require the use of the *2011 Edition Reference Tables for Physical Setting/Earth Science*.

Base your answers to questions 51 through 53 on the data table below, on the graph below, and on your knowledge of Earth science. The data table shows the projected percentages of radioactive isotope *X* remaining and its disintegration product *Z* forming over 6.5 billion years. The graph shows the disintegration of radioactive isotope *X*.

Radioactive Isotope X (%)	Disintegration Product Z (%)	Time (billion years)
100	0	0
50	50	1.3
25	75	2.6
12.5	87.5	3.9
6.25	93.75	5.2
3.125	96.875	6.5

51 On the grid *below*, construct a line graph by plotting the percentages of disintegration product Z forming over 6.5 billion years. Connect *all six plots* with a line. The percentages of radioactive isotope X have already been plotted. [1]

52 Identify radioactive isotope X. [1]

_____

53 Calculate the amount, in grams, of an original 300-gram sample of radioactive isotope X remaining after 3.9 billion years. [1]

_____ g

Base your answers to questions 54 through 57 on the weather map below and on your knowledge of Earth science. The map shows the location of a low-pressure system over New York State during summer. Isobar values are recorded in millibars. The darker shading indicates areas of precipitation. Some New York State locations are indicated.

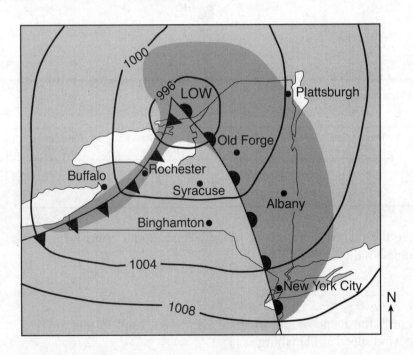

54 Describe the change in air pressure that will most likely occur at Rochester by the time that the cold front has reached Syracuse. Then describe what will most likely happen to the amount of cloud cover in Rochester with this change in air pressure and location of the cold front. [1]

Change in air pressure: _____

Amount of cloud cover: _____

55 The station model below represents the weather conditions at Buffalo, New York, at the time that this map was prepared.

Buffalo, New York

In the table *below*, record the weather data represented on this station model. [1]

Air Temperature (°F)	Barometric Pressure (mb)	Wind Direction from the	Wind Speed (knots)

56 State the relative humidity at Albany when the air temperature is equal to the dewpoint. [1]

_____ %

57 Identify the name of the weather instrument used to measure the wind speed at Plattsburgh. [1]

_____

Base your answers to questions 58 through 61 on the diagram and passage below and on your knowledge of Earth science. The diagram represents the orbits of Earth, Comet Tempel-Tuttle, and planet X, another planet in our solar system. Arrows on each orbit represent the direction of movement.

Orbit of Comet Tempel-Tuttle

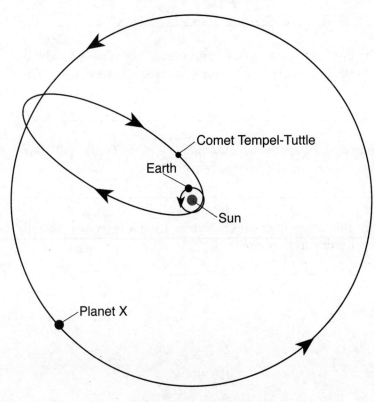

(Not drawn to scale)

Comet Tempel-Tuttle

Comet Tempel-Tuttle orbits our Sun and is responsible for the Leonid meteor shower event observed from Earth. This meteor shower occurs every year in November and is visible in the night sky as Earth passes through the debris left in space by this comet. The debris from the comet produces meteors that are smaller than a grain of sand, which enter Earth's atmosphere and burn up in the mesosphere temperature zone. Comet Tempel-Tuttle's orbital distance from the Sun ranges from about 145 million kilometers at its closest approach to 2900 million kilometers at its farthest distance. Its two most recent closest approaches to the Sun occurred in 1965 and one revolution later in 1998.

58 Identify the name of the object located at one of the foci of the elliptical orbit of Comet Tempel-Tuttle. [1]

_____

59 Identify the solar system planet represented by planet X, which orbits near Comet Tempel-Tuttle's farthest distance from the Sun. [1]

_____

60 Determine the year in which Comet Tempel-Tuttle will next make its closest approach to the Sun. [1]

_____

61 Identify the force that causes debris from the comet to fall through Earth's atmosphere. [1]

_____

Base your answers to questions 62 through 65 on the diagram below and on your knowledge of Earth science. The diagram represents the apparent path of the Sun across the sky as seen by an observer on Earth's surface on June 21. Points A, B, C, and D represent positions of the Sun at different times of the day. The angle of *Polaris* above the horizon as seen in the nighttime sky is indicated.

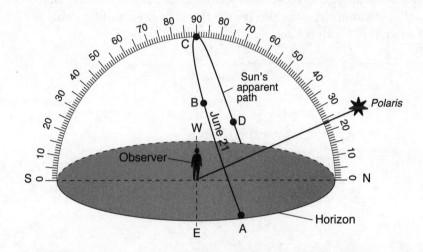

62 Describe the changes in the length of the observer's shadow as the Sun appears to move from position A to position D. [1]

_____

_____

63 Describe *one* piece of evidence from the diagram that supports the inference that the observer is located at the Tropic of Cancer. [1]

_____

_____

64 State the number of daylight hours at this location on September 23. [1]

_____ h

65 The Sun's apparent movement through the sky is caused by Earth's rotation. Identify the device that was used to first demonstrate that Earth rotates. [1]

_____

PART C

Answer all questions in this part.

Directions (66–85): Record your answers in the spaces provided. Some questions may require the use of the *2011 Edition Reference Tables for Physical Setting/Earth Science*.

Base your answers to questions 66 through 68 on the models below and on your knowledge of Earth science. The models represent cutaway views of four planets in our solar system, showing their inferred interior struc-tures. Each planet is shown in relation to the size of Earth.

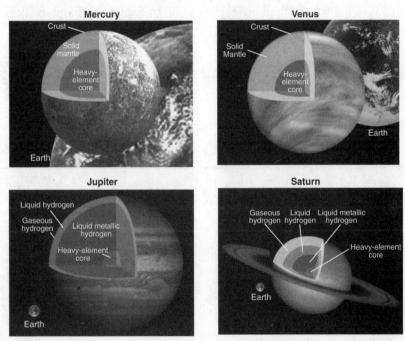

Seeds, Michael and Backman, Dana. 2011. *The Solar System*.

66 Determine how many times larger Jupiter's equatorial diameter is, compared to Earth's equatorial diameter. [1]

_____ **times larger**

67 Explain why Jupiter appears brighter in the night sky than Mercury, despite Jupiter's greater distance from Earth. [1]

_____

_____

68 Identify *two* terrestrial planets shown in the models. Explain why they are considered terrestrial planets. [1]

Terrestrial planets: _____ and _____

Explanation: _____

_____

Base your answers to questions 69 through 72 on the diagram below and on your knowledge of Earth science. The diagram represents several streams converging and eventually flowing into a lake. Points *X* and *Y* indicate locations on either side of a meander in the stream. Points *A* and *B* indicate locations in the streams where the stream discharge was measured in cubic meters per second. The circled region labeled *C* represents a depositional feature.

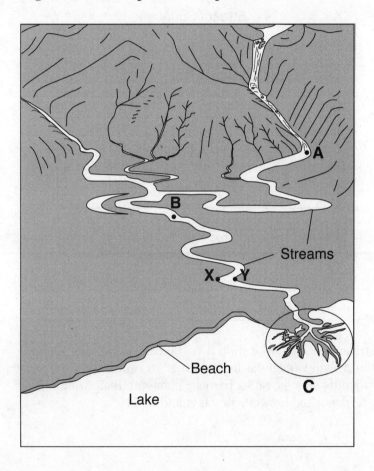

69 Identify the name of the depositional feature labeled C. [1]

_____

70 In the box *below*, draw a cross-sectional view of the general shape of the stream bottom from point X to point Y. [1]

71 Describe how the size and shape of sediments will most likely change as they are transported downstream. [1]

Size: _____

Shape: _____

72 The stream velocity at location A is 100 centimeters per second, and the stream velocity at location B is 10 centimeters per second. Identify *one* possible particle diameter that would most likely be deposited between points A and B. [1]

_____ cm

Base your answers to questions 73 through 76 on the map and graphs below and on your knowledge of Earth science. The map shows the locations of two cities, Hastings, Nebraska, and Riverhead, New York. The graphs show average monthly air temperatures for Hastings and Riverhead.

73 Explain why Hastings has a greater annual temperature range than Riverhead. [1]

_____

_____

74 Explain why the angle of insolation at both locations is approximately the same at solar noon on any given day. [1]

_____

_____

75 Identify the planetary wind belt that primarily influences the climates of *both* Hastings and Riverhead. [1]

_____

76 Name the ocean current that most likely has the greatest effect on the climate of Riverhead. [1]

_____ **Current**

_____

Base your answers to questions 77 through 80 on the block diagram below and on your knowledge of Earth science. The block diagram represents a region of sedimentary rock that has been intruded by magma, which has since solidified. Points *X* and *Y* identify locations at the boundary between the igneous intrusion and surrounding sedimentary rock layers. Letters *A* and *B* represent specific rock units. Letter *C* represents rock formed from the lava flow of the nearby volcano. The rock layers have *not* been overturned.

Adapted from www.brocku.ca/earthsciences

77 Identify *two* processes that formed the sedimentary rock layers represented in the diagram. [1]

   1: _____ 2: _____

78 Describe *one* piece of evidence shown in the diagram that indicates that rock unit A is younger than rock unit B. [1]

   _____
   _____

79 Explain why the igneous rock that formed at location C is composed of crystals less than 1 millimeter in size. [1]

   _____
   _____

80 State the names of *two different* metamorphic rocks that are most likely found in the zone of contact metamorphism at locations X and Y. [1]

   Location X: _____
   Location Y: _____

Base your answers to questions 81 through 85 on the passage and map below, the field map below, and on your knowledge of Earth science. The map shows the location of Crater Lake in Oregon in the western United States. The field map shows lake depths and some isolines in Crater Lake recorded in meters. Line *AB* and line *CD* are reference lines. Letter *X* represents a location on the lake bottom.

Crater Lake

Crater Lake is the deepest lake in the United States. The lake formed in the crater at the top of volcanic Mount Mazama after it exploded in a violent eruption approximately 7700 years ago. The rim of the crater is approximately 2300 meters (7500 feet) above sea level and is mostly composed of the rock andesite. The average yearly air temperature at the lake is 38°F, and snowfall often occurs from October through June. Hydrothermal activity (heating of the water) is ongoing under the lake, indicating that this region is still volcanically active.

81 On the field map *above*, draw the 500-meter depth isoline. [1]

82 On the grid *below*, construct a profile along line *AB* by plotting the lake depth of each isoline that crosses line *AB*. Connect *all* plots with a line to complete the profile. [1]

83 Determine the gradient, in meters per kilometer, between points C and D. [1]

_____ m/km

84 Determine *one* possible depth, in meters, of Crater Lake at location X. [1]

_____ meters

85 Line YZ on the diagram below represents the mineral composition of an andesitic rock taken from the bottom of Crater Lake.

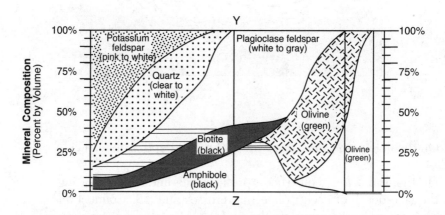

Identify the percent by volume of each of the three minerals in this andesitic rock. [1]

Plagioclase feldspar: _____ %

Biotite: _____ %

Amphibole: _____ %

Answers June 2019
Physical Setting/Earth Science

Answer Key

PART A

1. 2	8. 3	15. 4	22. 1	29. 1
2. 1	9. 4	16. 2	23. 3	30. 3
3. 2	10. 2	17. 1	24. 4	31. 4
4. 1	11. 3	18. 4	25. 2	32. 1
5. 4	12. 2	19. 4	26. 2	33. 3
6. 1	13. 3	20. 3	27. 1	34. 4
7. 2	14. 1	21. 2	28. 4	35. 2

PART B-1

36. 1	39. 4	42. 2	45. 3	48. 3
37. 1	40. 4	43. 4	46. 1	49. 1
38. 3	41. 2	44. 3	47. 4	50. 2

PART B–2 and **PART C**. *See* **Answers Explained**.

Answers Explained

PART A

1. **2**

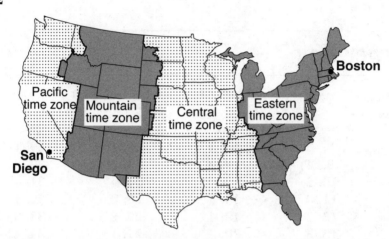

Earth rotates 15° per hour. So time zones were set up around a series of meridians spaced at 15° intervals starting at the prime meridian, which passes through Greenwich, England. Earth rotates from west to east. So for every time zone you travel west of the prime meridian, the clocks are set one hour earlier. According to the map, San Diego is located within the shaded region labeled "Pacific time zone" and Boston is located within the shaded region labeled "Eastern time zone." Note that the Pacific time zone is three time zones to the west of the Eastern Time Zone. Therefore, clocks in the Pacific time zone are set 3 hours earlier than those in the Eastern Time Zone. Thus, when it is 11 a.m. in San Diego, it is 2 p.m. in Boston.

2. **1** Find the Electromagnetic Spectrum chart in the *Reference Tables for Physical Setting/Earth Science*. Note that visible light at the red end of the spectrum has a longer wavelength than visible light at the blue end of the spectrum. Thus, a shift toward the longer wavelength end of the spectrum is a shift toward the red end of the spectrum. If a source of electromagnetic waves is moving away from an observer at the same time as the source is emitting light of a particular wavelength, fewer wave crests will reach the eye of the observer each second. The eye will interpret this as meaning that the light has a longer wavelength than it actually has. In other words, the light will appear shifted toward the red end of

the spectrum. It is given that the two elements were detected in a distant star that was moving away from Earth. Thus, all of the elements in the distant star are moving away from Earth. Therefore, the entire set of spectral lines would shift toward the red end of the spectrum.

3. **2**

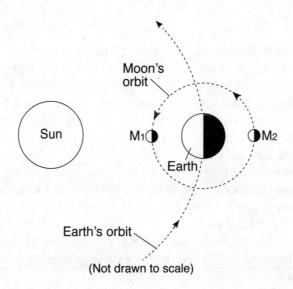

(Not drawn to scale)

A solar eclipse occurs when the Moon passes directly between the Sun and Earth, causing the Moon to cast a shadow on Earth. Note that when the Moon is in position M_1, it is directly between the Sun and Earth, the Moon's shadow falls on Earth and a solar eclipse occurs.

A lunar eclipse occurs when the Moon moves into Earth's shadow. When Earth's shadow falls on the Moon, Earth blocks sunlight from reaching the Moon. Earth's shadow is located on the side of Earth opposite the Sun. Therefore, in order for a lunar eclipse to occur, the Moon must be on the side of Earth directly opposite the Sun. Note that when the Moon is in position M_2, it is on the side of Earth directly opposite the Sun, causing Earth's shadow to fall on the Moon and a lunar eclipse to occur.

Only table (2) correctly matches each type of eclipse with the orbital position of the Moon and the cause of each eclipse.

4. **1**

Find the Geologic History of New York State chart in the *Reference Tables for Physical Setting/Earth Science*. In the column labeled "Era," find the statement "Estimated time of origin of Earth and solar system." Trace left to the scale labeled "Million years ago," and note that the origin of Earth occurred about 4600 million years ago. Thus, the entire length of the timeline from the point labeled "Origin of Earth" to the present day represents 4600 million years.

In the column labeled "Life on Earth," locate "Humans, mastodonts, mammoths." From humans, trace left to the time scale labeled "Million years ago." Note that humans first appeared on Earth less than 1.8 million years ago.

The time humans have existed (1.8 million years) is but a tiny fraction of the total time Earth has existed (1.8 ÷ 4600 = 0.00039). Thus, the timeline should show that humans first appeared on Earth at a time very close to the present day. Therefore, lettered dot *A* best indicates the geologic time when humans first appeared on Earth.

5. **4** Observations indicate that the universe is expanding. If the motion of galaxies is traced back in time, there is a point about 13–14 billion years ago at which the galaxies were very close to each other. Thus, we have a model in which the universe started out with all its matter and energy in a very small volume and then expanded outward in all directions. This motion is similar to an explosion, hence the name for this model—the Big Bang theory. According to this theory, the universe was initially smaller, denser, much hotter, and filled with a uniform glowing fog of hydrogen plasma. As the universe exploded outward in all directions from its origin point, both the plasma and the radiation filling it spread out and grew cooler. As the universe cooled, some of the matter condensed to form stars and their solar systems. Thus, the universe is older than our solar system.

The universe is everything that we can perceive. It includes all of the stars, gas clouds, galaxies, and other objects that we can detect by the radiation they emit. The universe contains billions of galaxies, each of which consists of hundreds of billions of stars. Our solar system consists of just a single star and all of the objects (such as planets) that orbit that star. Thus, the universe is larger than our solar system and contains more stars than the universe.

Thus compared to the solar system, the universe is older, larger, and contains more stars.

6. **1** As Earth revolves around the Sun, the side of Earth facing the Sun experiences day and the side facing away from the Sun experiences night. Since the stars are only visible at night, the portion of the universe whose stars (and the constellations they form) are visible to an observer on Earth varies cyclically as Earth revolves around the Sun. Thus, the motion that allows an observer on Earth to view different constellations throughout the year is Earth orbiting the Sun.

WRONG CHOICES EXPLAINED

(2) The idea that the stars in constellations orbit around Earth is part of the geocentric theory, which has been proven to be incorrect. Furthermore, Earth is too small and too far from the stars in constellations for Earth to exert enough gravitational attraction to hold the stars in orbit around Earth. Therefore, this motion does not occur and cannot allow an observer on Earth to view different constellations throughout the year.

(3) Earth orbits the Sun, not the stars in constellations. The stars in constellations are too far away to exert enough gravitational attraction on Earth to hold it in orbit. Therefore, this motion does not occur and cannot allow an observer on Earth to view different constellations throughout the year.

(4) Objects orbit a body that is both massive enough and close enough to exert a gravitational attraction strong enough to hold the objects in orbit. The Sun has too little mass and is too far from the stars in constellations to exert a strong enough gravitational attraction to hold them in orbit. For example, *Sirius*, in the constellation Canis Major, is located about 8.6 light-years from the Sun and is about twice as massive as the Sun. Therefore, this motion does not occur and cannot allow an observer on Earth to view different constellations throughout the year.

7. **2** Find the Tectonic Plates map in the *Reference Tables for Physical Setting/Earth Science*. In the key, note the symbol for a "Mantle hot spot." On the Tectonic Plates map, locate the Tasman Hot Spot off the southeast coast of Australia. Draw a horizontal line through the center of the Tasman Hot Spot, and extend it to intersect with the latitude scales along the left and right edges of the map. Note the value at these intersections—36°. Recall that 0° on this scale represents the equator. Latitudes above the 0° mark are north of the equator, and those below the 0° mark are south of the equator. Therefore, the Tasman Hot spot is located at about 36° S latitude. Now draw a vertical line through the Tasman Hot Spot, and extend it to intersect with the longitude scales along the top and bottom edges of the map. Note the value at these intersections—160°. Recall that 0° on the longitude scale represents the prime meridian. Longitudes to the left of the prime meridian from 0° to 180° are west longitudes; longitudes to the right of the prime meridian from 0° to 180° are east longitudes. Note that this particular map

is not centered on the prime meridian! The east longitudes to the right of the prime meridian wrap around and continue from the left edge of the map. Therefore, although the Tasman Hot is located to the left of the prime meridian on the map, it is beyond the 180° longitude line. In fact, the Tasman Hot Spot lies between 0° and 180° to the right of the prime meridian. Therefore, the Tasman Hot Spot is located at 160° E longitude. When taken together, the approximate location of the Tasman Hot Spot is 36° S 160° E.

8. **3** Find the Selected Properties of Earth's Atmosphere chart in the *Reference Tables for Physical Setting/Earth Science*. Locate 10–20 miles on the "Altitude" scale. Note that miles (mi) are listed on the right-hand side of the scale. Trace right to the section labeled "Temperature Zones," and note that this altitude range corresponds to the atmospheric temperature zone labeled "Stratosphere." Thus, most ozone is located in the temperature zone of the atmosphere known as the stratosphere.

9. **4** Planetary winds tend to blow in a straight line from regions of high pressure toward regions of low pressure. As these winds are blowing, Earth is spinning on its axis. This rotation causes the winds to appear to be turning toward the right in the Northern Hemisphere and toward the left in the Southern Hemisphere. This phenomenon is called the Coriolis effect. Thus, the Coriolis effect occurs as a result of the spinning of Earth on its axis.

WRONG CHOICES EXPLAINED

(1) The Coriolis effect does not change even though Earth's distance from the Sun varies. Therefore, it is unlikely that the Coriolis effect is the direct result of Earth's distance from the Sun.

(2) The inclination of Earth's axis causes unequal heating of the planet's surface. This causes differences in air temperature and therefore differences in air pressure. However, these differences in air pressure cause winds to blow directly from regions of high pressure to regions of low pressure, not cause them to be deflected.

(3) Earth's orbiting around the Sun causes seasonal changes in temperature and can lead to planetary regions of high and low pressure. However if Earth did not rotate, the planetary winds would blow in a straight line from regions of high pressure to regions of low pressure regions without being deflected.

10. **2** Find the Relative Humidity (%) chart in the *Reference Tables for Physical Setting/Earth Science*. Find the "Dry-Bulb Temperature (°C)" scale along the left side of the chart, and locate the row labeled "18." Trace this row to the right to the cell containing the value "64." From this cell, trace vertically upward to the

"Difference Between Wet-Bulb and Dry-Bulb Temperatures (°C)" scale along the top of the chart. Note the value listed there is "4." Thus, when the dry-bulb temperature is 18°C and the relative humidity is 64%, the difference between the wet-bulb and dry-bulb temperatures is 4°C.

Now find the Dewpoint (°C) chart in the *Reference Tables for Physical Setting/ Earth Science*. Locate the column headed "4" in the "Difference Between Wet-Bulb and Dry-Bulb Temperatures (°C)" scale along the top of the chart. Locate the row corresponding to 18°C on the "Dry-Bulb Temperature (°C)" scale along the left side of the chart. Trace to the right from the row labeled "18" until you intersect the column headed "4." At this intersection, note the value "11." Thus, when the dry bulb temperature is 18°C and the relative humidity is 64%, the dewpoint is 11°C.

11. **3** The process by which liquid water changes into water vapor is called evaporation, or vaporization. Find the Properties of Water chart in the *Reference Tables for Physical Setting/Earth Science*. Note that the heat gained during vaporization is 2260 J/g. Thus, in order for one gram of liquid water at its boiling point to change into water vapor, 2260 Joules of heat is gained from the surrounding environment.

12. **2**

By definition, the uppermost level of Earth material that is saturated by groundwater is known as the water table. The water table marks the boundary between the zone of saturation (dark gray shaded area) in which water fills all the pore spaces in the Earth material and the zone of aeration (light gray shaded area) in which air fills all the pore spaces. Thus, the dashed line labeled X, which marks the upper boundary of the zone of saturation, is the water table.

WRONG CHOICES EXPLAINED

(1) A watershed is the area drained by a stream and its tributaries, not the upper level of Earth material that is saturated with water.

(3) and (4) Note the arrows labeled "Infiltration" passing through the light gray shaded area of the ground above the dashed line labeled X. Infiltration is the downward movement of liquid water into and through interconnected openings among soil particles due to gravity. An Earth material through which water can infiltrate is said to be permeable. It is given that dashed line X represents the upper level of Earth material that is saturated by groundwater. In order for the Earth material to become saturated, its pore spaces must be interconnected so that the water can pass through the Earth material to fill empty pores. A substance through which water can pass is said to be permeable. Thus, both the Earth material above and below dashed line X are permeable. Therefore, it is unlikely that dashed line X indicates either impermeable bedrock or impermeable soil.

13. **3** Find the Geologic History of New York State chart in the *Reference Tables for Physical Setting/Earth Science.* Locate Mastodont, Condor, *Eospirifer,* and *Coelophysis* among the drawings of index fossils along the bottom of the chart. Note that mastodont corresponds to the circled letter "O," condor to "S," *Eospirifer* to "Y," and *Coelophysis* to "L." Now find the column labeled "Time Distribution of Fossils." Note the bold gray lines labeled with names and circled letters. The names are of types of fossil organisms. The circled letters are keyed to the illustrations of index fossils printed along the bottom of the chart and are placed on the lines to indicate the approximate time of existence of each specific important fossil. Locate the circled letters "O," "S," "Y," and "L." Note that "O" lies on the gray shaded line labeled "mammals." "S" is on the line labeled "birds." "Y" is on the line labeled "brachiopods." "L" is on the line labeled "dinosaurs." Thus, a mastodont was a mammal, a condor was a bird, an *Eospirifer* was a brachiopod, and a *Coelophysis* a dinosaur.

From the pictures of the index fossils, note that a mastodont is shown as a hairy elephant, a condor as a large winged bird, *Coelophysis* as a dinosaur that walked upright on its hind legs, and an *Eospirifer* as a shelled organism. Present-day elephants live on land, so it is reasonable to infer that an elephant-like mammal such as mastodont also lived on land. Therefore, mastodont fossils would not indicate a marine environment. Condors still exist today and live in mountainous regions, so it is reasonable to infer that condors in the past also lived in such regions. Therefore, condor fossils would not indicate a marine environment. Marine dinosaurs had flippers, not legs. Thus, *Coelophysis* was most likely a dinosaur that lived on land and *Coelophysis* fossils would not indicate a marine environment. However, present-day brachiopods are shelled organisms similar in appearance to clams or scallops and live in shallow marine environments. It is reasonable to infer that in the past, brachiopods also lived in such environments. Thus, the presence of the index fossil *Eospirifer* in sedimentary surface bedrock would most likely indicate that a marine environment once existed in the region where the sediments were deposited.

14. **1**

Cold fronts marks the boundary along which a cold air mass is advancing against a warmer air mass. The cold air mass is denser than the warmer air ahead of it. So the cold air mass pushes against and under the warm air mass like a wedge, forcing the warmer air upward. The result is a boundary that curves sharply upward in the direction of the cold air. Note that in diagram A, cold air is advancing against warm air and that the warm air is being forced upward. Note, too, that the boundary between the cold and warm air curves sharply upward toward the cold air. Thus, diagram A represents a cold front.

A warm front marks the boundary along which a warm air mass is advancing against a cold air mass. The warmer air mass is less dense than the colder air ahead of it. So as the warmer air pushes against the cooler air ahead of it, the warmer air is forced up and over the cooler air, forming a gently sloping boundary. Note that in diagram B, warm air is advancing against cool air and moving up and over the cool air, forming a gently sloping boundary. Thus, diagram B represents a warm front.

When a fast-moving cold front overtakes a slow-moving warm front, the cold front pushes completely under the warm air until it collides with the cool air ahead of the warm air and lifts the warm air entirely aloft, forming an occluded front. Note that in diagram C, cold air has pushed completely under the warm air and has reached the cool air ahead, lifting the warm air entirely aloft. Thus, diagram C most likely represents an occluded front.

A stationary front forms along the boundary between a warm air mass and a cold air mass when neither air mass moves appreciably in either direction. A stationary front slowly takes on the shape of a warm front as the denser cold air slowly slides beneath the less dense, warmer air. Note that none of the diagrams show a frontal boundary in which neither warm nor cold air is advancing. Thus, none of the diagrams shows a stationary front.

Therefore, the table that best matches each letter with the type of frontal boundary it represents is shown in Choice 1—*A* is a cold front, *B* is a warm front, and *C* is an occluded front.

15. **4**

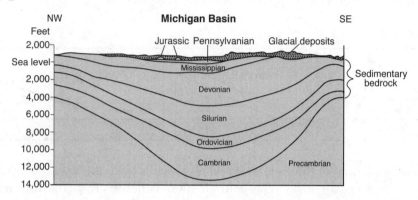

It is given that the sedimentary bedrock formed from sediments deposited in an ancient depositional basin. Basins are circular structures that arch downward, forming a bowl-like depression. Basins are formed by gentle downward bending, or downwarping, of rock due to compression forces acting on the crust. Note that the ancient Precambrian bedrock forms such a depression. As sediments were washed into this low-lying area, they accumulated in layers as shown in the cross section. Thus, the process that most likely caused the formation of the Michigan Basin is downwarping.

WRONG CHOICES EXPLAINED

(1) Uplift would cause the surface to be raised, not lowered. The uplifted surface would become a site of erosion, not a depositional environment. Therefore, it is unlikely that uplift caused the formation of the Michigan Basin

(2) The cross section shows no fault lines and no evidence of rock layers being displaced in either the sedimentary bedrock or the Precambrian bedrock. Therefore, there is no evidence that faulting caused the formation of the Michigan Basin.

(3) The glacial deposits at the surface are the result of erosion and deposition, not metamorphism. Note the rock layers labeled "Sedimentary bedrock" in the cross section. Sedimentary rocks form by the compaction and cementation of sediments, not metamorphism. There is no evidence in the cross section that the Precambrian bedrock has undergone metamorphism. Therefore, there is no evidence that metamorphism caused the formation of the Michigan Basin.

16. **2**

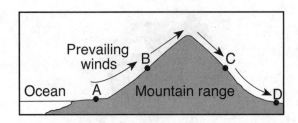

The cross section shows prevailing winds forcing air to rise along the windward side of the mountain range. As the air rises to higher elevations, air pressure decreases, causing the air to expand and cool. When the air cools to its dewpoint temperature, water vapor begins to condense, forming clouds. If enough water vapor condenses and if large water droplets form, precipitation will occur on the windward side. However, precipitation on the windward side is much more likely at location B than at location A, because at location A, the air has not yet been forced to rise much above sea level and has not yet been cooled to its dewpoint.

As the drier air sinks on the leeward side of the mountains, air pressure increases, causing the air to be compressed and warm. The air is warmed above its dewpoint, and any available moisture evaporates. Therefore, clouds and precipitation are much less likely to form on the leeward side of the mountain (locations C and D).

Thus, the location that would most likely have cloud cover and precipitation is B.

17. **1** All other factors being equal, dark-colored materials absorb more insolation than light-colored materials. Due to the high specific heat of water, land surfaces increase in temperature more than water surfaces when insolation strikes. Thus of the choices given, the darkest-colored land surface is the blacktop parking lot. Therefore, the blacktop parking lot will absorb the most insolation.

18. **4** Greenhouse gases allow sunlight to enter the atmosphere and strike Earth's surface but absorb the infrared radiation reradiated by Earth's surface, trapping the heat in the atmosphere and resulting in global warming. Some of these greenhouse gases occur in nature, such as water vapor, carbon dioxide, methane, ozone, and nitrous oxide. Others are exclusively human-made, such as the chlorofluorocarbons used in aerosol sprays. Thus in addition to water vapor and carbon dioxide, only choice (4)—methane—contains another major greenhouse gases found in Earth's atmosphere.

19. **4**

Note the relative position of the state of Washington on the North American continent. Find the Geologic History of New York State chart in the *Reference Tables for Physical Setting/Earth Science*. In the column labeled "Inferred Positions of Earth's Landmasses," locate the diagram in which the position of the state of Washington on the North American continent is closest to the equator. Note that this diagram is labeled "458 million years ago." On the time scale labeled "Million years ago" to the right of the column labeled "Epoch," locate the point corresponding to "458." From this point, trace left to the column labeled "Period," and note that 458 million years ago corresponds to the Ordovician Period. Therefore, the geologic period during which the region of Washington State was closest to the equator is the Ordovician.

20. **3** Along the eastern sides of many ocean basins, the global trade winds blow parallel to the coast and warm surface waters are carried offshore, producing strong coastal upwelling. The nutrient-rich water brought to the surface by this upwelling makes these coastal areas some of the ocean's richest fishing grounds. El Niño originally referred to the changes in surface currents along the coasts of Peru and Chile. Every year, usually in December, the trade winds subside, the surface ocean currents flowing offshore slow, upwelling slacks off, and the water gets warmer. Local fishermen have known about this for centuries since it marks the end of the peak fishing season. Since the change in currents comes around Christmas, they called it El Niño—"The Christ Child." Every few years, however, the change is greater than normal. The surface water gets much warmer, upwelling ceases altogether, and the surface ocean current reverses to inshore. The warming of the surface waters during an El Niño causes a large region of low pressure to form over the southeastern Pacific Ocean. This, in turn, causes changes in the jet stream winds, which profoundly affect global weather patterns. On average, during the last ten El Niño climate events, changes to the jet stream winds have resulted in warmer, drier winters in New York State. Thus, the climate event that occurs when surface waters in the eastern equatorial area of the Pacific Ocean become warmer than normal and may cause a warm, dry winter in New York State is an El Niño.

WRONG CHOICES EXPLAINED

(1) Air mass formation occurs all the time, not just when the eastern equatorial Pacific surface waters are warmer than usual.

(2) The Doppler effect refers to the change in the perceived wavelength of sound or light emitted by a source that is moving toward or away from an observer, not a climate event.

(4) A monsoon is a seasonally reversing system of surface winds caused by temperature differences between land and ocean, not the warmer than normal surface water in the eastern equatorial Pacific Ocean.

21. **2** Find the Geologic History of New York State chart in the *Reference Tables for Physical Setting/Earth Science*, and locate the column labeled "NY Rock Record." The thick black or shaded lines indicate that rocks or sediments from that time period exist in New York State. If there is no line, there is no rock or sediment of that age in New York State. Trace the areas where there is no line in the "NY Rock Record" column to the left to the column labeled "Period." Note that New York State contains no rocks corresponding to any part of the Neogene, Paleogene, and Permian Periods. Thus in addition to the Permian, two other entire geologic periods for which there is no sediment or bedrock in the New York State rock record are the Paleogene and Neogene Periods.

22. **1**

Find the Scheme for Sedimentary Rock Identification in the *Reference Tables for Physical Setting/Earth Science*. In the column labeled "Map Symbol," note the symbols that correspond to those shown in the outcrops. From each map symbol used in the outcrops, trace left to the column labeled "Rock Name." Note the name of the rock associated with each symbol.

Rock layers can sometimes be correlated on the basis of distinct similarities in physical characteristics, such as composition, color, thickness, and fossil remains. Based on composition, thickness, and sequence, the limestone, shale, siltstone sequence in the two outcrops can be correlated. Thus, limestone rock layer D correlates with limestone layer B; shale rock layer C correlates with the shale layer above B in Outcrop 2; and the siltstone rock layer above rock layer C correlates with siltstone layer A.

Note that all of the rock layers in the two outcrops are horizontal and do not appear to have been overturned. In fact, the question specifically states that the rock layers have *not* been overturned. In such cases, the principle of superposition states that the uppermost layer is the youngest. Therefore in outcrop 1, rock layer C is younger than rock layer D. Rock layers D and B correlate, so they are the same age. In outcrop 2, rock layer A is younger than rock layer B. Rock layer A is also younger than the shale layer beneath it that correlates with rock layer C. Therefore, rock layer A is younger than rock layer C. Thus, A is the youngest lettered rock layer.

23. **3**

Stream drainage patterns refer to the patterns formed by the system of streams in an area as they flow down slopes and join with other streams. Water flows from higher elevations to lower elevations. Therefore, water flows along either side of the tops of the long, parallel ridges shown in the block diagram into the valleys between the ridges, forming a stream drainage pattern as shown in map (3).

WRONG CHOICES EXPLAINED

(1) Drainage diagram (1) shows a series of concentric, circular streams. This type of drainage pattern forms where there is series of concentric circular ridges and valleys. The block diagram in the question shows a series of long, parallel ridges and valleys, not concentric circular ridges and valleys.

(2) On a flat plain, the slope is gentle. Streams flow downhill slowly and typically form a dendritic stream drainage pattern. In a dendritic stream drainage pattern, there are numerous small streams that join together to form larger streams in a treelike pattern as shown in drainage pattern (2). The block diagram in the question shows a series of long, parallel ridges and valleys, not a flat plain.

(4) Drainage diagram (4) shows a series of streams flowing outward from a central area. Water flows from higher elevations to lower elevations. Therefore, the central area in this drainage diagram must be at a higher elevation than the surrounding areas. This is typically the case on rounded surfaces, such as conical hills or domes. The block diagram in the question shows a series of long, parallel ridges and valleys, not a conical hill or dome.

24. **4** Water is important to many weathering processes. Many chemical weathering processes require water. Water is a reactant in hydration and carbonation. Additionally, water provides a medium in which acid reactions can occur. Chemical reactions also occur faster at warm temperatures than at cool temperatures. Warm, rainy climates also support increased biological activity, ranging from plant action to production of humic acid as plant matter decomposes. Therefore, a warm, rainy climate will result in more pronounced weathering than will a cool, dry climate. An increase in rainfall will also result in an increase in erosion. Abundant rainfall and warm temperatures also promote the growth of a protective layer of vegetation that protects soil from rapid runoff and erosion, leading to rounded slopes. Thus compared to a landscape that develops in a cool, dry climate, a landscape that develops in a warm, rainy climate will most likely weather and erode faster, so the landforms are more rounded.

25. **2** Find the Inferred Properties of Earth's Interior chart in the *Reference Tables for Physical Setting/Earth Science*. In the cross section at the top of the chart, locate "Mid-Atlantic Ridge." Note that the Mid-Atlantic Ridge lies within the section labeled "Atlantic Ocean." Therefore, the crustal bedrock at the Mid-Atlantic Ridge is considered oceanic crust. Locate the "Density (g/cm^3)" scale along the right side of the cross section. Note that oceanic crust is listed as basaltic and has a density of 3.0 g/cm^3. Thus, Earth's crustal bedrock at the Mid-Atlantic Ridge is composed mostly of basalt, with a density of 3.0 g/cm^3.

26. **2** When an earthquake occurs, *P*-waves and *S*-waves start moving outward from the focus at the same time. However since these waves travel at different speeds, they do not arrive at a seismic station at the same time. It is given that the first *P*-wave traveled 4000 kilometers from the epicenter of the earthquake to the seismic station. Find the Earthquake *P*-Wave and *S*-Wave Travel Time graph in

the *Reference Tables for Physical Setting/Earth Science*. Locate 4000 kilometers (4.0×10^3 km) along the horizontal "Epicenter Distance ($\times 10^3$ km)" axis at the bottom of the graph. From this point, trace vertically until you intersect the bold line labeled "P." From this intersection, trace horizontally to the left to the "Travel Time (min)" axis. Read the *P*-wave travel time—7 minutes. If the first *P*-wave takes 7 minutes to travel the 4000 kilometers from the epicenter to the seismic station and arrives at 7:10:00 p.m., it must have left the epicenter 7 minutes before 7:10:00 p.m., or 7:03:00 p.m. Thus, the earthquake occurred at 7:03:00 p.m.

Now that you know the origin time of the earthquake, you can use this procedure to determine the travel time of the *S*-wave and its arrival time. From 4000 kilometers on the horizontal "Epicenter Distance ($\times 10^3$ km)" axis, trace vertically until you intersect the bold line labeled "S." From this intersection, trace horizontally to the left to the "Travel Time (min)" axis. Read the *S*-wave travel time—12 minutes 40 seconds. If the first *S*-wave takes 12 minutes 40 seconds to travel the 4000 kilometers from the epicenter to the seismic station and the earthquake started out at 7:03:00 p.m., it would arrive at the seismic station at 7:15:40 p.m.

Therefore, other information that could be determined from this recording was that the earthquake occurred at approximately 7:03.00 p.m., and the *S*-wave arrived at 7:15:40 p.m.

27. **1** Find the Average Chemical Composition of Earth's Crust, Hydrosphere, and Troposphere chart in the *Reference Tables for Physical Setting/Earth Science*. In the section labeled "Crust," locate the column headed "Percent by mass." According to the chart, the two elements that make up the largest percent by mass of Earth's crust are oxygen (46.10%) and silicon (28.20%). These percentages are best represented by the graph in Choice (1).

28. **4**

Note that the elongated hill in the aerial photograph has a teardrop shape and slopes downward from the wider end toward the narrower end. It is given that the hill formed by glacial deposition. An elongated, teardrop-shaped hill formed by glacial deposition is called a drumlin. Drumlins are formed when the ice of a glacier slides over previously deposited piles of sediment, causing them to become elongated into a teardrop shape. Thus, the elongated hill in the aerial photograph is best identified as a drumlin.

WRONG CHOICES EXPLAINED

(1) Kettles are bowl-shaped depressions formed when buried blocks of glacial ice melt. They are not elongated, teardrop-shaped hills.

(2) Moraine is material transported by a glacier and then deposited, typically as ridges at the glacier's edges or end. A drumlin may have started out as a pile of sediment deposited by a glacier (moraine), but then it was reshaped as the glacier advanced over it once again.

(3) A sand dune is a mound of sand deposited by wind. It is not a feature formed by glacial deposition.

29. **1**

www.alicehenderson.com

When a glacier moves down a valley, it alters the valley's shape. Since ice does not meander back and forth as readily as water does, the glacier often breaks through obstacles to form a straighter path and a flat bottom. The ice also cuts back the walls of the valley, making them steeper. The result is a wide, straight, U-shaped valley with a flat bottom and steep sides. As the glacier moves along, loose rock may freeze into the ice at the bottom of the glacier. As the glacier moves over bedrock, these rocks scrape parallel grooves or scratches, called striations, into the bedrock beneath the glacier. Therefore, the result of the glacial action of the valley glacier shown in the photograph will be a U-shaped valley with scratched and grooved bedrock.

30. **3** Find the Scheme for Sedimentary Rock Identification in the *Reference Tables for Physical Setting/Earth Science.* In the column labeled "Texture," locate the term "bioclastic." Note that "bioclastic" occurs in two rows. Trace these rows to the right to the column labeled "Composition." Note that bioclastic sedimentary rocks may be composed of calcite or carbon. Find the Properties of Common Minerals table in the *Reference Tables for Physical Setting/Earth Science.* In the column labeled "Mineral Name," locate calcite. Trace this row to the left to the column labeled "Composition," and note the formula for calcite – $CaCO_3$. Calcite is a compound consisting of elements represented by the symbols Ca, C, and O. In the "Chemical symbols" key at the bottom of the chart, note that "C" is the symbol for the element carbon. Therefore, the element always found in bioclastic rocks is carbon.

31. **4** Soil is the accumulation of loose, weathered material that covers much of Earth's land surface. The main component is weathered rock. However, a true soil also contains water, air, bacteria, and decayed plant and animal material (humus).

Since weathered rock is the main component of soil, one of the most important factors controlling the development of soil is the type of rock that was weathered. Therefore, bedrock composition is a major factor controlling the development of soil in a given location. Another important factor controlling the development of soil is the weathering process that converts the bedrock to soil. The single most important factor affecting weathering is climate. Climate is typically expressed in terms of two factors: temperature and precipitation. Both of these factors influence the type and rate of weathering. Warm climates favor chemical weathering; cold climates favor physical weathering. In both cases, the more moisture that is present, the more pronounced the weathering. Therefore, two major factors that control the development of soil in a given location are bedrock composition and climate.

32. **1**

Layers of rock are generally deposited in an unbroken sequence. However if forces within Earth uplift rocks, deposition ceases. Erosion may wear away layers of rock before the land surface is low enough for another layer to be deposited. The uplift may cause rock layers to tilt or fold. The result is an unconformity, which is a break or gap in the sequence of a series of rock layers. Thus, the rocks above an unconformity are quite a bit younger than those below it. The erosional surface marking the gap between the older and younger rocks is generally drawn as a wavy line.

On the cross section, note that the Cambrian and Silurian limestones are separated by a wavy line. This wavy line represents an unconformity. Note, too, that below the wavy line the rock layers are folded and that above the wavy line the layers are flat.

Find the Geologic History of New York State chart in the *Reference Tables for Physical Setting/Earth Science*. In the column labeled "Period," locate the Cambrian and the Silurian. Trace right to the time scale labeled "Million years ago." Note that the Cambrian Period ended 488 million years ago and that the Silurian Period began 444 million years ago. So the Silurian limestone above the wavy line is more that 40 million years younger than the Cambrian limestone below the wavy line. In other words, there is a gap of more than 40 million years between the two limestone layers.

During those more than 40 million years, a number of things most likely happened. First, it can be inferred that uplift occurred because sediments stopped being deposited in an unbroken sequence. When rocks are uplifted, deposition stops because the surface is no longer at a low point where sediments settle to the ground. The folding in the lower rock layers indicates that forces acting on the crust were great enough to bend and fold rock. The folding pushed rock upward like a piece of paper on a desk when you push the two ends toward each other. Once uplifted, the rock would be exposed to weathering and be broken down into sediments. Erosion would then carry away the sediments, and the land surface would be worn down until it resembled the wavy line of the erosional surface.

Find the Scheme for Sedimentary Rock Identification in the *Reference Tables for Physical Setting/Earth Science*. In the column labeled "Rock Name," locate "Limestone." Trace left to the column labeled "Comments," and note that limestone is made up of "Precipitates of biologic origin or cemented shell fragments." This indicates that limestone forms in water, typically in shallow seas. In order for the Silurian limestone to be deposited, the land surface had to subside and become a seafloor. Thus, the last geologic event that occurred before the Silurian limestone was deposited was subsidence.

When taken together, the sequence of geologic events that occurred after the formation of the Cambrian limestone and before the formation of the Silurian limestone was uplift, weathering, erosion, and subsidence.

33. **3** Find the Scheme for Igneous Rock Identification in the *Reference Tables for Physical Setting/Earth Science*. In the chart at the top of the scheme, locate the "Granite." Then trace vertically down to the graph labeled "Mineral Composition (relative by volume)." Note that this column intersects the region labeled "Quartz (clear to white)." Thus, granite contains quartz.

Find the Scheme for Metamorphic Rock Identification in the *Reference Tables for Physical Setting/Earth Science*. In the column labeled "Rock Name," locate "Gneiss." Trace the row to the left to the column labeled "Composition." Note that gneiss contains quartz. Therefore, the two rocks that contain quartz are granite and gneiss.

WRONG CHOICES EXPLAINED

(1) Find the Scheme for Igneous Rock Identification in the *Reference Tables for Physical Setting/Earth Science*. In the chart at the top of the scheme, locate "Gabbro." Then trace vertically down to the graph labeled "Mineral Composition (relative by volume)." Note that this column does not intersect the region labeled "Quartz (clear to white)." Thus, gabbro does not contain quartz.

(2) Find the Scheme for Igneous Rock Identification in the *Reference Tables for Physical Setting/Earth Science*. In the chart at the top of the scheme, locate "Dunite." Then trace vertically down to the graph labeled "Mineral Composition (relative by volume)." Note that this column does not intersect the region labeled "Quartz (clear to white)." Thus, dunite does not contain quartz.

(4) Find the Scheme for Igneous Rock Identification in the *Reference Tables for Physical Setting/Earth Science*. In the chart at the top of the scheme, locate "Scoria." Then trace vertically down to the graph labeled "Mineral Composition (relative by volume)." Note that this column does not intersect the region labeled "Quartz (clear to white)." Thus, scoria does not contain quartz.

34. **4** Find the Properties of Common Minerals table in the *Reference Tables for Physical Setting/Earth Science*. In the column labeled "Fracture," note that only seven of the minerals on the chart display fracture. From each of these minerals, trace right to the column labeled "Composition," and check if the formula listed is one of the answer choices. Note that only Choice (4), $(Fe,Mg)_2SiO_4$, is a mineral that usually exhibits fracture, not cleavage.

35. **2** Find the Properties of Common Minerals table in the *Reference Tables for Physical Setting/Earth Science.* In the column labeled "Uses," locate "pencil lead, lubricants." From this phrase, trace right to the column labeled "Mineral Name." Note that the name of the mineral used to make pencil lead is graphite.

PART B-1

36. **1**

Locate the first low tide on Dec. 23. Trace vertically downward to the horizontal axis, and note that it occurred at 3:00 a.m. Now locate the first low tide on Dec. 24, and note that it occurred at 4:00 a.m., about 25 hours later. Next, locate the first low tide on Dec. 25, and note that it occurred at 5:00 a.m., about 25 hours after the first low tide on Dec. 24. Thus, the pattern is that the first low tide for each day shown on the graph occurs about 25 hours after the first low tide the day before. Therefore, if the pattern continues, the first low tide on Dec. 26 occurred at 6:00 a.m., 25 hours after the first low tide on Dec. 25.

37. **1** Note that the line representing water levels at Kings Park rises and falls in a repeating pattern. In other words, the changes in water level are cyclic. Note, too, that all of these changes repeat in a specific period of time. Therefore, they are predictable. For example, the first low tide on Dec. 23 occurs at 3:00 a.m., the first low tide on Dec. 24 occurs at 4:00 a.m., and the first low tide on Dec. 25 occurs at 5:00 a.m. Each of the first low tides are 25 hours apart. Thus, the changing water levels at Kings Park show a pattern that is cyclic and predictable.

38. **3**

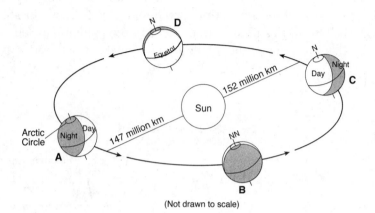

(Not drawn to scale)

It is given that aphelion is Earth's farthest position from the Sun and occurs around July 4. July 4 occurs during the summer season (June 21–September 21) in the Northern Hemisphere. Thus, Earth is at aphelion during the summer in the Northern Hemisphere.

39. **4** On the diagram, note the point in Earth's orbit at which Earth is at aphelion, farthest from the Sun (152 million km). The summer solstice occurs on or about June 21, shortly before Earth reaches aphelion around July 4. Thus, location C corresponds to the summer solstice. The winter solstice occurs on about December 21, shortly before Earth reaches perihelion, closest to the Sun (147 million km). Thus, location A corresponds to the winter solstice. The fall equinox occurs about halfway between the summer solstice and the winter solstice. Note the arrows indicating the direction in which Earth revolves around the Sun in its orbit. Thus, location D, which is halfway between locations C and A, corresponds to the fall equinox. Therefore, location B corresponds to the spring equinox.

The location at which the Sun's vertical rays strike Earth's surface varies in an annual cycle with the seasons. For example, on June 21, Earth's Northern Hemisphere is tilted farthest toward the Sun and the Sun's vertical rays strike Earth's surface at 23.5° N latitude (Tropic of Cancer). At the equinoxes in March and September, Earth is tilted neither toward nor away from the Sun, and the Sun's vertical rays strike Earth's surface at the equator. On December 21, Earth's Southern Hemisphere is tilted farthest toward the Sun and the Sun's vertical rays strike Earth's surface at 23.5° S latitude (Tropic of Capricorn). Therefore, the Sun's vertical ray is moving from the equator southward to the Tropic of Capricorn between September 21 (fall equinox) and December 21 (winter solstice), that is, between positions D and A.

40. **4** As explained in the answer to question 39, position A corresponds to the winter solstice in the Northern Hemisphere. In moving from position A back to position A, Earth goes from one winter solstice to the next winter solstice, a period of one year. One year is roughly 365 days. Earth completes one rotation in 24 hours, or one day. Thus, in one year Earth, rotates about 365 times. Therefore, as Earth moves in its orbit from position A back to position A, it rotates 365 times.

41. **2** Earth's rotational axis is tilted at an angle of 23.5° from a perpendicular to Earth's orbital plane.

42. **2**

Most storm systems are associated with low-pressure centers. Air moves from regions of higher pressure toward regions of lower pressure. In a low-pressure system, the pressure gradient causes the air to move inward toward the low-pressure center. At the same time, the Coriolis Effect causes the moving air to be deflected to the right in the Northern Hemisphere. The combination of these two motions results in winds that blow inward in a counterclockwise pattern.

43. **4** Isobars are lines connecting points of equal air pressure. The spacing of isobars indicates the nature of the pressure gradient. The more closely spaced the isobars are on a weather map, the greater the pressure gradient and the higher the wind speed. Thus, the best evidence on a weather map to indicate high-speed winds near the center of the storm system would most likely be isobars drawn close together.

WRONG CHOICES EXPLAINED

(1) Clouds may cover the sky at many different wind speeds or even if there is no wind at all. Therefore, cloud cover of 100% would provide no information about wind speed.

(2) The type of precipitation is simply the form of moisture falling to the ground. Indicating the type of precipitation would provide no information about wind speed.

(3) High winds can occur at a wide range of temperature and de point values. Therefore, indicating temperature and dewpoint values would provide no information about wind speed.

44. **3** Storm systems are moved by planetary winds and by the polar front jet stream. So the storm track will typically be in the direction of the planetary winds and polar front jet stream. Find the Tectonic Plates map in the *Reference Tables for Physical Setting/Earth Science*. Note that the region of North America covered by this storm system lies between 30° N and 60° N latitude. Now find the Planetary Wind and Moisture Belts in the Troposphere chart in the *Reference Tables for Physical Setting/Earth Science*. Note that between 30° N and 60° N latitude, the planetary winds blow from the southwest toward the northeast. Therefore as this storm system follows a normal storm track, it will most likely move toward the northeast.

45. **3**

Note that the plate boundary at point A is shown as a solid line with arrows on either side of the line pointing in opposite directions parallel to the line. Find the Tectonic Plates map in the *Reference Tables for Physical Setting/Earth Science*. In the key at the bottom of the map, note that this symbol corresponds to a transform boundary. Therefore, the plate boundary represented at point A is a transform plate boundary.

WRONG CHOICES EXPLAINED

(1) Find the Tectonic Plates map in the *Reference Tables for Physical Setting/Earth Science*. In the key at the bottom of the map, locate the symbol representing a divergent plate boundary. Note that the line and arrows at point A do not correspond to this symbol.

(2) Find the Tectonic Plates map in the *Reference Tables for Physical Setting/Earth Science*. In the key at the bottom of the map, locate the symbol representing a convergent plate boundary. Note that the solid line and arrows at point A do not correspond to this symbol.

(4) Find the Tectonic Plates map in the *Reference Tables for Physical Setting/Earth Science*. In the key at the bottom of the map, Note that symbol for a complex plate boundary is a dashed line, not the solid line and arrows shown at point A.

46. **1** The segment of plate X shown on the map lies just north of about 40° N latitude between 20° E and 40° E longitude. Find the Tectonic Plates map in the *Reference Tables for Physical Setting/Earth Science*. Locate the region just north of 40° N latitude, between 20° E and 40° E longitude. Note that this region corresponds to the Eurasian Plate.

The segment of plate Y shown on the map lies north of about 30° N latitude between 20° E and 35° E longitude. Find the Tectonic Plates map in the *Reference Tables for Physical Setting/Earth Science*. Locate 40° N latitude, between 20° E and 35° E longitude, just west of the Arabian Plate. Note that this region corresponds to the African Plate. (Note that Africa is split in this map projection; one-half is near the right edge and one-half is near the left edge of the map. The Arabian Plate is shown next to the part of Africa near the left edge of the map.) Therefore, plate X is the Eurasian Plate, and plate Y is the African Plate.

47. **4** The two New York State index fossils most closely related to the corals that were radioactively dated in this study are themselves corals. Find the Geologic History of New York State chart in the *Reference Tables for Physical Setting/Earth Science*. In the column labeled "Time Distribution of Fossils (including important fossils of New York)," locate the vertical bar labeled "Corals." Note that

the circled letters representing index fossils on this bar are "T," "U," and "V." Thus, the index fossils at the bottom of the chart corresponding to the circled letters "T," "U," and "V" are corals. Locate the drawings of "T," "U," and "V" at the bottom of the chart, and note their names: "T"—*Lichenaria*, "U"—*Cystiphyllum*, and "V"—*Pleurodictyum*. Thus, the two New York State index fossils most closely related to the corals that were radioactively dated in this study are choice (4)—*Lichenaria* and *Pleurodictyum*.

48. **3** Tsunamis may be only a few meters high, but they have very long wavelengths and travel much more rapidly than do ordinary ocean waves. Tsunamis have been clocked moving faster than 500 km/hr with wavelengths as great as 200 kilometers. When tsunamis reach shallow water, they are slowed by friction with the ocean bottom and begin to pile up as large waves. In bays and narrow channels, their high speed and long wavelength may be funneled into huge breaking waves more than 20 m high. The force exerted by such a huge, fast-moving mass of water can do extensive damage and cause widespread coastal flooding. The best thing to do if a tsunami approaches is to evacuate to higher ground. Therefore, the activity that could best prepare residents along the Mediterranean coast to reduce loss of human life during a future tsunami is to plan evacuation routes to higher ground.

WRONG CHOICES EXPLAINED

(1) Boarding up windows will do little to reduce the loss of human life because buildings are often completely inundated or washed away during a tsunami.

(2) Removing heavy objects from walls will do little to reduce the loss of human life because this will do nothing to prevent water from entering the room and drowning the inhabitants.

(4) A basement is likely the worst place to be during a tsunami because water will flood into the basement and drown the inhabitants.

49. **1**

www.nationalgeographic.com

Note that the feature shown in the photograph is located above the surrounding land, and that there is no vegetation covering the nearby land surface. The main agent of erosion in dry climates is wind. Sandblasting is the process by which sand carried by the wind collides with and wears away solid objects. When the sand particles collide with the solid object, the impact causes small pieces of the object to break off. This can cause the object being hit with particles to become smooth or worn. Thus, the agent of erosion that is most likely sandblasting this rock formation is wind.

50. **2** It is given that the photograph shows a sandstone erosional feature. Therefore, the rock making up this feature is sandstone. Find the Scheme for Sedimentary Rock Identification in the *Reference Tables for Physical Setting/ Earth Science*. In the column labeled "Rock Name," locate "Sandstone." Then trace left to the column labeled "Grain Size." Note that sandstone is made up of sand grains that range from 0.006 cm–0.02 cm.

PART B-2

51.

Radioactive Isotope X (%)	Disintegration Product Z (%)	Time (billion years)
100	0	0
50	50	1.3
25	75	2.6
12.5	87.5	3.9
6.25	93.75	5.2
3.125	96.875	6.5

To construct a graph that shows the percentages of disintegration product Z forming over 6.5 billion years, use the data in the table to plot the points for disintegration product Z and time in billion years as follows. The table states that at 0 billion years, there is 0% of Z. To plot these coordinates, locate "0" on the horizontal scale labeled "Time (billion years)" and locate "0" on the vertical scale labeled "Percentage (%)." Plot the point of intersection. Next, the table states that at 1.3 billion years, there is 50% of Z. Locate "1.3" on the horizontal scale labeled "Time (billion years)." Trace upward along the "1.3" line until you intersect the line corresponding to "50" on the vertical scale labeled "Percentage (%)." Plot the point of intersection. Repeat this process for each of the remaining coordinates: (2.6, 75), (3.9, 87.5), (5.2, 93.75), and (6.5, 96.875). Then connect the plotted intersection points with a line as shown in the graph below.

One credit is allowed **if the centers of *all six* plots are within or touch the circles shown and are correctly connected with a line that passes within or touches each circle.**

Note: Credit is allowed if the student-drawn line does *not* pass through the student plots but is still within or touches the circles.

52. Half-life is the time required for one-half (50%) of the atoms of a radioactive isotope to decay into a stable disintegration product. According to the table, every 1.3 billion years, half of the remaining radioactive isotope X decays into disintegration product Z. Therefore, the half-life of radioactive isotope X is 1.3 billion years. Find the Radioactive Decay Data chart in the *Reference Tables for Physical Setting/Earth Science*. In the column labeled "Half-Life (years)," locate 1.3 billion (1.3×10^9). Trace left to the column labeled "Radioactive Isotope," and note that potassium-40 has a half-life of 1.3×10^9 (1.3 billion) years. Thus, radioactive isotope X is potassium-40.

One credit is allowed for an acceptable response. Acceptable responses include but are not limited to:

- **potassium-40**
- **^{40}K**
- **K-40**

53. On the table, locate "3.9" in the column labeled "Time (billion years)." Trace left to the column labeled "Radioactive Isotope X (%)." Note that after 3.9 billion years, 12.5% of radioactive isotope X will remain. Therefore, after 3.9 billion years, 12.5% of the original 300-gram sample of radioactive isotope X, or 37.5 grams, will remain ($300 \times 0.125 = 37.5$).

One credit is allowed for **37.5 g.**

54.

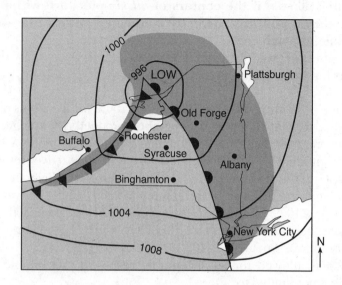

Find the Key to Weather Map Symbols in the *Reference Tables for Physical Setting/Earth Science*. In the section labeled "Fronts," note that the symbol for a cold front is a bold line with triangles. Locate the symbol corresponding to a cold front on the map, and note that it is located very near Rochester. A cold front marks the boundary where cold air is advancing against warmer air. The side of the front on which the symbols are drawn indicates the direction in which the cold air is advancing. Thus, the air mass west of the cold front is a cold air mass and the air mass east of the cold front is a warm air mass. Therefore, when the cold front reaches Syracuse, it will be well past Rochester and Rochester will be located in a cold air mass, not in a warm air mass. Cold air is denser than warm air, so cold air exerts a higher air pressure. Therefore by the time the cold front has reached Syracuse, the air pressure at Rochester will have increased.

Note the dark shaded area along the cold front indicating areas of precipitation. By the time the cold front reaches Syracuse, this area will be well past Rochester. Cold air masses contain less moisture than do warm air masses, so clouds are less likely to form with cold air masses. Therefore, the cloud cover in Rochester will most likely decrease.

One credit is allowed if both responses are correct. Acceptable responses include but are not limited to:
Change in air pressure:

- **slight decrease, then a steady increase**
- **generally increasing/rising**
- **lower to higher**
- **greater**

Amount of cloud cover:

- **It decreased.**
- **lower percent**
- **clear/0%**
- **There are fewer clouds.**
- **little cloud cover**

55. Find the Key to Weather Map Symbols in the *Reference Tables for Physical Setting/Earth Science*. In the section labeled "Station Model Explanation," note how the values for air temperature, barometric pressure, wind direction, and wind speed are represented on a station model. Note the station model for Buffalo, New York. The value corresponding to air temperature is "72," which is 72°F. The value corresponding to barometric pressure is "010," which is 1001.0 mb. (Air pressure rarely rises above 1050.0 mb or drops below 950.0 mb. Therefore, the general rule is to place a 9 in front of the digits on the station model if they are greater than 500 and to place a 10 in front of the digits if they are less than 500 and then insert a decimal between the last two digits.) The line representing wind direction extends to the northwest, corresponding to a NW wind direction. There is one whole feather and one half feather on the wind direction line, corresponding to a wind speed of 15 knots.

One credit is allowed if ***all four*** weather variables are correct as shown below.

Air Temperature (°F)	Barometric Pressure (mb)	Wind Direction from the	Wind Speed (knots)

Note: Credit is allowed if the student places the correct units in the boxes as part of their response. Do *not* allow credit for a barometric pressure of 010, as this is the format used on a station model, not the actual barometric pressure.

56. When air temperature cools to the dewpoint, water vapor in the air begins to condense and the relative humidity is 100%. Therefore, the relative humidity at Albany when the air temperature is equal to the dew point is 100%.

One credit is allowed for **100%**.

57. One of the most common instruments used to measure wind speed is an anemometer. A simple anemometer has cups mounted on a shaft that cause the shaft to spin when wind blows against them. The faster the wind blows, the faster the cups spin the shaft and the higher the number that registers on the anemometer's scale. Small handheld anemometers, called wind gauges, contain small fan blades that spin in the wind. There are also a variety of other instruments that can be used to measure wind speed. A wind sock is a cone-shaped tube made of fabric that resembles a giant sock but is open at both ends. As wind speed increases, the sock rises like a flag in a breeze. The higher the wind speed, the closer the sock flies to horizontal. Some handheld digital wind speed meters contain a metal wire that is heated by an electric current. The wire is cooled by air flow, and the meter instantly calculates wind speed. Doppler radar can also be used to determine wind speed. For Doppler radar, atmospheric objects moving inbound (toward the radar) produce a positive shift in frequency of the radar signal. Objects moving away from the radar (outbound) produce a negative shift in frequency. This change in frequency can then be analyzed to determine wind speed.

One credit is allowed. Acceptable responses include but are not limited to:

- **anemometer**
- **wind sock**
- **wind speed meter**
- **wind gauge**
- **Doppler radar/radar**

58.

Orbit of Comet Tempel-Tuttle

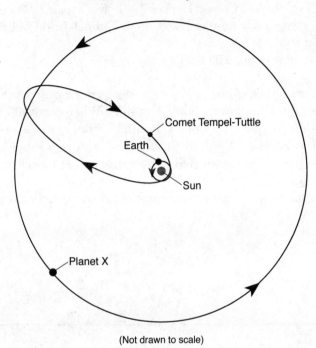

(Not drawn to scale)

A comet in an elliptical orbit is orbiting the foci of that ellipse. The reading passage states that Comet Tempel-Tuttle orbits our Sun. Therefore, our Sun is located at one of the foci of the elliptical orbit of Comet Tempel-Tuttle.

One credit is allowed for the **Sun**.

59. The diagram shows that the orbit of planet X is slightly less than the farthest distance of Comet Tempel-Tuttle from the Sun. The reading passage states that Comet Tempel-Tuttle's orbital distance from the Sun is "2900 million kilometers at its farthest distance." Find the Solar System Data chart in the *Reference Tables for Physical Setting/Earth Science*. In the column labeled "Mean Distance from Sun (million km)," locate the value closest to but less than 2900 million kilometers, which is 2871.0. From this value, trace left to the column labeled "Celestial Object" Note that Uranus is located 2871.0 million kilometers from the Sun. Therefore, the solar system planet represented by planet X is Uranus.

One credit is allowed for **Uranus**.

60. The reading passage states that Comet Tempel-Tuttle's "two most recent closest approaches to the Sun occurred in 1965 and one revolution later in 1998." Thus, the period of revolution of Comet Tempel-Tuttle is 33 years (1998 − 1965 = 33). Therefore, the next closest approach of Comet Tempel-Tuttle will occur 33 years after 1998, or in 2031.

One credit is allowed for **2031.**

61. All matter exerts a mutual gravitational attraction on all other matter. Thus, matter has a tendency to draw together over time. When particles of matter, such as the debris left in space by Comet Tempel-Tuttle, come near Earth, those particles are drawn toward Earth by this force of mutual gravitational attraction. Therefore, the force that causes debris from the comet to fall through Earth's atmosphere is gravity.

One credit is allowed for **gravity *or* gravitation *or* gravitational pull.**

62.

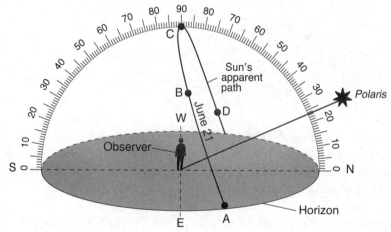

At sunrise at position A, the Sun would be low in the sky to an observer. The low angle of the Sun would cause a long shadow to be cast by the observer. At position B, the Sun would be higher in the sky and the shadow cast by the observer would be shorter. At position C, the Sun would be at an even higher in the sky and the shadow cast by the observer would be even shorter than at position B. At position D, the Sun would again be low in the sky and the observer would cast a long shadow. Thus, from position A to position D, the length of the observer's shadow will decrease from position A to position C and then increase from position C to position D.

One credit is allowed for an acceptable response. Acceptable responses include but are not limited to:

- **The length of the observer's shadow will decrease from *A* to *C*, then increase from *C* to *D*.**
- **The shadow will get shorter, then longer.**
- **The shadow was longest at *A*, became shortest at *C*, then became longer at *D*.**
- **decreases then increases**

63. The diagram shows that the altitude of *Polaris* at the observer's position is 23.5°. The altitude of *Polaris* is the same as an observer's latitude north of the equator. Therefore, the observer is located at 23.5° N latitude. Find the Surface Ocean Currents map in the *Reference Tables for Physical Setting/Earth Science*. Along the right edge of the map, note that 23.5° N is labeled "Tropic of Cancer." Thus, one piece of evidence that supports the inference that the observer is located at the Tropic of Cancer is the altitude of *Polaris*.

From the altitude of *Polaris*, it is clear that the observer is located in the Northern Hemisphere. It is given that the diagram represents the Sun's path on June 21, which is the summer solstice in the Northern Hemisphere. The diagram shows that at its highest point (noon), the altitude of the Sun is 90°. In other words, the Sun is directly overhead the observer. On June 21, the Sun is directly overhead (altitude 90°) at the Tropic of Cancer, which is located at latitude 23.5° N.

One credit is allowed for an acceptable response. Acceptable responses include but are not limited to:

- **The altitude of *Polaris* is at 23.5° or $23\frac{1}{2}°$.**
- **The noon Sun for June 21 is directly overhead.**
- **The highest Sun position is at 90° (zenith) on June 21.**
- **the angle/altitude of *Polaris***

Note: Credit is *not* allowed for $23\frac{1}{2}°$ N or for 23.5° North, because a compass direction should not be included with an altitude value.

64. September 23 is the fall equinox. When Earth is at the equinox position, day and night are equal in length. Thus, an observer would experience 12 hours of daylight and 12 hours of darkness. Therefore, on September 23, the number of daylight hours at this location will be 12 hours.

One credit is allowed for **12 h.**

65. In 1851, French physicist Jean Foucault suspended a heavy iron ball on a long steel wire from the top of the dome of the Pantheon in Paris. As this pendulum swung back and forth, it appeared to change direction slowly, passing over different lines on the floor until eventually coming full circle to its original position after 24 hours. Since Foucault knew that the path of a freely swinging pendulum would not change on its own, he concluded that the apparent shift in the direction of swing of the pendulum was due to the floor (Earth's surface) rotating beneath the pendulum. Thus, the apparent change in direction of swing of a Foucault pendulum provides evidence that Earth is rotating. Therefore, the device that was used to first demonstrate that Earth rotates was the Foucault pendulum.

One credit is allowed for **Foucault pendulum *or* pendulum.**

PART C

66.

Seeds, Michael and Backman, Dana. 2011. *The Solar System.*

Find the Solar System Data chart in the *Reference Tables for Physical Setting/Earth Science*. In the column labeled "Celestial Object," locate Earth and Jupiter. Trace these rows right to the column labeled "Equatorial Diameter (km)." Note that Earth has an equatorial diameter of 12,756 km and

Jupiter has an equatorial diameter of 142,984 km. Therefore, compared to Earth's equatorial diameter, Jupiter's equatorial diameter is 11.2 times larger (142,984 ÷ 12,756 = 11.2).

One credit is allowed for any value from **11 to 12 times larger.**

67. Planets are visible from Earth by the light they reflect from the Sun. Planets vary greatly in apparent brightness, or how bright they appear in the sky. Planets may appear bright because they reflect more light from the Sun (greater albedo) or because they are closer to Earth. Larger planets reflect more light than do smaller planets. Find the Solar System Data chart in the *Reference Tables for Physical Setting/Earth Science*.

In the column labeled "Celestial Object," locate Mercury and Jupiter. Trace these rows right to the column labeled "Equatorial Diameter (km)." Note that Mercury has an equatorial diameter of 4,879 km and Jupiter has an equatorial diameter of 142,984 km. Therefore, the reason why Jupiter appears brighter in the night sky than Mercury, despite Jupiter's greater distance from Earth, is that Jupiter is so much larger than Mercury and reflects more sunlight than Mercury.

One credit is allowed for an acceptable response. Acceptable responses include but are not limited to:

- **Jupiter is much larger than Mercury.**
- **Larger objects appear brighter than smaller objects.**
- **Jupiter reflects more light/has a greater albedo.**

68. The planets can be divided by mass and density into the terrestrial (Earthlike) and the Jovian (Jupiter-like) planets. The terrestrial planets are small, dense, and rocky. They are composed mostly of metal silicates and iron. The terrestrial planets include the four innermost planets: Mercury, Venus, Earth, and Mars. The Jovian planets are large, low in density, and gaseous. They are composed mainly of hydrogen and helium. The Jovian planets include the outer planets: Jupiter, Saturn, Uranus, and Neptune. The models show three of the terrestrial planets: Mercury, Earth, and Venus.

One credit is allowed for *two* correct terrestrial planets and a correct explanation. Acceptable responses include but are not limited to:

Terrestrial planets:

- **Mercury**
- **Venus**
- **Earth**

Explanations:

- **Terrestrial planets are small and rocky.**
- **Their interior layers are not gaseous.**
- **Terrestrial planet densities are greater than Jovian planet densities.**
- **They are closer to the Sun.**
- **They have solid crusts.**

Note: Do *not* accept Mars as the terrestrial planet because it is *not* shown in the models.

69.

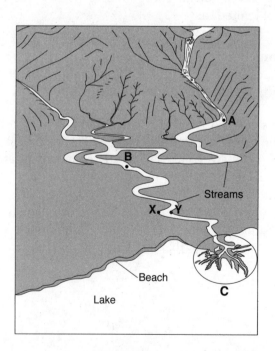

The diagram shows a large, fan-shaped deposit where a stream flows into the lake. When a stream flows into a standing body of water, the stream slows down and most of the stream's sediment is deposited. The resulting landform is a large, flat, fan-shaped pile of sediment at the mouth of the stream called a delta. Thus, depositional feature labeled C is a delta.

One credit is allowed for **delta.**

70. Note that in the diagram, points X and Y cut across a section of the stream where the stream channel curves in a meander. When a stream curves, or meanders, water velocity is higher along the outside of the curve and lower along the inside of the curve because the water is forced to cover a greater distance along the outside of the curve. Since stream erosion is greatest where the velocity of the water is highest, the outside of the curve of a meandering channel experiences more erosion than the inside of the curve. In the diagram, point X is located at the outside of a curve in the stream, which is where the velocity of the water and erosion are the greatest. Point Y is located at the inside of a curve in the stream, which is where the velocity of the water and erosion are the least. Therefore, the streambed will be more deeply eroded (the stream channel will be deeper) near point X than it is near point Y. Draw a line from X to Y on the cross section in your answer booklet showing that the stream channel is deepest near point X.

One credit is allowed if **the student's line is drawn from point X to point Y and shows that the stream channel is deepest near side X.**

Example of a 1-credit response:

71. As sediments are carried along by the moving water in a stream, they bounce off of and rub against one another, a process called abrasion. When these sediments collide, smaller pieces break off sharp corners and edges on their surface, causing the sediments to become smaller and more rounded. The sediments are also abraded as they bounce and roll along the bottom of the streambed. Therefore, as sediments are transported downstream, their size will become smaller and their shape more rounded.

One credit is allowed for *two* correct responses. Acceptable responses include but are not limited to:

Size:

- **becomes smaller**
- **decrease**

Shape:

- **becomes more rounded**
- **less angular**
- **Sharp edges are worn down.**

Note: The term "smooth" is *not* accepted for the shape of the sediments, because smooth does not describe a shape but a texture.

72. Find the Relationship of Transported Particle Size to Water Velocity graph in the *Reference Tables for Physical Setting/Earth Science*. Along the horizontal axis "Stream Velocity (cm/s)," locate a stream velocity of 100 cm/s. Trace upward to the bold curve and then horizontally to the axis labeled "Particle Diameter (cm)." Note that water traveling at 100 cm/s can move particles with diameters of up to 2.5 cm, which are in the size range labeled "Pebbles." At that velocity, the stream can also carry all particles of smaller diameter. Next, locate a stream velocity of 10 cm/s along the horizontal axis "Stream Velocity (cm/s)," and trace upward to the bold curve. Note that water traveling at 10 cm/s can move particles with diameters of up to about 0.18 cm, which are also in the size range labeled "Pebbles." Therefore, when the stream's velocity decreases from 100 to 10 centimeters per second, particles with diameters greater than 0.2 cm will be deposited. Thus, clay, silt, sand, and smaller pebbles will stay in transport; some pebbles (those between 0.18 and 2.5 cm in diameter) are deposited.

One credit is allowed for **any value from 0.18 cm to 2.5 cm.**

73.

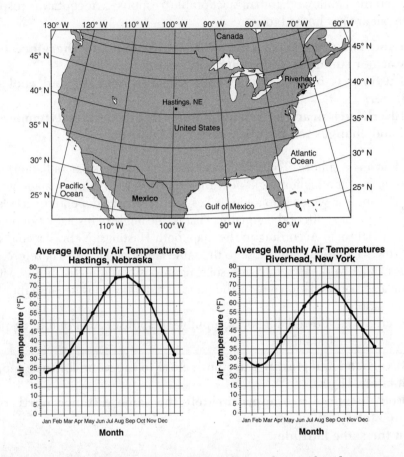

According to the map, Hastings, Nebraska, and Riverhead, New York, are located at nearly the same latitude. Therefore, both cities receive roughly the same intensity of insolation. However, land surfaces increase in temperature more than water surfaces when insolation strikes them. Land surfaces also cool off more rapidly than water surfaces. As a result, air temperatures are warmer in the summer and cooler in the winter over landmasses than they are over oceans at the same latitude. Therefore, large bodies of water tend to moderate the temperatures of nearby landmasses by warming them in winter and cooling them in summer. Thus, Hastings, Nebraska, most likely has a greater difference between maximum and minimum temperatures than Riverhead, New York, because Riverhead is located near a large body of water that moderates its temperatures. Therefore, Hastings has a greater annual temperature range than Riverhead because Hastings is located inland where land surfaces heat up and cool down faster. In contrast, Riverhead is near a large body of water that moderates its temperatures.

One credit is allowed for an acceptable response. Acceptable responses include but are not limited to:

- **Land has a lower specific heat than water and therefore causes warmer summers and cooler winters for Hastings.**
- **Hastings is inland, where land surfaces heat up and cool down faster.**
- **Riverhead is near a large body of water that moderates its temperature.**
- **Land changes temperature faster than water.**

74. Earth is a sphere. Therefore, insolation (incoming solar radiation) strikes Earth's surface at a higher angle near the equator and at a lower angle near the poles. The angle at which insolation strikes Earth's surface depends on a location's distance north or south of the equator. That is, the angle of insolation depends on a location's latitude. According to the map, both Hastings, Nebraska, and Riverhead, New York, are located close to the same latitude (41° N). Therefore, when Hastings and Riverhead rotate into solar noon position directly in line with the Sun, the angle of insolation will be the same in both locations because they are at the same latitude.

One credit is allowed for an acceptable response. Acceptable responses include but are not limited to:

- **Riverhead and Hastings are both close to the same latitude (approximately 41° N).**
- **Both locations are approximately the same distance north of the equator.**
- **at the same latitude**

75. Find the Planetary Wind and Moisture Belts in the Troposphere chart in the *Reference Tables for Physical Setting/Earth Science*. Note that the prevailing wind belt in the region between 30° N and 60° N is labeled "S.W. Winds." These winds are commonly called the prevailing southwesterlies or westerlies. Hastings, Nebraska, and Riverhead, New York, are both located close to the same latitude—41° N. Therefore, Hastings and Riverhead are located in this planetary wind belt. Thus, the planetary wind belt that primarily influences the climates of both Hastings and Riverhead are the S.W. winds.

One credit is allowed for an acceptable response. Acceptable responses include but are not limited to:

- **prevailing southwesterlies *or* westerlies**
- **S.W. Winds**

76. The map shows that Riverhead, New York, is located along the east coast of North America at about 41° N latitude. Find the Surface Ocean Currents map in the *Reference Tables for Physical Setting/Earth Science*. Locate the position corresponding to 41° N latitude along the east coast of North America. Note that the surface ocean current nearest this location is labeled the "Gulf Stream C." Therefore, the ocean current that most likely has the greatest effect on the climate of Riverhead is the Gulf Stream Current.

One credit is allowed for **Gulf Stream Current.**

77.

Adapted from www.brocku.ca/earthsciences

All sedimentary rock forms from sediments: either rock fragments (clastic sediments) or mineral crystals (precipitates or evaporites). Clastic sediments are formed when rock is uplifted, weathered, eroded, and then deposited. Sediments that are precipitates form when chemical reactions occur in water, forming insoluble substances that crystallize and then sink to the bottom. Sediments that are evaporites form when seawater evaporates. Removing water from a solution (dewatering) but not removing the dissolved minerals causes the solution to become more concentrated. When the solution of dissolved minerals in the remaining water becomes saturated, dissolved minerals start to crystallize out of the solution, and those crystals sink to the bottom.

Find the Rock Cycle in Earth's Crust diagram in the *Reference Tables for Physical Setting/Earth Science*. Locate the oval labeled "Sediments." Follow the arrow leading to the box labeled "Sedimentary Rock." Note the processes listed along the arrow: deposition, burial, compaction, and cementation. These processes are collectively called lithification, because they convert sediment to rock.

One credit is allowed for *two* correct responses. Acceptable responses include but are not limited to:

- **compaction**
- **cementation**
- **lithification**
- **deposition/sedimentation**
- **burial**
- **dewatering**
- **evaporation of water**
- **precipitation from seawater**
- **weathering**
- **erosion**
- **uplift**

78. According to the key, rock unit A is igneous rock. Note that rock unit A cuts across rock unit B and that the boundary between rock units A and B is marked with a symbol representing contact metamorphism. An igneous intrusion is younger than any rock layer or structure that it cuts across. In order to be contact metamorphosed, a rock layer must have already existed when the igneous intrusion occurred. Also note that rock unit A extends above rock unit B. If rock unit B formed after rock unit A, rock unit B would cover rock unit A. Therefore, rock unit A is younger than the rock unit B because it cuts across rock unit B, caused rock unit B to undergo contact metamorphism, and extends above rock unit B.

One credit is allowed for an acceptable response. Acceptable responses include but are not limited to:

- **Rock unit A cuts across (cross-cutting) rock unit B.**
- **There is contact metamorphism in rock unit B.**
- **Rock unit A intrudes into rock unit B.**
- **Parts of A extend above B.**

79. According to the key, both rock unit A and rock unit C are igneous rock. Note that rock unit C is an extension of rock unit A and has flowed out onto Earth's surface. Also note the broad, cone-shaped mound where rock unit A emerges at the surface—a volcano. Thus, rock unit C is an extrusive igneous rock that flowed out of a volcano as lava and then cooled and solidified at Earth's surface. When molten rock emerges onto Earth's surface, it is suddenly exposed to an environment that is hundreds or even thousands of degrees cooler than its temperature, so the molten rock cools rapidly. During rapid cooling, mineral

crystals may not have any time to form, in which case the solid formed is a noncrystalline glassy material. Alternatively, the mineral crystals may not able to grow very large before the rock solidifies.

Find the Scheme for Igneous Rock Identification in the *Reference Tables for Physical Setting/Earth Science*. In the chart at the top of the scheme, locate the section labeled "Extrusive (Volcanic)." Trace right to the column labeled "Crystal Size." Note that extrusive rocks are either noncrystalline or have crystals that are less than 1 mm in size. Thus, the igneous rock that formed at location C is composed of crystals less than 1 mm in size because the lava cooled quickly at Earth's surface.

One credit is allowed for an acceptable response. Acceptable responses include but are not limited to:

- **The rock at C has a fine texture because the lava cooled quickly.**
- **Rock C crystallized/solidified on Earth's surface.**
- **Extrusive igneous rock doesn't have time to grow large crystals.**
- **The environment of formation is extrusive/volcanic.**

80. Find the Scheme for Sedimentary Rock Identification in the *Reference Tables for Physical Setting/Earth Science*. In the column labeled "Map Symbol," locate the symbols corresponding to those of the sedimentary rocks in the block diagram at locations X and Y. Trace left to the column labeled "Rock Name." Note that the sedimentary rock layer at location X corresponds to limestone and the sedimentary rock layer at Y corresponds to sandstone.

Find the Scheme for Metamorphic Rock Identification in the *Reference Tables for Physical Setting/Earth Science*. In the column labeled "Type of Metamorphism," locate the row labeled "Contact (heat)." Trace right to the column labeled "Comments," and note the statement "Various rocks changed by heat from nearby magma/lava." Now trace right to the column labeled "Rock Name," and note that hornfels is a metamorphic rock that forms from various rocks by contact metamorphism. Thus, hornfels may form at both locations X and Y. In the column labeled "Type of Metamorphism," locate the row labeled "Regional or Contact." Trace right to the column labeled "Comments," and note the statements "Metamorphism of quartz sandstone" and "Metamorphism of limestone or dolostone." From these rows, trace right to the column labeled "Rock Name." Note that metamorphism of sandstone (location Y) forms quartzite and metamorphism of limestone or dolostone (location X) forms marble. Thus, metamorphic rocks that are most likely found in the zone of contact metamorphism at location X include marble and hornfels and at location Y include quartzite and hornfels.

One credit is allowed if *both* responses are correct. Acceptable responses include:

Location X:

- **marble**
- **hornfels**

Location Y:

- **quartzite**
- **hornfels**

Note: Credit is *not* allowed if hornfels is used for *both* locations X and Y.

81.

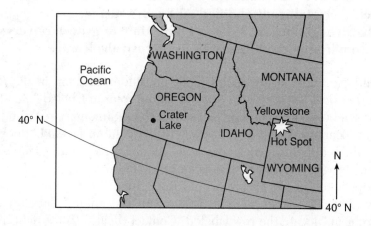

An isoline connects points of equal field value, and the field values shown on this map are depth. The 500-meter depth isoline connects all points that are located 500 meters below the lake surface. To construct the 500-meter depth isoline, connect all points that are at a depth of 500 meters with a smooth line. Several points on the map show a value of "500," however, these are not the only points on the map with that value. Begin with the point marked "500" to the lower left of point X. To the right of this point there is a point labeled "550" and a depth isoline labeled "400." It is logical to infer that between the depths "550" and "400," there is a point with the depth 500. Therefore, draw the isoline to curve south to pass between the 400-m isoline and the point labeled "550." Then curve north to reach the point labeled "500" to the lower right of point X. Now trace the isoline to the north between the points labeled "490" and "550." Then curve the isoline sharply to the south around the outside of the "550" and roughly parallel the 400-meter isoline. At the next point labeled "550" and extend the isoline between the "550" and the nearby "450." Then extend the isoline south between the "450" and the "570." Continue extending the line to the south between the points labeled "460" and "560" to the next point labeled "500." From there, extend the isoline west between the points labeled "450" and "560." Then turn north to pass between the points labeled "450" and "565." Then curve to the northwest between the point labeled "540" and the 400-meter isoline to the next point labeled "500." From here, extend the isoline west between the points labeled "460" and "530" to the next point labeled "500." From there, extend the line east between the points labeled "450" and "540" to connect with the point labeled "500" at which you began. A correctly drawn 500-meter depth isoline is shown below.

One credit is allowed if **the 500-meter depth isoline is correctly drawn. If additional isolines are drawn, all isolines must be correct to receive credit.**

Example of a 1-credit response:

Note: The isoline must touch *all five* points for the 500-meter depth.

82. To construct a profile of the depth of Crater Lake along line *AB*, proceed as follows. Place the straight edge of a piece of scrap paper along the solid line connecting point *A* to point *B*. Mark the edge of the paper at points *A* and *B* and

wherever the paper intersects a depth isoline. Wherever the paper intersects a depth isoline, label the mark with the depth of the isoline as shown below.

Now place this paper along the lower edge of the grid provided in your answer booklet so that points *A* and *B* on the paper align with points *A* and *B* on the grid. Then at each point where a depth isoline crosses the edge of the paper, draw a dot on the grid at the appropriate depth. Finally, connect all of the dots in a smooth curve to form the finished profile as shown below. Note that along the profile, the bottom of the lake lies between the 400-foot isoline and the 500-foot isoline. Therefore, the low point of the profile line must extend below 400 feet but not below 500 feet.

One credit is allowed if *all five* plots are located within or touch the rectangles shown and are correctly connected with a line that passes within or touches each rectangle.

83. Find the Equations chart of the *Reference Tables for Physical Setting/Earth Science*, and note the equation for gradient:

$$\text{Gradient} = \frac{\text{change in field value}}{\text{distance}}$$

The map shown is a lake depth map. The field value is depth. Note that point C is located directly on the 0-meter depth isoline and that point D is located directly on the 400-meter depth isoline. Thus, the change in field value (depth) from C to D is 400 meters.

On a piece of scrap paper, mark off the distance along the straight line between points C and D. Compare the distance marked off on the scrap paper to the scale printed beneath the map to determine the distance between points C and D. The distance is about 1.5 kilometers. Substitute these values in the equation and solve:

$$\text{Gradient} = \frac{400 \text{ m}}{1.5 \text{ km}} = 266.6 \text{ m/km}$$

Thus, the gradient between location A and location B is approximately 267 m/km.

One credit is allowed for **any value from 250 to 286 m/km *or* –250 to –286 m/km.**

84. Note that the interval between isolines on the map is 100 meters. The map shows that X is located within the isoline labeled "200." The concentric isolines surrounding the 200-meter isoline decrease in value as you approach location X. Therefore, the depth within the 200-meter isoline is less than 200 meters. Since no additional isoline exists within the 200-meter isoline and the interval between isolines is 100 meters, the depth of location X is less than 200 meters but more than 100 meters.

One credit is allowed for **any value greater than 100 meters but less than 200 meters**.

85.

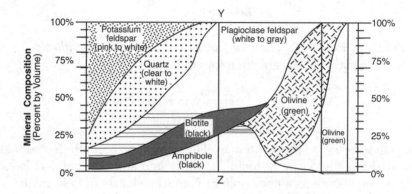

According to the diagram, line YZ extends through the regions of three minerals on the chart: amphibole, biotite, and plagioclase feldspar. The total distance from Z to Y on the line corresponds to 100% on the scale along the left edge of the chart labeled "Mineral Composition (Percentage by Volume)." The distance along line YZ that extends through each of the three mineral regions corresponds to the percentage by volume of each of those minerals in the rock represented by line YZ. To determine the percentage of each of these three minerals in the rock, first mark off the distance along line YZ that extends through each region. Then compare each distance to the scale labeled "Mineral Composition (Percent by Volume)" along the left side of the diagram. For example using a piece of scrap paper, mark off the distance that line YZ passes through the region labeled "Amphibole (black)." Now place the scrap paper so that one of these marks aligns with 0% on the scale labeled "Mineral Composition (Percentage by Volume)." Read the value corresponding to the other mark. Note that percentage of line YZ that passes through the amphibole region corresponds to about 27% on the percentage by volume scale. Thus, the rock represented by line YZ is 27% amphibole by volume. Repeat this procedure for each of the other two mineral regions through which line YZ passes on the diagram. Note that the rock represented by line YZ also contains about 16% biotite and 57% plagioclase feldspar by volume.

One credit is allowed if **all three** percentages are within the ranges shown below.

Plagioclase feldspar: **any value from 54% to 60%**
Biotite: **any value from 13% to 19%**
Amphibole: **any value from 24% to 30%**

SELF-ANALYSIS CHART June 2019

Topic	Question Numbers (Total)	Wrong Answers (x)	Grade
Standards 1, 2, 6, and 7: Skills and Application			
Skills			
Standard 1 Analysis, Inquiry, and Design	4, 5, 7–11, 13, 14, 19, 21, 25–27, 30, 32–37, 40, 43, 45–47, 49–53, 55–68, 71–77, 79–81, 83, 85		$\frac{100(57-x)}{57} = \%$
Standard 2 Information Systems			
Standard 6 Interconnectedness, Common Themes	1–4, 7, 12, 14–16, 19, 21–23, 27–29, 32, 36–46, 50–52, 54–56, 59, 62–64, 66–71, 73–85		$\frac{100(57-x)}{57} = \%$
Standard 7 Interdisciplinary Problem Solving			
Standard 4: The Physical Setting/Earth Science			
Astronomy			
The Solar System (MU 1.1a, b; 1.2d)	3, 38–40, 58–61, 64		$\frac{100(9-x)}{9} = \%$
Earth Motions and Their Effects (MU 1.1c, d, e, f, g, h, i)	1, 6, 9, 36, 37, 41, 62, 63, 65		$\frac{100(9-x)}{9} = \%$
Stellar Astronomy (MU 1.2b)			
Origin of Earth's Atmosphere, Hydrosphere, and Lithosphere (MU 1.2e, f, h)			
Theories of the Origin of the Universe and Solar System (MU 1.2a,c)	2, 5, 66–68		$\frac{100(5-x)}{5} = \%$

SELF-ANALYSIS CHART June 2019

Topic	Question Numbers (Total)	Wrong Answers (x)	Grade
Meteorology			
Energy Sources for Earth Systems (MU 2.1a, b)	8		$\frac{100(1-x)}{1} = \%$
Weather (MU 2.1c, d, e, f, g, h)	10, 14, 42–44, 54–57		$\frac{100(9-x)}{9} = \%$
Insolation and Seasonal Changes (MU 2.1i; 2.2a, b)	17, 74		$\frac{100(2-x)}{2} = \%$
The Water Cycle and Climates (MU 1.2g; 2.2c, d)	11, 12, 16, 18, 20, 73, 75, 76		$\frac{100(8-x)}{8} = \%$
Geology			
Minerals and Rocks (MU 3.1a, b, c)	30, 33–35, 50, 77, 79, 80, 85		$\frac{100(9-x)}{9} = \%$
Weathering, Erosion, and Deposition (MU 2.1s, t, u, v, w)	28, 29, 31, 49, 69, 70–72		$\frac{100(8-x)}{8} = \%$
Plate Tectonics and Earth's Interior (MU 2.1j, k, l, m, n, o)	7, 15, 19, 25–27, 45, 46, 48		$\frac{100(9-x)}{9} = \%$
Geologic History (MU 1.2i, j)	4, 13, 21, 22, 32, 47, 51–53, 78		$\frac{100(10-x)}{10} = \%$
Topographic Maps and Landscapes (MU 2.1p, q, r)	23, 24, 81–84		$\frac{100(6-x)}{6} = \%$
ESRT *2011 Edition Reference Tables for Physical Setting/Earth Science*	4, 5, 7, 8, 10, 11, 13, 14, 19, 21, 25–27, 30, 33–35, 40, 45–47, 50, 52, 53, 55, 56, 59, 63, 66–68, 72, 75–77, 79, 80, 83, 85		$\frac{100(39-x)}{39} = \%$

To further pinpoint your weak areas, use the Topic Outline in the front of the book.
MU = Major Understanding (see Topic Outline)

Examination August 2019
Physical Setting/Earth Science

PART A
Answer all questions in this part.

Directions (1–35): For *each* statement or question, choose the word or expression that, of those given, best completes the statement or answers the question. Some questions may require the use of the *2011 Edition Reference Tables for Physical Setting/Earth Science*. Record your answers in the space provided.

1 Most asteroids in our solar system are located in the asteroid belt between the orbits of Mars and Jupiter. What is the approximate distance from the Sun to the asteroid belt?

(1) 129 million km
(2) 210 million km
(3) 403 million km
(4) 1103 million km

2 The table below shows tide data for a location on the north shore of Long Island, in New York State.

Day	Tide	Time
Tuesday	High Tide	12:11 a.m.
	Low Tide	6:23 a.m.
	High Tide	12:36 p.m.
	Low Tide	6:49 p.m.
Wednesday	High Tide	1:02 a.m.
	Low Tide	7:15 a.m.
	High Tide	1:27 p.m.

Based on these data, what is the most likely time of the next high tide?

(1) 1:53 a.m.
(2) 1:53 p.m.
(3) 7:40 a.m.
(4) 7:40 p.m.

3 Which planet takes longer to complete one rotation on its axis than it does to complete one orbit of the Sun?

   (1) Mercury
   (2) Venus
   (3) Earth
   (4) Mars

   3 _____

4 Jupiter is a Jovian planet. Compared to the terrestrial planets, Jupiter has a

   (1) shorter period of revolution
   (2) shorter distance to the Sun
   (3) greater density
   (4) greater mass

   4 _____

5 The entire constellation of Orion is visible in the night sky in January to an observer in New York State. Which statement explains why this constellation is *not* visible in the night sky to this observer in June?

   (1) Earth rotates on its axis.
   (2) Earth revolves around the Sun.
   (3) The constellation Orion orbits the Sun.
   (4) The tilt of Earth's axis changes throughout the year.

   5 _____

6 The approximate age of the universe is estimated to be

   (1) 4.6 million years
   (2) 4.6 billion years
   (3) 13.8 million years
   (4) 13.8 billion years

   6 _____

7 The photograph below shows the apparent path of the Sun photographed at 20-minute intervals and combined into one photograph.

Baker, Robert H. *An Introduction to Astronomy*

Which motion is responsible for the apparent path of the Sun shown in the photograph?

(1) Earth's rotation
(2) Earth's revolution
(3) Sun's rotation
(4) Sun's revolution

7 ____

8 The mass extinction of the dinosaurs, approximately 65.5 million years ago, is inferred by most scientists to have been caused by

(1) a large energy surge from the surface of the Sun
(2) the occurrence of a major ice age
(3) an impact event occurring on Earth's surface
(4) earthquakes occurring along crustal plate boundaries

8 ____

9 The diagram below represents Earth as viewed from space. Letter A represents the angle of tilt of Earth's rotational axis. Line XY is perpendicular to the plane of Earth's orbit.

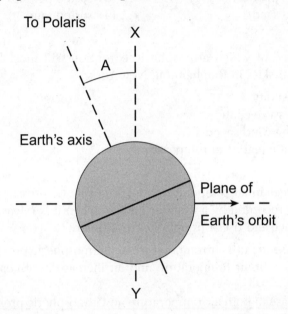

What is the value of the angle represented by letter A?
(1) 15.0°
(2) 23.5°
(3) 66.5°
(4) 90.0°

10 In the spring, when snow cover on the land melts, the water will most likely infiltrate Earth's surface where that land surface is

(1) still frozen with a steep slope
(2) still frozen with a gentle slope
(3) no longer frozen with a steep slope
(4) no longer frozen with a gentle slope

11 Water is returned from Earth's surface to the atmosphere by

(1) condensation and transpiration
(2) condensation and precipitation
(3) evaporation and transpiration
(4) evaporation and precipitation

12 Which type of sediment sample normally has the greatest permeability rate?
  (1) unsorted pebbles
  (2) unsorted sand
  (3) sorted pebbles
  (4) sorted sand   12 _____

13 The presence of which atmospheric condition will most likely result in clear skies in Binghamton, New York?
  (1) high humidity
  (2) high-pressure center
  (3) increasing wind speed
  (4) dewpoint equal to air temperature   13 _____

14 Which changes in air temperature and atmospheric pressure will normally be recorded by a weather balloon when it is released at Earth's surface and rises through the troposphere?
  (1) a decrease in both air temperature and atmospheric pressure
  (2) a decrease in air temperature and an increase in atmospheric pressure
  (3) an increase in both air temperature and atmospheric pressure
  (4) an increase in air temperature and a decrease in atmospheric pressure   14 _____

15 Which gas is most effective in absorbing incoming harmful ultraviolet radiation in Earth's stratosphere before that radiation reaches Earth's surface?
  (1) nitrogen
  (2) hydrogen
  (3) oxygen
  (4) ozone   15 _____

16 The major cause of monsoon rains in India and southeast Asia is seasonal
  (1) shifts in the prevailing wind belts
  (2) shifts in ocean currents
  (3) changes in the energy radiated from the Sun
  (4) changes in worldwide atmospheric temperatures   16 _____

17 The cross section below represents the windward and leeward sides of a mountain. The arrows show the direction of air movement over the mountain. Points X and Y represent locations on Earth's surface.

Compared to the temperature and water vapor content of the air at location X, the temperature and water vapor content at location Y are most likely

(1) warmer and wetter
(2) warmer and drier
(3) cooler and wetter
(4) cooler and drier   17 ___

18 Which surface ocean current transports cool water to lower latitudes?

(1) Gulf Stream
(2) Peru Current
(3) West Greenland Current
(4) East Australia Current   18 ___

19 Which land surface characteristics produce the greatest amount of absorption of insolation at Earth's surface?

(1) dark color and rough texture
(2) dark color and smooth texture
(3) light color and rough texture
(4) light color and smooth texture   19 ___

20 The planetary surface winds and air currents near Earth's equator are usually

(1) converging and sinking
(2) diverging and sinking
(3) converging and rising
(4) diverging and rising   20 _____

21 Oxygen in Earth's early atmosphere was first produced during the Precambrian from

(1) cyanobacteria in Earth's oceans
(2) volcanic activity along plate boundaries
(3) the absorption of sunlight by plants
(4) evaporation of ocean water   21 _____

22 Which life-form existed on Earth for the *shortest* period of time?

(1) dinosaurs   (3) ammonoids
(2) trilobites   (4) placoderm fish   22 _____

23 During the late Silurian epoch, very salty seas extended from New York State to Michigan. These environmental conditions resulted in the formation of halite layers. At the same time, *Eurypterus remipes* lived in nearby environments. Both the halite layers and *Eurypterus remipes* fossils can now be used to identify a specific geologic time interval because both formed over a

(1) large geographic area and in a short geologic time
(2) large geographic area and in a long geologic time
(3) small geographic area and in a short geologic time
(4) small geographic area and in a long geologic time   23 _____

24 If it takes a *P*-wave five minutes to travel from the epicenter of an earthquake to a seismic station, approximately how long will it take an *S*-wave to travel that same distance?

(1) 15 minutes   (3) 9 minutes
(2) 12 minutes   (4) 4 minutes   24 _____

25 Earthquake S-waves do *not* pass through which two interior Earth layers?

(1) rigid mantle and asthenosphere
(2) asthenosphere and stiffer mantle
(3) stiffer mantle and outer core
(4) outer core and inner core   25 _____

26 The cross section below represents the bedrock structure of a section of Earth's crust. Letters A through H represent rock units. Line XY represents a fault. The rock layers have *not* been overturned.

When did faulting along line XY occur?

(1) after the intrusion of rock unit A
(2) after the deposition of rock units B, C, and D
(3) after the formation of rock units E and F
(4) before the formation of rock units G and H   26 _____

27 The cross section below represents rock units within Earth's crust.

The age of the granite that makes up the inclusions is most likely

(1) older than the sandstone, but the same age as the granite bedrock
(2) older than the sandstone and the granite bedrock
(3) younger than the sandstone, but the same age as the granite bedrock
(4) younger than the sandstone and the granite bedrock

28 A portion of the Generalized Landscape Regions of New York State map below shows the location of the Catskill Aqueduct that flows into the Kensico Reservoir, which supplies the residents of New York City with drinkable water.

Which sequence shows the order of landscape regions that are crossed as water flows through the Catskill Aqueduct?

(1) Allegheny Plateau, Hudson-Mohawk Lowlands, Taconic Mountains, Newark Lowlands
(2) Allegheny Plateau, Hudson-Mohawk Lowlands, Hudson Highlands, Manhattan Prong
(3) Atlantic Coastal Plain, Newark Lowlands, Hudson Highlands, Hudson-Mohawk Lowlands
(4) Atlantic Coastal Plain, Manhattan Prong, Hudson Highlands, Allegheny Plateau

29 The map below shows the land area in New York State drained by the Oswego River and its tributaries.

The land area drained by the Oswego River and its tributaries is called a

(1) delta
(2) watershed
(3) water table
(4) floodplain

30 The photograph below shows a portion of the Genesee River in western New York. Letters A, B, C, and D are locations in the river.

At which location would deposition of sediments most likely be greater than erosion?

(1) A
(2) B
(3) C
(4) D   30 _____

31 The rate of soil development in tropical areas is usually greater than the rate of soil development in arctic areas because tropical areas have

(1) less chemical weathering and a scarcity of living organisms
(2) less chemical weathering and an abundance of living organisms
(3) more chemical weathering and a scarcity of living organisms
(4) more chemical weathering and an abundance of living organisms   31 _____

32 New York State's Finger Lakes exist today because

(1) U-shaped valleys were dammed by glacial sediments
(2) V-shaped valleys are being eroded by streams
(3) a drop in sea level occurred, leaving the lakes
(4) a rise in sea level occurred, flooding the region   32 _____

33 Which igneous rock could physically weather to beach sand that contains the minerals pyroxene, plagioclase feldspar, and olivine?

(1) dunite
(2) granite
(3) peridotite
(4) gabbro

33 ____

34 Which two physical properties of graphite make it a good mineral for use in pencils?

(1) luster and fracture
(2) cleavage and color
(3) hardness and streak
(4) greasy feel and composition

34 ____

35 Which mineral is produced when two atoms of iron chemically combine with three atoms of oxygen?

(1) garnet
(2) pyrite
(3) magnetite
(4) hematite

35 ____

PART B–1

Answer all questions in this part.

Directions (36–50): For *each* statement or question, choose the word or expression that, of those given, best completes the statement or answers the question. Some questions may require the use of the *2011 Edition Reference Tables for Physical Setting/Earth Science*. Record your answers in the space provided.

Base your answers to questions 36 through 38 on the map below and on your knowledge of Earth science. The map shows the locations of Jamestown, Watertown, and Kingston in New York State.

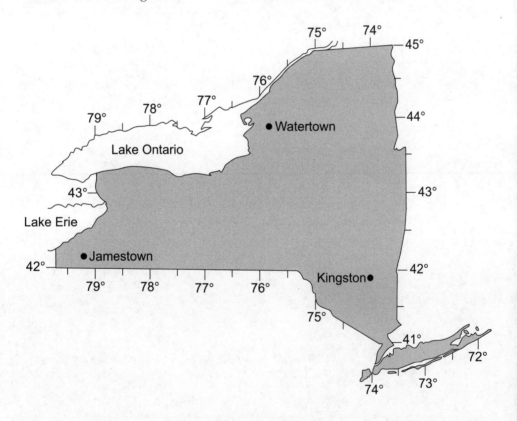

36 Compared to the time of sunrise in Kingston, the time of sunrise in Jamestown would be approximately

(1) 1 hour earlier
(2) 1 hour later
(3) 20 minutes earlier
(4) 20 minutes later

36 _____

37 An observer in Watertown, New York, would see the star *Polaris* at an altitude of approximately
 (1) 44°
 (2) 45°
 (3) 75°
 (4) 76°   37 _____

38 Earth's system of longitude is based on which motion?
 (1) Earth's rotation
 (2) the Sun's rotation
 (3) Earth's revolution around the Sun
 (4) the Sun's revolution around Earth   38 _____

Base your answers to questions 39 through 41 on the passage and photograph below and on your knowledge of Earth science. The photograph shows a portion of the Patagonia Marble Caves found in South America.

Patagonia Marble Caves

The Patagonia Marble Caves of South America are found on the shores of General Carrera Lake at a location of 46.5° S 72° W. Most of the water in the lake comes from the melting of nearby glaciers. Many small particles carried by the glacier remain suspended in the meltwater that fills this lake, causing a distinct blue color. Over the last 6200 years, the water of the lake has been weathering and eroding the marble bedrock found along the shores and within the lake itself. The marble dissolved faster at the water surface, where the moving water is interacting with the marble bedrock, producing countless caves, mazes, columns, and tunnels in the marble.

http://www.dont-complain.com/2015/01/17/marble-caves-chile/

39 The map below shows four locations in South America, labeled A, B, C, and D.

Which lettered point on the map best represents the location of the Patagonia Marble Caves?

(1) A
(2) B
(3) C
(4) D

40 Which two processes formed the marble bedrock shown in the photograph?

(1) heat and pressure
(2) compaction and cementation
(3) melting and solidification
(4) precipitation and evaporation   40 ____

41 Most marble is composed primarily of

(1) quartz and/or potassium feldspar
(2) calcite and/or dolomite
(3) halite and/or olivine
(4) pyroxene and/or plagioclase feldspar   41 ____

Base your answers to questions 42 through 44 on the photographs below and on your knowledge of Earth science. One photograph shows a digital device that recorded several weather variables. The second photograph shows two weather instruments, labeled A and B.

Adapted from: https://www.pce-instruments.com/us/measuring-instruments/

42 The barometric pressure was recorded on the digital device in millibars. What is the equivalent air pressure in inches of mercury (Hg)?

(1) 29.59 in of Hg
(2) 29.60 in of Hg
(3) 29.62 in of Hg
(4) 29.65 in of Hg

42 _____

43 Based on the outside air temperature and relative humidity shown on the digital device, what is the approximate dewpoint for the time shown on the device?

(1) 8°C
(2) 12°C
(3) 18°C
(4) 31°C

43 _____

44 Which table correctly identifies weather instruments A and B and the weather variable that each measures?

	Weather Instrument	Weather Variable Measured
A	wind vane	wind direction
B	anemometer	wind speed

(1)

	Weather Instrument	Weather Variable Measured
A	wind vane	wind speed
B	anemometer	wind direction

(3)

	Weather Instrument	Weather Variable Measured
A	anemometer	wind speed
B	wind vane	wind direction

(2)

	Weather Instrument	Weather Variable Measured
A	anemometer	wind direction
B	wind vane	wind speed

(4)

44 _____

Base your answers to questions 45 through 47 on the diagrams below and on your knowledge of Earth science. Diagram A represents a Foucault pendulum set up at a particular location on Earth. The line on the floor marked 0 is the path of the pendulum when it was first set in motion. Diagram B represents the curving of planetary winds located in two areas of Earth's surface due to the Coriolis effect.

Diagram A: Foucault's Pendulum

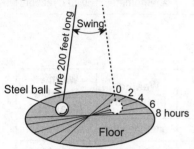

Diagram B: Coriolis Effect on Planetary Winds

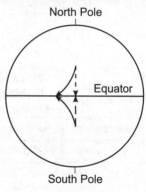

(Not drawn to scale)

Key
- - -> Original direction of wind
——> Deflected path of wind

45 The apparent shift in the direction of swing of a Foucault pendulum and the curving of winds by the Coriolis effect are both evidence of Earth's

(1) tilt (3) rotation
(2) shape (4) revolution   45 _____

46 In diagram A, the apparent shift of the path of the pendulum is shown at two-hour intervals for the first eight hours of pendulum motion. During the eight hours, the pendulum's path was displaced 60 degrees from where it started. How many degrees did the apparent path of the pendulum change each hour (h)?

(1) 7.5°/h (3) 60°/h
(2) 15°/h (4) 120°/h   46 _____

47 In diagram *B*, what are the names of the planetary wind belts formed by the deflected winds shown in the Northern and Southern hemispheres?

Hemisphere	Wind Belt
Northern	Southeast winds
Southern	Northwest winds

(1)

Hemisphere	Wind Belt
Northern	Southwest winds
Southern	Northwest winds

(3)

Hemisphere	Wind Belt
Northern	Northeast winds
Southern	Southeast winds

(2)

Hemisphere	Wind Belt
Northern	Northwest winds
Southern	Southeast winds

(4)

47 _____

Base your answers to questions 48 through 50 on the cross section below and on your knowledge of Earth science. The cross section shows the boundary between the Juan de Fuca Plate and the North American Plate.

(Not drawn to scale)

48 In which diagram do the arrows best represent the relative motion of the upper mantle at this plate boundary?

(1)

(3)

(2)

(4)

48 _____

49 Compared to the crust of the North American Plate, the crust of the Juan de Fuca Plate is
(1) thicker and less dense
(2) thicker and more dense
(3) thinner and less dense
(4) thinner and more dense

49 _____

50 The boundary between the asthenosphere and the stiffer mantle is located below the Earth's surface at a depth of approximately
(1) 100 km   (3) 2500 km
(2) 700 km   (4) 3000 km

50 _____

PART B–2

Answer all questions in this part.

Directions (51–65): Record your answers in the spaces provided. Some questions may require the use of the *2011 Edition Reference Tables for Physical Setting/Earth Science*.

Base your answers to questions 51 through 53 on the passage below, the cross section below, and on your knowledge of Earth science. The passage describes the discovery of ocean floor magnetism. The cross section represents a pattern of normal and reverse polarity of the magnetic field preserved in the igneous bedrock of the oceanic crust west of the Mid-Atlantic Ridge. The magnetic polarity pattern of the bedrock on the east side of the ridge has been left blank.

Ocean Floor Magnetism

Scientists in the early 1960s were surprised to find there was a pattern in the ocean floor magnetism preserved in the bedrock of the Atlantic Ocean floor. They found that the magnetism in the bedrock was arranged in an orderly pattern parallel to the Mid-Atlantic Ridge. This mountain ridge, often marked by earthquakes and volcanic eruptions, runs roughly north-south. Earth is currently in a period of normal polarity. However, the magnetic record preserved in the rocks indicates that Earth's magnetic poles have reversed positions many times in the past. Since the initial discovery of this ocean floor magnetism, similar magnetic patterns have also been found parallel to the mid-ocean ridges in all of the other oceans.

51 Complete the diagram *below* by shading in the pattern of normal polarity on the east side of the Mid-Atlantic Ridge center. Assume that the rate of plate movement was constant on both sides of the ridge center. Your answer must show the correct width and placement of each normal polarity section. [1]

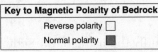

52 Identify the type of tectonic plate boundary at the mid-ocean ridges where these magnetic patterns were produced. [1]

_____ **plate boundary**

53 On the set of axes *below*, draw a line to represent the relationship between the distance from the Mid-Atlantic Ridge and the age of the ocean floor bedrock. [1]

Base your answers to questions 54 through 56 on the photograph below and on your knowledge of Earth science. The photograph is of a rock sample composed of pebbles that have been cemented together.

www.geolsoc.org.uk

54 Identify the name of the sedimentary rock. [1]

55 Identify the total possible range of particle diameters, in centimeters, for a particle to be classified as a pebble. [1]

The range is from _____ **cm** to _____ **cm**.

56 Describe the evidence from the photograph that supports the inference that the particles in this rock were most likely transported by running water. [1]

_____

_____

_____

Base your answers to questions 57 through 59 on the diagrams below and on your knowledge of Earth science. The diagrams represent a model of the disintegration of a sample of the radioactive isotope carbon-14 (^{14}C).

57 Determine the number of carbon-14 atoms that would most likely remain at the end of the second half-life. [1]

_____

58 Identify the stable disintegration product that is produced when carbon-14 decays. [1]

_____

59 Identify the radioactive isotope that has a half-life that is approximately the same as the estimated age of Earth, by using the Radioactive Decay Data table in the *Earth Science Reference Tables*. [1]

_____

_____

Base your answers to questions 60 through 62 on the diagrams below and on your knowledge of Earth science. The diagrams represent drill core samples from two different locations (I and II). A drill core is a cylinder of rock material removed from the bedrock. Letters A through J represent different rock layers. Some layers contain index fossils. The rock layers shown have *not* been overturned.

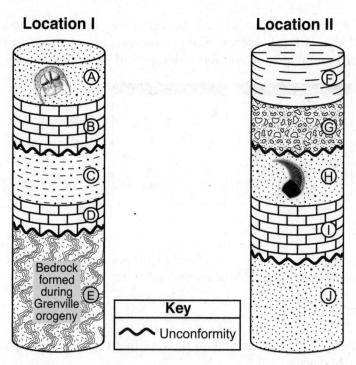

60 Identify the geologic period when rock layer A was deposited. [1]

_____ **Period**

61 List the relative ages of rock layers *D*, *E*, *F*, and *G* by listing the letters from oldest to youngest. [1]

_____ → _____ → _____ → _____

Oldest ———————————————————→ Youngest

62 Identify *two* processes that formed the unconformities in these drill cores. [1]

Process 1: _____

Process 2: _____

Base your answers to questions 63 through 65 on the graph below and on your knowledge of Earth science. The graph shows the discharge of the Delaware River at Barryville, New York, for a one-week period during March 2004.

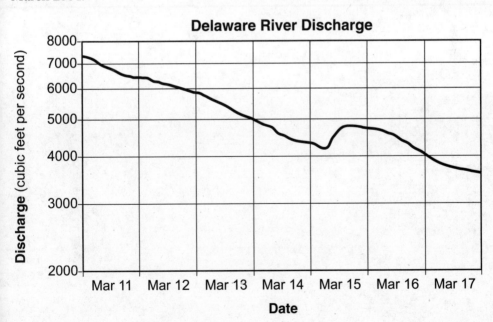

63 Identify the date shown on the graph when the river was most likely carrying the greatest amount of sediment. [1]

_____

64 State *one* possible cause for the increase in stream discharge on March 15. [1]

_____

65 The Delaware River flows out of the Catskills and into the Atlantic Ocean. Identify the general compass direction toward which the Delaware River flows along the Pennsylvania–New York State border. [1]

_____

PART C

Answer all questions in this part.

Directions (66–85): Record your answers in the spaces provided. Some questions may require the use of the *2011 Edition Reference Tables for Physical Setting/Earth Science*.

Base your answers to questions 66 through 68 on the topographic map below and on your knowledge of Earth science. The map shows an eroded drumlin at Chimney Bluffs State Park along the shoreline of Lake Ontario east of Rochester, New York. Lines *AB* and *CD* are reference lines. Elevations are shown in feet.

66 Calculate the gradient along the line between points *A* and *B* in feet per mile. [1]

_____ ft/mile

67 On the grid *below*, construct a topographic profile along line *CD* by plotting the elevation of each contour line that crosses line *CD*. The elevations for points *C* and *D* have been plotted on the grid. Connect *all nine* plots with a line from point *C* to point *D* to complete the profile. [1]

68 Describe *one* piece of evidence represented by the contour lines on the map that indicates the north side of Chimney Bluffs is steep. [1]

_____

_____

Base your answers to questions 69 through 71 on the weather map below and on your knowledge of Earth science. The map shows the center of a low-pressure system (**L**). The location for Albany, New York, is shown.

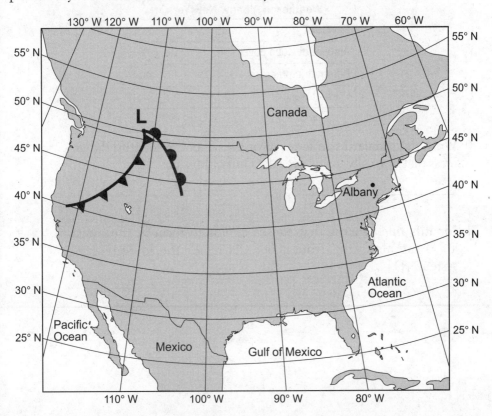

69 The station model below represents the weather conditions for Albany, New York.

Complete the table *below* by indicating the weather conditions represented on the station model for Albany, New York. [1]

Weather in Albany, New York	
Dewpoint	°F
Amount of cloud cover	%
Barometric pressure	mb
Present weather	

70 Write the standard two-letter air-mass symbol to identify the air mass that normally forms over the Gulf of Mexico. [1]

_____

71 Identify *one* process that causes clouds to form in the moist air rising along the frontal boundaries of the low-pressure system. [1]

_____

Base your answers to questions 72 through 74 on the passage and data table below and on your knowledge of Earth science. The data table shows the monthly average high air temperatures, in degrees Fahrenheit (°F), and monthly average snowfall, in inches (in), at the summit (top) of Mount Washington in New Hampshire.

Mount Washington

Mount Washington, located in the state of New Hampshire, is one of the highest mountains east of the Mississippi River. This mountain, as well as many mountains across upstate New York and the northeast, has deep-cut glacial valleys formed by continental glaciers approximately 1.7 to 1.8 million years ago. Hurricane-force wind gusts are observed at the summit of Mount Washington on the average of 110 days per year, including a record wind speed of 231 miles per hour. It also receives very high levels of snow, averaging 282 inches (23.5 feet) of snow per year.

Mount Washington Monthly Average High Air Temperatures and Average Snowfall

Month	Jan	Feb	Mar	Apr	May	Jun	Jul	Aug	Sep	Oct	Nov	Dec
Average High Air Temperature (°F)	14	15	21	30	41	50	54	53	47	36	28	18
Average Snowfall (in)	44	40	45	36	12	1	Trace	0.1	2	18	38	46

72 On the grid *below*, construct a line graph by plotting the average high air temperatures for each month listed in the data table. Connect *all twelve* plots with a line. [1]

73 Identify the climate factor that best explains why Mount Washington has relatively low air temperatures throughout the year. [1]

_____

74 State the name of the geologic epoch when continental glaciers last carved the valleys around Mount Washington and across upstate New York. [1]

_____ **Epoch**

Base your answers to questions 75 through 78 on the table below and on your knowledge of Earth science. The table shows the velocities, in kilometers per second (km/s), for several galaxies, represented by letters A, B, C, D, and E, that are moving away from Earth. The vast majority of stars and galaxies in the universe are moving away from our solar system. Scientific evidence indicates that the farther away a galaxy is, the faster it is moving away.

Velocities of Galaxies Moving Away From Earth	
Galaxy	Velocity (km/s)
A	61,000
B	15,000
C	1200
D	39,000
E	22,000

75 List the galaxies in order from closest to Earth to farthest from Earth. [1]

Closest to Earth: _____

_____

_____

_____

Farthest from Earth: _____

76 Identify the name of the theory for the formation of the universe that scientists developed after observing that most galaxies are moving away from each other. [1]

_____

77 Identify the evidence scientists use to determine that a galaxy is moving away from Earth. [1]

_____

_____

78 A star in one of these galaxies has a surface temperature of 8000 K and a luminosity of 10. Identify the stage and color of this star. [1]

Stage: _____

Color: _____

Base your answers to questions 79 through 81 on the graph below and on your knowledge of Earth science. The graph shows the altitude of the Sun for a 24-hour period on a certain day of the year at four different latitudes.

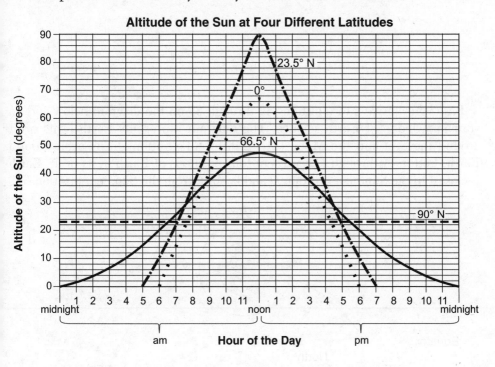

79 Identify the date represented by the data shown on the graph. [1]

_____

80 Identify the latitude that has the greatest intensity of insolation at approximately noon on this date, and describe the evidence shown on the graph to support your answer. [1]

Latitude: _____

Evidence: _____

_____

81 Identify the compass direction toward which an observer's shadow would extend if the observer is located exactly at 90° N at noon on this date. [1]

_____

Base your answers to questions 82 through 85 on the diagram below and on your knowledge of Earth science. The diagram represents Earth in its orbit around the Sun and the Moon (M) in different positions in its orbit around Earth. Letters A through D represent four positions of Earth in its orbit. Earth is closest to the Sun near position D and farthest from the Sun near position B.

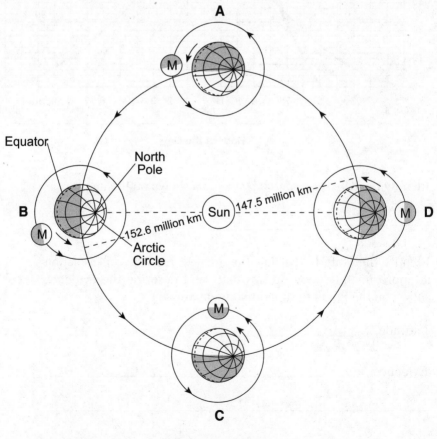

(Not drawn to scale)

82 The photograph below shows the surface of the Moon as seen from Earth during the full moon phase.

Identify the lettered position of Earth and the Moon when this full moon phase can be observed, and state the name of the eclipse that could occur at this position. [1]

Lettered position: _____

Name of eclipse: _____

83 Identify the Northern Hemisphere season that is beginning at position *A*. [1]

_____

84 Explain why the gravitational attraction between the Sun and Earth increases as Earth moves from position *C* to position *D*. [1]

_____

_____

85 Describe *one* piece of evidence shown by the diagram that indicates this is a heliocentric model. [1]

_____

_____

Answers August 2019
Physical Setting/Earth Science

Answer Key

PART A

1. 3	8. 3	15. 4	22. 4	29. 2
2. 1	9. 2	16. 1	23. 1	30. 3
3. 2	10. 4	17. 2	24. 3	31. 4
4. 4	11. 3	18. 2	25. 4	32. 1
5. 2	12. 3	19. 1	26. 3	33. 4
6. 4	13. 2	20. 3	27. 1	34. 3
7. 1	14. 1	21. 1	28. 2	35. 4

PART B–1

36. 4	39. 3	42. 3	45. 3	48. 4
37. 1	40. 1	43. 3	46. 1	49. 4
38. 1	41. 2	44. 1	47. 2	50. 2

PART B–2 and **PART C**. *See* **Answers Explained**.

Answers Explained

PART A

1. **3** It is given that the asteroid belt is between the orbits of Mars and Jupiter. Find the Solar System Data table in the *Reference Tables for Physical Setting/Earth Science*. Locate both "Mars" and "Jupiter" in the column labeled "Celestial Object." Follow both rows to the right, and note the column labeled "Mean Distance from Sun (million km)." Note that Mars and Jupiter orbit the Sun at distances of 227.9 and 778.4 million kilometers, respectively. Therefore, most asteroids are located between 227.9 and 778.4 million kilometers from the Sun. Of the choices given, only choice (3) 403 million km is within this range.

2. **1** Ocean tides are cyclic and predictable. By using subtraction, it can be determined that the time from the last high tide on Tuesday until the next high tide Wednesday morning is 12 hours 26 minutes (e.g., 12:36 p.m. − 1:02 a.m. = 12:26). Therefore, the next high tide after the last high tide on Wednesday will occur about 12 hours and 26 minutes after 1:27 p.m., or at 1:53 a.m.

3. **2** Find the Solar System Data chart in the *Reference Tables for Physical Setting/Earth Science*. Locate "Venus" in the column labeled "Celestial Object." Follow the row for Venus to the right, and note in the column labeled "Period of Revolution (d=days) (y=years)" the value 224.7 days. Continue following the row for Venus to the right. In the column labeled "Period of Rotation at the Equator," note the value 243 days. Thus, Venus takes more time to complete one rotation on its axis than to complete one revolution around the Sun.

WRONG CHOICES EXPLAINED:

(1) Find the Solar System Data chart in the *Reference Tables for Physical Setting/Earth Science*. Locate "Mercury" in the column labeled "Celestial Object." Note that Mercury's period of revolution is 88 days and its period of rotation is 59 days. Thus, it takes *less* time for Mercury to complete one rotation on its axis than to complete one revolution around the Sun.

(3) Find the Solar System Data chart in the *Reference Tables for Physical Setting/Earth Science*. Locate "Earth" in the column labeled "Celestial Object." Note that Earth's period of revolution is 365.26 days and its period of rotation is 23 hours 56 minutes 4 seconds. Thus, it takes *less* time for Earth to complete one rotation on its axis than to complete one revolution around the Sun.

(4) Find the Solar System Data chart in the *Reference Tables for Physical Setting/Earth Science*. Locate "Mars" in the column labeled "Celestial Object." Note that the period of revolution for Mars is 687 days and its period of rotation is 24 hours 37 minutes 23 seconds. Thus, it takes *less* time for Mars to complete one rotation on its axis than to complete one revolution around the Sun.

4. **4** The planets can be divided by mass and density into the terrestrial (Earth-like) and Jovian (Jupiter-like) planets. The terrestrial planets include the four innermost planets: Mercury, Venus, Earth, and Mars. The Jovian planets include the four outermost planets: Jupiter, Saturn, Neptune, and Uranus. Find the Solar System Data chart in the *Reference Tables for Physical Setting/Earth Science*. Locate the column labeled "Mass (Earth = 1)." Note that the four terrestrial planets range in mass from 0.06 to 1.00, while Jupiter has a mass of 317.83. Thus, compared to the terrestrial planets, Jupiter has a greater mass.

WRONG CHOICES EXPLAINED:
(1) Find the Solar System Data chart in the *Reference Tables for Physical Setting/Earth Science*. Locate the column labeled "Period of Revolution (d=days) (y=years)." Note that the periods of revolution for the terrestrial planets range from 88 days to 687 days. Note that the period of revolution of Jupiter is 11.9 years. Thus, compared to the terrestrial planets, Jupiter has a longer period of revolution.

(2) Find the Solar System Data chart in the *Reference Tables for Physical Setting/Earth Science*. Locate the column labeled "Mean Distance from Sun (million km)." Note that the distance from the Sun for the terrestrial planets ranges from 57.9 to 227.9 million km. Note that Jupiter's distance from the Sun is 778.4 million km. Thus compared to the terrestrial planets, Jupiter has a longer distance to the Sun.

(3) Find the Solar System Data table in the *Reference Tables for Physical Setting/Earth Science*. Locate the column labeled "Density (g/cm^3)." Note that the densities of the terrestrial planets range from 3.9 g/cm^3 to 5.5 g/cm^3 while the density of Jupiter is 1.3 g/cm^3. Thus, compared to the terrestrial planets, Jupiter has a lower density.

5. **2** At different times of the year, Earth is at different points along its orbit, and the night side of Earth faces different portions of the universe. Since stars are visible only at night, only constellations in the direction of Earth's night side are visible from Earth. Since the constellation Orion is visible in New York State at night in January, the night side of Earth faces Orion in January. Five months later, in June, Earth has revolved to a position in its orbit nearly opposite the Sun from its January position. Thus, the night side of Earth faces away from Orion in June.

Orion is not visible in New York State in June because the Sun lies between Earth and Orion.

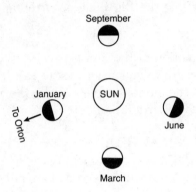

Therefore, the best explanation for why this constellation is not visible in the night sky to this observer in June is that Earth revolves around the Sun.

WRONG CHOICES EXPLAINED:
(1) Earth rotates on its axis once every day. Earth's daily rotation causes a daily apparent motion of constellations, not a seasonal change in constellations visible from Earth.
(3) The Sun is located too far from the stars in constellations to exert enough gravitational attraction to hold the constellations in orbit and cause them to revolve around the Sun.
(4) The side of Earth that is in darkness depends on only Earth's position in relation to the Sun. In whatever direction Earth is tilted, night always occurs on the side of the planet located directly opposite the Sun. Thus, the tilt of Earth's axis does not affect the direction Earth faces at night (nor does it affect the constellations that are visible in that direction).

6. **4** According to the Big Bang theory, the universe started out with all of its matter and energy in a small volume and then expanded outward in all directions, a motion very much like an explosion. The explosion associated with the Big Bang theory and the formation of the universe occurred *before* the formation of Earth and the solar system. In other words, the universe is older than either Earth or the solar system. Find the Geologic History of New York State chart in the *Reference Tables for Physical Setting/Earth Science*. Locate the column labeled "Eon." Note that the entry "Estimated time of origin of Earth and solar system" corresponds to 4600 million years ago (4.6 billion years ago) on the time scale. Thus, the explosion associated with the Big Bang theory took place

more than 4.6 billion years ago. Based on the current expansion rate of the universe, scientists extrapolate back to when the universe was all concentrated in a small volume and infer that the Big Bang occurred approximately 13.8 billion years ago.

7. **1** This apparent motion of the Sun across the sky is the result of Earth's rotation on its axis. As Earth rotates from west to east, the Sun appears to move from east to west. Thus, the motion responsible for the apparent path of the Sun shown in the photograph is Earth's rotation.

WRONG CHOICES EXPLAINED:
(2) Earth takes about 365 days (one year) to complete one 360-degree revolution around the Sun, which is about 1 degree of arc per day. The Sun has clearly moved through more than 1 degree of arc during the time shown in the photograph. Thus, it is unlikely that Earth's revolution around the Sun is responsible for the apparent path of the Sun shown in the photograph.

(3) In order for the Sun to change position in the sky, it must change position relative to an observer on Earth. The Sun's rotation on its axis does not change its position relative to an observer on Earth. The Sun could remain stationary to an observer on Earth and still rotate. Therefore, the apparent path of the Sun through the sky cannot be caused by the Sun's rotation on its axis.

(4) The Sun does not revolve around Earth. Therefore, this motion could not cause the apparent path of the Sun shown in the photograph.

8. **3** Evidence suggests that an impact event was responsible for the extinction of many life-forms at the end of the Cretaceous period approximately 65.5 million years ago. The dividing line between the Cretaceous and the Tertiary is a thin layer of clay that has been identified in sediments worldwide. This boundary clay contains numerous indications of a massive impact event: levels of iridium higher than those found in Earth's crust but similar to those found in meteorites, shocked quartz grains, melt spherules, soot from the widespread forest fires ignited by the impact, as well as evidence of large waves such as would be caused by an ocean impact. The impact of asteroids or large meteorites would generate a massive shock wave and spew large amounts of dust high into the atmosphere. The resulting decrease in sunlight would cause global temperatures to decrease markedly and have devastating effects on photosynthetic life and all of the life-forms that depend on them for food. Thus, some scientists have inferred that the mass extinction of the dinosaurs approximately 65.5 million years ago was caused by an impact event occurring on Earth's surface.

WRONG CHOICES EXPLAINED:

(1) Measurements of the Sun's energy output indicate that the Sun is relatively stable. Although there are fluctuations associated with the sunspot cycle, these do not occur only at times of mass extinctions. Thus, there does not appear to be a cause and effect relationship between large energy surges from the surface of the Sun and mass extinctions.

(2) About five major ice ages have occurred during Earth's history. The most recent, the Quaternary, began about 2.8 million years ago. The ice age before that, the Karoo, occurred between 360 and 260 million years ago. Thus, the mass extinction of dinosaurs about 65.5 million years ago did not occur during a major ice age. Therefore, there is no basis to infer that the mass extinction of dinosaurs was caused by the occurrence of a major ice age.

(4) Earthquakes are unlikely to have caused mass extinctions, because they occur almost continuously, not solely at times of mass extinctions.

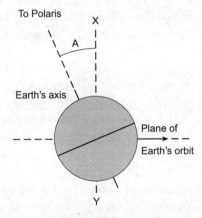

9. **2** It is given that line *XY* is perpendicular to the plane of Earth's orbit. Earth's axis of rotation is tilted 23.5° from a line perpendicular to the plane of its orbit. Thus, the value of the angle represented by letter *A* is 23.5°.

10. **4** The rate at which meltwater can infiltrate soil depends on a number of factors. One factor is the surface slope. The gentler the slope, the more slowly the water will flow downhill and the more time the water has to infiltrate before it runs off. Another factor is the permeability of the surface. If the land surface is still frozen, ice fills the pore spaces between particles, making the surface impermeable. However, if the surface is no longer frozen, the spaces between particles are not blocked by ice, allowing water to infiltrate. Thus, meltwater will most likely infiltrate where the land surface is no longer frozen with a gentle slope.

11. **3** Water is returned to Earth's atmosphere in the form of water vapor. To form water vapor, water must change from either the solid or the liquid phase to the gaseous phase. Since most of Earth's surface is liquid water, most water vapor enters the atmosphere by changing from a liquid to a gas, a process called evaporation. Water vapor also enters the atmosphere directly from the leaves of plants that cover much of Earth's land surfaces, a process called transpiration. Thus, water is returned from Earth's surface to the atmosphere by evaporation and transpiration.

WRONG CHOICES EXPLAINED:
(1) Condensation refers to the change from the gaseous to the liquid phase of matter. Thus, condensation does not return water vapor to the atmosphere.

(2) Condensation refers to the change from the gaseous to the liquid phase of matter. Precipitation is condensed moisture that falls to the ground. Neither process returns water from Earth's surface to the atmosphere.

(4) Precipitation is condensed moisture that falls to the ground. This process returns water from the atmosphere to Earth's surface, not from Earth's surface to the atmosphere.

12. **3** Permeability is the rate at which water infiltrates (trickles through) sediment. The smaller the sediment particles, the smaller the empty spaces are between the sediment particles. Small spaces constrict the flow of water through soil. Large sediment particles have larger spaces between them, which allow water to flow through the sediment more rapidly. Therefore, as sediment particle size increases, permeability rate also increases. Find the Relationship of Transported Particle Size to Water Velocity graph in the *Reference Tables for Physical Setting/Earth Science*. Note the dashed lines labeled with the size ranges corresponding to clay, silt, sand, pebbles, cobbles, and boulders. Note that pebbles are larger in diameter than sand. Thus, pebbles have a greater permeability rate than sand. Sorting also affects permeability rate. Unsorted pebbles have smaller spaces through which water can infiltrate than do sorted pebbles. In unsorted pebbles, smaller pebbles fill the spaces between larger pebbles. Therefore, sorted pebbles have the greatest permeability rate.

13. **2** High-pressure centers are usually associated with air sinking toward the surface. When air sinks, it goes through a warming and drying process. Therefore, the relative humidity of the air decreases, and condensation and cloud formation become less likely. Thus, high-pressure centers are usually associated with clear skies.

WRONG CHOICES EXPLAINED:

(1) Clouds form when air is cooled to its dewpoint and water vapor condenses into water droplets. The closer the air temperature is to the dewpoint, the higher the relative humidity is. Therefore, cloudy skies are associated with higher relative humidity than are clear skies.

(3) Most clouds and precipitation occur at or around frontal boundaries. Frontal boundaries mark the transition from higher pressure cool air to lower pressure warm air. The greatest pressure differences, and thus the highest wind speeds, occur at or near these boundaries. Thus, increasing wind speed is associated with frontal boundaries and cloudy skies, not with clear skies.

(4) When the air temperature reaches the dewpoint, moisture begins to condense, clouds form, and precipitation results. Thus, when the dewpoint equals the air temperature, cloudy skies are the most likely result, not clear skies.

14. **1** Find the Selected Properties of Earth's Atmosphere chart in the *Reference Tables for Physical Setting/Earth Science*. Locate the vertical axis labeled "Altitude." Note that an altitude of "0" corresponds to ground level. Now find "Troposphere" in the portion of the chart labeled "Temperature Zones." Note the dashed line labeled "Tropopause" marking the top of the troposphere. On the graph labeled "Temperature (°C)," note that from ground level to the tropopause, temperature decreases. On the graph labeled "Pressure (atm)," note that from ground level to the tropopause, atmospheric pressure decreases. Thus, as a weather balloon released at Earth's surface rises through the troposphere, there will be a decrease in both air temperature and atmospheric pressure.

15. **4** The ozone layer is a portion of the upper stratosphere that contains high concentrations of the gas ozone. One property of ozone is that it absorbs certain wavelengths of ultraviolet radiation. The short wavelength and high energy of ultraviolet radiation enables it to damage cells. Therefore, the gas most effective in absorbing incoming harmful ultraviolet radiation in Earth's stratosphere before that radiation reaches Earth's surface is ozone.

16. **1** A monsoon is a seasonally reversing system of surface winds caused by temperature differences between continents and neighboring oceans. A monsoon is a large-scale version of the type of *daily* changes in circulation we see in land and sea breezes. In a monsoon, we see *seasonal* changes in wind direction between a large landmass and surrounding oceans.

For example during the summer, the large landmass of Asia warms up and a low-pressure center develops in the southern part of the continent. This low-pressure center results in winds that blow inland from the surrounding oceans.

These moist summer monsoon winds bring clouds and heavy rains. During the winter, intense cooling produces a mass of very cold, very dry air over the northern part of the continent. The result is a high-pressure center in the northern part of the continent that causes the winds to reverse direction and blow from the continental interior outward to the coast. Although these winter monsoon winds may warm up as they descend toward the coast, they pick up little moisture as they blow over the land and bring clear, dry weather. Thus, the major cause of monsoon rains in India and Southeast Asia is seasonal shifts in the prevailing wind belts.

WRONG CHOICES EXPLAINED:

(2) A monsoon is a seasonally reversing system of surface winds caused by temperature differences between land and ocean, not shifts in ocean currents.

(3) Measurements of the Sun's energy output indicate that it is relatively stable. Although fluctuations are associated with the sunspot cycle, these do not occur on a seasonal basis and are therefore unlikely to be the major cause of monsoons.

(4) Monsoons are regional, not worldwide, occurrences. Changes in worldwide temperatures would affect all parts of Earth, not just certain regions.

17. **2** The cross-section shows prevailing winds forcing air to rise along the windward side of the mountain range. As the air rises to higher elevations, air pressure and temperature decrease, causing the air to expand and cool. When the air cools to its dewpoint temperature, water vapor begins to condense, forming clouds. If enough water vapor condenses and if large water droplets form, precipitation will occur on the windward side. Thus, location X will be cool and wet.

As the air sinks on the leeward side of the mountains, air pressure increases, causing the air to be compressed and, therefore, warm. The air is warmed above its dewpoint, and any available moisture evaporates. Therefore, clouds and precipitation are much less likely to form on the leeward side of the mountain, and location Y will be warm and dry. Therefore, compared to location X, the temperature and water vapor content at location Y are most likely warmer and drier.

18. **2** Find the Surface Ocean Currents map in the *Reference Tables for Physical Setting/Earth Science*. In the key, note the symbols for warm and cool ocean currents. Locate the Gulf Stream, Peru Current, West Greenland Current, and East Australia Current. Note that of these, the Peru Current and West Greenland Current are cool currents. Note that the Peru Current transports cool water toward the equator and the West Greenland Current transports cool water toward the pole. Therefore, the surface ocean current that transports cool water to lower latitudes is the Peru Current.

19. **1** Color is a fairly good indicator of whether a substance absorbs more light than it reflects. The lighter the color of a substance, the more light is being reflected by it; the darker the color, the more light is absorbed by it.

If a surface is smooth, light is more likely to be immediately reflected away from the surface than if it is rough. The irregularities on a rough surface can cause some of the light to hit the surface several times before leaving. Each time the light strikes the surface, a little more of its energy is absorbed. Thus, the greatest amount of absorption of insolation at Earth's surface is produced by a dark color and rough texture.

20. **3** Note that the question refers to *surface* winds and air currents. Find the Planetary Wind and Moisture Belts in the Troposphere chart in the *Reference Tables for Physical Setting/Earth Science*. Note the arrows showing air movements at the surface near the equator. As you can see, the surface winds north and south of the equator are both blowing toward the equator. In other words, the surface winds are converging at the equator. The arrows also show air currents at the equator moving upward from the surface toward the tropopause, or rising. Thus, the planetary surface winds and air currents near Earth's equator are usually converging and rising.

21. **1** Earth's early atmosphere most likely formed by outgassing, which was the release of water and trace gases trapped in rocks in the interior of the hot, young Earth. Comets that collided with Earth were another likely source of gases. Comets are rich in ices of water, carbon dioxide, carbon monoxide, ammonia, and organic compounds. Thus, Earth's early atmosphere contained virtually no free oxygen. However, over time, life-forms evolved that were able to carry out photosynthesis. Photosynthetic life-forms use energy from sunlight to convert carbon dioxide and water into carbohydrates (food) and oxygen. Therefore, once photosynthetic life evolved, these organisms released oxygen into the environment.

Find the Geologic History of New York State chart in the *Reference Tables for Physical Setting/Earth Science*. In the column labeled "Era," locate the statements "Oceanic oxygen produced by cyanobacteria combines with iron, forming iron oxide layers on ocean floor" and "Oceanic oxygen begins to enter the atmosphere." Cyanobacteria, or blue-green algae, are single-celled organisms that produce gaseous oxygen as a by-product of photosynthesis. Oxygen is highly reactive and combines with substances dissolved in seawater. Some of these substances, such as carbonates, were soluble and remained in solution in the oceans. Others, such as iron oxides, were insoluble and settled out of the oceans to form layers of iron oxide on the ocean floor. Eventually, oxygen levels grew high enough in the oceans that excess oxygen began to escape into the atmosphere. As photosynthetic organisms multiplied over time, photosynthesis steadily increased. Over millions of years, the oxygen released by these photosynthetic organisms steadily increased the oxygen in Earth's atmosphere to its present level. Thus, scientists infer that oxygen in Earth's early history was first produced during the Precambrian from cyanobacteria in Earth's oceans.

WRONG CHOICES EXPLAINED:

(2) Volcanic activity along plate boundaries occurred throughout Earth's history and was occurring long before oxygen was first produced in Earth's atmosphere. Therefore, it is unlikely that volcanic activity first produced the oxygen in Earth's atmosphere.

(3) Find the Geologic History of New York State chart in the *Reference Tables for Physical Setting/Earth Science*, and locate the "Life on Earth" column. Note in this column that the earliest land plants and animals appeared during the Silurian, well after oxygen in Earth's early atmosphere was first produced during the Precambrian. Therefore, it is unlikely that the absorption of sunlight by plants first produced the oxygen in Earth's atmosphere.

(4) The evaporation of ocean water involves the change from liquid water to water vapor. Water vapor is not oxygen. Furthermore, oceans existed and ocean water evaporated long before oxygen in Earth's early atmosphere was produced. Therefore, it is unlikely that evaporation of ocean water first produced the oxygen in Earth's atmosphere.

22. **4** Find the Geologic History of New York State chart in the *Reference Tables for Physical Setting/Earth Science*, and locate the column labeled "Time Distribution of Fossils." Note the vertical gray bars labeled with the names of different groups of organisms. The vertical gray bars indicate the range of time during which each group of organisms existed on Earth. The bottom of the gray bar indicates when the group first appeared on Earth, and the top of the bar indicates when the group became extinct or, if it extends to the top of the column,

that the group still exists. Locate the bars labeled "dinosaurs," "trilobites," "ammonoids," and "placoderm fish." Note that of these four bars, the bar labeled "placoderm fish" is the shortest. Therefore, of these four choices, the life-form that existed on Earth for the shortest period of time was placoderm fish.

23. **1** Find the Geologic History of New York State chart in the *Reference Tables for Physical Setting/Earth Science*. Note that *Eurypterus* is fossil "**M**" in the illustrations of specific important index fossils printed along the bottom of the chart. In the column labeled "Time Distribution of Fossils," locate the circled letter "**M**" along the black line labeled "eurypterids." Trace horizontally to the left to the column labeled "Period," and note that the approximate time of the existence of *Eurypterus* was during the Silurian. Index fossils are remains of organisms that had distinctive body features, were common and abundant, and had a broad, even worldwide range yet existed for only a short period of time. Their distinctive body features and broad distribution make index fossils easy to find in widely separated rock layers. Their short existence pinpoints the time period during which the rock layer was formed. Thus, organisms that later became good index fossils lived over a wide geographic area and existed for a short time. Evaporites, such as halite, that formed during a short period of time when very salty inland seas existed in an area allows one to roughly date the age of the rocks. Abundance over a wide geographic area also allows rocks in widely separated places to be correlated. Thus, both the halite layers and the *Eurypterus remipes* fossils can now be used to identify a specific geologic time interval because both formed over a large geographic area and in a short geologic time.

24. **3** Find the Earthquake *P*-Wave and *S*-Wave Travel Time graph in the *Reference Tables for Physical Setting/Earth Science*. On the vertical axis labeled "Travel Time (min)," locate "5." From this point, trace right until you intersect the bold curve labeled "*P*." From this intersection, trace downward vertically to the horizontal axis labeled "Epicenter Distance ($\times 10^3$ km)." Note that you intersect the axis at "2600 km." Thus, a *P*-wave takes 5 minutes to travel 2600 km. Now, trace vertically upward from 2600 km until you intersect the bold curve labeled "*S*." From this intersection, trace horizontally to the left to the vertical axis labeled "Travel Time (min)." Note that an *S*-wave takes 9 minutes to travel 2600 km. Thus, an *S*-wave takes 9 minutes to travel the same distance a *P*-wave travels in 5 minutes.

25. **4** *P*-waves are compression waves that compress and expand rock in the direction of wave travel. Solids, liquids, and gases are all compressible; therefore, *P*-waves can be transmitted through solids, liquids, and gases. On the other hand, *S*-waves are transverse waves that twist rock back and forth, deforming their shape

in a direction perpendicular to that of wave travel (laterally). S-waves can be transmitted only through solids. S-waves cannot travel through liquids and gases, because when liquids and gases are deformed laterally, they do not return to their original shape. Find the Inferred Properties of Earth's Interior in the *Reference Tables for Physical Setting/Earth Science*. Locate the outer core in the cross section at the top of the diagram. From the outer core, trace downward between the dashed lines to the "Temperature (°C)" graph, and note that the actual temperature of the outer core is higher than its melting temperature. Thus, the outer core is a liquid. Note, too, that the outer core is the only place on the "Temperature (°C)" graph where this is the case, so the outer core is the only layer of Earth's interior that is a liquid. Therefore, S-waves from an earthquake cannot pass through the liquid outer core and as a result will also not reach the inner core. Thus, earthquake S-waves do not pass through the outer core and the inner core.

26. **3** Note that the fault along line *XY* has displaced rock units *E*, *F*, *G*, and *H*. Any rock unit displaced by a fault is older than the fault because it already had to exist in order to be displaced. Thus, the faulting along *XY* occurred after the formation of rock units *E*, *F*, *G*, and *H*. Therefore, the faulting along line *XY* occurred after the formation of rock units *E* and *F*.

WRONG CHOICES EXPLAINED:

(1) Note that the igneous rock intrusion *A* cuts across the fault and is not displaced by the fault. If the intrusion had existed before faulting occurred, the fault would cut across the intrusion and the intrusion would have been displaced by the fault. Therefore, the faulting along line *XY* occurred before, not after, the intrusion of rock unit *A*.

(2) Note that rock units *B* and *C* have not been displaced by faulting along line *XY*. Thus, the faulting along line *XY* occurred before rock units *B* and *C* were deposited.

(4) Note that the fault along line *XY* has displaced rock units *G* and *H*. Any rock unit displaced by a fault is older than the fault because it already had to exist in order to be displaced. Thus, the faulting along line *XY* occurred after the formation of rock units *G* and *H*.

27. **1** An inclusion is a solid fragment, liquid globule, or pocket of gas enclosed in a mineral or rock. The geologic principle of inclusions states that the inclusions in a rock are older than the rock itself. Thus, the granite inclusions are older than the sandstone. Note that the granite inclusions lie near the granite bedrock beneath the sandstone. Therefore, it is reasonable to infer that the inclusions shown near the bottom of the sandstone are pieces of the granite bedrock broken loose by weathering before the sandstone was deposited. Thus, the age of the granite that makes up the inclusions is most likely older than the sandstone, but the same age as the granite bedrock.

28. **2** It is given that the Catskill Aqueduct flows into the Kensico Reservoir. Find the Generalized Landscape Regions of New York State map in the *Reference Tables for Physical Setting/Earth Science*. Locate the region corresponding to the location of the Catskill Aqueduct. Note that the Catskills are part of the

Allegheny Plateau. Therefore, the landscape regions that are crossed as water flows through the Catskill Aqueduct, in order, are the Allegheny Plateau, Hudson-Mohawk Lowlands, Hudson Highlands, and Manhattan Prong.

29. **2** By definition, a watershed is the area drained by a stream and its tributaries. A river is a stream. The question states that the map represents the land area drained by the Oswego River and its tributaries. Therefore, the land area drained by the Oswego River and its tributaries is called a watershed.

WRONG CHOICES EXPLAINED:

(1) A delta is a large, flat, fan-shaped pile of sediment at the mouth of a stream, not the land area drained by a river and its tributaries.

(3) A water table is the upper surface of groundwater filling underground pore spaces, not the land area drained by a river and its tributaries.

(4) A floodplain is a low plain adjacent to a stream that is likely to be flooded if a stream overflows, not the area drained by a river and its tributaries.

30. **3** Stream erosion is greatest where water velocity is greatest. When the stream curves, or meanders, velocity is greatest along the outside of the curve and lowest along the inside of the curve. Water velocity is greatest along the outside of the curve because the water is forced to cover a greater distance along the outside of the curve. Along a straight section of a stream's channel, velocity is greatest in the center of the stream, which is farthest from the friction of the banks or bed. Since stream deposition is greatest where the velocity of the water is slowest, the inside of the curve of a meandering channel experiences the greatest amount of deposition. Of the four locations on the Genesee River, only location *C* is on the inside curve of a meander. Therefore, deposition of sediments would most likely be greater than erosion at location *C*.

31. **4** Soil is the accumulation of loose, weathered material that covers much of Earth's land surface. The main component of soil is weathered rock. However, a true soil also contains water, air, bacteria, and decayed plant and animal material (humus).

Since weathered rock is the main component of soil, one of the most important factors controlling the development of soil is the multitude of weathering processes that convert the bedrock to soil. The single most important factor affecting weathering processes is climate. Climate is typically expressed in terms of two factors: temperature and precipitation. Both of these factors influence the type and rate of weathering. Warm climates favor chemical weathering; cold climates favor physical weathering. In both cases, the more moisture that is present, the more pronounced is the weathering. Tropical climates tend to have high temperatures and abundant rainfall. The combination of heat and water accelerates chemical weathering, leading to the rapid breakdown of rock. Living organisms can cause both physical and chemical weathering of rock. For example, tree roots can grow into crevices in rock and pry the rock apart, causing mechanical breaking apart of the rock. Moss and fungi can grow on the surface of rock. Their roots produce a weak acid that can chemically break down the rock. Ants and earthworms bring rock fragments to the surface, where the fragments are exposed to air and water that further weather them. Arctic areas tend to have cold temperatures and low rainfall. These two factors result in living organisms being less abundant in arctic areas than in tropical areas. Therefore, the rate of soil development is usually greater in tropical areas than in arctic areas because tropical areas have more chemical weathering and an abundance of living organisms.

32. **1** During the Pleistocene, huge continental glaciers formed in arctic and subarctic regions and flowed outward (south), merging with smaller valley glaciers that had already formed in mountainous regions. These processes covered most of New York State with a layer of ice thousands of feet thick. Nearly all of New York State displays evidence of glaciation. Glacial lakes, such as the Finger Lakes, formed when the glaciers retreated and meltwater filled depressions whose openings were blocked by moraines (piles of sediment) deposited by the glaciers. Since glaciers often flow through existing valleys, the Finger Lakes' basins are probably old stream valleys that were widened and deepened by the glacier into U-shaped valleys and later filled with meltwater when the ends of the valleys were dammed by moraines. Thus, New York State's Finger Lakes exist today because U-shaped valleys were dammed by glacial sediments.

WRONG CHOICES EXPLAINED:

(2) If V-shaped valleys are being eroded by streams, the water is flowing in the stream, not forming a lake.

(3) If the Finger Lakes were bodies of water stranded by falling sea levels, the water in the lakes would be saltwater, not freshwater.

(4) If the Finger Lakes were bodies of water created by rising sea levels flooding the region, the water in the lakes would be saltwater, not freshwater.

33. **4** Physical weathering breaks down rocks without changing their chemical composition. Therefore, the chemical composition of the rock fragments (beach sand) is the same as the composition of the rock from which they formed. Find the Scheme for Igneous Rock Identification in the *Reference Tables for Physical Setting/Earth Science*. On the graph labeled "Mineral Composition (relative by volume)," locate the regions labeled pyroxene, plagioclase feldspar, and olivine. Find the section of the graph in which a vertical line would pass through the regions for all three of these minerals. From this section, trace upward to the chart in the scheme labeled "Igneous Rocks." Note that it intersects the rocks gabbro, diabase, basalt, vesicular basalt, scoria, and basaltic glass. Thus of the choices given, the igneous rock that could physically weather into beach sand that contains the minerals pyroxene, plagioclase feldspar, and olivine is gabbro.

WRONG CHOICES EXPLAINED:

(1) Find the Scheme for Igneous Rock Identification in the *Reference Tables for Physical Setting/Earth Science*. In the part of the Scheme labeled "Igneous Rocks," locate dunite. From there, trace vertically downward to the graph labeled "Mineral Composition (relative by volume)." Note that dunite contains only the mineral olivine. Therefore, dunite could not physically weather to form beach sand that contained pyroxene or plagioclase feldspar.

(2) Find the Scheme for Igneous Rock Identification in the *Reference Tables for Physical Setting/Earth Science*. In the part of the Scheme labeled "Igneous Rocks," locate granite. From there, trace vertically downward to the graph labeled "Mineral Composition (relative by volume)." Note that granite contains no pyroxene or olivine. Therefore, granite could not physically weather to form beach sand that contained pyroxene or olivine.

(3) Find the Scheme for Igneous Rock Identification in the *Reference Tables for Physical Setting/Earth Science*. In the part of the Scheme labeled "Igneous Rocks," locate peridotite. From there, trace vertically downward to the graph labeled "Mineral Composition (relative by volume)." Note that peridotite contains no plagioclase feldspar. Therefore, peridotite could not physically weather to form beach sand that contained plagioclase feldspar.

34. **3** Graphite is used in pencils because of its ability to leave dark marks on paper. The dark marks are the powdered graphite left behind on the paper. To leave a streak on paper, the graphite must be very soft. In order for the streak to be easily visible, it must be dark in color. Find the Properties of Common Minerals table in the *Reference Tables for Physical Setting/Earth Science*. In the column labeled "Mineral Name," locate graphite. From graphite, trace left to the column labeled "Hardness," and note that graphite has a hardness of 1–2. Thus, graphite is among the softest of minerals. Now trace right to the column labeled "Distinguishing Characteristics," and note that graphite has a black streak. Thus, the two physical properties of graphite that make it a good mineral to use in pencils are hardness and streak.

35. **4** Find the Properties of Common Minerals table in the *Reference Tables for Physical Setting/Earth Science*. At the bottom of the table, note that the chemical symbol for iron is "Fe" and the chemical symbol for oxygen is "O." Thus, the chemical formula for a mineral that is a chemical combination of two atoms of iron and three atoms of oxygen is written as Fe_2O_3. Now find the column labeled "Composition," and locate Fe_2O_3. From there, trace right to the column labeled "Mineral Name." Note that the mineral with the chemical formula Fe_2O_3 is hematite. Thus, the mineral produced when two atoms of iron chemically combine with three atoms of oxygen is hematite.

WRONG CHOICES EXPLAINED:
(1) Find the Properties of Common Minerals table in the *Reference Tables for Physical Setting/Earth Science*. In the column labeled "Mineral Name," locate garnet. From garnet, trace left to the column labeled "Composition." Note that the composition of garnet is $Fe_3Al_2Si_3O_{12}$. In other words, garnet is formed by chemically combining three atoms of iron, two atoms of aluminum, three atoms of silicon, and twelve atoms of oxygen, not two atoms of iron and three atoms of oxygen.

(2) Find the Properties of Common Minerals table in the *Reference Tables for Physical Setting/Earth Science*. In the column labeled "Mineral Name," locate pyrite. From pyrite, trace left to the column labeled "Composition." Note that the composition of pyrite is FeS_2. In other words, pyrite is formed by chemically combining one atom of iron and two atoms of sulfur, not two atoms of iron and three atoms of oxygen.

(3) Find the Properties of Common Minerals table in the *Reference Tables for Physical Setting/Earth Science*. In the column labeled "Mineral Name," locate magnetite. From magnetite, trace left to the column labeled "Composition." Note that the composition of magnetite is Fe_3O_4. In other words, magnetite is formed by chemically combining three atoms of iron and four atoms of oxygen, not two atoms of iron and three atoms of oxygen.

PART B–1

36. **4** Earth rotates through 360° of longitude in 24 hours, or 15° of longitude per hour (360/24 = 15). Thus, every 15° of longitude that separates two locations corresponds to a difference of 1 hour (60 minutes) in solar time. Find Jamestown on the map and note that it is located at 79.2° W longitude. Find Kingston on the map and note that it is located at 74° W longitude. Thus, Kingston and Jamestown are separated by 5.2° of longitude. A 5.2° difference in longitude is equivalent to a difference in solar time of 0.35 hours (5.2°/15°/h = 0.35 h), or about 21 minutes (60 min/h × 0.35 h = 21 min). Thus, Jamestown and Kingston will experience sunrise 21 minutes apart. As Earth rotates from west to east, Kingston in the eastern part of New York State will be carried from darkness into daylight (experience sunrise) before Jamestown in the western part of New York State. Therefore, compared to the time of sunrise in Kingston, the time of sunrise in Jamestown will be approximately 20 minutes later.

37. **1** Find Watertown on the map, and note that it is located at 44° N latitude. The altitude of *Polaris* is the same as an observer's latitude north of the equator. Therefore, an observer in Watertown, New York, would see the star *Polaris* at an altitude of approximately 44°.

38. **1** Earth's axis and poles provide the reference points for lines of longitude. Lines of longitude are great circles that pass through the north and south poles and that cross the equator at 90°. These great circles are also called meridians. The longitude of a point is the angular distance between two meridians: a fixed reference meridian and the meridian passing through the point. Since the angular distance in any full circle is 360° and the angular distance covered in one Earth rotation is 360°, it was logical to link Earth's rotation to the longitude coordinate system. In one rotation, Earth moves through 360° of angular distance and 360° of longitude. Thus, Earth's system of longitude is based on Earth's rotation.

39. **3** The reading passage states "The Patagonia Marble Caves . . . at a location of 46.5° S 72° W." Locate the vertical longitude lines labeled "70° W" and "80° W." Longitude 72° W is located between longitudes 70° W and 80° W, that is, to the left of the 70° W longitude line. Locate the latitude lines labeled "40° S" and "50° S." Latitude 46.5° S is located between latitudes 40° S and 50° S, that is, below the 40° S latitude line. Therefore, the lettered point on the map that best represents the location of the Patagonia Marble Caves is left of the 70° W line and below the 40° S line, or *C*.

40. **1** Find the Scheme for Metamorphic Rock Identification in the *Reference Tables for Physical Setting/Earth Science.* In the column labeled "Rock Name," locate marble. Trace this row to the left to the "Comments" column, and note that the metamorphism of limestone or dolostone forms marble. Metamorphism is the change in the mineral composition or texture of a rock due to heat, pressure, or chemical activity without the rock undergoing melting into liquid magma. Therefore, two processes that formed the marble bedrock shown in the photograph are heat and pressure.

WRONG CHOICES EXPLAINED:

(2) Compaction and cementation are processes that transform loose particles of sediment into solid sedimentary rocks, not processes that change an existing rock into a metamorphic rock such as marble.

(3) Melting and solidification are processes that lead to the formation of a new igneous rock, not a metamorphic rock such as marble.

(4) Precipitation and evaporation are processes involving the movement of water out of or into the atmosphere, not processes that result in the formation of a metamorphic rock such as marble.

41. **2** Find the Scheme for Metamorphic Rock Identification in the *Reference Tables for Physical Setting/Earth Science*. In the column labeled "Rock Name," locate marble. Trace this row to the left to the column labeled "Composition," and note that marble is composed of calcite and/or dolomite. Thus, most marble is composed primarily of calcite and/or dolomite.

42. **3** The barometric pressure shown on the digital device is 1003.1 mb. Find the Pressure scale in the *Reference Tables for Physical Setting/Earth Science*. Locate "1003.1" on the left side of the scale labeled "millibars (mb)." Trace right to the corresponding scale labeled "inches (in of Hg*)." Note that 1003.1 millibars corresponds to 29.62 inches of Hg. Note at the bottom of the scale that "*Hg = mercury." Thus, 1003.1 millibars corresponds to 29.62 inches of mercury.

43. **3** According to the digital device, the outside air temperature is 30.0°C and the outside relative humidity is 49%. Find the Relative Humidity (%) chart in the *Reference Tables for Physical Setting/Earth Science*. Find the "Dry-Bulb Temperature (°C)" scale along the left side of the chart, and locate the row labeled "30." Trace this row to the right to the cell containing the value "49." From this cell, trace vertically upward to the "Difference Between Wet-Bulb and Dry-Bulb Temperatures (°C)" scale along the top of the chart. Note the value listed there is "8." Thus, when the dry-bulb temperature is 30°C and the relative humidity is 49%, the difference between the wet-bulb and dry-bulb temperatures is 8°C.

Now find the Dewpoint (°C) chart in the *Reference Tables for Physical Setting/Earth Science*. Locate the column headed "8" in the "Difference Between Wet-Bulb and Dry-Bulb Temperatures (°C)" scale along the top of the chart.

Locate the row corresponding to 30°C on the "Dry-Bulb Temperature (°C)" scale along the left side of the chart. Trace to the right from the row labeled "30" until you intersect the column headed "8." At this intersection, note the value of "18." Thus, when the dry-bulb temperature is 30°C and the relative humidity is 49%, the dewpoint is 18°C.

Weather Instruments

44. **1** In the photograph, weather instrument A consists of an arrow mounted on a rod in such a way that it can rotate freely. Wind exerts more force on the wide tail of the arrow than on the narrow head of the arrow. This causes the arrow to swing so that its head points in the direction from which wind is blowing, which is then registered on the digital device. This instrument is called a wind vane. Thus, the weather variable measured by a wind vane is wind direction.

Weather instrument B consists of a series of cups mounted on a shaft. The cups cause the shaft to spin when wind blows against them. The faster the wind blows, the faster the cups spin the shaft and the higher the wind speed that registers on the digital device. This instrument is called an anemometer. Thus, the weather variable measured by an anemometer is wind speed.

Therefore, A is a wind vane that measures wind direction, and B is an anemometer that measures wind speed.

45. **3** In 1851, the French physicist Jean Foucault devised a conclusive proof of Earth's rotation by means of a pendulum. He suspended a heavy iron ball on a long steel wire from the top of the dome of the Pantheon in Paris. As the pendulum swung back and forth, it appeared to change direction slowly, passing over different lines on the floor until eventually coming full circle to its original position after 24 hours. Since he knew that the path of a freely swinging pendulum would not change on its own, Foucault concluded that the apparent shift in the direction

of swing of the pendulum was due to the floor (Earth's surface) rotating beneath the pendulum. Thus, a Foucault pendulum is evidence of Earth's rotation.

Gaspard Coriolis successfully predicted the behavior of objects moving near a rotating Earth. In a stationary system, the behavior of fluids, such as the atmosphere, would be to move directly in a straight line from regions of high pressure to regions of low pressure. However, that is not what occurs on Earth. Large-scale movements of the atmosphere appear to be deflected from a straight line and instead follow a curved path. Since the curving path of winds would not occur on a stationary Earth, this apparent deflection is considered the result of Earth's rotation.

Thus, the apparent shift in the direction of swing of a Foucault pendulum and the curving of winds by the Coriolis effect are both evidence of Earth's rotation.

WRONG CHOICES EXPLAINED:

(1) If Earth's axis was not tilted but if Earth still rotated, there would still be a Coriolis effect. Thus, the Coriolis effect is not evidence of the tilt of Earth's axis.

(2) The shape of the surface beneath a freely swinging pendulum (whether flat or spherical) has no effect on the pendulum's direction of swing because there is no contact between the pendulum and the surface. Thus, a Foucault pendulum provides no evidence of Earth's shape.

(4) Since Earth's axis remains pointed in the same direction (parallel to itself) as Earth orbits the Sun, the direction of swing of a Foucault pendulum would remain unchanged in relation to Earth as the planet revolves around the Sun. Thus, the apparent shift in the direction of swing of a Foucault pendulum does not provide evidence of Earth's revolution. Furthermore, if Earth only rotated but did not revolve around the Sun, the curving path of winds would occur. Thus, the Coriolis effect is not considered evidence of Earth's revolution.

46. **1** Find the Equations chart in the *Reference Tables for Physical Setting/ Earth Science*. Locate the equation for rate of change:

$$\text{rate of change} = \frac{\text{change in value}}{\text{time}}$$

It is given that the pendulum's path was displaced 60 degrees during eight hours. Substitute these values into the rate of change equation and solve:

$$\text{rate of change} = \frac{60°}{8 \text{ h}} = 7.5°/\text{h}$$

Thus, the apparent path of the pendulum changed 7.5°/h.

Diagram B: Coriolis Effect on Planetary Winds

(Not drawn to scale)

Key
- - → Original direction of wind
—→ Deflected path of wind

47. **2** In diagram *B*, the Northern Hemisphere is the half of Earth extending from the equator to the North Pole; the Southern Hemisphere is the half of Earth extending from the equator to the South Pole.

By convention, when north is toward the top of a diagram and south is toward the bottom, east is to the right and west is to the left. Thus, in diagram *B*, the deflected path of wind in the Northern Hemisphere is from the north to the west, and the deflected path of wind in the Southern Hemisphere is from the south to the west. Winds are named by the direction *from* which they blow, not the direction *toward* which they blow. Therefore, to an observer in the path of these winds, the deflected winds in the Northern Hemisphere *come from* the northeast and the deflected winds in the Southern Hemisphere *come from* the southeast. Now find the Planetary Wind and Moisture Belts in the Troposphere chart in the *Reference Tables for Physical Setting/Earth Science*. Note that the planetary wind belt just north of the equator is labeled "N.E. Winds," and the planetary wind belt just south of the equator is labeled "S.E. Winds." Thus, the names of the planetary wind belts formed by deflected winds shown in the Northern and Southern Hemispheres is best shown by the table in choice (2).

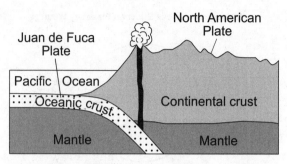

(Not drawn to scale)

48. **4** Find the Tectonic Plates map in the *Reference Tables for Physical Setting/Earth Science*, and locate the Juan de Fuca Plate. According to the key at the bottom of the map, the symbols along the eastern edge of the Juan de Fuca Plate correspond to a convergent plate boundary (subduction zone). Note that the arrows in the key indicate that the plates on either side of a convergent boundary are moving toward one another. Convection currents in the mantle cause lithospheric plates to move. Therefore, the relative motion of the upper mantle on either side of the boundary is toward the boundary. This relative motion is best shown by the diagram in choice (4).

49. **4** According to the diagram, the Juan de Fuca Plate consists of oceanic crust, and the North American Plate consists of continental crust. By visual inspection of the diagram, it can be seen that the Juan de Fuca Plate is thinner than the North American Plate. Find the Inferred Properties of Earth's Interior chart in the *Reference Tables for Physical Setting/Earth Science*. Locate the section labeled "Density (g/cm^3)" along the right-hand side of the cross section. Note that continental crust has a density of 2.7 g/cm^3 and that oceanic crust has a density of 3.0 g/cm^3. Thus, oceanic crust is denser than continental crust. Therefore, compared to the crust of the North American Plate, the Juan de Fuca Plate is thinner and more dense.

50. **2** Find the Inferred Properties of Earth's Interior chart in the *Reference Tables for Physical Setting/Earth Science*. Locate the asthenosphere and stiffer mantle in the cross section of Earth's interior at the top of the chart. Locate the dashed line corresponding to the boundary between the asthenosphere and the stiffer mantle. Follow this dashed line downward until it intersects the horizontal scale labeled "Depth (km)." Note that the dashed line intersects the scale at approximately 700 km. Thus, the boundary between the asthenosphere and the stiffer mantle is located below Earth's surface at a depth of approximately 700 km.

PART B–2

51. When lava is extruded in the rifts in mid-ocean ridges, crystals of minerals affected by magnetism align with Earth's magnetic field while the lava cools. When solidified, the rock contains a record of Earth's magnetic field locked in its crystals. Studies of ancient rocks show that Earth's magnetic field has reversed many times in the past. Research ships carrying sensitive instruments that can measure small changes in magnetism discovered that magnetism in rocks on the ocean floors reveals a striking pattern of strips of normal and reversed magnetic fields. The pattern on one side of the ridge is the mirror image of the pattern on the other side of the ridge. This suggested that rocks were forming at the ridges and then moving sideways away from the ridges. Thus, to complete the diagram, shade the pattern of normal polarity on the east side of the Mid-Atlantic Ridge center so that the width and placement of each normal polarity section is the mirror image of the pattern on the west side of the ridge, as shown below.

One credit is allowed if **the width and placement of the shaded areas are correctly indicated, as shown below.**

52. Find the Tectonic Plates map in the *Reference Tables for Physical Setting/Earth Science*. Note the symbol representing most mid-ocean ridges. In the key, note that the symbol corresponding to the one representing most mid-ocean ridges is a divergent plate boundary. Also note the statement "usually broken by transform faults along mid-ocean ridges." Thus, the type of tectonic plate boundary at the mid-ocean ridges where these magnetic patterns were produced is a divergent plate boundary.

One credit is allowed for **divergent/diverging**.

53. Find the Tectonic Plates map in the *Reference Tables for Physical Setting/ Earth Science*, and locate the Mid-Atlantic Ridge. Note that according to the key, the Mid-Atlantic Ridge is a divergent boundary. Along divergent boundaries, adjacent plates move apart and open rifts through which magma can rise to the surface and solidify to form new ocean floor. As the plates continue to move apart, this process is repeated, constantly creating new ocean floor and pushing the older floor on either side of the mid-ocean ridge away from the ridge. Thus, the age of ocean floor bedrock increases with distance away from the mid-ocean ridge. Therefore, the line representing the relationship between the distance from the Mid-Atlantic Ridge and the age of the ocean floor bedrock should form a U-shape or a V-shape, with the youngest age at the Mid-Atlantic Ridge and the oldest ages farthest from the ridge as shown below.

One credit is allowed for **a U-shape or a V-shape, with the youngest age at the Mid-Atlantic Ridge and the oldest ages farthest from the ridge**.

Examples of 1-credit responses:

Note: One credit is allowed even if the oldest ages are *not* exactly at the same "oceanic bedrock age" level.

www.geolsoc.org.uk

54. It is given that the rock in the photograph is composed of pebbles that have been cemented together. Note that the pebbles in the photograph of the rock are rounded in shape.

Find the Scheme for Sedimentary Rock Identification in the *Reference Tables for Physical Setting/Earth Science*. In the "Grain Size" column, note the entry "pebbles, cobbles, and/or boulders embedded in sand, silt, and/or clay." Trace this double row right to the "Comments" column, and note the single row corresponding to "rounded fragments." Trace this row right to the "Rock Name" column, and note that a sedimentary rock composed of rounded pebbles is "conglomerate." Trace left to the "Texture" column, and note that conglomerate has a "clastic (fragmental)" texture. Clastic sedimentary rocks are composed of fragments of rock or sediments. Find the Rock Cycle in Earth's Crust diagram in the *Reference Tables for Physical Setting/Earth Science*. Note along the arrow leading from "Sediments" to "Sedimentary Rock" the label "compaction and/or cementation." Thus, a rock sample composed of rounded pebbles that have been cemented together is conglomerate.

One credit is allowed for **conglomerate**.

55. Find the Relationship of Transported Particle Size to Stream Velocity graph in the *Reference Tables for Physical Setting/Earth Science*. Along the right side of the graph, locate the region labeled "pebbles" among the listed particle size names. Note that the dashed line marking the upper limit of the pebble range is labeled "6.4" and the dashed line marking the lower limit of the pebble range is labeled "0.2." Also note that these dashed lines intersect the vertical axis labeled "Particle Diameter (cm)." Therefore, these values represent centimeters. Particles in the range of 0.2 cm to 6.4 cm in diameter are classified as pebbles. Thus, the total possible range of particle diameters, in centimeters, is 0.2 cm to 6.4 cm.

One credit is allowed for **0.2 cm** to **6.4 cm**.

56. Note that the particles in the rock mostly have a smooth, rounded shape. As sediments are carried along by running water, they bounce off of and rub against one another. These collisions cause smaller pieces to break off of sharp corners and edges on the surface of the sediments, particularly at corners that protrude—a process called abrasion. The sediments are also abraded as they bounce, roll, and scrape against the bottom of the streambed. As a result, sediments eroded by running water change in size and shape, becoming smaller, smoother, and more rounded. Therefore, the smooth, rounded shape of the pebbles in the conglomerate in the photograph supports the inference that the particles in this rock were likely transported by running water.

One credit is allowed. Acceptable responses include but are not limited to:

- **Rounded pebbles usually indicate transportation by running water.**
- **Sediments tumbling or bouncing in a stream produce the round shapes.**
- **The particles are smooth and have round shapes.**

57. According to the key, carbon-14 atoms are shaded on the diagram. Count the number of shaded particles in the original sample and note that it contains sixteen carbon-14 atoms. Count the number of shaded particles in the sample after one half-life and note that it contains eight carbon-14 atoms. The decay of a radioactive isotope occurs at a statistically predictable rate known as its half-life. Half-life is the time required for one-half of the unstable radioactive isotope (such as carbon-14) to change into a stable disintegration product. Every half-life, one-half of the radioactive isotope atoms in the sample decay, forming atoms of stable disintegration product. Thus, at the end of the second half-life, one-half of the eight carbon-14 atoms that remained in the sample after the first half-life will decay, leaving four carbon-14 atoms.

One credit is allowed for **4 *or* four**.

58. Find the Radioactive Decay Data chart in the *Reference Tables for Physical Setting/Earth Science*. Locate "Carbon-14" in the "Radioactive Isotope" column. Trace right to the column labeled "Disintegration" and note that $^{14}C \rightarrow {}^{14}N$. Find the Properties of Common Minerals table in the *Reference Tables for Physical Setting/Earth Science*. In the key labeled "*Chemical symbols" at the bottom of the chart, note that "C = carbon" and "N = nitrogen." Thus, carbon-14 disintegrates into the stable disintegration product nitrogen-14.

One credit is allowed. Acceptable responses include but are not limited to:

- **nitrogen-14**
- ^{14}N
- $^{14}C \rightarrow {}^{14}N$

59. Find the Geologic History of New York State chart in the *Reference Tables for Physical Setting/Earth Science*. In the column labeled "Era," find the statement "Estimated time of origin of Earth and solar system." Trace left to the scale labeled "Million years ago" and note that the origin of Earth occurred about 4600 million years ago (4.6 billion years ago).

Locate the Radioactive Decay Data chart in the *Reference Tables for Physical Setting/Earth Science*. Note that uranium-238 (^{238}U) has a half-life of 4.5×10^9 years; that is, 4.5 billion years. Thus, the radioactive isotope that has a half-life that is approximately the same as the estimated age of Earth is uranium-238 (^{238}U).

One credit is allowed. Acceptable responses include, but are not limited to:

- **uranium-238**
- ^{238}U

Note: The response "uranium" alone is *not* accepted because there are different isotopes of uranium.

60. Index fossils are remains of organisms that had distinctive body features, were common and abundant, and had a broad, even worldwide range yet existed for only a short period of time. Their distinctive body features and broad distribution make index fossils easy to find in widely separated rock layers. Their short existence pinpoints the time period during which the rock layer was formed.

Find the Geologic History of New York State chart in the *Reference Tables for Physical Setting/Earth Science*. Locate the diagrams of index fossils along the bottom of the chart. Note that the index fossil that most closely resembles the one shown in rock layer A is "B–*Cryptolithus*." In the column labeled "Time Distribution of Fossils," locate the circled letter "**B**" and note that it lies on the gray bar labeled "trilobites." Note the explanation at the top of the column, "The center of each lettered circle indicates the approximate time of existence of a specific index fossil." From the center of the circled letter "**B**" representing *Cryptolithus*, trace left to the column labeled "Period." Note that *Cryptolithus* lived during the Ordovician Period. Therefore, the geologic period when rock layer A was deposited was the Ordovician Period.

One credit is allowed for **Ordovician Period**.

Note: "Middle Ordovician" is *not* accepted because that is an epoch.

61. It is given that the rock layers at these locations have not been overturned. The principle of superposition states that the bottom layer of a sedimentary series is the oldest, unless it was overturned or had older rock thrust over it, because the bottom layer was deposited first. Similarly, a rock layer is younger than those below it.

As explained in answer 60, rock layer A formed during the Ordovician Period. Based on superposition, rock layer D is older than rock layer A but younger than rock layer E. Find the Geologic History of New York State chart in the *Reference Tables for Physical Setting/Earth Science*.

Now locate the index fossil shown at location II in rock layer *H* in the diagrams of index fossils along the bottom of the chart. Note that the index fossil in rock layer *H* most closely resembles "**D**–*Valcouroceras*." In the column labeled "Time Distribution of Fossils," locate the circled letter "**D**". From the center of the circled letter "**D**" representing *Valcouroceras*, trace left to the column labeled "Period." Note that *Valcouroceras* lived during the Ordovician Period. Therefore, rock layer *H* formed during the Ordovician Period. Thus, rock layer *H* and rock layer *A* were both formed during the Ordovician Period, or are roughly the same age, and allow location I to be correlated with location II.

By superposition, rock layer *G* is younger than rock layer *H*, and rock layer *F* is younger than rock layer *G*. Rock layer *D* is older than rock layer *A* and, by correlation, older than rock layers *F* and *G*. Finally, by superposition, rock layer *E* is older than rock layer *D*. Thus from oldest to youngest, the relative ages of rock layers *D*, *E*, *F*, and *G* are: *E-D-G-F*.

One credit is allowed for **the correct order shown below**.

62. Layers of rock are generally deposited in an unbroken sequence. However, if forces within Earth cause rocks to be uplifted, deposition ceases. Weathering and erosion may wear away layers of rock before the land surface is low enough for another layer to be deposited. The uplift may cause rock layers to tilt or fold. The result is an unconformity, which is a break or gap in the sequence of a series of rock layers. Thus, the rocks above an unconformity are quite a bit younger than those below it. Thus, it is reasonable to infer that the rock units beneath the lower of the two unconformities were uplifted until they emerged from their depositional environment. Then their surface was weathered and eroded to form the unconformity. After that, the rock units subsided or were worn down until they were low enough for deposition to begin once again. Then several layers were deposited atop the unconformity, each burying the layers below it. Finally, the whole process occurred again, forming the upper unconformity and uppermost layers. Thus, the formation of the unconformities involved many processes, including uplift, emergence, weathering, erosion, subsidence, deposition, and burial. Any two of these processes should be listed as your answer.

One credit is allowed for *two* acceptable processes. Acceptable responses include but are not limited to:

- **uplift/emergence**
- **weathering**
- **erosion**

- submergence/subsidence/sinking
- deposition/sedimentation/precipitation
- burial

63. Stream discharge is the volume of water that flows through a stream in a given amount of time, measured in units like cubic feet per second. In general, the faster water is moving, the greater its carrying power and the more sediment the water can carry. All other things being equal, velocity is highest in streams with the highest discharge (rate of water flow). Thus, the greatest amount of sediment will be transported by the Delaware River when its discharge is the greatest. On the graph labeled "Delaware River Discharge," locate the highest point on the line. From that point, trace vertically downward to the horizontal axis labeled "Date" and note that the greatest discharge occurred on March 11.

Thus, the date when the river was most likely carrying the greatest amount of sediment was March 11.

One credit is allowed for **March 11**.

64. During a heavy rainfall or during a sudden snowmelt in the spring, the rate at which water builds up on the surface of the land exceeds the rate at which it can seep into the ground. Water that accumulates at the surface but does not seep into the ground moves downhill as runoff and adds to the volume of water in a stream. Therefore, the increase in stream discharge on March 15 was most likely due to rain or snowmelt that increased runoff into the stream, causing an increase in the volume of water in the stream.

One credit is allowed. Acceptable responses include but are not limited to:

- **precipitation**
- **rain shower**
- **snowmelt**

- **increased runoff**
- **flooding**
- **increase in the volume of water in the stream**

Note: Credit is *not* allowed for "velocity of water" or "increase in water velocity" because water velocity is a result of an increase in water volume or discharge.

65. It is given that the Delaware River flows out of the Catskills and into the Atlantic Ocean. First, find the Generalized Landscape Regions of New York State map in the *Reference Tables for Physical Setting/Earth Science,* and locate "The Catskills." Next, find the Generalized Bedrock Geology of New York State map in the *Reference Tables for Physical Setting/Earth Science.* Locate the region corresponding to the Catskills and the Delaware River along the Pennsylvania–New York State border. Note the location of the Atlantic Ocean to the southeast of the Catskills. Thus, the general compass direction toward which the Delaware River flows *along the New York State border* is toward the southeast. (Note that the Delaware River turns where the Pennsylvania–New York State border meets the New Jersey border and flows toward the south along the New Jersey border.)

One credit is allowed for **southeast/SE *or* south/S *or* south southeast/SSE**.

PART C

66. Find the Equations section in the *Reference Tables for Physical Setting/ Earth Science*, and note the equation for gradient. To determine the gradient, proceed as follows:

   a. Write the equation for determining the gradient:

$$\text{gradient} = \frac{\text{change in field value}}{\text{distance}}$$

   b. Note that point A lies on the 300-foot contour line and point B lies on the 370-foot contour line. Align the edge of a piece of scrap paper with points A and B on the map, and mark off the distance from point A to point B. Compare this to the map scale below the map, and note that points A and B are 0.1 miles apart. Substitute the elevations of points A and B and the distance between them in the equation and solve:

$$\text{gradient} = \frac{370 \text{ ft} - 300 \text{ ft}}{0.1 \text{ mile}} = \frac{70 \text{ ft}}{0.1 \text{ mile}} = 700 \text{ ft/mile}$$

   c. Label the answer with the correct units.

$$\text{Gradient} = 700 \text{ ft/mile}$$

One credit is allowed for any value from **636** to **778 ft/mile**.

67. To construct a topographic profile along line *CD*, proceed as follows. Place the straight edge of a piece of scrap paper along the solid line connecting point *C* to point *D*. Mark the edge of the paper at points *C* and *D* and wherever the paper intersects a contour line. Wherever the paper intersects a contour line, label the mark with the elevation of that contour line as shown below.

Now place this scrap paper along the lower edge of the grid provided, so points *C* and *D* on the paper align with points *C* and *D* on the grid. Then at each point where a contour line crossed the edge of the paper, draw a dot on the grid at the appropriate elevation.

Finally, connect all of the dots in a smooth curve to form the finished profile, as shown below.

One credit is allowed if the centers of *all seven* plots are within or touch the rectangles shown and *all nine* plots are correctly connected with a line, from C to D that passes within or touches each rectangle.

Note: One credit is allowed if the line misses a plot, but is still within or touches the rectangle.

68. On a topographic map, contour lines connect points of equal elevation. The distance between one contour line and the next indicates the distance over which the elevation changes. The closer together the contour lines are on a map, the shorter the distance one must travel in order to change elevation, that is, the steeper the slope. Thus, the closely spaced contour lines on the north side of Chimney Bluffs indicate that the north side of Chimney Bluffs has the greatest change in elevation over the shortest distance and, therefore, the steepest slope.

One credit is allowed. Acceptable responses include but are not limited to:

- **The contour lines are close together on the north side of Chimney Bluffs.**
- **The spacing between the contour lines is small.**
- **There is a great change in elevation in a short distance on the north side.**

69. Find the Key to Weather Map Symbols in the *Reference Tables for Physical Setting/Earth Science* and locate the section labeled "Station Model Explanation." Note the positions in which the values for dewpoint, amount of cloud cover, barometric pressure, and present weather are placed on a station model.

Note that dewpoint (°F) is shown to the lower left of the station model. Thus, the dewpoint in Albany, New York, is 64°F.

Note that the amount of cloud cover is represented by how much of the circle at the center of the station model is shaded. The entire circle at the center of the station model is shaded. Thus, the cloud cover at Albany, New York, is 100%.

Note that barometric pressure is shown at the upper right of the station model and that the three digits listed represent the last three digits of the air pressure in millibars. Since air pressure rarely rises above 1050.0 mb and rarely drops below 950.0 mb, the general rule is to place a 9 in front of the three digits on the station model if they are greater than 500 and place a 10 in front of the three digits if they are less than 500. Then place a decimal point between the last two digits. According to the station model, the barometric pressure is represented by the digits 009. Since the three digits are less than 500, place a 10 in front of the digits and a decimal point between the last two digits. Therefore, the barometric pressure at the weather station is 1000.9 mb.

Finally, note that present weather is indicated with a symbol. According to the "Present Weather" key, the symbol shown on the station model represents haze. Therefore, the present weather at Albany, New York, is haze. Record these values in the table provided, as shown below.

One credit is allowed if ***all four*** **weather conditions are correctly recorded as shown below.**

Weather in Albany, New York	
Dewpoint	64°F
Amount of cloud cover	100%
Barometric pressure	1000.9 mb
Present weather	Haze

70. The characteristics of an air mass are the result of the geographic region over which it formed, or its *source region*. Air resting on or moving very slowly over a region tends to take on the characteristics of that region. For example, air that sits over the Gulf of Mexico is resting on warm water. The air is warmed by contact with the water, and evaporation causes much humidity to enter the air. As a result, the air becomes warm and moist, just like the Gulf of Mexico. An air mass that forms over water is called a maritime air mass. A warm air mass is called a tropical air mass. Therefore, the air mass that normally forms over the Gulf of Mexico is a maritime tropical air mass.

Find the Key to Weather Map Symbols in the *Reference Tables for Physical Setting/Earth Science* and locate the section labeled "Air Masses." Note that mT is the symbol for a maritime tropical air mass.

One credit is allowed for **mT**.

Notes: Credit is allowed for using either uppercase or lowercase letters. However, credit is *not* allowed if the air mass letters are reversed, such as Tm.

For students who used the Spanish edition, either exclusively or in conjunction with the English edition of the exam, credit is allowed for the correct two-letter air mass symbol as it appears in either the English or Spanish *Reference Tables for Physical Setting/Earth Science*.

71. Low-pressure centers are usually associated with warm, moist air that is rising along frontal boundaries. As warm, moist air rises, it expands, which causes adiabatic cooling. As the air cools, the air temperature approaches the dewpoint temperature. When the air is cooled to its dewpoint or below, the water vapor in the air condenses. Clouds form when water vapor in the atmosphere changes back into tiny droplets of liquid water by the process of condensation or forms ice crystals by the process of deposition (sublimation).

One credit is allowed. Acceptable responses include but are not limited to:

- **condensation**
- **cooling**
- **expansion**
- **deposition/phase change directly from water vapor to ice**

Month	Jan	Feb	Mar	Apr	May	Jun	Jul	Aug	Sep	Oct	Nov	Dec
Average High Air Temperature (°F)	14	15	21	30	41	50	54	53	47	36	28	18
Average Snowfall (in)	44	40	45	36	12	1	Trace	0.1	2	18	38	46

72. To construct a graph of the average high air temperatures for each month, do the following. In the data table, note that during the month of January, the "Average High Air Temperature (°F)" is 14. On the grid provided, locate the intersection of January on the horizontal axis labeled "Months" and 14 on the vertical axis labeled "Air Temperature (°F)." Draw an **X** at this intersection. Repeat this process for each of the remaining months and their corresponding average high air temperatures listed in the table given in the question. When all of the **X**s have been plotted, connect them with a smooth line. A graph with all points correctly plotted and connected with a smooth line is shown below.

One credit is allowed if **the centers of *all twelve* plots are within or touch the circles shown and are correctly connected with a line that passes within or touches each circle**.

Note: Credit is allowed if the student-drawn line does not pass through the student plot points but is still within or touches the circles. Credit is *not* allowed if the student makes any attempt to graph average snowfall on the Monthly Average High Temperatures graph because the vertical axis is in degrees Fahrenheit, not inches.

73. It is given that Mount Washington is one of the highest mountains east of the Mississippi. Temperature is a major climate factor. Mountains are, by definition, regions of high elevation. Find the Selected Properties of Earth's Atmosphere chart in the *Reference Tables for Physical Setting/Earth Science*. Locate the graph labeled "Temperature (°C)." Note that temperature decreases rapidly with increasing altitude (elevation) in the troposphere. Elevation is a climate factor because there is a decrease in temperature with an increase in elevation. The fact that temperature decreases about 1°C for every 100-meter rise in elevation explains why high mountains may have tropical vegetation at their bases but permanent ice and snow at their peaks. Thus, the climate factor that best explains why Mount Washington has low air temperatures throughout the year is elevation (altitude above sea level).

One credit is allowed. Acceptable responses include but are not limited to:

- **elevation**
- **height above sea level**
- **high altitude**

Note: Credit is *not* allowed for "mountain" or "top of a mountain" because this identifies the location from which the data were taken but does not identify the climate factor affecting this mountain that causes low air temperatures.

74. According to the reading passage, Mount Washington "has deep-cut glacial valleys formed by continental glaciers approximately 1.7 to 1.8 million years ago." Find the Geologic History of New York State chart in the *Reference Tables for Physical Setting/Earth Science*. Locate the time scale (millions of years ago) between the "Epoch" and "Life on Earth" columns. On this scale, locate 1.7 to 1.8 million years ago, and trace horizontally left to the "Epoch" column. Note that this time interval corresponds to the Pleistocene Epoch.

One credit is allowed for **Pleistocene Epoch**.

Velocities of Galaxies Moving Away From Earth	
Galaxy	Velocity (km/s)
A	61,000
B	15,000
C	1200
D	39,000
E	22,000

75. It is given that the farther away a galaxy is, the faster it is moving away. Therefore, to list the galaxies in order from closest to Earth to farthest from Earth, they should be listed in order from lowest to highest velocities at which they are moving away from Earth. According to the table, this is: C-1200, B-15,000, E-22,000, D-39,000, and A-61,000.

One credit is allowed for **a correct list as shown below**.

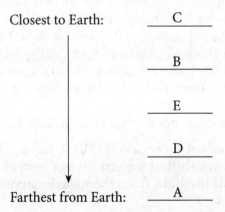

Note: One credit is allowed if the correct velocities are substituted for the letters.

76. The best explanation for the observation that all galaxies are moving away from each other is that the universe is expanding in all directions. If the motion of galaxies is then traced back in time, there is a point about 13 to 14 billion years ago at which the galaxies were very close to each other. Thus, we have a model in which the universe started out with all its matter and energy in a very small volume and then expanded outward in all directions. This motion is similar to an explosion, hence the name for this model—the Big Bang theory.

One credit is allowed for **the Big Bang theory** *or* **the Big Bang**.

77. When light emitted by the atoms in a star's core passes through the star's cooler outer layers, certain wavelengths of light are absorbed. The result is a pattern of dark spectral lines in the spectrum of light emitted by the star. Spectral lines are highly atom specific and can be used to identify the chemical element that produced the spectrum. When compared with the spectral lines of elements on Earth, the spectral lines of elements in distant stars have the same pattern, but that pattern is shifted from its normal position. By observing the pattern of spectral lines in light emitted by galaxies, scientists are able to determine that a galaxy is moving toward or away from Earth in the following manner. If a source of electromagnetic waves is moving *away* from an observer at the same time as the source is emitting light of a particular wavelength, fewer wave crests will reach the eye of the observer each second. The eye will interpret this as meaning that the light has a longer wavelength than it actually has, and the spectral lines will be shifted toward the longer wavelength end of the spectrum. If the source is moving toward the observer, the opposite will occur. This is called the Doppler effect.

Find the Electromagnetic Spectrum chart in the *Reference Tables for Physical Setting/Earth Science*. Locate the section of the chart labeled "Visible light." Note that light at the red end of the spectrum has a longer wavelength than light at the blue end of the spectrum. Thus, if the wavelengths of light appear longer than they actually are, they will be shifted toward the red end of the spectrum. When the wavelengths of light (and spectral lines) from distant galaxies are observed, they are *all* lengthened and shifted toward the red end of the spectrum, or red-shifted, indicating that *all* of these galaxies are all moving away from us.

One credit is allowed. Acceptable responses include but are not limited to:

- **The Doppler effect shows a red shift in light.**
- **Spectral lines are shifted toward the red end of the spectrum.**
- **Wavelengths of the light from the galaxies appear to be longer.**
- **Doppler effect**
- **Red shift**

Note: Credit is *not* allowed for "cosmic background radiation" because even though this is evidence for the Big Bang event, it is not used to determine that a galaxy is moving away from Earth.

78. Find the Characteristics of Stars chart in the *Reference Tables for Physical Setting/Earth Science*. Locate "10" on the scale labeled "Luminosity (Rate at which a star emits energy relative to the Sun)" along the left side of the chart and "8000" on the scale labeled "Surface Temperature (K)" along the bottom of the chart. Trace right from luminosity 10 and upward from temperature 8000K, and

locate the point where these two values intersect. Note that the intersection falls within the region labeled "Main Sequence (Early stage)." Now from "8000" on the scale labeled "Surface Temperature (K)" along the bottom of the chart, trace downward to the scale labeled "Color." Note that 8000K corresponds to a star with a white color.

One credit is allowed if *both* responses are acceptable.
Stage: **main sequence** *or* **early stage**
Color: **white**

79. According to the graph, the altitude of the Sun at 23.5° N is 90°. On June 21, the summer solstice in the Northern Hemisphere, the Sun is directly overhead (altitude 90°) at the Tropic of Cancer, which is located at latitude 23.5° N. Thus, the date represented by the data shown on the graph is June 21.

One credit is allowed. Acceptable responses include but are not limited to:

- **June 20**
- **June 21**
- **June 22**
- **summer solstice**

80. As the angle of insolation (altitude of the Sun) increases from 0° toward 90°, the same amount of insolation is spread over smaller and smaller areas. Insolation is most intense when it strikes a surface at 90° (that is, perpendicular to a surface) because the insolation is concentrated in the smallest possible area. Therefore, as the noontime altitude of the Sun (angle of insolation) increases, the intensity of insolation increases. According to the graph, the altitude of the Sun is 90° at latitude 23.5° N (Tropic of Cancer). Therefore, the latitude that has the greatest intensity of insolation at approximately noon on this date is 23.5° N.

One credit is allowed for **23.5° N** *or* **Tropic of Cancer and acceptable evidence**. Acceptable evidence includes but is not limited to:

- **The graph shows the altitude of the Sun is greatest at 23.5° N.**
- **Sun is highest in the sky.**
- **The Sun's altitude is 90°.**
- **The Sun is directly overhead.**

81. According to the graph, if an observer is located at 90° N, the altitude of the Sun would be 23.5° and the observer would cast a shadow. Latitude 90° N corresponds to the North Pole. From the North Pole, all directions are to the south. Therefore, at 90° N at noon on this date, an observer's shadow would extend to the south.

One credit is allowed for **south**.

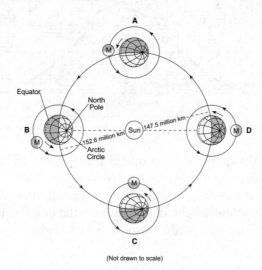

82. An observer on Earth can only see that portion of the Moon that is illuminated by the Sun's rays and only from the side of Earth that is in darkness. In the diagram, the part of the Moon that is *not* shaded represents the illuminated portion of the Moon. According to the diagram, when the Moon is in position *D*, its entire illuminated portion faces Earth's nighttime side and the full moon phase would be observed.

When the moon is at position *D*, Earth is directly between the Moon and the Sun. When this occurs, the Moon can move through Earth's shadow, and a lunar eclipse can be viewed. Thus, a lunar eclipse could occur when the Moon is at position *D*.

One credit is allowed for **lettered position *D*** and **lunar eclipse**.

83. When sunlight strikes the spherical Earth, exactly one-half of Earth's sphere is illuminated and experiences daylight and exactly one-half is in darkness. The boundary between daylight and darkness forms a great circle that bisects Earth's sphere. On an equinox, the Sun's direct rays are at the equator, and the boundary between daylight and darkness passes through both the North and South Poles as it bisects Earth's spherical surface. Thus, position A represents an equinox. Note the arrows on Earth's orbit indicate its direction of motion. From position A, Earth would orbit toward position B. On the first day of summer (summer solstice) in the Northern Hemisphere, Earth's axis of rotation is tilted farthest toward the Sun. The position in the diagram at which Earth's axis of rotation is tilted farthest toward the Sun is position B. Therefore, position B represents the first day of summer. Therefore, position A represents the equinox that occurs before the first day of summer, or the spring equinox. The season between the spring equinox and the summer solstice is spring. Therefore, the Northern Hemisphere season that is beginning at position A is spring.

One credit is allowed for **spring**.

84. It is given that Earth is closest to the Sun near position D. Therefore, as Earth moves from position C to position D, the distance between Earth and the Sun decreases.

The force of gravitational attraction between any two objects is directly proportional to their masses and indirectly proportional to the square of the distance between their centers. Newton expressed this relationship in a simple mathematical expression:

$$F \propto \frac{m_1 \cdot m_2}{d^2}$$

Therefore, the force of gravitational attraction between the two objects increases as the distance between them decreases. Thus, the gravitational attraction between the Sun and Earth increases as Earth moves from position C to position D because the distance between the Sun and Earth is decreasing.

One credit is allowed. Acceptable responses include but are not limited to:

- **Earth is getting closer to the Sun.**
- **The distance between Earth and the Sun is decreasing.**
- **Earth's orbit is elliptical, and position D is closer to the Sun.**

85. A geocentric (*geo* = "Earth," *centric* = "centered") model of the solar system is one in which Earth is motionless at the center of the solar system and all other objects revolve around Earth. In the heliocentric (*helios* = "Sun," *centric* = "centered") model, the Sun is near the center of the solar system and all other objects in the system revolve around the Sun. Note that in the diagram, Earth orbits the Sun. Note, too, the arrows indicating that Earth rotates (is not motionless as in the geocentric model). Therefore, this diagram is a heliocentric model because it shows the Sun near the center of the diagram with a rotating Earth orbiting the Sun.

One credit is allowed. Acceptable responses include but are not limited to:

- **The Sun is near the center of the diagram.**
- **Earth orbits the Sun.**
- **The small arrows by Earth indicate that Earth rotates.**

SELF-ANALYSIS CHART August 2019

Topic	Question Numbers (Total)	Wrong Answers (x)	Grade
Standards 1, 2, 6, and 7: Skills and Application			
Skills			
Standard 1 Analysis, Inquiry, and Design	1–5, 8, 10, 12, 14, 18, 20–25, 28, 33–36, 39–44, 46–50, 52–55, 57–62, 64–74, 76–81, 84		$\dfrac{100(60-x)}{60} = \%$
Standard 2 Information Systems			
Standard 6 Interconnectedness, Common Themes	2, 5, 7, 9, 10, 15, 17, 20, 23, 25–30, 36–39, 44–52, 54–56, 60–64, 67, 69–71, 75, 80, 82–85		$\dfrac{100(46-x)}{46} = \%$
Standard 7 Interdisciplinary Problem Solving			
Standard 4: The Physical Setting/Earth Science			
Astronomy			
The Solar System (MU 1.1a, b; 1.2d)	1–3, 8, 82–85		$\dfrac{100(8-x)}{8} = \%$
Earth Motions and Their Effects (MU 1.1c, d, e, f, g, h, i)	5, 7, 9, 36–39, 45, 46, 79, 81		$\dfrac{100(11-x)}{11} = \%$
Stellar Astronomy (MU 1.2b)	78		$\dfrac{100(1-x)}{1} = \%$
Origin of Earth's Atmosphere, Hydrosphere, and Lithosphere (MU 1.2e, f, h)	21		$\dfrac{100(1-x)}{1} = \%$
Theories of the Origin of the Universe and Solar System (MU 1.2a,c)	4, 6, 75–77		$\dfrac{100(5-x)}{5} = \%$

SELF-ANALYSIS CHART August 2019

Topic	Question Numbers (Total)	Wrong Answers (x)	Grade
Meteorology Energy Sources for Earth Systems (MU 2.1a, b)			
Weather (MU 2.1c, d, e, f, g, h)	13–15, 42–44, 69–71		$\dfrac{100\,(9-x)}{9} = \%$
Insolation and Seasonal Changes (MU 2.1i; 2.2a,b)	15, 16, 18–20, 47, 80		$\dfrac{100\,(7-x)}{7} = \%$
The Water Cycle and Climates (MU 1.2g; 2.2c,d)	10–12, 17, 72, 73		$\dfrac{100\,(6-x)}{6} = \%$
Geology Minerals and Rocks (MU 3.1a, b, c)	33–35, 40, 41, 54		$\dfrac{100\,(6-x)}{6} = \%$
Weathering, Erosion, and Deposition (MU 2.1s, t, u, v, w)	29–32, 55, 56, 63–65		$\dfrac{100\,(9-x)}{9} = \%$
Plate Tectonics and Earth's Interior (MU 2.1j, k, l, m, n, o)	24, 25, 48–53		$\dfrac{100\,(8-x)}{8} = \%$
Geologic History (MU 1.2 i, j)	22, 23, 26, 27, 57–62, 74		$\dfrac{100\,(11-x)}{11} = \%$
Topographic Maps and Landscapes (MU 2.1p, q, r)	66–68		$\dfrac{100\,(3-x)}{3} = \%$
ESRT 2011 Edition Reference Tables for Physical Setting/Earth Science	1, 3, 4, 14, 18, 20–25, 28, 33–36, 40–44, 47–50, 52, 54, 55, 58–60, 64–66, 69–71, 74, 78		$\dfrac{100\,(39-x)}{39} = \%$

To further pinpoint your weak areas, use the Topic Outline in the front of the book.
MU = Major Understanding (see Topic Outline)

NOTES

NOTES

NOTES

NOTES

NOTES

NOTES

NOTES

Let's Review Regents:

Earth Science— Physical Setting

Revised Edition

Edward J. Denecke, Jr., B.A., M.A.
Formerly, William H. Carr J.H.S. 194Q
Whitestone, New York

© Copyright 2021, 2020, 2018, 2017, 2016, 2015, 2014, 2013, 2012, 2011, 2010, 2009, 2008, 2007, 2006, 2005, 2004, 2003, 2002, 2001, 2000, 1995 by Kaplan, Inc., d/b/a Barron's Educational Series

All rights reserved.
No part of this book may be reproduced or distributed in any form or by any means without the written permission of the copyright owner.

Published by Kaplan, Inc., d/b/a Barron's Educational Series
1515 W. Cypress Creek Road
Fort Lauderdale, FL 33309
www.barronseduc.com

ISBN: 978-1-5062-6464-6

10 9 8 7

Kaplan, Inc., d/b/a Barron's Educational Series print books are available at special quantity discounts to use for sales promotions, employee premiums, or educational purposes. For more information or to purchase books, please call the Simon & Schuster special sales department at 866-506-1949.

TABLE OF CONTENTS

PREFACE v

CAUTION! DON'T USE THIS BOOK UNTIL YOU READ THIS vii

TOPIC ONE	**Astronomy**	**1**
UNIT ONE	**From a Geocentric to a Heliocentric Universe**	**1**

Chapter 1:	Early Astronomy and the Geocentric Model	1
Chapter 2:	The Development of the Heliocentric Model	19
Chapter 3:	Heliocentric Earth Motions and Their Effects	41
Chapter 4:	Earth's Coordinate System and Mapping	69
Chapter 5:	Our System of Time	97

UNIT TWO	**Modern Astronomy**	**121**

Chapter 6:	Stars, Their Origin and Evolution	121
Chapter 7:	The Solar System	143
Chapter 8:	Theories of the Origin of the Universe	173

UNIT THREE	**Earth's History**	**187**

Chapter 9:	The Origins of Earth and Its Moon	187
Chapter 10:	The Origin and Structure of Earth's Atmosphere	199
Chapter 11:	The Origin and Nature of Earth's Hydrosphere	211
Chapter 12:	The Origin and History of Life on Earth	237

TOPIC TWO	**Geology**	**275**
UNIT FOUR	**Earth Materials**	**275**

Chapter 13:	Minerals	275
Chapter 14:	Rocks	299

UNIT FIVE	**The Dynamic Earth**	**333**

Chapter 15:	Earthquakes and Earth's Interior	333
Chapter 16:	Volcanoes and Earth's Internal Heat	365
Chapter 17:	Plate Tectonics	381

Table of Contents

UNIT SIX — Weathering, Erosion, and Deposition — 419

Chapter 18:	Weathering and Soil Formation	419
Chapter 19:	Erosion	441
Chapter 20:	Deposition	483

TOPIC THREE — Meteorology — 527

UNIT SEVEN — The Atmosphere, Weather, and Climate — 527

Chapter 21:	The Atmosphere	527
Chapter 22:	Weather	559
Chapter 23:	Climate	613

Appendix A: Answers to Review Questions — 639
Appendix B: Reference Tables for Physical Setting/Earth Science — 695

Glossary of Earth Science Terms — 711

June 2019 Exam — 729

Answer Key — 768

Index — 769

Preface

To the Student:

"On the surface, islands may seem separate, but underneath they are all connected."

Earth science is not the study of isolated facts but rather the development of a deep understanding and appreciation of the interconnectedness of Earth phenomena, processes, and systems. In Earth science, the key to success is thorough understanding and the ability to demonstrate what you understand. This book is a concise text and review aid in which the author has tried to make Earth science as understandable as possible to you, the student, by incorporating the following features:

- Explanations of concepts and understandings are detailed, yet simply and clearly stated. They are designed to help you grasp the "how" and "why" of an idea, rather than just stating the idea.
- Important terms are printed in **boldface type** where they are defined in the text.
- The illustrations are designed to make difficult ideas easier to understand. Many of the illustrations are similar to those used in Regents examination questions in order to familiarize you with the types of diagrams you will be asked to analyze and interpret.
- Each chapter ends with a wide range of review questions from previous Regents examinations, including constructed response and extended constructed response questions.
- A glossary and a complete index make it easy for you to find the definition of a specific term and the pages in the book where a topic is covered.
- A full-length Regents examination provides you with the opportunity to assess your understanding and to test your Earth science knowledge and skills before taking the Regents examination.

Earth science offers the challenge and excitement of new theories, new discoveries, and new problems to be solved. It is a science in which sweeping new theories are being tested and applied to puzzling new observations. It is a science in which revolutionary advances in human knowledge of Earth and the other planets of our solar system are being made almost daily. I hope that studying Earth science will fill you with wonder and delight at the complexities of our planet Earth.

Preface

To the Teacher:

Let's Review Regents: Earth Science—Physical Setting is a concise text and review aid for courses based on the New York State *Physical Setting/ Earth Science Core Curriculum*, a comprehensive course of study in Earth science on the secondary level. However, the material in this book provides such comprehensive coverage of topics in Earth science that it can be used as a review text to supplement virtually any secondary course in Earth science taught in the United States, using any major textbook.

This edition reflects the content of the New York State *Physical Setting/ Earth Science Core Curriculum*. It is organized into three major topics: astronomy, geology, and meteorology. Each topic addresses a key idea of Standard 4: The Physical Setting/Earth Science, and is divided into units based upon the performance indicators for that key idea. Each unit is subdivided into chapters that deal with groups of related major understandings underlying the performance indicator. Specific skills identified in Standards 1, 2, 6, and 7 are introduced with the appropriate major understanding. Figures and text follow the updated 2011 edition of the *Physical Setting/ Earth Science Reference Tables*.

The review questions in this edition have been chosen from previous Regents examinations to reflect content consistent with the *Physical Setting/Earth Science Core Curriculum* and suitability for use with the reference tables. Constructed response and extended constructed response practice questions are included at the end of each chapter.

I wish to thank my wife, Gerry, for her infinite patience and my children, Meredith, Abigail, and Benjamin, for their loving support during the preparation of this manuscript.

CAUTION!
Don't Use This Book Until You Read This

You are taking Earth science and want to get good grades on your exams. Well, before you even look at the Earth science subject matter in this book, let's talk about how to use this book. Many students don't really know how to read a textbook or a review book. Suppose you get a homework assignment such as "Read pages 73–82 in the text, and answer questions 1–5 on page 83." What will you do? Typical students turn to page 73 and start reading, sentence by sentence, paragraph by paragraph. Once in a while they may stop to see how much more they have to read. When they finally reach the last sentence on page 82, they consider that they have "studied" the material and then turn to page 83 and try to answer the questions. This simple procedure doesn't require a lot of time or effort. But just reading the text once is a really weak approach to studying. So what is a better way to study?

Understand How the Text Is Organized

First, it is important to realize that textbooks are not written like novels. Textbooks are not meant to be read straight through; instead, they are structured to guide in-depth study. If you understand how a textbook is arranged, you can organize your reading into small, useful blocks.

Let's Review Regents is based upon the New York State *Physical Setting/Earth Science Core Curriculum* and is organized as follows:

ORGANIZATION

Physical Setting/Earth Science Core Curriculum	Let's Review Regents: Earth Science—Physical Setting
STANDARDS tell what you are expected to know and be able to do.	
KEY IDEAS tell you the really important ideas relating to the standard.	**TOPICS** are based upon the major disciplines in earth science—astronomy, geology, and meteorology—that are addressed in the key ideas of the standards.
PERFORMANCE INDICATORS describe what you should be able to do in order to show that you understand the key ideas.	**UNITS** are based upon the elements of the performance indicators for each key idea. For example, Unit 2: Modern Astronomy addresses Performance Indicator 1.2—Describe current theories about the origin of the universe and solar system.

Caution!

MAJOR UNDERSTANDINGS list what you need to know in order to do the things described in the performance indicators.	**CHAPTERS** address groups of related major understandings underlying each performance indicator. For example, Chapter 12: The Origin and History of Life on Earth addresses Major Understandings 1.2h–j.
	SECTIONS break chapters into manageable reading blocks to help you better understand what you read. For example, Chapter 18: Weathering and Soil Formation has such sections as Physical Weathering, Chemical Weathering, and The Products of Weathering.

Know How to Spell Success: S-Q-3R

Before you read, you need to realize that studying means not just reading, but also *thinking deeply* about what you have read. One good approach is to take notes while reading and then study your notes before a test. A better approach is to read each chapter subsection once, read it again and highlight the important points, and then study those points after you finish reading each chapter.

One of the best approaches to studying is called **S-Q-3R**, which stands for **S**urvey, **Q**uestion, **R**ead, **R**ecite, and **R**eview. Here's how it works:

1. You **Survey** the chapter by reading through the section and subsection headings to get a mental map of the material.
2. You ask **Questions** about the material by turning each heading into a question.
3. Then you **Read** the text, a subsection at a time, with the purpose of answering these questions.
4. Next, you **Recite** by jotting down brief notes about what you have read, making an outline or a graphic organizer, or writing a summary. (Note: In this case the word *recite* doesn't mean to "speak publicly"; it means "to re-cite, or cite again." To *cite* is to quote, or mention. Here, *recite* means to list or itemize important ideas you have read.)
5. Finally, you **Review** by rereading your notes and answering questions about the material. They may be questions that you pose to yourself or that have appeared on prior tests.

This five-step approach will take more time than just reading through, but the reward for the extra time you spend will be better grades. Therefore, plan to have enough time to study before you sit down for a session with this book.

Know What's on the Test

The key to success in preparing for any test is to know what will be expected of you so that you can review and practice beforehand. The New York State Physical Setting/Earth Science Regents Examination has four parts:

Part A—Multiple-Choice. In a multiple-choice question you are given several choices from which to select the one that best answers the questions or completes the statement. Many practice questions of this type from previous Regents examinations are included at the end of each chapter of this book. Part A of the exam focuses on Earth science content from Standard 4.

Part B—Multiple-Choice and Constructed Response. In a constructed response question there is no list of choices from which to select an answer; rather, you are required to provide the answer. Constructed response questions can test skills ranging from constructing graphs or topographic maps to formulating hypotheses, evaluating experimental designs, and drawing conclusions based upon data. Practice questions of this type are also included at the end of each chapter. In Part B you will be asked to demonstrate skills identified in Standards 1, 2, 6, and 7 in the context of Earth science.

Part C—Extended Constructed Response. The constructed response questions require more time (5–10 minutes per item) and effort on your part to answer. Questions in Part C require you to apply your Earth science knowledge and skills to real-world problems and applications. You may be asked to produce short essays, design controlled experiments, predict outcomes, or analyze the risks and benefits of various solutions to a problem.

Part D—Laboratory Performance Tasks. These tasks test your laboratory skills. You will take this part of the exam sometime during the 2 weeks before the written Regents. Laboratory performance tasks involve skills such as using instruments (e.g., rulers, external protractors, triple beam balances, graduated cylinders, stopwatches), observing properties of Earth materials, performing calculations, and collecting and analyzing data.

Note: The following description represents that information the State Education Department has stated may be shared with students before taking the performance part of the examination. You should be familiar with the skills being assessed because you have used them in laboratory activities throughout the year. However, you will not be allowed to practice the entire test or any of the individual stations before this performance component is administered.

Caution!

Station 1 ... *Mineral and Rock Identification*
The student determines the properties of a mineral and identifies that mineral using a flowchart. Then the student classifies two different rock samples and states a reason for each classification based on observed characteristics.

Station 2 ... *Locating an Epicenter*
The student determines the location of an earthquake epicenter using various types of data that were recorded at three seismic stations.

Station 3 ... *Constructing and Analyzing an Asteroid's Elliptical Orbit*
The student constructs a model of an asteroid's elliptical orbit and compares the eccentricity of the orbit with that of a given planet.

Extensive analyses of questions of all types can be found in the companion volume to this book, ***Barron's Regents Exams and Answers: Earth Science—Physical Setting***. Together, these review books will help you prepare for the New York State Physical Setting/Earth Science Regents Examination by clearly explaining what you should know and be able to do in order to perform well on the exam and by providing you with practice questions from prior exams that are thoroughly explained.

Unit One

FROM A GEOCENTRIC TO A HELIOCENTRIC UNIVERSE

CHAPTER 1

EARLY ASTRONOMY AND THE GEOCENTRIC MODEL

> **KEY IDEAS** People have observed the stars for thousands of years, using them to find direction, note the passage of time, and express human values and traditions. To an observer on Earth, it appears that Earth stands still and everything else moves around it. Thus, in trying to make sense of how the universe works, it was logical for early astronomers to start with those apparent truths. To comprehend our modern view of the universe, it is helpful to begin by understanding these first attempts to explain the universe in terms of what can be seen from our vantage point on Earth. As technology has progressed, so has our understanding of celestial objects and events.

KEY OBJECTIVES
Upon completion of this chapter, you will be able to:

- Explain the meaning of the term celestial object.
- Compare and contrast "apparent" and "real" motion.
- Explain how the celestial sphere model of the sky accounts for the motions of celestial objects.
- Explain how Earth's rotation makes it appear that the Sun, the Moon, and the stars are moving around Earth once a day.
- Locate *Polaris* in the night sky.

OBSERVING THE SKY

If you kept a list of things observed in the sky, it might include birds, smoke, clouds, rainbows, halos, lightning, stars, the Moon, the Sun, and comets. One of the first ideas that might occur to you is that the sky has depth. Some things in it appear closer, and some appear farther away. Why? Perspective! From everyday experiences you know that closer objects block your view of more distant objects. For example, if you hold your hand in front of your

Early Astronomy and the Geocentric Model

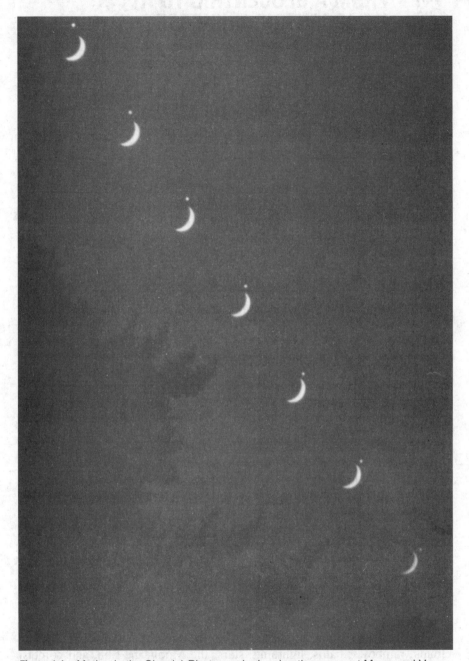

Figure 1.1 Motion in the Sky. (a) Photograph showing the crescent Moon and Venus setting. Exposures were made every 8 minutes, showing the changes in position of these celestial objects over time. Note the motion of the Moon relative to Venus.

(b) A time exposure taken with a camera aimed at *Polaris* over Mt. Fuji, latitude 35° N. Note the circular star trails. Source: Fuji Under Polaris by Victor Porof.

eyes, you cannot see a more distant tree. Therefore, if a bird flying by blocks your view of a cloud, you logically conclude that the bird is closer to you than the cloud. Then you see a cloud move "in front of" the Sun, and you conclude that the cloud is closer to you than the Sun. Or perhaps you see a solar eclipse, and you conclude that the Moon is closer to you than the Sun. In this way, all of the objects on your list could be put in order of distance from an observer.

Careful observation also leads you to realize that many of these closer objects or phenomena are associated with the atmosphere. You feel a wind and see it moving the clouds. You see a rainbow in the spray of a waterfall or in a distant rain shower and realize that it is caused by the interplay of sunlight and tiny droplets of water in the air. You see lightning flash between a cloud and the ground. In this way, you can classify the things seen in the sky into two groups: those that are part of, or occur in, the atmosphere, and those that are beyond the atmosphere.

Celestial Objects

Celestial objects are objects that can be seen in the sky that are not associated with Earth's atmosphere. The most numerous of the celestial objects are the stars. To an observer on Earth, stars are simply points of light that vary in size, brightness, and color. The Sun, the Moon, the planets, and comets are also examples of celestial objects. Clouds, rainbows, halos, and other phenomena seen in the sky that are part of, or occur in, Earth's atmosphere are *not* considered celestial objects.

Celestial Motion

If you observe celestial objects for even a short while, it is clear that they change position in the sky over time. You have probably noticed the Sun in different places in the sky at different times of the day. Thus, it seems that the Sun is "moving." Similarly, if you observe the Moon and stars carefully, you find that they, too, are seen in different places in the sky at different times of the night. Try going out on a moonlit night and noting the Moon's position at 7:00 P.M. If you go out again at 10:00 P.M., you'll notice that the Moon has changed position in the sky. The same is true of stars. When you observe the sky, you find that every celestial object changes position over time, or is in motion. See Figure 1.1 on pages 2 and 3.

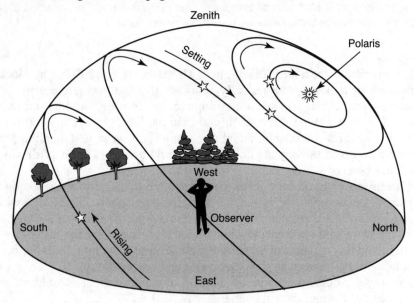

Figure 1.2 The Apparent Motion of Celestial Objects to an Observer in New York State.

If you keep track of this motion, you discover something curious. The motion of celestial objects is not random. They don't all move in different directions at different speeds. Instead, with very few exceptions, every single one of these thousands of objects appears to move in the same general direction—*from east to west*. And if you measure the rates at which all of these celestial objects are moving, you discover something even more curious—with few exceptions (such as the Moon), *they appear to move at the same rate!*

Careful records of this motion reveal that all celestial objects appear to move across the sky from east to west along a path that is an arc, or part of a circle. Since celestial objects appear to follow a circular path at a constant rate of 15 degrees per hour, or one complete circle every day (24 hr/day × 15°/hr = 360°/day), this motion is called **apparent daily motion**.

Figure 1.3 Constellations Are Imaginary Patterns of Stars.

In the Northern Hemisphere, all the circles formed by completing the arcs along which celestial objects move are centered very near the star *Polaris*. The apparent circular motion of celestial objects causes them to come into view from below the eastern horizon and to sink from view beneath the western horizon (that is, to rise in the east and set in the west). See Figure 1.2.

Early observers noted that the positions of celestial objects change in a daily and yearly cyclic pattern. They discovered that understanding these patterns of motion was very useful. Since the positions of celestial objects change with time and location, such changes can be used to *determine time* and to *find one's position on Earth*. Since the distribution of stars is random, these observers devised **constellations**, imaginary patterns of stars, to help them keep track of the changing positions of celestial objects. See Figure 1.3.

"Apparent" Versus "Real" Motion

So far, we have used the word *apparent* when referring to celestial motions because the motion of an object is always judged with respect to some other object or point. The idea of absolute motion or rest is misleading because there are several possible reasons why an object may *appear* to an observer to be moving. One possibility is that the observer is standing still and the *object* is moving. Another possibility is that the object is standing still and *the observer* is moving. A third possibility is that *both the observer and the object* are moving, but one is moving faster, or in a different direction, than the other. This is the case when you are in a car speeding down a highway;

as you look out of the car window, trees along the side of the road seem to whiz by. Of course, your brain tells you that the trees are rooted to the ground and that they only seem to whiz by because you are riding in a car. But to your eyes alone, you are not moving; the image of the trees is moving from one side of your window to the other. Now think about driving past a person walking along the sidewalk. The person is moving, but also seems to whiz by your window. Now think of a person sitting next to you in your car. To you, would that person look as though he or she was moving?

By now you should realize that the problem of determining which of the two is moving, the object or the observer, is not always easy to solve. If the signs that tell the body it is moving are removed, an observer may not realize that he or she is in motion. (Do you really feel as if you are moving at 400 miles per hour when watching a movie in an airplane cruising in level flight at that speed?) Without signs telling the observer's body that he or she is moving, any object seen changing position will be interpreted as a moving object by the observer.

THE CELESTIAL SPHERE

Early observers reasoned that when they looked at the sky they were standing still because their senses gave them no signs that they were moving. They *felt* as if they were standing still. Therefore, they interpreted the changing positions of celestial objects to mean that the *celestial objects* were moving. They visualized all celestial objects as revolving around a motionless Earth.

One effect of apparent daily motion is that the sky appears to move as if it were a single object. Here's a simple analogy. If a yellow bus with the words School Bus painted on its side drives past you, the words and letters don't end up looking like this SchoolBus just because the bus is moving forward. Even though they are moving forward, all of the letters in the two words stay in a fixed pattern because they are part of a single object—the bus. In much the same way, the stars in the sky stay in a fixed pattern even as you observe them moving through the sky.

It is not surprising, then, that early observers imagined that the sky was a single object—a huge dome. This "sky model" envisioned an observer as standing on a flat, circular disk representing Earth's surface and imagined the sky as a dome arching over the observer's head. The circumference of the flat disk was the horizon where an observer would see the sky meet Earth's surface in all directions. Since the "dome" of the sky was in motion, and new parts would come into view as others dropped out of sight, these observers imagined that the dome extended beyond the horizon. As they followed through on this model, they realized that, if the dome were extended far enough, it would form a hollow ball, or sphere, surrounding Earth. They imagined a huge "sky ball," or **celestial sphere**, slowly spinning around a motionless Earth. See Figure 1.4. To these observers the Sun, Moon, and stars were either holes in the celestial sphere or objects attached to it.

The "celestial sphere" was a nice model because it accounted for many observations. It explained why objects appeared, arced across the sky, disappeared, and then reappeared the next day. Imagine it as a ball tied to a rope and swung in a circle around your head. First the ball arcs across your line of sight as you swing it in front of you, next it disappears as it swings around behind you, and then it reappears as it swings around in front of you again. This model explained why all of the celestial objects moved in the same direction at the same speed. It also explained why the stars remained in fixed positions relative to one another. This Earth-centered, or **geocentric**, model of the universe was used successfully for thousands of years to explain most observations of celestial objects.

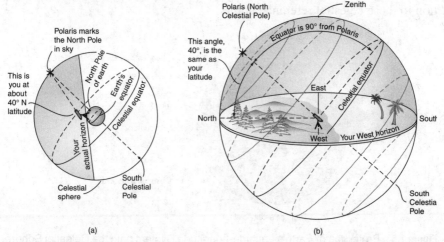

Figure 1.4 The Celestial Sphere, an Imaginary Sphere Surrounding Earth. The most you see at any one time is half of this sphere. Certain reference points on the celestial sphere are defined in relation to reference points on Earth. The celestial poles lie directly over Earth's poles; the celestial equator lies over Earth's equator midway between the celestial poles. Other points are defined by their positions in relation to the observer: the zenith is a point directly above the observer, the celestial meridian is the circle that runs through the celestial poles and the zenith. As Earth rotates from west to east, all objects in the sky appear to move from east to west, revolving around the north celestial pole. (a) View from a spot outside the celestial sphere. (b) Observer's view.

Even though we now know that the motion of celestial objects is due to Earth's rotation, it is still sometimes useful, when discussing objects in the sky, to think of them as part of a sphere surrounding Earth. The most that an observer would see at any one time would be half of this sphere; but we still refer to this imaginary half-sphere, or dome, visible over our heads as the celestial sphere. The circle formed by the intersection of the celestial sphere and the ground is called the **horizon**. The point on the celestial sphere that is right over an observer's head at any given time is the **zenith**. The imaginary circle that passes through the north and south points on the horizon and through the zenith is the **celestial meridian**.

A Simple Celestial Coordinate System

A useful coordinate system for locating objects on the celestial sphere can be set up by projecting Earth's Equator and poles onto the sky. As shown in Figure 1.5, Earth's Equator, North Pole, and South Pole correspond to a "celestial equator" and "north and south celestial poles" on the celestial sphere. Celestial objects can be located in the sky by their positions in relation to these celestial reference points.

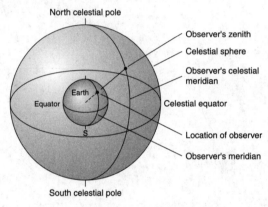

Figure 1.5 Projection of Earth's Latitude-Longitude System onto the Celestial Sphere.

The star *Polaris* is located very close to the north celestial pole, making it a convenient reference point for determining the north-south positions of celestial objects in the Northern Hemisphere. *Polaris* can be located by following the "pointer stars," *Dubhe* and *Merak*, in the bowl of the Big Dipper in the constellation Ursa Major. See Figure 1.6.

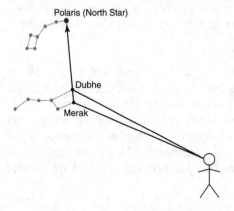

Figure 1.6 The "Pointer Stars," *Dubhe* and *Merak*, in the Bowl of the Big Dipper. Use these two stars to find the North Star, *Polaris*, and also to judge angular distances; they are about 5° apart.

A convenient reference point for determining the east-west positions of objects on the celestial sphere is the Sun. Objects to the west of the Sun on the celestial sphere will "rise" before the Sun and "set" before it. Likewise, objects to the east of the Sun trail behind it and will "rise" after the Sun and "set" after it. See Figure 1.7.

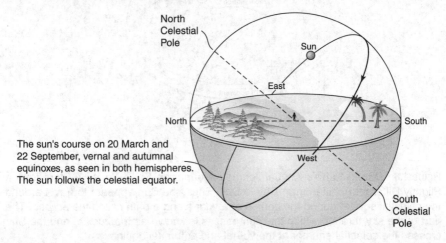

The sun's course on 20 March and 22 September, vernal and autumnal equinoxes, as seen in both hemispheres. The sun follows the celestial equator.

Figure 1.7 The Sun's Path on March 20 and September 22, the Vernal and Autumnal Equinoxes. The Sun follows the celestial equator.

The Sun's Path

Each day, because of Earth's rotation, the Sun moves along an imaginary path on the celestial sphere. Over the course of a year, however, it also follows an imaginary path on the celestial sphere. As you can see in Figure 1.8, the apparent position of the Sun with respect to the background stars change continuously as Earth orbits the Sun. The nighttime side of Earth is always opposite the Sun, so the background stars seen at night also change continuously as Earth orbits the Sun. When Earth has made one complete revolution in its orbit, the Sun will return to its starting point against the background stars. In other words, the Sun traces out a closed path on the celestial sphere once a year. The apparent path of the Sun through the stars on the celestial sphere over the course of the year is called the **ecliptic**. Since Earth's axis of rotation is tilted 23½° to the plane of its orbit, the ecliptic is tilted 23½° with respect to the celestial equator.

The ecliptic is important because the Sun, the Moon, and the planets are always found near it. As we will see later, this occurs because all of these objects in our solar system lie nearly in the same plane.

Early Astronomy and the Geocentric Model

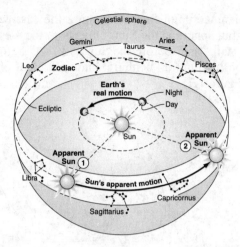

Figure 1.8 During Earth's Annual Journey Around the Sun, We View Stars from a Slightly Different Position from Day to Day. Thus, the Sun appears to travel around the celestial sphere during the course of a year along a path called the *ecliptic*. The part of the sky through which the Sun passes is known as the *zodiac*, and the Sun crosses the celestial equator at the vernal and autumnal equinoxes.

The Problem of Planets

There were, however, some problems with the geocentric model. Early astronomers also observed that certain points of light changed position with respect to the background of stars in the sky. They called these points of light *planets*, from the Greek word for "wanderer."

Astronomers working before the invention of the telescope and before anyone understood the present structure of the solar system counted seven such wanderers or planets: Mercury, Venus, Mars, Jupiter, Saturn, the Moon, and the Sun. This list differs from our modern list of planets in several ways:

- Earth is missing, because no one realized that the points of light wandering in the sky and the Earth on which these observers stood were in any way alike.
- The Sun and the Moon were classified as planets because they wandered on the celestial sphere, just like Mars and Jupiter and the other planets.
- Uranus and Neptune are missing because they were not discovered until the telescope made them easily visible. Uranus, which is barely visible to the naked eye, was discovered in 1781. Neptune, which can't be seen at all without a telescope, was discovered in 1846.

Planets differ from stars in a number of ways. As already mentioned, the relative positions of stars on the celestial sphere are fixed, while planets move relative to the stars. Stars can be seen anywhere on the celestial sphere; planets are always found near the ecliptic (that imaginary yearly path of the

Sun on the celestial sphere). Stars appear to "twinkle," but the brighter planets do not. Even through a telescope, stars appear as points of light, while the larger and nearer planets appear as disks.

These observed differences between planets and stars, particularly the "wandering" of planets on the celestial sphere, attracted a lot of attention from early astronomers. Their attempts to explain these differences ultimately led to the development of a new model of the universe.

MULTIPLE-CHOICE QUESTIONS

In each case, write the number of the word or expression that best answers the question or completes the statement.

1. Which of the following is *not* a celestial object?
   (1) the Sun
   (2) the Moon
   (3) a rainbow
   (4) a star

2. As viewed from Earth, most stars appear to move across the sky each night because
   (1) Earth revolves around the Sun
   (2) Earth rotates on its axis
   (3) stars orbit around Earth
   (4) stars revolve around the center of the galaxy

3. Which real motion causes the Sun to appear to rise in the east and set in the west?
   (1) the Sun's revolution
   (2) the Sun's rotation
   (3) Earth's revolution
   (4) Earth's rotation

Base your answers to questions 4 and 5 on the time-exposure photograph shown below. The photograph was taken by aiming a camera at a portion of the night sky above a New York State location and leaving the camera's shutter open for a period of time to record star trails.

4. Which celestial object is shown in the photograph near the center of the star trails?
   (1) the Sun
   (2) the Moon
   (3) *Sirius*
   (4) *Polaris*

5. During the time exposure of the photograph, the stars appear to have moved through an arc of 120°. How many hours did this time exposure take?
   (1) 5 h
   (2) 8 h
   (3) 12 h
   (4) 15 h

6. How many degrees does the Sun appear to move across the sky in four hours?
   (1) 60°
   (2) 45°
   (3) 15°
   (4) 4°

Base your answers to questions 7 through 11 on your knowledge of Earth science and on the diagram, which represents observations of the apparent paths of the Sun in New York State on the dates indicated.

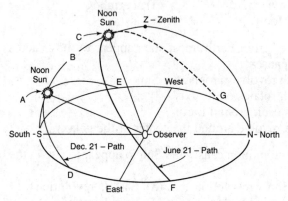

7. On the basis of the diagram, which statement is true?
   (1) The Sun passes through the zenith on December 21.
   (2) The Sun rises due east and sets due west on December 21.
   (3) The Sun passes through the zenith on June 21.
   (4) The Sun rises north of east and sets north of west on June 21.

8. Which statement about the Sun's path is true?
   (1) The Sun's path varies with the seasons.
   (2) The midpoint of the Sun's path is the zenith.
   (3) The angle of the Sun's path to the horizon is greatest on December 21.
   (4) The Sun's path on certain days of the year is shown by line *SZN*.

9. Which arc represents a part of the observer's horizon?
   (1) *DAE*
   (2) *FCG*
   (3) *SBZN*
   (4) *DSEG*

10. On which date will the noon sun be nearest to position *B*?
    (1) September 21
    (2) November 21
    (3) December 21
    (4) January 21

11. Which arc represents part of the observer's celestial meridian?
    (1) *SBZ* (3) *SDF*
    (2) *DFN* (4) *GCF*

12. An observer on Earth measures the angle of sight between Venus and the setting Sun

(Not drawn to scale)

Which statement best describes and explains the apparent motion of Venus over the next few hours?
(1) Venus will set 1 hour after the Sun because Earth rotates at 45° per hour.
(2) Venus will set 2 hours after the Sun because Venus orbits Earth faster than the Sun orbits Earth.
(3) Venus will set 3 hours after the Sun because Earth rotates at 15° per hour.
(4) Venus will set 4 hours after the Sun because Venus orbits Earth slower than the Sun orbits Earth.

13. The constellation Pisces changes position during a night as shown in the diagram below.

Which motion is mainly responsible for this change in position?
(1) revolution of Earth around the Sun
(2) rotation of Earth on its axis
(3) revolution of Pisces around the Sun
(4) revolution of Pisces on its axis

14. The diagram below represents a portion of the constellation Ursa Minor. The star *Polaris* is identified.

Ursa Minor can be seen by an observer in New York State during all four seasons because Ursa Minor is located almost directly
(1) above Earth's equator
(2) above Earth's North Pole
(3) overhead in New York State
(4) between Earth and the center of the Milky Way

Base your answers to questions 15 and 16 on the map of the night sky below, which represents the apparent locations of some of the constellations that are visible to an observer at approximately 40° N latitude at 9 P.M. in April. The point directly above the observer is labeled zenith.

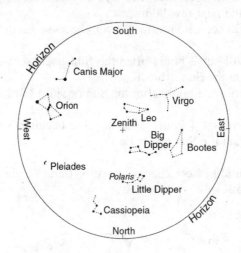

15. Which map best illustrates the apparent path of Virgo during the next 4 hours?

16. Which motion causes the constellation Leo to no longer be visible to an observer at 40° N in October?
    (1) spin of the constellation on its axis
    (2) revolution of the constellation around the Sun
    (3) spin of Earth on its axis
    (4) revolution of Earth around the Sun

17. At a location in the Northern Hemisphere, a camera was placed outside at night with the lens pointing straight up. The shutter was left open for four hours, resulting in the star trails shown below.

At which latitude were these star trails observed?
(1) 1° N   (3) 60° N
(2) 30° N   (4) 90° N

Early Astronomy and the Geocentric Model

CONSTRUCTED RESPONSE QUESTIONS

Base your answers to questions 18 through 20 on the diagram below and on your knowledge of Earth science. The diagram represents a time-exposure photograph taken by aiming a camera at *Polaris* in the night sky and leaving the shutter open for a period of time to record star trails. The angular arcs (star trails) show the apparent motions of some stars.

18. Identify the motion of Earth that causes these stars to appear to move in a circular path. [1]

19. Determine the number of hours it took to record the star trails labeled on the diagram. [1]

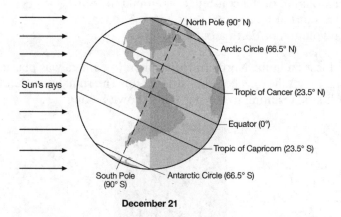

December 21

(Not drawn to scale)

20. The diagram above represents Earth as viewed from space. The dashed line indicates Earth's axis. Some latitudes are labeled. On the diagram, draw an arrow that points from the North Pole toward *Polaris*. [1]

Base your answer to question 21 on the diagram below, which shows the Sun's apparent path as viewed by an observer in New York State on March 21.

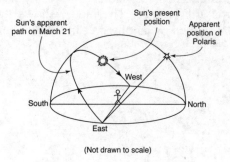

(Not drawn to scale)

21. At approximately what hour of the day would the Sun be in the position shown in the diagram? [1]

Base your answers to questions 22 through 24 on diagram 1 and on diagram 2, which show some constellations in the night sky viewed by a group of students. Diagram 1 below shows the positions of the constellations at 9:00 p.m. Diagram 2 shows their positions two hours later.

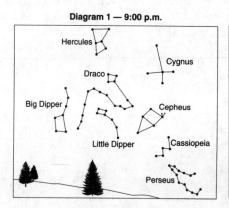

22. Circle *Polaris* on diagram 2. [1]

23. In which compass direction were the students facing? [1]

24. Describe the apparent direction of movement of the constellations Hercules and Perseus during the two hours between student observations. [1]

17

Early Astronomy and the Geocentric Model

EXTENDED CONSTRUCTED RESPONSE QUESTIONS

Base your answers to questions 25 and 26 on the sky model below and on your knowledge of Earth science. The model shows the Sun's apparent path through the sky as seen by an observer in the Northern Hemisphere on June 21.

25. Describe the evidence, shown in the sky model, which indicates that the observer is not located at the North Pole. [1]

26. Identify the cause of the apparent daily motion of the Sun through the sky. [1]

Base your answers to questions 27 through 29 on the diagram below and on your knowledge of Earth science. The diagram is a model of the sky (celestial sphere) for an observer at 50° N latitude. The Sun's apparent path on June 21 is shown. Point A is a position along the Sun's apparent path. Angular distances above the horizon are indicated.

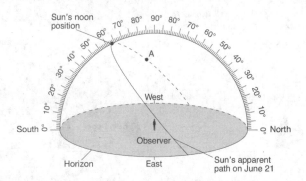

27. On the celestial sphere diagram, place an **X** on the Sun's apparent path on June 21 to show the Sun's position when the observer's shadow would be the longest. [1]

28. The Sun travels 45° in its apparent path between the noon position and point A. Identify the time when the Sun is at point A. Include A.M. or P.M. with your answer. [1]

29. Describe the general relationship between the length of the Sun's apparent path and the duration of daylight. [1]

CHAPTER 2
THE DEVELOPMENT OF THE HELIOCENTRIC MODEL

> **KEY IDEAS** Modern astronomy traces its beginning to the publication in May 1543 by Nicolaus Copernicus of a new **heliocentric**, or Sun-centered, model of the universe. Although Aristarchus of Samos had proposed a Sun-centered model almost 1,800 years earlier, the idea that Earth is moving at great speed had been dismissed as obvious nonsense since no one could feel any motion. Copernicus discarded the idea of a stationary Earth and argued that Earth and the planets circle the Sun. His logical and mathematical arguments paved the way for further investigations. The shift from an Earth-centered to a Sun-centered model was revolutionary and has evolved into our current concept of the universe.

KEY OBJECTIVES
Upon completion of this chapter, you will be able to:
- Compare and contrast the geocentric and heliocentric models of the universe.
- Describe the investigations that led scientists to understand that most of the observed motions of celestial objects are the result of Earth's motion around the Sun.
- Explain how gravity influences the motions of celestial objects.
- Determine the gravitational force between two objects, given their masses and the distance between their centers.
- Analyze the relationships among a planet's distance from the Sun, gravitational force, period of revolution, and speed of revolution.

EARLY MODELS: ARISTOTLE AND PTOLEMY

Ancient Greek thinkers, particularly Aristotle, set a pattern of belief that persisted for 2,000 years—the universe had a large, stationary Earth at its center; and the Sun, the Moon, and the stars were arranged around Earth in a perfect sphere, with all of these bodies orbiting Earth in perfect circles at constant speeds. The Egyptian astronomer Ptolemy refined this concept into an elegant mathematical model of circular motions that enabled astronomers to predict the positions of celestial objects fairly accurately and could account for many of the "problem" observations that plagued Aristotle's model.

Aristotle's Geocentric Universe

Aristotle, a Greek philosopher who lived from 384 B.C. to 322 B.C., wrote about and taught many subjects, including history, philosophy, drama, poetry, and ethics. His wide-ranging knowledge and insight earned him a prominent place among the great thinkers of antiquity.

Aristotle's was a common sense view of the universe. He understood the celestial sphere model and its ability to explain most casual observations, such as the apparent movements of celestial bodies. As records of careful measurements were kept over time, however, some problems arose. The Sun doesn't follow the same path through the sky all year long. The Moon changes position relative to the stars from night to night. Five (actually, nine) "stars," out of the thousands seen in the sky, don't stay in fixed positions relative to the others, but "wander" around in the sky. These moving objects, as explained in Chapter 1, came to be called **planets**, from *planetes*, the Greek word for "wanderer." Aristotle realized that a one-sphere model couldn't explain these "problem" observations, so he revised the model.

Spheres Within Spheres

Aristotle reasoned that, if some objects move differently, they must be on different celestial spheres! Aristotle explained the "problem" observations by proposing a universe consisting of eight crystalline (i.e., transparent) spheres nesting one inside the other like a set of Russian dolls, with Earth at the very center. The Sun, Moon, stars, and planets were fixed to the surface of separate spheres, which rotated around the unmoving Earth. All motions of the spheres were perfect circles. By having the spheres spinning at slightly different rates and at slightly different angles in relation to one another, most of the "problem" observations could be accounted for. Either the spheres moved because they were self-propelled, or, as was thought more likely, their motion was initiated by a supernatural being. See Figure 2.1.

Common Sense

In Aristotle's model, Earth too was a sphere, the perfect shape, as could be seen when its shadow was visible against the Moon during an eclipse. Common sense indicated that Earth wasn't moving because no motion could be felt, but Aristotle believed there was other evidence as well. If Earth moved, objects falling in a straight line should fall to the side of points directly beneath them.

According to Aristotle, the natural state of things on Earth was to be at rest. Natural motion on Earth was toward its center. Unlike the perfect circular motion of the spheres, the motion of objects on Earth was imperfect straight-line motion. The spheres were perfectly clear and were composed of ether, a substance that could not be changed or destroyed.

Since Aristotle's universe has Earth at its center, it is called a **geocentric**, or Earth-centered, model of the universe.

Figure 2.1 Aristotle's Model of the Universe. Crystalline spheres were nested one inside the other, with Earth at the center. The spheres and their attached stars and planets rotated around Earth.

Ptolemy's Geocentric Model

There were some obvious problems with Aristotle's view of the universe. The most obvious was visible to the naked eye. There were times when the planets changed course in the sky; for example, at times Mars would stop and then move backward, a phenomenon called **retrograde motion**. Since the crystal spheres of the Aristotelian universe could not stop or change direction, this observation could not be explained until the second century A.D., when Claudius Ptolemaeus, usually referred to as Ptolemy, proposed an ingenious theory.

Ptolemy, an Egyptian, lived and worked in the Greek settlement at Alexandria in about A.D. 140. There he studied mathematics and astronomy and developed a model of the universe based upon Aristotle's teachings. The details of his model are carefully spelled out in his great book, *Almagest*.

The Development of the Heliocentric Model

Explaining Retrograde Motion

Ptolemy's view was that each planet was fixed to a small sphere that was in turn fixed to a larger sphere. The smaller sphere and its attached planet turned at the same time that the larger sphere turned. As a result there could be times when, to an observer on Earth, the planet appeared to be moving backward. Ptolemy called the circular motions of the larger spheres *deferents* and the motions of the smaller spheres *epicycles*. He placed Earth's sphere off the center of its deferent. See Figure 2.2.

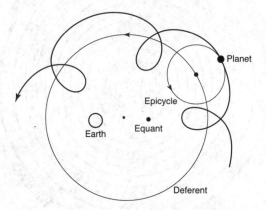

Figure 2.2 Ptolemy's Universe. Ptolemy added epicycles to Aristotle's model to explain retrograde motion and changes in apparent diameter.

With Ptolemy's ingenious modifications, Aristotle's universe could explain all casual, naked-eye observations of the universe. For 1,000 years astronomers studied and preserved Ptolemy's work, making no changes in his basic theory. It became part of the accepted thinking of the time. See Figure 2.3.

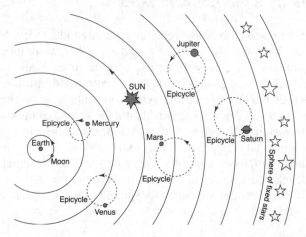

Figure 2.3 Ptolemy's Geocentric Model of the Universe. This model was accepted for well over 1,000 years.

Problems with Predictions

At first, the Ptolemaic system was able to predict the motions of celestial objects with a fair degree of accuracy. However, as the centuries passed, the differences between what the Ptolemaic system predicted and what was actually observed grew so large they could not be ignored. At first, earlier astronomers blamed these discrepancies on poor instruments or inaccurate observations. Arabian and, later, European astronomers corrected the system, recalculated constants, and even added new epicycles. King Alfonso X of Castile paid for the last great correction of the Ptolemaic model. Ten years of observations and calculations were then published as the Alfonsine Tables. By the 1500s, however, the Alfonsine Tables were also inaccurate, often being off by as much as 2°, which is four times the angular diameter of the moon—a significant error.

THE HELIOCENTRIC MODEL

Copernicus

At about the same time that astronomers were struggling with the inaccurate Alfonsine Tables, there was a serious need for calendar reform. By the beginning of the 1500s, the Julian calendar was off by about 11 days. Easter, a major church holiday, was particularly hard to determine. Both the Hebrew calendar, which was based upon the Moon, and the Julian calendar, which was based upon the Sun, had to be used to calculate the phase of the moon, upon which the date of Easter depended. A secretary of Pope Sixtus IV asked Nicolaus Copernicus, a priest-mathematician from Poland (see Figure 2.4), to examine the problem of calendar reform.

Figure 2.4 Nicolaus Copernicus.

Copernicus recognized that any calendar reform would have to resolve the relationship between the Sun and the Moon. After much study of the problem, Copernicus proposed a mathematically elegant solution in which he suggested a **heliocentric**, or Sun-centered, universe with a moving Earth.

In 1514, he distributed a brief manuscript outlining his ideas, but was discreet because he recognized the potential dangers of questioning church dogma. Not until his death in 1543 was his full argument in favor of a Sun-centered system published. Even then, he avoided heresy charges by crediting classical Greek sources with the idea, thus implying that the concept did not originate with him.

In Copernicus' model of a heliocentric universe, the center of the universe was a point near the Sun. Earth orbited the Sun and spun once a day on its axis. See Figure 2.5.

The Development of the Heliocentric Model

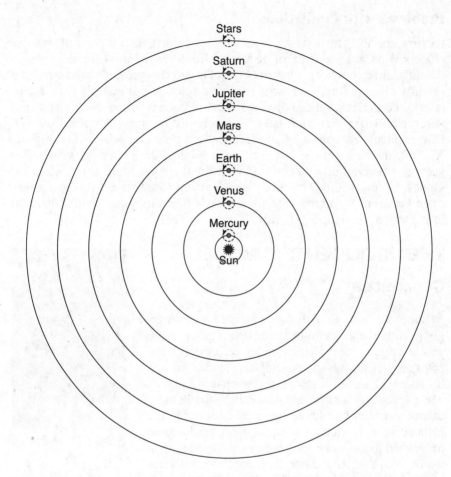

Figure 2.5 The Copernican Heliocentric Universe. Copernicus proposed a Sun-centered model in which all planets and stars moved in perfect circles around the Sun.

Copernicus reasoned that retrograde motion occurs because Earth moves faster in its orbit than do planets farther from the Sun. Earth and the other planets all move continuously in their orbits around the Sun, but Earth moves toward an outer planet in one part of its orbit and then passes it and moves away from it. However, planets moving in perfect circles around the Sun could not explain all of the observed details of their motions, and in the end Copernicus, too, resorted to epicycles and did no better at predicting the positions of celestial objects than Ptolemy.

While Copernicus' system was also erroneous, his idea that the universe was heliocentric, or Sun-centered, was correct and gradually gained acceptance. Probably the most important reasons why his theory was eventually accepted were the revolutionary mood of the world in his lifetime and the simple, forthright way in which his model explained retrograde motion. See Figure 2.6.

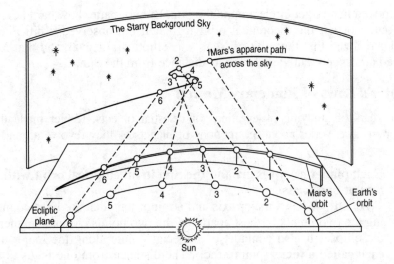

Figure 2.6 Copernicus' Simple, Forthright Explanation of Retrograde Motion. Both Earth and Mars move in a continuous path, but the inner planet (Earth) covers more of its orbit in the same time period, changing its point of view toward the outer planet (Mars).

Contributions of Tycho Brahe and Johannes Kepler

Tycho Brahe: Precision Observer

Shortly after Copernicus died, a Danish nobleman named Tycho Brahe became interested in astronomy. After observing that the Alfonsine Tables were nearly a month off in predicting a conjunction of Jupiter and Saturn, and observing a "new star" produced by a supernova, that is, the explosion of a very large star, Tycho questioned the Ptolemaic system of a perfect, unchanging heaven in a small book he wrote. His book was widely read, and the King of Denmark gave him funds to build a world-class astronomical observatory. Telescopes had not yet been invented, so Tycho devised many ingenious devices for measuring celestial motions precisely. When the King of Denmark died, Tycho fell out of favor and accepted a position as court astronomer to the Holy Roman Emperor in Prague, taking with him all of the data from the observatory in Denmark. In Prague, the emperor commissioned him to publish a revision of the Alfonsine Tables. Tycho hired several young mathematicians to help him with his task.

Johannes Kepler: Orbits Are Ellipses, Not Circles

One of Tycho's young assistants was Johannes Kepler. Shortly after beginning the project commissioned by the emperor, Tycho died unexpectedly. Before he died, however, he recommended Kepler to take over his position. As court astronomer, Kepler spent six years trying to work out the orbit of the planet Mars, using Ptolemy's system of the planet moving in a small circle

The Development of the Heliocentric Model

that moved in a larger circle around the Sun. But no matter how hard he tried, he could not get the theoretical orbit to match the observed orbit. Finally Kepler realized that the orbit of Mars was elliptical, or oval, and that Mars moved at a speed that varied with its distance from the Sun.

Kepler's Laws of Planetary Motion

After years of studying observations of celestial objects, Kepler made three important discoveries about the motions of planets as they revolve around the Sun.

1. **Each planet revolves around the Sun in an elliptical orbit with the Sun at one focus.**

    An **ellipse** has a major axis and a minor axis that are lines connecting the two points farthest apart and the two points closest together on the ellipse. It also contains two special points along the major axis, each called a **focus** (plural, foci). The distance from one focus to any point on the ellipse and back to the other focus is always the same.

    As a result it is very easy to draw an ellipse using two tacks and a loop of string. Press the tacks into a board, loop the string around them, and place a pencil in the loop on page 26. Keep the string taut; then, as you move the pencil, it will trace out the shape of an ellipse. See Figure 2.7.

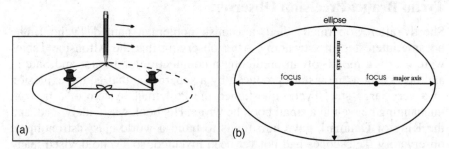

Figure 2.7 (a) The way to draw an ellipse (b) The main parts of an ellipse

The closer together the foci, the more nearly circular the ellipse. The farther apart the foci, the flatter the ellipse. The flatness of an ellipse is called its **eccentricity**. Eccentricity is expressed as the ratio between the distance between the foci and the length of the major axis:

A perfect circle would have an eccentricity of 0; a straight line has an eccentricity of 1.

$$\text{Eccentricity} = \frac{\text{distance between foci}}{\text{length of major axis}}$$

Since the orbit of each planet is an ellipse, the distance from each planet to the Sun varies during its orbit. See Figure 2.8. For example, Earth's distance to the Sun varies from 147×10^6 kilometers on January 3 at its closest (perihelion) to 152×10^6 kilometers on July 6 at its farthest (aphelion). The difference between these distances, 5×10^6 kilometers, is the distance between the foci of Earth's elliptical orbit. This measurement is very small compared to the length of the major axis, 299×10^6 kilometers, so the eccentricity of Earth's orbit is very small (0.17), indicating that the orbit is very nearly a circle. Although many illustrations show Earth's orbit around the Sun in a perspective view that exaggerates its eccentricity, if viewed from directly overhead the orbit would appear very nearly circular. (In this connection, it is interesting to note that Ptolemy's system of circular orbits *almost* worked because Earth's orbit is *almost* a circle. However, that slight difference from a perfect circle was enough to throw Ptolemy's system into question over time.)

As the distance between Earth and the Sun changes, the apparent diameter of the Sun changes in a cyclic manner. When Earth is closest to the Sun on January 4, the Sun has its greatest apparent diameter. On July 4, when the Sun is farthest away, it has its smallest apparent diameter.

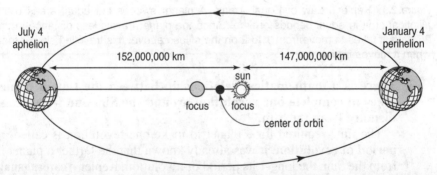

Figure 2.8 View of Earth's Elliptical Orbit with the Sun at One Focus. Earth is closest to the Sun at perihelion and farthest away at aphelion. Distances are approximate.

2. **The planets do not move at a constant velocity.**
   In Figure 2.9, the elliptical shape of a planet's orbit is exaggerated to show variation in velocity more clearly. The times the planet takes to move from 1 to 2, from 3 to 4, and from 8 to 9 are all equal. However, if you look at the diagram carefully, you can see that the distance from 1 to 2 is less than the distance from 8 to 9 even though the distances were covered in the same time. This means that the planet is moving fastest when it is closest to the Sun and slowest when it is farthest from the Sun.

The Development of the Heliocentric Model

Kepler did not know *why* this was so; he only determined that the variation did occur. We now know that this cyclic changing velocity of the planets as they move around the Sun is due to changes in the gravitational force between a planet and the Sun as distance changes. As a planet approaches the Sun (distance decreases), gravitational force increases and, since it is acting in the same direction in which the planet is moving, causes the planet to move faster. As a planet moves away from the Sun (distance increases), gravitational force decreases and, since it is now acting opposite to the direction in which the planet is moving, causes the planet to slow down.

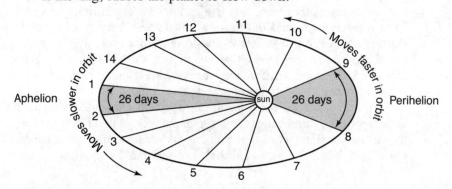

Figure 2.9 Kepler's Law of Equal Areas. A planet sweeps out equal areas of its elliptical orbit in equal periods of time. Since the planet travels less distance in the 26 days it takes to move from 1 to 2 on the ellipse above, it is traveling slower than when it moves from 8 to 9.

3. **There is a mathematical relationship between the time a planet takes to complete one revolution around the Sun and its average distance from the Sun.**

   The time required for a planet to make one revolution is called its **period of revolution**. It was already known that the farther a planet is from the Sun, the longer its period of revolution. Kepler's careful analysis showed that there is a mathematical relationship between the two factors. See the table accompanying Figure 2.10. A planet's period of revolution *squared* is proportional to its distance from the Sun *cubed*:

   $$T^2 \propto R^3$$

   If we measure the period of revolution in units of Earth-years, and let Earth's average distance from the Sun equal 1 unit of distance, called an *astronomical unit* (AU), the relationship simplifies. Then T^2 becomes $(1)^2$, R^3 becomes $(1)^3$, and, since $(1)^2 = 1$ and $(1)^3 = 1$:

   $$T^2 = R^3$$

   This relationship for the planets in our solar system can be plotted as a curve on a graph. See Figure 2.10.

28

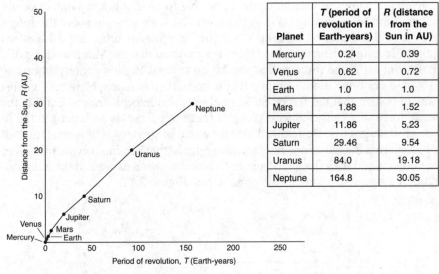

Figure 2.10 Relationship between Period of Revolution and Distance from the Sun

Galileo and Newton: Improving the Heliocentric Model

Galileo: Observations That Challenged Aristotle's Geocentric Universe

Shortly after Kepler's works were published, Galileo Galilei, an Italian astronomer and physicist, turned his telescope on the heavens and made several discoveries that further undermined the Ptolemaic system. His discovery of imperfections on the Moon's surface, spots on the surface of the Sun, and moons circling Jupiter challenged Aristotle's view of the heavens as perfect and unchanging. He saw Venus go through a full set of phases, which was not possible according to the Ptolemaic system of epicycles. Galileo also studied motion and solved Copernicus' falling-object problem. He argued that, if Earth is in motion, so are all objects on it. Therefore, an object that is dropped moves sideways at the same speed as Earth and falls on a spot directly beneath its point of release. Galileo became an outspoken champion of the heliocentric model.

Newton: Explaining Motion

Eleven months after Galileo died in 1642, Isaac Newton was born in England. Newton brought the discoveries of Copernicus, Galileo, and Kepler together. Using a few key concepts (mass, momentum, acceleration, and force), three laws of motion (inertia, the dependence of acceleration on force and mass, and action and reaction), and the law of universal gravitation, Newton was able to explain both the motions of objects on Earth and the distant motions of celestial objects.

The Development of the Heliocentric Model

Newton's laws of motion made it possible to predict how an object would move if the forces acting on it were known. Newton thought about the forces that would be needed to keep a satellite moving in orbit around another object. In considering the Moon, Newton realized that the Moon would circle Earth only if some force pulled the Moon toward Earth's center; otherwise it would continue moving in a straight line off into space. Newton's genius was to realize that the force that keeps the Moon in orbit around Earth is the same force that causes objects close to Earth (e.g., apples on a tree) to fall to the ground: gravity. He realized that the force of **gravity** is universal, that all objects are attracted to one another with a force that depends on the masses of the objects and their distance from each other. Newton expressed this relationship in a simple mathematical formula. See Figure 2.11.

$$F = -G\frac{m_1 m_2}{d^2}$$

Figure 2.11 Newton's Law of Gravity. In this equation, F is the force of gravity acting between two masses, G is the gravitational constant, m_1 and m_2 are the masses, and d is the distance between them.

Gravity and Orbital Motion

How does gravity keep satellites moving in a curved orbit? Imagine that a cannonball is shot out of a cannon aimed horizontally. If there were no gravity, inertia would cause the cannonball to fly off horizontally until some force stopped it. However, with gravity pulling the cannonball downward toward Earth's center, as the cannonball is flying horizontally its path curves downward and eventually the cannonball strikes Earth's surface. If a more powerful charge is used in the cannon, the cannonball will travel farther horizontally before it strikes Earth. With a sufficiently powerful charge, the cannonball would travel far enough horizontally that, as its path curved downward because of gravity, Earth's surface would curve away because of its spherical shape, and the cannonball would never strike Earth's surface. Instead, gravity would cause the cannonball to fall downward at the same rate that Earth's surface curves away from it, and it would fall unendingly in a circular path around Earth—it would be in orbit. See Figure 2.12.

Figure 2.12 Orbital Velocity. As a fired cannonball travels through the air, it is drawn downward by gravity in a curved path until it strikes Earth's surface. If it is fired with more energy, it will travel farther in a curved path before crashing into Earth's surface. If it is fired with enough energy, its path will curve downward at the same rate that Earth curves away from its path, and it will go into orbit.

Similarly, as Newton explained, it is a combination of two forces that keeps the planets moving in their curved paths around the Sun. The combination of a planet's forward motion and its motion toward the Sun due to gravity results in circular motion—the planet's orbit around the Sun. See Figure 2.13.

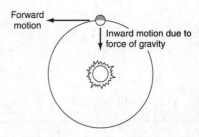

Figure 2.13 The Two Motions That Explain Why a Planet Travels in Orbit.

Now let's return to our cannonball analogy. For an orbit just above Earth's surface, the cannonball would have to be shot out of the cannon at 7.9×10^3 meters per second, or about 18,000 miles per hour. If the cannonball was propelled at a higher speed, it would travel farther outward before being pulled back by gravity, and the orbit would be elliptical instead of circular. If the cannonball was shot out at a velocity equal to or greater than 11.2×10^3 meters per second, or about 25,000 miles per hour, it would be able to escape Earth's gravity and fly out of orbit. See Figure 2.14.

Figure 2.14 Circular Orbit, Elliptical Orbit, and Escape Velocity.

The Development of the Heliocentric Model

With Newton's explanation of the causes of motion, the heliocentric theory began to firmly displace the geocentric theory as the generally accepted model of the universe. Our modern view of planetary motions in the solar system is based upon the heliocentric model.

MULTIPLE-CHOICE QUESTIONS

For each case, write the number of the word or expression that best answers the question or completes the statement.

1. The diagram below represents a simple geocentric model. Which object is represented by the letter **X**?

(Not drawn to scale)

(1) Earth
(2) Sun
(3) Moon
(4) *Polaris*

2. Which object orbits Earth in both the Earth-centered (geocentric) and Sun-centered (heliocentric) models of our solar system?
(1) the Moon
(2) Venus
(3) the Sun
(4) *Polaris*

3. Which diagram best represents the motions of celestial objects in a heliocentric model?

4. The modern heliocentric model of planetary motion states that the planets travel around
   (1) the Sun in slightly elliptical orbits
   (2) the Sun in circular orbits
   (3) Earth in slightly elliptical orbits
   (4) Earth in circular orbits

5. Which characteristic of the planets in our solar system increases as the distance from the sun increases?
   (1) equatorial diameter
   (2) eccentricity of orbit
   (3) period of rotation
   (4) period of revolution

6. The symbols below represent two planets.

   ⑤ represents a planet with a mass 5 times Earth's mass.

   ⑨ represents a planet with a mass 9 times Earth's mass.

   Which combination of planet masses and distances produces the greatest gravitational force between the planets?

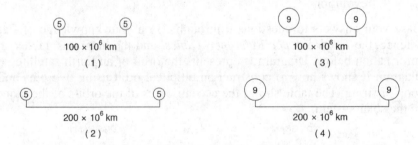

7. The diagram shows Earth (*E*) in orbit about the Sun. If the gravitational force between Earth and the Sun were suddenly eliminated, toward which position would Earth then move?
   (1) 1
   (2) 2
   (3) 3
   (4) 4

(Not drawn to scale)

The Development of the Heliocentric Model

8. The diagram below represents planets *A* and *B*, of equal mass, revolving around a star.

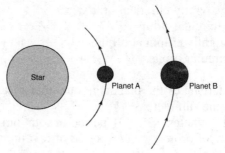

Compared to planet *A*, planet *B* has a
(1) weaker gravitational attraction to the star and a shorter period of revolution
(2) weaker gravitational attraction to the star and a longer period of revolution
(3) stronger gravitational attraction to the star and a shorter period of revolution
(4) stronger gravitational attraction to the star and a longer period of revolution

Base your answers to questions 9 through 13 on your knowledge of Earth science, the *Earth Science Reference Tables*, and the diagrams, tables, and information below. Diagram I represents the orbit of an Earth satellite, and diagram II shows how to construct an elliptical orbit using two pins and a loop of string. The table shows the eccentricities of the orbits of the planets in the solar system.

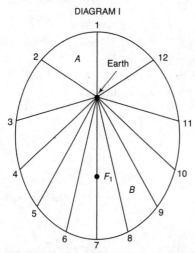

The satellite was at position 1 precisely at midnight on the first day. It arrived at position 2 the next midnight, at 3 the next, and so on.

Planet	Eccentricity of Orbit
Mercury	0.206
Venus	0.007
Earth	0.017
Mars	0.093
Jupiter	0.048
Saturn	0.056
Uranus	0.047
Neptune	0.008

9. At which position represented in diagram I would the gravitational attraction between the Earth and the satellite be greatest?
   (1) 1   (2) 7   (3) 3   (4) 11

10. According to the table, the orbit of which planet would most closely resemble a circle?
    (1) Mercury   (2) Venus   (3) Saturn   (4) Mars

11. What is the approximate eccentricity of the satellite's orbit?
    (1) 0.31   (2) 0.40   (3) 0.70   (4) 2.5

12. The Earth satellite takes 24 hours to move between each numbered position on the orbit. How does the orbital speed of the satellite in section A of its orbit (between positions 1 and 2) compare to its orbital speed in section B (between positions 8 and 9)?
    (1) It is moving faster in section A than in section B.
    (2) It is moving slower in section A than in section B.
    (3) Its speed in section A is equal to its speed in section B.
    (4) It is speeding up in section A and slowing down in section B.

Note that question 13 has only three choices.

13. If the pins in diagram II were placed closer together, the eccentricity of the ellipse being constructed would
    (1) decrease   (2) increase   (3) remain the same

Base your answers to questions 14 and 15 on the diagram below, which represents the current locations of two planets, *A* and *B*, orbiting a star. Letter **X** indicates a position in the orbit of planet *A*. Numbers 1 through 4 indicate positions in the orbit of planet *B*.

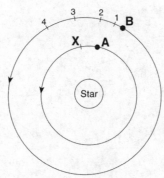

(Not drawn to scale)

Unit One GEO- TO HELIOCENTRIC

The Development of the Heliocentric Model

14. As planet *A* moves in orbit from its current location to position **X**, planet *B* most likely moves in orbit from its current location to position
    (1) 1 (3) 3
    (2) 2 (4) 4

15. If the diagram represents our solar system and planet *B* is Venus, which planet is represented by planet *A*?
    (1) Mercury (3) Earth
    (2) Jupiter (4) Mars

16. The graph below shows the varying amount of gravitational attraction between the Sun and an asteroid in our solar system. Letters *A*, *B*, *C*, and *D* indicate four positions in the asteroid's orbit.

Which diagram best represents the positions of the asteroid in its orbit around the Sun? [Note: The diagrams are not drawn to scale.]

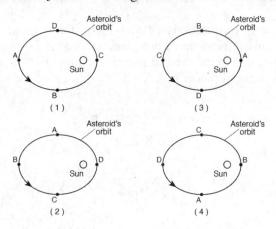

CONSTRUCTED RESPONSE QUESTIONS

17. Listed below are statements of several theories and observations, some of which may have served as partial bases for Newton's Law of Universal Gravitation. For each statement, (a)–(e), write the number preceding the name of the scientist, chosen from the list below, who first made the observation or proposed the theory.

(1) Brahe, (2) Newton, (3) Copernicus, (4) Ptolemy, (5) Galileo, (6) Aristotle, (7) Kepler

(a) The Sun is the center of our solar system. [1]
(b) Retrograde motions by planets can be explained by the use of epicycles. [1]
(c) The true orbits of planets are ellipses. [1]
(d) There is always gravitational force between two masses. [1]
(e) Falling bodies accelerate at a rate independent of their masses. [1]

Base your answers to questions 18 through 21 on the diagram below, which shows the heliocentric model of a part of our solar system. The planets closest to the Sun are shown. Point B is a location on Earth's equator.

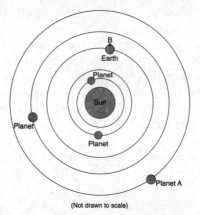

(Not drawn to scale)

18. State the name of planet A.

19. Explain why location B experiences both day and night in a 24-hour period.

20. On the graph to the right, draw a line to show the general relationship between a planet's distance from the Sun and the planet's period of revolution.

The Development of the Heliocentric Model

21. Identify one feature of the geocentric model of our solar system that differs from the heliocentric model shown.

Base your answers to questions 22 through 24 on the diagram below, which represents a model of Earth's orbit. Earth is closest to the Sun at one point in its orbit (perihelion) and farthest from the Sun at another point in its orbit (aphelion). The Sun and point *B* represent the foci of this orbit.

(Not drawn to scale)

22. Explain why Earth's orbit is considered to be elliptical.

23. Describe the change that takes place in the gravitational attraction between Earth and the Sun as Earth moves from perihelion to aphelion and back to perihelion during one year.

24. Describe how the shape of Earth's orbit would differ if the Sun and focus *B* were farther apart.

Base your answers to questions 25 and 26 on the diagram of the ellipse below.

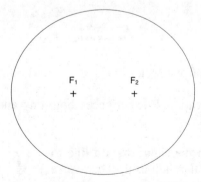

25. Calculate the eccentricity of the ellipse to the *nearest thousandth*. [1]

26. State how the eccentricity of the given ellipse compares to the eccentricity of the orbit of Mars. [1]

EXTENDED CONSTRUCTED RESPONSE QUESTIONS

Base your answers to questions 27 and 28 on the data table below, which lists the apparent diameter of the Sun, measured in minutes and seconds of a degree, as it appears to an observer in New York State. (Apparent diameter is how large an object appears to an observer.)

Apparent Diameter of the Sun During the Year

Date	Apparent Diameter (' = minutes " = seconds)
January 1	32'32"
February 10	32'25"
March 20	32'07"
April 20	31'50"
May 30	31'33"
June 30	31'28"
August 10	31'34"
September 20	31'51"
November 10	32'18"
December 30	32'32"

27. On the grid below, graph the data shown on the table by marking with a dot the apparent diameter of the Sun for *each* date listed and connecting the dots with a smooth, curved line. [2]

28. Explain why the apparent diameter of the Sun changes throughout the year as Earth revolves around the Sun. [1]

The Development of the Heliocentric Model

Base your answers to questions 29 and 30 on the diagram below, which shows Earth's orbit and the orbit of a comet within our solar system.

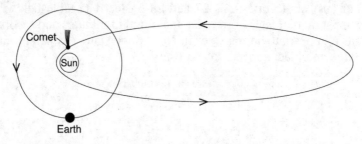

(Not drawn to scale)

29. Explain how this comet's orbit illustrates the heliocentric model of our solar system.

30. Explain why the time required for one revolution of the comet is more than the time required for one revolution of Earth.

Base your answers to questions 31 and 32 on the diagram below, which represents an asteroid's elliptical orbit around the Sun. The dashed line is the major axis of the ellipse.

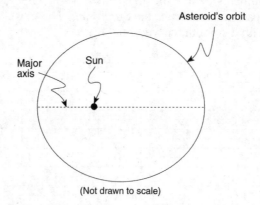

(Not drawn to scale)

31. Place a circle, **O**, on the orbital path where the velocity of the asteroid would be the least.

32. The Sun is located at one focal point of the orbit. Place an **X** on the diagram at the location of the second focal point.

CHAPTER 3
HELIOCENTRIC EARTH MOTIONS AND THEIR EFFECTS

KEY IDEAS The shift from a geocentric to a heliocentric model of the universe changed the way most people saw themselves in relation to the physical universe. The heliocentric model required the apparently immobile Earth to spin completely around on its axis once a day and the universe to be far larger than anyone had imagined; worst of all, Earth lost its position at the center of the universe and became just one of nine planets that orbit the Sun, a typical star in a vast and ancient universe.

However, the heliocentric model has greatly improved human understanding of complex phenomena, such as variations in day length, seasons, ocean tides, eclipses, phases of the Moon, the apparent motion of planets, and the annual traverse of the constellations.

KEY OBJECTIVES
Upon completion of this chapter, you will be able to:

- Explain how the Foucault pendulum and the Coriolis effect provide evidence of Earth's rotation.
- Describe Earth's rate of rotation, the orientation of Earth's axis of rotation with respect to the plane of its orbit, and the effects of Earth's changing position with regard to the Sun.
- Explain how seasonal changes in the apparent positions of constellations provide evidence of Earth's revolution.
- Interpret diagrams of Earth's orbit to determine how the Sun's apparent path through the sky and the relative position of the noon Sun vary with the seasons, and how the length of daylight varies throughout the year at different locations.
- Explain how the changing relative positions of Earth, the Moon, and the Sun account for the phases of the Moon, tides, and eclipses.

EARTH MOTIONS

Rotation

In the modern heliocentric model, Earth rotates at a rate of 15° per hour, or one complete rotation every 24 hours. **Rotation** is a motion in which every part of an object is moving in a circular path around a central line called the **axis of rotation**, like an ice skater spinning around a line through his or her body. Earth's axis of rotation is a line passing through the North Pole, Earth's center, and the South Pole. Earth's axis of rotation is almost directly aligned with the star *Polaris* and is tilted at an angle of 23½° from a perpendicular to the plane passing through the centers of Earth and the Sun. See Figure 3.1.

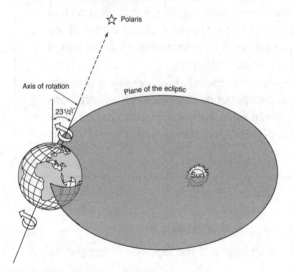

Figure 3.1 Rotation. Earth rotates from west to east once every 24 hours around an axis that runs through the poles.

Evidence of Rotation

The heliocentric theory requires Earth to rotate on its axis once every 24 hours. It is one thing to be able to successfully explain away a problem such as the lack of deflection of a falling object; it is quite another to actually *prove* Earth's rotation. The arguments in favor of rotation may be quite reasonable, but reasonableness is not scientific proof. A French physicist, Jean Foucault, and a French mathematician, Gaspard Coriolis, provided this proof.

The Foucault Pendulum

Jean Foucault used an ingeniously simple method to prove Earth's rotation. Foucault knew that gravity pulls a pendulum only toward Earth's center. It does not act laterally to change the plane of the pendulum's swing. Therefore, if set freely swinging in the absence of any lateral forces, a pendulum should swing in a fixed plane. Foucault further reasoned that the heavier he made the weight of the pendulum, the more force would be required to push it out of its plane of motion, making it unlikely that small breezes would deflect the pendulum. Finally, he realized that, if he made the pendulum really long, it would have a long period and would swing for many hours without needing a push to get it going again (which would introduce a possibility of exerting a lateral force). Foucault argued that, if Earth were motionless, such a pendulum would swing in a plane whose direction would not change. However, if Earth rotated, it would rotate beneath the pendulum and would change position relative to the swinging pendulum. The result would be a pendulum whose plane of swing would appear to rotate relative to the ground in a direction opposite to that of Earth's rotation.

Foucault's idea is easiest to understand if you visualize a pendulum swinging over the North Pole. As the pendulum swings, Earth moves beneath it. In 24 hours the plane of the swinging pendulum will appear to make one rotation. At the Equator, however, the plane of the pendulum will not appear to change. In between, the period of rotation of swinging pendulums will vary from 24 hours at the poles to infinity at the Equator. At 40° N, a pendulum takes 37 hours to make one rotation. See Figure 3.2.

In 1851, Foucault suspended a freely swinging pendulum from the inside of the dome of the Pantheon building in Paris. The high dome allowed the use of a very long (60-meter) pendulum. Foucault fastened the pendulum's heavy weight to one side of the room with a thin cord and sealed all entrances to eliminate possible drafts that could exert a lateral force on the pendulum. He then burned the cord to set the pendulum swinging smoothly in a fixed plane. Through windows, observers were able to watch the pendulum rotate slowly in a direction opposite to that of Earth's rotation, thus proving that Earth rotates.

Figure 3.2 The Foucault Pendulum.

The Coriolis Effect

Gaspard Coriolis first described the behavior of objects moving in a rotating frame of reference. His work successfully predicted the behavior of objects moving near the rotating Earth. In a stationary system, fluids such as the atmosphere or the oceans move directly in a straight line from regions of high pressure to regions of low pressure. However, that behavior does not occur on Earth. Large-scale movements of both the atmosphere and the oceans follow curved paths.

Study the following diagrams in the *Reference Tables for Physical Setting/Earth Science* (Appendix B): Surface Ocean Currents (page 698) and Planetary Wind and Moisture Belts in the Troposphere (page 708). Note that ocean currents form large circular patterns, called *gyres*, that flow clockwise in the Northern Hemisphere and counterclockwise in the Southern

Hemisphere. Note, too, that planetary winds follow paths that curve to the right in the Northern Hemisphere and to the left in the Southern Hemisphere. Such motions should not occur if Earth is stationary, but are precisely what would be expected if Earth is rotating. For a full description of why this occurs, refer to the section on the Coriolis effect in Chapter 22. Since the curving paths of ocean currents and weather systems would not occur on a stationary Earth, their existence is considered proof of Earth's rotation.

Revolution

As Earth rotates, it revolves around the Sun once every 365¼ days. **Revolution** is the motion of one body around another body in a path called an *orbit*. The revolving body is termed a **satellite** of the body it orbits. The body that a satellite orbits is its *primary*. Thus, Earth is a satellite of the Sun, its primary. The plane of Earth's orbit is called the *ecliptic* because eclipses occur when Earth, the Moon, and the Sun align in this plane.

Evidence of Revolution

Parallelism of Earth's Axis of Rotation

As stated previously, Earth's axis of rotation is tilted 23½° from a perpendicular to the plane of its orbit. Spinning like a top, Earth holds its axis fixed in space as it moves around the Sun. Another way to describe this is to say that Earth's axis of rotation at any point in Earth's orbit is parallel to its axis at any other point. As a result, the Northern Hemisphere is tipped toward the Sun in June and away from the Sun in December.

Figure 3.3 Revolution. (a) The constant orientation of Earth's axis as the planet orbits the Sun.

(b) Earth revolves around the Sun once every 365.25 days. During each season, different parts of the universe are visible at night.

Annual Traverse of the Constellations

As Earth revolves around the Sun, the side of Earth facing the Sun experiences day and the side facing away experiences night. Since the stars are visible only at night, the portion of the universe whose stars are visible to an observer on Earth varies cyclically as Earth revolves around the Sun. See Figure 3.3.

Precession

Precession is the very slow change in the direction of Earth's axis of rotation. Earth's axis sweeps around in a cone-shaped path like the wobbling of a spinning top. Almost 26,000 years are required for each sweep, causing Earth's North Pole to move slowly with respect to *Polaris*. Precession is caused by the gravitational pulls of the Sun and Moon. See Figure 3.4.

Since Earth's precession takes place over such a long period of time, it has little effect during a human lifetime. During your lifetime, Earth's axis will point slightly closer to *Polaris*. It will point closest to *Polaris* around the year A.D. 2100 and will then begin to move away from *Polaris*. In 13,000 years, Earth's axis will point at the star *Vega* instead of *Polaris*.

(a)

Figure 3.4 Precession. Precession causes Earth's axis to wobble like a top. (a) The force of gravity (*A*) on the spinning top tries to tip it over. The result is precession. If the top were not spinning, it would simply topple over. (b) The net force of the Moon (*B*) on Earth's tidal bulge tries to pull the planet's equatorial plane into the ecliptic plane. Again, the result is precession. Like the top, if Earth were not rotating, its equator would correspond to the ecliptic.

(b)

Heliocentric Earth Motions and Their Effects

EFFECTS OF EARTH'S MOTIONS

The value of any model lies in its ability to explain past and current observations and to predict future observations. A revolving and rotating planet with its spin axis tilted at 23½° to a line perpendicular to its orbital plane fits all terrestrial and celestial observations.

The Sun's Path

The Sun's apparent path through the sky from sunrise to sunset is an arc, like the paths of all other celestial objects, and is the result of Earth's rotating on its axis. As Earth (and any observer on Earth's surface) rotates from west to east, the Sun appears to move from east to west. See Figure 3.5.

Figure 3.5 (a) Direction in which a Single Observer Will See the Sun at Three Different Times of the Day. As Earth rotates from west to east, an observer on Earth's surface rotates along with Earth and the direction in which that observer sees the Sun changes. At sunrise (6 A.M.), the observer sees the Sun toward the east; at noon, toward the south; and at sunset (6 P.M.), that same observer will have rotated to yet another position and see the Sun toward the west. (b) To the Rotating Observer, the Sun Appears to Move Through the Sky from East to West.

At 6 A.M., an observer at sunrise would see the Sun to the east and low in the sky, at the horizon. At 9 A.M., Earth has rotated 45° from west to east and an observer would see the Sun to the southeast and higher up, above the horizon. At 12 noon, Earth has rotated another 45° from west to east and an observer would see the Sun to the south and at its highest point above the horizon, its noon position. At 3 P.M., Earth has rotated a further 45° and an observer would see the Sun to the southwest and lower above the horizon than at noon. At 6 P.M., Earth has rotated another 45° and an observer would see the Sun set to the west, low in the sky and at the horizon.

Note that, as Earth rotated through 180° from west to east, an observer saw the Sun move from east to southeast to south to southwest to west. The Sun's path through the sky began at sunrise at the eastern horizon at 6 A.M., arced upward, reaching a high point at noon, and then arced downward, ending at sunset at the western horizon at 6 P.M. What the observer saw as Earth rotated is summarized by path *B* in Figure 3.6.

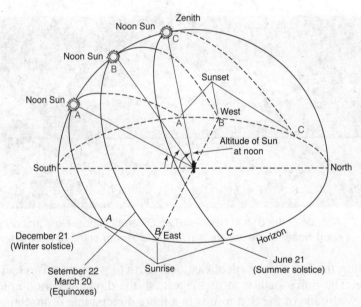

Figure 3.6 The Sun's Apparent Path. An observer at 42° N latitude would see the Sun move along different paths across the sky on different dates.

Changes in the Sun's Path

The length and the position of the Sun's path vary with the seasons and latitude. Sunlight that reaches Earth's surface is called **insolation**, short for *in*coming *sol*ar radi*ation*. The length of the Sun's path determines the length of time that insolation reaches Earth's surface, or the **duration of insolation**. The longer the Sun's path, the greater the duration of insolation, that is, the more hours of daylight. The position of the Sun's path determines the angle at which sunlight strikes Earth's surface, or **angle of insolation**. The closer to perpendicular the angle of insolation, the greater the intensity of the insolation, that is, the more it warms Earth's surface. The position of the Sun's path also determines the altitude of the Sun at noon. See Figure 3.6.

Changes with the Seasons

On different days of the year, an observer will see the Sun follow different paths through the sky. This is so because the Sun's position relative to an observer changes as Earth orbits the Sun. The points of sunrise and sunset vary, as does the altitude of the Sun at noon. See Figure 3.7.

In December, the position of sunrise to an observer in the United States is south of east and the position of sunset is south of west. The length of the Sun's path is shortest, so the daylight period is shortest. Because of Earth's tilted axis of rotation, the noon Sun is at its lowest altitude of the year. The combination of a short daylight period and a low angle of insolation causes low temperatures.

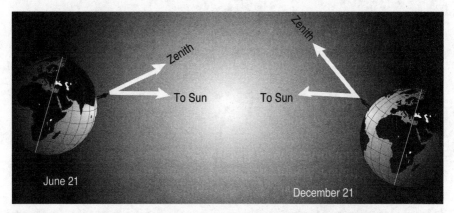

Figure 3.7 Altitude of the Sun at Noon on Two Different Dates. Note that on June 21 the Sun is seen closer to the zenith; that is, it is closer to being directly overhead.

In June, the Sun rises north of east, sets north of west, and rises to a higher altitude. The Sun's path is at its longest, so the daylight period is longest. The high altitude of the Sun results in a more direct angle of insolation. The combination of a long daylight period and a more direct angle of insolation results in high temperatures.

At the equinoxes in March and September, the Sun rises due east and sets due west. The length of the Sun's path results in exactly equal periods of daylight and darkness. The Sun's altitude at noon is halfway between its highest and lowest points. The length of day and the angle of insolation are intermediate between the two extremes; therefore, the temperatures are moderate.

From December to June, the Sun's path gets longer each day and the altitude of the Sun at noon increases. On June 21, the altitude of the Sun at noon stops increasing. Therefore, this date is called the **summer solstice**, meaning summer "Sun stop." From June to December, the Sun's path gets shorter each day and the altitude of the Sun at noon decreases. On December 21, the altitude of the Sun at noon stops decreasing. Therefore, this date is called the **winter solstice**, meaning winter "Sun stop." March 21 and September 21, when the daylight and darkness periods are equal, are called, respectively, the **spring equinox** and the **fall equinox**, meaning spring and fall "equal night." This cyclic pattern of change repeats in an annual cycle.

Changes with Latitude

Now consider what two observers at different latitudes will see on a given day as Earth rotates. As you can see in Figure 3.8, the observer near the Equator will see the Sun at a higher altitude at noon than the observer in New York State on the same day.

The altitude of the noon Sun at any location can be determined quite easily if you know the latitude of that location and the latitude at which the Sun is directly overhead on that day. To find the altitude of the Sun at noon, find the difference between the latitude of the location and the latitude at which the Sun is directly overhead. Then subtract this number of degrees from 90°.

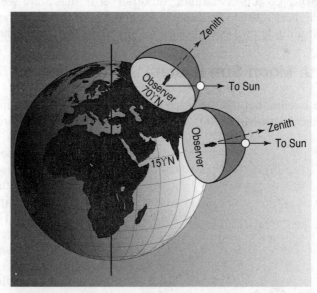

Figure 3.8 Altitude of the Sun at Different Latitudes. Observers at different latitudes will see the noon Sun at different altitudes on the same date.

Example:
On March 21, the Sun is directly overhead at the Equator, 0° latitude. On that day, what is the altitude of the Sun at noon in New York City?

$$\text{Latitude of NYC} - \text{latitude where Sun is directly overhead} = \text{difference in latitude}$$

$$41° \text{N} - 0° = 41°$$

$$90° - \text{difference in latitude} = \text{altitude of Sun at noon in NYC}$$

$$90° - 41° = 49°$$

The altitude of the Sun in New York City on March 21 is 49°.

Since the farthest Earth tips toward the Sun is 23½°, the Sun is never directly overhead north of 23½° N (Tropic of Cancer) or south of 23½° S (Tropic of Capricorn). There is always a difference between the latitude of a location in the continental United States and the latitude at which the Sun is directly overhead, and therefore the altitude of the Sun at noon is always less than 90° in the United States.

Because of New York's latitude, an observer in that State sees the noon Sun at its highest altitude on June 21, at about 73° above the horizon, and at its lowest altitude on December 21, at only about 23° above the horizon.

The Earth-Moon-Sun System

Few of the events that can be observed in the sky are as striking as those involving the Earth-Moon-Sun system. Phases of the Moon are probably the most familiar phenomena, but others involving this system include tides and eclipses of the Sun and Moon.

As Earth revolves around the Sun, the Moon circles Earth in an elliptical orbit with a period of 27.32 days. The Moon's orbit is tilted at an angle of about 5° from the plane of Earth's orbit around the Sun. The Moon moves rapidly in its orbit, covering 13° every day. As a result, each day

One revolution of the moon

One rotation of the moon

Figure 3.9 Revolution and Rotation of the Moon. The Moon completes one rotation in the same time it makes one revolution. Therefore, the same side of the Moon is always facing Earth.

its position against the backdrop of stars changes by 13°, or about 26 times its apparent diameter. The Moon also rotates on its axis once every 27.32 days. Thus, the same side of the Moon always faces Earth. See Figure 3.9.

Phases of the Moon

The Moon does not produce visible light of its own; the Moon is visible only because light from the Sun is reflected from its surface. An observer on Earth is able to see only that portion of the Moon that is illuminated by the Sun. As the Moon moves around Earth, different portions of the side of the Moon facing Earth are illuminated by sunlight, and the Moon passes through a cycle of **phases**. See Figure 3.10.

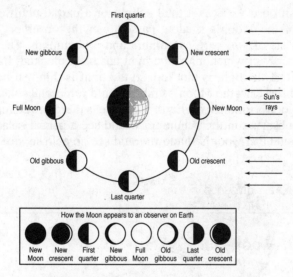

Figure 3.10 The Phases of the Moon.

Although the Moon makes one revolution in 27.32 days, it takes 29.5 days to go through a complete cycle of phases. Why the extra two days? Let's start at new Moon, when the Moon is between Earth and the Sun. At the same time that the Moon revolves around Earth, Earth is revolving around the Sun at a rate of about 1° per day. Thus, by the time the Moon has completed one revolution, the Earth's position in relation to the Sun is not the same as it was when the Moon started its revolution. In 27 days the Earth has moved 27° in its orbit. Moving at about 13° per day, the Moon takes about 2 days to catch up to Earth and align with it and the Sun in a new Moon phase. The word *month* has its origin in "Moon-th," which referred to this 29.5-day cycle of phases.

Eclipses of the Sun and Moon

A **solar eclipse** occurs when the Moon passes directly between Earth and the Sun, casting a shadow on Earth and blocking our view of the Sun. Both the Moon and the Earth are illuminated by the Sun and cast shadows in space. As seen in Figure 3.11, the shadows cast by Earth and the Moon are extremely long and narrow.

Figure 3.11 The Moon's Shadow. The 5½° tilt of the Moon's orbit and the small size of the Moon's shadow makes it easy for the shadow to miss Earth and not cause an eclipse.

Whether an observer sees a total eclipse or a partial eclipse depends on which part of the Moon's shadow passes over the observer. The Moon's shadow consists of two parts, an umbra and a penumbra. The **umbra** is a region of shadow in which all of the light has been blocked. In the **penumbra** only part of the light is blocked, so the light is dimmed but not totally absent. An observer in the Moon's umbra would see a total solar eclipse. The path of the Moon's umbra over Earth's surface is therefore called the **path of totality**. An observer in the penumbra would see a partial solar eclipse. An observer outside the Moon's shadow would see no eclipse. See Figure 3.12.

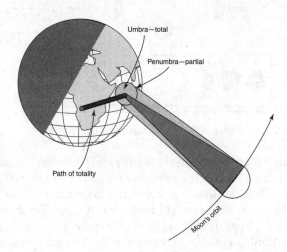

Figure 3.12 A Solar Eclipse. Observers in the umbra experience a total eclipse, while those in the penumbra experience a partial eclipse.

The umbra (i.e., the dark, inner part) of the Moon's shadow barely reaches Earth, and the small, circular shadow that the Moon casts on Earth's surface is never more than 269 kilometers in diameter. The 5° tilt of the Moon's orbit, together with the small size of the shadow that can reach Earth, makes it easy for the shadow to miss Earth at full and new Moon. Thus, total eclipses of the

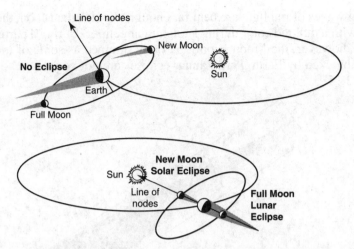

Figure 3.13 Eclipses of the Sun and Moon. Solar and lunar eclipses can occur when the Moon intersects the ecliptic in new-moon or full-moon position. If the Moon is above or below the ecliptic during these phases, no eclipse occurs.

Sun are rare. For an eclipse to occur, the plane of the Moon's orbit must intersect the shadows being cast by both Earth and the Moon. See Figure 3.13.

Since the Moon's orbit is elliptical, the Moon's distance from Earth varies. If the Moon lines up directly with the Sun at a point in its orbit when the Moon is farthest away from Earth, the umbra of the Moon's shadow does not reach Earth's surface. If the umbra does not reach the surface, there is no total solar eclipse. However, a type of partial eclipse called an **annular eclipse** occurs under these conditions. In an annular eclipse the Moon is too far from Earth to completely block our view of the Sun. During the eclipse a narrow ring, or annulus, of light is visible around the edges of the Moon. See Figure 3.14.

The eclipse viewed from space — The eclipse viewed from Earth

Figure 3.14 An Annular Eclipse and the Way It Appears When Viewed from Space and from Earth.

A **lunar eclipse** occurs when the full Moon moves through Earth's shadow. If the Moon moves into Earth's umbra, a total lunar eclipse is seen. If the Moon moves into Earth's penumbra, a partial lunar eclipse is seen. When the Moon is totally in Earth's umbra, it does not completely disappear from view. While no direct sunlight reaches the Moon, some light that is bent as it passes through Earth's atmosphere reaches the Moon. Since only the long waves of red light are bent far enough to reach the Moon, the Moon glows with a dull red color during a total lunar eclipse. During a partial lunar eclipse, however, the Moon is only partially dimmed as some of the Sun's light is blocked by Earth. Partial lunar eclipses are not very impressive. See Figure 3.15.

Figure 3.15 A Lunar Eclipse. If the moon passes only through the Earth's penumbra, a partial eclipse of the Moon is observed. If the Moon passes through Earth's umbra, a total eclipse of the Moon is observed.

Tides

Every day you feel the mutual attraction of gravity between Earth and your body pulling you downward with a force called your *weight*. But Earth's gravity is not the only gravity acting on you. While the Moon is farther away from you than Earth's center and has less mass than Earth, it still exerts a measurable force on you and everything else on Earth.

The part of Earth's surface that faces the Moon is about 6,000 kilometers closer to the Moon than is Earth's center. Therefore, the force of gravity the Moon exerts on this surface is stronger than the gravity exerted on Earth's center. Although Earth's surface is solid, it is not absolutely rigid. The Moon's gravity causes Earth's surface to flex outward, forming a bulge several inches high. As the Moon moves around Earth, the bulge moves across the surface as it remains beneath the Moon.

An inches-high bulge in the bedrock spread over half of Earth's surface is barely noticeable. Water is a fluid, however, and can move much more readily than rock in response to the Moon's gravity. Water in the oceans attracted by the Moon's gravity flows into a bulge of water on the side of Earth facing the Moon. A bulge of water also forms on Earth's far side. The Moon pulls on Earth's center more strongly than on Earth's far side, thus attracting Earth away from the oceans on the far side. The water flows into this space, creating a bulge. The water from the area between these bulges flows into them, creating a deep region and a shallow region in the ocean waters, or **tides**.

As Earth rotates on its axis, the positions of the tidal bulges remain fixed in line with the Moon. As the rotating Earth carries a location into a tidal bulge, the water deepens and the tide rises on the beach. As the location rotates out of the tidal bulge, the water becomes shallower and the tide falls. Since there are two bulges on opposite sides of Earth, the tide rises and falls twice a day.

The Sun also produces tidal bulges in Earth's surface and oceans. At new Moon and full Moon, the Sun's tidal bulges and the Moon's tidal bulges align with one another and combine. The result is very high and very low tides, called **spring tides** because they "spring so high," not because they happen during the spring season. See Figure 3.16a. Spring tides occur at every new and full Moon whatever the season. During the first- and third-quarter phases of the Moon, the tidal bulges of the Sun and Moon are at right angles to one another and very nearly cancel each other. The result is **neap tides**, in which there is very little difference between high and low tides. See Figure 3.16b.

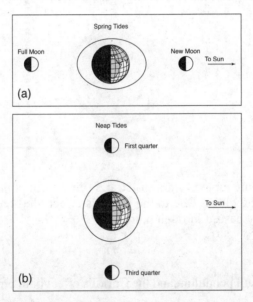

Figure 3.16 Spring and Neap Tides. (a) When the Sun and Moon pull in the same direction, their tidal forces combine and tidal bulges on Earth are larger. Spring tides occur at new Moon or full Moon. (b) When the Sun and Moon pull at right angles, their tidal forces do not combine and tidal bulges are much smaller. Neap tides occur at first- and third-quarter Moon.

Heliocentric Earth Motions and Their Effects

MULTIPLE-CHOICE QUESTIONS

In each case, write the number of the word or expression that best answers the question or completes the statement.

1. Which observation provides the best evidence that Earth rotates?
   (1) The position of the planets among the stars changes during the year.
   (2) The location of the constellations in relationship to *Polaris* changes from month to month.
   (3) The length of the shadow cast by a flagpole at noontime changes from season to season.
   (4) The direction of swing of a freely swinging pendulum changes during the day.

2. The diagram below represents a globe that is spinning to represent Earth rotating. The globe is spinning in the direction indicated by the arrow. Points $A, B, C, D, X,$ and Y are locations on the globe.

   A student attempted to draw a straight line from point X to point Y on the spinning globe. Due to the Coriolis effect, the student's drawn line most likely passed through point
   (1) A (3) C
   (2) B (4) D

3. The Foucault pendulum and the Coriolis effect both provide evidence of Earth's
   (1) revolution   (3) tilted axis
   (2) rotation   (4) elliptical orbit

4. As Earth travels in its orbit, Earth's axis
   (1) remains parallel to itself at all Earth positions
   (2) remains aligned with the Sun's axis
   (3) is perpendicular to the Moon's axis
   (4) is pointing toward the center of the Milky Way

5. Which motion causes some constellations to be visible in New York State only during winter nights and other constellations to be visible only during summer nights?
   (1) Stars in constellations revolve around Earth.
   (2) Stars in constellations revolve around the Sun.
   (3) Earth revolves around the Sun.
   (4) Earth rotates on its axis.

6. The diagram below represents some constellations and one position of Earth in its orbit around the Sun. These constellations are visible to an observer on Earth at different times of the year.

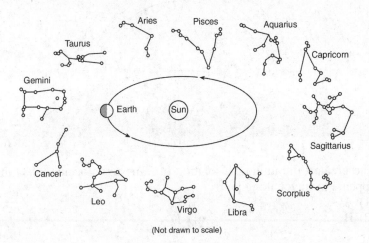

(Not drawn to scale)

When Earth is located in the orbital position shown, two constellations that are both visible to an observer on Earth at midnight are
   (1) Libra and Virgo
   (2) Gemini and Taurus
   (3) Aquarius and Capricorn
   (4) Cancer and Sagittarius

Heliocentric Earth Motions and Their Effects

Base your answers to questions 7 through 9 on the diagram below and on your knowledge of Earth science. The diagram represents four apparent paths of the Sun, labeled A, B, C, and D, observed in Jamestown, New York. The June 21 and December 21 sunrise and sunset positions are indicated. Letter **S** identifies the Sun's position on path C at a specific time of day. Compass directions are indicated along the horizon.

7. The greatest duration of insolation in Jamestown occurs when the Sun appears to travel along path
   (1) A
   (2) B
   (3) C
   (4) D

8. At what time of day is the Sun at position S?
   (1) 6 A.M.
   (2) 9 A.M.
   (3) 3 P.M.
   (4) 6 P.M.

9. When the Sun appears to travel along path D at Jamestown, which latitude on Earth receives the most direct rays from the Sun?
   (1) 42° N
   (2) 23.5° N
   (3) 0°
   (4) 23.5° S

10. A cycle of Moon phases can be seen from Earth because the
    (1) Moon's distance from Earth changes at a predictable rate
    (2) Moon's axis is tilted
    (3) Moon spins on its axis
    (4) Moon revolves around Earth

Base your answers to questions 11 through 13 on the diagram below, which shows Earth in orbit around the Sun and the Moon in orbit around Earth. M_1, M_2, M_3, and M_4 indicate positions of the Moon in its orbit. Letter *A* indicates a location on Earth's surface.

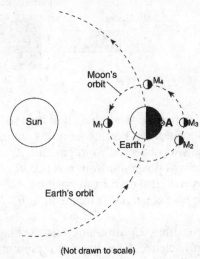

(Not drawn to scale)

11. An observer at location *A* on Earth views the Moon when it is at position M_3. Which phase of the Moon will the observer see?

12. At which Moon position could a solar eclipse be seen from Earth?
    (1) M_1 (3) M_3
    (2) M_2 (4) M_4

13. An observer at location *A* noticed that the apparent size of the Moon varied slightly from month to month when the Moon was at position M_4 in its orbit. Which statement best explains this variation in the apparent size of the Moon?
    (1) The Moon expands in summer and contracts in winter.
    (2) The Moon shows complete cycles of phases throughout the year.
    (3) The Moon's period of rotation is equal to its period of revolution.
    (4) The Moon's distance from Earth varies in a cyclic manner.

14. The diagram below shows the relative positions of the Sun, the Moon, and Earth when an eclipse was observed from Earth. Positions A and B are locations on Earth's surface.

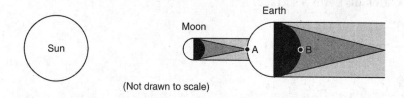

(Not drawn to scale)

Which statement correctly describes the type of eclipse that was occurring and the position on Earth where this eclipse was observed?
(1) A lunar eclipse was observed from position A.
(2) A lunar eclipse was observed from position B.
(3) A solar eclipse was observed from position A.
(4) A solar eclipse was observed from position B.

Base your answers to questions 15 through 18 on the diagram below and on your knowledge of Earth science. The diagram represents the Moon at four positions, labeled A, B, C, and D, in its orbit around Earth. The position of the full-Moon phase is labeled.

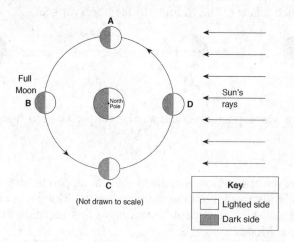

15. Approximately how many days (d) does it take for the Moon to move from the phase shown at position A to the full-Moon phase?
(1) 7.4 d   (3) 27.3 d
(2) 14.7 d   (4) 29.5 d

16. Which phase of the Moon could be observed from New York State when the Moon is at position C?

   (1)   (2)   (3)   (4)

17. The same side of the Moon always faces Earth because the Moon's period of revolution
    (1) is longer than the Moon's period of rotation
    (2) equals the Moon's period of rotation
    (3) is longer than Earth's period of rotation
    (4) equals Earth's period of rotation

18. Solar and lunar eclipses rarely happen during a cycle of phases because the
    (1) Moon's orbit is circular and Earth's orbit is elliptical
    (2) Moon's orbit is elliptical and Earth's orbit is elliptical
    (3) plane of the Moon's orbit is different from the plane of Earth's orbit
    (4) plane of the Moon's orbit is the same as the plane of Earth's orbit

Base your answers to questions 19 through 21 on the graph below and on your knowledge of Earth science. The graph shows the tidal range (the difference between the highest tide and the lowest tide) recorded in Minas Basin, Nova Scotia, during November 2007. The phase of the Moon on selected days is shown above the graph. The dates that the Moon was farthest from Earth (apogee) and closest to Earth (perigee) are indicated under the graph.

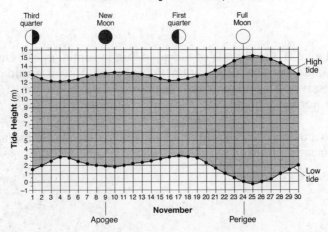

19. The tidal range on November 8 was approximately
    (1) 11 m   (3) 13 m
    (2) 2 m   (4) 15 m

Heliocentric Earth Motions and Their Effects

20. The highest high tides and the lowest low tides occurred when the Moon was near
    (1) apogee and a new-Moon phase
    (2) apogee and a full-Moon phase
    (3) perigee and a new-Moon phase
    (4) perigee and a full-Moon phase

21. The next first-quarter Moon after November 17 occurred closest to
    (1) December 9   (3) December 17
    (2) December 14   (4) December 24

CONSTRUCTED RESPONSE QUESTIONS

Base your answers to questions 22 and 23 on the diagram below, which represents the sky above an observer in Elmira, New York. Angular distances above the horizon are indicated. The Sun's apparent path for December 21 is shown.

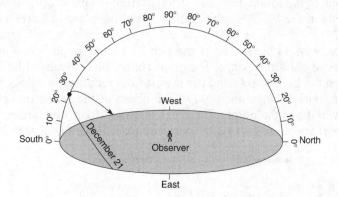

22. On March 21, the altitude of the noon Sun in Elmira is 48°. On the diagram, draw the Sun's apparent path for March 21 as it would appear to the observer. Be sure your path begins and ends at the correct positions on the horizon and indicates the correct altitude of the Sun at noon. [1]

23. On what date of the year does the maximum duration of insolation usually occur at Elmira? [1]

Base your answers to questions 24 and 25 on the diagram below, which shows the locations of high and low tides on Earth at a particular time.

(Not drawn to scale)

24. Identify the force that causes ocean tides on Earth. [1]

25. Approximately how many hours will pass between high tide and the following low tide? [1]

Base your answers to questions 26 through 29 on the data table below. The data table shows the latitude of several cities in the Northern Hemisphere and the duration of daylight on a particular day.

Data Table

City	Latitude (°N)	Duration of Daylight (hr)
Panama City, Panama	9	11.6
Mexico City, Mexico	19	11.0
Tampa, Florida	28	10.4
Memphis, Tennessee	35	9.8
Winnipeg, Canada	50	8.1
Churchill, Canada	59	6.3
Fairbanks, Alaska	65	3.7

26. On the grid, plot with an **X** the duration of daylight for each city shown in the data table. Connect your **X**s with a smooth, curved line. [1]

27. Based on the data table, state the relationship between latitude and the duration of daylight. [1]

28. Use your graph to determine the latitude at which the Sun sets 7 hours after it rises. [1]

29. The data were recorded for the first day of a certain season in the Northern Hemisphere. State the name of this season. [1]

Base your answers to questions 30 through 32 on the diagram below, which shows Earth as viewed from space on December 21. Some latitudes are labeled.

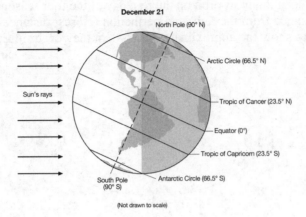

30. On the diagram, place an **X** at a location on Earth's surface where the Sun was directly overhead at some time on December 21. [1]

31. State one factor, other than the tilt of Earth's axis, that causes seasons to change on Earth. [1]

32. At which latitude is *Polaris* observed at an altitude of 66.5°? [1]

Heliocentric Earth Motions and Their Effects

EXTENDED CONSTRUCTED RESPONSE QUESTIONS

Base your answers to questions 33 through 37 on the diagram below. The diagram shows Earth revolving around the Sun. Letters A, B, C, and D represent Earth's location in its orbit on the first day of the four seasons. Aphelion (farthest distance from the Sun) and perihelion (closest distance to the Sun) are labeled to show the approximate times when they occur in Earth's orbit.

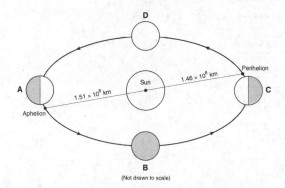

33. On the diagram, draw a line through Earth at location *A* to represent Earth's tilted axis on the first day of summer in the Northern Hemisphere. Label the North Pole end of the axis. [1]

34. On the diagram, draw an arrow on Earth at location *D* to show the direction of Earth's rotation. Extend the arrow from one side of Earth to the other side of Earth. [1]

35. Approximately how many days does it take Earth to travel from location *B* to location *C*? [1]

36. Explain why the gravitational attraction between the Sun and Earth decreases as Earth travels from location *D* to location *A*. [1]

37. Explain why an observer in New York State sees some different constellations in the night sky when Earth is at location *A* compared to when Earth is at location *C*. [1]

Base your answers to questions 38 through 42 on the diagram below, which represents eight positions of the Moon in its orbit around Earth.

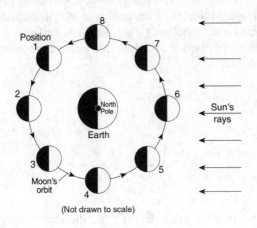

(Not drawn to scale)

38. On the diagram, circle the position of the Moon where a solar eclipse is possible. [1]

39. On the diagram below, shade the portion of the Moon that is in darkness to show the phase of the Moon at position 3, as viewed from New York State. [1]

40. Using the terms *rotation* and *revolution*, explain why the same side of the Moon always faces Earth. [1]

41. Explain why the Moon's gravity has a greater effect on Earth's ocean tides than the Sun's gravity. [1]

42. The table below shows times of ocean tides on March 4 for a city on the Atlantic coast of the United States.

Ocean Tides on March 4

Tide	Time
high	12:00 a.m.
low	6:13 a.m.
high	12:26 p.m.

Determine the time when the next low tide occurred. Include A.M. or P.M. in your answer, if needed. [1]

Heliocentric Earth Motions and Their Effects

Base your answers to questions 43 and 44 on the diagram below, which represents an exaggerated model of the shape of Earth's orbit, and on your knowledge of Earth science. The positions of Earth in its orbit on December 21 and June 21 are indicated. The positions of perihelion (when Earth is closest to the Sun) and aphelion (when Earth is farthest from the Sun) are also indicated. Both perihelion and aphelion occur approximately two weeks after the dates shown.

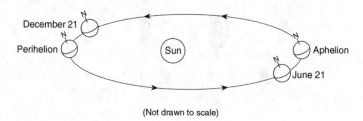

(Not drawn to scale)

43. How many months after Earth's perihelion position does Earth's aphelion position occur? [1]

44. Explain why warm summer temperatures occur in New York State when Earth is at aphelion. [1]

Base your answers to questions 45 through 47 on the diagram below and on your knowledge of Earth science. The diagram represents the Sun's apparent path on the equinoxes and the longest and shortest days of the year for a location in New York State. Points X, Y, and Z represent the solar noon positions along daily Sun paths X, Y, and Z.

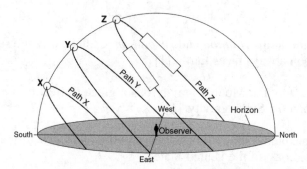

45. On the diagram above, draw one arrow in each box on path Z to indicate the Sun's apparent direction of movement along path Z. [1]

46. State one possible date of the year represented by each apparent path of the Sun. [1]

47. State the rate, in degrees per hour, that the Sun appears to travel along path X from sunrise to sunset. [1]

CHAPTER 4
EARTH'S COORDINATE SYSTEM AND MAPPING

KEY IDEAS Very early on, a variety of evidence led astronomers to conclude that Earth is a sphere. However, if only the shape of Earth is known, celestial observation can be used only to determine relative position—for example, 60° south of the North Pole. But how far is that from other locations: 1,000 kilometers? 10,000 kilometers? To determine actual distances between places on Earth's surface, the planet's size must also be determined. Only then can true-to-scale maps be constructed.

The size and shape of Earth can be readily determined by combining Earth-based measurements with simple observations of the sky. Observations show that Earth is a very slightly oblate sphere with an equatorial diameter of 12,757 kilometers.

The ability to locate and map positions on Earth's surface is essential to a wide variety of human activities ranging from engineering to urban planning to national defense. Topographic maps represent the three-dimensional shape of Earth's surface on a two-dimensional surface. A field is a region of space with a measurable quantity at every point. Field maps can be drawn to represent any quantity that varies in a region of space.

KEY OBJECTIVES
Upon completion of this chapter, you will be able to:

- Describe Earth's shape and explain how it can be determined by simple observations.
- Explain how Earth's size can be calculated using Earth-based measurements and simple observations of the sky.
- Use the latitude and longitude coordinate system to locate points on Earth's surface.
- Construct field maps and calculate gradient within a field.
- Read and interpret a topographic map.

THE NATURE OF EARTH'S SHAPE AND SURFACE

In order to map Earth's surface, its size and shape must be known. As stated above, Earth's size and shape can be readily determined by combining Earth-based measurements with simple observations of the sky. Let's begin by considering what we currently know about Earth's shape, and then examine how we arrived at this knowledge.

Earth's Shape

Earth's shape is very nearly a perfect sphere. A perfect sphere has exactly the same diameter when measured in any direction. Actual measurements of Earth's dimensions deviate slightly from this ideal. The polar diameter, or diameter measured from the North Pole through the center of Earth to the South Pole, is 12,714 kilometers. The equatorial diameter, or diameter measured from a point at the Equator through the center of Earth to the Equator, is 12,756 kilometers. Thus, Earth's spherical shape "bulges" very slightly at the Equator and is very slightly "flattened" at the poles; this shape is called an **oblate** (flattened) **spheroid**.

Roundness

Earth's actual shape, however, is so close to being a perfect sphere that the eye cannot detect its oblateness. When viewed from space, Earth appears perfectly round. Any cross section of Earth looks like a perfect circle. To gain an idea of how close Earth is to being perfectly round, let's suppose that we made a scale model of Earth—a globe. If we used a scale of 1 centimeter = 1,000 kilometers, the globe would have a polar diameter of 12.714 centimeters and an equatorial diameter of 12.756 centimeters—a little bigger than a softball. The difference in diameters would be 0.042 centimeter, or less than half a millimeter! We would need a micrometer to measure this difference in diameters.

Look carefully at the circle in Figure 4.1. Its polar diameter is 4.238 centimeters, and its equatorial diameter is 4.252 centimeters. It was drawn exactly one-third the size of the globe described above so that it would fit on this page. Can you tell that it is not perfectly round?

Earth's oblateness is the result of forces produced by Earth's rotation on its axis. Just as a loose skirt will swirl outward if the wearer spins around, planet Earth "swirls" outward

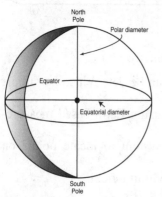

Figure 4.1 Schematic Diagram of Earth Showing Polar and Equatorial Diameters.

when it rotates. However, since Earth is much stiffer than a skirt, the distance it moves outward is much less.

Smoothness

In addition to being almost perfectly round, Earth is also very smooth. Compared with its diameter, the irregularities on Earth's surface (mountains and ocean basins) are relatively small. Mount Everest rises 8,848 meters, or about 8.8 kilometers, above sea level. That is about 7/10,000 of Earth's diameter (8.8 km/12,756 km). If you were to draw Mount Everest to scale on the circle in Figure 4.1, it would protrude from the surface less than 1/20 of a millimeter. Mark off a millimeter on a blank piece of paper, and then try to make dots small enough so that 20 will fit between the lines. If you put one of those dots on the outside of the circle, it would stick out of the circle as much as Mount Everest protrudes from Earth. As you can see, the dot barely affects the smoothness of the circle.

When you consider that most of the protrusions and indentations on the Earth's surface are much smaller than Mount Everest, you can see why Earth appears smooth. The only reason that Earth's mountains and valleys seem so large to us is that, compared to Earth, we are tiny. If you were to draw humans to scale on the model globe described in the preceding section, you would need a microscope to see them.

Evidence of Earth's Shape

Earth's true shape is a question that has captured the minds and imaginations of humans for thousands of years. Contrary to the popular belief that Columbus was among the first to believe that Earth is a sphere, Aristotle reached this conclusion in the third century B.C.

Ships and Eclipses

Aristotle based his conclusion upon simple observations of the sky and Earth-based measurements. This use of observation to support his ideas, rather than merely stating them as theories, places Aristotle among the earliest scientific thinkers. First, Aristotle observed that Earth's shadow during a lunar eclipse is definitely round. Although this could also happen if Earth was a cone or a cylinder, there was additional evidence supporting a spherical shape. Travelers reported that, as they went farther north or south, some stars appeared lower and lower in the sky. Furthermore, Aristotle noted that ships disappeared bow-first over the horizon, no matter in which direction they sailed. These observations could occur only if the ships were constantly moving along a curved surface that changed the angle at which they were viewed by an observer.

Today, a number of observations provide additional evidence of Earth's shape.

Photographs Taken from Space

The most direct evidence of Earth's shape consists of photographs taken from space. These photographs show that Earth is indeed spherical. When very precise photographs are taken, they can be analyzed and precise measurements of Earth's image in the photographs can be made. These measurements indicate that the Earth's polar and equatorial diameters are indeed slightly different, confirming that its shape is an oblate spheroid.

Observations of the Altitude of Polaris (the North Star)

Observations of *Polaris* provide us with information that can be interpreted, using simple geometry, to provide evidence of Earth's spherical shape and its oblateness. *Polaris* is a distant star that is located almost exactly over Earth's North Pole and is almost in line with its axis of rotation. *Polaris* is very far away from Earth, literally trillions of Earth-diameters distant. If it were observed from two points farthest apart on Earth's surface, the directions in which it was seen would vary by only a minute fraction of a degree. See Figure 4.2.

Figure 4.2 Direction to *Polaris*.

The direction to *Polaris* from all points on Earth is not measurably different.

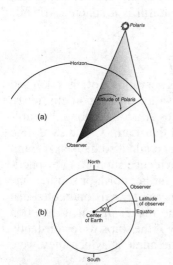

Figure 4.3 (a) Altitude of *Polaris* (b) Latitude of an Observer.

The **altitude** of *Polaris* is the angle between the star and the horizon with the observer at the vertex, as shown in Figure 4.3a. The latitude of an observer is the angle of the observer north or south of the Equator; see Figure 4.3b.

If Earth were a flat disk, the altitude of *Polaris* would be almost exactly the same at all locations and at all times. See Figure 4.4a.

It is true that, to a fixed observer, as Earth rotates, the altitude of *Polaris* does not appear to change, as shown in Figure 4.4b. However, if the observer travels north or south of the original location, the altitude does change, a result that would be expected if Earth is spherical. See Figure 4.4c.

If Earth were a perfect sphere, the altitude of *Polaris* would be the same as an observer's latitude and the distance traveled to change the altitude of *Polaris* by 1° would always be the same. See Figure 4.5.

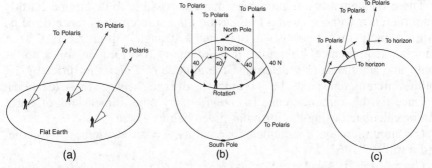

Figure 4.4 The Altitude of *Polaris* on a Flat and on a Spherical Earth. (a) If Earth were a flat disk, the altitude of *Polaris* would always be almost exactly the same at any location and at all times. (b) To a fixed observer on a spherical Earth, the altitude of *Polaris* does not appear to change as Earth rotates. (c) However, if the observer travels north or south, the altitude does change, a result that would be expected if Earth is spherical.

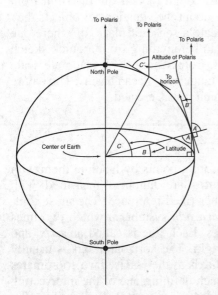

Figure 4.5 Relationship of the Altitude of *Polaris* to the Latitude of an Observer. By simple geometry it can be shown that the altitude of *Polaris* is the same as an observer's latitude. The sum of the angles of a triangle is equal to 180°. Similarly, since a straight line has a measure of 180°, ∠A + ∠B + 90° = 180°. Also, ∠A = ∠A' because they are alternate interior angles. Therefore, ∠B (latitude) = ∠B' (altitude of *Polaris*).

Determining Earth's Size

Modern measuring instruments based upon lasers allow us to measure Earth with great precision. However, Earth's size was estimated quite accurately more than 2,000 years ago. The principle is simple: if Earth is a sphere, then its circumference is a circle and all of the mathematical relationships between the various parts of a circle hold true for Earth.

These mathematical relationships make it possible to determine Earth's circumference without having to measure the entire circumference directly. If it is a circle, then it consists of 360° and a 1° angle will intersect 1/360 of Earth's circumference. Therefore, if observers at two locations on a north-south line simultaneously determine the altitude of *Polaris* to differ by 1°, Earth's circumference must be 360 times the distance between those locations.

Once Earth's circumference is known, many other dimensions of Earth can be calculated using the formulas given below, where

C = circumference, d = diameter, r = radius, A = surface area, V = volume, and π (pi) = 3.1462:

$$C = \pi d, \, d = 2r, \, A = 4\pi r^2, \, V = \frac{4}{3}\pi r^3$$

MAPPING EARTH'S SURFACE

Maps

A **map** is a model of Earth's surface. Although models are different from the real thing, they are tools for learning about the situations or objects they represent. A map is meant to communicate a sense of place, of where one point is in relation to another. A map can be anything from a quick sketch showing a friend how to get to the park from school to an elaborate scale model of Earth complete with mountain ranges and ocean basins. The nature of a map depends on the purpose for which it was created.

Coordinate Systems

A **coordinate system** is a method of locating points by labeling them with numbers called coordinates. **Coordinates** are numbers measured with respect to a system of lines or some other fixed reference. The most commonly used coordinate system is the *Cartesian* system, in which two lines on a flat surface intersect at right angles. Each line is called an **axis**, and the point where they intersect is the **origin**. The horizontal axis is usually referred to as the x-axis, and the vertical axis as the y-axis. Two coordinates describe any point on the flat surface, each defining one of the intersecting lines. For example, a fixed point can be located on a graph by describing the two lines $x = 4$ and $y = 3$. As shown in Figure 4.6a, these two lines intersect at a single point. By convention, coordinates in this system are written with the x value first and the y value second—(x, y), so the coordinates of the point shown are $(4, 3)$. The coordinate axes, in terms of which position is specified, form the **frame of reference** of a coordinate system.

Many coordinate systems can be devised to locate points on a surface. Another common example is the *polar* coordinate system, in which the frame of reference consists of a line and a point. Each point has coordinates (r, θ), which refer to a line called the **axis** and a point on that line called the **pole**. The *r coordinate* is the distance from the pole to the point. The θ *coordinate* is the angle between the axis and the line formed by joining the point and the pole measured counterclockwise. Thus, the coordinates of point P_1 are (4, 30) and those of point P_2 are (3, 240). See Figure 4.6b.

(a) The Cartesian coordinate system
(b) The polar coordinate system

Figure 4.6 Two Common Coordinate Systems. (a) The Cartesian coordinate system (b) The polar coordinate system

The Latitude-Longitude Coordinate System

In order to describe the position of any point on Earth's spherical surface, a coordinate system has been set up that uses two coordinates known as latitude and longitude. The latitude-longitude system consists of two sets of lines that cross each other at right angles. *Latitude* lines (called *parallels*) run in an east-west direction, and *longitude* lines (called *meridians*) run in a north-south direction. This type of system and the words *latitude* and *longitude* were already being used to describe such lines at the time of the Egyptian astronomer Claudius Ptolemy in A.D. 150.

Parallels

Since Earth is a sphere, east-west lines on its surface, such as the Equator, actually form circles. The circles formed by lines of latitude are called **parallels** because, if you drew a series of east-west lines, the circles formed would all be parallel to one another. See Figure 4.7a. Parallels lie in planes that are at right angles to Earth's axis of rotation. See Figure 4.7b.

To create a coordinate system on a globe or map, it is necessary to have a frame of reference, or fixed reference lines. The fixed reference line for latitude in this system is the **Equator**, an east-west line midway between the North and South Poles. Parallels are described by their angular distances north or south of the Equator as measured from the center of Earth. See Figure 4.7c.

Latitude is defined as *positive* for locations north of the Equator and *negative* for locations south of the Equator. Of course, latitude may also be given as **North (N)** or **South (S)** latitude. Thus, latitude is 0° at the Equator, and may be written as +90°, or 90° N, at the North Pole and –90°, or 90° S, at the South Pole. The latitude of New York City is about +41°30′ or 41°30′ N. Notice that the farther a location is from the Equator, the smaller the circle and the shorter its length. The east-west line at 60° N is only half as long as the Equator.

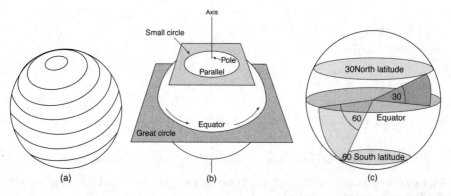

Figure 4.7 (a) Parallels are east-west lines on Earth's surface. (b) The Equator and all parallels lie in planes that are at right angles to the axis of rotation. (c) Latitude is a measure of the angle between the plane of the Equator and a line joining the center of Earth with a point on its surface.

Great Circles and Meridians

Earth's axis and poles also provide the reference points for lines of longitude. To understand longitude, you must first grasp the concept of a great circle. A **great circle** is a line drawn around a sphere to form a circle whose plane passes through the center of the sphere. If the plane of a circle doesn't pass through the center of the sphere, a small circle is formed. See Figure 4.8. An infinite number of great circles can be drawn on a sphere. The Equator, however, is the *only* great circle whose plane is at right angles to Earth's axis.

One interesting property of great circles is that they all cut a sphere exactly in half (e.g., the Equator divides Earth into the Northern and Southern hemispheres). Another is that the arc of a great circle connecting any two points on a sphere's surface is the shortest

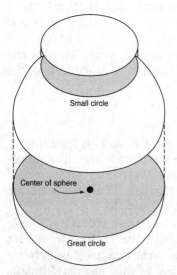

Figure 4.8 A Great Circle and a Small Circle.

distance between those points! This is why airliners travel along "great circle routes."

On Earth, any great circle that passes through both the North and the South poles is called a **meridian**. Earth's axis of rotation lies in the plane of every meridian. See Figure 4.9. **Meridians of longitude** go only halfway around Earth, from pole to pole. Although a meridian of longitude is actually only half of a meridian, it is often referred to simply as a meridian.

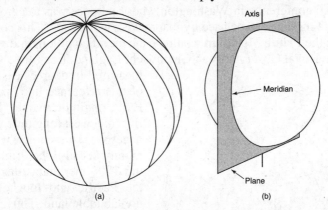

Figure 4.9 (a) Meridians are great circles that pass through both poles. (b) The plane of a meridian passes through Earth's axis.

The **longitude** of a point is the angular distance between two meridians: a fixed reference meridian and the meridian passing though the point. It is measured from the center of Earth east or west along the Equator between the fixed reference meridian of longitude and the meridian of longitude passing through the point. See Figure 4.10.

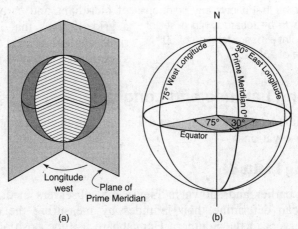

Figure 4.10 (a) Longitude can be thought of as the angle between the planes of two meridians. (b) Longitude is measured from the center of Earth as an angle east or west of the Prime Meridian.

Unlike the Equator, which is the *only* line halfway between the poles, the reference meridian, or *Prime Meridian*, is an arbitrarily chosen line and could be almost anywhere. Over the years, on different maps, it *has* been located in different places. Mapmakers usually began numbering the lines of longitude at whichever meridian passed through the site of their national observatory. As recently as 1881, 14 different prime meridians were being used on topographic survey maps. In 1884 the United States hosted an international conference in Washington whose purpose was to agree upon a single prime meridian to be used by mapmakers worldwide. This conference agreed that the **Prime Meridian** would be the meridian of longitude that runs through the Royal Observatory in Greenwich, England. Longitude would be measured in degrees east or west of this reference line up to 180°. East longitude would be positive, and west longitude would be negative. This choice was made mainly because the Prime's twin, the 180° meridian, runs through the Pacific and touches almost no habitable land. Thus, the 180° meridian is a good choice for an international date line, that is, a line marking the transition from one date to the next.

Any meridian will cross the Equator and all other parallels at right angles. Therefore, once reference lines are chosen, a system of meridians and parallels forms a grid of lines that intersect at right angles—a neat coordinate system! See Figure 4.11.

Figure 4.11 The geographic grid of parallels and meridians that allows any point on Earth's surface to be located with a set of coordinates. Point *P* has a latitude of 50° N and a longitude of 75° W.

Determining Latitude and Longitude

Both the latitude and the longitude of a location can be determined by making simple observations.

Determining Latitude

As described earlier and shown in Figure 4.5, observers in the Northern Hemisphere can determine their latitudes by measuring the altitude of *Polaris*. Observers in the Southern Hemisphere use the positions of other stars and make corrections for their deviation from Earth's axis of rotation.

Determining Longitude

The longitude of a particular location is determined by observing the time and the position of the Sun. Local noon is the time at which the Sun is at its highest altitude of the day. At that moment, a line drawn from the Sun to the center of Earth will pass through that location's meridian. A few moments later, Earth will have rotated from east to west and another meridian will align with the Sun and experience local noon. Thus, different longitudes will experience local noon at different times.

Since Earth makes one complete 360° rotation every 24 hours, the relationship between time and degrees rotated can be calculated. For example, in 1 hour Earth will rotate 360°/24, or 15°. Now, suppose you were at a location and the time was exactly local noon. One hour later, Earth would have rotated 15° and a location 15° of longitude away would be experiencing local noon. Two hours later, Earth would have rotated 30° and a location 30° of longitude away would experience local noon. By comparing the difference in time between local noons at any two locations, you can easily calculate their difference in longitude. Every hour of difference equals 15° of longitude, but this doesn't tell you a location's actual longitude, only its longitude relative to another point.

Suppose, however, that one of the times you knew was the time at which local noon occurred in *Greenwich*! Then you would know the difference in longitude at your location from *zero* longitude, and that would be your actual longitude. Therefore, in order to measure longitude at a particular location you need to know two times—local noon and the time of local noon in Greenwich. Each hour of difference between local time and Greenwich time equals 15° of longitude. Since Earth rotates from west to east, locations east of Greenwich will experience local noon earlier than Greenwich, while those west of Greenwich will experience local noon later than Greenwich. For example, if Old Forge, New York, experiences local noon about 5 hours later than Greenwich, England, its longitude is 5 × 15°, or 75°. Since local noon in Old Forge occurs after local noon in Greenwich, it is west longitude.

To measure longitude, then, you need a radio broadcast or a chronometer (a watch set to Greenwich time) to tell you what time it is in Greenwich when you experience local noon. You can then look up the time of local noon in Greenwich on that day in an astronomical table and, using the difference in times, calculate longitude. The need for such clocks drove the development of ever more precise timekeeping devices, culminating with today's atomic clocks.

Field Maps

A **field** is a region of space that has a measurable quantity at every point. Some examples of field quantities are temperature, pressure, magnetism, gravity, and elevation. Field maps can be used to represent any quantity that varies in a region of space.

Isolines

One way to represent field quantities on a two-dimensional field map is to use isolines. **Isolines** connect points of equal field values. For example, a temperature field map shows lines connecting points of equal temperature, or isotherms. See Figure 4.12.

Figure 4.12 Temperature Field Map.

Gradient

Within a field, field values change as you move from place to place. The rate at which a field value changes is called its **gradient**. Gradient can be calculated as follows:

$$\text{Gradient} = \frac{\text{amount of change in field value}}{\text{distance through which change occurs}}$$

Example:

Figure 4.13 shows pollutant levels measured in Lake Meredith. Isolines connect points of equal pollutant concentration, measured in parts per billion (ppb). Abby's Island and Ben's Island are 2 kilometers apart, and the pollutant concentration between them changes from 30 to 50 parts per billion. The rate of change, or gradient, is:

$$\text{Gradient} = \frac{50 \text{ ppb} - 30 \text{ ppb}}{2 \text{ km}}$$

$$= \frac{20 \text{ ppb}}{2 \text{ km}}$$

$$= 10 \text{ ppb/km}$$

The gradient is 10 parts per billion per kilometer.

Fields seldom remain unchanged over time. A field map shows a field at a particular point in time, the time at which its field values were measured. For example, the temperature field over the United States changes drastically from July to December. Also, Earth's magnetic field has changed position many times since Earth formed. Therefore, field maps need to be updated periodically in order to represent current field conditions. On weather maps, which represent rapidly changing fields, values are updated as often as once an hour.

Figure 4.13 Meredith Lake Containing Abby's Island and Ben's Island.

Topographic Maps: Elevation Field Maps

Topographic maps are scale models of a part of Earth's surface that show its three-dimensional shape in two dimensions. A topographic map is actually a type of field in which the field value is elevation above sea level of points on Earth's surface.

Contour Lines

Isolines that connect points of equal elevation are called **contour lines** because they represent the shapes, or contours, of Earth's surface. A contour line shows the shape that would be formed if the land surface were sliced by a horizontal plane at a particular elevation above sea level. An unvarying vertical distance called the **contour interval** separates successive contour lines. See Figure 4.14.

Contour lines around an enclosed depression have **hachure marks** pointing into the depression in order to distinguish them from small hills. See Figure 4.15.

Figure 4.14 Three-Dimensional Mountain Sliced by Horizontal Plane, Showing Contour Line.

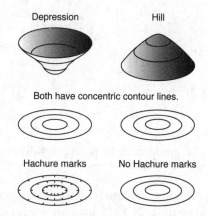

Figure 4.15 Depression Contours Versus Regular Contours.

Map Symbols

Topographic maps provide a view of the ground as seen from vertically above. Surface features are represented by map symbols. An extensive list of these symbols is shown in Figure 4.16. Symbols are blue for bodies of water, black and red for human-made structures, and brown for contour lines and other relief symbols.

Map Scale

A *map scale* is the ratio between the distance shown on a map and the actual distance on the ground. For example, if the map scale is expressed as 1:100,000, then 1 unit of length on the map is equal to 100,000 of the same units on the ground. Both numbers in the ratio have the same units, but those units can be anything. For example, the map scale indicates that 1 centimeter on the map equals 100,000 centimeters on the ground. But it also means that 1 inch on the map equals 100,000 inches on the ground. On most topographic maps, a graphic scale such as the one shown in Figure 4.17 is used to represent the map scale. The graphic scale can be used as a ruler to measure distances on the map.

Map Direction

The convention used in most topographic maps is that the top of the map is north. However, since north is determined with a compass, and geographic north differs from magnetic north, most topographic maps include an arrow showing each. The map in Figure 4.18 shows magnetic north with an arrow labeled "MN" and geographic North with an arrow labeled "GN."

Figure 4.16 Selected Topographic Map Symbols.

Figure 4.17 A Graphic Map Scale. The ratio 1:62,500 means that 1 unit of distance on the map equals 62,500 units of distance on the ground. Thus, 1 inch on the map is 62,500 inches on the ground, which is slightly less than 1 mile (1 mi = 63,360 in.). One centimeter on the map equals 62,500 centimeters on the ground (0.625 kilometer). The graphic scale is printed so that each unit is precisely the correct size for the distance indicated. (In this case 1 mile on the scale measures 1 inch in length.) The first unit on the graphic scale is subdivided so that it can be used to make more precise measurements.

Figure 4.18 Topographic Map. Source: *Exploring Earth Science*, Walter A. Thurber and Robert E. Kilburn, Allyn and Bacon, 1965.

Map Profiles

It is often useful to construct a profile from a topographic map. A **map profile** shows what a cross section of land looks like between two points. Figure 4.19 shows how a profile can be constructed.

1. The two points between which the profile is to be drawn are chosen, and a line is drawn to connect them.
2. The edge of a piece of paper is placed along the line, and the edge of the paper is marked wherever it intersects a contour line.

Figure 4.19 Steps in Constructing a Map Profile. Source: *Exploring Earth Science*, Walter A. Thurber and Robert E. Kilburn, Allyn and Bacon, 1965.

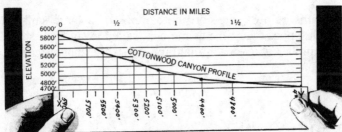

3. The paper is moved to a piece of graph paper on which a vertical scale has been marked to match the contour lines intersected. At each point where a contour line crosses the edge of the paper, a line of the appropriate height is drawn on the graph paper.
4. The endpoints of the lines are connected in a smooth line to form the finished profile.

Conventions Used in Topographic Maps

When topographic maps are drawn or read, the following conventions apply:

- All points on a contour line have the same elevation.
- **Index contours** are wide (bold). Elevation values are printed in several places along these lines to help the user determine elevations.
- All contour lines are closed, but they may run off the map.
- Two contour lines of different elevations may not cross each other.
- Contour lines may merge at a cliff or waterfall.
- The spacing of contour lines indicates the nature of the slope. The closer the contour lines are spaced, the steeper the slope. Even spacing indicates a uniform slope.
- Where contour lines cross a stream, they always form a V whose apex points up the valley.
- Where contour lines cross a ridge between valleys, they often form a V whose apex points down the valleys.

MULTIPLE-CHOICE QUESTIONS

In each case, write the number of the word or expression that best answers the question or completes the statement.

1. Which diagram most accurately shows the cross-sectional shape of Earth?

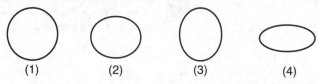

2. The diagrams below represent photographs of a large sailboat taken through telescopes over time as the boat sailed away from shore out to sea. The number above each diagram shows the magnification of the telescope lens.

   Which statement best explains the apparent sinking of this sailboat?
   (1) The sailboat is moving around the curved surface of Earth.
   (2) The sailboat appears smaller as it moves farther away.
   (3) The change in density of the atmosphere is causing refraction of light rays.
   (4) The tide is causing an increase in the depth of the ocean.

3. At sea level, which location would be farthest from the center of Earth?
   (1) the North Pole   (3) 45° N
   (2) 23½° S   (4) the Equator

4. The best evidence of Earth's nearly spherical shape is obtained through
   (1) telescopic observations of other planets
   (2) photographs of Earth from an orbiting satellite
   (3) observations of the Sun's altitude made during the day
   (4) observations of the Moon made during lunar eclipses

5. New York State's highest peak, Mt. Marcy, is located at approximately
   (1) 44°10′ N 74°05′ W
   (2) 44°05′ N 73°55′ W
   (3) 73°55′ N 44°10′ W
   (4) 74°05′ N 44°05′ W

6. Since Denver's longitude is 105° W and Utica's longitude is 75° W, sunrise in Denver occurs
   (1) 2 hours earlier
   (2) 2 hours later
   (3) 3 hours earlier
   (4) 3 hours later

7. Which New York State location has surface bedrock that has been subjected to very intense regional metamorphism?
   (1) 41°00′ N 72°15′ W
   (2) 42°30′ N 75°00′ W
   (3) 44°00′ N 76°00′ W
   (4) 44°30′ N 74°00′ W

8. Which New York State city is located at 42°39′ N, 73°45′ W?
   (1) Buffalo
   (2) Albany
   (3) Ithaca
   (4) Plattsburgh

9. *Polaris* is used as a celestial reference point for Earth's latitude system because *Polaris*
   (1) always rises at sunset and sets at sunrise
   (2) is located over Earth's axis of rotation
   (3) can be seen from any place on Earth
   (4) is a very bright star

10. From Utica, New York, *Polaris* is observed at an altitude of approximately
    (1) 43°
    (2) 47°
    (3) 75°
    (4) 90°

The diagrams below illustrate systems that can be used to determine position on a sphere. Refer to them to answer questions 11 and 12.

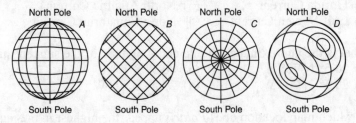

11. Systems of line like those illustrated above are called
    (1) latitude systems
    (2) coordinate systems
    (3) great circle systems
    (4) axis systems

12. Which of the systems illustrated is most like the latitude-longitude system used on Earth?
    (1) A
    (2) B
    (3) C
    (4) D

Base your answers to questions 13 through 15 on the passage and map below. The map shows sections of the Atlantic Ocean, the Caribbean Sea, and the Gulf of Mexico.

Shipwreck

In 1641, the crew of the ship *Concepcion* used the Sun and stars for navigation. The crew thought that the ship was just north of Puerto Rico, but ocean currents had carried them off course. The ship hit a coral reef and sank off the coast of the Dominican Republic. The **X** on the map marks the location of the sunken ship.

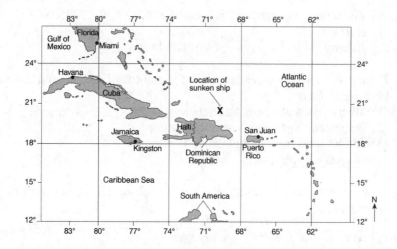

13. The *Concepcion* was carried off course to the northwest by an ocean current flowing from the
    (1) Florida Current
    (2) Gulf Stream Current
    (3) North Atlantic Current
    (4) North Equatorial Current

14. What is the approximate latitude and longitude of the sunken ship?
    (1) 20.5° N 70° E
    (2) 20.5° N 70° W
    (3) 20.5° S 70° E
    (4) 20.5° S 70° W

15. At which map location does *Polaris* appear the highest in the nighttime sky?
    (1) Miami, Florida
    (2) Kingston, Jamaica
    (3) Havana, Cuba
    (4) San Juan, Puerto Rico

16. When the time of day for a certain ship at sea is 12 noon, the time of day at the Prime Meridian (0° longitude) is 5 P.M. What is the ship's longitude?
    (1) 45° W   (3) 75° W
    (2) 45° E   (4) 75° E

17. Which statement is *not* true about fields?
    (1) They do not change with time.
    (2) They are often illustrated by the use of isolines.
    (3) Any one place in a field has a measurable value at a specific time.
    (4) Gradients indicate the degree of change from place to place in a field.

18. A topographic map is a two-dimensional model that uses contour lines to represent points of equal
    (1) barometric pressure   (3) elevation above sea level
    (2) temperature gradient   (4) magnetic force

Base your answers to questions 19 through 22 on the topographic map below. Elevations are in feet. Points *A* and *B* are locations on the map.

19. Toward which direction does the Green River flow?
    (1) northeast   (3) southeast
    (2) northwest   (4) southwest

20. What is the gradient along the straight line between points *A* and *B*?
    (1) 10 ft/mi (3) 25 ft/mi
    (2) 20 ft/mi (4) 35 ft/mi

21. Which graph best represents the profile along line *AB*?

    (1)

    (3)

    (2)

    (4)

22. What evidence can be used to determine that the land surface in the northeast corner of the map is relatively flat?
    (1) a rapidly flowing river
    (2) a large region covered by water
    (3) the dark contour line labeled 300
    (4) the absence of many contour lines

23. A topographic map and an *incorrectly* constructed profile from point *A* to point *H* on the map are shown below.

    What mistake was made in the construction of this profile?
    (1) using a contour interval of 10 feet
    (2) plotting points *A* through *H* the same distance apart horizontally
    (3) drawing a curved line instead of a straight line from point *B* to point *C*
    (4) increasing the elevation from point *F* to point *H*

CONSTRUCTED RESPONSE QUESTIONS

Base your answers to questions 24 through 26 on the diagram below, which represents a north polar view of Earth on a specific day of the year. Solar times at selected longitude lines are shown. Letter *A* represents a location on Earth's surface.

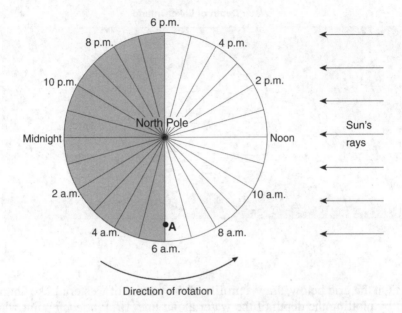

24. How many degrees apart are the longitude lines shown in the diagram? [1]

25. State the altitude of *Polaris* as seen by an observer at the North Pole. [1]

26. How many hours of daylight would an observer at location *A* experience on this day? [1]

Base your answers to questions 27 through 30 on the field map below and on your knowledge of Earth science. The map shows the depth of Lake Ontario. Isoline values indicate water depth, in feet. Points A, B, and C represent locations on the shoreline of Lake Ontario. Points D and E represent locations on the bottom of the lake.

27. On the grid below, draw a profile of the bottom of western Lake Ontario by plotting the depth of the water along line AB. Plot *each* point where an isoline showing depth is crossed by line AB. Connect the plots with a line, starting at A and ending at B, to complete the profile. [1]

28. Calculate the gradient of the lake bottom between point C and point D. Label your answer with the correct units. [1]

29. What is a possible depth of the water at location E? [1]

30. What evidence shown on the map indicates that the southern section of the bottom of Lake Ontario has the steepest slope? [1]

Base your answers to questions 31 and 32 on the diagrams below. The top diagram shows a depression and hill on a gently sloping area. The bottom diagram is a topographic map of the same area. Points *A*, *X*, and *Y* are locations on Earth's surface. A dashed line connects points *X* and *Y*. Elevation is indicated in feet.

31. What is a possible elevation of point *A*? [1]

32. On the grid below, construct a topographic profile along line *XY* by plotting a point for the elevation of *each* contour line that crosses line *XY*. Points *X* and *Y* have already been plotted on the grid. Connect the points with a smooth, curved line to complete the profile. [2]

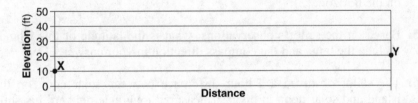

EXTENDED CONSTRUCTED RESPONSE QUESTIONS

Base your answers to questions 33 through 36 on the diagram provided below, which shows observations made by a sailor who left his ship and landed on a small deserted island on June 21. The diagram represents the apparent path of the Sun and the position of *Polaris*, as observed by the sailor on this island.

Sailor's Observations on the Deserted Island

33. On the diagram above, draw an arrow on the June 21 path of the Sun to show the Sun's direction of apparent movement from sunrise to sunset. [1]

34. The sailor was still on the island on September 23. On the diagram above, draw the Sun's apparent path for September 23, as it would have appeared to the sailor. Be sure your September 23 path indicates the correct altitude of the noon Sun and begins and ends at the correct points on the horizon. [2]

35. Based on the sailor's observations, what is the latitude of this island? Include the units and the compass direction in your answer. [1]

36. The sailor observed a 1-hour difference between solar noon on the island and solar noon at his last measured longitude onboard his ship. How many degrees of longitude is the island from the sailor's last measured longitude onboard his ship? [1]

Base your answers to questions 37 through 42 on the topographic map below and on your knowledge of Earth science. Dashed lines separate the map into sections I, II, III, and IV. Letters A through E represent locations on Earth's surface. The points in section I represent elevations in feet.

37. On the map, complete the 600-ft, 700-ft, and 800-ft contour lines in section I. Extend the lines to the edge of the map. [1]

38. On the map, draw a line showing the most likely path of a second creek that begins at location E and flows into Spruce Creek. [1]

Earth's Coordinate System and Mapping

39. Describe how the topography within section II is different from the topography within section IV. [1]

40. What is a possible elevation of location *E*? [1]

41. Calculate the gradient between locations *A* and *B*. [1]

42. On the grid below, construct a topographic profile along line *CD* by plotting the elevation of *each* contour line that crosses line *CD*. Connect all the plots with a line to complete the profile. [1]

CHAPTER 5

OUR SYSTEM OF TIME

> **KEY IDEAS** For thousands of years, people have used the changing positions of celestial objects to note the passage of time. The frame of reference for our system of time has changed over the years; however, it is still based upon the changing positions of celestial objects due to Earth's motions. Earth's rotation provides a basis for the day and our system of local time; meridians of longitude are the basis of time zones. The revolution of the Moon is the basis for the month, and the revolution of Earth around the Sun provides the basis for the year and our current calendar system.

KEY OBJECTIVES
Upon completion of this chapter, you will be able to:

- Explain how Earth's rotation forms the basis for the day and our system of local time.
- Compare and contrast the solar day and the sidereal day.
- Explain the difference between a synodic month and a sidereal month.
- Describe how meridians of longitude are used to define time zones.
- Analyze the seasonal changes in the Sun's apparent path through the sky.

THE DAY

The seemingly endless cycle of daylight and darkness was probably the earliest timekeeping unit—the day. It is the most basic of all natural cycles. Few things influence our lives as deeply as the change from light to dark, from day to night, from being asleep and being awake. The problem with the day is where to begin and end it and how to divide it up.

The earliest way of expressing "time of day" was simply to point to a place in the sky where the Sun is found at that time. Here is where the Sun rises; over there is where it sets. Morning, noon, and afternoon can be easily determined simply by looking at which side of the sky the Sun is in. Selecting a specific landmark with which the Sun aligns, such as a hut, a tree, or a distant mountain peak, can further refine this system. Using such markers, it is not difficult to divide daytime into 6, 9, or even 12 divisions.

Our System of Time

Dividing the Day into Hours

Timekeeping became a bit more precise with the simple invention of a stick stuck vertically into the ground. In sunlight, a vertical stick casts a shadow. As the Sun changes position, the shadow also changes position. The path of the Sun through the sky could be recorded as the line traced on the ground by the tip of the stick's shadow. This line could then be divided into equal units that, in turn, could be used to divide daytime into uniform time periods. The next small step was to realize that stars could also be sighted along the tip of a vertical stick and their motions tracked at night. In turn, these motions could then be used to divide nighttime into uniform time periods.

More than 3,000 years ago, the Egyptian priests of the Sun god Ra decided to divide the day into 12 hours of daytime and 12 hours of night. They chose 12 because they believed this number was mystical and symbolized the gods. The problem they faced was that most days did not have daylight and darkness periods of equal length. In New York State, for example, daylight varies from fewer than 9 hours in the winter to more than 15 hours in the summer. Consequently, dividing daylight into 12 equal time periods would yield a longer "hour" in the summer and a shorter "hour" in the winter. Eventually, the Egyptians discovered that there are only 2 days, the equinoxes, during which daylight and darkness are of equal length, so daylight and darkness were merged into a single 24-hour day with days and nights that

Figure 5.1 Changes in the Sunset Position along the Horizon During an 8-Month Period, Moving South from Position A to the Winter Solstice (B), Then North Again in C through the Spring Equinox (E) to the summer solstice (H).
Source: *Astronomy: The Cosmic Journey*, William K. Hartman, Wadsworth, 1987.

vary in length. But the Egyptians still had to measure the length of that 24-hour day.

You might think that measuring the length of a day is easy—simply pick a fixed point during the day, and then measure the time that elapses until that same point in the day is reached again. Then divide this time into 24 equal "hours." Great idea, but the question now becomes, At what point should you start measuring the day?

Suppose you decide to use sunrise to sunrise, or sunset to sunset, as your fixed time interval. You soon discover that the time of sunrise is not fixed. During part of the year the Sun rises a bit later each day, and during another part of the year it rises a bit earlier each day. See Table 5.1.

TABLE 5.1
TIME OF SUNRISE AND SUNSET DURING A TYPICAL YEAR AT MASSENA, NEW YORK (44°56′ N, 74°51′ W)

Date	Sunrise (A.M.) (Eastern Standard Time)	Sunset (P.M.) (Eastern Standard Time)
January 21	7:30	4:52
February 21	6:52	5.35
March 21	6:00	6:14
April 21	5:04	6:53
May 21	4:24	7:29
June 21	4:13	7:50
July 21	4:34	7:38
August 21	5:09	6:55
September 21	5:46	5:58
October 21	6:24	5:04
November 21	7:06	4:25
December 21	7:35	4:21

You also find that the positions of sunrise and sunset also change. See Figure 5.1 for changes in the sunset position over an 8-month period. If the times *and* the positions of sunrise and sunset vary, how, then, did people find a fixed point from which to measure the length of an entire day? The answer lies in the apparent path of the Sun. Look at Figure 5.2, which shows the Sun's apparent path through the sky on three different days of the year. Notice that,

Figure 5.2 The Apparent Path of the Sun as Viewed by an Observer in New York State Varies from Season to Season, but the Sun Always Crosses the Observer's Meridian at Noon.

Our System of Time

although the positions of sunrise and sunset and the length of the Sun's path vary, the midpoint of all three paths occurs when the Sun crosses the line running due north-south through the observer's zenith.

You recall from Chapter 4 that this imaginary line is the observer's meridian. Also notice that, when the Sun crosses the observer's meridian, it is at its highest point in the sky for that day and a vertical stick will cast the shortest shadow. The moment when the Sun is at its highest point in the sky is called **solar noon**. No matter how the number of daylight hours varies, solar noon always occurs exactly in the middle of the daylight period. Also, the Sun is always seen in the same direction at solar noon.

The Solar Day

The position of solar noon can be marked using two fixed points, one on the horizon and one nearby, such that they line up with the Sun at solar noon. (Since the Sun crosses the observer's meridian at solar noon, a line connecting these markers will run due north-south; a shadow cast by a vertical stick will also lie along a north-south line—a handy way of determining direction without a compass.) Whenever the Sun aligns with these two markers, it is solar noon. The time Earth takes to rotate from one solar noon until the next is called a **solar day**; that is, a solar day is the time from when the Sun crosses an observer's meridian until the Sun next crosses the observer's meridian.

Figure 5.3 At Precisely the Same Instant, a Person in New York City Experiences Solar Noon While in Tokyo, Japan, the Time is 2:00 A.M.

Local Time

So far, we have talked about the day in a very general way, with solar noon occurring when the Sun crosses an observer's meridian. However, an observer's meridian depends on his or her longitude, meaning that each observer has his or her own time scale based upon his or her location. A careful look at Figure 5.3 should make it clear that, while someone in New York City is experiencing solar noon, someone in Tokyo, Japan, is asleep in bed at 2 A.M. Our solar noon in New York (~75° W) occurs some 3 hours earlier than solar noon in California (~120° W)—a 45°

difference in longitude—because Earth takes 24 hours to spin through 360°, rotating through 15° every hour.

In the past, every locality set its clocks to its own solar noon. Clocks in Syracuse would be set to a different local time from those in Buffalo. As soon as people began traveling long distances rapidly, by stagecoach, train, or steamboat, however, local timekeeping created severe problems for companies that wanted to set up arrival and departure timetables.

Time Zones

In 1883 and 1884, the countries of the world agreed to set up a series of time zones. In a **time zone** the same time is used throughout the zone, instead of the local time at each place within the zone.

Since Earth rotates 15° per hour, time zones were set up around a series of meridians spaced at 15° intervals starting at the Prime Meridian, which, you have learned, passes through Greenwich, England. Time zones officially cover 7½° on either side of these meridians, so places between longitude 7½° W and 7½° E all set their clocks to **Greenwich Mean Time** (GMT). The next zone to the west extends from longitude 7½° W to 22½° W, and all clocks in that zone are set 1 hour behind GMT. New York State falls in the time zone surrounding the 75° W meridian, which extends from 67½° W to 82½° W. This zone sets its clocks 5 hours behind GMT.

To avoid inconveniences, the boundaries are shifted where they pass over land so that small countries or whole states fall within the same time zone. Figure 5.4 shows how the time zone boundaries in the United States have been shifted to follow state lines.

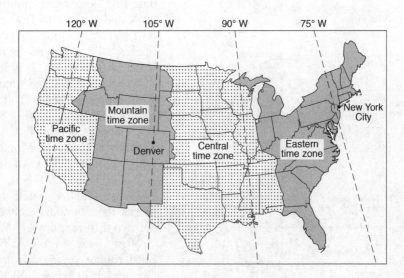

Figure 5.4 The Time Zones of the United States.

The International Date Line

As you travel west of Greenwich, England, you pass through 12 time zones before reaching longitude 180°, ending up 12 hours *behind* GMT. Likewise, as you travel east of Greenwich, England, you also pass through 12 time zones to reach longitude 180°, but now you are 12 hours *ahead* of GMT. How can you be 12 hours behind and 12 hours ahead of GMT at the same location? The answer is that the international date line is located here, at 180° longitude, and marks the boundary between 2 consecutive days; as you cross the date line from east to west, you advance 1 day. The time of day doesn't change, only the date.

To understand how the date line works, consider that the change from one day to the next is made at midnight. Look at Figure 5.5a. At Greenwich, England, it is solar noon on Monday, and the international date line is at midnight. Point A is at 1:00 A.M. Monday, and point B is at 11:00 P.M. Monday. Now (see Figure 5.5b), Earth rotates for 1 hour, and Greenwich goes from noon on Monday to 1:00 P.M. on Monday afternoon. Point A is now at 2:00 A.M. Monday and point B goes from 11:00 Monday to midnight, but the date line goes from midnight on Monday night to 1:00 P.M. on *Tuesday* morning. The date line is still 12 hours behind GMT (1:00 P.M. − 12 hr = 1:00 A.M.) and at the same time is 12 hours ahead of GMT (1:00 P.M. + 12 hr = 1:00 A.M.). But 12 hours behind 1:00 P.M. Monday is 1:00 A.M. Monday, while 12 hours *ahead* of 1:00 P.M. GMT is 1:00 A.M. on *Tuesday*. Thus, the date line marks the boundary between the two different days.

Now (see Figure 5.5c), Earth rotates yet another hour. Greenwich goes to 2:00 P.M. Monday, point A is at 3:00 A.M. Monday and the date line

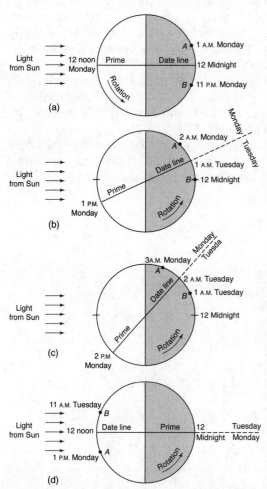

Figure 5.5 Understanding the International Date Line.

is at 2:00 A.M. Tuesday morning, but now point *B* is at 1:00 A.M. *Tuesday*. At this time it is Monday at point *A*, but Tuesday at point *B*. Let's continue this all the way to noon on Tuesday at the international date line (see Figure 5.5d). Now Greenwich is at midnight, finally entering Tuesday, and point *A* is at 1:00 P.M. Monday, but point *B* is at 11:00 A.M. Tuesday. The dividing line between Monday and Tuesday is the international date line. East of the date line it is Tuesday; west of the date line it is Monday.

This sequence continues until the date line once again rotates to the midnight position, when the boundary changes from Monday/Tuesday to Tuesday/Wednesday. With each successive rotation of Earth, the boundary advances another day.

THE MONTH

Just as the concept of a day arose from the natural cycle of light and darkness, the concept of a month originated in the natural cycle of lunar phases. (See Chapter 4.) The Moon, with its changes in shape, brightness, and position, has always fascinated humans. Indeed, there is evidence that the Moon has been used from the very earliest days of human civilization as a measuring tool for time. A Cro-Magnon eagle bone found in Le Placard, France, with notches cut at regular intervals and marked into groups, suggests that the Moon was the first celestial object used by humans for timekeeping.

When people first began to use arithmetic to count time, it was not difficult to determine that the Moon went through its full cycle of phases in 29 or 30 days.

Synodic and Sidereal Months

We now know that we see the phases of the Moon because the Moon revolves around Earth in an elliptical orbit that is tilted at an angle of 5½° to the plane of the ecliptic. (Refer to Figure 3.11.) We also know that the Moon goes through its cycle of phases, from one new Moon until the next, in precisely 29.530588 mean solar days, or 29 days, 12 hours, 44 minutes, and 3 seconds. This 29½-day month, from one new Moon phase until the next, is called a **synodic month**.

Like a solar day, the synodic month is based upon the Moon's position in relation to the Sun. New Moon occurs when the Moon is aligned between the Sun and Earth. Because Earth moves in its own gigantic orbit around the Sun while the Moon is making its much smaller orbit of Earth, the Moon has to travel more than 360° around Earth to align again with the Sun. See Figure 5.6.

If, instead of the Sun, we use alignment with a distant star, we find that the Moon completes a 360° orbit of Earth every 27.32166 mean solar days, or 27 days, 7 hours, 43 minutes, and 11 seconds. This 27⅓-day month, based upon alignment with a distant star and representing a 360° orbit of Earth, is called a **sidereal month**.

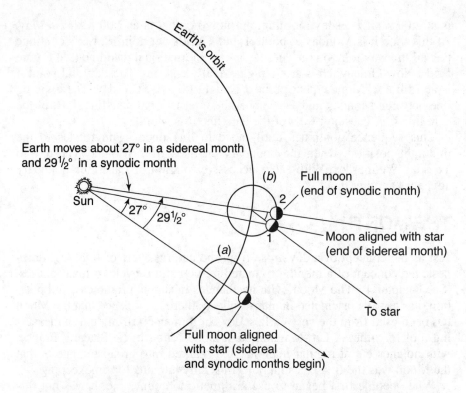

Figure 5.6 Difference in Length of Sidereal and Synodic Months. (a) The full Moon is observed to be near some bright star. (b) Roughly 27⅓ days later, at position 1, the Moon is again near the same bright star, completing a *sidereal month*. But Earth and the Moon have moved together around the Sun during that time. The Moon must revolve for 2⅙ more days before the Moon, Earth, and the Sun are again lined up so that the Moon is full again (position 2), completing a cycle of phases, or a **synodic month**. Thus, the time between full Moons, or the synodic month, is 27⅓ + 2⅙ = 29½ days.

Since the phases of the Moon are so universally visible, the 29½-day synodic month has become the basis for all calendar months. However, lunar months will not fit neatly into a solar year without fractional division, nor will days fit exactly into lunar months. Since the solar year is 365.242199 mean solar days long, you can't get exactly a whole number of days into a solar year either. How, then, can the year be divided into months and weeks? A convenient week will have a number of days that will divide equally into 365. Yet, no matter how you look at it, the week is an artificial unit of time and could have been 5, 6, or even 10 days long. For example, there could have been a 73-week year made up of 5-day weeks for a total of 365 days, because 5 is one of the few practical numbers that divides into 365.

The subdivision of the month into 7-day weeks can be traced back to the Genesis story of creation. It coincides roughly with dividing the month into four equal parts: new Moon to first quarter, first quarter to full Moon, full Moon to last quarter, and last quarter back to new Moon. And 7 days works nicely because 7 × 52 = 364, very close to the actual length of a year. But the 7-day week really became firmly entrenched because, in the Genesis story, God created the world in 7 days, which gave this number a sacred significance that survives to this day.

THE YEAR

The concept of a year originated in the natural cycle of seasons. In the Middle East, that cradle of civilization, there were probably two seasons—a hot, dry summer season and a cool, wet winter season. The Egyptians had a built-in time marker in the annual flooding of the Nile River during the wet season. But seasonal changes in temperature and rainfall are highly variable. Sometimes the Nile floods earlier in the season, sometimes later. And who in New York State hasn't experienced occasional stretches of balmy weather in March, only to encounter freezing temperatures and a late spring snowstorm in April? Thus, a more reliable indicator of the time of year is needed. Here again, the observation of celestial objects provided a valuable tool, and the Sun takes center stage.

Seasons and the Sun's Path

Because of Earth's rotation, the Sun appears to move in an arc across the sky from east to west. The daily path that the Sun traces through the sky goes through a cycle of changes with the seasons. Also, although the Sun always rises to the east of an observer and sets to the west, the exact positions of sunrise and sunset on the horizon change cyclically with the seasons too. Three factors combine to produce these cyclic changes: Earth's *revolution* around the Sun, the *tilt* of Earth's axis of rotation, and the **parallelism of Earth's axis** of rotation (the position of the axis at any given time is always parallel to its position at any other time). Figure 5.7 shows how these factors result in important changes in Earth's position relative to the Sun on key dates in the seasonal cycle of changes.

The Solstices

As Earth orbits the Sun, it also rotates on its axis. At one end of its axis is the North Pole, and at the other end the South Pole. Earth's axis of rotation is tilted at an angle of 23½°, so that during one part of its orbit the North Pole is leaning away from the Sun and the South Pole is leaning toward the Sun. The day on which the North Pole leans farthest *away from* the Sun is December 21, and this day is called the **winter solstice**. As you can see in Figure 5.8a, because of Earth's spherical shape only one parallel, the Tropic of Capricorn receives the Sun's direct rays (the direct rays strike Earth perpendicular to its surface).

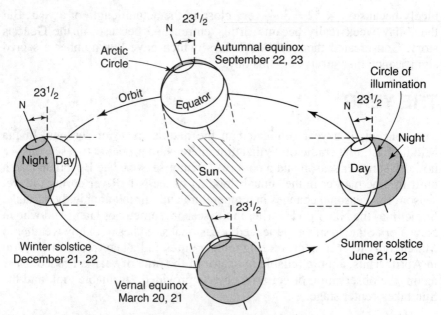

Figure 5.7 Earth's Changing Position in Relation to the Sun.

On December 21, the direct rays of the Sun reach farthest south of the Equator—$23\frac{1}{2}°$ S—and this parallel is given a special name, the Tropic of Capricorn. However, when the Sun's rays strike the Tropic of Capricorn at 90°, they are striking New York City, which is 64.5° of latitude away, at 90° − 64.5°, or only 25.5°—the lowest angle of the year. At positions north of New York City, the Sun's rays strike at an even smaller angle. Therefore, the shadow cast by a stick at noon in New York State is at its longest on December 21.

As Earth orbits the Sun, its axis remains tilted at this angle, so that, when it reaches the *other* side of the Sun, it is parallel to its previous position and the South Pole now leans away from the Sun while the North Pole leans toward it. The day on which the North Pole leans farthest *toward* the Sun is June 21, and this day is called the **summer solstice**. On this day, the direct rays of the Sun reach farthest north of the Equator—$23\frac{1}{2}°$ N—and this parallel is given a special name, the Tropic of Cancer. Locations north of the Tropic of Cancer never receive the Sun's direct rays, but on June 21 the rays are the most nearly direct of the year. Therefore, the shadow cast by a stick at solar noon in New York State is at its shortest, and the shadow falls due north.

The Equinoxes

Midway between these ends of Earth's orbit, however, neither the North nor the South Pole leans toward the Sun. To be sure, the poles are still tilted, but to the side, rather than toward or away from the Sun. In fact, at these midpoints, Earth's shadow cuts precisely through the poles. (Refer to the vernal and autumnal equinoxes shown in Figure 5.7.) In other words, the edge of Earth's shadow forms a great circle, cutting Earth precisely in half. And that great circle runs through the poles—it is a meridian! As Earth rotates, every point on its surface spends exactly one-half rotation in the light and one-half rotation in Earth's shadow; day and night are equal in length. These midpoints occur on March 21 and September 23; and these dates are called, respectively, the spring and fall **equinoxes** (*equinox* is the Latin word for "equal night"). On these dates, the direct rays of the Sun strikes the Equator. See Figure 5.8b.

The Sun's Changing Path

Figure 5.8 (a) Altitude of the noon Sun at the winter solstice (b) Altitude of the noon Sun at the equinoxes

At each of the positions shown in Figure 5.7, an observer on Earth's surface would see the Sun follow a different path as it rose, arced through the sky, and set. Figure 5.9 shows the changing path of the Sun to an observer at 42° N as Earth orbits the Sun.

Path *A* is the Sun's path on December 21, the winter solstice. Note that on this date the Sun rises at a point south of east on the horizon and sets south of west, and the arc traced by the Sun through the sky is the shortest. The shorter the Sun's daily path through the sky, the shorter the period of daylight, so December 21 is the shortest day of the year—less than 9 hours in New York State. The altitude of the Sun at noon is also at its lowest for the year, so sunlight strikes Earth's surface indirectly. Indirect sunlight is less intense than direct sunlight and therefore has less of a warming effect. Indirect sunlight, coupled with a short daylight period, is the cause of the winter season's lower temperatures.

Path *B* is the Sun's path on March 21 and September 23, the equinoxes. After the winter solstice, the sunrise and sunset points begin to migrate northward a bit each day, and the arc traced by the Sun slowly lengthens. When the Sun reaches path *B* on March 21, sunrise is due east, sunset is due west, and the daylight period is exactly 12 hours long; this is the spring equinox. After the spring equinox, the points of sunrise and sunset keep migrating northward, and the Sun's path keeps lengthening, until June 21, the summer solstice.

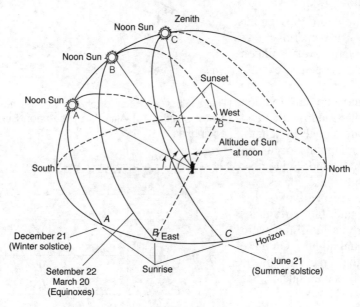

Figure 5.9 The Changing Path of the Sun to an Observer at 42° N.

Path *C* is the Sun's path on June 21, the summer solstice. On this date, the Sun rises farthest north of east and sets farthest north of west on the horizon. The Sun's path is the longest of the year; therefore, the daylight period is also the longest of the year—more than 15 hours in New York State! The noon Sun also reaches its greatest altitude of the year, and sunlight strikes Earth's surface more directly and intensely than at any other time of the year. More direct, intense sunlight, coupled with longer daylight hours, is the cause of the summer season's higher temperatures.

After June 21, the points of sunrise and sunset begin to migrate southward each day, and the arc traced by the Sun shortens. By September 23, the Sun is back to path *B*. Once again the Sun rises due east and sets due west, and the daylight period is exactly 12 hours in length. After the fall equinox the Sun's path continues its steady southward migration. Each day it rises and sets farther south and traces a shorter arc through the sky until the cycle is complete on December 21, roughly 365 days later. Then the cycle begins anew, moving from path *A* to *B* to *C* and back through *B* to *A* again.

Measuring the Year

Such cyclic seasonal changes in the Sun's path and sunrise-sunset position have long been used by cultures worldwide as fixed points to measure the length of a year and to subdivide it. In principle, the average number of days between successive recurrences of an annual event will yield an estimate for the length of a year.

One method is to use the length of a shadow at noon to find the winter and summer solstices. The Chinese used the number of days between two such solstices to calculate the length of a year.

Between the tropics of Cancer and Capricorn, there is always at least one day each year when the Sun is directly overhead at noon and there are no shadows at noontime. The Incas of Peru used this fact to figure out the length of a year. They noted the days when the Sun could be seen at noon through a skinny vertical tube pointing at zenith, and counted the average number of days between recurrences.

Another method is to note the positions of sunrise and sunset on the horizon. At the solstices, the Sun rises and sets at its northernmost and southernmost positions. Of course, for this method to work, the exact same position must be used for observations and the exact point on the horizon where the sun rises and sets must be carefully marked. It is thought that the circular arrangement of the huge rocks at Stonehenge may have been used for this purpose.

Table 5.2 shows the length of the mean solar year determined by these methods in many different cultures over time — with an amazing degree of accuracy!

TABLE 5.2
ESTIMATED LENGTH OF THE MEAN SOLAR YEAR

Place	Date	Estimated Length of Year (mean solar days)	Error
Babylon	700 B.C.	365.24579	0.00344
Egypt	150 B.C.	365.2466	0.0043
China	8 B.C.	365.25016	0.0078
India	A.D. 476	366.2589	0.0167
Mexico	700	365.2420	−0.0001
Arabia	900	365.24056	−0.00170
Samarkand	1400	365.24223	0.00030
Europe	1500	365.24222	0.00034
China	1620	365.24219	−0.00003
Europe	1600	365.24222	−0.00003
Modern	2000	365.242199	

Our System of Time

MULTIPLE-CHOICE QUESTIONS

In each case, write the number of the word or expression that best answers the question or completes the statement.

1. The length of an Earth day is determined by the time required for approximately one
   (1) Earth rotation
   (2) Earth revolution
   (3) Sun rotation
   (4) Sun revolution

2. The length of an Earth year is based on Earth's
   (1) rotation of 15°/hr
   (2) revolution of 15°/hr
   (3) rotation of approximately 1°/day
   (4) revolution of approximately 1°/day

3. The time required for the Moon to show a complete cycle of phases when viewed from Earth is approximately
   (1) 1 day
   (2) 1 week
   (3) 1 month
   (4) 1 year

4. On June 21, where will the Sun appear to rise for an observer located in New York State?
   (1) due west
   (2) due east
   (3) north of due east
   (4) south of due east

5. When does local solar noon occur for an observer in New York State?
   (1) when the clock reads 12 noon
   (2) when the Sun reaches its maximum altitude
   (3) when the Sun is directly overhead
   (4) when the Sun is on the Prime Meridian

6. A student in New York State looked toward the eastern horizon to observe sunrise at three different times during the year. The student drew the following diagram that shows the positions of sunrise, *A*, *B*, and *C*, during this one-year period.

   Which list correctly pairs the location of sunrise to the time of the year?
   (1) *A*—June 21   *B*—March 21   *C*—December 21
   (2) *A*—December 21 *B*—March 21   *C*—June 21
   (3) *A*—March 21   *B*—June 21   *C*—December 21
   (4) *A*—June 21   *B*—December 21   *C*—March 21

7. The map below shows a portion of the Middle East. Points *A*, *B*, *C*, *D*, and *X* are locations on Earth's surface.

When it is 10:00 A.M. solar time at location *X*, at which location is 11:00 A.M. solar time being observed?
(1) *A*   (3) *C*
(2) *B*   (4) *D*

8. The diagram below shows Earth and the Moon in four locations during their orbits. Arrows *A* through *D* represent different motions of Earth, the Moon, and the Sun.

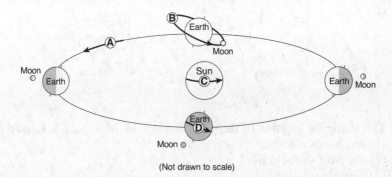

(Not drawn to scale)

Which arrow represents a rate of movement of approximately 1° per day?
(1) *A*   (3) *C*
(2) *B*   (4) *D*

Our System of Time

Base your answers to questions 9 through 11 on the diagram below and on your knowledge of Earth science. The diagram shows a pin perpendicular to a card. The card was placed outdoors in the sunlight on a horizontal surface. The positions of the pin's shadow on the card were recorded several times on March 21 by an observer in New York State.

(Drawn to scale)

9. Which diagram best represents the length of the pin's shadow at 2 P.M. on March 21?

10. The changing location of the pin's shadow on March 21 is caused by
    (1) the Sun's rotation
    (2) the Sun's revolution
    (3) Earth's rotation
    (4) Earth's revolution

11. On June 21, the card and pin were placed in the same position as they were on March 21. The diagram below shows the positions of the pin's shadow.

(Drawn to scale)

Which statement best explains the decreased length of each shadow on June 21?
(1) The Sun's apparent path varies with the seasons.
(2) The Sun's distance from Earth varies with the seasons.
(3) The intensity of insolation is lower on June 21.
(4) The duration of insolation is shorter on June 21.

Base your answers to questions 12 through 15 on the diagram and data table below. The diagram represents the Sun's apparent paths as viewed by an observer located at 50° N latitude on June 21 and March 21. The data table shows the Sun's maximum altitude for the same two dates of the year. The Sun's maximum altitude for December 21 has been left blank.

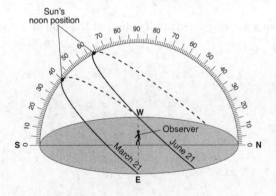

Data Table

Date	Sun's Maximum Altitude
June 21	63.5°
March 21	40°
December 21	

12. Which value should be placed in the data table for the Sun's maximum altitude on December 21?
(1) 16.5°
(2) 23.5°
(3) 40°
(4) 90°

13. Which graph best represents the relationship between the time of day and the length of a shadow cast by the observer on March 21?

14. Which statement best compares the intensity and angle of insolation at noon on March 21 and June 21?
    (1) The intensity and angle of insolation are greatest on March 21.
    (2) The intensity and angle of insolation are greatest on June 21.
    (3) The intensity of insolation is greatest on June 21, and the angle of insolation is greatest on March 21.
    (4) The intensity of insolation is greatest on March 21, and the angle of insolation is greatest on June 21.

15. Which diagram represents the approximate location of the Sun at 3 P.M. on March 21?

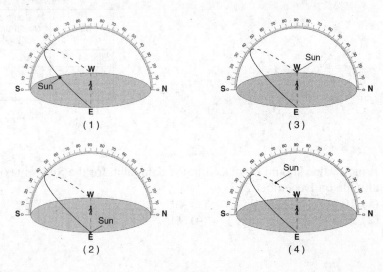

Base your answers to questions 16 through 20 on the diagram below, which represents Earth in relation to the Moon and Sun during the spring equinox as viewed from a point in space above the North Pole.

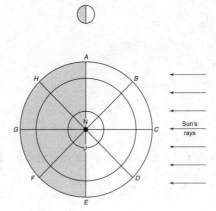

16. The Sun's vertical rays are striking Earth at this time at latitude
(1) 0°
(2) 23½° N
(3) 23½° S
(4) 66½° N

17. If it is 12 noon at point C, the time at F is
(1) 3 A.M.   (3) 6 P.M.
(2) 6 A.M.   (4) 9 P.M.

18. Sunrise on Earth is occurring at the point on the diagram labeled
(1) A (3) E
(2) B (4) G

19. If the Prime Meridian is at position B, the longitude at E is
(1) 90° E   (3) 135° E
(2) 90° W   (4) 135° W

20. In 7 days the Moon will have moved into a position nearly opposite the meridian indicated by letter
(1) B (3) G
(2) E (4) H

115

21. The diagram below indicates regions of daylight and darkness on Earth on the first day of summer in the Northern Hemisphere. Four latitudes are labeled A, B, C, and D.

At which latitude is the Sun above the horizon for the *least* number of hours on the day shown?
(1) A
(2) B
(3) C
(4) D

22. Since Denver's longitude is 105° W and Utica's longitude is 75° W, sunrise in Denver occurs
(1) 2 hours earlier
(2) 2 hours later
(3) 3 hours earlier
(4) 3 hours later

Base your answers to questions 23 and 24 on the United States time zone map shown below. The dashed lines represent meridians (lines of longitude).

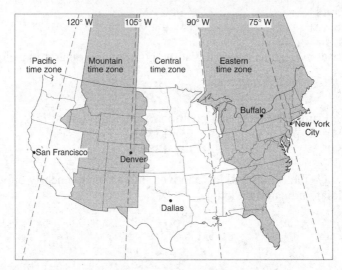

23. If the time in Buffalo, New York, is 5 A.M., what time would it be in San Francisco, California?
(1) 8 A.M.
(2) 2 A.M.
(3) 3 A.M.
(4) 4 A.M.

24. The basis for the time difference between adjoining time zones is Earth's
    (1) 1° per hour rate of revolution
    (2) 1° per hour rate of rotation
    (3) 15° per hour rate of revolution
    (4) 15° per hour rate of rotation

Base your answers to questions 25 through 27 on the diagram below, which represents an exaggerated view of Earth revolving around the Sun. Letters *A, B, C,* and *D* represent Earth's location in its orbit on the first day of each of the four seasons.

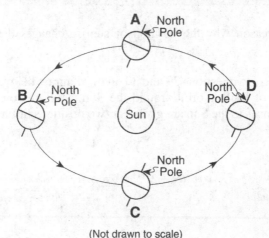

(Not drawn to scale)

25. Which location in Earth's orbit represents the first day of fall (autumn) for an observer in New York State?
    (1) *A* (3) *C*
    (2) *B* (4) *D*

26. Earth's rate of revolution around the Sun is approximately
    (1) 1° per day   (3) 15° per hour
    (2) 360° per day   (4) 23.5° per hour

27. Which observation provides the best evidence that Earth revolves around the Sun?
    (1) Stars seen from Earth appear to circle *Polaris*.
    (2) Earth's planetary winds are deflected by the Coriolis effect.
    (3) The change from high ocean tide to low ocean tide is a repeating pattern.
    (4) Different star constellations are seen from Earth at different times of the year.

Our System of Time

CONSTRUCTED RESPONSE QUESTIONS

28. The diagram below shows the position of sunrise along the horizon for a period of time from September 10 until December 21 as seen by an observer near Binghamton, New York.

State one reason why the position of sunrise changes during this time period. [1]

Base your answers to questions 29 and 30 on the diagram below. The diagram shows a model of Earth's orbit around the Sun. Two motions of Earth are indicated. Distances to the Sun are given for two positions of Earth in its orbit.

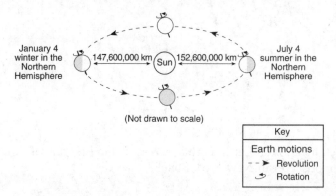

29. On the diagram above, place an **X** on Earth's orbit to indicate Earth's position on May 21. [1]

30. Explain why New York State experiences summer when Earth is at its greatest distance from the Sun. [1]

Extended Constructed Response Questions

Base your answers to questions 31 and 32 on the data table below, which shows the azimuths of sunrise and sunset on August 2 observed at four different latitudes. Azimuth is the compass direction measured, in degrees, along the horizon, starting from north.

Data Table

Latitude	Azimuths of Sunrise and Sunset	Letter Code
30° N	sunrise 69°	A
	sunset 291°	B
40° N	sunrise 66°	C
	sunset 294°	D
50° N	sunrise 61°	E
	sunset 299°	F
60° N	sunrise 51°	G
	sunset 309°	H

31. On the outer edge of the azimuth circle below, mark with an **X** the positions of sunrise and sunset for *each* latitude shown in the data table. Write the correct letter code beside each **X**. The positions of sunrise and sunset for 30° N have been plotted and labeled with letters *A* and *B*. [2]

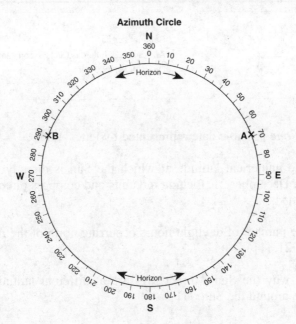

Azimuth Circle

119

Our System of Time

32. State the relationship at sunrise between the latitude and the azimuth. [1]

Base your answers to questions 33 through 36 on the table below and on your knowledge of Earth science. The table provides information about sunlight received on four dates of a certain year. Letter A represents a date. The arrows indicate the Sun's direct rays.

Date	Position of Earth Relative to the Sun's Rays	Seasonal Event
Sept 23		Fall equinox: Equal day and night Sun on the horizon at poles Direct ray at equator
Dec 21		Winter solstice: Area north of Arctic Circle in constant darkness
A		Spring equinox: Equal day and night Sun on the horizon at poles Direct ray at equator
June 21		Summer solstice: Area south of Antarctic Circle in constant darkness Direct ray at 23.5° N

33. Identify *one* possible date represented by letter A. [1]

34. State the numerical latitude at which the Sun is directly overhead at noon on December 21. Include the units and compass direction in your answer. [1]

35. State the number of daylight hours occurring north of the Arctic Circle on June 21. [1]

36. Explain why the Sun's direct rays are at different latitudes as Earth revolves around the Sun. [1]

Unit Two
MODERN ASTRONOMY

Chapter 6
STARS, THEIR ORIGIN AND EVOLUTION

KEY IDEAS Humans perceive the universe by the radiation it emits. Through much of recorded history, our understanding of the universe was based upon observations made with the unaided eye. This was a great limitation, for our eyes perceive only a small fraction of the radiation emitted by the universe—radiation with wavelengths in the range called *visible light*. Technological advances have greatly extended the scope of human perception and have led to the observations upon which our current theories of the universe are based. The vast majority of observable objects in the universe are stars. A star is a hot, luminous, gaseous celestial body. Stars form when gravity causes clouds of matter to contract until nuclear fusion of light elements into heavier ones occurs. Fusion releases great amounts of energy over millions of years.

Stars vary in size, temperature, and age. The majority of stars, including the Sun, fall into the main sequence when plotted on the Hertzsprung-Russell diagram according to luminosity and spectral class. Stars undergo a series of changes as they age.

A galaxy is a system of stars, cosmic dust, and gas held together by gravitation. A galaxy typically contains billions of stars and may be thousands of light-years in diameter, and the universe contains billions of such galaxies. Our Sun is a medium-sized star within a spiral galaxy of stars known as the Milky Way.

KEY OBJECTIVES
Upon completion of this chapter, you will be able to:

- Give examples of how progress in technology has increased our understanding of celestial objects.
- Distinguish among stars, galaxies, and nebulae.
- Describe the Hertzsprung-Russell diagram, and explain the relationship of a star's mass to its luminosity and temperature.
- List the three main steps leading to the birth of a star and the stages in the life cycle of a star such as our Sun.

HOW HUMANS PERCEIVE THE UNIVERSE

When an observer "sees" a star or any other object, the lens in his or her eye has focused light emanating from that object to form an image on the eye's retina. However, the human eye has limitations. The light entering the eye is limited to what passes through the eye's rather small lens. If the light intensity is too low, it doesn't trigger the nerve cells in the retina and we don't "see" the light. If the intensity is too great, the nerves are overwhelmed and signal the pupil to become smaller, allowing even less light to enter the eye.

TELESCOPES

A telescope uses a large lens to form a real image and then uses a second lens as a magnifying glass to examine the image. In this way, your eye sees an enlarged, close-up image. Such a combination of two lenses, one to form a real image and one to magnify the real image, is called a **refracting telescope**. See Figure 6.1a. If, on the other hand, a mirror is used to create the real image, which is then magnified, the instrument is a **reflecting telescope**. See Figure 6.1b. The lens or mirror that forms the real image in a telescope is called the *objective*. The magnifier used to examine the image is the *eyepiece*.

Figure 6.1 (a) A simple refracting telescope (b) A Newtonian focus reflecting telescope
Source: *Discovering Astronomy*, Robert Chapman, W. H. Freeman, 1978.

The larger the lens or mirror is in a telescope, the more light it gathers and the larger, brighter, and more detailed the image. The telescope on Mount Palomar has a lens that is 5 meters in diameter and has a light-gathering power that is 1 million times greater than the tiny 5-millimeter lens in your eye! Since telescopes can gather more light than the human eye, they allow us to see objects too dim to be visible with the naked eye. When under magnification, telescopes let us see details in the images not otherwise visible. Some telescopes can even "see" light that is not visible to our eyes. They use sensors that detect electromagnetic radiation invisible to the human eye and then convert that radiation into images our eyes can see. All of these technological advances have added tremendously to our store of observational evidence about the universe.

SPECTROSCOPES

Another valuable tool of the modern astronomer is the spectroscope. In 1666, Isaac Newton demonstrated that white light is actually a combination of all colors. He passed light through a triangular piece of glass, or prism, and found that the prism spread the light out into a rainbow of colors, or spectrum. See Figure 6.2. The colors of the spectrum are, in order from longest to shortest wavelength, red, orange, yellow, green, blue, indigo, and violet. This sequence can be remembered by using the name **Roy G. Biv**. Each letter in the name is the first letter of a color, and the letters are in correct order.

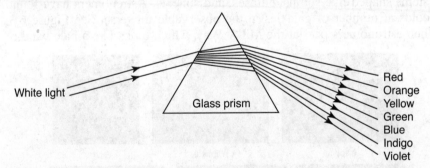

Figure 6.2 A Beam of White Light Entering a Prism Is Dispersed into a Spectrum.

In a spectroscope, astronomers pass light from a star through a prism before that light enters a telescope. Then the enlarged, close-up image of the spectrum is examined. When burned as a gas in the laboratory, each chemical element produces a unique pattern of bright spectral lines, which, like a fingerprint, identifies the element. By analyzing the specific wavelengths of light emitted by a star, astronomers are able to determine the chemical elements present in that star.

SOME OBSERVATIONS OF THE UNIVERSE MADE WITH INSTRUMENTS

By using instruments such as telescopes and spectroscopes, astronomers can see more of the universe, and obtain more detailed information about it, than ever before. Let's consider some of the observations made with these instruments that have led to current theories about the universe.

Stars

First, there are *a lot* more stars in the universe than the 1,500 or so that can be seen on a really dark night with the naked eye. Since telescopes gather more light than do eyes, stars too dim to be seen with the eye alone are visible through a telescope. Patches of sky that look empty to the naked eye are revealed to be filled with stars when viewed through a telescope. As ever-larger telescopes have been built, ever-dimmer stars have been detected. The number of observable stars has grown from thousands to hundreds of billions.

Galaxies

The high resolution of telescopes has revealed much about the structure of the universe. Some objects, which to the unaided eye looked like dim smudges of light, resolve into clusters of billions of stars, called *galaxies*, in systems shaped like spirals, ellipses, and spheres. Astronomers have found a colossal number of galaxies in the observable universe. See Figure 6.3. When astronomers peer at the Milky Way, which looks like a pale band of

Figure 6.3 Galaxies Are Systems Containing Billions of Stars. The universe is filled with galaxies of all shapes and sizes. Sources: Elliptical (M 89) © Canada-France-Hawaii Telescope Corporation; Elliptical (NGC 5866) from NASA, ESA, and The Hubble Heritage Team (STScI/AURA); Spiral (Andromeda Galaxy M31) from NASA/JPL/California Institute of Technology; Spiral (The Pinwheel Galaxy, M101) from ESA/Hubble; Barred spiral (NGC 1300) from The Hubble Heritage Team, ESA, and NASA; Irregular (NGC 1569) from NASA, ESA, Hubble Heritage (STScI/AURA)

Figure 6.4 Nebulae. (a) Orion nebula. Source: NASA (b) Horsehead nebula. Source: © Canada-France-Hawaii Telescope Corporation

light to the naked eye, they see that it actually consists of light from countless millions of stars. They realized from the distribution of stars that the Milky Way is a galaxy and that our Sun is one of the stars in the Milky Way galaxy.

Nebulae

Nebulae are clouds of gas and dust in the universe. The Orion nebula is a cloud of gas and dust lit by a group of stars within it. See Figure 6.4a. This nebula is dense enough to scatter starlight, and thus we can see it. By spectral analysis, the composition of the dust and gases in the cloud can be determined. When astronomers analyzed invisible radiation emitted by the cloud, they found that it extends out several times the area of the region we can see by visible light.

Sometimes dust is visible in a negative sort of way, as in the Horsehead nebula. There, a dust cloud is so thick that it blocks visible light, forming a black region (shaped like a horse's head) in front of the nebula. See Figure 6.4b. The dust actually absorbs starlight and reradiates it at longer wavelengths. To our eyes, the dust appears dark; to an infrared detector, it glows brightly.

Observations such as these indicate that there is a lot of "dark" matter in the universe, that is, matter that is not visible to the unaided eye.

THE SIZE AND STRUCTURE OF THE UNIVERSE

Based on observations made with a wide variety of instruments, astronomers have determined that the observable universe is incredibly vast and filled with energy and matter. Distances in the observable universe are so large that they are often measured in light-years. A light-year is the distance light travels in one Earth year, or about 9½ trillion kilometers. Our current estimates are that the universe is 93 billion light-years in diameter.

Observations have indicated that matter is not evenly spaced throughout the universe but occurs mostly in clusters. The largest clusters of matter are often galaxies. The universe contains billions of galaxies. Some galaxies are estimated to be more than 1 million light-years across. Our Milky Way galaxy is relatively small, at only about 100,000 to 150,000 light-years in diameter. The clouds of dust and gas that comprise nebulae are also very large. Estimates of the size of some of these dust- and gas-filled regions of space reach as high as 2 million light-years in diameter.

Each galaxy consists of billions of stars. Most stars have formed solar systems, but the majority of those solar systems are different from our own. Stars and their solar systems are much smaller than the galaxies. The diameters of most stars can be measured in light-seconds or light-minutes, solar systems in a few light-years.

CHARACTERISTICS OF STARS

Stars vary in many characteristics, including brightness, color, and temperature. A *star* is a hot, luminous, gaseous celestial body. Stars are distant, blazing suns moving through space at different distances from Earth. We see stars by the light they emit. On a clear, dark night you can see about 2,000 stars with the unaided eye. With a telescope you can see billions. Stars differ from each other in size, temperature, and age.

Stellar Brightness

Go outside on any clear night, and you will see that stars vary greatly in *apparent brightness*. Some appear very bright in the sky, while others are so dim you can barely make them out. You might think that the brighter stars appear that way because they are giving off more light, but the explanation is not that simple. Stars may appear bright because they emit more light, *or* because they are closer to Earth. The farther away a source of light is located, the dimmer it appears to an observer. Therefore, astronomers distinguish between a star's apparent brightness and its *luminosity*, or the actual amount of light the star shines into space each second. Since the Sun is the nearest and best known star, the luminosity of other stars is often stated in terms of the Sun's luminosity, which is 3.85×10^{28} watts (the equivalent of 3,850 billion trillion 100-watt light bulbs all shining together). The brightest stars are more than 1 million times as luminous as the Sun, while the dimmest are only 1/10,000 (0.0001) as luminous.

Star Color and Temperature

If you look carefully at stars, you discover that many show definite shades of color. *Betelgeuse* is reddish. Another bright star, *Arcturus*, is a pale orange-red, while *Vega*, a prominent star in the summer sky, shows a definite blue tint. Why the different colors?

The temperature of a star is related to the average speed of the particles of which it is composed. We say "average" speed because some of the particles are moving faster and some are moving slower. The speed at which matter moves determines the wavelength of the energy it emits. The faster it is moving, the shorter the wavelength of the energy it gives off. Thus, some of the particles in a star emit short-wavelength radiation while others emit long-wavelength radiation. The result is that a star gives off radiation that is a mix of many different wavelengths.

A *radiation curve* shows the amount of radiation a hot object gives off at different wavelengths. The more radiation a star emits at a particular wavelength, the higher the curve at that wavelength. The area under the curve indicates the total amount of energy being radiated by the star.

Figure 6.5 shows the radiation curves for a theoretical star at two different temperatures. As you can see, the higher the temperature, the shorter the peak wavelength of the energy emitted. Note, too, that the total amount of energy emitted by the star (the area under the curve) increases with temperature. The relationship between the total energy emitted by a star over all wavelengths and its temperature is directly proportional to its surface area. Thus, if we know the surface temperature of a star, we can calculate the amount of energy being emitted by each square meter of its surface, or its *luminosity*.

Figure 6.5 The Radiation Curves of a Theoretical Object at Two Different Temperatures. Note that, as temperature increases, the peak radiation shifts.

Figure 6.6 shows the actual radiation curve of the Sun and the radiation curve of a theoretical star at a temperature of 6,000 K. Note how closely the two curves match. For this reason, scientists believe the Sun has a surface temperature of about 6,000 K. Note, too, that the Sun radiates most intensely in the yellow range, but remember that the *peak* wavelength emitted by a star is not the *only* wavelength it is emitting. Stars emit a mix of wavelengths, so we perceive their light as white tinted the color of the peak wavelength, or as pastel shades of color. Hence, the Sun appears white with a decidedly yellow hue.

Figure 6.6 The Actual Radiation Curve of the Sun Closely Matches That of a Theoretical Object at 5777 K.

The physical conditions in the surface of a star, such as the Sun, are such that theoretical models can be used to derive information about stars. By analyzing the amount of energy emitted at each wavelength in a star's spectrum, astronomers can determine the peak wavelength, and thus the temperature, of the star. Also, the pattern of spectral lines reveals which elements are in the star, and measurements of the intensities of various bright lines in the spectrum show how much of each element is present.

The Hertzsprung-Russell Diagram

Early in the twentieth century, the relationship between the luminosities and temperatures of stars was discovered independently by two astronomers: Ejnar Hertzsprung of Denmark and Henry N. Russell of the United States. The ***Hertzsprung-Russell diagram*** (or H-R diagram) is a plot of luminosity versus surface temperature for the stars. See Figure 6.7. Every point on the H-R diagram represents a star. The star's temperature is read along the horizontal axis and its luminosity along the vertical axis. When several thousand stars are chosen at random and plotted on an H-R diagram, the dots do not scatter randomly over the graph. Instead, they fall into definite regions, forming patterns that show a meaningful relationship between a star's luminosity and its temperature.

Roughly 90 percent of all stars fall on a diagonal line across the diagram, called the ***main sequence***, which runs from the upper left (very hot, luminous blue supergiants) to the lower right (cool, dim red dwarfs). Why do stars follow this pattern? As was mentioned earlier, an increase in the temperature of a star results not only in a color shift in the radiation it emits, but also an increase in the total energy it gives off per unit of surface area. This increase in total energy emitted is seen as an increase in the star's brightness.

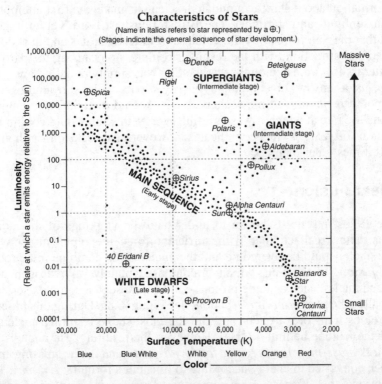

Figure 6.7 The Hertzsprung-Russell Diagram. When luminosity is plotted versus temperature, stars fall into distinct regions in patterns that represent the relationship between luminosity and temperature. Source: The State Education Department, *Earth Science Reference Tables*, 2011 Edition (Albany, New York; The University of the State of New York).

Most of the remaining 10 percent of stars fall above and to the right of the main sequence (cool, bright supergiants or red supergiants) or to the lower left (hot, dim white dwarfs). How can a star be cool yet bright, or hot yet dim? Consider, for example, 4,000°C stars. Since they all have the same surface temperature, they emit the same amount of energy per square meter of surface area. Thus, for a star above the main sequence, such as *Aldebaran*, to be more luminous than a red dwarf at the same temperature on the main sequence, it must have more square meters to radiate; in other words, it must be larger. Similarly, for 10,000°C stars that fall below the main sequence, such as white dwarfs, to be dimmer, they must be smaller than main sequence stars of the same temperature.

THE ORIGIN AND EVOLUTION OF STARS

Early Ideas

As information about stars accumulated, scientists made the first attempts to explain sunlight (and starlight). The British physicist Lord Kelvin hypothesized that the Sun formed by the collapse of interstellar gas and dust due to gravitational attraction, with the resultant compression causing the particles of matter to heat up. He thought that the Sun was simply radiating away this energy, just as any hot object radiates energy into its cooler surroundings, but his hypothesis ran into a time-scale problem. According to its size, temperature, and the rate at which it was emitting energy, the Sun would cool in tens of millions of years. Evidence indicates, however, that rocks on Earth are at least 10 times older than that.

Nuclear Fusion

By the 1930s, physicists began to understand the workings of the atomic nucleus. They realized that with sufficient force the repulsion between atomic nuclei could be overcome and the nuclei could combine in a process called *fusion*. In the Sun, which is composed mainly of hydrogen, four hydrogen nuclei (each a single proton) combine through a complex series of steps called the *proton-proton reaction*. See Figure 6.8a. During this process, two protons transform to neutrons and the most stable resulting nucleus is that of the isotope helium-4 (^{4}He). However, the resulting ^{4}He nucleus *has 0.7% less mass than the original four hydrogen nuclei*. The missing mass has been converted to energy according to Einstein's formula: $E = mc^2$. This formula says that, if mass, m, is converted to energy, the amount of energy created, E, is equal to m times the speed of light, c, squared. See Figure 6.8b.

If all of the hydrogen in the Sun were converted into helium, enough energy would be created to keep the Sun shining at its present rate for 100 billion years. However, as the ratio of hydrogen to helium in the Sun changes, so will its structure. After about 10 billion years the Sun will begin to undergo profound changes. But assuming that the Sun is about as old as the 4–5 billion-year-old Earth, it is only about halfway through its lifetime—so don't panic.

With star lifetimes measured in billions of years, it is not surprising that the stars we see today are pretty much the same as those observed by early astronomers. However, with all of written human history encompassing little more than 5,000 years, how can astronomers hope to learn anything about the birth, evolution, and death of a star? Luckily, stars did not all form at the same

Figure 6.8 (a) The proton-proton reaction, one of several nuclear reactions that result in the fusion of hydrogen into helium in stars, releasing energy. (b) One atom of helium has less mass than the four atoms of hydrogen from which it formed. The mass is not lost, but is converted into energy. The amount of energy produced per unit of mass is described by Einstein's famous equation, $E = mc^2$.

time, nor do they all go through their life cycle at the same rate. Therefore, it is possible to find stars at all stages of development in the universe and to piece together their life cycles from this evidence.

From Nebula to Protostar

The formation of a star begins in the immense interstellar (between stars) clouds of dust and gas, or *nebulae*, scattered throughout the universe. Inside these nebulae, which are composed mainly of hydrogen, stars are born through the process of gravitational collapse. However, the temperature of the cloud determines whether or not stars will form. If the temperature is too high, the atoms in the cloud move around too quickly to coalesce under the influence of gravity.

The first step in star formation is fragmentation of the cloud, in which cloudlets of matter form from denser portions of the cloud. See Figure 6.9.

The force of gravity pulls the dust and gas in toward the center of a cloudlet. As the particles of matter come closer together, mutual gravitational attraction strengthens, and the matter contracts and becomes even denser. As this process continues, gravitational attraction strengthens, attracting more and more dust and gas. As more and more matter accretes, or gradually accumulates and comes together, gravitational contraction causes temperatures and pressures to rise. Also, the cloudlet further contracts, rotates faster (as an ice skater spins faster as the arms are drawn inward toward the body), and becomes flattened out into a disk.

Figure 6.9 A Nebula Condenses into Cloudlets of Matter in the First Stage of Star Formation.

Heat flows from the hot center of the condensing cloudlet to its cooler surface. At first, the heat generated in the condensing cloudlet is radiated away into the infinite reservoir of the universe around it. As the cloudlet thickens, however, the outer layers begin to absorb radiation. From this point on, the heat generated by its collapse can no longer escape the thickening cloudlet. Now the cloudlet heats up, and its internal pressure increases as the heat energy is shared among its gas and dust particles, causing their velocities to increase. Eventually, the energy of collisions between particles bouncing them apart balances the gravitational energy drawing them together. The rate of compression slows and stops, temperatures reach the point where the whole mass glows, and fragmentation ceases. The cloudlet has now become a hot, relatively dense, disk-shaped region called a ***protostar***. The protostar is approximately the size of a solar system, and its surface temperature is about 4,000 K. See Figure 6.10.

The balance between the outward pressure of very hot gases and the inward pull of gravity should prevent the collapse of a protostar. The interplay of magnetic fields and charged particles within the cloudlet, however, permits randomly formed clumps of matter poor in charged particles to collapse into the core, causing it to become increasingly dense and hot.

A Star Is Born: Mass Is Destiny

As the core of a protostar collapses, the dust and gas collide more and more vigorously, and in each collision some of the energy of motion is converted into heat. Temperatures at the center of the core reach tens of millions of degrees, and pressures rise to billions of atmospheres. When the temperature in the center of the protostar reaches 10 million K, nuclear fusion reactions are triggered, and a star is born.

What happens next depends upon the mass and composition of the protostar. Stars that have about the same mass and chemical composition go through the same stages of evolution in about the same time. High-mass stars evolve the fastest, while stars of very low mass take the longest time to evolve.

Red Dwarfs

Low-mass protostars have correspondingly low gravity. Their cores balance at a slow rate of fusion and low temperatures. With less than a third the mass of the Sun, these stars are smaller and cooler and therefore appear red rather than yellow; thus they are called red dwarfs. Red dwarfs fuse hydrogen at such a slow rate that they can remain stable for hundreds of millions of years. The smallest red dwarfs may last for trillions of years before all of their hydrogen is consumed.

Figure 6.10 Three Stages in the Evolution of a Protostar. (a) An interstellar gas cloud begins to contract because of its own gravity. (b) A central condensation forms, and the cloud rotates faster and flattens. (c) The star forms in the cloud center, surrounded by a rotating disk of gas.

Sun-Class Stars

Mid-sized stars such as the Sun have enough mass to escape the dead-end fate of a red dwarf. They are larger, have higher gravity, and, once fusion has begun, stabilize at a higher temperature. Although they contain more hydrogen than dwarfs, their higher temperature causes fusion to occur at a faster rate, exhausting their supply of hydrogen in less than 50 billion years. As the star's hydrogen is converted into helium, the core becomes mostly helium; eventually the star is composed of a helium core surrounded by a shell of hydrogen gas. Without fusion of hydrogen in the core to support it, inward pressure of gravity from the surrounding shell causes the helium core to contract. Heat from the contraction triggers fusion in the hydrogen remaining in the surrounding shell. This fusion, together with heat from the still contracting core, causes the shell to puff out to perhaps 100 times its former size. The surface cools because of the expansion and now appears red. The star has moved off the main sequence and is now a red giant. See Figure 6.11.

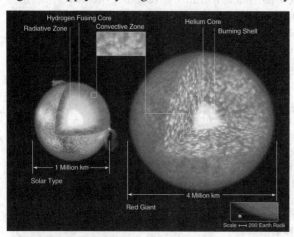

Figure 6.11 (a) At birth, a Sun-class star is mainly hydrogen, which fuses into helium, releasing energy. (b) After several billion years, a star consumes most of its hydrogen and develops a helium core. Heat from contraction of the helium core triggers fusion in the outer shell of hydrogen, causing the star to "puff up" into a red giant. Source: European Southern Observatory (ESO).

After about 100 million years, the core of a red giant contracts and heats up until it reaches the incredible temperature of 100,000,000 K, at which point helium begins to fuse into heavier elements, including carbon. In less than 100 million years this process ends, leaving a hot carbon core. But with insufficient gravity to fuse carbon, the star dies, its envelope of gases drifts off, and its hot, highly compressed core remains as a white dwarf.

Short-Lived Giants

Stars of at least three solar masses shine hot, bright, and blue. At the very upper end of the main sequence are hot supergiants with 30 times the Sun's mass and 100,000 times its brightness. To shine 100,000 times brighter than the Sun, these stars must fuse hydrogen 100,000 times faster, but they have only 30 times the Sun's mass. Thus, a supergiant should last for a time period that is 100,000 ÷ 30, or roughly 3,000, times less than the period of time the Sun should exist. Thus, a typical supergiant exists for only about 3 million years.

MULTIPLE-CHOICE QUESTIONS

In each case, write the word or expression that best answers the question or completes the statement.

1. Compared to stars viewed with the unaided eye, stars viewed with telescopes appear
   (1) larger and brighter
   (2) larger and dimmer
   (3) smaller and brighter
   (4) smaller and dimmer

2. When viewed by the unaided eye, the Milky Way looks like a pale band of light. When viewed through a telescope, the Milky Way
   (1) looks like a pale band of light
   (2) appears to be a glowing nebula
   (3) is seen to consist of billions of stars
   (4) appears to be the tail of a large comet

3. Which instrument is used to study the compositions of stars?
   (1) sextant
   (2) spectroscope
   (3) seismograph
   (4) chronometer

4. Which color of the visible spectrum has the *shortest* wavelength?
   (1) violet
   (2) blue
   (3) yellow
   (4) red

5. An observer viewing the sky through a telescope sees a fuzzy, glowing region in the constellation Orion. The region has an irregular shape, and some stars seem to be shining through it. The observer is most likely viewing a
   (1) planet
   (2) comet
   (3) meteor
   (4) nebula

6. In which list are celestial features correctly shown in order of increasing size?
   (1) galaxy, solar system, universe, planet
   (2) solar system, galaxy, planet, universe
   (3) planet, solar system, galaxy, universe
   (4) universe, galaxy, solar system, planet

7. Which of the following statements best describes the difference between a galaxy and a nebula?
   (1) A galaxy consists of stars; a nebula consists of dust and gas.
   (2) There are two types of nebula but only one type of galaxy.
   (3) A galaxy always emits light; a nebula never emits light.
   (4) A galaxy consists of matter; a nebula consists of energy.

Stars, Their Origin and Evolution

8. A star differs from a planet in that a star
   (1) has a fixed orbit
   (2) is self-luminous
   (3) revolves about the Sun
   (4) shines by reflected light

9. If three identical 100-watt light bulbs were placed at distances of 1 meter, 10 meters, and 100 meters, respectively, from an observer, each would seem to have a different brightness. Applied to stars, this concept is called
   (1) density
   (2) magnitude
   (3) volume
   (4) twinkling

10. Which factor does *not* directly affect the apparent brightness of *Polaris*?
    (1) its motion about the north celestial pole
    (2) its distance from Earth
    (3) its mass
    (4) its temperature

11. Which object in space emits light because it releases energy produced by nuclear fusion?
    (1) Earth's Moon
    (2) Halley's comet
    (3) Venus
    (4) *Polaris*

12. Which sequence of stars is listed in order of increasing luminosity?
    (1) *Spica, Rigel, Deneb, Betelgeuse*
    (2) *Polaris, Deneb, 40 Eridani B, Proxima Centauri*
    (3) *Barnard's Star, Alpha Centauri, Rigel, Spica*
    (4) *Procyon B, Sun, Sirius, Betelgeuse*

13. Which star is more massive than our Sun but has a lower surface temperature?
    (1) *40 Eridani B*
    (2) *Sirius*
    (3) *Aldebaran*
    (4) *Barnard's Star*

14. Which characteristics best describe the star *Betelgeuse*?
    (1) reddish orange with low luminosity and high surface temperature
    (2) reddish orange with high luminosity and low surface temperature
    (3) blue-white with low luminosity and low surface temperature
    (4) blue-white with high luminosity and high surface temperature

15. Compared to the luminosity and surface temperature of red main sequence stars, blue main sequence stars are
    (1) less luminous and have a lower surface temperature
    (2) less luminous and have a higher surface temperature
    (3) more luminous and have a lower surface temperature
    (4) more luminous and have a higher surface temperature

16. To an observer on Earth, the Sun appears brighter than the star *Rigel* because the Sun is
    (1) hotter than *Rigel*
    (2) more luminous than *Rigel*
    (3) closer than *Rigel*
    (4) larger than *Rigel*

17. Compared to other groups of stars, the group that has relatively low luminosities and relatively low temperatures is the
    (1) red dwarfs
    (2) white dwarfs
    (3) red giants
    (4) blue supergiants

Base your answers to questions 18 through 21 on the diagram below and on your knowledge of Earth science. The diagram represents two possible sequences in the evolution of stars.

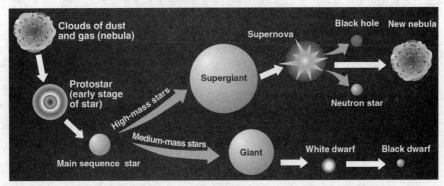

(Not drawn to scale)

18. What causes clouds of dust and gas to form a protostar?
    (1) magnetism
    (2) gravitational attraction
    (3) expansion of matter
    (4) cosmic background radiation

19. Which property primarily determines whether a giant star or a supergiant star will form?
    (1) mass
    (2) color
    (3) shape
    (4) composition

20. Which table includes data that are characteristic of the surface temperature and luminosity of some white dwarf stars?

Surface Temperature	5000 K
Luminosity	100

(1)

Surface Temperature	10,000 K
Luminosity	100

(3)

Surface Temperature	5000 K
Luminosity	0.001

(2)

Surface Temperature	10,000 K
Luminosity	0.001

(4)

21. Which process generates the energy that is released by stars?
    (1) nuclear fusion
    (2) thermal conduction
    (3) convection currents
    (4) radioactive decay

CONSTRUCTED RESPONSE QUESTIONS

Base your answers to questions 22 through 24 on the Characteristics of Stars graph in the *Reference Tables for Physical Setting/Earth Science*.

22. Describe the relationship between temperature and luminosity of main sequence stars. [1]

23. In which group of stars would a star with a temperature of 5,000°C and a luminosity of approximately 100 times that of the Sun be classified? [1]

24. Complete the table below by identifying the color and classification of the star *Procyon B*. The data for the Sun have been completed as an example. [1]

Star	Color	Classification
Sun	yellow	main sequence
Procyon B		

Base your answers to questions 25 through 27 on the passage below and on your knowledge of stars and galaxies.

Stars

Stars can be classified according to their properties, such as diameter, mass, luminosity, and temperature. Some stars are so large that the orbits of the planets in our solar system would easily fit inside them.

Stars are grouped together in galaxies covering vast distances. Galaxies contain from 100 billion to over 300 billion stars. Astronomers have discovered billions of galaxies in the universe.

25. Arrange the terms *galaxy*, *star*, and *universe* in order from largest to smallest. [1]

26. Complete the table below by placing an **X** in the boxes that indicate the temperature and luminosity of each star compared to our Sun. [1]

	Temperature		Luminosity	
Stars	Hotter	Cooler	Brighter	Dimmer
Procyon B				
Barnard's Star				
Rigel				

27. The star *Betelgeuse* is farther from Earth than the star *Aldebaran*. Explain why *Betelgeuse* appears brighter or more luminous than *Aldebaran*. [1]

Base your answers to questions 28 through 30 on the graph below, which shows the early formation of main sequence stars of different masses (M). The arrows represent temperature and luminosity changes as each star becomes part of the main sequence. The time needed for each star to develop into a main sequence star is shown on the main sequence line.

28. Describe the relationship between the original mass of a star and the length of time necessary for it to become a main sequence star. [1]

29. Describe the change in luminosity of a star that has an original mass of 0.5 M as it progresses to a main sequence star. [1]

30. Identify the force that causes the accumulation of matter that forms the stars. [1]

EXTENDED CONSTRUCTED RESPONSE QUESTIONS

Base your answers to questions 31 through 33 on the table below, which lists some information about *Barnard's Star*.

Barnard's Star

Distance from Sun	• 6.0 light-years* • currently moving toward the Sun (and Earth) and will get as close as 3.8 light-years in approximately 11,000 years
Characteristics of Barnard's Star	• less than 17 percent of the Sun's mass • approximately 20 percent of the Sun's diameter • age thought to be between 11 and 12 billion years old and may last another 40 billion years • no planets observed orbiting Barnard's Star

*A light-year is the distance light travels in one year.

31. Compared to the surface temperature and luminosity of the Sun, describe the relative surface temperature and the relative luminosity of *Barnard's Star*. [1]

32. List *Barnard's Star*, the Sun, and the universe in order by age from oldest to youngest. [1]

33. If a planet with the same mass as Earth were discovered orbiting *Barnard's Star* at the same distance that Earth is orbiting the Sun, why would there be less gravitational attraction between this new planet and *Barnard's Star* than there is between Earth and the Sun? [1]

Base your answers to questions 34 and 35 on the flowchart below and on your knowledge of Earth science. The flowchart shows the evolution of stars.

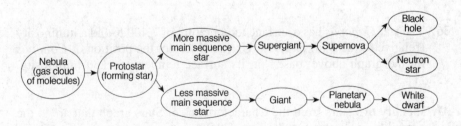

34. Identify the force responsible for the contraction of a nebula (a gas cloud of molecules) to form a protostar. [1]

35. Describe how the diameter and luminosity of a main sequence star change as the star becomes either a giant or a supergiant. [1]

Stars, Their Origin and Evolution

Base your answers to questions 36 through 39 on the Characteristics of Stars graph below and on your knowledge of Earth science.

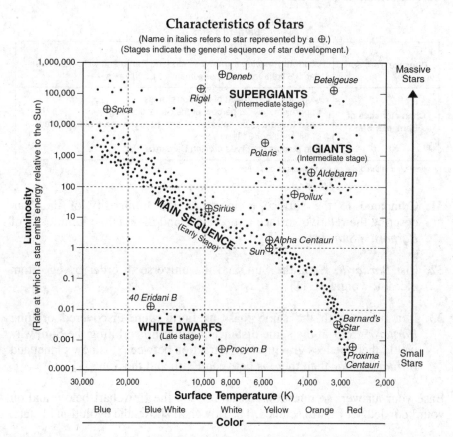

36. The star *Canopus* has a surface temperature of 7,400 K and a luminosity (relative to the Sun) of 1,413. Use an **X** to plot the position of *Canopus* on the graph above, based on its surface temperature and luminosity. [1]

37. Identify *two* stars from the Characteristics of Stars graph that are at the same life-cycle stage as the Sun. [1]

38. Describe *one* characteristic of the star Spica that causes it to have a greater luminosity than *Barnard's Star*. [1]

39. Describe how the relative surface temperature and the relative luminosity of *Aldebaran* would change if it collapses and becomes a white dwarf like *Procyon B*. [1]

Chapter 7
THE SOLAR SYSTEM

KEY IDEAS Our solar system formed about 5 billion years ago from a giant cloud of gas and debris. The solar system consists of nine large planets, their satellites, and a variety of smaller objects, ranging from asteroids, meteors, and comets to tiny particles of dust and gas, all in orbit around a central star, our Sun.

During the formation of the solar system, gravity caused Earth and the other planets to become layered according to the density of the materials of which they were composed. The distance of each planet from the Sun was a key factor in determining the planet's characteristics. Powerful emissions from the Sun drove off most of the nearby gases, leaving behind the small, dense, rocky terrestrial planets. The more distant Jovian planets had sufficient gravity to retain much of the gas that surrounded them, and thus evolved into the large, low-density, gaseous planets.

Formation of the solar system left numerous smaller objects, such as asteroids, comets, and meteors, orbiting the Sun. From time to time, these objects collide with planets. Impact craters from such collisions have been identified in Earth's crust, and impact events have been correlated with mass extinctions and global climate change on Earth.

KEY OBJECTIVES
Upon completion of this chapter, you will be able to:

- List the members of the solar system.
- Describe the nebular theory of the formation of the solar system.
- Compare and contrast the characteristics of the nine large planets, distinguishing between the terrestrial planets and the Jovian planets.
- Describe the other components of our solar system, such as asteroids, meteors, and comets.
- Explain the evidence linking impact events with mass extinctions and global climate change on Earth.

The Solar System

THE SOLAR SYSTEM'S PLACE IN THE UNIVERSE

We currently know that Earth is one of several planets that orbit a star—the Sun. The Sun is millions of times closer to Earth than is any other star. Light travels at a finite speed of 300,000 kilometers per second. Light from the Sun reaches Earth in less than 10 *minutes*, but light from the next closest star takes several *years* to get to Earth. The light from very distant stars takes several billion years to reach Earth. When we look at distant stars, we see them as they were when the light we now see left each star. Therefore, when we look at distant stars, we look back in time.

The universe is so large that the distance light travels in a year, or a light-year, is used to measure its distances. A light-year is a distance of about 9½ trillion kilometers. Our fastest rockets would take thousands of years to reach the nearest star beyond the Sun. Compared with the vast distances between stars, the distances between the Sun and its planets are small. Most astronomers estimate the universe to be about 15 billion light-years in radius, and it is therefore thought to be about 15 billion years old.

Stars are not scattered evenly throughout the universe; gravity has drawn them together in huge clumps called galaxies. A **galaxy** is a system of hundreds of billions of stars. The universe contains many billions of galaxies, each, in turn, containing billions of stars. Our solar system is located near the edge of a disk-shaped galaxy of stars called the **Milky Way galaxy**, which gets its name from the faint white band of its stars that can be seen from Earth on a clear, dark night. (See Figure 7.1.) The Milky Way galaxy contains more than 100 billion stars revolving in huge orbits around the center of the galaxy.

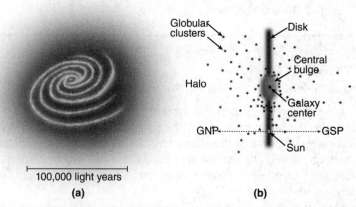

Figure 7.1 The Milky Way Galaxy. Viewed face on (a) and edge on (b), the shapes and locations of the nucleus, disk, and halo can be seen. Note the position of the Sun, about two-thirds of the way out from the nucleus in the Orion arm of the disk.

THE ORIGIN AND EVOLUTION OF THE SOLAR SYSTEM

Our solar system consists of the Sun, the eight planets, fifty or so moons, and thousands of asteroids, meteors, and countless comets that orbit the Sun. On the whole, though, the solar system *is* the Sun, for the Sun contains 99.9 percent of all the mass of the whole system. It is unlikely that the origin of a family of objects (the planets, moons, etc.) representing such a tiny fraction of the solar system as a whole is not closely linked to the formation of the Sun. Any theory of the formation of the solar system has to account for observational evidence, such as the following:

1. More than 99 percent of the mass of the solar system is contained in the Sun.
2. All the planets move around the Sun in the same direction and in roughly the same plane.
3. All of the planets (except Venus and Uranus) spin in the same direction, close to the plane of the Sun's equator; and most of the moons also spin in the same direction as their planets, close to the plane of those planets' equators.
4. The planets can be divided by mass and density into the terrestrial and Jovian planets. The terrestrial planets are composed mostly of metal silicates and iron; the Jovian planets, mostly of hydrogen and helium.
5. The planets exhibit a fairly regular spacing of orbits.
6. All the solid bodies of the solar system that have been measured—Earth, the Moon, and meteorites—are roughly 4.6 billion years old, and formed within about 0.1 billion years of each other, that is, at about the same time.
7. Some meteorites, aside from their loss of volatile (easily vaporized) elements such as rare gases, are virtually identical in composition to the Sun.

The question is, Did the solar system form as part of the natural process of star formation or as some later event? As early as the eighteenth century, two theories were proposed to account for the formation of the solar system: the catastrophic theory and the nebular theory.

The *catastrophic theory* suggests that early in the Sun's life there was a collision or near collision between the Sun and a passing celestial body. Such an event would cause a solar upheaval, resulting in streamers and globs of gaseous solar material being thrown off into space. As this material condensed, it would form a great many small bodies, all revolving around the Sun. Some of these would be close enough to each other to aggregate by mutual gravitational attraction to form planets. Others would be left revolving around the Sun as asteroids, planetoids, and comets. However, the catastrophic theory requires an improbable event and does not adequately explain the even spacing of planets and their varying compositions.

The Solar System

In recent years, with the aid of computer modeling, the *nebular theory* has gained more widespread support. The modern **nebular theory** proposes that the solar system formed from an interstellar cloud of dust and gas, enriched in heavier elements from earlier supernovae, which condensed into a disk-shaped protostar. The source of planet formation was the disk of dust and gas surrounding the protostar. Turbulence in the dust and gas of the spinning disk caused it to fragment and sort itself out into concentric rings according to the mass and the speed at which the material in it was revolving. See Figure 7.2.

Figure 7.2 The Formation of the Solar System from a Solar Nebula to Our Present Solar System.

Most collisions between celestial bodies tend to cause material to fall inward and end up in the protostar. Some of the dust particles, however, stick together upon collision to form grains. This process of sticking, or accretion, continues, slowly forming larger and larger bodies called **planetesimals**. Computer simulations indicate that over a period of millions to tens of millions of years, Earth-sized bodies can be formed in this way.

The nature of the dust that accretes depends upon its position in the disk, and it changes over time. The temperature of the disk decreases with movement outward from the core. Where temperatures are lower than 1,500 K, rocky and metallic grains can survive. This region, between 0.5 and 5 astronomical units from the Sun (recall that an AU, or astronomical unit, is the average distance between Earth and the Sun: about 150 million kilometers), is where the terrestrial planets are found today. Beyond about 5 astronomical units, the disk was cold enough so that water ice could condense and survive. Thus, today, in the outer solar system bodies made of rock and ice—the moons of the gas giants—are visible. The Jovian planets probably began forming with the accretion of rock and ice. But, as the body grew larger, it reached a point where it gravitationally attracted gas (mostly hydrogen) from the surrounding solar nebula. As the gas built up around the growing planet, the planet's gravity strengthened, attracting yet more gas, rock, and ice. As the giant planets formed, they produced disks of gas and dust of their own, out of which their moons formed. (The formation of Earth's Moon, which is not much smaller than Earth itself, must have occurred in a different way, which will be discussed later.) Computer simulations suggest that Jupiter and Saturn could have formed in this way in just a few million years. Uranus and Neptune, in the colder outer reaches of the disk, would have taken longer, perhaps 10 million years. As the planets drew material from the solar nebula, it dissipated and planet formation ceased as the planets literally ran out of gas and dust to draw upon.

That the terrestrial planets did not acquire huge gaseous envelopes may be due to a number of reasons. The surrounding gas may have been too hot—its molecules moving too vigorously—to form a stable envelope. The solar wind, or stream of charged particles and radiation emanating from the protostar, may have stripped the planets of their hydrogen envelopes as the solar nebula dissipated. Or perhaps the lack of water ice resulted in a slower rate of accretion, so that by the time the terrestrial planets were big enough to attract gas the solar nebula had already dissipated.

Beyond the orbit of Neptune, extending out to about 1,000 astronomical units, lies the **Kuiper belt**—a region of smaller orbiting objects left over from planet formation. Kuiper belt objects range in size from tiny grains to minor planets hundreds of kilometers in diameter. The dwarf planet Pluto and its moon Charon are believed to be Kuiper objects that strayed close enough to the Sun to be drawn into a planetary orbit. See Figure 7.3.

Figure 7.3 The Kuiper Belt. This is a region of objects surrounding the solar system and consisting of particles, ranging in size from dust to planetesimals, that are remnants of solar system formation. It extends from the orbit of Neptune out to about 1,000 AU.

Even farther out lies a huge, shell-like cloud of icy material called the **Oort cloud**, after the Dutch astronomer Jan Oort, who proposed its existence in 1950. The Oort cloud is thought to surround the solar system completely and extend to a distance of 100,000 astronomical units (more than 3,000 times the Sun–Neptune distance). See Figure 7.4. The very low temperatures in the Oort cloud allow gases such as methane, nitrogen, and carbon monoxide to form ices that accrete, along with water and dust, and preserve a record of the composition of the original solar nebula. It is believed that the clumps of icy material are visible as comets originate within the Oort cloud. The gravitational influence of passing stars causes some of these orbiting ice balls to fall toward the inner reaches of the solar system. Drawn inward by the Sun's gravity, and perturbed by the gravity of major planets, some settle into short-term orbits; one example is Halley's comet, which orbits the Sun once every 76 years. Other, shorter period comets include Tempel 2 (5.3 years) and Encke's comet (3.3 years). Some comets, however, have longer periods, so long that astronomers cannot measure them and may be taken by surprise. One such comet, the Daylight comet of 1910, was probably the brightest seen in the twentieth century. See Figure 7.4.

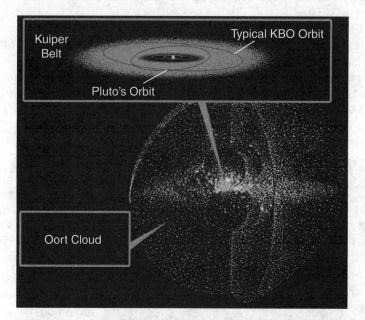

Figure 7.4 The Oort Cloud. A giant cloud of comets is believed to surround the solar system. Stars passing the Oort cloud tear comets from their orbits, sending some off into space and others into long-period orbits around the Sun.

Although the disk-shaped protostars can be observed in molecular clouds such as the Orion nebula, planetary formation around stars has never been observed. Of the billions of stars in the universe, only ten or so planets orbiting other stars have been definitively identified. Thus, much of the current thinking about planetary formation is based upon computer modeling and is therefore subject to revision as new observational data are obtained.

THE STRUCTURE OF THE SOLAR SYSTEM

The solar system is defined as the Sun, the planets that orbit the Sun, the satellites of those planets, and many small interplanetary bodies such as asteroids and comets. See Figure 7.5. The **Sun** is a star, which is composed of gases and emits electromagnetic radiation produced by nuclear reactions in its interior. **Planets** are objects that are large enough that their gravity has pulled them into a round shape. Their gravity is also strong enough that they have swept up all nearby objects and debris and now orbit in a clear path around the Sun. **Satellites** are solid bodies that orbit planets.

Figure 7.5 The Solar System. Note that in addition to the eight planets and their moons, the solar system includes the asteroid belt, dwarf planets such as Pluto, and comets that circle the Sun in highly eccentric orbits.

Beginning at the center, the major planets of the solar system are Mercury, Venus, Earth, Mars, Jupiter, Saturn, Uranus, and Neptune. A common memory aid for this sequence is **My Very Educated Mother Just Served Us Nachos.** Some basic information about the bodies in the solar system is summarized in Table 7.1, the Solar System Data chart of the *Reference Tables for Physical Setting/Earth Science*. The major planets of the solar system are divided into two groups based upon their size and composition: the four innermost, small, dense **terrestrial planets** and the four large, much less dense outermost **Jovian planets**.

The Solar System

In addition to the eight major planets, several smaller dwarf planets orbit the Sun. **Dwarf planets** have enough gravity that they are rounded, or nearly round in shape. However, they have not swept up everything near their path and may orbit in a zone that still has many other objects in it. Currently, there are three known dwarf planets—Ceres, Pluto, and Eris. All are less than half the size of the planet Mercury and at best have only a trace of an atmosphere.

TABLE 7.1 SOLAR SYSTEM DATA

Celestial Object	Mean Distance from the Sun (million km)	Period of Revolution (d = days) (y = years)	Period of Rotation at Equator	Eccentricity of Orbit	Equatorial Diameter (km)	Mass (Earth = 1)	Density (g/cm^3)
SUN	—	—	27 d	—	1,392,000	333,000.00	1.4
MERCURY	57.9	88 d	59 d	0.206	4,879	0.06	5.4
VENUS	108.2	224.7 d	243 d	0.007	12,104	0.82	5.2
EARTH	149.6	365.26 d	23 h 56 min 4 s	0.017	12,756	1.00	5.5
MARS	227.9	687 d	24 h 37 min 23 s	0.093	6,794	0.11	3.9
JUPITER	778.4	11.9 y	9 h 50 min 30 s	0.048	142,984	317.83	1.3
SATURN	1,426.7	29.5 y	10 h 14 min	0.054	120,536	95.16	0.7
URANUS	2,871.0	84.0 y	17 h 14 min	0.047	51,118	14.54	1.3
NEPTUNE	4,498.3	164.8 y	16 h	0.009	49,528	17.15	1.8
EARTH'S MOON	149.6 (0.386 from Earth)	27.3 d	27.3 d	0.055	3,476	0.01	3.3

Source: The State Education Department, *Earth Science Reference Tables*, 2011 Edition (Albany, New York; The University of the State of New York).

The Terrestrial Planets

The four planets closest to the Sun are Mercury, Venus, Earth, and Mars. These "inner" planets are terrestrial; that is, they resemble Earth in size and rocky composition. They also have about the same density as Earth.

Mercury

Mercury is one of the five planets that can be seen from Earth with the unaided eye. Mercury is very difficult to spot, however, because it is never more than 28° from the Sun, whose glare usually hides it. Through a telescope from Earth, Mercury, like the Moon and Venus, can be seen to exhibit phases, but only dark, fuzzy features can be detected on its surface.

Mercury revolves around the Sun once every 88 days and rotates once every 58.65 days. Because it is the closest planet to the Sun, it receives the most intense radiation. The long rotation period causes the side of Mercury facing the Sun to receive nonstop sunlight for a long period of time. At the point where the Sun is directly overhead of Mercury, the surface temperature reaches 700 K—hot enough to melt lead. At the same time, the side facing away from the Sun is in darkness for long periods and cools to 100 K. Like the Moon, Mercury has no atmosphere because its small gravitational field was unable to hold on to any gases. The lack of an atmosphere allows debris from space to strike the surface of Mercury unhindered. A fly-by of Mercury by the Mariner 10 spacecraft showed a cratered surface remarkably like that of the Moon.

Venus

Venus is one of the brightest objects seen in the sky; only the Sun and the Moon are brighter. Since its orbit is larger than Mercury's, Venus may be seen as far as 47° from the Sun. This feature, together with its brightness, makes it one of the most obvious of all celestial objects.

Venus is almost identical in size to Earth. Like Mercury, Venus can be seen to go through a series of phases. However, Venus rotates in a direction opposite to that of Earth and most other planets.

A telescope can detect little about the surface of Venus because the planet's dense atmosphere, consisting mostly of carbon dioxide, hides the surface from view. This thick, cloudy atmosphere produces a greenhouse effect that traps heat; several Venera probes soft-landed on Venus by the Soviet Union recorded temperatures near 750 K and atmospheric pressures 90 times greater than Earth's. Spectroscopic studies indicate the presence of sulfuric acid, hydrochloric acid, and hydrofluoric acid. In February 1974, the Mariner 10 spacecraft came within 5,800 kilometers of Venus. As it flew by, cameras equipped with special filters and films recorded pictures by ultraviolet light. These photographs revealed details of Venus's atmosphere unseen in visible light, such as cloud motions indicating wind speeds of 100 meters per second (200 mph) in the upper atmosphere.

Earth

Earth is the third planet from the Sun and is the largest of the inner planets. Earth's atmosphere is rich in oxygen and nitrogen. From space, many details of Earth's surface can be seen; more than 70 percent is covered with liquid water. Earth has one natural satellite, the Moon.

Mars

Mars is the fourth planet from the Sun. In the sky, it appears as a reddish star. When Mars is viewed through a telescope, its reddish brown, desert-like surface and polar "ice caps" can be seen. Mars's axis is tilted 24° to its orbit (similar to Earth's 23½° tilt); but since its year is nearly twice as long as Earth's, its seasons are longer. The polar "ice caps," consisting mostly of carbon dioxide with some water ice, melt and refreeze as the seasons pass.

Beginning in 1965, a series of Mariner spacecraft passed Mars and photographed its surface, which is pockmarked with craters much like the Moon's. The photographs also revealed long, sinuous channels that look just like dry riverbeds. Most analysts think these channels were cut by liquid water, but have no idea where the water is now. The Viking probes that landed on Mars found a rocky surface with volcanoes (the largest is four times the size of Mount Everest) and vast, flat plains, but no traces of life. Mars atmosphere is very thin, with less than 1 percent of the surface pressure of Earth. It is composed mainly of carbon dioxide with traces of water vapor, argon, ozone, oxygen, carbon monoxide, and hydrogen.

Mars has two natural satellites, the moons Phobos and Deimos. These moons are tiny; Phobos is only 25 kilometers, and Deimos is just 15 kilometers, in diameter. However, they orbit much closer to the surface than our Moon does to Earth and would therefore appear large—about one-half the size of our Moon—to an observer on Mars. Phobos revolves around Mars in just under 8 hours, faster than the planet rotates, so Phobos's rising and setting are the result of its motion, not the rotation of Mars.

The Jovian Planets

All the more distant Jovian planets—Jupiter, Saturn, Uranus, and Neptune—are gas giants. Although they have more mass than the terrestrial planets, they are less dense. Gas giants have thick atmospheres (hence their name), composed largely of gaseous hydrogen compounds such as water (H_2O), methane (CH_4), and ammonia (NH_3) surrounding a small rocky or liquid core. Despite their large size, the gas giants rotate very rapidly, thereby causing a distinct equatorial bulge.

Jupiter

Jupiter is the most massive planet in the solar system, containing 70 percent of all the mass outside of the Sun (which is still only one-thousandth of the mass of the Sun). In spite of its large mass, Jupiter is less dense than Earth because its volume is 1,300 times that of Earth.

Through even a small telescope, Jupiter can be seen as a disk crossed by narrow, parallel bright and dark bands. Through more powerful telescopes, Jupiter's great red spot is visible. In 1973 and 1974, Pioneer 10 and 11 flew by Jupiter and took many pictures, which were sent back to Earth. They revealed Jupiter's surface in more detail than had ever been seen.

The visible features of Jupiter's disk are the tops of clouds in the deep atmosphere. A mottled appearance suggests the presence of convective cells. The current thinking is that heated gases from deep within the atmosphere rise and cool, causing clouds to condense and reflect sunlight, thereby forming bright spots. The clouds then spread out to the north and south, sink, and clear up, thus appearing darker. The rapid rotation of Jupiter causes these cloudy and clear regions to swirl together into parallel bands. The great red spot is thought to be the eye of an immense, hurricanelike storm in Jupiter's atmosphere.

Jupiter's interior structure was inferred from gravity measurements made by Pioneer spacecraft. The core is probably a small, rocky ball similar in composition to Earth. The rest of the planet is mostly hydrogen in two distinct layers: an inner layer of liquid hydrogen under such tremendous pressures that it acts like a metal, and an outer layer of liquid hydrogen that acts as hydrogen does on Earth. This outer layer changes gradually from pure hydrogen to hydrogen compounds such as water, ammonia, and ammonium hydrosulfide in the atmosphere.

Jupiter has many natural satellites because of its powerful gravitational field. Four bright moons of Jupiter were first discovered by Galileo. Over the years, ever more powerful telescopes were brought to bear on Jupiter, and by 1975 fourteen moons had been discovered. The four Galilean moons, Io, Europa, Ganymede, and Callisto, are all about the same size as Earth's Moon; the rest are much smaller. Voyager 1 discovered a faint ring around Jupiter, and photographs showed that one moon, Io, has active volcanoes that cover its surface with red and yellow sulfur compounds. Pictures also revealed that Europa has a smooth, ice-covered surface and that Ganymede and Callisto are covered with craters like those on Earth's Moon.

Saturn

Saturn resembles Jupiter in composition and structure, but its cloud markings show less contrast. The surface of Saturn consists mostly of bands of yellowish and tan clouds. The density of Saturn, 0.7 gram per cubic centimeter, is the lowest of any planet, and is, in fact, less than that of water. If a large enough body of water could be found, Saturn would float in it!

Saturn is best known for its ring system. When Galileo first saw Saturn through a telescope in 1610, he drew it as a blurry object with another blurry object on either side and thought it was a triple planet. In 1655, however, Christian Huygens, a Danish physicist and astronomer, discovered that a ring system surrounds the planet. The rings' dimensions are remarkable; they stretch for 274,000 kilometers from tip to tip but are barely 100 meters thick! In fact, the rings are so thin that observers from Earth lose sight of them when they appear edge-on as Earth passes even with their plane.

Saturn's rings are composed of countless particles ranging in size from that of a golf ball to that of a house. Spectroscopic studies have proved that these particles are frozen water or are covered by frozen water. The rings are separated by several large gaps and thousands of finer divisions. The reason that the gaps exist is poorly understood, but one theory is that gravitational effects are responsible. Where did the ring particles come from? Some possibilities are that the particles condensed from gas as Saturn formed, that they are fragments of a satellite that was blown apart by a collision with a comet or an asteroid, and that they are the remains of a comet or asteroid torn apart by tidal forces while passing very close to Saturn.

Like Jupiter, Saturn has an extensive system of natural satellites. It has at least 17 moons, including Titan, which is almost half the size of Earth, and many smaller moons clustered near the rings. Titan is so large that it has an atmosphere of its own, consisting mainly of nitrogen with traces of ethane, acetylene, ethylene, and hydrogen cyanide.

Uranus

The English astronomer William Herschel first discovered Uranus in 1781. This planet is so far from Earth that it cannot be seen with the unaided eye and a telescope cannot resolve any markings on it; only a faint greenish color is visible. In 1986 Voyager 2 flew by Uranus and revealed that its atmosphere has an almost featureless blue haze overlying deeper clouds. Voyager 2 recorded a minimum temperature of 51 K (–368°F) and a composition matching that of Jupiter and Saturn. In 1977, evidence of rings around Uranus had been observed. Voyager 2 obtained the first clear pictures of those rings, showing them to be much narrower than those of Jupiter or Saturn.

Uranus's rotation, like that of Venus, is retrograde, and its axis of rotation, pointing toward the Sun, is almost in line with the plane of its orbit. Uranus has five major moons, as well as ten smaller moons only recently discovered by Voyager. The larger moons are composed of ice and black soil. All of the moons are heavily cratered, and cracks and canyons can be seen on several.

Neptune

The discovery of Uranus led astronomers to search for other planets. By 1800, observation of Uranus had revealed that it has irregular motions that run counter to Kepler's laws. Some scientists thought that this phenomenon was caused by a breakdown of gravitation at great distances from the Sun, but others guessed correctly that Uranus was being pulled from its theoretical orbit by an even further planet. Using laborious calculations, astronomers set out to predict where the new planet should be located in order to produce the effects observed on Uranus. In 1846, within 12 hours of beginning a search based upon the calculations of French astronomer Urbain Leverrier, two German astronomers discovered Neptune.

Neptune's atmosphere has been found to contain hydrogen, helium, and some methane, which gives the planet its bluish color. It also has faint cloud patterns resembling the bands of Jupiter and Saturn. Like the other gas giants, Neptune rotates rapidly and has a distinct equatorial bulge. The temperature of its atmosphere is 60 K, warmer than expected for a body so far from the Sun. Its high temperature suggests that Neptune may have an internal source of heat.

Neptune has eight known natural satellites. The largest moon, Triton, revolves around the planet in a direction opposite to that of all other satellites in the solar system. The second largest moon, Nereid, revolves in the normal direction, but its orbit is highly eccentric.

Interplanetary Bodies

In addition to the planets, many smaller bodies orbit the Sun and are therefore also considered part of the solar system. These objects include asteroids, comets, meteors, and meteoroids.

Asteroids

Asteroids are solid bodies having no atmosphere that orbit the Sun. They are tiny planets with well-determined orbits. More than 2,000 asteroids have been found in the solar system; most of them orbit in the gap between Mars and Jupiter. Some asteroids move in very elliptical orbits. The asteroid Icarus comes closer to the Sun than any other asteroid in the solar system.

The composition of asteroids has been studied by examining the wavelength of light that reflects from their surface. If asteroids were perfect mirrors, the light that reflects from them would be identical to the sunlight that strikes their surface. Minerals reflect light of different wavelengths in different ways, however, so that the light that reflects from an asteroid's surface differs from sunlight. By analyzing light reflected from asteroids, astronomers can infer their composition. There are two basic types of asteroids, those with surface reflections characteristic of metals and silicate minerals and those with reflections characteristic of carbonaceous chondrites (stony with a high carbon content).

Meteoroids

In addition to the asteroids between Mars and Jupiter, chunks of matter orbit the Sun in orbits that cross those of the planets. When one of these chunks of matter hits Earth's atmosphere at very high speed, it vaporizes because of friction with the air. As it streaks through the air and vaporizes, it emits light and to an observer on Earth appears as a streak of light. The chunk of matter is a **meteoroid**, the glowing object that streaks through the sky is called a **meteor**, and any of the matter that survives to strike the ground is a **meteorite**. Meteoroids vary in size from sand-sized particles to chunks weighing many tons. Meteor showers occur when Earth crosses the path of a clump or stream of meteoroids. See Figure 7.6.

Meteorites tell us a great deal about the solar system. There are three main types of meteorites: stony, stony-iron, and iron. Stony meteorites are like Earth's crust, whereas iron meteorites resemble its core. Stony-iron meteorites are like the iron-rich material of Earth's deep mantle. One explanation for this similarity between meteorites and portions of Earth is that a process called **gravitational differentiation** took place while the solar system formed. The premise is that the solar system originally had a uniform composition. As gravity drew clumps of the original matter together, however, the material separated into layers because lighter elements do not experience as strong a gravitational attraction as heavier elements. This separation resulted in Earth's layered interior and also layered interiors for all other bodies in the solar system.

The Solar System

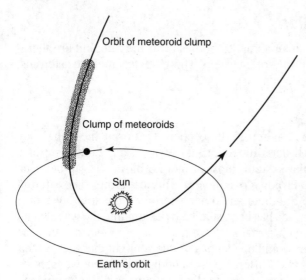

Figure 7.6 Meteor Shower.

The solar system is thought to have originally contained many large asteroids that collided frequently. With each collision, material was broken off and ejected, some into highly eccentric orbits. If these large asteroids had a layered structure, they would be expected to break into fragments with different compositions reflecting the different layers. See Figure 7.7. For this reason, iron meteorites are considered examples of the composition of Earth's core.

Comets

Comets are another type of object orbiting the sun. Comets, like planets, move in elliptical orbits, but their orbital ellipses are highly elongated. Some comets vary from being 1 astronomical unit from the Sun at perihelion to

Figure 7.7 Production of Meteoroids. Collisions between large asteroids with layered interiors gave rise to the three types of meteorites found on Earth. Source: Adapted from Formation of Meteorites by Fernando de Gorocica.

more than 1,000 astronomical units at aphelion. Kepler's law of equal areas tells us that, although such a comet moves very fast while it is near the Sun, it must move very slowly when it is such a huge distance away. Therefore, some comets take as long as 2 million years to make one orbit of the Sun. Comets are generally divided into long-period and short-period comets. Short-period comets orbit the sun in 200 years or less; long-period comets require more than 200 years to complete an orbit.

Comets are thought to have a solid **nucleus** consisting of meteoroid particles embedded in ice. When the nucleus approaches the Sun and heats up, the ice sublimes into a cloud of gas, called a **coma**, around the nucleus. As the comet gets closer to the Sun, the coma increases in size as more gas sublimes, and particles emitted from the Sun collide with the gas molecules and push them out of the coma, forming a **tail**. Under the influence of this solar "wind," the tail always streams away from the Sun. It does *not* stream out along the path of the comet like a contrail out of a jet engine. See Figure 7.8.

Figure 7.8 Changes That Take Place in a Comet as Its Orbit Carries It Past the Sun.
Source: *Astronomy: The Cosmic Journey*, William K. Hartman, Wadsworth, 1987.

The Solar System

Impact Events

From time to time, the orbits of asteroids, comets, and meteors are disturbed by collisions or by the gravity of other objects. Their new orbits may then place them on a collision path with a planet such as Earth. When a high-speed object (a comet, asteroid, or meteoroid) collides with a solid surface (a planet or moon), the collision creates a distinctive bowl-shaped hole called an **impact crater**.

The impact of the speeding object sends out shock waves that spread outward and downward into the ground, and the object's energy of motion is quickly changed to heat. The enormous pressures and heat of the shock wave shatter the ground. Close to the impact, rock is vaporized and melted. Farther away, it is pulverized. The shock waves also raise a rim around the crater and spew material off the sides. When the ground bounces back from the shock wave, it heaves up a central peak in the crater. See Figure 7.9a. These unique characteristics distinguish impact craters from other bowl-shaped holes, such as volcanic craters and sinkholes.

Photographs of planets and their moons reveal surfaces pockmarked with thousands upon thousands of craters. This evidence shows that impact events are not at all unusual, though they have decreased over time as gravity has swept most debris out of the inner solar system. On Earth, few impact craters have been found because they are quickly erased by water and geologic activity. Of the roughly 200 impact craters discovered on Earth, one of the largest and best known is Meteor Crater, near Flagstaff, Arizona. See Figure 7.9b.

The energy released by the impact of a high-speed object depends upon its mass and the speed of impact. Typical impact speeds on the Moon are about 40 kilometers per second, or about 100,000 miles per hour. Impact speeds

Figure 7.9 (a) Formation of an impact crater. (b) Meteor crater near Flagstaff, Arizona.

would be even higher on Earth because of its stronger gravity. The energy released during a collision between Earth and a 1-kilometer asteroid moving at 100,000 miles per hour would be equivalent to setting off the world's entire nuclear arsenal at once—a dozen times over! Such a huge release of energy has the capability to transform oceans and atmospheres and to destroy life on a planetary scale.

The impact on Earth of an asteroid or comet fragment roughly 10 kilometers in diameter would have widespread disastrous effects. Such an impact would gouge a crater more than 100 kilometers in diameter. It would blow a temporary hole the same size in the atmosphere and hurl dust into the upper atmosphere. The dust blown into the stratosphere would darken Earth for months with drastic effects on global climate. When this dust settled, it would fall on land and sea alike, carrying with it the chemical signature of the asteroid. Rock near the site would be "shock-heated," and the shock wave and molten debris thrown out of the crater would knock down trees across thousands of kilometers of land. An ocean impact would create huge waves that would submerge land for hundreds of miles around the impact site.

There is much evidence that an impact event was responsible for the sudden extinction of many life-forms at the end of the Cretaceous period. At that time, about 65 million years ago, 15 percent of shallow-water *families* of organisms became extinct, including 80 percent of all shallow-water invertebrate species. The dinosaurs, too, disappeared around this time. The dividing line between the Cretaceous and the Paleogene, called the K/Pg boundary (K is the symbol commonly used by geologists for the Cretaceous), is a thin layer of clay that has been identified in sediments worldwide. K/Pg boundary sediments contain numerous indications of a massive impact event. The clay has an abundance of platinum-group elements—iridium, osmium, gold, platinum, etc.—that is more similar to that found in meteorites than that in Earth's crust. Iridium, in particular, is more abundant than in normal crustal rocks. Other features associated with an impact that are found in this thin boundary of clay include shocked quartz grains, melt spherules (droplets formed from molten rock), graphite and other evidence of burning, and evidence of large waves such as would be caused by an ocean impact.

Solar System Exploration

Exploration of the solar system has yielded information about the origin of Earth and the solar system. Exploration of other planets will help scientists to understand how or whether these planets can benefit humans. In the future, colonization of other planets may be possible and/or necessary. Asteroids may become a valuable source of metals and mineral resources.

The Solar System

MULTIPLE-CHOICE QUESTIONS

In each case, write the number of the word or expression that best answers the question or completes the statement.

1. The diagram below represents a side view of the Milky Way galaxy.

(Not drawn to scale)

At approximately which position is Earth's solar system?
(1) *A* (3) *C*
(2) *B* (4) *D*

2. The Milky Way galaxy is best described as
   (1) a type of solar system
   (2) a constellation visible to everyone on Earth
   (3) a region in space between the orbits of Mars and Jupiter
   (4) a spiral-shaped formation composed of billions of stars

Base your answers to questions 3 through 6 on the diagram below. The diagram represents the inferred stages in the formation of our solar system. Stage 1 shows a contracting gas cloud. The remaining stages show the gas cloud flattening into a spinning disk as planets formed around our Sun.

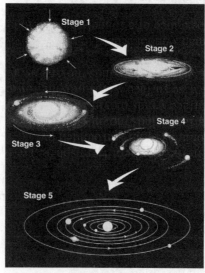

(Not drawn to scale)

160

3. Which force was mostly responsible for the contraction of the gas cloud?
(1) friction
(2) gravity
(3) magnetism
(4) inertia

4. Which process was occurring during some of these stages that resulted in the formation of heavier elements from lighter elements?
(1) conduction
(2) radiation
(3) radioactive decay
(4) nuclear fusion

5. Approximately how long ago did stage 4 end and stage 5 begin?
(1) 1 billion years
(2) 5 billion years
(3) 20 billion years
(4) 100 billion years

6. Compared to the terrestrial planets, the Jovian planets in stage 5 have
(1) larger diameters
(2) higher densities
(3) shorter periods of revolution
(4) longer periods of rotation

7. Which terms describe the motion of most objects in our solar system?
(1) noncyclic and unpredictable
(2) noncyclic and predictable
(3) cyclic and unpredictable
(4) cyclic and predictable

8. Which bar graph best represents the equatorial diameters of the eight planets of our solar system?

The Solar System

9. Compared to the terrestrial planets, the Jovian planets are
   (1) larger and less dense
   (2) smaller and more dense
   (3) closer to the Sun and less rocky
   (4) farther from the Sun and more rocky

10. Which characteristic of the planets in our solar system increases as the distance from the Sun increases?
    (1) equatorial diameter
    (2) eccentricity of orbit
    (3) period of rotation
    (4) period of revolution

11. Which planet has completed less than one orbit of the Sun in the last 100 years?
    (1) Mars
    (2) Mercury
    (3) Neptune
    (4) Uranus

12. Which planet's day (period of rotation) is longer than its year (period of revolution)?
    (1) Mercury
    (2) Venus
    (3) Jupiter
    (4) Saturn

13. The diagram below shows the relative positions of Earth and Mars in their orbits on a particular date during the winter of 2007.

(Not drawn to scale)

Which diagram correctly shows the locations of Earth and Mars on the same date during the winter of 2008?

14. Why is the surface of Mercury covered with meteor impact craters, while Earth's surface has relatively few craters?
    (1) Mercury is larger than Earth, so it gets hit with more meteors.
    (2) Mercury is an older planet, so it has a longer history of meteor impacts.
    (3) Earth's less dense water surface attracts fewer meteors.
    (4) Earth's hydrosphere and atmosphere destroyed or buried most meteor impact sites.

15. The surface of Venus is much hotter than expected, considering its distance from the Sun. Which statement best explains this fact?
    (1) Venus has many active volcanoes.
    (2) Venus has a slow rate of rotation.
    (3) The clouds of Venus are highly reflective.
    (4) The atmosphere of Venus contains a high percentage of carbon dioxide.

16. The solar system object in the photograph below is 56 kilometers long.

    The object in the photograph is most likely
    (1) an asteroid   (3) Earth's Moon
    (2) Neptune   (4) Mercury

The Solar System

Base your answer to question 17 on the diagram below. This diagram shows a portion of the solar system.

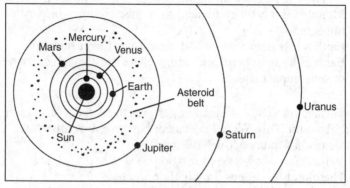

(Not drawn to scale)

17. What is the average distance, in millions of kilometers, from the Sun to the asteroid belt?
    (1) 129
    (2) 189
    (3) 503
    (4) 857

18. A person observes that a bright object streaks across the nighttime sky in a few seconds. What is this object most likely to be?
    (1) a comet
    (2) a meteor
    (3) an aurora
    (4) an orbiting satellite

Base your answers to questions 19 and 20 on the diagrams below. The diagrams represent the events that occur when a large meteor, such as the one believed to have caused the extinction of many organisms, impacts Earth's surface. Diagram A shows the meteor just before impact. Diagram B represents the crater forming, along with the vapor and ejecta (the fragmented rock and dust) thrown into the atmosphere.

Diagram A: Before Impact

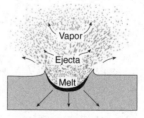

Diagram B: During Impact

19. Which statement best explains how the global climate would most likely be affected after this large meteor impact?
    (1) Large quantities of ejecta in the atmosphere would block insolation and lower global temperatures.
    (2) An increase in vapor and ejecta would allow radiation to escape Earth's atmosphere and lower global temperatures.
    (3) Ejecta settling in thick layers would increase the absorption of insolation by Earth's surface and raise global temperatures.
    (4) Forest fires produced from the vapor and ejecta would raise global temperatures.

20. Many meteors are believed to be fragments of celestial objects normally found between the orbits of Mars and Jupiter. These objects are classified as
    (1) stars   (3) planets
    (2) asteroids   (4) moons

Base your answers to questions 21 through 25 on the diagram below and on your knowledge of Earth science. The diagram represents the orbital paths of the four Jovian planets and Halley's comet around the Sun. Halley's comet has a revolution period of 76 years. In 1986, Halley's comet was at perihelion, its closest point to the Sun. Letters A, B, C, and D represent locations of Halley's comet in its orbit. Location D represents Halley's comet at aphelion, its farthest point from the Sun. The comet's tail is shown at perihelion and at locations B and C.

21. Based on the pattern shown above, which diagram best represents the correct position of the comet's tail at location A relative to the Sun?

The Solar System

22. Compared to the orbit of the Jovian planets, the orbit of Halley's comet is
    (1) less elliptical, with a shorter distance between its foci
    (2) less elliptical, with a greater distance between its foci
    (3) more elliptical, with a shorter distance between its foci
    (4) more elliptical, with a greater distance between its foci

23. Compared to the velocity of Jupiter in its orbit, the velocity of Halley's comet is
    (1) always less
    (2) always greater
    (3) always the same
    (4) sometimes less and sometimes greater

24. This diagram of our solar system represents a
    (1) geocentric model with the Sun near the center
    (2) geocentric model with Earth near the center
    (3) heliocentric model with the Sun near the center
    (4) heliocentric model with Earth near the center

25. Which sequence lists the Jovian planets in order of increasing mass?
    (1) Jupiter, Saturn, Neptune, Uranus
    (2) Uranus, Neptune, Saturn, Jupiter
    (3) Jupiter, Saturn, Uranus, Neptune
    (4) Neptune, Uranus, Saturn, Jupiter

CONSTRUCTED RESPONSE QUESTIONS

Base your answers to questions 26 through 30 on the side-view model of the solar system in your answer booklet and on your knowledge of Earth science. The planets are shown in their relative order of distance from the Sun. Letter *A* indicates one of the planets.

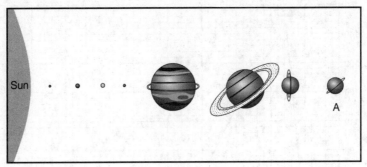

(Not drawn to scale)

26. The center of the asteroid belt is approximately 503 million kilometers from the Sun. On the drawing above, draw an **X** on the model between two planets to indicate the center of the asteroid belt. [1]

27. State the period of rotation at the equator of planet *A*. Label your answer with the correct units. [1]

28. How many million years ago did Earth and the solar system form? [1]

29. Calculate how many times larger the equatorial diameter of the Sun is than the equatorial diameter of Venus. [1]

30. Identify the process that occurs within the Sun that converts mass into large amounts of energy. [1]

The Solar System

Base your answers to questions 31 through 34 on the graph below and on your knowledge of Earth science. The graph shows planet equatorial diameters and planet mean distances from the Sun. Neptune is *not* shown.

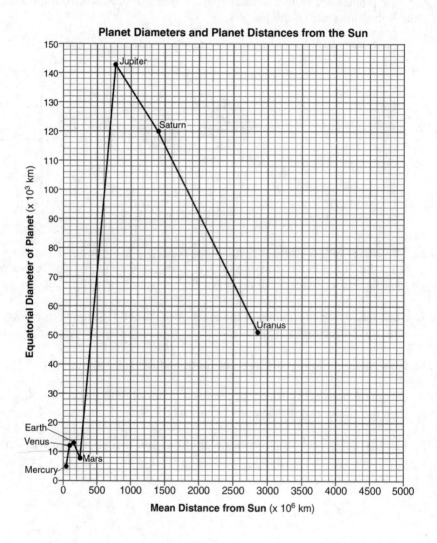

31. Place an **X** on the graph to indicate where Neptune would be plotted, based on its mean distance from the Sun and its equatorial diameter. [1]

32. The diagram below represents Earth drawn to a scale of 1 cm = 2,000 km. Centimeter markings along the equatorial diameter of Earth are also shown on the diagram. On the diagram, shade in the space between the centimeter markings to represent the equatorial diameter of Earth's Moon at this same scale. [1]

Scale: 1 cm = 2000 km

33. Compared to the periods of revolution and periods of rotation of the terrestrial planets, how are the periods of revolution and periods of rotation for the Jovian planets different? [1]

34. The center of the asteroid belt is approximately 404 million kilometers from the Sun. State the name of the planet that is closest to the center of the asteroid belt. [1]

The Solar System

EXTENDED CONSTRUCTED RESPONSE QUESTIONS

Base your answers to questions 35 through 38 on the table below and on your knowledge of Earth science. The table lists the average surface temperature, in kelvins, and the average orbital velocity, in kilometers per second, of each planet of our solar system.

Data Table

Planet	Average Surface Temperature (K)	Average Orbital Velocity (km/s)
Mercury	440	47.87
Venus	737	35.00
Earth	288	29.78
Mars	208	24.13
Jupiter	163	13.07
Saturn	133	9.69
Uranus	78	6.81
Neptune	73	5.43

35. On the grid *below*, construct a bar graph to represent the average surface temperature for each planet. [1]

36. Approximately 97% of Venus's atmosphere is carbon dioxide. Describe how carbon dioxide contributes to the unusually high average surface temperature of Venus. [1]

37. Use the set of axes below to draw a line that represents the general relationship between the mean distances of planets from the Sun and the average orbital velocities of the planets. [1]

38. The orbital velocity of Earth is sometimes faster and sometimes slower than its average orbital velocity. Explain why the orbital velocity of Earth varies in a cyclic pattern. [1]

Base your answers to questions 39 through 41 on the diagram below and on your knowledge of Earth science. The diagram represents the present position of our solar system in a side view of the Milky Way galaxy. The distance across the Milky Way galaxy is measured in light-years.

39. What is the distance, in light-years, from the center of the Milky Way galaxy to our solar system? [1]

40. Galaxies are classified based on their shape. What is the shape of the Milky Way galaxy when viewed from directly above? [1]

41. List the following astronomical features, in order of relative size, from smallest to largest. [1]

Sun
Jupiter
Milky Way galaxy
Universe
Our solar system

CHAPTER 8

THEORIES OF THE ORIGIN OF THE UNIVERSE

> **KEY IDEAS** The universe is vast and is estimated to be over 10 billion years old. The current theory is that the universe was created by an explosion called the *big bang*. Evidence for this theory includes the red-shifted spectra of distant galaxies and cosmic background radiation. There are several possible models for the future of the universe, depending upon its total mass.

KEY OBJECTIVES
Upon completion of this chapter, you will be able to:

- Identify the underlying assumptions and limitations of cosmology.
- Describe the observed structure of the universe.
- Identify evidence that the universe is expanding.
- Describe the big bang theory and the evidence for it.
- Compare and contrast the future of the universe according to the open and closed models of the universe.

WHAT IS THE UNIVERSE?

The universe, because of its seemingly orderly arrangement, is sometimes called the *cosmos*. **Cosmology** is the study of the nature of the universe as a whole. It is the scientific inquiry into the origin, evolution, and fate of the universe. By making assumptions that are not at odds with the observable universe, cosmologists create models, or theories, that attempt to describe the universe, its origin, and its future. The model is then used to make predictions until an observation is found that contradicts it, whereupon the model is modified or discarded in favor of a new model.

Astronomers define the universe in three different ways. The *observable* universe is all that we can see. It includes all the stars, gas clouds, galaxies, and other objects that we can detect by the radiation they emit. As new and more powerful instruments are developed to detect radiation, more objects can be perceived, and the observable universe grows. The *entire* universe is everything we can see, plus everything else that may be there. We can infer things about the entire universe only on the basis of what we know of the observable universe. We're not even sure that there is an entire universe!

Theories of the Origin of the Universe

The *physical* universe is that part of the entire universe that can be described by the laws of physics. It extends a little beyond the observable universe because it includes objects that we cannot see, but that can be detected by their effects on objects we can see.

The Origin and Evolution of the Universe

Cosmologists assume that the laws of physics are identical throughout the universe and that the universe has the same appearance to all observers no matter where they are located. This assumption means that the universe cannot have an edge; if it did, it would appear different to an observer near the edge than to an observer near the center. The geometry of space must be such that all observers see themselves as being at the center. Cosmologists believe that the only motion that can occur in such geometry is expansion or contraction of the universe.

All this not only sounds mind-boggling; it actually is! The problem with thinking about the universe is that there are too few facts of which we can be certain. Therefore, we must make assumptions about the universe to make progress in cosmology. False assumptions, however, can lead us to nonsensical results! Consider a well-known example, Olber's paradox, shown in Figure 8.1.

Figure 8.1 Olber's Paradox. If the universe is infinitely large, infinitely old, and uniformly filled with stars, any line of sight from Earth should intersect a star. This assumption predicts that nights should not be dark; instead, the sky should glow with starlight.

The discrepancy between what Olber's paradox theorizes and what we actually observe tells us that one of his assumptions is wrong. We now know that the light from many distant stars moving away from us is red-shifted into the infrared range, which is not visible to human eyes. Also, there are many stars that do not radiate energy in the range of visible light. If we can't see light, it can't make the sky glow. Furthermore, while the universe may be infinitely large, the observable universe is only 16–20 billion light-years in diameter. Light from distances beyond 20 billion light-years has not yet reached us.

The Structure of the Universe

The distribution of matter, such as stars, dust, and gas, throughout the universe is not uniform; there appears to be a high degree of clustering. On a very large scale, the universe seems to be made up of clusters of objects. Each of these objects is a cluster of smaller objects, which are clusters of yet smaller objects, which are clusters of even smaller ones. See Figure 8.2.

Figure 8.2 Clustering in the Universe.

The largest clusters detected so far are clusters of clusters of galaxies, called *superclusters*. Superclusters contain from 5 to 40 clusters of galaxies and are mind-boggling in size. An average supercluster is about 100 million light-years across; a **light-year**, you will recall, is the distance light travels in 1 year at the speed of 300,000 kilometers per second.

Clusters of galaxies are "smaller" objects—only a few million light-years across. Galaxies, which are clusters of stars, are even smaller—only about 100,000 light-years in diameter. But consider that an average galaxy contains about a million million stars, and the average star is 100 or so times the diameter of the Earth and millions of times its volume—not bad for a "small" object! Finally, each star is a cluster of atoms. And what are atoms but clusters of protons, neutrons, and electrons?

Despite the almost unimaginable number of stars, galaxies, and clouds of dust and gas that the universe contains, it is still mostly empty space. Vast distances separate the clusters of matter that can be observed.

The Expanding Universe

Observations of objects in the universe indicate that it is expanding. Other galaxies are moving away from us, and the most distant galaxies are racing away the fastest. See Figure 8.3. The evidence of this motion away from Earth comes from the light reaching us from distant galaxies, which has a lower frequency than would be expected. The explanation for this shift lies in a phenomenon called the *Doppler effect*.

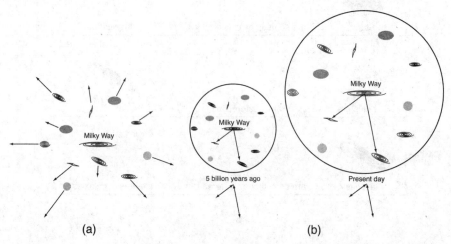

Figure 8.3 (a) All galaxies are observed to be moving away from our Milky Way Galaxy (arrows indicate speeds). (b) Expansion of the Universe explains why all galaxies are moving away from us.

The **Doppler effect** is the shift of a spectrum line away from its normal wavelength caused by motion of the source toward or away from the observer. See Figure 8.4. If the source is approaching the observer, there is a blue-shift toward shorter wavelengths of light. If the source is moving away from the observer, there is a red-shift toward longer wavelengths of light.

Why does this occur? Light consists of electromagnetic waves. The human eye distinguishes between one wave and another by wavelength, that is, the distance between one crest of a wave and the next. Now, suppose a source and an observer are not moving in relation to each other. The source is emitting waves of a particular wavelength; let's say 500 nanometers (billionths of a meter) at the speed of light. The observer will perceive a certain number of crests per second, and the light will be "seen" as a particular color, in this case yellow. Now, if the source moves toward the observer at the same time that it is emitting the waves, more crests will reach the eye of the observer per second. This is the same result that the eye would experience if the light waves had a shorter wavelength. Therefore, the eye will interpret the light as being a different color, let's say blue, even though the source is emitting yellow light. If the source is moving away from the observer, just the opposite happens. Fewer crests per second reach the eye, and the yellow light is interpreted as having a longer wavelength, for example, that of red light.

Figure 8.4 The Doppler Effect.

The amount of shift is proportional to the speed at which the source is approaching or receding from the observer. Therefore:

$$\frac{\text{shift in wavelength}}{\text{normal wavelength}} = \frac{\text{approach or recession speed}}{\text{speed of light}}$$

When the light from distant galaxies is observed, all the spectra are seen to be red-shifted, indicating that the galaxies are moving away from us. See Figure 8.5. Furthermore, the farther away the galaxy, the more it is red-shifted, indicating that distant galaxies are moving away from Earth faster than nearer galaxies. The only explanation for such an observation is that the universe is expanding in all directions.

Figure 8.5 Actual Photos of Red-Shifted Dark H and K Lines of Gaseous Calcium in the Spectra of Galaxies Moving Away from Earth at Different Velocities.

Theories of the Origin of the Universe

The Big Bang Theory

Since the universe appears to be expanding, it must have been smaller at some time in the past. If the galaxies could be traced back in time, there would be a point at which they were all very close to one another. According to the observed expansion rate, this point occurred between 10 and 20 billion years ago. Thus, we have a model in which the universe started out with all of its matter in a small volume and then expanded outward in all directions, much like an explosion. For this reason, this model of the universe is called the **big bang theory**. See Figure 8.6.

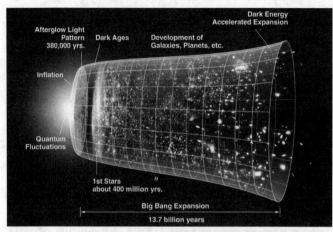

Figure 8.6 The Big Bang. During the big bang, matter was thrown out uniformly in all directions. Thereafter, gravity caused the matter to coalesce into clumps—galaxies. Near us we see galaxies, farther away we see young galaxies (dots), and at a great distance we see radiation coming from the hot clouds of the big bang explosion.

The big bang was, however, quite different from any explosion on Earth. An explosion on Earth can be observed from a distance; it comes from a specific point in space at a particular time. But the big bang did not occur at some pre-existing point in space. Instead, the entire universe, including space and time, originated in the big bang. Before it occurred, there was no place from which an observer could stand and watch the explosion, and there was no time as we know it. With the big bang, both space and time came into existence simultaneously. What we perceive as movement of galaxies away from us is due to the expansion of space, and what we perceive as the elapsing of time is the expansion of time. Thus, galaxies are not hurtling through space; the space itself is expanding and carrying the galaxies apart in the process.

After the big bang, energy and matter spread out as the universe expanded, so temperature and density dropped. The rapidly expanding matter cooled and fragmented into clouds of gas that coalesced into clusters of galaxies 10 to 20 billion years ago. When an observer looks at a distant galaxy today, she is not seeing it as it is now, but as it was long ago when the light now entering her eye first left the galaxy. The red-shift in the light seen coming from distant galaxies is due to the expansion of the universe.

The most distant galaxies are so far away that the light reaching us today left them soon after the big bang occurred. Light from beyond these galaxies is so ancient it comes from the original hot clouds of the big bang. The radiation from the original big bang is even more red-shifted than the light from distant galaxies. In fact, it has such a large red-shift that the radiation reaching us from the original big bang arrives at Earth in the range of long-wave infrared or short microwave radio waves. When radiation is red-shifted, its wavelength makes it appear that the source is much cooler than it actually is. Thus, the radiation given off by the original big bang, when red-shifted, appears like radiation given off by an object with a temperature of 2.7 K (about 2.7° above absolute zero—very cold!). Since we are looking at the big bang from within the sphere of the expanding shock wave (see Figure 8.6c), that radiation should be coming at us from *all* directions, even from what may appear to be "empty space." Thus, space should be aglow with long-wavelength infrared and microwave radiation. Although predicted in the 1940s, this **cosmic background radiation** glow was not detected until the mid-1960s by Arno Penzias and Robert Wilson at Bell Laboratories in New Jersey. For their discovery of the big bang's primordial background radiation, they received the Nobel Prize in 1978.

The Future of the Universe

The future of the universe depends upon the total amount of matter in the universe. Gravity is a strong and pervasive force. If the density of matter in the universe is greater than a certain critical amount, gravity will be able to slow the expansion of the universe and slowly turn expansion into contraction. The problem is that measurements of the amount of matter that exists in the universe vary so much that none of the possibilities described below can be ruled out. One difficulty is that the matter we can see, that is, the matter that emits light, represents only a tiny fraction of all the matter in the universe. We can perceive matter only by the electromagnetic radiation it emits, and the amount and type of radiation emitted depends upon the temperature of the matter. The "dark" matter, which we cannot see because it doesn't emit electromagnetic radiation, will determine the fate of the universe.

The Open Universe

If the density of matter falls short of the critical amount, there won't be enough matter to exert the gravitational pull needed to stop the expansion. The universe will continue to expand and cool. Although stars will continue to form for a while, the amount of matter available for star formation will dwindle until no new stars form. Eventually, all of the mass in the universe will be contained in cold, dead planets and burned out stars. The universe will end as ever more widely scattered matter in a black void containing no light or heat.

If the universe has exactly the critical density of matter, expansion will continue, but will slow down at an ever-decreasing rate. In this case the fate of the universe will be similar to that of an ever-expanding universe, except that at some point it will stop moving entirely. See Figure 8.7.

Figure 8.7 Stages of the Open Universe.

The Closed Universe

If the density of matter in the universe even slightly exceeds the critical density, then, as stated previously, the matter will exert enough gravitational attraction to stop the expansion and start a contraction. Ultimately, all of the matter would again collapse back into a small volume—the big crunch. And what then? One theory is that the universe is oscillating and the big crunch would give rise to another big bang. Another theory is that the big crunch would form an enormous black hole that would consume space and time, just as the big bang produced them. See Figure 8.8.

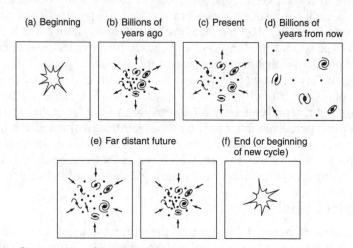

Figure 8.8 Stages of the Closed Universe.

Will we ever know which model is correct? Perhaps, but even then we will encounter new questions and continue to search for answers.

Multiple-Choice Questions

In each case, write the number of the word or expression that best answers the question or completes the statement.

1. Observations of the universe indicate that the distribution of matter in the universe is
   (1) equally concentrated everywhere
   (2) concentrated in clusters of various sizes
   (3) most concentrated in the Milky Way galaxy
   (4) constantly increasing in concentration

2. The diagram below represents the bright-line spectrum for an element.

   Violet   Red

   The spectrum of the same element observed in the light from a distant star is shown below.

   Violet   Red

   The shift in the spectral lines indicates that the star is moving
   (1) toward Earth
   (2) away from Earth
   (3) in an elliptical orbit around the Sun
   (4) in a circular orbit around the Sun

3. The red shift of light from most galaxies is evidence that
   (1) most galaxies are moving away from Earth
   (2) a majority of stars in most galaxies are red giants
   (3) light slows down as it nears Earth
   (4) red light travels faster than other colors of light

4. The theory that the universe is expanding is supported by data from the
   (1) nuclear decay of radioactive materials
   (2) nuclear fusion of radioactive materials
   (3) blue shift of light from distant galaxies
   (4) red shift of light from distant galaxies

5. If the universe appears to be expanding, which conclusion is most logical concerning the universe?
   (1) The universe must have been smaller at some time in the past.
   (2) The universe must have been larger at some time in the past.
   (3) The density of the universe is increasing.
   (4) The temperature of the universe is increasing.

6. According to the Big Bang theory, which graph best represents the relationship between time and the size of the universe from the beginning of the universe to the present?

Base your answers to questions 7 through 11 on the passage below and on your knowledge of Earth science.

Cosmic Microwave Background Radiation

In the 1920s, Edwin Hubble's discovery of a pattern in the red shift of light from galaxies moving away from Earth led to the theory of an expanding universe. This expansion implies that the universe was smaller, denser, and hotter in the past. In the 1940s, scientists predicted that heat (identified as cosmic microwave background radiation) left over from the Big Bang would fill the universe. In the 1960s, satellite probes found that cosmic microwave background radiation fills the universe uniformly in every direction, and indicated a temperature of about 3 kelvins (K). This radiation has been cooling as the universe has been expanding.

7. Scientists infer that the universe began approximately
   (1) 1.0 billion years ago
   (2) 3.3 billion years ago
   (3) 8.2 billion years ago
   (4) 13.7 billion years ago

8. Which graph best shows the relationship of the size of the universe to the temperature indicated by the cosmic microwave background radiation?

9. Cosmic microwave background radiation is classified as a form of electromagnetic energy because it
   (1) travels in waves through space
   (2) moves faster than the speed of light
   (3) is visible to humans
   (4) moves due to particle collisions

10. The current temperature indicated by the cosmic microwave background radiation is
    (1) higher than the temperature at which water boils
    (2) between the temperature at which water boils and room temperature
    (3) between room temperature and the temperature at which water freezes
    (4) lower than the temperature at which water freezes

11. The time line below represents time from the present to 20 billion years ago. Letters *A*, *B*, *C*, and *D* represent specific times.

    Which letter on the time line best represents the time when scientists estimate that the Big Bang occurred?
    (1) *A*   (3) *C*
    (2) *B*   (4) *D*

12. The observable universe is estimated to be roughly 16–20 billion years old. Which statement best describes why a galaxy located 25 billion light-years from Earth may not be visible to an observer on Earth?
    (1) Galaxies 25 billion light-years away would emit no visible light.
    (2) Light from beyond 20 billion light-years has not yet reached Earth.
    (3) Light from beyond 20 billion light-years passed our galaxy before Earth existed.
    (4) No galaxies are located farther than 5 billion light-years from Earth.

Constructed Response Questions

Base your answers to questions 13 and 14 on the calendar model shown below of the inferred history of the universe and on your knowledge of Earth science. The 12-month time line begins with the big bang on January 1 and continues to the present time, which is represented by midnight on December 31. Several inferred events and the relative times of their occurrence have been placed in the appropriate locations on the time line.

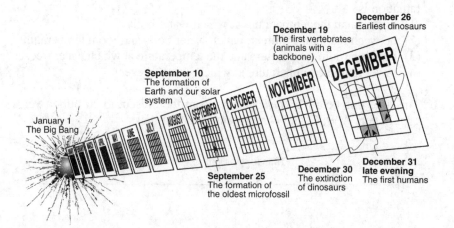

13. State one piece of evidence used by scientists to support the theory that the big bang event occurred. [1]

14. How many million years of Earth's geologic history elapsed between the event that occurred on September 10 and the event that occurred on September 25 in this model? [1]

15. State two observations successfully explained by the big bang theory. [2]

16. The critical density of the universe is the minimum average density of the matter required for the force of gravity to stop the expansion of the universe without reversing it. Infer the future of the universe for the following two conditions:
    (a) density of the universe > critical density
    (b) density of the universe < critical density

EXTENDED CONSTRUCTED RESPONSE QUESTIONS

Base your answers to questions 17 through 19 on the data table below, which shows some galaxies, their distances from Earth, and the velocities at which they are moving away from Earth.

Name of Galaxy	Distance (million light-years)	Velocity (thousand km/s)
Virgo	70	1.2
Ursa Major 1	900	15
Leo	1100	19
Bootes	2300	40
Hydra	3600	61

One light-year = distance light travels in one year

17. On the grid below, use an **X** to plot the distance and velocity for each galaxy from the data table to show the relationship between each galaxy's distance from Earth and the velocity at which it is moving away from Earth. Connect the **X**s with a smooth line. [1]

18. State the general relationship between a galaxy's distance from Earth and the velocity at which the galaxy is moving away from Earth. [1]

19. Another galaxy is traveling away from Earth at a velocity of 70 thousand kilometers per second. Estimate that galaxy's distance from Earth in millions of light-years. [1]

Theories of the Origin of the Universe

Base your answers to questions 20 and 21 on your knowledge of Earth science and on the newspaper article shown below, written by Paul Recer and printed in the *Times Union* on October 9, 1998.

Astronomers peer closer to big bang

WASHINGTON—The faintest and most distant objects ever sighted—galaxies of stars more than 12 billion light years away—have been detected by an infrared camera on the Hubble Space Telescope.

The sighting penetrates for the first time to within about one billion light years of the very beginning of the universe, astronomers said, and shows that even at that very early time there already were galaxies with huge families of stars.

"We are seeing farther than ever before," said Rodger I. Thompson, a University of Arizona astronomer and the principal researcher in the study.

Thompson and his team focused an infrared instrument on the Hubble on a narrow patch of the sky that had been previously photographed in visible light. The instrument detected about 100 galaxies that were not seen in the visible light and 10 of these were at extreme distance.

He said the galaxies are seen as they were when the universe was only about 5 percent of its present age. Astronomers generally believe the universe began with a massive explosion, called the "big bang," that occurred about 13 billion years ago.

Since the big bang, astronomers believe that galaxies are moving rapidly away from each other, spreading out and becoming more distant.

20. The big bang theory is widely believed by astronomers to explain the beginning of the universe. Why does the light from distant galaxies support the big bang theory? [1]

21. Compare the age of Earth and our solar system to the age of these distant galaxies of stars. [1]

Unit Three
EARTH'S HISTORY

CHAPTER 9
THE ORIGINS OF EARTH AND ITS MOON

> **KEY IDEAS** Earth formed as colliding matter accreted. A giant cloud of gas and debris contracted to form our solar system, with Earth reaching its present size about 4.5 billion years ago. The process of accretion heated the young Earth above the melting point of silicate minerals, creating a vast molten region extending downward from the surface. Within this molten region, denser materials sank toward the center and less dense materials floated to the surface, creating Earth's layered internal structure.
>
> Internal convection currents are responsible for Earth's magnetic field and the motion of crustal plates. Earth's Moon probably formed when collision of Earth with another large object caused a chunk of molten Earth material to spin off into a circular orbit.

KEY OBJECTIVES
Upon completion of this chapter, you will be able to:

- Describe the process of accretion by which Earth is thought to have formed.
- Explain how gravity caused Earth to become layered according to the densities of the materials of which it is composed.
- Describe the giant impact theory of the origin of the Moon.

THE ORIGIN OF EARTH

A Fiery Birth

Earth and the other inner, or terrestrial, planets accreted by collisions. In this process of accretion, matter collided with the planets and was held to them by gravity. The matter gradually accumulated, causing the planets to grow in size. When another body impacts a planet, its energy of motion (kinetic energy) is converted to heat. The larger and denser the planet, the stronger the gravitational force exerted on the incoming body, and the greater the energy of the impact. Therefore, of the terrestrial planets, Earth was heated the most, and Mercury the least. The temperature reached depends on the rate of impacts as compared to the rate at which heat is radiated away.

The heat produced by accretion caused Earth, Venus, and perhaps Mars to reach internal temperatures above the melting points of silicate minerals. Thus, the earliest Earth had a vast molten region, extending from the surface partway into the interior. Heat flowed from the hot surface toward the cooler center *and* outward to space. Shortly after accretion brought Earth to its present size, roughly 4.5 billion years ago, interior temperatures became high enough to partially melt the mixed solids of silicates, as well as iron. Once molten, elements that had been locked into crystalline structures were free to recombine with other, compatible elements.

The Layered Earth

Substances in the molten Earth then underwent ***differentiation***, a process by which denser materials (principally iron) sank toward the center, forming the core, while less dense materials (e.g., sodium and potassium ions) floated to the top. As a rock falling off a cliff releases gravitational potential energy as it falls, so the iron that fell toward the core released gravitational energy, creating extra heat. About 32 percent of the mass of our planet is iron, and iron is 50 percent denser than silicates, so the heat generated by formation of the core was substantial and fueled even more melting and differentiation. While this process was going on, about 4.5 billion years ago, an event that resulted in the formation of the Moon occurred, as discussed later in this chapter.

Eventually, Earth's gravity attracted most of the bodies in space surrounding it. With the surrounding space swept clean, the accretion rate fell off. In its molten state, Earth lost heat by convection. With little or no incoming energy from impacts, and removal of heat by convection, Earth cooled, forming a solid crust. The end result of differentiation and cooling was a planet with an internal structure that is layered like an onion, with the least dense materials at the surface and the densest ones at the center.

As you can see in Figure 9.1, there are three main zones within Earth: the crust, the mantle, and the core. The core is composed of mainly iron and nickel and extends from the center of Earth out to a depth of 2,900 kilometers. It consists of two parts: a solid inner core and a fluid outer core. Over the core, out to a depth of about 70 kilometers, lies the mantle. The mantle is rather plastic in nature and is composed mainly of dark, dense silicates. It is subdivided into several layers based upon density, composition, and fluidity. Overlying the mantle is the solid crust, also subdivided by density and composition. Ranging from 0 to 70 kilometers in thickness, it is chemically diverse but contains a lot of lighter, less dense silicates than those in the mantle.

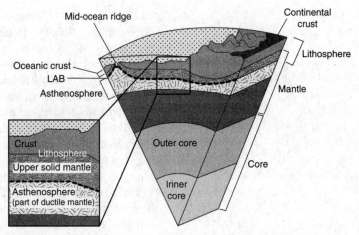

Figure 9.1 Cross Section Showing Earth's Layered Interior.

Earth also experiences radioactive heating. Among the elements present in the developing Earth were uranium, thorium, and potassium, each of which has radioactive isotopes. These elements also have large atomic radii and therefore floated to the top and became concentrated in crustal rocks, mainly granite. Decay of the radioactive isotopes in the crust and mantle generates roughly 30 trillion watts of power a year. Today, between radioactive decay and heat transferred to the mantle by the core, there is enough heating to soften mantle rock and allow convective flow to remove heat. The patterns of motion, however, are complex. Simple convection currents in the mantle are interrupted by plumes of hot material driven by heat from the core, which convects separately.

The Earth as a Magnet

The iron core was able to create a magnetic field. Magnetic fields in the solar nebula magnetized some of the rocks and iron grains that later accreted, forming Earth. Thus, some iron in the early Earth was magnetic. But iron is also a conductor of electricity. Convection in the molten core caused the electrically conductive iron to move in the presence of the magnetic field. This induced an electric current, which in turn created a stronger magnetic field. Further convection of the iron in this field induced an even larger current, thereby creating an even stronger magnetic field. Fueled by heat escaping from the core, this self-perpetuating process, called a *magnetic dynamo*, resulted in Earth's magnetic field. See Figure 9.2.

Earth's magnetic field, or **magnetosphere**, extends out into space about four Earth radii. Blasted by the solar wind's steady stream of high-energy charged particles, the magnetosphere sticks out like a tail on the side of Earth facing away from the Sun. See Figure 9.3.

The Origins of Earth and Its Moon

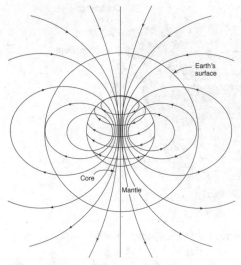

Figure 9.2 Earth's Magnetic Field. Convection motion of the molten iron core induces a magnetic field.

High-energy charged particles can be deadly. Penetrating organisms like a bullet, they can disrupt and destroy living tissue. Fortunately for life on Earth, moving charged particles have a magnetic field of their own, and many of the deadly particles of the solar wind are deflected by the magnetosphere. Also, some charged particles are trapped by Earth's magnetic field and move around rapidly inside two doughnut-shaped regions of the magnetosphere known as the Van Allen belts. See Figure 9.3.

High-energy charged particles moving through the atmosphere stimulate the atoms and ions of the atmosphere to radiate light, producing auroras. The Aurora Borealis, or northern lights, and Aurora Australis, or southern lights, are shimmering bands of light that shine in the night sky. Charged particles tend to be funneled down to Earth near the poles by the magnetosphere, and auroras are seen mainly in Earth's Arctic and Antarctic regions. See Figure 9.4.

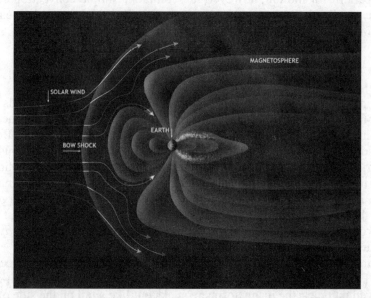

Figure 9.3 Earth's Magnetosphere and Van Allen Belts. The solar wind distorts Earth's magnetosphere. The Van Allen belts are two doughnut-shaped regions of charged particles trapped by the magnetosphere that surround Earth.

Figure 9.4 (a) Charged particles funneled into the upper atmosphere near the poles by lines of magnetic force surrounding Earth cause shining lights known as auroras.
Source: *The Earth Sciences*, Arthur N. Strahler, Harper & Row, 1971. Library of Congress Catalog Card Number 78-127335.
(b) An Aurora Borealis photographed during a winter night in Alaska.

THE ORIGIN OF THE MOON

Early Ideas

Until the mid-1980s, the origin of the Moon was a difficult issue for astronomers because the Moon is unusually large compared to Earth and moves in a circular orbit. A variety of theories, ranging from capture of a passing body by Earth's gravity, to ejection of a chunk of matter by a rapidly spinning Earth, to formation in place at the same time as Earth, had been suggested. Attempts to model the third theory, formation in place, with computers presents problems. The Moon's density is consistent with little or no iron content. Capture of the Moon, while possible, is unlikely, and capture into a circular orbit is even more unlikely. Finally, there are problems with the physical plausibility of a molten Earth spinning rapidly enough to throw off the Moon.

Apollo's Clues and a New Theory

Rock and dust samples brought back from the Moon by the Apollo missions in the 1980s virtually eliminated all of the early theories. Although the Moon's small size limits possible geologic activity, Moon rocks are more similar to Earth's mantle than to meteorites like those that would have accreted to form the Moon. Moon rocks are also richer in silicates and poorer in metals and volatile elements than Earth rocks. Simplistically put, Moon rocks could be made by heating rocks from Earth's mantle until they vaporize, and then recondensing the less volatile materials.

This geochemical mystery has prompted planetary scientists to consider a new theory—that the Moon is the result of a *giant impact*, a huge collision between Earth and another planet-sized body.

How likely is such an impact? Early in the formation of the solar system, conditions were right for this kind of collision. At first, accreting planetesimals would have been small and would have moved in roughly circular orbits. Collisions would have been gentle, largely resulting in accretion. As planets grew larger, however, close passes of these bodies would have sent them into elliptical orbits, which, in turn, would increase collision speeds. By the time the terrestrial planets had formed, catastrophic collisions with solar-system debris traveling in high-speed elliptical orbits would have been likely. Small bodies hitting big ones would vaporize and melt, adding to the larger body. Big bodies hitting other big bodies would have had disastrous results. It was probably a giant impact that knocked Uranus on its side and spun out a disk from which its moons formed.

The Giant Impact

Computer simulations show that, if a planet as big as Mars, or slightly bigger, struck Earth, a big chunk of Earth's mantle, but very little iron core, would have been spun off. Also, part of the material ejected would have entered a circular orbit, forming a ring around Earth. See Figure 9.5.

The rest of the ejected material would have reaccreted into Earth (computer simulations of the event indicate that the iron core of the planetesimal would have merged with Earth's in a matter of hours) or been lost into orbit around the Sun. The material ejected would have left a sizable hole, but the spinning of the still malleable Earth would quickly have restored the planet's normal shape.

Most of the material ejected into space would have been vaporized, with only the least volatile materials remaining solid. Much of the volatile material (water and less dense ions) would be lost. This hypothesis is consistent with the lower density minerals with which rocks from the lunar highlands are enriched. Since little of Earth's core was ejected, little iron would be present. Material in the ring then collected to form the Moon.

For the first billion years of its life, the newly formed Moon was bombarded by meteorites. They created great craters and melted the surface that is now the lunar crust. About a billion years into the Moon's life, accretion and radioactive decay produced enough heat to cause melting and differentiation in the upper 500 kilometers or so. Volcanoes erupted, pouring out huge floods of basalt, which now form the maria. By about 3 billion years ago, the Moon had cooled enough that volcanic activity ceased. With the exception of a few major impacts and small lava flows, the Moon has remained virtually unchanged since then. Because it has no water, atmosphere, or weather, no weathering, erosion, or deposition occurs on the Moon.

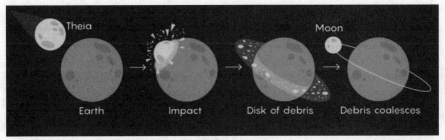

Figure 9.5 Formation of the Moon by a Giant Impact.

When did this sequence of events take place? The oldest rocks recovered from the Moon date to 4.4–4.5 billion years ago, so it certainly didn't occur any later. This age also sets a time by which Earth's core must have formed. Formation of the core had to occur before the impact that formed the Moon, because the Moon is poor in iron. Most likely, the Moon formed early in Earth's history, close to or before 4.5 billion years ago.

MULTIPLE-CHOICE QUESTIONS

In each case, write the number of the word or expression that best answers the question or completes the statement.

1. Earth formed by accretion as a giant cloud of gas and debris contracted to form our solar system. During the process of accretion, the energy of motion of bodies that collided with Earth was converted into
   (1) magnetic force
   (2) heat energy
   (3) electrical force
   (4) matter

2. Of the terrestrial planets, Earth was heated the most by collisions during accretion because
   (1) collisions with Earth produced greater impact energy because of Earth's large mass and high density
   (2) Earth was closest to the Sun and received the most intense solar radiation
   (3) bodies that collided with Earth were hotter than bodies that collided with other planets
   (4) bodies that collided with Earth were heated by friction with Earth's magnetic field

3. What caused the interior of Earth to separate into layers?
   (1) a decrease in the rate of rotation of Earth
   (2) the gravitational pull on materials of varying densities
   (3) variations in heating by the Sun due to Earth's tilt
   (4) collisions with meteors and comets

4. Compared to Earth's crust, Earth's core is believed to be
   (1) less dense, cooler, and composed of more iron
   (2) less dense, hotter, and composed of less iron
   (3) more dense, hotter, and composed of more iron
   (4) more dense, cooler, and composed of less iron

5. The density of Earth's crust is
   (1) less than the density of the outer core but greater than the density of the mantle
   (2) greater than the density of the outer core but less than the density of the mantle
   (3) less than the density of both the outer core and the mantle
   (4) greater than the density of both the outer core and the mantle

6. Earth's magnetic field is likely a result of
   (1) convection currents in Earth's mantle
   (2) convection currents in Earth's core
   (3) a high concentration of iron in Earth's crust
   (4) high-energy particles in the solar wind

7. Much of Earth's internal heat is produced by
   (1) decay of radioactive elements
   (2) fusion of hydrogen into helium
   (3) friction between Earth's surface and atmosphere
   (4) energy absorbed from sunlight

8. Analysis of rock and dust samples brought back from the Moon by the Apollo missions indicates that Moon rocks
   (1) have a composition more similar to that of Earth's mantle than to meteorites
   (2) contain elements never before identified on Earth
   (3) have a composition very different from that of Earth rocks
   (4) have a high concentration of iron, similar to that of Earth's core

9. Which statement about the age of Earth rocks and the age of Moon rocks is true?
   (1) Moon rocks are much younger than Earth rocks.
   (2) Moon rocks are much older than Earth rocks.
   (3) Moon rocks are about the same age as Earth rocks.
   (4) Moon rocks cannot be dated because they contain no radioactive elements.

10. Which of the following models of the Moon's origin is currently considered most likely?
    (1) The Moon was spun off by a rapidly spinning, molten Earth.
    (2) The Moon was a passing body that was captured by Earth's gravity.
    (3) The Moon formed in place at the same time that Earth formed.
    (4) The Moon was ejected from a molten Earth by a giant impact.

11. Measurements of radioactivity in the rocks of Earth's crust currently place the age of Earth closest to
    (1) 100,000 years   (3) 4.6 billion years
    (2) 1.0 million years   (4) 16 billion years

12. Displays of auroras (northern lights) are believed to be caused by
    (1) radiation of infrared rays from Earth
    (2) reflection of the midnight Sun on northern snowfields
    (3) gases glowing under the impact of electrically charged particles
    (4) radioactive dust in the upper atmosphere

13. Craters and layers of basalt on the Moon's surface probably formed as a result of
    (1) meteorite impacts and the heating of surface rock due to accretion
    (2) recent volcanic eruptions
    (3) wrinkling of the Moon's molten surface as it cooled and contracted
    (4) bubbles of gas rising through the molten rock just beneath the Moon's surface

Base your answers to questions 14 and 15 on the data table below. The data table provides information about the Moon, based on current scientific theories.

Information About the Moon

Subject	Current Scientific Theories
Origin of the Moon	Formed from material thrown from a still-liquid Earth following the impact of a giant object 4.5 billion years ago
Craters	Largest craters resulted from an intense bombardment by rock objects around 3.9 billion years ago
Presence of water	Mostly dry, but water brought in by the impact of comets may be trapped in very cold places at the poles
Age of rocks in terrae highlands	Most are older than 4.1 billion years; highland anorthosites (igneous rocks composed almost totally of feldspar) are dated at 4.4 billion years
Age of rocks in maria plains	Varies widely from 2 billion to 4.3 billion years
Composition of terrae highlands	Wide variety of rock types, but all contain more aluminum than rocks of maria plains
Composition of maria plains	Wide variety of basalts
Composition of mantle	Varying amounts of mostly olivine and pyroxene

14. Which statement is supported by the information in the table?
    (1) The Moon was once a comet.
    (2) The Moon once had saltwater oceans.
    (3) Earth is 4.5 billion years older than the Moon.
    (4) Earth was liquid rock when the Moon was formed.

15. Which Moon feature is an impact structure?
    (1) crater
    (2) maria plain
    (3) terrae highland
    (4) mantle

16. In which parts of Earth's interior would melted or partially melted material be found?
    (1) stiffer mantle and inner core
    (2) stiffer mantle and outer core
    (3) crust and inner core
    (4) asthenosphere and outer core

CONSTRUCTED RESPONSE QUESTIONS

17. Explain the role of gravity in the process of differentiation that created Earth's layered internal structure. [2]

18. State one reason why, in light of current knowledge, it is unlikely that the Moon is a captured passing body. [2]

19. Construct a labeled diagram of Earth's magnetic field. [2]

EXTENDED CONSTRUCTED RESPONSE QUESTIONS

20. Base your answers to parts (a) through (c) on the *Earth Science Reference Tables* and the diagram below, which shows Earth's internal structure and its three main layers.

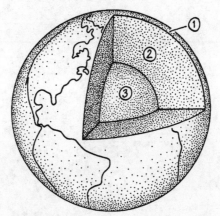

    (a) Identify the three main layers of Earth's interior. [3]
    (b) State the thickness of each of the three main layers of Earth's interior. [3]
    (c) Describe the process by which Earth's layered internal structure is thought to have been formed. [2]

21. The photographs below show the Moon and Earth as viewed from space. It is inferred that Earth had many impact craters similar to those shown on the Moon.

(Not drawn to scale)

The Moon has many more impact craters visible on its surface than Earth has on its surface. State *two* reasons that Earth has so few visible impact craters. [2]

CHAPTER 10

THE ORIGIN AND STRUCTURE OF EARTH'S ATMOSPHERE

> **KEY IDEAS** Earth is surrounded by an atmosphere that extends several hundred kilometers out into space. Earth's early atmosphere formed as a result of the outgassing of water vapor, carbon dioxide, nitrogen, and other gases, in lesser amounts, from its interior. The evolution of life caused dramatic changes in the composition of Earth's atmosphere. Free oxygen did not form in the atmosphere until oxygen-producing organisms evolved. Earth's current atmosphere has a layered structure, with each layer differing in composition and temperature from other layers.

KEY OBJECTIVES
Upon completion of this chapter, you will be able to:

- Describe the process of outgassing that formed Earth's early atmosphere.
- Explain how the evolution of life led to the development of Earth's current oxygen-rich atmosphere.
- Compare and contrast Earth's early reducing atmosphere and current oxidizing atmosphere.
- Describe the layered structure of Earth's current atmosphere.

THE ORIGIN OF EARTH'S ATMOSPHERE

Earth's Lost Atmosphere

In an early stage, while still accreting and forming a core, Earth's atmosphere was a cloud of silicate vapor. Then, as accretion and core formation stopped, Earth cooled and the silicate vapor condensed, forming molten and solid rock. Any hydrogen and other gases left from the solar nebula would quickly have been swept away by the vigorous solar wind, leaving primordial Earth with no atmosphere.

Formation of Earth's Present Atmosphere

Outgassing

Where, then, did the gases of our present atmosphere originate? One likely source was **outgassing**, the release, from the interior of the hot, young Earth, of water and trace gases trapped in rocks. (Recall that the original matter from which Earth accreted was mostly bits of rock and water ice.) Compounds of hydrogen, carbon, oxygen, and nitrogen, such as carbon dioxide, ammonia, methane, and a large amount of water, worked their way to the surface of the molten Earth. Even after Earth had cooled and formed a solid crust, outgassing during volcanic eruptions would have continued. In addition to ash and lava, gases such as hydrogen sulfide, carbon dioxide, and water vapor are still being released during volcanic eruptions today. Table 10.1 compares, in terms of composition, the gases found in Earth's atmosphere and hydrosphere with those given off during volcanic eruptions and emitted by other sources such as hot springs, geysers, and fumaroles.

TABLE 10.1 COMPARISON OF THE COMPOSITIONS OF GASES FROM INTERNAL EARTH SOURCES WITH THOSE OF EARTH'S ATMOSPHERE AND HYDROSPHERE*

Gas	Some Gases of Earth's Hydrosphere and Atmosphere	Gases in Hot Springs, Fumaroles, and Geysers	Volcanic gases from Basaltic Lava of Mauna Loa and Kilauea
Water vapor, H_2O	92.8	99.4	57.8
Total carbon, as CO_2	5.1	0.33	23.5
Sulfur, S_2	0.13	0.03	12.6
Nitrogen, N_2	0.24	0.05	5.7
Argon, A	Trace	Trace	0.3
Chlorine, Cl_2	1.7	0.12	0.1
Fluorine, F_2	Trace	0.03	—
Hydrogen, H_2	0.07	0.05	0.04

*Data from W. W. Rubey (1952), *Geol. Soc. Amer., Bull.*, vol. 62, p. 1137. Figures in table represent percentages by weight.

Cometary Impacts

Another likely source of gases was comets and other bodies from farther out in the forming solar system. The outer solar system is cold enough to condense many volatile materials, and comets are rich in ices of water, carbon dioxide, carbon monoxide, ammonia, and organic compounds. Today, hundreds of Earth-masses of comets orbit the Sun. According to computer models, there were even more comets in the early solar system and they commonly intersected the orbits of Venus, Earth, and Mars. Collisions of comets with planets would have released the comets' icy gases into the atmospheres of the planets they struck. Hundreds of millions of years of cometary collisions could have added large amounts of material to Earth's atmosphere. As stated earlier, in addition to water, comets would have provided carbon dioxide, carbon monoxide, methane, ammonia, nitrogen, and other gases. Carbon dioxide was also released from rocks in Earth's mantle, so this gas probably dominated Earth's early atmosphere after water condensed out.

Virtually no molecular oxygen is found in comets, very little exists on Venus and Mars, and it was absent from Earth's early atmosphere. Evidence of the absence of oxygen comes from ancient rocks containing minerals that would have been unstable in an oxygen atmosphere.

Development of Earth's Atmosphere

There were a number of stages in the development of Earth's present, oxygen-rich atmosphere. Once liquid water had condensed on Earth's surface, the atmosphere began to change.

A Reducing Atmosphere

Carbon dioxide is very soluble in water, and was probably removed from the atmosphere by dissolving in the ocean and then combining with a substance such as calcium to form limestone. Oxygen began to be formed photochemically; water absorbs ultraviolet radiation from the Sun, and the water molecule (H_2O) breaks down into hydrogen gas and oxygen gas. The light hydrogen can escape from the atmosphere, so it does not recombine. The remaining oxygen forms molecular oxygen (O_2) and ozone (O_3). However, the weathering of minerals and reactions with volcanic gases bind the oxygen almost as quickly as it is formed, so that this element remained very scarce, less than one hundred-millionth of present atmospheric levels. Oxygen levels this low formed a **reducing atmosphere**, that is, an atmosphere in which elements and compounds that would combine with oxygen in today's atmosphere remained stable (e.g., iron would not form rust).

An Oxidizing Atmosphere

Once photosynthetic life (such as chlorophyll-containing life-forms) appeared and spread around the planet, however, there was a new source of oxygen. See Figure 10.1. As these organisms multiplied over time, photosynthesis steadily increased. Geological evidence suggests that between 2.2 and 2.0 billion years ago the oxygen level in the atmosphere skyrocketed to roughly one-hundredth of its present level—more than a millionfold increase. Oxygen levels grew high enough to form an **oxidizing atmosphere**, one in which oxidation of elements and compounds occurred but that was still too low to sustain aerobic respiration (the oxidation of food to produce energy). However, at this concentration, enough molecular oxygen, which strays into the upper atmosphere and is chemically changed into ozone by ultraviolet rays, is formed to sustain an ozone shield that blocks incoming ultraviolet rays.

Figure 10.1 Possible Algaelike Fossils Found in Rocks More than 3.1 Billion Years Old.
Source: *The Earth Sciences*, Arthur N. Strahler, Harper & Row, 1971.

As the oxygen level of the atmosphere continued to increase, and ozone shielded organisms from damaging ultraviolet rays, photosynthetic organisms proliferated. But these primitive photosynthesizers were anaerobic; their body metabolism did not use oxygen. In fact, anaerobic organisms cannot survive if oxygen levels rise too high. Oxygen is a waste product of anaerobic photosynthesizers, just as urea is a waste product of humans. Most organisms can tolerate waste products in moderate amounts, but high levels are toxic (which is why humans with damaged kidneys undergo dialysis to remove urea from their blood). At first this limited oxygen levels; as a population of anaerobic photosynthesizers would increase, oxygen levels would rise to toxic levels. This, in turn, would cause their population (and oxygen production) to decrease until oxygen levels were once again tolerable. By about 1.7 billion years ago, however, organisms had developed higher and higher oxygen tolerances, so production of this element surged again.

An Aerobic Atmosphere

Eventually, Earth formed an **aerobic atmosphere**, one with sufficient oxygen to support life-forms that metabolize oxygen. Over time, aerobic organisms (life-forms that use oxygen for body metabolism) evolved to take advantage of this oxygen-rich atmosphere. The number and vigor of photosynthetic life-forms (e.g., algae, green plants) increased. Photosynthesis has increased along with them, but is almost balanced by the consumption of oxygen by these new organisms during respiration. Photochemical oxygen production, as well as loss by weathering and volcanic activity, is now insignificant compared with photosynthesis and respiration. In this new balance, oxygen production slightly exceeds loss, and within the past billion years atmospheric oxygen has steadily risen to its current level. See Figure 10.2.

Figure 10.2 The Geologic and Climatological History of the Development of Earth's Atmosphere.

THE STRUCTURE OF EARTH'S ATMOSPHERE

The Composition of Air

The atmosphere, which now has a total mass of about 5,000 trillion tons, is held in place by Earth's gravity and extends several hundred kilometers into space. The **atmosphere** consists mostly of gases but also contains water, ice, dust, and other particles. This mixture of gases and other substances is called **air**.

Dry air is composed mainly of nitrogen and oxygen, with traces of other gases as well. On average, air today contains about 78 percent nitrogen and 21 percent oxygen; the remaining 1 percent consists mostly of argon, with traces of carbon dioxide and other gases. The water vapor content of air may range from 0 percent over deserts and polar ice caps to as much as 4 percent in a tropical jungle. See Table 10.2. Air also contains varying amounts of water vapor, dust, and chemicals released by industry and microorganisms.

TABLE 10.2 THE COMPOSITION OF AIR IN THE TROPOSPHERE

Gas	Percent by Volume in Dry Air
Nitrogen	78
Oxygen	21
Argon	Almost 1
Neon	Trace
Helium	Trace
Krypton	Trace
Xenon	Trace
Hydrogen	Trace
Ozone	Trace
Carbon dioxide	Trace
Nitrous oxide	Trace
Methane	Trace

Layers of the Atmosphere

With movement upward through the atmosphere, the temperature changes. The atmosphere is divided into four distinct layers according to temperature: the **troposphere**, **stratosphere**, **mesosphere**, and **thermosphere**.

Figure 10.3 shows the changes in temperature that occur in the atmosphere. Notice that, near the ground, temperature decreases with altitude. Then, at an altitude of about 11 kilometers, the air stops getting colder and starts to become warmer. This change in behavior signals that the boundary from one layer to another—in this case from troposphere to stratosphere—has been crossed. The boundary is called the **tropopause** because, at the top of the troposphere, the temperature stops decreasing, or pauses. You can see from Figure 10.3 that such a change happens twice more, at the stratopause and the mesopause.

These changes in temperature are the result of differences in the compositions of the different atmospheric layers. The upper stratosphere contains the gas ozone. Ozone absorbs ultraviolet light from the Sun, thereby warming the stratosphere. Once past the ozone, the temperature decreases in the mesosphere. In the thermosphere, gases and charged particles absorb ultraviolet light and other wavelengths of energy and turn them into heat energy. Once again, temperatures rise. In fact, the amount of energy absorbed by this outermost layer is so great that temperatures exceed the boiling point of water. Some gas molecules absorb so much energy they lose or gain electrons and become ions, that is, charged particles. These ions in the thermosphere form a layer, the ionosphere, that can reflect radio waves.

Figure 10.3 Selected Properties of Earth's Atmosphere. Source: The State Education Department, *Earth Science Reference Tables*, 2011. (Albany, New York; The University of the State of New York).

As you can see from the atmospheric pressure and water vapor sections of Figure 10.3, most of the air in the atmosphere and all of the water vapor is confined to the troposphere. Air exerts pressure in proportion to the amount of air that is present in the atmosphere. At the tropopause air exerts only one-tenth of the pressure that it does at sea level. This means that there is only about one-tenth as much air! At the stratopause there is only about one-thousandth of the air that is present at sea level, and by the mesopause less than one ten-thousandth the air. The atmosphere extends for hundreds of kilometers above the lithosphere and hydrosphere, but most of the air is confined to the 8–10 kilometers closest to their surface.

MULTIPLE-CHOICE QUESTIONS

In each case, write the number of the word or expression that best answers the question or completes the statement.

1. Scientists infer that most of Earth's earliest atmosphere was produced by
   (1) a collision with a giant gas cloud
   (2) capturing gases from a nearby planet
   (3) vaporizing comets that impacted Earth's surface
   (4) the escape of gases from Earth's molten surface

2. Earth's early atmosphere formed during the Early Archean Era. Which gas was generally absent from the atmosphere at that time?
   (1) water vapor   (3) nitrogen
   (2) carbon dioxide   (4) oxygen

3. The diagram below shows four different chemical materials escaping from the interior of early Earth.

   Which material contributed *least* to the early composition of the atmosphere?
   (1) SiO_2 (3) N_2
   (2) H_2O (4) CO_2

4. Most scientists believe Earth's Early Archean atmosphere was formed primarily by gases released from
   (1) stream erosion
   (2) chemical weathering
   (3) volcanic eruptions
   (4) plant transpiration

5. The accumulation of water vapor, carbon dioxide, and nitrogen in Earth's early atmosphere approximately 4 billion years ago resulted mainly from
   (1) outgassing from Earth's interior
   (2) radioactive decay
   (3) photosynthesis by the earliest land plants
   (4) convection currents in Earth's outer core

6. Scientists infer that oxygen in Earth's atmosphere did *not* exist in large quantities until after
   (1) the first multicellular, soft-bodied marine organisms appeared on Earth
   (2) the initial opening of the Atlantic Ocean
   (3) the first sexually reproducing organisms appeared on Earth
   (4) photosynthetic cyanobacteria evolved in Earth's oceans

7. Earth's atmosphere consists
   (1) mostly of oxygen, argon, carbon dioxide, and water vapor
   (2) entirely of ozone, nitrogen, and water vapor
   (3) of gases that cannot be compressed
   (4) of a mixture of gases, liquid droplets, and minute solid particles

8. What is the average air pressure exerted by Earth's atmosphere at sea level, expressed in millibars and inches of mercury?
   (1) 1,013.25 mb and 29.92 in. of Hg
   (2) 29.92 mb and 1,013.25 in. of Hg
   (3) 1,012.65 mb and 29.91 in. of Hg
   (4) 29.91 mb and 1,012.65 in. of Hg

9. The atmosphere is bound to Earth by
   (1) magnetic fields
   (2) atmospheric pressure
   (3) the force of gravity
   (4) chemical bonding at the interface

10. Which statement most accurately describes Earth's atmosphere?
    (1) The atmosphere is layered, with each layer possessing distinct characteristics.
    (2) The atmosphere is a shell of gases surrounding most of Earth.
    (3) The atmosphere's altitude is less than the depth of the ocean.
    (4) The atmosphere is more dense than the hydrosphere but less dense than the lithosphere.

11. In which two temperature zones of the atmosphere does the temperature increase with increasing altitude?
    (1) troposphere and stratosphere
    (2) troposphere and mesosphere
    (3) stratosphere and thermosphere
    (4) mesosphere and thermosphere

12. Which atmospheric temperature zone is located between 8 and 32 miles above Earth's surface and contains an abundance of ozone?
    (1) troposphere   (3) mesosphere
    (2) stratosphere   (4) thermosphere

13. At which position would the surrounding air temperature most likely be warmest?
    (1) mesosphere   (3) thermosphere
    (2) stratosphere   (4) troposphere

14. As the altitude increases within Earth's stratosphere, air temperature generally
    (1) decreases, only
    (2) increases, only
    (3) decreases, then increases
    (4) increases, then decreases

15. At an altitude of 95 miles above Earth's surface, nearly 100% of the incoming energy from the Sun can be detected. At 55 miles above Earth's surface, most incoming X-ray radiation and some incoming ultraviolet radiation can no longer be detected. This missing radiation was most likely
    (1) absorbed in the thermosphere
    (2) absorbed in the mesosphere
    (3) reflected by the stratosphere
    (4) reflected by the troposphere

16. Which gas in Earth's upper atmosphere is beneficial to humans because it absorbs large amounts of ultraviolet radiation?
    (1) water vapor   (3) nitrogen
    (2) methane   (4) ozone

17. The altitude of the ozone layer near the South Pole is 20 kilometers above sea level. Which temperature zone of the atmosphere contains this ozone layer?
    (1) troposphere   (3) mesosphere
    (2) stratosphere   (4) thermosphere

18. Scientists are concerned about the decrease in ozone in the upper atmosphere primarily because ozone protects life on Earth by absorbing certain wavelengths of
    (1) X-ray radiation
    (2) ultraviolet radiation
    (3) infrared radiation
    (4) microwave radiation

CONSTRUCTED RESPONSE QUESTIONS

Base your answers to questions 19 through 22 on the passage below.

Earth's Early Atmosphere

Early in Earth's history, the molten outer layers of Earth released gases to form an early atmosphere. Cooling and solidification of that molten surface formed the early lithosphere approximately 4.4 billion years ago. Around 3.3 billion years ago, photosynthetic organisms appeared on Earth and removed large amounts of carbon dioxide from the atmosphere, which allowed Earth to cool even faster. In addition, they introduced oxygen into Earth's atmosphere, as a by-product of photosynthesis. Much of the first oxygen that was produced reacted with natural Earth elements in the lithosphere, such as iron, and produced new varieties of rocks and minerals. Eventually, photosynthetic organisms produced enough oxygen so that it began to accumulate in Earth's atmosphere. About 450 million years ago, enough oxygen was in the atmosphere to allow for the development of an ozone layer 30 to 50 kilometers above Earth's surface. This layer was thick enough to protect organisms developing on land from the ultraviolet radiation from the Sun.

19. State *one* reason why the first rocks on Earth were most likely igneous in origin. [1]

20. Identify *one* mineral with a red-brown streak that formed when oxygen in Earth's early atmosphere combined with iron. [1]

21. Identify the temperature zone of the atmosphere in which the ozone layer developed. [1]

22. Complete the pie graph below to show the percent by volume of nitrogen and oxygen gases currently found in Earth's troposphere. Label each section of the graph with the name of the gas. The percentage of other gases is shown. [1]

EXTENDED CONSTRUCTED RESPONSE QUESTIONS

Base your answers to questions 23 through 25 on the passage below and on your knowledge of Earth science.

Ozone in Earth's Atmosphere

Ozone is a special form of oxygen. Unlike the oxygen we breathe, which is composed of two atoms of oxygen, ozone is composed of three atoms of oxygen. A concentrated ozone layer between 10 and 30 miles above Earth's surface absorbs some of the harmful ultraviolet radiation coming from the Sun. The amount of ultraviolet light reaching Earth's surface is directly related to the angle of incoming solar radiation. The greater the Sun's angle of insolation, the greater the amount of ultraviolet light that reaches Earth's surface. If the ozone layer were completely destroyed, the ultraviolet light reaching Earth's surface would most likely increase human health problems, such as skin cancer and eye damage.

23. State the name of the temperature zone of Earth's atmosphere where the concentrated layer of ozone gas exists. [1]

24. Explain how the concentrated ozone layer above Earth's surface is beneficial to humans. [1]

25. Assuming clear atmospheric conditions, on what day of the year do people in New York State most likely receive the most ultraviolet radiation from the Sun? [1]

Base your answers to questions 26 and 27 on the table below and on your knowledge of Earth science. The table shows air temperatures and air pressures recorded by a weather balloon rising over Buffalo, New York.

Altitude Above Sea Level (m)	Air Temperature (°C)	Air Pressure (mb)
300	16.0	973
600	16.5	937
900	15.5	904
1,200	13.0	871
1,500	12.0	842
1,800	10.0	809
2,100	7.5	778
2,400	5.0	750
2,700	2.5	721

26. On the grid provided below, construct a graph of altitude above sea level and air temperature by following the directions below.
    (a) Plot an **X** for the air temperature recorded at *each* altitude shown on the table. [1]
    (b) Connect the **X**s with a solid line. [1]

27. State the relationship shown in the table between altitude above sea level and air pressure recorded by the rising weather balloon. [1]

CHAPTER 11
THE ORIGIN AND NATURE OF EARTH'S HYDROSPHERE

KEY IDEAS Approximately 70 percent of Earth's surface is covered by a relatively thin layer of water. Earth's oceans formed as a result of precipitation over millions of years. The presence of early oceans is indicated by sedimentary rocks of marine origin, dating back about 4 billion years.

Earth has continuously been recycling water ever since the outgassing of water early in the planet's history. This constant recirculation of water at and near Earth's surface is described by the water (hydrologic) cycle. Water leaves Earth's surface and enters the atmosphere by evaporation or transpiration from plants. Water is returned from the atmosphere to Earth's surface by precipitation.

A portion of the precipitation becomes runoff over the land or infiltrates into the ground and is stored in the soil or groundwater below the water table. Soil capillarity influences these processes. Porosity, permeability, and water retention also affect runoff and infiltration. The amount of precipitation that seeps into the ground or runs off is influenced by climate, slope of the land, soil, rock type, vegetation, land use, and degree of saturation.

KEY OBJECTIVES
Upon completion of this chapter, you will be able to:

- Describe the origin of Earth's hydrosphere.
- Identify the forms in which water exists on Earth.
- Analyze the water cycle, and describe the transfer of water and energy into and out of the atmosphere at each step of the cycle.
- Explain why fresh water is a limited but renewable resource.
- Give examples of how factors such as porosity, permeability, water retention, and soil capillarity affect infiltration and runoff.
- Describe the formation of the water table and some methods by which groundwater is obtained for use at Earth's surface.

THE ORIGIN OF THE HYDROSPHERE

Earth is often referred to as "the water planet." Most people know that the oceans are full of water, and think water is commonplace because there is so much of it around. On a cosmic scale, however, water is not so common. Earth is the *only* known planet that has liquid water on its surface! This has resulted from a unique combination of Earth's temperature and the unusual properties of water. Water not only covers Earth's surface, but also makes life as we know it possible.

Let us now consider the origin and nature of Earth's oceans and the properties of water that make it such an unusual substance.

The Origin of Earth's Water

The origin of Earth's water is directly linked to the formation of Earth. Meteorites called Type I carbonaceous chondrites provide a clue to the probable origin of Earth's water. They contain 5 percent carbon compounds and 20 percent water tied up in mineral compounds. These meteorites are believed to be debris left over from the formation of the solar system. As such, they represent a type of material thought to have formed while the planets were condensing from a cloud of dust and hydrogen gas. See Table 11.1.

TABLE 11.1 COMPOSITION OF EARTH AND CHONDRITE METEORITES*

Rank	Element	Symbol	Earth Average (% by wt.)	Average of Chondrites (% by wt.)
1	Iron	Fe	34.6 ⎫	27.2 ⎫
2	Oxygen	O	29.5 ⎬ 92.0	33.2 ⎬ 91.8
3	Silicon	Si	15.2 ⎪	17.1 ⎪
4	Magnesium	Mg	12.7 ⎭	14.3 ⎭
5	Nickel	Ni	2.4	1.6
6	Sulfur	S	1.9	1.9
7	Calcium	Ca	1.1	1.3
8	Aluminum	Al	1.1	1.2
9	Sodium	Na	0.57	0.64
10	Chromium	Cr	0.26	0.29
11	Manganese	Mn	0.22	0.25
12	Cobalt	Co	0.13	0.09
13	Phosphorus	P	0.10	0.11
14	Potassium	K	0.07	0.08
15	Titanium	Ti	0.05	0.06

Source: *The Earth Sciences*, Arthur N. Strahler, Harper & Row, 1971.

*Data from *Principles of Geochemistry*, 3rd Ed., Brian Mason, John Wiley, 1966.

As Earth cooled and condensed from a cloud of dust and gas, hydrogen reacted with oxygen to form water, which in turn reacted with crystals of solids that were forming. The products of these reactions were silicate minerals that contain water and oxidized iron. Therefore, the original form in which water existed on Earth was as part of solid compounds—rocks. The common clay minerals found in soil are examples of this type of compound.

As these substances formed, they were drawn inward by gravity and heated. We know that, as rocks are heated above 500°C, the water in their molecules is released in a process called *degassing*. When vented to Earth's surface, the water is released into the atmosphere, a process called **outgassing**, and upon further cooling condenses to form liquid water. Much evidence supports the idea that much of Earth's water was released to the surface soon after it formed. Even today, however, the process continues on a small scale through degassing associated with volcanic eruptions.

From an Ocean of Gas to an Ocean of Water

Outgassing from Earth's interior and collisions of comets with Earth added large amounts of water vapor to the developing atmosphere. Consider that a typical comet contains about 10^{15} (one million billion) kilograms of water. Since all of Earth's oceans hold roughly 10^{21} (one trillion billion) kilograms of water, only a million or so comet collisions could have created Earth's entire water supply ($10^6 \times 10^{15} = 10^{21}$). This is not an unlikely number of collisions, considering the estimated number of existing comets, which are but a fraction of those that crowded the early solar system.

As Earth cooled, water vapor in the atmosphere would have condensed and fallen as torrential rains that ran off high places and filled low-lying basins. At first, striking a still warm Earth, water would vaporize quickly and return to the atmosphere. But as the planet continued to cool, liquid water would remain for longer and longer periods in the basins of Earth's surface. Eventually, the oceans and lakes were permanently filled with water that returned to the atmosphere only because of evaporation by the Sun, as is the case today.

This reservoir of liquid water blanketing Earth is called the **hydrosphere**. The hydrosphere, however, is very thin. If you dipped a basketball in water and then shook the water off, the thin film of water adhering to its surface would be a good model of the hydrosphere. See Figure 11.1.

Outgassing of water during volcanic eruptions still occurs today, and so do collisions with icy comets. (Consider the 1994 collision of Comet Shoemaker-Levy with Jupiter.) As water continues to work its way to Earth's surface and to be carried in by comets, the oceans may gradually increase in volume until, in a few billion years, the hydrosphere covers the whole planet.

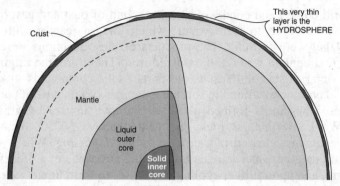

Figure 11.1 The Hydrosphere. The thicknesses of the crust and hydrosphere are greatly exaggerated in this diagram. The hydrosphere is actually much thinner than shown here.

The Origin of Seawater

How did seawater become salty? The general assumption is that the ocean was originally a huge freshwater lake. Over the billions of years that the ocean has existed, soluble ions freed by chemical weathering have been carried into the ocean by rivers, creating the saline waters of today. We know, for example, that the extremely salty Dead Sea and Great Salt Lake have resulted because all types of substances entered them, but only water vapor left them. But at the current rate at which streams add salts to the ocean, it could have reached its present salinity in roughly 100 million years. This time span falls far short of the current estimates, based upon radioactive decay rates and other data, of 3.5–4 billion years for the age of the ocean. Why the discrepancy?

One answer is that the salinity of the ocean does not increase at a constant rate, and probably decreases from time to time. For example, the end of a glacial period would flood the oceans with melting ice, decreasing salinity worldwide. Shallow seas that become landlocked and then evaporate would remove substantial amounts of salt from the ocean. (Salt deposits near Moab, Utah, that formed when a vast inland sea evaporated are as much as 4,000 feet thick!)

Another answer is that the rate at which water has been added by outgassing has not been uniform or continuous. Thus, it appears that salinity, along with sea level, has risen and fallen with the freezing and melting of ice and the removal of huge amounts of salt by evaporation of landlocked seas.

Seawater Today

Seawater is a complex mixture of dissolved inorganic matter, dissolved organic matter, dissolved gases, and particulate matter. **Salinity** is the total amount of dissolved material in seawater. Salinity is measured in parts per thousand (‰) by weight in 1 kilogram of seawater.

Inorganic Matter

By weight, seawater is about 96.5 percent pure water and about 3.5 percent dissolved inorganic substances. See Figure 11.2. Six elements—chlorine, sodium, magnesium, sulfur, calcium, and potassium—make up more than 90 percent of the materials dissolved in seawater. The minor elements include strontium, bromine, and boron. The elements in seawater are almost always present as part of chemical compounds, most of which are salts. Although the salinity of seawater varies from 33 percent to 37 percent, depending upon where in the ocean the sample is taken, the ratio of the major dissolved substances remains about the same.

Figure 11.2 Materials Dissolved in Seawater. Source: *Introduction to Oceanography*, 3rd Ed., David A. Ross, Prentice-Hall, 1982.

Organic Matter

The dissolved organic matter in seawater comes mainly from wastes excreted by organisms and from the decaying bodies of dead organisms. It is usually present in small amounts (0–6 parts per million) and is highly variable. Dissolved organic matter includes organic carbon, carbohydrates, proteins, amino acids, organic acids, and vitamins. Organic compounds containing nitrogen and phosphorus oxidize to form nitrates and phosphates.

Dissolved Gases

Nitrogen, oxygen, and carbon dioxide are the major gases dissolved in seawater. Dissolved gases should not be confused with gaseous elements that are part of compounds. For example, water is a compound of hydrogen and oxygen, both of which are gases. However, the compound is not a gas. The oxygen that is part of the water molecule is not a dissolved gas; it is bound in the molecule and can't be "breathed in" and used for respiration. Other gases dissolved in seawater include helium, neon, argon, krypton, and xenon. With the exception of oxygen and carbon dioxide, the gases in seawater usually enter it from the atmosphere.

Seawater has two sources of oxygen: the atmosphere and the plants that live in the ocean. Oxygen gas dissolves directly into seawater from the atmosphere. Photosynthesis by plants living in the upper layers of the ocean (mainly phytoplankton) releases oxygen into seawater.

Particulates

Particulates are solid particles of material that do not dissolve in seawater, but are suspended or settling through it. Aside from living organisms, particulate matter consists mainly of fine-grained minerals and decaying organic material. Particulate matter in seawater varies greatly depending on ocean currents, winds, and nearness to rivers and streams.

WATER: A UNIQUE SUBSTANCE

Earth's hydrosphere is the result of a unique combination of Earth's temperature and the properties of water. Water has unusual properties that make it different from any other substance on Earth. These properties enable liquid water to exist on Earth's surface and are responsible for its chemical and physical behavior. If water had properties typical of a substance with its molecular size and weight, Earth would be a very different place indeed.

The Structure of Water

A water molecule consists of one relatively large oxygen atom combined with two smaller hydrogen atoms. The result is a somewhat lopsided molecule that creates an unequal distribution of electrical charges. The hydrogen side of the molecule is slightly positive; the oxygen side, slightly negative. This unusual structure causes water molecules to be attracted to one another and "stick" together. The attractions between molecules of liquid water are called **hydrogen bonds**. Hydrogen bonds are weak, but they make water behave differently from any other substance on Earth. See Figure 11.3.

Figure 11.3 The Structure of Water Molecules Causes Hydrogen Bonding. The two ends of a water molecule have opposite electrical charges that attract each other as do opposite poles of a magnet. The weak attractions between the slightly positive end of one water molecule and the slightly negative end of another water molecule are known as hydrogen bonds.

Although weak, hydrogen bonds hold water molecules together more strongly than is the case in substances lacking such bonds. Thus, liquid water has a high **surface tension**; to push liquid water molecules apart and to penetrate the surface of liquid water, a relatively strong force must be exerted to overcome the hydrogen bonds. Water striders, long-legged bugs that move about on the surface of water, spread their mass over several legs, so that no one leg exerts enough force to break through the surface.

The same "stickiness" that holds water molecules together also makes them adhere to the surfaces of other substances. For example, when water seeps through the ground, some of it "sticks" to the soil particles so that their surface stays "wet" even after most of the water has drained through them. The water that adheres to the surface of soil particles is an important source of water for plant life.

Water's "stickiness" is also responsible for *capillarity*, the tendency for water's surface to rise in a narrow opening. The water rises because its attraction to the walls of the opening pulls it upward, and surface tension drags the water surface between the walls upward along with it. Capillarity is an important process in the movement of water through porous rocks and soil.

The States of Water

Matter can exist in three different physical forms—liquid, gas, and solid—called the **states**, or **phases**, of matter. Water is the only substance on Earth that occurs naturally in all three phases of matter. Since the majority of the hydrosphere is in the liquid state, we'll begin our discussion with water as a liquid.

Liquid Water

In the liquid phase, molecules of water are in constant motion. Because of hydrogen bonds individual water molecules cluster together in clumps of two to eight molecules. Although weak, hydrogen bonds hold water molecules together more strongly than is the case in substances lacking such bonds.

The average speed of the molecules in a substance is reflected in its temperature; the faster the molecules are moving, the higher the temperature. The addition of heat to water makes its molecules move faster, and thus raises its temperature. However, some of the heat added to the water goes into breaking hydrogen bonds instead of promoting molecular motion. Therefore, a large amount of heat energy is needed to get water molecules to move faster, and thus increase the temperature of water. This heat energy is not lost; it is stored *in* the water. Thus, liquid water holds more heat than do substances without hydrogen bonds; and its ability to hold heat, or **specific heat capacity**, is high. In fact, water has one of the highest specific heat capacities of any naturally occurring substance.

Water Vapor

If a water molecule moves fast enough, it can break free of the molecules around it. This process, known as **evaporation**, marks the change from the liquid phase of water to the gaseous or vapor phase. See Figure 11.4. The molecules in water vapor are so fast moving, and move so much farther apart than in the liquid, that they are not held together by hydrogen bonds. As temperature rises, and liquid molecules move faster and faster, more and more escape, so the rate of evaporation increases. When the temperature reaches 100°C, nearly all of the molecules break free of their neighbors and try to enter the vapor state all at once; in other words, the water boils.

Once water begins to boil, added heat goes into breaking hydrogen bonds between molecules instead of increasing molecular speed. Therefore, while water is changing phase, it does not change temperature. The amount of heat required to evaporate a substance is called the **latent heat of vaporization**.

Figure 11.4 Evaporation Takes Place When Molecules Break Free of the Hydrogen Bonds of the Liquid and Enter the Gaseous State.

The word *latent* means "hidden," and in this term refers to the added heat that is not evident as a temperature change. As long as there is boiling water to be evaporated, the temperature of the liquid remains constant at 100°C. Water absorbs a large amount of heat when it evaporates; in other words, it has a high latent heat of vaporization.

At temperatures lower than the boiling point, only the most energetic, fastest moving molecules can break free of the hydrogen bonds and become a gas. Since the fastest moving molecules leave the liquid, those that are left behind are slower moving. As a result, the average speed of the molecules in the liquid, and thus its temperature, decreases. This effect is known as **evaporative cooling**. Our bodies use evaporative cooling to regulate body temperature—when sweat evaporates, it cools the skin.

As liquid water cools, the molecules move slower, pack closer together, and take up less space. In other words, the volume of the water decreases. Since the same mass of water is occupying less volume, the water becomes denser. Thus, liquid water gets denser as it gets colder, but only down to a temperature of about 4°C.

Water as a Solid

As water continues to cool, the molecules move more slowly and rebound from one another more weakly. Eventually, the hydrogen bonds between the molecules suppress these motions and begin to hold the molecules in a fixed, three-dimensional pattern. This process, known as **freezing**, marks the change from the liquid phase to the solid phase. Solids in which atoms or molecules are arranged in such a regular, fixed pattern are called **crystals**. The solid, crystalline form of water is, of course, ice.

In ice, the hydrogen bonds of water hold the molecules in a hexagonal, or six-sided, pattern. As shown in Figure 11.5, the molecules of water are spaced farther apart in ice than in liquid water. What happens might be compared to putting up a tent. When disassembled, the tent poles fit into a small bag; but when they are connected together, they form a large, open framework. Similarly, when water freezes, it expands. Since the same mass of water occupies more volume in ice than in liquid water, ice is less dense than liquid water and therefore floats. It is extraordinarily unusual for the solid phase of a substance to be less dense than the liquid phase. This unique property of water has important consequences for aquatic life. As ice forms in a body of water, it floats and forms an insulating blanket that keeps the subsurface water from cooling rapidly in cold weather. Organisms can live in the liquid water below the surface ice.

If ice were denser than liquid water, it would sink as it formed and build up at the bottom. In colder climates, lakes and even some parts of the oceans would freeze solid. During the summer, only a thin surface layer would thaw, and these bodies of water would be much less hospitable to life.

The Origin and Nature of Earth's Hydrosphere

Figure 11.5 The Crystalline Structure of Ice. In ice, hydrogen bonding holds water molecules in a fixed structure in which they are spaced farther apart than in liquid water.

A large amount of heat is required to melt ice. Water molecules in ice crystals vibrate, but they do not move around freely. For melting to occur, the hydrogen bonds holding ice crystals together must first be broken. As heat energy is added to ice, the molecules vibrate faster and eventually break the hydrogen bonds holding the crystal together; then the ice melts. The molecules then move around as well as vibrate. Since a lot of energy is needed to break the hydrogen bonds holding ice crystals together, ice melts at a much higher temperature than do substances that do not have hydrogen bonds. Similarly, since hydrogen bonds help pull water molecules together into crystals, water does not have to be cooled as much before freezing. Thus, water melts and freezes at a much higher temperature than do substances that lack hydrogen bonds. If water did not have hydrogen bonds, the change from solid to liquid (or liquid to solid) would occur at about –90°C instead of 0°C.

As ice crystals are melting, added heat goes into breaking hydrogen bonds instead of increasing molecular speed. Therefore, while water is changing phase it does not change temperature. The amount of heat required to melt a substance is called the **latent heat of fusion**. Here, again, the word *latent* means "hidden," and in this term refers to the added heat not evident as a temperature change. As long as there is ice left to be melted, the temperature of the ice-water mixture remains constant at 0°C.

Water: The Universal Solvent

Water is often called the universal solvent because it can dissolve more things than any other naturally occurring substance. Water is particularly good at dissolving molecules held together by the electrical attraction between oppositely charged particles. Common table salt, or sodium chloride, is such a molecule. It is composed of positively charged sodium (Na^+) and negatively charged chlorine (Cl^-). These sodium and chlorine ions have much stronger charges than the slightly charged ends of a water molecule. When a salt crystal is placed in water, the strong charges of the sodium and chlorine ions attract water molecules. Water's positive end is attracted to (or will attract) the negative chlorine; likewise, water's negative end will be attracted to the positive sodium. Thus, water molecules cluster around both ions in a salt molecule. As water molecules cluster around the ions, they cancel some of the attraction between the sodium and the chlorine, and the two move farther apart. Then more water molecules can come between them until the attraction is eliminated. Finally, the sodium and chlorine ions in the salt crystal pull apart, and the crystal **dissolves**. Once the sodium and chlorine ions separate, water molecules surround them completely, insulating them from each other and preventing them from recombining to form a crystal. See Figure 11.6.

Figure 11.6 Water Molecules Cluster Around the Electrically Charged Sodium and Chlorine Ions in a Salt Crystal. This weakens the bonds between the ions, causing them to separate, and the crystal dissolves.

THE WATER CYCLE

The *water cycle* is the process by which Earth's water moves into and out of the atmosphere. See Figure 11.7. Let's begin by describing this process in the oceans that cover nearly three-quarters of Earth's surface.

1. Each day, insolation (*in*coming *sol*ar radi*ation*) reaching the surface of the oceans provides energy to the water molecules at the surface. As the molecules heat up, trillions of tons of water change from liquid to gas, or *evaporate*. The resulting water vapor enters the atmosphere, where it is circulated by winds and carried upward by rising warm air.

The Origin and Nature of Earth's Hydrosphere

2. When air is cooled to its dewpoint by expansion during rising or contact with cooler surfaces, the water vapor changes from gas back to liquid, or *condenses*, forming tiny water droplets. These tiny water droplets suspended in air form clouds.
3. When the water droplets become too large to remain suspended, they fall back to the surface as *precipitation*. Precipitation may fall on the oceans or the land. Water that falls on the oceans has completed its cycle and may evaporate back into the atmosphere, beginning the entire process anew.
4. A number of things may happen to precipitation that falls on land. It may become *runoff*, water that flows downhill into rivers and streams that eventually carry it back into an ocean or other body of water. Or, it may seep into the ground to become **groundwater**, water that has infiltrated open spaces in the outer part of Earth's crust. Groundwater may move slowly back to the oceans through the ground, but most of it seeps into rivers and streams. It is then carried back to the oceans.
5. Some of the water may evaporate again to be carried farther over land, or be blown back over the oceans.
6. Once the water is back in the oceans, the cycle begins all over again.

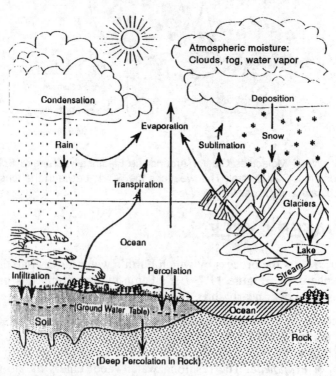

Figure 11.7 The Water Cycle. Source: Permission from Thomas McGuire.

There are many variations in the water cycle. Sometimes the water that evaporates over the oceans condenses there and falls directly back into the oceans as rain. Water falling on land may evaporate almost immediately, and in some cases precipitation evaporates before it ever reaches the ground.

Plants play a role in the water cycle as well. The roots of plants absorb water that has seeped into the soil. Then the water is transported to their leaves and released back into the atmosphere by a process called **transpiration**. In an area of abundant vegetation, such as a forest, transpiration returns more than one-third of precipitation back to the atmosphere as water vapor.

Other variations are possible, but all are part of the endless water cycle that receives its energy from the Sun. The water cycle continually renews our supply of fresh water. Each day an estimated 15 trillion liters of water in the form of rain or snow fall on the United States alone. Earth is a closed system, and water consumed today will eventually be recycled for use by a future generation.

In theory, Earth's water will never be exhausted. If, however, persistent, long-lasting pollutants enter the basins and ground from which Earth's water supply is drawn, water that has been cleansed by evaporation will be polluted as soon as it condenses and falls back into these polluted areas. Consider what will happen if the ground is contaminated with nuclear waste, which remains radioactive for tens or even hundreds of thousands of years. For the entire time the material remains radioactive, any water that infiltrates that ground will immediately become unusable.

GROUNDWATER

Porosity

Earth's surface is not completely solid; rather, it is filled with empty spaces, or **pores**. When precipitation falls on the surface, some of it **infiltrates**, or seeps through openings in the surface and into pore spaces. Pore spaces are usually filled with air unless water, oil, or natural gas has forced the air out. The amount of open space in a material is its **porosity**. Porosity is expressed as a percentage of the total volume of a given substance. For example, if half of the volume of a sample of rock or sediment is open space, the specimen is said to have a porosity of 50 percent. Earth materials differ greatly in porosity. Factors that affect porosity are shown in Figure 11.8. Loose sediments such as sand, gravel, and clay usually have the highest porosity.

Porosity can be calculated by measuring the amount of water required to fill the pores in a substance and then setting up a ratio of the volume of water to the total volume of the sample being tested.

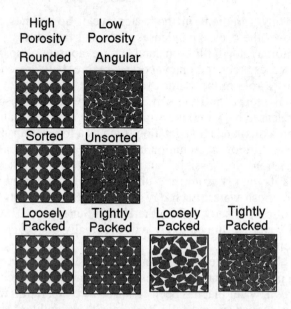

Figure 11.8 Factors That Affect the Porosity of a Material.

Example:
Water is poured into a 100-cubic-centimeter sample of sand. When 35 milliliters of water have been poured in, the water is just even with the surface of the sample. What is the porosity of the sand? [Note: 1 ml = 1 cm^3]

$$\text{Porosity} = \frac{\text{volume of pore space}}{\text{total volume of sample}} \times 100\%$$

$$= \frac{35 \text{ cm}^3}{100 \text{ cm}^3} \times 100\% = 35\%$$

The porosity of the sand is 35%.

Permeability

Permeability is the rate at which water can infiltrate a material. A substance through which water passes rapidly is said to be **permeable**; a substance through which water cannot flow is **impermeable**. The permeability of a substance depends on two factors: the size of its pores and the degree to which they are interconnected. See Figure 11.9. Small pores constrict the flow of water; and if water cannot get from one pore to another, it cannot flow through a substance.

Knowing the permeability of a material is very useful. From it, the rate at which rainwater will sink into the ground can be determined, as well as the speed and direction in which sewage in a septic tank will flow. Also, the rate at which a well will refill with water after some has been pumped out can be determined.

Figure 11.9 Isolated and Interconnected Pore Spaces. Permeable materials have interconnected pore spaces.

The Water Table

When rain falls on land or when snow that has fallen melts, gravity pulls the water down into the ground through interconnected pore spaces. Some of the water clings to particles near the surface, forming a moist layer called the **soil moisture zone**. This is the layer from which plants obtain the water they need to survive. Most of the water that infiltrates continues to trickle downward through the ground until it reaches an impermeable substance. Then the water begins to fill all the pore spaces from the impermeable material upward. The region of permeable ground in which all of the pore spaces are filled with water is the **zone of saturation**.

The top of the zone of saturation, known as the **water table**, is the boundary between pores filled with water and pores filled with air. Just above the water table is a narrow region, called the **capillary fringe**, in which narrow pore spaces are filled with water drawn upward by capillary action. The area above the capillary fringe, where most of the pore spaces are filled with air, is called the **zone of aeration**. See Figure 11.10.

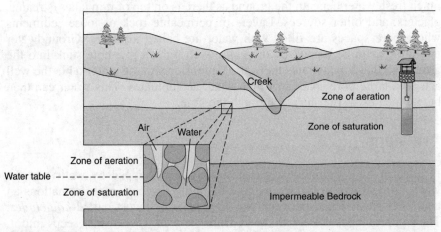

Figure 11.10 Zones of Soil Water and Ground Water. The water table is the boundary between the zone in which pores are filled with air and the zone in which pores are filled with water.

Groundwater can leave the ground in several ways. It can be drawn up by **capillary action** through tiny interconnected pore spaces and reach the soil moisture zone, where plant roots remove it. Or, instead, it may evaporate and slowly diffuse out of the ground through the zone of aeration, or it may seep through the ground sideways and slowly move out of the area. If the surface of the ground dips beneath the level of the water table, water can seep out and run off, forming streams, or collect in depressions, forming springs, ponds, or lakes. When the water table in a depression is just above the surface, a swamp may form. See Figure 11.11.

Figure 11.11 Variations in the Subsurface Position of the Water Table. The water table roughly follows the contours of Earth's surface. Where the surface dips beneath the water table, water seeps out of the ground and can collect in depressions, forming a body of water such as a lake or pond. If the water seeping out runs off, it may form a stream. Impermeable layers in hills can create a perched water table, which may seep out of a hillside as a spring.

Wells

Groundwater is an important source of fresh water. There is 37 times as much fresh water beneath the ground as there is on top of it in lakes, streams, glaciers, and other sources. Layers of permeable rock or loose sediments whose pore spaces are filled with water are called **aquifers**. Groundwater in aquifers can be tapped by digging a well, which is a hole sunk into the ground so that it penetrates the water table. Beneath the water table the well acts as a large pore space in which water accumulates. This water can then be pumped up to the surface through the hole.

Artesians

In artesian rock formations, groundwater can be tapped by wells and the water brought to the surface without pumping. In these rock formations an aquifer is sandwiched between two impermeable layers called *aquicludes*. The aquifer slopes downward away from the area where water seeps into it. The aquicludes act like the walls of a pipe, confining the water as it seeps downslope, so that water pressure builds up in the aquifer. If the upper aquiclude is pierced by a well, this pressure may push the water up to the surface or even higher. The amount of pressure in an artesian well depends on the difference in elevation between the well and the water table in the aquifer. See Figure 11.12. Artesian wells get their name from Artois, a former province of France, where the first well of this type was dug.

Figure 11.12 An Artesian Formation. The water rises in an artesian well because of hydrostatic pressure created by the weight of the water in the aquifer.

MULTIPLE-CHOICE QUESTIONS

In each case, write the number of the word or expression that best answers the question or completes the statement.

1. Large amounts of water vapor were added to Earth's developing atmosphere by
   (1) outgassing and collisions with comets
   (2) photosynthesis by early life-forms
   (3) dust and hydrogen gas
   (4) electrical discharges in the atmosphere

2. Approximately what fraction of Earth's surface do oceans now cover?
   (1) ¼   (3) ½
   (2) ⅓   (4) ¾

3. What is the major source of the dissolved minerals that affect the salinity of ocean water?
   (1) ice sheets   (3) continental erosion
   (2) industrial pollutants (4) tropical storms

4. Which profile when drawn to true scale most accurately shows the relationship between the ocean width of 1,500 kilometers and a maximum depth of 8 kilometers?

5. A lake without outlets may increase in salinity if the inflow of fresh water
   (1) is equal to or less than the amount of water lost by evaporation
   (2) is greater than the amount of water lost by evaporation
   (3) carries no dissolved substances into the lake
   (4) is used to power turbines in hydroelectric plants

6. Water is a good solvent because
   (1) a water molecule contains two hydrogen atoms and one oxygen atom
   (2) water is held together by strong molecular bonds
   (3) the two ends of a water molecule have different electrical charges
   (4) water is the most common compound on Earth

7. Liquid water can store more heat energy than an equal amount of any other naturally occurring substance because liquid water
   (1) covers 71% of Earth's surface
   (2) has its greatest density at 4°C
   (3) has the higher specific heat
   (4) can be changed into a solid or a gas

8. During which phase change will the greatest amount of energy be absorbed by 1 gram of water?
   (1) melting   (3) evaporation
   (2) freezing   (4) condensation

9. When 1 gram of liquid water at 0° Celsius freezes to form ice, how many total joules of heat are lost by the water?
   (1) 2.11   (3) 334
   (2) 4.18   (4) 2260

10. When water vapor condenses, how much heat energy will be released into the atmosphere?
    (1) 2,260 joules/gram
    (2) 334 joules/gram
    (3) 4.18 joules/gram
    (4) 2.11 joules/gram

11. What best explains why, in early spring, ice remains longer on Lake Erie than on the surrounding land areas when the air temperature is above freezing?
    (1) Water has a higher specific heat than land.
    (2) Energy is needed for water to evaporate.
    (3) Cool winds from the surrounding land cool the ice on the lake.
    (4) Air temperature does not affect water temperature.

12. The flowchart below shows part of Earth's water cycle. The question marks indicate a part of the flowchart that has been deliberately left blank.

    (1) condensation   (3) evaporation
    (2) deposition   (4) infiltration

13. Which processes of the water cycle return water vapor directly to the atmosphere?
    (1) evaporation and transpiration
    (2) infiltration and capillarity
    (3) freezing and precipitation
    (4) water retention and runoff

The Origin and Nature of Earth's Hydrosphere

Base your answers to questions 14 through 17 on the diagram below, which shows a model of the water cycle. Letters *A* through *F* represent some processes of the water cycle. Letter *X* indicates the top of the underground zone that is saturated with water.

The Water Cycle

(Not drawn to scale)

14. Which process is represented by letter *F*?
  (1) capillarity
  (2) infiltration
  (3) condensation
  (4) vaporization

15. What does letter *X* represent?
  (1) the water table
  (2) a floodplain
  (3) sea level
  (4) impermeable rock

16. If the surface soil is saturated and precipitation increases, there will be
  (1) a decrease in the amount of groundwater
  (2) a decrease in the surface elevation of the lake
  (3) an increase in the rate of capillarity
  (4) an increase in the amount of runoff

17. The processes of transpiration and evaporation are represented by letters
  (1) *A* and *B*
  (2) *B* and *E*
  (3) *C* and *A*
  (4) *D* and *E*

Base your answers to questions 18 through 21 on the diagram below, which shows four tubes containing 500 milliliters of sediment labeled *A*, *B*, *C*, and *D*. Each tube contains well-sorted, loosely packed particles of uniform shape and size and is open at the top. The classification of the sediment in each tube is labeled.

(Not drawn to scale)

18. Water will be able to infiltrate each of these sediment samples if the sediment is
  (1) saturated and impermeable
  (2) saturated and permeable
  (3) unsaturated and impermeable
  (4) unsaturated and permeable

19. Water was poured into each tube of sediment. The time it took for the water to infiltrate to the bottom was recorded, in seconds. Which data table best represents the recorded results?

Tubes	Infiltration Time (s)
A	5.2
B	3.4
C	2.8
D	2.3

(1)

Tubes	Infiltration Time (s)
A	2.4
B	2.9
C	3.6
D	3.8

(3)

Tubes	Infiltration Time (s)
A	3.2
B	3.3
C	3.2
D	3.3

(2)

Tubes	Infiltration Time (s)
A	3.0
B	5.8
C	6.1
D	2.8

(4)

20. Each tube is filled with water to the top of the sediments, and the tube is covered with a fine screen. The tubes are then tipped upside down so the water can drain. In which tube would the sediment retain the most water?
    (1) A
    (2) B
    (3) C
    (4) D

21. Which graph best represents the relationship between soil particle size and the rate at which water infiltrates permeable soil?

Base your answers to questions 22 and 23 on the cross section below, which represents part of Earth's water cycle. Letters A, B, C, and D represent processes that occur during the cycle. The level of the water table and the extent of the zone of saturation are shown.

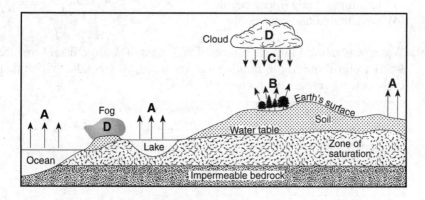

22. Which two letters represent processes in the water cycle that usually cause a lowering of the water table?
    (1) A and B
    (2) A and C
    (3) B and D
    (4) C and D

23. What are two water cycle processes not represented by arrows in this cross section?
    (1) transpiration and condensation
    (2) evaporation and melting
    (3) precipitation and freezing
    (4) runoff and infiltration

24. Which diagram best illustrates the condition of soil below the water table?

25. Which processes are most likely to cause a rise in the water table?
(1) runoff and erosion
(2) precipitation and infiltration
(3) deposition and burial
(4) solidification and condensation

26. During a rainstorm, when soil becomes saturated, the amount of infiltration
(1) decreases and runoff decreases
(2) decreases and runoff increases
(3) increases and runoff decreases
(4) increases and runoff increases

CONSTRUCTED RESPONSE QUESTIONS

Base your answers to questions 27 through 29 on the diagram below, which shows the changes in the molecular structure of water that occur as it changes temperature.

27. According to the diagram, in which phase would water have the greatest density? Explain your answer. [1]

28. What would happen to the spacing between molecules of water as a gas if the temperature was increased to 300°C? [1]

29. Describe how the volume of a particular sample of water changes as the sample changes from water into ice. [1]

Base your answers to questions 30 through 32 on the diagram below and on your knowledge of Earth science. The diagram represents a portion of a stream and its surrounding bedrock. The arrows represent the movement of water molecules by the processes of the water cycle. The water table is indicated by a dashed line. Letter *A* represents a water cycle process occurring at a specific location. Letter *d* represents the distance between the water table and the land surface.

30. Identify water cycle process *A*, which produces cloud droplets. [1]

31. Describe the soil permeability and the land surface slope that allow the most infiltration of rainwater and the least runoff. [1]

32. Explain why the distance, *d*, from the water table to the land surface would decrease after several days of heavy rainfall. [1]

Base your answers to questions 33 through 35 on the block diagram below, which represents a house in New York State with a well that supplies water for people. A truck is spreading salt near a gasoline station to melt the snow on the road. Two soil zones are labeled on the diagram.

(Not drawn to scale)

33. Place an **X** on the block diagram to indicate the location of the water table. [1]

34. Identify one process that occurred in rising, moist air that caused the clouds to form at this location. [1]

35. Explain why, in winter, most of the meltwater produced by salting the road will not infiltrate the soil. [1]

EXTENDED CONSTRUCTED RESPONSE QUESTIONS

Base your answers to questions 36 and 37 on the diagram below, which shows some processes in the water cycle.

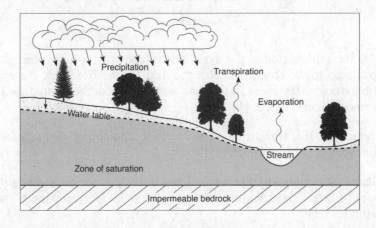

36. State the relationship between the amount of precipitation in this area and the height of the water table above the impermeable bedrock. [1]

37. Describe *one* change that would cause more water to evaporate from this stream. [1]

Base your answers to questions 38 through 40 on the data table below. Six identical cylinders, *A* through *F*, were filled with equal volumes of sorted spherical particles. The data table shows the particle diameters, in centimeters, and the amount of time, in seconds, for water to flow equal distances through each cylinder.

Data Table

Cylinder	Particle Diameter (cm)	Flow Time (s)
A	0.07	51
B	0.08	39
C	0.10	25
D	0.14	13
E	0.16	10
F	0.18	8

38. Use the information in the data table to construct a line graph. On the grid next to the data table, plot the data for the flow time for each of the particle sizes given in the data table. Connect the plotted data with a smooth, curved line. [1]

39. Determine the flow time in a cylinder containing particles with a diameter of 0.13 centimeter. [1]

40. State *one* reason why the water flows faster through the cylinders containing larger particles than through the cylinders containing smaller particles. [1]

CHAPTER 12

THE ORIGIN AND HISTORY OF LIFE ON EARTH

KEY IDEAS All of Earth's present-day life forms have evolved from common ancestors reaching back about three billion years to the simplest one-celled organisms. Before that time, simple molecules may have formed complex organic molecules that slowly evolved into cells capable of replicating themselves. The history of life on Earth has been pieced together from geological, anatomical, and molecular evidence. The geological evidence has come mainly from the sequence of changes seen in fossils found in successive layers of rock that have formed over more than a billion years. Human existence has been very brief compared to the expanse of geologic time.

Geologic history can be reconstructed by observing sequences of rock types and fossils to correlate bedrock at various locations. The characteristics of rocks indicate the processes by which they formed and the environments in which these processes took place. Fossils preserved in rocks provide information about past environmental conditions.

Geologists have divided Earth history into time units based upon the fossil record. Age relationships among bodies of rocks can be determined using principles of original horizontality, superposition, inclusions, cross-cutting relationships, contact metamorphism, and unconformities. The presence of volcanic ash layers, index fossils, and meteoritic debris can provide additional information. The regular rate of nuclear decay (half-life time period) of radioactive isotopes allows geologists to determine the absolute ages of materials found in some rocks.

KEY OBJECTIVES
Upon completion of this chapter, you will be able to:

- Identify the characteristics of life and two ideas about how it may have originated.
- Explain how a study of the fossil record shows that life-forms have evolved through geologic time.

- Describe how fossils provide evidence of past environments.
- Determine the relative ages of a series of rock layers and any igneous intrusions, faults, folds, or fossils they may contain, based upon the principle of superposition.
- Explain the significance of index fossils and volcanic ash deposits in correlating widely separated rock layers.
- Interpret the geologic time scale.
- Determine the absolute age of a rock, given the relative amounts of a radioisotope and decay product in the rock, and the half-life of the radioisotope.

THE GEOLOGIC RECORD OF LIFE'S HISTORY

Until the nineteenth century, nearly everyone believed that Earth was only a few thousand years old. Scientific evidence, though, indicates that some of the rocks on Earth's surface are several *billion* years old. Observations of patterns in rock layers and the location of various kinds of fossils allow inferences concerning the relative ages of rocks and the events that formed them. The absolute age of a rock can be determined from the relative amounts of a radioisotope and its decay products in the rock. The rock and fossil record reveals that life-forms originated early in Earth's history and that they and the environments in which they lived have changed over time.

How Can the Order in Which Geological Events Occurred Be Determined?

Much of what is known about Earth's history, and that of life, has been learned by studying rocks. Rock layers contain traces of events that occurred in the past and provide clues to the origins of the layers and the environments in which they formed. With careful logic, inferences can be drawn about Earth's past from data obtained from rocks. For example, the presence in a rock layer of sorted sand grains that are rounded implies that the grains were carried by running water. Also, the size of the grains may indicate how fast the water that deposited them was moving. If a rock layer contains shells, the type of shell may show that the rock formed in a lake, not an ocean. Shells can also provide clues to the depth and temperature of the water. Microscopic pollen grains in the rock could identify plants living around the lake when the rock formed. Still other rock layers may contain evidence of glaciers, deserts, or volcanic eruptions.

Such inferences are based upon an assumption proposed by James Hutton in 1795—uniformitarianism. **Uniformitarianism** is the idea that Earth's features, such as mountains, valleys, and rock layers, have formed gradually by processes still underway, not by instantaneous creation. It assumes that Earth has always behaved much as it does now; that water, for example, ran downhill in streams carrying sand and silt 10 million years ago just as it does today. This idea can be summarized in the statement "The present is the key to the past."

According to uniformitarianism, every rock layer contains a record of a short part of Earth's long history. By working out the age of each layer in a series, geologists can determine the order in which the layers formed and can arrange the events they represent in the order in which these events occurred. In this way a history of Earth can be pieced together.

There are two ways of expressing the ages of Earth materials or geologic events, relative age and actual age. **Relative age** is the age of a rock, fossil, geologic feature, or event relative to that of another rock, fossil, geologic feature, or event. Thus, relative age is not an exact age in years; it tells you only that one thing is older or younger than another. If you say that you are older than a first grader but younger than your parents, you are giving your relative age.

Absolute age is the age of a rock, fossil, geologic feature, or event given in units of time, usually years. If you state, "I am 15 years old," you are giving your actual age. The absolute age of Earth materials and events is usually determined by analyzing radioisotopes in objects.

Determining Relative Age

Two key ideas in determining relative age are the law of original horizontality and the principle of superposition. The **law of original horizontality** states that sediments deposited in water form flat, level layers parallel to Earth's surface.

The **principle of superposition** is the concept that the bottom layer of a series of sedimentary layers is the oldest, unless it has been overturned or older rock has been thrust over it. Each layer forms atop the one that is already there. (Layers of sediment do not form in midair, nor do they hover halfway between the ocean surface and the ocean bottom!) Therefore, each layer is younger than the one under it and older than the one on top of it. The relative ages of any two layers can be determined based upon which layer is on top and which is on the bottom. See Figure 12.1.

Figure 12.1 Series of Rock Layers Showing Oldest to Youngest.

The principle of superposition must be used with care, however, because there are events that can disturb the positions of rock layers. Forces within Earth may tilt, fold, or fault the layers. Older layers may be pushed on top of younger ones. In such cases, one must work out the original positions of the rock layers before applying the principle of superposition.

In general, a rock layer is older than any joint, fault, or fold that appears in it; the rock had to already exist in order to be folded or faulted. By unfolding or unfaulting the rock layers, one can determine the positions of the layers before they were disturbed, and can then apply the principle of superposition to determine their relative ages.

For example, let's look at Figure 12.2. If we were to judge only by the position of the rock layers in the core sample at (3), we would conclude that layer C is the youngest and A is the oldest because C is on top and A is on the bottom. If we look at the entire rock structure, however, we see that the three layers of rock have been folded and overturned. By unfolding the layers, we place the layers in their original position and see that layer A is the *youngest* while C is the oldest.

Figure 12.2 The Original Position of Rock Layers Must Be Determined Before Applying the Principle of Superposition. Originally horizontal rock layers (1), may be folded by compression (2), causing older rock layers to appear above younger rock layers (3). Source: *Macmillan Earth Science*, Eric Danielson and Edward J. Denecke Jr., Macmillan Publishing Co., 1989.

The same is true of faults, that is, fractures along which the rocks on either side have moved. If rock layers fracture and then move along that fracture, they are displaced. Any rock displaced by a fault is older than the fault. During faulting, underlying rock layers may be pushed up so that they are found on top of younger rock. See Figure 12.3. Again, to obtain true relative ages, one must work backward to the position the rock layers were in before the fault offset them.

Igneous intrusions or extrusions are often found associated with other types of rock. Igneous intrusions form when molten rock forces its way into preexisting rock, cools, and hardens. Thus, an intrusion is younger than any rock it cuts through. See Figure 12.4. Extrusions form during volcanic eruptions when molten rock flows out onto Earth's surface as lava and hardens,

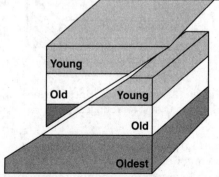

Figure 12.3 Fault, Showing Older Rocks Pushed Atop Younger Ones.

Figure 12.4 Five Rock Layers Before and After Intrusion of Magma. Layer 0, which formed from the cooling magma, is younger than the five rock layers into which it was intruded.

or is blown into the atmosphere and settles on the ground, forming a blanket of volcanic rock particles. Thus, extrusions are younger than the rocks beneath them, but older than any layer that may form over them. Therefore, if an igneous body is found in rock, one must first determine whether it is an intrusion or an extrusion before the relative ages of the layers can be established.

Unconformities

Layers of rock are generally deposited in an unbroken sequence. However, if forces within Earth uplift rocks, deposition ceases. Erosion may wear away many layers of rock before the land is low enough for another layer to be deposited. The result is an **unconformity**, a break or gap in the sequence of a series of rock layers. Thus, the rocks above an unconformity are quite a bit younger than those below it.

There are three types of unconformity. See Figure 12.5. **Angular unconformities** form when rock layers are tilted or folded before being eroded. When new layers are deposited, they form horizontally and the layers below the unconformity are at an angle to those above it. **Disconformities** are irregular erosional surfaces between parallel layers of rock. Disconformities occur when deposition stops and layers are eroded, but no tilting or folding occurs. These surfaces are not easy to discern and are often found when fossils of very different ages are discovered in adjacent layers. **Nonconformities** are places where sedimentary layers lie on top of igneous or metamorphic rocks.

How Can Rocks and Geologic Events in One Place Be Matched with Those in Other Locations?

Correlation

With care and logic the relative ages of rock layers in any particular location can be worked out. However, this information provides only a small part of the overall picture of Earth's history, and unconformities may mean that even more details are missing. To broaden the scope of the picture, rocks in one place need to be correlated with rocks in other locations. The process of **correlation** involves determining that rock layers in different areas are the same age. In this way, rocks in one location may fill in gaps in the record in another location. Correlation also allows the relative ages of rocks in widely separated outcrops to be determined.

Walking the Outcrop

Rock layers can sometimes be followed from one location to another by walking the outcrop. See Figure 12.6. *Walking the outcrop* means physically tracing the layers from one place to another. In this way, rock layers in two different places, such as two adjacent mountains or ridges, can sometimes be correlated. Unfortunately, rock layers are rarely continuously visible for any distance. Most rocks are hidden beneath regolith, the loose fragments of rock and soil that blanket Earth's surface. Rock outcrops poke out here and there like islands. As a result, geologists are often faced with the task of correlating rocks in widely separated outcrops.

Figure 12.5 Angular Unconformity, Disconformity, and Nonconformity.

Figure 12.6 Walking the Outcrop.

Matching Physical Characteristics

Rocks can sometimes be correlated on the basis of distinct similarities in physical characteristics such as composition, color, thickness, and fossil remains. Distinctive rocks or sequences of rock types can also be used to correlate rocks.

For example, Alfred Wegener used this type of correlation as part of his evidence for continental drift. He stated: "Igneous rocks in Brazil and Africa [have] no less than five parallels: 1) the older granite, 2) the younger granite, 3) alkali-rich rocks, 4) volcanic Jurassic rock and intrusive dolerite, and 5) kimberlite, alnoite, etc. . . . The last of the rock groups (kimberlite, alnoite) is best known since both in Brazil and South Africa the beds yield the famous diamond finds. In both these regions the peculiar type of stratification known as 'pipes' occurs. There are white diamonds in Brazil in Minas Geraes State and in South Africa north of the Orange River only."

This type of correlation must be done with care, though, because different formations can look almost identical.

GRAPTOLITES
395-450 million years ago

Didymograptus

Monograptus

Ammonoids
144-208 million years ago

x½

Manticoceros (ammonoid)

TRILOBITES
500-600 million years ago

Serrodiscus (agnostic)

X2

Elliptocephala (otenellid)

Cryptolithus X2

Phacops

Phacops (enrolled)

Figure 12.7 Graptolites, Trilobites, and Ammonoids Labeled with Ages. Source: The New York State Museum. Educational Leaflet #28, *Geology of New York, A Simplified Account*, Second Edition. Printed with permission of New York State Museum, Albany, N.Y.

Index Fossils

Index fossils are remains of organisms that had distinctive body features, were abundant, and had a broad, even worldwide range, yet existed only for a short period of time. The best index fossils include swimming or floating organisms that evolved rapidly and were distributed widely, such as graptolites, trilobites, and ammonoids. See Figure 12.7. Their distinctive body shapes and broad distribution make index fossils easy to find in widely separated rock layers (see Figure 12.8). Also, their short existence pinpoints the time period during which those rock layers were formed.

Figure 12.8 Correlation by Fossils of Strata in Several Locations.

Key Beds

Key beds are well-defined, easily identifiable layers or formations that have distinctive characteristics or fossil content that allows them to be used in correlation. Key beds can be readily identified and are not easily confused with other layers. Materials such as volcanic ash layers that are rapidly deposited over a wide area make excellent key beds.

One well-known key bed is the iridium-rich layer of rock discovered in Italy, Denmark, and many other places around the world. Iridium is an extremely rare element on Earth, but not in meteorites. The iridium-rich layer is thought to have formed 66 million years ago when a meteorite impact created a dust cloud that encircled Earth and then slowly settled to the ground.

How Can Earth's Geologic History Be Sequenced from the Fossil and Rock Record?

The Geologic Column

Even though the rock record in any one place is incomplete, correlation has made it possible to piece together a fairly complete record of Earth's geologic history by examining rocks in different locations. By the nineteenth century, geologists had correlated rocks worldwide into a single sequence called the **geologic column**. Rocks are still being added to the geologic column as more outcrops are mapped and described.

The Geologic Time Scale

Geologists then divided Earth's history into a sequence of time units called the **geologic time scale**. *Geologic time* is the entire time that Earth has existed. The geologic time scale divides geologic time into units and subunits based upon the fossil record. Geologic time is divided into eons, eons are subdivided into eras, eras are subdivided into periods, and periods are subdivided into epochs. At first, these time units were based only upon the relative ages of fossils in rocks of the geologic column, but radioisotope dating has enabled geologists to assign actual ages to the time units of the geologic time scale as shown in Figure 12.9.

Eons

Eons are the largest time units on the geologic time scale. The oldest is the *Archean* (Greek "ancient"); it covers the time from Earth's formation to the appearance of multicelled organisms. Archean rocks are the oldest known on Earth and contain microscopic fossils of single-celled, bacteria-like organisms. The *Proterozoic* (Greek "earlier life") is the next oldest eon; its rocks contain traces of multicelled organisms that had no preservable hard parts. The most recent eon is the *Phanerozoic* (Greek "visible life"), which provides an abundant fossil record of preserved hard parts.

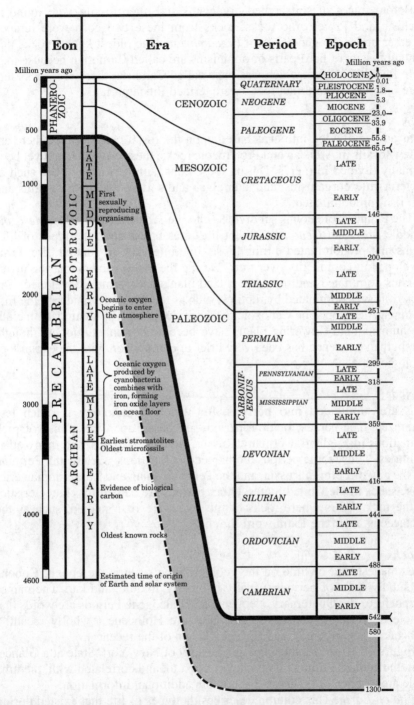

Figure 12.9 The Geologic Time Scale. Source: The State Education Department, *Earth Science Reference Tables*, 2011 ed. (Albany, New York; The University of the State of New York).

Before traces of soft-bodied organisms and microfossils were found in Archean and Proterozoic rocks, rocks from these two eons were lumped together under the general term *Precambrian*. The oldest known rocks that contain fossils of hard parts of organisms are called Cambrian because they were found in Wales, which was once called Cambria; therefore, rocks that were older than Cambrian rocks were termed Precambrian.

Eras

Eons are subdivided into **eras** based upon the fossil record. Since there are so few fossils in Archean and Proterozoic rocks, these eons have not yet been formally divided into eras. During these eons, simple forms of life such as bacteria, microorganisms, and later algae and soft animals (e.g., worms and jellyfish) predominated.

The Phanerozoic eon is subdivided into the *Paleozoic* (old life), *Mesozoic* (middle life), and *Cenozoic* (recent life) eras based upon the types of life-forms that predominated during those time intervals. In general, life-forms developed in complexity over time. During the Paleozoic era, marine invertebrates dominated and the earliest land plants and animals developed. The Mesozoic was dominated by reptiles, such as the dinosaurs, and saw the first mammals develop. The Cenozoic has been dominated by mammals, including humans, and flowering plants have become dominant. Notice, though, that human existence has been very brief in comparison with the expanse of geologic time.

Periods

Eras are subdivided into **periods**, for which terminology is much less organized. The names for time periods are based upon the names of rock formations in England, Germany, Russia, the United States, and many other countries. Thus, some periods are named for locations, such as the Permian for the city of Perm in Russia, and the Pennsylvanian and Mississippian after those states in the United States. Other periods are named for characteristics of the rock layers where rocks of this age were first studied, such as the Cretaceous, from the Latin word for chalk.

Epochs

The smallest unit of time on the geologic time scale is the **epoch**. **Epochs** are subdivisions of periods, usually into early, middle, and late. The names of epochs of the Quaternary, Neogene, and Paleogene Periods are words that denote degrees of recentness; examples are Holocene ("wholly recent"), Miocene ("less recent"), and Eocene ("dawn of the recent").

Figure 12.10 shows the Geologic History of New York State at a Glance. In it, the geologic time scale is shown. Next to, and correlated with, the time scale are a series of columns that contain additional information.

Life on Earth. This column describes the types of life that existed during the different time periods based upon the fossil record.

Rock Record in N.Y. This column shows the time periods for which there is a rock and fossil record in New York State. The thick black line indicates that rock or fossil evidence from that time period exists in New York State. If there is no line, there is no rock or fossil record for that time. These gaps reflect unconformities in the rocks of New York State.

Time Distribution of Fossils (Including Important Fossils of New York). This column contains bold black lines labeled with names and circled letters. The names identify types of fossil organisms. The circled letters are keyed to the illustrations of specific index fossils printed along the top of the chart, and are placed on the line to indicate the approximate time of existence of each specific index fossil. For example, find A at the bottom of the line labeled "Trilobites." Now look to the left, and note that organism A lived at the end of the Early epoch of the Cambrian period.

Important Geologic Events in New York. This column describes in more detail how the plate tectonic events shown in the preceding column affected New York State. It describes the origin of well-known features of New York State, such as the Palisades Sill, which was intruded during the late Triassic, and the Adirondacks, which were uplifted during the Pliocene.

Inferred Position of Earth's Landmasses. This column shows the changing positions of landmasses due to plate movements. North America is shown in black to highlight its movements. The latitude and longitude lines show the positions of landmasses in relation to the Equator and poles, as well as to each other. Notice that during the Devonian and Mississippian periods, North America was centered on the Equator, a location that explains how coal beds formed from tropical rain forests.

How Can the Absolute Age of a Rock Be Determined?

Early attempts at determining the age of Earth on the basis of uniformitarianism were inaccurate at best. For example, careful measurement of the accumulation of sediments in a shallow sea over objects of known age, such as shipwrecks, shows that about 15,000 to 30,000 years are needed to form a layer of sediment 1 meter thick. Layers of sedimentary rock exist that are more than 2,000 meters thick. If the sediment accumulated at a rate of 1 meter in 20,000 years, then the layers would have taken 2,000 meters × 20,000 years per meter or *40 million years to form*! If this method is applied to the entire geologic column, an estimate for the age of Earth can be obtained.

Another early method used to estimate Earth's age was based upon the salt content of the oceans. Salt is carried into the oceans by rivers, so scientists measured the amounts of salt in all the oceans. Then they measured the amounts of salt all the world's rivers are carrying into the oceans. Finally they calculated how long the rivers would take to carry in all the salt now in the oceans.

The Origin and History of Life on Earth

The Geologic History of New York State at a Glance Chart

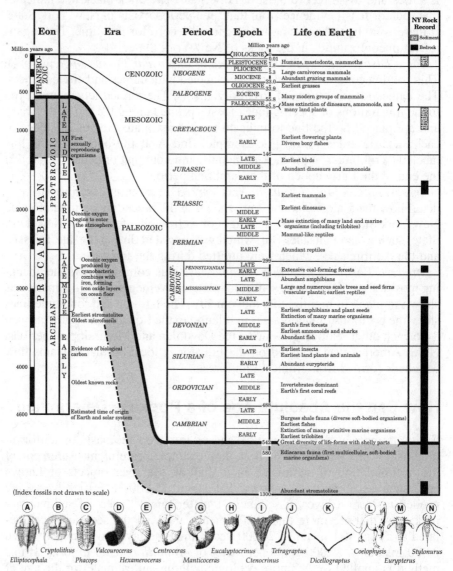

Figure 12.10 Geologic History of New York State. Source: The State Education Department, *Earth Science Reference Tables*, 2011 ed. (Albany, New York; The University of the State of New York).

Based on such attempts, the age of the Earth was estimated at several hundred million years. We now know that this figure is much too low. Dating by those early methods was not reliable because they depended too heavily on rates that vary widely from place to place and through time. What was needed was a way to measure time by a process that does not vary, a process that runs continuously through time and leaves a record with no gaps in it. The discovery of radioactivity by A. H. Becquerel in 1896 provided the needed process.

All elements are made up of tiny bits of matter called *atoms*. Most elements have a number of isotopes. **Isotopes** are varieties of the same element whose atoms differ slightly in mass. For example, most carbon atoms have a

The Origin and History of Life on Earth

mass of twelve units. This isotope is called carbon-12. Some carbon atoms, however, have a mass of fourteen units. This isotope is carbon-14.

The most common isotopes of elements are stable, meaning their atoms do not change. However, some isotopes are unstable. In a process called **radioactive decay**, the atoms of unstable isotopes break apart. During this process the atoms go through a series of changes. They give off energy and some of the small particles that make up their nucleus. Finally, they form a stable isotope of a new element that is not radioactive. This new substance is called the decay product. For example, uranium-238, a radioactive isotope, decays slowly into the stable decay product lead-206.

Radioactive decay takes place at a steady, constant rate. It is not affected by outside factors such as changes in temperature, pressure, or chemical state. The minerals in certain types of rocks contain radioactive isotopes that start to decay when the rock forms. Therefore, the decay process may be used as a clock to determine the actual ages of these rocks.

TABLE 12.1 RADIOACTIVE DECAY DATA

RADIOACTIVE ISOTOPE	DISINTEGRATION	HALF-LIFE (years)
Carbon-14	$^{14}C \rightarrow {}^{14}N$	5.7×10^3
Potassium-40	$^{40}K \nearrow {}^{40}Ar \searrow {}^{40}Ca$	1.3×10^9
Uranium-238	$^{238}U \rightarrow {}^{206}Pb$	4.5×10^9
Rubidium-87	$^{87}Rb \rightarrow {}^{87}Sr$	4.9×10^{10}

Source: The State Education Department, *Earth Science Reference Tables*, 2011 ed. (Albany, New York; The University of the State of New York).

The decay of a radioactive isotope occurs at a statistically predictable decay rate known as its half-life. The **half-life** of a radioisotope is the time required for one-half of the unstable radioisotope to change into a stable decay product. Table 12.1 lists the half-lives and decay products of four commonly used radioisotopes.

If the half-life of a radioisotope is known, the age of a rock can be determined from the relative amounts of the radioisotope and its decay product in the rock. Every half-life, the percentage of decay product will increase and the percentage of radioisotope will decrease. Thus, the ratio of radioisotope to decay product indicates how many half-lives have gone by. This information, in turn, reveals how many years the decay process has been going on. In this manner, geologists are able to find the absolute ages of rocks. Figure 12.11 shows how the relative amounts of radioisotope and its decay product change as each half-life elapses.

Example:

A rock contains 50 grams of potassium-40 and 50 grams of its decay product argon-40. How old is the rock?

Since the entire decay product was originally radioactive isotope, the rock originally contained 100 grams of potassium-40 (the 50 g of the radioisotope that still exist plus the 50 g that decayed into argon-40). Thus, exactly one-half of the original radioisotope has decayed; the rock is one half-life old. From Table 13.1 we know that the half-life of potassium-40 is 1.3×10^9 years, or 1.3 billion years.

The rock is 1.3 billion years old.

There are two reasons why a certain isotope may be used to date a rock. It may be used because it is present in the minerals of that rock, or it may be used because its half-life is long or short. Isotopes with long half-lives are used to date very old rocks; those with short half-lives, to date younger objects. For example, potassium-40 can be used to date rocks between 100,000 and 4.6 billion years old. On the other hand, carbon-14 is best for dating objects between 100 and 50,000 years old. Table 12.2 gives data for six radioisotopes used in dating.

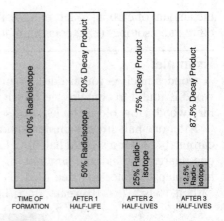

Figure 12.11 Ratios of Isotope and Decay Product after One, Two, and Three Half-Lives.

TABLE 12.2 SIX RADIOISOTOPES USED IN DATING*

Isotopes				
Parent	Decay Product	Half-life (years)	Dating Range (years)	Minerals or Other Materials That Can Be Dated
Uranium-238	Lead-206	4.5 billion	10 million–4.6 billion	Zircon
Uranium-235	Lead-207	710 billion	10 million–4.6 billion	Uraninite
Uranium-232	Lead-208	14 billion	10 million–4.6 billion	
Potassium-40	Argon-40 Calcium-40	1.3 billion	50,000–4.6 billion	Muscovite, biotite, hornblende
Rubidium-87	Strontium-87	47 billion	10 million–4.6 billion	Muscovite, biotite, potassium feldspar
Carbon-14	Nitrogen-14	5,730 ± 30	100–70,000	Wood, peat, grain, charcoal, bone, tissue, cloth, shell, stalactites, glacier ice, ocean water

Carbon-14 is especially useful because it can be used to date the remains of living things. All living things contain carbon, and some of that carbon is carbon-14, which is present in air, water, and food. As long as an organism is alive, the amount of carbon-14 in its body remains constant. Whatever decays is quickly replaced from its surroundings. However, when the organism dies, the carbon-14 that decays is not replaced. The longer the organism has been dead, the less carbon-14 remains. Carbon-14 can be used to date fossils in which the original material remains unchanged, such as wood, bones, and shells. However, this radioisotope has a short half-life. After about 50,000 years

The Origin and History of Life on Earth

the amount of carbon-14 left in a fossil is too small to be measured. Therefore, other radioisotopes (e.g., potassium-40) must be used to date older rocks.

Example:

A wood branch is found buried beneath layers of sediment in a lake. The wood is found to contain one-fourth of the carbon-14 present in contemporary wood. How old is the layer of sediment in which the wood is found?

After one half-life, a sample would contain one-half the normal amount of carbon-14. After a second half-life, half of that would have decayed, leaving one-quarter the normal amount of carbon-14. Therefore, two half-lives have elapsed since the tree died. Table 12.1 indicates that the half-life of carbon-14 is 5.7×10^3 years, or 5,700 years.

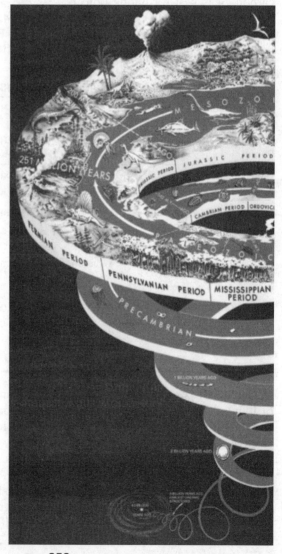

Figure 12.12 Geologic Time Scale on Earth.

Since two half-lives have elapsed, the wood and therefore the layer of sediment are 5,700 × 2 or 11,400 years old.

How Can the Rock Record and Fossil Evidence Reveal Changes in Past Life and Ancient Environments?

Many thousands of layers of sedimentary rock contain evidence of the long history of Earth and the changing life-forms whose fossils are found in the rocks. Fossils show us that a great variety of plants and animals lived on Earth in the past. Fossils can be compared to one another and to living organisms by their similarities and differences. Some fossil organisms are similar to existing organisms, but many are quite different. A study of the fossil record reveals that more recently deposited rock layers are more likely to contain fossils resembling existing life-forms and that most life-forms of the geologic past have become extinct.

Evolution

The history of life on Earth depicted by the fossil record is one of change. The fossil record shows that, since its beginning, many new life-forms have appeared and most old forms have disappeared. It reveals also that life-forms have changed gradually over time, or evolved, so that the current life-forms differ significantly from the earliest ones.

Scientists have traced many sequences (inferred from ages of rock layers) of anatomical forms over time and have observed the accumulation of differences from one generation to the next. These researchers are convinced that **biological evolution**, that is, the development of existing life-forms from earlier, different ones, is what has led to species as different from one another as algae are from whales.

Natural Selection

The theory of natural selection provides a scientific explanation for the evolution of life-forms seen in the fossil record. **Natural selection** proposes the following mechanism for evolution:

1. Individual organisms of the same species have different characteristics, and sometimes these differences give one organism an advantage in surviving and reproducing.
2. Offspring that inherit the advantage are more likely to survive and reproduce.
3. Over time the proportion of individuals with the advantageous characteristic will increase.

In this way, natural selection leads to organisms that are well suited to their environments. Changes in the environment can affect the survival value of some inherited characteristics. Then the characteristics that are advantageous for survival may change, so changes in the environment can lead to changes in organisms. Small differences between parents and offspring can accumulate over many generations so that descendants may become very different from their ancestors.

Evolution by natural selection does not imply long-term progress toward a goal, or have a set direction. Nor does evolution always occur gradually. Much recent information indicates that evolution occurs in spurts brought on by sudden, large-scale changes in the environment. A good analogy for evolution is the growth of a hedge. Some branches exist from the beginning of the hedge's life with little or no change. Other branches grow and then die out. Still others grow a little, and then branch apart repeatedly. The end result is a complex network of large and small branches.

On the basis of the fossil record, life on Earth is thought to have begun about 4 billion years ago as simple, one-celled organisms. During the first 2 billion years, only these one-celled organisms existed. About 1 billion years ago, cells with nuclei began to develop. Since then, increasingly complex multicellular organisms have evolved.

One important reason to preserve life on Earth is that evolution builds on what exists. The more variety there is at the present time, the even greater variety can exist in the future. Human behavior that results in the extinction of organisms decreases the variety of life-forms on Earth and may have far-reaching effects in the future.

Ancient Environments

Rocks in which fossils are found provide evidence of ancient environments. Many fossils are clear indicators of particular types of environment. For example, fossil coral that is almost identical to existing corals can be found in some rocks. Corals that exist today can live only in shallow, tropical seas. Therefore, we can infer that rocks containing fossil coral existed in a

Figure 12.13 Fossil-Bearing Rock Layers with Illustrations of Environments Inferred from Fossil Evidence.

similar environment. Oak and birch trees inhabit moist, temperate climates. Abundant oak and birch pollen in a sedimentary layer implies that, when the rock layer formed, the climate was moist and temperate. The fossils in coal are the remains of tropical rain forest plants and animals, so layers of coal imply a tropical rain forest origin. Inferences of this type, together with the inferences that can be drawn from the composition of a given sedimentary layer, enable geologists to piece together a model of the environment that existed when that particular rock layer formed. See Figure 12.13.

MULTIPLE-CHOICE QUESTIONS

In each case, write the number of the word or expression that best answers the question or completes the statement.

1. It is difficult to state with certainty the way in which life on Earth originated because
   (1) Earth's original life-forms left no fossil record
   (2) no rocks date back to the time when life originated
   (3) the age of rocks formed at the time when life originated cannot be determined
   (4) the process of radioactive dating destroys the fossils found in rocks

2. Earth's age is most accurately determined by the
   (1) thickness of sedimentary strata
   (2) salinity of the oceans
   (3) study of fossils
   (4) radioactive dating of rock masses

3. Evidence suggests that the geologic processes of the past
   (1) were similar to those of the present
   (2) were different from those of the present
   (3) occurred at a faster rate than those of the present
   (4) occurred at a slower rate than those of the present

4. Much of the evidence for the evolution of life-forms on Earth has been obtained by
   (1) studying the life spans of present-day animals
   (2) radioactive dating of metamorphic rock
   (3) correlating widespread igneous ash deposits
   (4) examining fossils preserved in the rock record

5. Which scientific principle states that younger rock layers are generally deposited on top of older rock layers?
   (1) superposition
   (2) evolution
   (3) original horizontality
   (4) inclusion

6. The bedrock cross section below contains rock formations *A*, *B*, *C*, and *D*. The rock formations have *not* been overturned.

   Which sequence represents the relative ages of these rock formations, from oldest to youngest?
   (1) $B \rightarrow A \rightarrow C \rightarrow D$
   (2) $B \rightarrow D \rightarrow C \rightarrow A$
   (3) $D \rightarrow C \rightarrow A \rightarrow B$
   (4) $D \rightarrow B \rightarrow A \rightarrow C$

7. Older layers of rock may be found on top of younger layers of rock as a result of
   (1) weathering processes
   (2) igneous extrusions
   (3) joints in the rock layers
   (4) overturning of rock layers

8. Which feature in a rock layer is older than the rock layer?
   (1) igneous intrusions
   (2) mineral veins
   (3) rock fragments
   (4) faults

9. Unconformities (buried erosional surfaces) are good evidence that
   (1) many life-forms have become extinct
   (2) the earliest life-forms lived in the sea
   (3) part of the geologic rock record is missing
   (4) metamorphic rocks have formed from sedimentary rocks

10. The presence of brachiopod, nautiloid, and coral fossils in the surface bedrock of a certain area indicates the area was once covered by
    (1) tropical vegetation
    (2) glacial deposits
    (3) volcanic ash
    (4) ocean water

11. Volcanic ash deposits found in the geologic record are most useful in correlating the age of rock layers if the volcanic ash was distributed over a
    (1) large area during a short period of time
    (2) large area during a long period of time
    (3) small area during a short period of time
    (4) small area during a long period of time

Base your answers to questions 12 through 14 on the geologic cross section below in which overturning has not occurred. Letters *A* through *H* represent rock layers.

12. Which sequence of events most likely caused the unconformity shown at the bottom of rock layer *B*?
    (1) folding → uplift → erosion → deposition
    (2) intrusion → erosion → folding → uplift
    (3) erosion → folding → deposition → intrusion
    (4) deposition → uplift → erosion → folding

13. The folding of rock layers *G* through *C* was most likely caused by
    (1) erosion of overlying sediments
    (2) contact metamorphism
    (3) the collision of lithospheric plates
    (4) the extrusion of igneous rock

14. Which two letters represent bedrock of the same age?
    (1) *A* and *E*
    (2) *B* and *D*
    (3) *F* and *G*
    (4) *D* and *H*

15. The best method for the correlation of sedimentary rock layers several hundred kilometers apart is by comparing the
    (1) index fossils in the layers
    (2) layers by walking the outcrop
    (3) thickness of the rock layers
    (4) color of the rock layers

16. Which characteristic is most useful in correlating Devonian-age sedimentary bedrock in New York State with Devonian-age sedimentary bedrock in other parts of the world?
    (1) color
    (2) index fossils
    (3) rock types
    (4) particle size

Base your answers to questions 17 through 20 on the cross sections of three rock outcrops, *A*, *B*, and *C*. Line *XY* represents a fault. Overturning has not occurred in the rock outcrops.

17. The volcanic ash layer is considered a good time marker for correlating rocks because the volcanic ash layer
    (1) has a dark color
    (2) can be dated using carbon-14
    (3) lacks fossils
    (4) was rapidly deposited over a wide area

18. Which sedimentary rock shown in the outcrops is the youngest?
    (1) black shale   (3) tan siltstone
    (2) conglomerate   (4) brown sandstone

19. What is the youngest geologic feature in the three bottom layers of outcrop *C*?
    (1) fault   (3) unconformity
    (2) igneous intrusion (4) zone of contact metamorphism

20. Which processes were primarily responsible for the formation of most of the rock in outcrop *A*?
    (1) melting and solidification
    (2) heating and compression
    (3) compaction and cementation
    (4) weathering and erosion

21. The division of Earth's geologic history into units of time called eons, eras, periods, and epochs is based on
    (1) absolute dating techniques   (3) climatic changes
    (2) fossil evidence   (4) seismic data

22. The diagram below represents three bedrock outcrops. The layers have *not* been overturned. Letters *A* through *E* identify different rock layers. Fossils found in the rock layers are shown.

Which fossil could be classified as an index fossil?

Base your answers to questions 23 through 26 on the cross sections below, which represent two bedrock outcrops 15 kilometers apart. The rock layers have been numbered for identification, and some contain the index fossil remains shown.

23. When these rocks were deposited as sediments, this area was most likely
    (1) under the ocean
    (2) a desert between high mountains
    (3) repeatedly covered by lava flows
    (4) glaciated several times

24. Both organisms that formed the fossils found in rock layers 3 and 4
    (1) lived during the same period of geologic time
    (2) lived in polar regions
    (3) are members of the same group of organisms
    (4) are still alive today

25. Evidence best indicates that rock layers 4 and 8 were deposited during the same geologic period because both layers
    (1) contain the same index fossil
    (2) are composed of glacial sediments
    (3) contain index fossils of the same age
    (4) are found in the same area

26. Which two types of organisms survived the mass extinction that occurred at the end of the Permian Period?
    (1) trilobites and nautiloids
    (2) corals and vascular plants
    (3) placoderm fish and graptolites
    (4) gastropods and eurypterids

27. One reason *Tetragraptus* is considered a good index fossil is that *Tetragraptus*
  (1) existed during a large part of the Paleozoic Era
  (2) has no living relatives found on Earth today
  (3) existed over a wide geographic area
  (4) has been found in New York State

28. According to the fossil record, which group of organisms has existed for the greatest length of time?
  (1) gastropods   (3) mammals
  (2) corals   (4) vascular plants

Base your answers to questions 29 through 32 on the geologic cross section below and on your knowledge of Earth science. The cross section represents rock and sediment layers, labeled *A* through *F*. Each layer contains fossil remains, which formed in different depositional environments. Some layers contain index fossils. The layers have *not* been overturned.

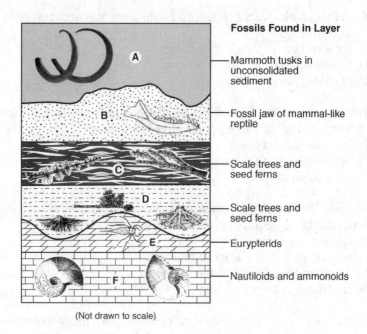

(Not drawn to scale)

29. Which pair of organisms existed when the unconsolidated sediment in layer *A* was deposited?
  (1) birds and trilobites
  (2) dinosaurs and mastodonts
  (3) ammonoids and grasses
  (4) humans and vascular plants

30. Which rock layer formed mainly from the compaction of plant remains?
    (1) *E*   (3) *C*
    (2) *B*   (4) *F*

31. During which geologic epoch was layer *F* deposited?
    (1) Late Devonian   (3) Early Devonian
    (2) Middle Devonian   (4) Late Silurian

32. The depositional environment during the time these layers and fossils were deposited
    (1) was consistently marine
    (2) was consistently terrestrial (land)
    (3) changed from marine to terrestrial (land)
    (4) changed from terrestrial (land) to marine

33. The graph below shows the radioactive decay of a 50-gram sample of a radioactive isotope.

    According to the graph, what is the half-life of this isotope?
    (1) 100 years   (3) 200 years
    (2) 150 years   (4) 300 years

34. Radioactive decay of ^{40}K atoms in an igneous rock has resulted in a ratio of 25 percent ^{40}K atoms to 75 percent ^{40}Ar and ^{40}Ca atoms. How many years old is this rock?
    (1) 0.3×10^9 y   (3) 2.6×10^9 y
    (2) 1.3×10^9 y   (4) 3.9×10^9 y

35. The table below shows the radioactive decay of carbon-14. Part of the table has been left blank.

Half-Life	Original Carbon-14 Remaining (%)	Number of Years
0	100	0
1	50	5,700
2	25	11,400
3		17,100
4		
5		

After 22,800 years, approximately what percentage of the original carbon-14 remains?
(1) 15%
(2) 12.5%
(3) 6.25%
(4) 3.125%

Base your answers to questions 36 and 37 on the graph below, which shows the generalized rate of decay of radioactive isotopes over 5 half-lives.

36. If the original mass of a radioactive isotope was 24 grams, how many grams would remain after 3 half-lives?
(1) 12
(2) 24
(3) 3
(4) 6

37. Which radioactive isotope takes the greatest amount of time to undergo the change shown on the graph?
(1) carbon-14
(2) potassium-40
(3) uranium-238
(4) rubidium-87

38. An igneous rock contains 10 grams of radioactive potassium-40 and a total of 10 grams of its decay products. During which geologic time interval was this rock most likely formed?
   (1) Middle Archean
   (2) Late Archean
   (3) Middle Proterozoic
   (4) Late Proterozoic

39. The graph below shows the extinction rate of organisms on Earth during the last 600 million years. Letters A through D represent mass extinctions.

Which letter indicates when dinosaurs became extinct?
   (1) A
   (2) B
   (3) C
   (4) D

40. According to the fossil record, which sequence correctly represents the evolution of life on Earth?
   (1) fish → amphibians → mammals → soft-bodied organisms
   (2) fish → soft-bodied organisms → mammals → amphibians
   (3) soft-bodied organisms → amphibians → fish → mammals
   (4) soft-bodied organisms → fish → amphibians → mammals

41. Which time line most accurately indicates when this sequence of events in Earth's history occurred?

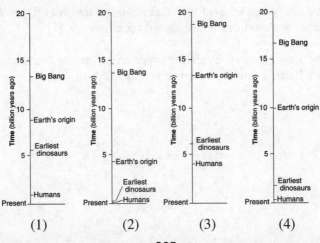

The Origin and History of Life on Earth

CONSTRUCTED RESPONSE QUESTIONS

Base your answers to questions 42 through 44 on the cross section below and on your knowledge of Earth science. The cross section represents a portion of Earth's crust. Letters A, B, C, and D indicate sedimentary rock layers that were originally formed from deposits in a sea. The rock layers have *not* been overturned.

(Not drawn to scale)

42. Geologic events V through Z are listed below.

   V. Metamorphism of some sedimentary rock
   W. Formation of sedimentary rock layers
   X. Tilting and erosion of sedimentary rock layers
   Y. Intrusion/extrusion of igneous rock
   Z. Erosion of some igneous rock

   List the letters V through Z to indicate the correct order of the geologic events, from oldest to youngest, that formed this portion of Earth's crust. [1]

43. Describe *one* piece of evidence that suggests that rock layer C formed in a deeper sea environment than did rock layer A. [1]

44. Describe *one* piece of evidence represented in the cross section that indicates Earth's crust has moved at this location. [1]

Base your answers to questions 45 through 48 on the cross sections below and on your knowledge of Earth science. The cross sections represent three bedrock outcrops, 1, 2, and 3, found several kilometers apart. The geologic time period when each sedimentary rock layer formed is shown. The symbols (☆, ○, X, □, and △) represent fossils of different types of organisms present in the rock layers.

Outcrop 1	
Permian	X
Pennsylvanian	X △
Mississippian	X
Devonian	○ □

Outcrop 2
Devonian ○
Silurian △
Ordovician □
Cambrian ☆

Outcrop 3	
Permian	△ X
Pennsylvanian	△ X
Devonian	○ △
Silurian	△

45. Draw the fossil symbol that represents the best index fossil. Describe *one* piece of evidence shown in the outcrops that indicates that this fossil has characteristics of a good index fossil. [1]

46. Write the outcrop number of the cross section that could be found in New York State. Describe the evidence that supports your answer. [1]

47. Explain why the geologic age of these rock layers could *not* be accurately dated using carbon-14. [1]

48. Explain why the index fossil *Coelophysis* is *not* preserved in any of the rock outcrops. [1]

Base your answers to questions 49 through 52 on the graph below and on your knowledge of Earth science. The graph shows the rate of decay of the radioactive isotope carbon-14 (^{14}C).

49. Complete the flowchart below by filling in the boxes to indicate the percentage of carbon-14 remaining and the time that has passed at the end of each half-life. [1]

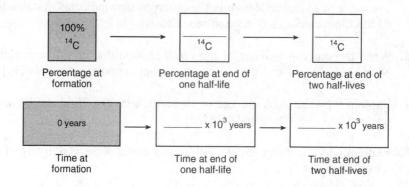

50. Identify the decay product formed by the disintegration of carbon-14. [1]

51. Explain why carbon-14 cannot be used to accurately determine the age of organic remains that are 1,000,000 years old. [1]

52. State the name of the radioactive isotope that has a half-life that is approximately the same as the estimated time of the origin of Earth. [1]

Base your answers to questions 53 and 54 on the cross section below and on your knowledge of Earth science. The unconformity is located at the boundary between Middle Proterozoic rock and Late Cambrian and Early Ordovician rock.

53. Identify *one* geologic process that occurred in this region that produced the unconformity in this outcrop. [1]

54. Identify by name the oldest New York State index fossil that could be found in the Early Ordovician bedrock. [1]

55. The table below shows information about Earth's geologic history. Letter *X* represents information that has been omitted.

Period	Million Years Ago	Index Fossil Found in Bedrock	Important Geologic Event
Triassic	251 to 200	*Coelophysis*	X

Identify *one* important geologic event that occurred in New York State that could be placed in the box at *X*. [1]

Base your answers to questions 56 through 58 on the cross section of part of Earth's crust in your answer booklet and on your knowledge of Earth science. On the cross section, some rock units are labeled with letters *A* through *I*. The rock units have *not* been overturned. Line *XY* represents a fault. Line *UV* represents an unconformity.

56. On the cross section, draw *two* arrows, one on each side of line *XY*, to show the direction of relative movement that has occurred along the fault. [1]

57. Write the letter of the oldest rock unit in the cross section. [1]

58. The table below shows the ages of the igneous rock units, determined by radioactive dating.

Rock Unit	D	G	H	B
Age (million years)	420	454	420	140

How many million years ago did rock unit *I* most likely form? [1]

EXTENDED CONSTRUCTED RESPONSE QUESTIONS

Base your answers to questions 59 through 64 on the geologic cross section shown below. Rock units A through H are shown. Several rock units contain fossils. Rock unit G was formed in a zone of contact metamorphism.

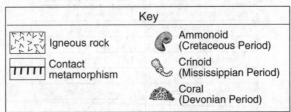

59. Place two **X**s on the cross section above to show the locations of two unconformities that formed at different times in geologic history. [1]

60. Identify two possible geologic periods during which the sediments that formed rock unit E could have been deposited. [1]

61. Describe the evidence shown in the cross section that indicates that rock unit C is younger than rock unit D. [1]

62. Identify the letter of the rock unit that was formed at the same time as igneous rock unit H. [1]

63. Identify one geologic period during which igneous intrusion H could have formed. [1]

64. Explain why the absolute age of the fossils shown in the cross section *cannot* be determined by using radioactive carbon-14. [1]

Base your answers to questions 65 through 68 on the passage and chart below, and on your knowledge of Earth science. The chart identifies some human species and the times when they are believed to have existed.

Human Species

Modern humans, *Homo sapiens*, appear to have evolved through several species of earlier members of the genus *Homo*. Each of these human species possessed specific features that made that species distinct. Many lived in (or at least have been discovered in) specific geographic areas, and existed for specific time ranges shown in the chart. In many cases, fossil remains are partial, often consisting of only teeth and skulls. Interpretation of human evolution continues to change with new discoveries.

Human Species Distributed Through Time

Human Species	Time of Existence from Fossil Evidence (million years ago)
Homo sapiens	0.25 to the present
Homo neanderthalensis	0.35 to 0.03
Homo rhodesiensis	0.6 to 0.1
Homo heidelbergensis	0.6 to 0.3
Homo mauritanicus	1.2 to 0.6
Homo erectus	1.5 to 0.2
Homo ergaster	1.8 to 1.25
Homo habilis	2.25 to 1.4

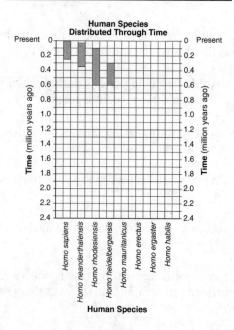

Human Species Distributed Through Time

65. Complete the graph next to the data table by drawing a bar to represent the time span that each human species existed. The bars for the first four species listed have already been drawn. [1]

66. Which human species shown in the chart was the first to exist? [1]

67. One species of the genus *Homo* could have evolved directly from another species of the genus *Homo* only if the other species:

    - existed before the new species appeared
    - did not become extinct before the new species appeared

    Identify *two* species of the genus *Homo* from which *Homo neanderthalensis* may have directly evolved. [1]

68. During which geologic epoch did the *Homo mauritanicus* species exist? [1]

Base your answers to questions 69 and 70 on the passage below and on your knowledge of Earth science.

Radiocarbon Dating

Radioactive carbon-14 (^{14}C), because of its short half-life, is used for the absolute dating of organic remains that are less than 70,000 years old. Carbon-14 is an isotope of carbon that is produced in Earth's upper atmosphere. High-energy cosmic rays from the Sun hit nitrogen-14 (^{14}N), producing radioactive ^{14}C. This ^{14}C is unstable and will eventually change back into ^{14}N through the process of radioactive decay. The proportions of ^{14}C and ordinary ^{12}C in Earth's atmosphere remain approximately constant.

Radioactive ^{14}C, just like ordinary ^{12}C, can combine with oxygen to make carbon dioxide. Plants use CO_2 during photosynthesis. The proportion of ^{14}C to ^{12}C in the cells and tissues of living plants is the same as the proportion of ^{14}C to ^{12}C in the atmosphere. After plants die, no new ^{14}C is taken in because there is no more photosynthesis. Meanwhile, the ^{14}C in the dead plant keeps changing back to ^{14}N, so there is less and less ^{14}C. The longer the plant has been dead, the less ^{14}C is found in the plant. The age of organic remains can be found by comparing how much ^{14}C is still in the organic remains to how much ^{14}C is in a living organism.

69. Radioactive ^{14}C was used to determine the geologic age of old wood preserved in a glacier. The amount of ^{14}C in the old wood is half the normal amount of ^{14}C currently found in the wood of living trees. What is the geologic age of the old wood? [1]

70. State *one* difference between dating with the radioactive isotope ^{14}C and dating with the radioactive isotope uranium-238 (^{238}U). [1]

Base your answers to questions 71 through 73 on the passage and diagram below and on your knowledge of Earth science. The diagram represents some of the Burgess shale community of organisms that existed together during part of the Cambrian Period. Thirteen different types of organisms are numbered in the diagram.

Burgess Shale Fossils

The Burgess shale fossil discovery revealed unique Cambrian life-forms, most of which were not present in the previously known fossil record. Normally, soft body parts of dead organisms are destroyed by scavengers and bacteria on the ocean floor. However, in the deep-water depositional environment of the Burgess shale, oxygen was lacking and organisms were buried rapidly, preserving the unique community seen in the diagram. The soft-bodied organisms had previously been unknown. The Burgess shale fossils were originally found in a layer of bedrock in southwestern Canada.

Adapted from: Briggs, et al., *The Fossils of the Burgess Shale*, Smithsonian Institution Press, 1994

71. During which epoch of the Cambrian Period were the Burgess shale organisms and sediments deposited? [1]

72. Explain why so many soft body parts of organisms were preserved in the Burgess shale. [1]

73. Identify the number of *one* organism in the diagram that is most likely a trilobite. [1]

274

Unit Four: EARTH MATERIALS

Chapter 13
MINERALS

KEY IDEAS The solid part of Earth is made up of rocks. Rocks, in turn, are composed of minerals, and most minerals are made of chemical elements. Earth consists of a great variety of minerals, whose properties depend on the history of how they were formed as well as the elements of which they are composed. Most properties of minerals can be explained in terms of the arrangement and properties of the atoms that compose them.

Minerals are made and remade on Earth's surface, in its oceans, and in the hot and high-pressure layers inside Earth. Observing and classifying minerals has helped us understand the great variety and complexity of Earth materials. Through the study of minerals, we have gained insight into Earth's historical development and its dynamics.

Minerals are important to us because many are sources of essential industrial materials, such as iron, copper, aluminum, and magnesium. The abundance of minerals ranges from almost unlimited to extremely rare. Many of the best sources, however, are depleted, making it more difficult and expensive to obtain those minerals. Also, the difficulty of extracting minerals from Earth's crust has important economic and environmental impacts. As limited resources, minerals must be used wisely.

KEY OBJECTIVES
Upon completion of this chapter, you will be able to:

- Explain how the physical properties of minerals are determined by their chemical compositions and crystal structures.
- Describe how minerals can be identified by well-defined physical and chemical properties, such as cleavage, fracture, color, specific gravity, hardness, streak, luster, crystal shape, and reaction with acid.
- Explain that chemical composition and physical properties determine how humans use minerals.
- Describe how minerals are formed inorganically by the process of crystallization as a result of specific environmental conditions. These include:

- Cooling and solidification of magma.
- Precipitation from water caused by such processes as evaporation, chemical reactions, and temperature changes.
- Rearrangement of atoms in existing minerals subjected to conditions of high temperature and pressure.

WHAT IS A MINERAL?

If you walk outside and pick up any earth material—a rock, sand, soil, gravel, or mud—you will hold minerals in your hand. Nearly all rocks are composed of one or more substances called *minerals*. When rocks are broken down into smaller pieces such as pebbles or sand, those smaller pieces are composed of the same minerals that were in the rock.

Nearly every single thing we use is made of, or contains, minerals. Minerals are essential to the life processes of plants and animals. Industry is equally dependent upon an abundant supply of minerals as raw materials. Without minerals, none of the devices and structures that surround us in our daily lives could be made.

Some minerals are rare and highly prized for their characteristics, such as gold for its conductivity, resistance to corrosion, and high luster, and diamonds for their beauty and hardness. Some minerals are raw materials from which industries manufacture the products that are the basis of a nation's wealth. Hematite, an iron ore, is needed to make the steel from which products ranging from cars to skyscrapers are manufactured. As such, minerals are of strategic importance and wars have been fought to gain control over mineral resources. Some minerals, however, are so abundant that they have little commercial value.

The word *mineral* means different things to different people. To a nutritionist, minerals are things to be eaten, along with proteins, carbohydrates, and vitamins. To a jeweler, a mineral is a stone to be cut or polished. To a geologist, a **mineral** is a naturally occurring, inorganic compound with a fixed chemical composition and an orderly internal arrangement of atoms.

Defining Characteristics of Minerals

Although there are many different minerals, they all share certain characteristics.

Minerals are naturally occurring. A **naturally occurring** material is formed as result of natural processes in or on Earth. It is not manufactured in a factory or synthesized in a laboratory. For example, most diamonds are formed naturally in Earth and are minerals. Synthetic diamonds made in the laboratory are not minerals.

Minerals are inorganic matter. **Inorganic** substances are not alive, never were alive, and do not come from living things. Thus, amber (a tree resin in which insects are often found embedded) and the fossil fuels coal, petroleum, and natural gas are not true minerals. They were formed from organic substances, animal or vegetable material, that once lived on Earth.

A chemical symbol or formula can be written for a mineral. Minerals are either elements or compounds. Both elements and compounds have definite chemical and physical properties.

Elements are substances that cannot be broken down into simpler substances by ordinary chemical means. Ninety-two different elements have been found to occur naturally on Earth, each with distinctly different physical and chemical properties. Elements consist of particles called atoms, and all atoms of an element have the same properties. One- or two-letter symbols are used to represent the atoms of elements. For example, the symbol for the element oxygen is O and the symbol for the element silicon is Si.

Table 13.1 shows the relative abundances of different elements in Earth's crust, hydrosphere, and troposphere. Notice that only eight elements make up more than 98 percent of Earth's crust, the two most common elements being oxygen and silicon.

TABLE 13.1 AVERAGE CHEMICAL COMPOSITION OF EARTH'S CRUST, HYDROSPHERE, AND TROPHOSPHERE

ELEMENT (symbol)	CRUST		HYDROSPHERE	TROPOSPHERE
	Percent by mass	Percent by volume	Percent by volume	Percent by volume
Oxygen (O)	46.10	94.04	33.0	21.0
Silicon (Si)	28.20	0.88		
Aluminum (Al)	8.23	0.48		
Iron (Fe)	5.63	0.49		
Calcium (Ca)	4.15	1.18		
Sodium (Na)	2.36	1.11		
Magnesium (Mg)	2.33	0.33		
Potassium (K)	2.09	1.42		
Nitrogen (N)				78.0
Hydrogen (H)			66.0	
Other	0.91	0.07	1.0	1.0

Source: The State Education Department, Earth Science Reference Tables, 2011 ed. (Albany, New York; The University of the State of New York)

Although some minerals found in Earth's crust are pure elements, such as native copper and silver, the elements in Earth's crust rarely exist by themselves. Most are chemically combined with other elements as compounds. *Compounds* consist of molecules, that is, groups of atoms joined together in a definite proportion. For example, the mineral calcite is a compound of the elements calcium, carbon, and oxygen (see Figure 13.1a). Every calcite molecule contains one atom of calcium, one atom of carbon, and three atoms of oxygen. To show the chemical

Figure 13.1 Molecular Models of Calcite and Quartz.

composition of a mineral, a formula can be written. The chemical formula for calcite is $CaCO_3$. The mineral quartz consists of molecules each containing one atom of silicon and two atoms of oxygen (Figure 13.1b). The chemical formula for quartz is SiO_2.

Every compound has distinct properties of its own. Thus, a mineral, which is either an element or a compound, has both a definite composition and distinct physical and chemical properties. Table 13.2 shows the chemical names and formulas of a few common minerals. The last column of the Properties of Common Minerals chart in the *Earth Science Reference Tables* has a more extensive list of the chemical compositions of common minerals.

TABLE 13.2 CHEMICAL NAMES AND FORMULAS OF SOME COMMON MINERALS

Mineral	Chemical Name	Chemical Formula
Calcite	Calcium carbonate	$CaCO_3$
Galena	Lead sulfide	PbS
Gypsum	Calcium sulfate-water	$CaSO_2 \cdot 2H_2O$
Olivine (fosterite)	Magnesium silicate	Mg_2SiO_4
Potassium feldspar	Potassium aluminum silicate	$KAlSi_3O_8$
Pyrite	Iron sulfide	FeS_2
Quartz	Silicon dioxide	SiO_2

Minerals have a crystalline form. The atoms or molecules of a mineral are the same throughout that mineral. When they are joined in fixed positions as a solid, a definite pattern is formed. A solid having a definite internal structural pattern is said to have a **crystalline** form. If the pattern is large enough to be seen with the unaided eye, the solid is called a **crystal**. The crystal form of a mineral determines its cleavage, or the way it splits or breaks, as well as many other properties. Mica, for example, splits into thin, flat sheets (see Figure 13.2).

Identifying Minerals

Mineralogists have examined Earth materials and identified more than 2,000 minerals. Minerals can be identified on the basis of well-defined physical and chemical properties. Several important properties commonly used in mineral identification are color, luster, streak, hardness, and cleavage, parting, and fracture. Other properties used to identify minerals include specific gravity, radioactivity, luminescence, and chemical, thermal, electrical, and magnetic properties, as well as elasticity and strength.

Figure 13.2 The Crystalline Pattern of Mica. Mica consists of sheets of tightly bonded silicon and oxygen atoms held together weakly by metallic ions.

Since no single property can be used to identify all minerals, mineral identification is usually a process of elimination. As each property is observed, it becomes evident what a mineral is not, rather than what it is. Step by step, the possibilities are narrowed down until the identity of the mineral is determined.

Color

Color is often the first property noticed about a mineral. When observing color, it is important to use a fresh surface of a mineral since exposed surfaces are often discolored by weathering. Color alone is an unreliable property by which to identify a mineral. Although some minerals are always the same color (e.g., sulfur is yellow), others may occur in a variety of colors. Thus quartz is found in many colors, some of which are rare (these specimens are highly prized as semiprecious gems). For example, amethyst is purple quartz, citrine is yellow quartz, and rose quartz is pink quartz. Therefore, quartz cannot be identified by color alone. Another reason why color is unreliable is that many minerals have almost the same color. For example, calcite, quartz, and halite all occur in white and transparent varieties and can look strikingly alike.

Luster

The **luster** of a mineral is the way light reflects from its surface. Luster can be metallic or nonmetallic. Nonmetallic lusters can be described as glassy, brilliant, greasy or oily, waxy, silky, pearly, or earthy.

Related to the luster of a mineral are its transparency and iridescence, or the play of colors in its interior or exterior.

Streak

Streak is the color of the fine powder left when a mineral is rubbed against a hard, rough surface. A piece of unglazed porcelain, called a *streak plate*, is usually used to create a streak. Although the color of a mineral may vary, the streak of a particular mineral is always the same color. For example, fluorite may range in color from green to blue; yet its streak is always white. This characteristic makes streak a useful property for identifying a mineral. It must be remembered, though, that a mineral's streak is not always the same color as the mineral because powders reflect light in a different way than large crystals. Pyrite crystals have a brassy, yellow color, but their streak is greenish, or brownish black! Most minerals have colorless or white streaks; therefore, a colored streak is very useful in identifying a mineral.

Hardness

Hardness is a mineral's resistance to being scratched. Talc is so soft that it can be scratched by a fingernail. At the other extreme, diamond is so hard that no other mineral can scratch it. The hardness of a mineral is usually stated in terms of Mohs scale (see Table 13.3). On this scale, ten typical minerals are arranged in order from the softest to the hardest.

Minerals

TABLE 13.3 MOHS SCALE

Mineral	Hardness	
Talc	1	SOFTEST
Gypsum	2	↑
Calcite	3	
Fluorite	4	
Apatite	5	
Orthoclase	6	
Quartz	7	
Topaz	8	
Corundum	9	↓
Diamond	10	HARDEST

To find the hardness of a mineral, you determine what minerals your sample can scratch and what minerals it cannot scratch. For example, tourmaline will scratch quartz or anything softer, but cannot scratch topaz or anything harder. The hardness of tourmaline, therefore, is between 7 and 8 (sometimes expressed as 7.5).

Cleavage, Parting, and Fracture

Some minerals break in ways that help to identify them. **Cleavage** (see Figure 13.3) is the tendency of a mineral to break parallel to atomic planes in its crystalline structure. *Parting* is the tendency to break along surfaces that follow a structural weakness caused by factors such as pressure, or along zones of different crystal types. These cleavage or parting surfaces often occur at very specific angles to one another and can be helpful in identifying a mineral.

Mineral	Cleavage	Appearance
Muscovite mica	Breaks in parallel sheets	
Pyroxene	Breaks along planes at 88° angle to one another in a prismatic shape	
Halite	Breaks along planes at 90° to one another in a cubic shape	
Calcite	Breaks along planes at a 75° angle to one another in a rhombohedral shape	

Figure 13.3 Some Common Minerals That Display Cleavage.

Some minerals have no planes of weakness, or are equally strong in all directions. They **fracture**; that is, they do not follow a particular direction when they break. Although fracture does not occur along smooth surfaces, it can display distinctive patterns that are very helpful in mineral identification (see Figure 13.4).

Patterns of Fracture	Description	Appearance
Conchoidal	Smooth, curved break that looks somewhat like the inside of a shell (e.g., quartz, olivine, gypsum)	
Fibrous or splintery	Fibers, as in asbestos Long, thin splinters (e.g., chrysotile, actinolite, tremolite, satin-spar calcite)	
Hackly	Jagged or sharp-edged surfaces (e.g., native metals such as copper, silver, iron, nickel)	
Uneven	Rough, irregular surfaces (e.g., sulfur, anhydrite, barite)	

Figure 13.4 Some Common Minerals That Display Fracture.

Specific Gravity

Every mineral has a certain density. If you have samples that are about the same size, you can compare the densities of two different minerals. When you hold one in each hand, the denser mineral will feel heavier than the less dense one. A more exact way to determine the relative densities of different minerals is to compare them all to one standard. **Specific gravity** is the ratio of the density of a mineral to the density of water. It tells us how much heavier than water a mineral is. For example, the density of galena is 7.5 grams per cubic centimeter; the density of water is 1 gram per cubic centimeter. Galena is 7.5 times denser than water and therefore has a specific gravity of 7.5.

Most rocks have a specific gravity of 2.5 to 3.5. Most people have a feeling for how heavy a chunk of mineral of a given size should be. Sometimes, though, when you pick up a sample of a mineral and toss it in your hand, it feels heavier than you expected. This feeling of unexpected weight is termed

Minerals

heft. You can usually judge by a mineral's heft that its specific gravity is greater than 4.0.

Chemical Tests

Two chemical tests are commonly used to identify minerals in the field. The acid test uses hydrochloric acid. Some minerals bubble when a drop of hydrochloric acid dropped on the mineral reacts with it. Calcite bubbles vigorously; dolomite, slowly.

The second chemical test is the taste test. Halite tastes the same as table salt, which it is, though not purified. **The taste test should be used with caution and only at your teacher's direction.**

Special Properties

Special properties that are displayed by some minerals and can be used to distinguish them from other minerals are summarized in Table 13.4.

TABLE 13.4 SOME SPECIAL PROPERTIES OF MINERALS

Property	Definition or Description
Magnetism	Mechanical force of attraction associated with moving electricity (magnetite)
Luminescence	Emission of light by a substance that has received energy or electromagnetic radiation of a different wavelength from an external stimulus
includes:	
phosphorescence	Light emitted after stimulus is removed (scheelite autunite)
fluorescence	Light emitted while stimulus is applied (fluorite, willemite)
triboluminescence	Light emitted when pressure is applied (fluorite, lepidolite)
thermoluminescence	Light emitted when heated (chlorophane fluorite)
Piezoelectricity	Electricity emitted when pressure is applied (quartz)
Pyroelectricity	Electricity emitted when heat is applied (tourmaline)
Flame color	Color of the flame when a mineral is intensely heated in a flame
Double refraction	Double image produced when an object is viewed through the mineral (Iceland spar calcite)
Play of colors	Series of colors produced when the angle of the light shining on the mineral is changed (labradorite, opal)
Chatoyancy and asterism	Appearance of being made of tiny parallel fibers (satin spar gypsum, cat's-eye chrysoberyl) Light scattered in a pattern that looks like a three- or six-rayed star (star rubies and sapphires)

Using the Properties of Common Minerals Chart to Identify Minerals

As mentioned earlier, mineral identification is a process of elimination. As each property of a mineral is observed, possibilities are narrowed down until the identity of the mineral is determined. The Properties of Common Minerals chart in the *Reference Tables for Physical Setting/Earth Science* is arranged to assist you in this process. From left to right the columns list properties in the order of their usefulness in narrowing down the possible identity of a mineral.

Example:
Which mineral contains iron, has a metallic luster, is hard, and has the same color and streak?
(1) biotite mica   (2) galena   (3) potassium feldspar   (4) magnetite

First, rearrange the properties given in the question in the order in which they appear in the Properties of Common Minerals chart in the *Earth Science Reference Tables*: metallic luster, hard, same color and streak, contains iron. Now, referring to the chart, follow these steps:

- Since the mineral has a metallic luster, all of the minerals with nonmetallic luster are eliminated. This leaves graphite, galena, magnetite, pyrite, and the earthy-red form of hematite (earthy is a nonmetallic luster).
- Since the mineral is hard, graphite and galena are eliminated. This leaves magnetite, pyrite, and the metallic silver form of hematite.
- Since the mineral has the same color and streak, pyrite (brassy yellow color, green-black streak) and hematite (metallic silver color, red-brown streak) are eliminated. This leaves magnetite.
- Since the mineral contains iron, check to see that magnetite contains iron. The last column lists the composition of magnetite as Fe_3O_4; and since the chemical symbol for iron is Fe, magnetite does, indeed, contain iron.

Thus, the mineral in the question is magnetite, choice 4.

Minerals as the Building Blocks of Rocks

Of the thousands of known minerals, about 100 are so common that they make up more than 95 percent of Earth's crust. Not surprisingly, these minerals are generally compounds of the most abundant elements in the crust. Nearly every rock contains one or more of these so-called *rock-forming minerals* (see Table 13.5). Of the rock-forming minerals, fewer than 20 comprise most of the rocks you are likely to find.

Minerals

TABLE 13.5 SOME MAJOR ROCK-FORMING MINERALS

Mineral or Mineral Group	Composed of These Elements
Olivines	Iron, magnesium, silicon, oxygen
Pyroxenes	Calcium, iron, magnesium, silicon, oxygen
Amphiboles	Sodium, calcium, aluminum, silicon, oxygen
Micas	Potassium, aluminum, iron, magnesium, silicon, oxygen, hydrogen
Feldspars	Potassium, sodium, calcium, aluminum, silicon, oxygen
Quartz	Silicon, oxygen
Hematite	Iron, oxygen
Magnetite	Iron, oxygen
Pyrites	Iron, sulfur
Gypsum	Calcium, sulfur, oxygen
Calcite	Calcium, carbon, oxygen
Dolomite	Calcium, magnesium, carbon, oxygen
Fluorite	Calcium, fluorine

HOW DO MINERALS DIFFER FROM EACH OTHER?

Minerals are grouped according to their chemical compositions. Major mineral groups include the silicates, sulfides, oxides, carbonates, and sulfates. Minor mineral groups include the nitrates, borates, chromates, phosphates, halides, hydroxides, and native elements.

Major Mineral Groups

Silicates

If we look at the chemical composition of Earth's crust (refer to Table 13.1), we see that oxygen is the most abundant element, by both mass and volume, and silicon is second, by mass. It is not surprising, then, that minerals composed mainly of silicon and oxygen are abundant and widespread. In all silicates, one ion (charged atom) of silicon is always joined to four ions of oxygen in the shape of a tiny **tetrahedron** (plural, tetrahedra), as shown in Figure 13.5. The bonds in this structure, which we will call the **silicon tetrahedron**, are very strong.

Silicon tetrahedra can join with positive ions of common elements in a wide variety of structures. Consider the next most abundant elements in Earth's crust after silicon and oxygen; aluminum, iron, calcium, sodium, magnesium, and potassium are all metals that form positive ions. When the positive ions of these elements join with negative silicon tetrahedra, a tremendous number of different silicates can be formed. Rings, chains, sheets, and frameworks are some of the different structures that silicon tetrahedra can form by joining with positive ions and sharing oxygen atoms.

An overhead view, as though looking through the large oxygen ions at the smaller silicon ion between them.

The natural position of the silicon tetrahedon. Notice that the silicon ion is nestled among the oxygen ions and is hidden from view.

An exploded view, showing the positions of the silicon ion and oxygen ions.

Figure 13.5 Three Views of a Silicon Tetrahedron. The silicon ion has a +4 charge (Si^{4+}), and each oxygen ion has a –2 charge (O^{2-}), so the net charge of the silicon tetrahedron is –4 $(SiO4)^{4-}$.

In the complex structures shown in Figure 13.6, many other elements and groups of elements, such as extra oxygen, water, aluminum and other metal ions, and hydroxide ions, promote electrical and mechanical balance so that these structures are stable.

The physical properties of minerals depend on the arrangement and bonding of their atoms, as is very clearly seen in the structures formed by silicon tetrahedra. Look at a typical sheet silicate, mica, in Figure 13.6. The tetrahedra all share oxygens and are tightly bound together. The metal ions between the sheets, however, are isolated from each other and are less tightly bound. They form a plane of weakness in the structure—a weakness that manifests itself in the tendency of mica to break into thin sheets. Quartz, on the other hand, is a framework silicate. All of the tetrahedra are bound tightly to each other, and no place in the structure is weaker than any other place. Therefore, quartz does not break apart easily; its structure resists outside forces, so it is hard and strong. When it does break, there are no planes of weakness, and the break is random and uneven.

This relationship between the arrangement and bonding of atoms and the physical properties of minerals extends to all mineral compounds. Knowing the structure of a mineral helps us to understand why that mineral displays certain physical properties.

Silicate minerals are so abundant and widespread that almost every rock contains one or more. Although there are more groups of nonsilicate minerals than there are of silicates, nonsilicates make up only a very small part of Earth's crust, and very few are important rock-forming minerals. A brief overview of four nonsilicate minerals follows.

Minerals

Independent Tetrahedra
e.g., olivine, zircon, garnet

Shared Oxygen

Ring Silicates
e.g., tourmaline, beryl

Shared Oxygen

Single-Chain Silicates
pyroxenes: e.g., enstatite, ferrosilite, augite, diopside, jadeite, spodumene

Shared Oxygen

Double-Chain Silicates
amphiboles: e.g., hornblendes, anthophyllite, tremolite, glaucophane

Shared Oxygen

Sheet Silicates
micas: e.g., muscovite, biotite, lepidolite; also, talc, chlorite, serpentine

All oxygens are shared

Framework Silicates
e.g., feldspars, quartz, nephaline, sodalites

Figure 13.6 Some Structures That Form with Silicon Tetrahedra.

Sulfides

Sulfides are compounds in which one or more ions of sulfur are combined with a metallic ion. Sulfide minerals usually display a metallic luster, and many are important ores of metals. Some examples of sulfides are galena (PbS), pyrite (FeS), marcasite (FeS_2), cinnabar (HgS), sphalerite (ZnS), and stibnite (Sb_2S_3).

Oxides

Oxides are compounds in which oxygen is joined with ions of other elements, usually metals. Common oxides include iron oxide (FeO), hematite (Fe_2O_3), magnetite (Fe_3O_4), corundum (Al_2O_3), ilmenite ($FeTiO_5$), chromite ($FeCr_2O_4$), and spinel ($MgAl_2O_4$). Magnetite is the only mineral attracted to a magnet. Iron oxides such as magnetite and hematite are mined to make iron and steel. Corundum is extremely hard and is widely used as an abrasive. When it forms with impurities, corundum produces red rubies and blue sapphires.

Carbonates and Sulfates

In oxides, oxygen forms a single ion and acts alone. In other minerals that contain oxygen, the oxygen ions join with other ions to form a group called a polyatomic ion. Two very common polyatomic ions are the **carbonate ion**, $(CO_3)^{2-}$, and the **sulfate ion**, $(SO_4)^{2-}$. Polyatomic ions are electrically charged and can combine with ions of the opposite charge to form mineral compounds.

When carbon dioxide dissolves in water, it often forms carbonate ions. These react with other ions in the water to form carbonate compounds that can precipitate out of the water or be left behind when the water evaporates. Some organisms take carbon dioxide from the environment and form, in their bodies as skeletons, compounds containing carbonate ions. In this way, carbonate minerals have accumulated into huge sedimentary rock formations and in some cases have been metamorphosed. Calcite (calcium carbonate) is the chief mineral in limestone and marble. Dolomite (calcium magnesium carbonate) is the chief mineral in dolostone.

Gypsum (hydrated calcium sulfate) is probably the best known sulfate mineral. It is mined for use in making wallboard, plaster, cement, fertilizer, and paint and as a filler in paper. Barite ($BaSO_4$) and celestite ($SrSO_4$) are important sources of barium and strontium, respectively, which are used in medicines and drilling.

Minor Mineral Groups

There are also other, less abundant ions that form minerals: nitrates (nitrogen and oxygen), borates (boron and oxygen), chromates (chromium and oxygen), phosphates (phosphorus and oxygen), and hydroxides (hydrogen and oxygen).

Halides are compounds of metallic ions with halogen elements such as chlorine, fluorine, and bromine. The only major rock-forming mineral in this group is halite (NaCl), a sedimentary mineral that forms when seawater evaporates.

Native elements are found as pure elements in Earth's crust. They include gold, silver, copper, and sulfur. These were among the first minerals to be mined by humans.

CONSERVATION OF EARTH'S RESOURCES

Minerals

The minerals in Earth are a treasure of almost unimaginable value. They help to fill many human needs, which, through time, have become greater. Minerals are used in making products ranging from steel to electric light bulbs. Unfortunately, minerals are a nonrenewable resource; that is, they can be used only once and then they are gone.

Imagine that every week you put a few cents of your allowance into a bank. For three years you do not take any money from the bank. Then you see a video game that you want, you open the bank, and you use the money to buy the game. In a few minutes, you have spent the money it took you three years to save. In the same way, it is easy to use up in a few years minerals that took millions of years to form. But while you will probably save more money in the future, mineral resources may not form again within your lifetime or even a thousand lifetimes. Every time a mineral is used, that much less remains. Thus, Earth's mineral resources will not last forever.

For this reason, people must conserve mineral resources, not just by saving these resources, but also by making sure that there is no waste in using them. Recycling and modernizing factories are parts of a good conservation policy. Another important part is research into better ways of production.

If conservation is practiced by everyone, Earth's mineral resources can be used to best advantage. Today there are recycling centers throughout the United States, and many communities have instituted mandatory recycling. Recycling centers process used aluminum cans, old bottles, scrap metals, and newspapers. These materials are then reused in the manufacture of aluminum, glass, steel, and paper products. You may have noticed on some cans a statement that they were made from recycled aluminum.

Fossil Fuels

At certain times in Earth's history, environmental conditions have enabled plant and marine organisms to grow at a rapid rate. Over millions of years, their bodies stored energy from sunlight that was converted into chemical bonds during photosynthesis. When these organisms completed their life cycles, their remains accumulated faster than decomposers could break them down and recycle them back into the environment. Over time, when these remains became buried by deposition, the pressure of overlying sediments and the increased temperatures at depth converted the layers of organic remains into coal beds and pools of petroleum. Since coal and petroleum are derived from organic remains and are used chiefly as fuels, they are called **fossil fuels**.

Fossil fuels are a vital source of energy and petrochemicals for the global economy. Transportation, urban development, industry, commerce, agriculture, and many other human activities depend on the amount and type of energy available. Most of the energy used today is obtained by burning fossil fuels. While coal was widely used in the past, petroleum and natural gas are now the fuels of choice because they are easier to obtain than coal, have many uses in industry, and are concentrated, easily portable sources of energy for cars, trucks, airplanes, and trains.

Unfortunately, the burning of fossil fuels releases into the atmosphere waste products that threaten the health of living things and have environmental risks associated with them. Current environmental conditions are not resulting in the large-scale accumulation of remains that can be converted into coal or petroleum. Therefore, for the foreseeable future, organic fossil fuels are a nonrenewable resource.

Global Implications of Resource Use and Distribution

The use and the distribution of Earth's resources have global political, financial, and social implications. Highly industrialized nations use more energy and mineral resources than less developed countries. The energy and natural resources we consume contribute to our high standard of living. This consumption, however, has led to a rapid depletion of Earth's natural resources and to mounting environmental crises, ranging from the devastation of marine life by oil-tanker spills to the breakdown of the ozone layer of the atmosphere by the chlorofluorocarbons (CFCs) used in refrigerators and air conditioners.

As developing nations industrialize and their urban centers grow, they demand more energy and natural resources, thereby entering into competition with all of the other industrialized nations for these resources. Often this competition has resulted in friction between nations and even in war. If we are to live in peace and maintain our environment, we must develop ways to slow the depletion of natural resources and to use them more efficiently.

The depletion of energy and other natural resources can be slowed by decisions to use efficient technologies, to conserve, and to recycle. These decisions can be implemented at many levels ranging from the national down to the personal. Government can restrict low-priority uses of materials, such as the use of petrochemicals to manufacture purely ornamental packaging that adds to our solid-waste disposal problems. Government can also set standards

Minerals

of technical efficiency, for example, mileage and emission requirements for automobiles or the use of insulation in home construction. Cars that are built with more power than their function warrants waste energy. (If the speed limit is 55 mph, why build cars with the power to do 100 mph?) However, such decisions often involve trade-offs of cost and social values. For example, when other, less damaging compounds are substituted for CFCs, new compressors are required, so older units must be replaced before they have worn out. Reduced emissions depend on catalytic converters filled with toxic heavy metals. The development of wise policies for conserving and managing natural resources is one of the great challenges now facing all nations.

MULTIPLE-CHOICE QUESTIONS

In each case, write the number of the word or expression that best answers the question or completes the statement.

1. Which group of elements is listed in increasing order based on the percent by mass in Earth's crust?
   (1) aluminum, iron, calcium
   (2) aluminum, silicon, magnesium
   (3) magnesium, iron, aluminum
   (4) magnesium, silicon, calcium

2. Oxygen is the most abundant element by volume in Earth's
   (1) inner core   (3) hydrosphere
   (2) troposphere   (4) crust

3. The diagram below represents a part of the crystal structure of the mineral kaolinite.

   Structure of Kaolinite

   An arrangement of atoms such as the one shown in the diagram determines a mineral's
   (1) age of formation   (3) physical properties
   (2) infiltration rate   (4) temperature of formation

4. The pie graph below shows the elements comprising Earth's crust in percent by mass.

Which element is represented by the letter *X*?
(1) silicon   (3) nitrogen
(2) lead   (4) hydrogen

5. The diagrams below show the crystal shapes of two minerals.

Quartz and halite have different crystal shapes primarily because
(1) light reflects from crystal surfaces
(2) energy is released during crystallization
(3) of impurities that produce surface variations
(4) of the internal arrangement of the atoms

6. A student created the table below by classifying six minerals into two groups, *A* and *B*, based on a single property.

Group A	Group B
olivine	pyrite
garnet	galena
calcite	graphite

Which property was used to classify these minerals?
(1) color   (3) chemical composition
(2) luster   (4) hardness

Minerals

7. Silicate minerals contain the elements silicon and oxygen. Which list contains only silicate minerals?
   (1) graphite, talc, and selenite gypsum
   (2) potassium feldspar, quartz, and amphibole
   (3) calcite, dolomite, and pyroxene
   (4) biotite mica, fluorite, and garnet

8. Which two minerals have cleavage planes at right angles?
   (1) biotite mica and muscovite mica
   (2) sulfur and amphibole
   (3) quartz and calcite
   (4) halite and pyroxene

9. Which two properties are most useful in distinguishing between galena and halite?
   (1) cleavage and color
   (2) luster and color
   (3) hardness and streak
   (4) streak and cleavage

10. What is the approximate density of a mineral with a mass of 262.2 grams that displaces 46 cubic centimeters of water?
    (1) 1.8 g/cm^3
    (2) 5.7 g/cm^3
    (3) 6.1 g/cm^3
    (4) 12.2 g/cm^3

11. Which mineral will scratch fluorite, galena, and pyroxene?
    (1) graphite
    (2) calcite
    (3) olivine
    (4) dolomite

12. The data table below gives characteristics of the gemstone peridot.

    Characteristics of Peridot

Luster	nonmetallic
Hardness	6.5
Color	green
Composition	$(Fe,Mg)_2SiO_4$

    Peridot is a form of the mineral
    (1) pyrite
    (2) pyroxene
    (3) olivine
    (4) garnet

13. The most abundant metallic element by mass in Earth's crust makes up 8.23% of the crust. Which group of minerals all normally contain this metallic element in their compositions?
    (1) garnet, calcite, pyrite, and galena
    (2) biotite mica, muscovite mica, fluorite, and halite
    (3) talc, quartz, graphite, and olivine
    (4) plagioclase feldspar, amphibole, pyroxene, and potassium feldspar

14. The minerals talc, muscovite mica, quartz, and olivine are similar because they
    (1) have the same hardness
    (2) are the same color
    (3) contain silicon and oxygen
    (4) break along cleavage planes

Base your answers to questions 15 and 16 on the data table below and on your knowledge of Earth science. The table provides information about four minerals, *A* through *D*.

Data Table

Mineral	Breakage	Hardness	Luster	Color
A	cleavage	2.5	metallic	silver
B	cleavage	2.5	nonmetallic	black
C	cleavage	3	nonmetallic	colorless
D	fracture	6.5	nonmetallic	green

15. The diagram below represents a sample of mineral *A*.

Mineral *A* is most likely
(1) garnet
(2) galena
(3) olivine
(4) halite

16. Which mineral can scratch *A*, *B*, and *C* but *cannot* scratch *D*?
    (1) talc
    (2) selenite gypsum
    (3) fluorite
    (4) quartz

Minerals

CONSTRUCTED RESPONSE QUESTIONS

Base your answers to questions 17 and 18 on the data table below, which shows the volume and mass of three different samples, A, B, and C, of the mineral pyrite.

Pyrite		
Sample	Volume (cm³)	Mass (g)
A	2.5	12.5
B	6.0	30.0
C	20.0	100.0

17. On the grid provided below, plot the data (volume and mass) for the three samples of pyrite and connect the points with a line. [2]

18. State the mass of a 10.0-cm³ sample of pyrite. [1]

19. State one reason why color is frequently less valuable than cleavage in mineral identification. [1]

20. In one or more complete sentences, explain why minerals are considered a nonrenewable resource. [1]

Base your answers to questions 21 and 22 on the hardness of the minerals talc, quartz, halite, sulfur, and fluorite.

21. On the grid below, construct a bar graph to represent the hardness of these minerals. [1]

22. Which mineral shown on the grid would be the best abrasive? State one reason for your choice. [1]

Base your answers to questions 23 through 25 on the chart below, which shows some physical properties of minerals and the definitions of these properties. The letters *A*, *B*, and *C* indicate parts of the chart that have been left blank. Letter *C* represents the name of a mineral.

23. Which physical property of a mineral is represented by letter *A*? [1]

24. State the definition represented by letter *B*. [1]

25. Identify *one* mineral that could be represented by letter *C*. [1]

Minerals

EXTENDED CONSTRUCTED RESPONSE QUESTIONS

Base your answers to questions 26 through 28 on the passage below.

Carbon

Carbon may be the most important element on our planet because it is the chemical building block of all living things. The element carbon is formed in dying stars and scattered when the stars explode. Our solar system formed from such star remnants. Pure carbon comes in several forms, which include the minerals graphite and diamond (hardness = 10), and the fossil fuels bituminous coal and anthracite coal. Almost all diamonds are mined from igneous rocks that originate at an approximate depth of 150 kilometers under immense pressure. Most graphite is formed through the metamorphism of organic material in rocks closer to Earth's surface.

26. Identify *two* uses for the mineral graphite. [1]

27. Explain why graphite and diamond have different properties. [1]

28. Complete the table below to show the properties of the minerals diamond and graphite. [1]

Property	Diamond	Graphite
color	variable	
luster	nonmetallic	
hardness		

Base your answers to questions 29 through 31 on the table below and on your knowledge of Earth science. The table shows the elements and their percent compositions by mass in the five minerals present in a rock sample.

Elements and Their Compositions by Mass in Five Minerals

Minerals Present in Rock Sample	Element (percent by mass)									
	Al	Ca	Fe	H	K	Mg	Na	O	Si	Ti
Amphibole	6.2	3.0	29.7	0.2	–	3.7	1.8	31.7	12.8	10.9
Plagioclase feldspar	9.7	–	–	–	14.2	–	–	46.3	29.8	–
Garnet	10.9	–	33.8	–	–	–	–	38.7	16.6	–
Muscovite mica	20.3	–	–	0.5	9.8	–	–	48.2	21.2	–
Quartz	–	–	–	–	–	–	–	53.2	46.8	–

29. Identify *one* use for the mineral garnet. [1]

30. Identify *one* mineral in this rock sample that can scratch the mineral olivine. [1]

31. All five of the minerals listed in the table are silicate minerals because they contain the elements silicon and oxygen. State the name of *one* other mineral found on the Properties of Common Minerals chart that is a silicate mineral. [1]

Base your answers to questions 32 through 36 on the passage and cross section below, which explain how some precious gemstones form. The cross section shows a portion of the ancient Tethys Sea, once located between the Indian-Australian Plate and the Eurasian Plate.

Precious Gemstones

Some precious gemstones are a form of the mineral corundum, which has a hardness of 9. Corundum is a rare mineral made up of closely packed aluminum and oxygen atoms, and its formula is Al_2O_3. If small amounts of chromium replace some of the aluminum atoms in corundum, a bright-red gemstone called a ruby is produced. If traces of titanium and iron replace some aluminum atoms, deep-blue sapphires can be produced.

Most of the world's ruby deposits are found in metamorphic rock that is located along the southern slope of the Himalayas, where plate tectonics played a part in ruby formation. Around 50 million years ago, the Tethys Sea was located between what is now India and Eurasia. Much of the Tethys Sea bottom was composed of limestone that contained the elements needed to make these precious gemstones. The Tethys Sea closed up as the Indian-Australian Plate pushed under the Eurasian Plate, creating the Himalayan Mountains. The limestone rock lining the seafloor underwent metamorphism as it was pushed deep into Earth by the Indian-Australian Plate. For the next 40 to 45 million years, as the Himalayas rose, rubies, sapphires, and other gemstones continued to form.

A Portion of the Tethys Sea 50 Million Years Ago

32. Which element replaces some of the aluminum atoms, causing the bright-red color of a ruby? [1]

33. State *one* physical property of rubies, other than a bright-red color, that makes them useful as gemstones in jewelry. [1]

34. Identify the metamorphic rock in which the rubies and sapphires that formed along the Himalayas are usually found. [1]

35. During which geologic epoch did the events shown in the cross section of the Tethys Sea occur? [1]

36. What type of tectonic plate boundary is shown in the cross section? [1]

CHAPTER 14
ROCKS

KEY IDEAS Rocks are the naturally formed, solid, coherent materials that comprise Earth. Most rocks are aggregates of minerals; that is, they are made of many grains stuck together. The grains may be mineral crystals or tiny bits of broken rock, or even solid parts of once-living organisms. The grains may be small or large, tightly or loosely bound together, and may all be made of one type of mineral or represent a variety of minerals.

Rocks are grouped into three "families" based upon how they were formed: igneous, metamorphic, and sedimentary. Igneous rocks form by the solidification of molten rock; metamorphic rocks form when heat, pressure, or chemical activity causes changes in existing rocks; and sedimentary rocks form by the chemical precipitation or the compaction and/or cementation of sediments. The composition and structure of a rock offer important clues as to its origin.

Of the three types of rock, no one type necessarily originates before another. Rock is constantly being created, destroyed, and altered in internal and external Earth processes. The schematic model of the possible sequences of processes by which any one rock may have been transformed into another is known as the rock cycle.

KEY OBJECTIVES
Upon completion of this chapter, you will be able to:

- Explain that rocks are usually composed of one or more minerals.
- Describe how rocks are classified by their origins, mineral contents, and textures.
- Explain how the characteristics of rocks indicate the processes by which they formed and the environments in which these processes took place.
- Describe how the properties of rocks determine their uses and also influence land usage by humans.

WHAT IS A ROCK?

A **rock** is a naturally formed, nonliving, firm, and coherent mass of solid Earth material. A pile of sand is not a rock because the grains are not bound together, so the sand is not firm and coherent. A tree is not a rock even though it is solid, because a tree is living. But coal, which has been compressed into a coherent mass of dead leaves, twigs, and other plant parts, *is* a rock. Most rocks, though, are aggregates of minerals.

There are three large "families" of rocks, each defined by the processes that form them. **Igneous** rocks (from the Latin *ignis*, meaning "fire") form by the cooling of molten rock. **Sedimentary** rocks form when sediments are compacted and/or cemented together, or by the precipitation of dissolved minerals. *Sediments* are small bits of matter deposited by water, ice, or wind. They can be fragments of rocks, shells, or the remains of plants or animals. **Metamorphic** rocks are formed when already existing rock is changed by great heat, great pressure, or chemical action. Let us now consider how each type of rock forms, and how the way in which it forms determines its characteristics.

TYPES OF ROCK

Igneous Rocks

Formation of Igneous Rocks

Igneous rocks form by the crystallization of a molten mixture of minerals and dissolved gases. This molten mixture is called **magma** while it is beneath Earth's surface and **lava** when it escapes onto the surface. Most magmas can be thought of as fiery soups of fast-moving ions, that is, electrically charged atoms or molecules. As the magma cools, these ions slow down, are attracted to one another, and settle into stable structures—molecules of mineral compounds joined in a structural array, or mineral crystals. See Figure 14.1.

The formation of mineral crystals is a process, not an instantaneous event. As magma or lava cools, the amount of time available for crystal structures to form determines how large they can become. Thus, the size of the mineral crystals in an igneous rock provides a clue to the speed at which the rock crystallized, and that in turn is a clue to its origin.

Figure 14.1 Cooling and Crystallization.

If cooling is very rapid, groups of atoms may solidify into a substance that lacks any distinctive structure—a glass. Such rapid cooling typically occurs at Earth's surface, where molten material comes into sudden contact with air or water at temperatures that are hundreds of degrees cooler than its temperature. Volcanic glasses, such as obsidian, can be observed forming when lava flows into the ocean and rapidly cools. From this type of observational data, we can infer the origin of glass found elsewhere, far from current volcanic activity or perhaps even buried beneath layers of other rock.

If cooling occurs more slowly, mineral crystals may begin to form, but may solidify while still microscopic in size. This, too, typically occurs at or near the surface of Earth. For example, the surface of lava, which may flow out of a fissure as a glowing liquid, can be observed to quickly cool and "skin over." This thin skin of rock shields the molten lava beneath it and allows the lava to cool more slowly. The lava inside may take days, weeks, or even months to solidify, thus allowing it to form crystals large enough to be seen with a microscope or magnifying glass.

Only very slow cooling allows the formation of crystals large enough to be visible to the unaided eye. Such slow cooling occurs underground, where the cooling magma is blanketed by the surrounding rock. Rock is a poor conductor of heat; and magma surrounded by rock, like hot coffee in a Thermos bottle, loses heat very slowly. Some underground pools of magma cool only a few degrees per century! Thousands of years may be required for the magma to totally crystallize. The result is large, visible crystals.

Characteristics of Igneous Rocks

Since most rocks are mixtures of minerals, rocks cannot be identified in the same way as minerals. A single rock may consist of minerals of several colors, hardnesses, and densities. For example, granite may contain white quartz with a hardness of 7, pink feldspar with a hardness of 6, and black mica with a hardness of 2–3. Thus, granite does not have a single characteristic color or hardness.

Two more general properties that are useful in identifying rocks are texture and mineral composition. **Texture** is the size, shape, and arrangement of the mineral crystals or grains in a rock. **Mineral composition** is simply the minerals that comprise the rock.

Igneous rocks are composed of randomly scattered, tightly interlocking mineral crystals. The main differences in the textures of igneous rocks involve the sizes and compositions of the mineral crystals. Most igneous rocks consist of a mixture of several of the following common minerals: quartz, feldspar, biotite, amphibole, pyroxene, and olivine.

Igneous rocks can be divided into two major types: intrusive and extrusive. Classification is based on whether the rocks formed from magma or lava. Magma and lava solidify into rock under very different conditions and thus produce rocks with different characteristics. Except for the volcanic glasses, all igneous rocks consist of tightly interlocking mineral crystals (see Figure 14.2). The texture of these crystals is a key to how a particular rock was formed.

Textures of Igneous Rocks

Intrusive Rocks

Figure 14.2 Tightly Interlocking Crystals of an Igneous Rock.

Magma, as previously defined, is molten rock inside Earth. Magma can move around by pushing its way into cracks or crevices in surrounding rock or by intruding into them. If the magma remains trapped underground and cools enough to solidify, an igneous rock will form. Igneous rock formed by the cooling of magma is called **intrusive**, or *plutonic*, rock.

As described earlier, magma generally cools slowly underground because it is blanketed by the surrounding rocks. Rocks are poor conductors of heat and do not allow the heat in the magma to escape very rapidly. When magma cools slowly, ions in the magma have time to align themselves in orderly structures called *crystals*. The slower the cooling, the more ions are able to align and the bigger the crystal. As a result, intrusive rocks generally have crystals large enough to be seen with the unaided eye. While you may not think a 1-millimeter crystal is large when you look at it, remember that millions upon millions of ions had to move into alignment with each other to reach that size. Rocks with large, visible crystals are said to have a coarse texture.

Granite, gabbro, and pegmatite are typical intrusive rocks.

Extrusive Rocks

When magma reaches the surface and pours out of Earth, it is called **lava**. Lava is usually pushed out, or extruded, from inside Earth in volcanoes or through large cracks in Earth's crust. Therefore, igneous rock formed from lava is called **extrusive**, or *volcanic*, rock.

Lava cools and solidifies into rock rapidly. When lava is extruded from a volcano or cracks in the crust, it is instantly exposed to an environment that may be 1,000°C colder than the lava's location just moments before. Since lava cools and solidifies rapidly, the ions in it have less time to align themselves into crystals, and the crystals that form are generally too small to be seen by the unaided eye. A magnifier or microscope may be needed to see the crystals in an extrusive rock. Rocks with crystals that can be seen only with magnification have a fine texture.

In some cases, magma cools so rapidly that no crystalline structures can form and the ions literally become frozen in place. When a rock of this type is examined with a microscope, no crystals can be seen. The physical characteristics of these unusual rocks are the same as those of glass, so they are called *volcanic glasses*. Their texture, in which no mineral crystals or grains are visible, is termed glassy. Obsidian is a common volcanic glass.

Sometimes, as lava is being extruded, gases dissolved in the lava form tiny bubbles. A glassy lava froth forms similar to the froth that overflows from a bottle of soda after it has been shaken and opened. In thick, viscous lava, the rock solidifies around the bubbles before the gas can completely escape, and a type of rock called *pumice* is formed. Since most of the glassy bubbles remain sealed, pumice will often float on water. In thinner, runnier lavas, the gas tends to bubble away rather than to form pumice. However, rapidly cooling surfaces may become pockmarked with small openings, or **vesicles**, made by the final escaping gas bubbles just as the lava solidified.

When lava is very forcibly extruded, as in an explosive volcanic eruption, a whole array of unusual rock materials form. These range from the thin, glassy strands ("Pele's hair") produced when very liquid lava is sprayed out, to large, streamlined globs ("volcanic bombs") that solidify while hurtling through the air. Such shapes are clues to the way in which the rock formed.

Basalt and rhyolite are typical extrusive rocks. The differences in the grain sizes of typical intrusive and extrusive rocks are shown in Figure 14.3.

(a) Basalt: Fine texture   **(b) Diabase**   **(c) Gabbro:** Coarse texture

Figure 14.3 Comparison of Grain Sizes in a Continuum from Extrusive (a) and Intrusive (b and c) Rocks. Basalt (a) forms at the surface. Diabase (b) may form as an intrusion near the surface as in a dike or sill. Gabbro (c) cools very slowly deep underground.

Porphyrys

In most igneous rocks the crystal grains are all roughly the same size, all large or all small. One type, however, has an unusual texture. In **porphyrys**, coarse mineral grains are embedded in a mass of fine mineral grains (see Figure 14.4). The isolated, large crystals are called *phenocrysts*; the surrounding fine-grained material is the *groundmass*. Porphyrys are thought to form in two stages: (1) a magma begins to solidify slowly at depth, and (2) the magma rises rapidly to the surface, where it finishes solidifying. The crystals formed at depth are large because the rate of cooling is slow. Nearer the surface, cooling is more rapid and the crystals formed are tiny.

Mineral Composition of Igneous Rocks

Figure 14.4 Porphyritic Texture—Large Phenocrysts of Quartz, Potassium Feldspar, and Plagioclase in Granite.

The mineral composition of an igneous rock is the product of the magma from which the rock solidifies. All common igneous rocks are mixtures of two or more of the following minerals: quartz, potassium feldspar, plagioclase feldspar, biotite mica, amphibole, pyroxene, and olivine. All are silicates, but they vary in composition, color, density, and other properties.

These silicate minerals are divided into two main groups: the **felsic** group, consisting of quartz and potassium and plagioclase feldspars, and the **mafic** group, consisting of silicates rich in magnesium and iron, such as olivine, pyroxene, amphibole, and biotite. The word *felsic* comes from *fel*dspar and *si*lica (the name of the compound SiO_2, of which quartz is composed). *Mafic* comes from *ma*gnesium and the symbol *Fe* for iron. Felsic lavas have a thick, pudding-like consistency. As a result, they clog up fissures, allowing pressure to build up and leading to explosive eruptions. Mafic lavas have a thinner, syrupy consistency. They tend to flow more freely and to erupt as a flowing river of lava or spurt out of fissures like a fountain. Also, felsic minerals are lighter in color and lower in density than mafic minerals. Felsic rocks are common in the continents, while mafic rocks are more frequently found in the ocean basins.

The Scheme for Igneous Rock Identification

There are more than 1,500 different igneous rocks, and the system for classifying them is very complex. Petrologists are trying to simplify the system. Figure 15.5 can be used to make a very basic classification of an igneous rock.

This scheme has two main parts: an upper half, in which some of the most common igneous rocks are arranged by characteristics, and a lower half, a diagram that shows the percent mineral compositions of these rocks.

In the upper half, rocks such as granite and rhyolite, which are composed of light-colored minerals, are on the left. Rocks such as basalt and gabbro, which are composed of dark-colored minerals, are on the right. Since dark-colored minerals tend to be denser than light-colored minerals, the dark-colored rocks on the right are denser than the light-colored rocks on the left. Glassy extrusive rocks (e.g., obsidian and pumice) are at the top. Fine-grained extrusive rocks (e.g., rhyolite and basalt) are in the middle. Coarse-grained, intrusive rocks (e.g., granite and gabbro) are on the bottom.

Figure 14.5 Scheme for Igneous Rock Identification. Source: The State Education Department, *Earth Science Reference Tables*, 2011 ed. (Albany, New York; The University of the State of New York).

In the lower half, the diagram shows the relative percents of the minerals of which the rocks are composed. As you can see, a particular type of rock can have a mineral composition that varies within a certain range. Look, for example, at the two different granites. Granite A consists of about 50% potassium feldspar, 30% quartz, 10% plagioclase feldspar, 7% biotite, and 3% amphibole. Granite B consists of only 10% potassium feldspar, 35% quartz, 25% plagioclase feldpsar, and 15% each of biotite and amphibole.

The following is a brief description of the rocks in Figure 14.5, starting at the top.

OBSIDIAN and BASALTIC GLASS form when lava cools so rapidly that crystals have no time to form before the lava solidifies. These rocks are very hard and brittle and break along shell-like depressions, forming sharp edges. Primitive peoples used them to make knives.

PUMICE hardens just as escaping gases whip the lava into a froth. The cooling rate is so fast that gases are trapped inside the frothy hardening rock.

Thus, pumice is full of holes and looks somewhat like a rocky sponge. In fact, pumice is often used to scrub surfaces clean. Formerly, sailors used pumice stones to scrub the decks of wooden ships. Because of the holes, most pumice is so light it floats on water.

VESICULAR ROCKS form when gas escaping from lava bubbles away into the atmosphere rather than being trapped inside and forming pumice. Vesicles are small openings formed by escaping gas just as the lava flow hardens. In many flows the vesicles later become filled with calcite, quartz, or some other mineral deposited by heated groundwater. The type of lava determines which vesicular rock forms. For example, a rhyolite lava will form vesicular rhyolite. In scoria, the gases escape from the just-hardening lava as large bubbles, forming a texture full of cavities. Large bubbles are more likely to form in a thin, fluid basalt lava. Thus, scoria has a basaltic composition.

RHYOLITE contains the same minerals as granite, but it is fine grained and much rarer than granite. It is formed in lava flows and can be found in volcanic regions.

ANDESITE is named for the Andes Mountains, where it was first studied. Andesites are fine grained and usually gray or green in color. They have the same composition as diorites.

BASALT is the most abundant extrusive rock. Most lava flows form basalt, which is dark colored and fine grained. The ocean floor is mostly basalt. Sometimes basalt rock is crushed and used in road and railroad beds.

GRANITE consists mainly of quartz and feldspars. It is light colored and coarse grained and is thought to form by very slow cooling deep within Earth. Granite is the most abundant of all the igneous rocks. Most of the crust beneath the continents is made of granite. Granite is commonly used as a building stone and in making monuments.

DIORITE is coarse grained and contains about equal amounts of light- and dark-colored minerals, giving it a "salt and pepper" appearance.

GABBRO is a dark, coarse-grained rock. It is a common intrusive rock in New England and other mountain areas. It is sometimes used as a building stone.

PERIDOTITE is composed almost entirely of two minerals: olivine and pyroxene. It occurs widely, mostly in small plutonic bodies. It is a dark-colored rock with a high density—3.3 grams per cubic centimeter.

DUNITE is composed almost entirely of olivine. It is very rare and ranges in color from yellow to dark green, with a texture like hard brown sugar. It is named for the Dun Mountains of New Zealand, one of the few places it is found.

PEGMATITE has abnormally large crystals. It forms from a mixture of minerals that have relatively low melting temperatures, and are therefore the last to solidify from a cooling magma. It forms from a magma rich in silica and consists largely of quartz, potassium feldspar, and muscovite and biotite mica. Individual minerals grains several meters long have been found in some pegmatites, and it is mined for muscovite mica, which can sometimes be found in sheets up to a meter across.

Sedimentary Rocks

Sediments

All sedimentary rocks are composed of sediments. **Sediments** can be broadly defined as solid fragments of material that have been transported and then deposited by air, water, or ice. Most sediments are fragments of rock, but bits and pieces of plants and animals, or even molecules dissolved in water, can also be sediments.

When deposited, sediments typically form loose, unconsolidated layers on Earth's surface. These layers may then be transformed by physical and chemical changes into sedimentary rock.

Lithification

Lithification is the conversion of loose sediments into coherent, solid rock by processes such as cementation, compaction, dessication, and crystallization. This conversion may occur while the sediment is being deposited, soon after deposition, or long after deposition.

Cementation is the binding together of sediment particles by substances such as clay, carbonates, and hydrates of iron. These substances literally act as cement to hold particles together, in a solid mass, when they *crystallize* or otherwise accumulate between the loose sediment particles.

Compaction is the reduction in volume or thickness of a sediment layer due to the increasing weight of overlying sediments that are continually being deposited or to the pressure produced by movements of Earth's crust. Compaction usually results in a decrease in pore space within the sediments as they become more tightly packed. This, in turn, commonly leads to *desiccation*, the drying out or drying up of water, as it is forced out of the pore spaces during compaction. As the water dries up, minerals dissolved in the water crystallize and help to hold the rock together.

Figure 14.6 illustrates the compaction and cementation processes.

Figure 14.6 Compaction and Cementation.

Most sedimentary rocks form as a result of the compression and cementing of sediments. In most cases these sediments are rock fragments that range in size from tiny particles, such as clay and silt, to larger particles, such as sand and pebbles.

Some sedimentary rocks, however, result from the evaporation of seawater or from organic processes. Seawater has many substances dissolved in it; most are salts. When seawater evaporates, the salts are left behind. They form salt crystals, which settle out of the water in layers. Rock salt forms in this way as seawater evaporates.

Some organisms, such as corals, take substances dissolved in seawater into their bodies and form skeletons. When these organisms die, their skeletons settle to the ocean floor and can build up in layers. Biogenic limestones are formed in this way.

Plant materials that build up in layers can change drastically during compaction. They become desiccated, and the substances from which they were composed break down. Since carbon is the chief component of organic materials, it is often the end product of the lithification of plant parts. Coal typically forms in this way.

Characteristics of Sedimentary Rocks

Sedimentary rocks have three unique characteristics:

They *form at or near the surface of Earth at normal temperatures and pressures*. As a result, the environment in which sedimentary rocks form is quite different from the environments in which most other types of rock have their origins. Therefore, the minerals contained in sedimentary rocks are often quite different from those found in igneous or metamorphic rocks.

They *form layers*. Water is the chief transporter of sediments on Earth's surface. While being transported by water, most sediments become rounded. Sediments deposited in water settle into horizontal layers of rounded particles. This layering is preserved during lithification, so most sedimentary rocks usually contain rounded grains cemented in layers. These successive layers, known as strata, or beds, preserve evidence of past surface conditions. The individual layers can be easily distinguished because they usually differ in some way, such as color, thickness, size of particles, mineral composition, or degree of cementation.

The *surfaces of layers in sedimentary rock often contain distinctive features* identical to those found on modern sediments. Examples are ripple marks created by waves in shallow water, polygon-shaped cracks formed by fine sediment that was dried out by the Sun, footprints of walking animals, and the tracks and trails of crawling animals such as snails.

When sand and coarser sediments are moved along by air or water, they often form streamlined mounds such as dunes or ripples. Within these mounds, the layers of sediment are inclined in a down-wind or down-current direction. As successive layers of dunes or ripples pass over the surface, they leave behind layers of inclined sediments, called *cross-bedded sediments*, as shown in Figure 14.7.

They *contain fossils*. Plants and animals often live in areas where sediments are being deposited. The remains of these plants and animals are often incorporated into layers of accumulating sediment and into the sedimentary rocks that form from them.

Figure 14.7 Diagram of Cross-Bedded Sediments.

Observations of current sediment layers and surfaces indicate that some characteristics enable us to infer which is the top, and which the bottom, of a series of sedimentary rock layers. This, in turn, allows us to infer the relative ages of the layers (bottom is oldest, and top is youngest). When a mixture of sediments settles in water, the larger particles reach the bottom first, and the smallest last. The result is graded bedding, as shown in Figure 14.8a. When graded bedding is preserved in a sedimentary rock, we can infer that the original top was where the smaller particles are located. Similarly, cross-bedded sediments (Figure 14.8b) allow us to infer not only top and bottom, but also the direction in which the wind or water current was moving when the beds were deposited. If a rock layer contains ripple marks, mud cracks, footprints, or trails, we can infer the top of the layer from their position. Also, when certain shells settle in water (Figure 14.8c), they end up with one side facing upward most of the time, so their position in a sedimentary rock again allows us to infer top and bottom.

(a) Graded Bedding **(b) Cross Bedding** **(c) Position of Fossils**

Figure 14.8 Diagram of Graded Bedding, Cross Bedding, and Fossil Position All Showing the Top of a Series of Layers. (a) Graded Bedding. Within each bed, particles are sorted and form layers that decrease in size from the bottom to the top of the bed. (b) Cross Bedding. Particles are well sorted and form inclined layers within each bed. The layers in the bed are steeply sloping at the top and gently sloping at the bottom. (c) Position of Fossils. The hard parts of many organisms that become fossilized have a preferred orientation after settling through water. For example, clam shells settle more often with the cupped side of the shell facing down.

Types of Sediment Particles

There are three major types of sediment particles: clastic, organic or biogenic, and chemical.

Rocks

Clastic sediments are rock or mineral fragments formed by the breakdown of rock due to weathering. The vast majority of sediments are of this type. These particles are grouped and named by size according to the Wentworth scale, as shown in Table 14.1.

TABLE 14.1 THE WENTWORTH SCALE

Particle-Size Range (diameter, cm)	Particle Name	Sediment Composed of That Particle
Greater than 25.6	Boulder	Boulder gravel
6.4–25.6	Cobble	Cobble gravel
0.2–6.4	Pebble	Pebble gravel
0.006–0.2	Sand	Sand
0.0004–0.006	Silt	Silt
Less than 0.0004	Clay	Clay

Organic, or **biogenic**, **sediments** are particles produced by the life activities of plants or animals. The most abundant organic sediment consists of the shells of aquatic animals. These may range from the large calcium carbonate shells of clams or snails to the microscopic silica skeletons of amoebas and diatoms. There are even some algae, which absorb calcium carbonate from seawater and incorporate it into their skeletons. When these organisms die, their shells and skeletons settle to the bottom and accumulate in layers of mud.

In warm seas, coral reefs form when huge populations of little animals called corals secrete calcium carbonate. Wave action smashes the reef into pieces, along with the shells of other animals living on the reef, producing a coarse sand or granule-sized sediment.

Organic sediments also include the woody tissue of plants (tree trunks, branches, twigs, and leaves), as well as spores and pollen. *Peat* is an organic sediment formed from plant material that accumulates in swamps, where stagnant water conditions prevent the oxidation of the plant material. Over time, buried peat may be transformed into coal.

The fatty or waxy tissues of land plants and the soft tissues of aquatic plants and animals do not form particles. However, this soft organic matter is usually mixed in with sediment particles and is considered to be the raw material from which petroleum forms underground.

Fossils in sedimentary rocks provide evidence of the environment in which they formed. The fossil of an organism that is clearly a fish, together with fossilized clam shells, indicates an aquatic environment. Pollen from citrus or palm trees in a rock is evidence of a warm terrestrial environment.

Chemical sediments consist of particles (crystals) that have crystallized out of solutions at or near Earth's surface. All the water on Earth's surface has some material dissolved in it. Seawater, for example, contains over 80 elements and hundreds of different compounds. A number of conditions can cause some of the substances dissolved in water to crystallize.

Let's use a simple analogy. Think of a solution as a container in which a substance is being held. A glass of juice is a good example. The solvent is like the glass, and the solute is like the juice. If there is more juice than can fit into the glass, the juice will overflow and spill to the ground. Now think of seawater. Water is the solvent, or glass; salts are the solute, or juice. Overflow is like crystallization; spilling to the ground is like precipitation.

In our example, there are two things that can change—the size of the glass and the amount of juice. Having more solvent is like having a bigger glass. Similarly, if more water is added, seawater can hold more salt. On the other hand, removing water is like having a smaller glass. If water is removed, less salt can be held.

Evaporation removes water from seawater. When the water in a solution evaporates, the solution becomes more and more concentrated. When it reaches saturation, the substance dissolved in the solution begins to crystallize out. Sediments made of particles that crystallized as a result of evaporation are called *evaporites*. Halite crystals are a common chemical sediment that forms along the edges of bodies of water undergoing extensive evaporation.

If you have ever tried to dissolve sugar in iced tea, you know that less sugar dissolves in cold tea than hot tea. The temperature of a solution affects the amount of material that can be dissolved in it. As the temperature of a solution falls, its ability to hold dissolved materials decreases. Cooling a solution is like making the glass in the preceding example smaller without removing any of the juice. The result is overflow. For example, when the mineral-laden water of a hot spring cools, calcite or even opal may precipitate from the water.

Particles may also crystallize out of a solution if more solute is added than the solution can hold. If you add juice to an already full glass, the glass will overflow. For example, aquatic plants and animals can increase the amount of carbon dioxide (CO_2) dissolved in the water around them and thereby cause calcium carbonate to crystallize out and sink to the bottom, or precipitate. Also, if a chemical reaction in a solution produces a substance that doesn't dissolve in water, that substance will also precipitate.

The Scheme for Sedimentary Rock Identification

Sedimentary rocks are classified as *fragmental*, *organic*, or *chemical*, depending on the sediments from which they lithified. As with igneous rocks, texture and mineral composition are the main properties used to classify and name sedimentary rocks, as shown in Figure 14.9. In sedimentary rocks, the mineral composition of the sediment particles tells us where the particles came from, and the texture provides clues about the processes that deposited the sediments.

In Figure 14.9, sedimentary rocks are divided into two main groups. In the upper part of the chart are "Inorganic Land-Derived Sedimentary Rocks," that is, rocks formed from clastic sediments. In the lower part are "Chemically and/or Organically Formed Sedimentary Rocks," that is, rocks formed from chemical or organic sediments.

Clastic sedimentary rocks are classified on the basis of grain size. Nonclastic rocks are identified primarily by composition and texture.

Rocks

INORGANIC LAND-DERIVED SEDIMENTARY ROCKS					
TEXTURE	GRAIN SIZE	COMPOSITION	COMMENTS	ROCK NAME	MAP SYMBOL
Clastic (fragmental)	Pebbles, cobbles, and/or boulders embedded in sand, silt, and/or clay	Mostly quartz, feldspar, and clay minerals; may contain fragments of other rocks and minerals	Rounded fragments	Conglomerate	
			Angular fragments	Breccia	
	Sand (0.006 to 0.2 cm)		Fine to coarse	Sandstone	
	Silt (0.0004 to 0.006 cm)		Very fine grain	Siltstone	
	Clay (less than 0.0004 cm)		Compact; may split easily	Shale	
CHEMICALLY AND/OR ORGANICALLY FORMED SEDIMENTARY ROCKS					
TEXTURE	GRAIN SIZE	COMPOSITION	COMMENTS	ROCK NAME	MAP SYMBOL
Crystalline	Fine to coarse crystals	Halite	Crystals from chemical precipitates and evaporites	Rock salt	
		Gypsum		Rock gypsum	
		Dolomite		Dolostone	
Crystalline or bioclastic	Microscopic to very coarse	Calcite	Precipitates of biologic origin or cemented shell fragments	Limestone	
Bioclastic		Carbon	Compacted plant remains	Bituminous coal	

Figure 14.9 Scheme for Sedimentary Rock Identification. Source: The State Education Department, *Earth Science Reference Tables*, 2011 ed. (Albany, New York; The University of the State of New York).

Metamorphic Rocks

Formation of Metamorphic Rocks

Metamorphic rocks get their name from the Greek words *meta*, meaning "change," and *morphs*, meaning "form." A metamorphic rock has literally been changed from its original form. A rock changes when it is exposed to an environment significantly different from the one in which it originally formed. In general, three factors cause metamorphic changes in rock: heat, pressure, and chemical activity.

Heat

Most minerals expand when heated, causing the atoms to move apart, and the bonds that hold them together to be stretched and weakened. If a mineral is heated enough, all of the bonds break, and the mineral melts. If a rock experiences such heating, a magma forms. In metamorphism, however, heating breaks *some* but not all of the bonds in the minerals of the rock. *Some* atoms break loose, migrate through the rock, and join with other atoms to form new minerals—metamorphic minerals—in the rock. The result is to change both the chemical composition and the structure of the rock, that is, to change the form of the rock. Such extreme heating of rock may be caused by deep burial (temperature increases with depth) or by contact with hot materials, such as magma.

Pressure

The effect of pressure on minerals is opposite to that of heat. Pressure forces the atoms in the mineral much closer together. The resulting stress on the bonds causes some of them to break. This, in turn, enables the atoms to rearrange into a more compact structure. The result is a significant change—a denser, harder rock.

Pressure exerted on rock may be due to deep burial or to movements in Earth's crust. Pressure due to deep burial is equal in all directions, but crustal movements may produce pressures that are greater in one direction than another, thereby deforming or flattening rocks. The effects on the particles in a rock are different in these two cases, as shown in Figure 14.10.

Figure 14.10 Equal Versus Directed Pressure on a Particle. Equal pressure (a) causes the particle to decrease in size, but the shape is unchanged. Unequal pressure (b) causes both a decrease in size and a change in shape.

Chemical Activity

A rock can also be changed when solutions rich in dissolved ions move through it. Such solutions are given off by cooling magmas; others come from metamorphism taking place deep underground. As these solutions move through pores in surrounding rock, they interact with minerals in the rock to form new minerals. Sometimes the water itself combines with minerals to form new ones. For example, olivine reacts with water to form talc or serpentine.

Metamorphism

The changes that occur in solid rock due to great temperature, great pressure, and chemical activity are known as **metamorphism**. Metamorphism occurs while rocks are still in the solid state. Changes produced by weathering, although they occur while rocks are solid, are not considered metamorphism. Neither are igneous processes such as the melting of rocks in high-temperature zones. True metamorphism occurs beneath Earth's surface at high temperatures and pressures. Under the conditions that produce metamorphism, ordinarily rigid, brittle rocks behave like soft modeling clay. The rocks can be bent and twisted without breaking. Rocks that originally had a clastic texture can recrystallize so that the metamorphic rock has a crystalline texture. Conversely, rocks that were originally crystalline can be crushed and broken into pieces and then relithified to form a clastic texture.

The preexisting rock from which a metamorphic rock forms is called the **parent rock**. By investigating the conditions that produce metamorphic changes, it is possible to infer the parent rock from the mineral compositions and structures of most metamorphic rocks.

Metamorphic rocks occur on a continuum from little alteration in the parent rock to major changes. Rarely are there sharp boundaries between parent rocks and metamorphic rocks; instead, gradations lead from one to the other. Limestone grades almost imperceptibly into marble. Shales grade into schists. Sandstones grade into quartzites. This characteristic has been helpful in determining the origins of certain metamorphic rocks. Metamorphic rocks are divided into groups based on the parent materials from which they were derived.

Types of Metamorphism
There are two major types of metamorphism: contact and regional. **Contact metamorphism** occurs when molten rock comes into contact with surrounding rocks. Heat given off by the cooling magma is conducted into the rock, where it causes changes in texture and mineral composition. Contact metamorphism is most intense at the contact between the magma and the surrounding rock, and decreases with distance from the magma until points are reached at which the effects of the heat are not felt. Generally, crystals are largest at the point of contact and become smaller with distance.

Regional metamorphism occurs over large areas and is generally associated with mountain building. Regional metamorphism takes its name from the large scale on which it occurs. For example, about 300 million years ago North America and Africa collided along a boundary between plates of Earth's crust. The result was the crushing of rocks from Alabama to New York and the formation of the Appalachian Mountains. During regional metamorphism, deeply buried sedimentary and igneous rocks are squeezed both by the weight of rock layers above and by movements of the crust, are intruded by igneous rocks, and are infused with chemically active fluids.

Characteristics of Metamorphic Rocks

When a rock is metamorphosed, nearly every characteristic can change, including texture and mineral composition. These changes can be so great that the metamorphic rock bears little, if any, resemblance to its original state. Therefore, to infer what a metamorphic rock once was, one needs to know what kinds of changes take place during metamorphism.

Changes in Texture
Crystalline Texture. During metamorphism the size, shape, and spacing of the grains (crystals or particles) in a rock are changed. Under both heat and pressure, the grains in a rock may partially melt and fuse together, forming larger grains (recrystallization); or, pressure may break brittle grains into smaller pieces, destroying the original texture of the rock. The heat and the pressure of metamorphism fuse any fragmental textured rock into a solid,

crystalline mass. As a result, most metamorphic rocks have a crystalline texture.

Increased Density. Pressure from deep burial closes pore spaces in sedimentary rocks and openings in igneous rocks. Pressure also compresses grains so that they are smaller and closer together. The rock becomes more compact. Its mass is forced into a smaller volume, and its density increases. Thus, the igneous or sedimentary rock is changed—it becomes metamorphic rock.

Foliation and Banding. A rock is said to be foliated when its crystals are arranged in layers or bands along which the rock breaks easily. These layers may be microscopic or thick enough to be easily seen. *Foliation* results when mineral grains recrystallize or are flattened under pressure. Even pressure compresses the grains so that they are smaller and closer together, but retain their original shape. Uneven pressure not only compresses grains but also changes their shape. Grains that may have been randomly oriented will form parallel layers (foliation) or become aligned when deformed under pressure.

Compression causes randomly dispersed crystals to become oriented in a direction perpendicular to the pressure. The oriented crystals form thin sheets or layers called *foliation*.

Migration of mineral crystals due to density differences produces *banding*—regions of light- and dark-colored minerals.

Figure 14.11 Random to Oriented Patterns Due to Pressure and Fusing of Grains, and Banding Caused by Density Separation.

Banding occurs when minerals of different densities recrystallize under pressure and separate into layers, like a mixture of oil and water. Since light-colored minerals tend to be less dense than dark ones, the layers that form are alternating light and dark.

Figure 14.11 shows the effects of pressure and recrystallization on rocks.

Distortion of Layers. When sedimentary rocks are metamorphosed, the layers in the rock may be distorted. Softened by the heat and squeezed by the pressure, the once-flat layers become twisted and contorted. Such distortion is commonly seen in large-scale formations of metamorphic rock.

Changes in Mineral Composition

The changes described above can occur without altering the original mineral composition of the rock. When a rock is metamorphosed, however, the original minerals often disappear and new minerals form in their place. The underlying reason for this change is lack of stability. Minerals that are stable in one environment are not stable in another. Sugar is stable at room temperature; but when heated strongly, it becomes unstable and breaks down into carbon and water. Clay minerals, such as kaolinite, provide another good example. Kaolinite forms by the weathering of feldspar. It has an open, sheetlike molecular structure that is stable at the temperatures and pressures normally found at Earth's surface. Under the heat and pressure associated with deep burial, however, it cannot survive, and forms micas that have a more compact, sheetlike molecular structure.

Classification of Metamorphic Rocks

Metamorphic rocks are classified according to mineral composition and texture, including foliation and banding.

An interesting question associated with metamorphic rocks is, When has a rock changed enough to be called metamorphic? Rocks do not change instantaneously from one type to another; the changes occur along a continuum from little alteration to major changes. Generally, if a change can be recognized in a rock, the rock is considered metamorphic. Table 14.2 lists some common rocks and the rocks they can become when metamorphosed.

TABLE 14.2 DERIVATIONS OF COMMON METAMORPHIC ROCKS

Parent Rock	Metamorphic Rock
Shale	Compaction and reorientation → slate
	Slate → micas and chlorite form → phyllite
	Phyllite → grains recrystallize and get coarser → schist
	Schist → banding occurs → gneiss
Shale	Contact with magma → granofels
Sandstone	Quartzite
Limestone or dolostone	Marble
Basalt	Hydration reactions → greenschist
	Greenschist → dehydration → amphibolite
	Amphibolite → banding occurs → gneiss

The Scheme for Metamorphic Rock Identification

Metamorphic rocks are classified according to their textures (including foliation and banding) and compositions. In Figure 14.12 metamorphic rocks are divided into two main groups, foliated and nonfoliated, based on texture. Then, within each of these groups, the rocks are classified by grain size.

If you look at the column headed "Composition," you will see that the foliates share a common composition, while each of the nonfoliates has a distinctly different composition. The majority of foliated rocks are derived from clastic sedimentary rocks rich in clay minerals, such as shale and siltstone. The "Comments" column tell you what the parent rock was in each case.

Notice that slate, schist, and gneiss really form a continuum. Slate forms first as shale is slightly metamorphosed. The flakelike micas align because of compression and form surfaces along which the rock will split. With further metamorphosis, alignment of the mica becomes even more pronounced and phyllite is formed. As metamorphosis continues, chlorite changes to biotite, foliation becomes stronger, grains become coarser because of recrystallization, and a schist is formed. Still further metamorphism causes the crystals to grow even larger, alternating layers of light and dark minerals (bands) form as crystals migrate because of density differences, and a gneiss forms. The larger, interlocking crystals of a gneiss cause these rocks to break less regularly than schists, but they will often break completely across layers. In the highly metamorphic environment in which gneisses form, rock is plastic and will often be contorted into the weirdly twisted patterns seen in gneiss.

TEXTURE		GRAIN SIZE	COMPOSITION	TYPE OF METAMORPHISM	COMMENTS	ROCK NAME	MAP SYMBOL
FOLIATED	MINERAL ALIGNMENT	Fine	MICA QUARTZ FELDSPAR AMPHIBOLE GARNET PYROXENE	Regional (Heat and pressure increases)	Low-grade metamorphism of shale	Slate	
		Fine to medium			Foliation surfaces shiny from microscopic mica crystals	Phyllite	
					Platy mica crystals visible from metamorphism of clay or feldspars	Schist	
	BAND- ING	Medium to coarse			High-grade metamorphism; mineral types segregated into bands	Gneiss	
NONFOLIATED		Fine	Carbon	Regional	Metamorphism of bituminous coal	Anthracite coal	
		Fine	Various minerals	Contact (heat)	Various rocks changed by heat from nearby magma/lava	Hornfels	
		Fine to coarse	Quartz	Regional or contact	Metamorphism of quartz sandstone	Quartzite	
			Calcite and/or dolomite		Metamorphism of limestone or dolostone	Marble	
		Coarse	Various minerals		Pebbles may be distorted or stretched	Metaconglomerate	

Figure 14.12 Scheme for Metamorphic Rock Identification. Source: The State Education Department, *Earth Science Reference Tables*, 2011 ed. (Albany, New York; The University of the State of New York).

WHAT IS THE ROCK CYCLE?

Where do rocks come from? You might say igneous rock comes from magma, sedimentary rock comes from sediments, and metamorphic rock comes from other rocks that have changed. But consider that magma is melted rock, sediments are broken rock, and there had to be an already-existing rock to be metamorphosed. Basically, then, rocks come from other rocks.

Rocks are constantly changing from one type to another in a never-ending **rock cycle**, shown in Figure 14.13. The outside circle shows the different forms in which rock matter can exist: magma, igneous rock, sediments, and so on. Arrows leading from one form to another are labeled with the changes that are taking place. Arrows inside the circle show some alternative paths in the rock cycle. For example, an igneous rock does not always break down into sediments. If buried, it may be metamorphosed; therefore, there is an arrow inside the circle leading from igneous rock to metamorphic rock.

Rocks

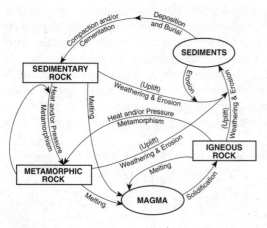

Figure 14.13 Rock Cycle in Earth's Crust. Source: The State Education Department, *Earth Science Reference Tables*, 2011 ed. (Albany, New York; The University of the State of New York).

Let's follow a rock through one possible cycle, beginning with magma. A magma solidifies and forms an igneous rock. The igneous rock is uplifted and exposed at the surface, where weathering breaks it down into sediments. Rain washes the sediments into a stream. The stream carries them to the ocean, where they are deposited and buried. Compaction and cementation produce a sedimentary rock. The sedimentary rock is exposed to heat and pressure, perhaps as two of the plates in Earth's crust collide, and a metamorphic rock is formed. If the metamorphic rock is forced deep enough into the crust, it melts and forms magma.

Notice that we have come full circle back to where we started. This is an example of the rock cycle. It is, however, only one of many possible paths. Except for the stray meteorite that reaches the surface, Earth is essentially a closed system. Rock doesn't enter or leave, but is constantly recycling from one form to another.

The rock cycle diagram has no beginning and no end, but of course the rock cycle really did start somewhere. There is much evidence that Earth was originally totally molten. No solid rock existed; there was only magma. Therefore, it is thought that the original rocks formed as this magma cooled, and the rock cycle began with igneous rocks.

MULTIPLE-CHOICE QUESTIONS

In each case, write the number of the word or expression that best answers the question or completes the statement.

1. All rocks contain
   (1) minerals
   (2) sediments
   (3) intergrown crystals
   (4) fossils

2. Rocks are classified as igneous, sedimentary, or metamorphic based primarily on their
   (1) texture
   (2) crystal or grain size
   (3) method of formation
   (4) mineral composition

3. Although more than 2,000 minerals have been identified, 90 percent of Earth's lithosphere is composed of 12 minerals: feldspar, quartz, mica, calcite, amphiboles, kaolinite, augite, garnet, magnetite, olivine, pyrite, and talc.
The best explanation for this fact is that most rocks
   (1) are monomineralic
   (2) are composed only of recrystallized minerals
   (3) have a number of minerals in common
   (4) have a 10% nonmineral composition

4. Which properties of a rock provide the most information about the environment in which the rock formed?
   (1) texture, composition, and structure
   (2) mass, composition, and color
   (3) density, volume, and mass
   (4) color, texture, and volume

5. Which processes lead directly to the formation of igneous rock?
   (1) weathering and erosion
   (2) compaction and cementation
   (3) heat and pressure
   (4) melting and solidification

6. The igneous rock gabbro most likely formed from molten material that cooled
   (1) rapidly at Earth's surface
   (2) slowly at Earth's surface
   (3) rapidly, deep underground
   (4) slowly, deep underground

7. Which igneous rock, when weathered, could produce sediment composed of the minerals potassium feldspar, quartz, and amphibole?
   (1) gabbro   (3) andesite
   (2) granite   (4) basalt

8. Which graph best shows the relationship between the compositions of different igneous rocks and their densities?

(1)

(2)

(3)

(4)

9. Which graph best represents the textures of granite, pegmatite, and rhyolite?

10. A nonvesicular rock is made entirely of green 2-millimeter-diameter crystals that have a hardness of 6.5 and show fracture, but *not* cleavage. The rock is most likely
    (1) shale
    (2) phyllite
    (3) dunite
    (4) schist

11. The graph below shows the relationship between the cooling time of magma and the size of the crystals produced.

Which graph correctly shows the relative positions of the igneous rocks granite, rhyolite, and pumice?

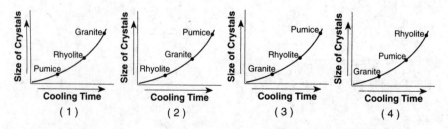

12. Which texture best describes an igneous rock that formed deep underground?
    (1) glassy
    (2) vesicular
    (3) fine grained
    (4) coarse grained

13. Obsidian's glassy texture indicates that it formed
    (1) slowly, deep below Earth's surface
    (2) slowly, on Earth's surface
    (3) quickly, deep below Earth's surface
    (4) quickly, on Earth's surface

14. The three statements below are observations of the same rock sample:
    - The rock has intergrown crystals from 2 to 3 millimeters in diameter.
    - The minerals in the rock are gray feldspar, green olivine, green pyroxene, and black amphibole.
    - There are no visible gas pockets in the rock.

    This rock sample is most likely
    (1) sandstone
    (2) gabbro
    (3) granite
    (4) phyllite

15. An extrusive igneous rock with a mineral composition of 35% quartz, 35% potassium feldspar, 15% plagioclase feldspar, 10% biotite, and 5% amphibole is called
    (1) rhyolite
    (2) granite
    (3) gabbro
    (4) basaltic glass

16. The diagram below shows magnified views of three stages of mineral crystal formation as molten material gradually cools.

    Stage 1   Stage 2   Stage 3

    Which rock normally forms when minerals crystallize in these stages?
    (1) shale
    (2) gneiss
    (3) gabbro
    (4) breccia

17. Which igneous rock has a vesicular texture and a felsic composition?
    (1) pumice
    (2) basalt
    (3) granite
    (4) scoria

18. Which mineral is most frequently found in both granitic continental crust and basaltic oceanic crust?
    (1) olivine
    (2) potassium feldspar
    (3) plagioclase feldspar
    (4) quartz

19. Where are Earth's sedimentary rocks generally found?
    (1) in regions of recent volcanic activity
    (2) deep within Earth's crust
    (3) along the midocean ridges
    (4) as a thin layer covering much of the continents

20. Rock layers showing ripple marks, cross bedding, and fossil shells were formed
    (1) from solidification of molten material
    (2) from deposits left by a continental ice sheet
    (3) by high temperature and pressure
    (4) by deposition of sediments in a shallow sea

21. A student classified the rock below as sedimentary.

    Which observation about the rock best supports this classification?
    (1) The rock is composed of several minerals.
    (2) The rock has a vesicular texture.
    (3) The rock contains fragments of other rocks.
    (4) The rock shows distorted and stretched pebbles.

22. Most of the sediment that is compacted and later forms shale bedrock is
    (1) clay
    (2) silt
    (3) sand
    (4) pebbles

23. One difference between a breccia rock and a conglomerate rock is that the particles in a breccia rock are
(1) more aligned
(2) more angular
(3) harder
(4) land derived

24. Which rock is sedimentary in origin and formed as a result of chemical processes?
(1) granite
(2) shale
(3) breccia
(4) dolostone

Base your answers to questions 25 through 27 on the drawings of six sedimentary rocks labeled *A* through *F*.

A Conglomerate B Breccia C Sandstone D Shale E Limestone F Rock salt

25. Most of the rocks shown were formed by
(1) volcanic eruptions and crystallization
(2) compaction and/or cementation
(3) heat and pressure
(4) melting and/or solidification

26. Which two rocks are composed primarily of quartz, feldspar, and clay minerals?
(1) rock salt and conglomerate
(2) rock salt and breccia
(3) sandstone and shale
(4) sandstone and limestone

27. Which table shows the rocks correctly classified by texture?

Texture	clastic	bioclastic	crystalline
Rock	A, B, C, D	E	F

(1)

Texture	clastic	bioclastic	crystalline
Rock	A, C	B, E	D, F

(3)

Texture	clastic	bioclastic	crystalline
Rock	A, B, C	D	E, F

(2)

Texture	clastic	bioclastic	crystalline
Rock	A, B, F	E	C, D

(4)

Rocks

Base your answers to questions 28 through 30 on the graph below and on your knowledge of Earth science. The graph shows the temperature, pressure, and depth environments for the formation of the three major rock types. Pressure is shown in kilobars (kb). Letters A through D identify different environmental conditions for rock formation.

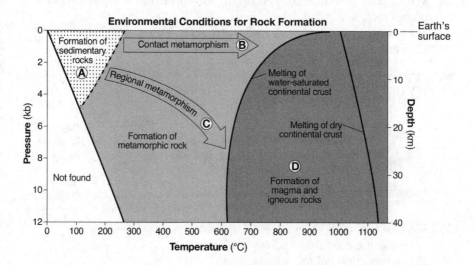

28. Which rock is most likely to form directly from rock material at a depth of 30 km and a temperature of 1000°C?
    (1) quartzite (3) shale
    (2) scoria (4) granite

29. Which letter represents the environmental conditions necessary to form gneiss?
    (1) A (3) C
    (2) B (4) D

30. At what pressure and temperature is sand most likely to be compacted into sandstone?
    (1) 2 kb and 150°C (3) 10 kb and 400°C
    (2) 6 kb and 200°C (4) 12 kb and 900°C

31. Metamorphic rocks result from the
    (1) erosion of rocks
    (2) recrystallization of rocks
    (3) cooling and solidification of molten magma
    (4) compression and cementation of soil particles

32. Which characteristics indicate that a rock has undergone metamorphic change?
    (1) The rock shows signs of being heavily weathered and forms the floor of a large valley.
    (2) The rock is composed of intergrown mineral crystals and shows signs of deformed fossils and structure.
    (3) The rock becomes less porous when exposed at the surface and is finely layered.
    (4) The rock contains a mixture of different-sized rounded grains or both felsic and mafic silicate minerals.

33. Which characteristic of rocks tends to increase as the rocks are metamorphosed?
    (1) density
    (2) volume
    (3) permeability
    (4) number of fossils present

34. Which two kinds of adjoining bedrock would most likely have a zone of contact metamorphism between them?
    (1) shale and conglomerate
    (2) shale and sandstone
    (3) limestone and sandstone
    (4) limestone and granite

35. Which statement is supported by information in the Rock Cycle in Earth's Crust diagram in the *Earth Science Reference Tables*?
    (1) Metamorphic rock results directly from melting and crystallization.
    (2) Sedimentary rock can be formed only from igneous rock.
    (3) Igneous rock always results from melting and solidification.
    (4) All sediments turn directly into sedimentary rock.

36. Which sequence of change in rock type occurs as shale is subjected to increasing heat and pressure?
    (1) shale → schist → phyllite → slate → gneiss
    (2) shale → slate → phyllite → schist → gneiss
    (3) shale → gneiss → phyllite → slate → schist
    (4) shale → gneiss → phyllite → schist → slate

37. Which rock is formed only by regional metamorphism?
    (1) slate
    (2) hornfels
    (3) dunite
    (4) marble

CONSTRUCTED RESPONSE QUESTIONS

Base your answers to questions 38 through 40 on the flowchart below and on your knowledge of Earth science. The flowchart shows the formation of some igneous rocks. The circled letters A, B, C, and D indicate parts of the flowchart that have not been labeled.

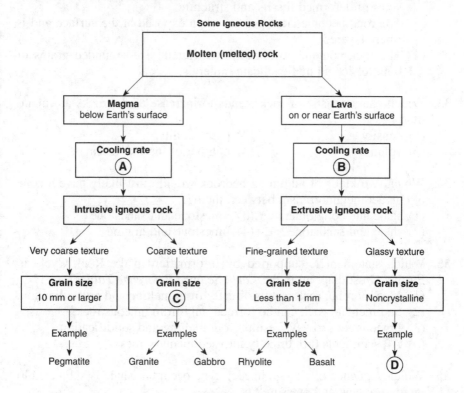

38. Contrast the rate of cooling at A that forms intrusive igneous rock with the rate of cooling at B that forms extrusive igneous rock. [1]

39. Give the numerical grain-size range that should be placed in the flowchart at C. Units must be included in your answer. [1]

40. State *one* igneous rock that could be placed in the flowchart at D. [1]

Base your answers to questions 41 through 43 on the photograph of a sample of gneiss below.

41. What observable characteristic could be used to identify this rock sample as gneiss? [1]

42. Identify *two* minerals found in gneiss that contain iron and magnesium. [1]

43. A dark-red mineral with a glassy luster was also observed in this gneiss sample. Identify the mineral and state *one* possible use for this mineral. [1]

Base your answers to questions 44 through 46 on the diagram of Bowen's Reaction Series below, which shows the sequence in which minerals crystallize as magma cools and forms different types of igneous rocks from the same magma. The arrow for each mineral represents the relative temperature range at which that mineral crystallizes.

Bowen's Reaction Series

Temperature Conditions	Minerals that Crystallize from Magma as the Magma Cools	Igneous Rock Type
High temperature (first to crystallize)	Olivine	Ultramafic (peridotite)
	Pyroxene / Amphibole / Plagioclase feldspar (More calcium rich)	Basaltic (basalt/gabbro)
Cooling magma ↓	Biotite mica / (More sodium rich)	Andesitic (andesite/diorite)
Low temperature (last to crystallize)	Muscovite mica / Quartz / Potassium feldspar	Granitic (rhyolite/granite)

44. According to Bowen's Reaction Series, how is the chemical composition of plagioclase feldspar found in basaltic rock different from the chemical composition of plagioclase feldspar found in granitic rock? [1]

45. Describe the temperature conditions shown in Bowen's Reaction Series that explain why olivine and quartz are *not* usually found in the same igneous rock type. [1]

46. Identify *one* similarity and *one* difference between the igneous rocks andesite and diorite. [1]

Base your answers to questions 47 through 49 on the magnified views shown below of the minerals found in an igneous rock and in a metamorphic rock. The millimeter scale indicates the size of the crystals shown in the magnified views.

47. Identify the environment of formation of this igneous rock based on the size of its intergrown crystals. [1]

48. Based on the minerals present, identify the relative color and density of this igneous rock compared to mafic igneous rocks with the same crystal size. [1]

49. Describe the texture shown by this metamorphic rock that indicates it could be schist. [1]

Base your answers to questions 50 through 52 on the data table below, which shows some characteristics of four rock samples, numbered 1 through 4. Some information has been left blank.

Data Table

Rock Sample Number	Composition	Grain Size	Texture	Rock Name
1	mostly clay minerals		clastic	shale
2	all mica	microscopic, fine	foliated with mineral alignment	
3	mica, quartz, feldspar, amphibole, garnet, pyroxene	medium to coarse	foliated with banding	gneiss
4	potassium feldspar, quartz, biotite, plagioclase feldspar, amphibole	5 mm		granite

50. State a possible grain size, in centimeters, for most of the particles found in sample 1. [1]

51. Write the rock name of sample 2. [1]

52. Write a term or phrase that correctly describes the texture of sample 4. [1]

Base your answers to questions 53 and 54 on the geologic cross section shown below and on your knowledge of Earth science.

53. State the name of the metamorphic rock at location *A*. [1]

54. Identify *one* characteristic that could be used to determine if the intrusive igneous rock has a mafic composition or a felsic composition. [1]

EXTENDED CONSTRUCTED RESPONSE QUESTIONS

Base your answers to questions 55 through 58 on the diagram in your answer booklet and on your knowledge of Earth science. The diagram represents several common rock-forming minerals and some of the igneous rocks in which they commonly occur. The minerals are divided into two groups, *A* and *B*. Dashed lines connect the diagram of diorite to the three minerals that are commonly part of diorite's composition.

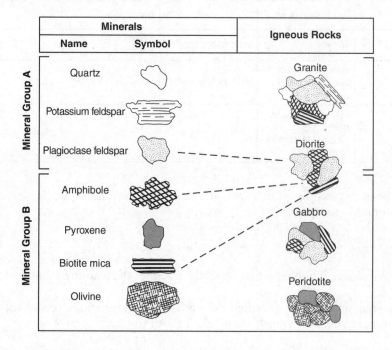

55. On the diagram, draw *five* lines to connect the diagram of granite to the symbols of the minerals that are commonly part of granite's composition. [1]

56. Describe *one* characteristic of the minerals in group *A* that makes them different from the minerals in group *B*. [1]

57. Based on the *Earth Science Reference Tables*, identify *one* other mineral found in some samples of diorite that is *not* shown in the diorite sample in the diagram. [1]

58. A sedimentary rock sample has the same basic mineral composition as granite. Describe one observable characteristic of the sedimentary rock that is different from granite. [1]

Base your answers to questions 59 through 62 on the passage below and on your knowledge of Earth science.

Dimension Stone: Granite

Dimension stone is any rock mined and cut for specific purposes, such as kitchen countertops, monuments, and the curbing along city streets. Examples of rock mined for use as dimension stone include limestone, marble, sandstone, and slate. The most important dimension stone is granite; however, not all dimension stone sold as granite is actually granite. Two examples of such rock sold as "granite" are syenite and anorthosite. Syenite is a crystalline, light-colored rock composed primarily of potassium feldspar, plagioclase feldspar, biotite, and amphibole, while anorthosite is composed almost entirely of plagioclase feldspar. Like actual granite, both syenite and anorthosite have large, interlocking crystals.

59. Explain why syenite is classified as a plutonic igneous rock. [1]

60. State *one* reason why anorthosite is likely to be white to gray in color. [1]

61. The igneous rock gabbro is sometimes sold as "black granite." Compared to the density and composition of granite, describe how the density and composition of gabbro are different. [1]

62. Identify *one* dimension stone mentioned in the passage that is composed primarily of calcite. [1]

Base your answers to questions 63 through 65 on the table and photograph below and on your knowledge of Earth science. The table shows the approximate mineral percent composition of an igneous rock. The photograph shows the true-scale crystal sizes in this igneous rock.

Mineral Name	Percentage of Mineral Present
plagioclase feldspar	55%
biotite	15%
amphibole	30%

centimeter

63. Identify *two* elements that are commonly found in all three minerals in the data table. [1]

64. Identify this igneous rock. [1]

65. Identify *two* processes that formed this rock. [1]

Unit Five
THE DYNAMIC EARTH

CHAPTER 15
EARTHQUAKES AND EARTH'S INTERIOR

> **KEY IDEAS** With current technology, we cannot drill more than a few kilometers into the solid Earth. How, then, can we know anything of the internal structure of Earth? Earthquakes provide the key to inferring this internal structure. Earthquake waves literally travel around and through Earth, because rock is an excellent medium for conducting wave motion. Networks of seismometers, which detect earthquake waves, blanket Earth's surface. The records of earthquakes detected by seismometers can be analyzed to determine precisely when and where an earthquake occurred. By analyzing records of many earthquakes, as well as earthquake-wave behavior, scientists have been able to piece together a picture of Earth's internal structure. One of the most important results of such analysis has been the inference of a core that is chemically different from the rest of Earth and that is divided into a solid inner core and a liquid outer core.

KEY OBJECTIVES
Upon completion of this chapter, you will be able to:
- Describe the causes and effects of earthquakes.
- Explain how seismic records can be analyzed to locate the epicenter of an earthquake.
- Explain how earthquake waves can be used to create and refine a model of Earth's interior.
- Compare and contrast the composition, density, and phase of matter in the inner core, outer core, mantle, and crust.
- Identify regions of high earthquake activity, and relate them to the structure of Earth's crust.

EARTHQUAKES

An **earthquake** is a sudden trembling of the ground. Seismologists, scientists who study earthquakes, estimate that over one million quakes occur each year—about one every second! During most earthquakes, the shaking is so slight it is barely noticeable to the senses. During minor earthquakes, the ground feels a bit like a railway station when the train rumbles in. Major earthquakes cause violent shaking and lurching of the ground. The ground may split open, buildings may topple, pipes and power lines may be broken, and roadways may collapse. A destructive earthquake is a catastrophe, something to be planned for and guarded against.

Causes of Earthquakes

The major cause of earthquakes is **faulting**, the sudden movement of rock along planes of weakness, called faults, in Earth's crust. However, some earthquakes are caused by volcanic eruptions, and explosions set off by humans are responsible for a few. Nuclear testing is usually detected by the earthquakes it causes. Recent research has shown that wells drilled to tap geothermal energy, pumping industrial wastewater into underground storage wells, and hydraulic fracturing (fracking) have triggered small earthquakes.

The **elastic rebound theory** explains the mechanism by which faulting causes earthquakes. At some places in Earth's crust, immense forces push or pull on the rock. Under this stress, rock can bend elastically, much as a wooden ruler can bend. However, if the stress increases, a point is reached where the rock can bend no further without breaking; it is at its elastic limit. If stressed beyond this point, the rock snaps and the two broken edges whip back, or rebound. Great masses of rock suddenly scrape past each other, and the shock of this wrenching action jars the crust and sets an earthquake in motion. The point where the rock breaks is called the **focus** of an earthquake. The focus may be just beneath the surface or hundreds of kilometers down.

As the rock rebounds from this deformation, it pushes against surrounding rock, causing it to deform. The energy of this back and forth motion, or vibration, is transmitted as a series of deformations spreading out through the rock in all directions. The motion of a change in a medium is termed *wave motion*. The deformations of rock that spread out from the focus of an earthquake are therefore called *earthquake waves*, or **seismic waves**. Waves transport energy from one place to another through the motion of a change in the medium. Seismic waves transport energy from the focus, through the surrounding rock, to Earth's surface. The earthquake is first felt at the **epicenter**, a point on the surface directly above the focus. See Figure 15.1.

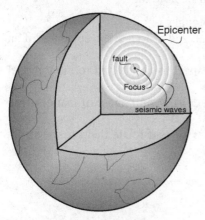

Figure 15.1 Fault, Focus, Epicenter, and Emanating Seismic Waves.

Earthquake Waves

To understand the behavior of earthquake waves, it is necessary to review a few basic concepts relating to wave behavior in general.

Wave Behavior

A wave in a rope is a simple analogy that will help you visualize what is going on when seismic waves travel through Earth. It will also help you to understand some of the inferences that can be drawn from the behavior of seismic waves.

If you give one end of a stretched rope a quick shake, a bump, or wave, travels down the rope at some speed. See Figure 15.2. If the rope is uniform and completely flexible, the wave keeps the same shape as it moves down the rope. This is what happens in an earthquake, when faulting gives the crust a quick shake that produces seismic waves.

As a wave travels down a rope, its forward part is moving upward and its rear part is moving downward. See Figure 15.3. The up and down motions of the string have kinetic energy. Work has to be done to produce the wave by pulling against the tension of the string, so the deformed string has potential energy. Thus, a wave contains both kinetic and potential energy as it travels along a rope. In like manner, seismic waves transfer energy as they travel through Earth. It is this energy that causes earthquake damage.

Figure 15.2 Wave Traveling Along a Rope.

Figure 15.3 Movement of Rope as Wave Passes Through It.

The speed of the wave depends on the properties of the rope—how dense it is and how tightly it is stretched. Pulses move slowly down a loose, heavy rope because this rope has a lot of inertia and responds slowly to the forces acting on it. On the other hand, waves move rapidly down a taut, light rope because this rope has a greater tendency to straighten out. The same is true of seismic waves moving through Earth. The denser and more rigid the Earth material, the faster seismic waves travel through them.

When a wave reaches the end of the rope, it may be **reflected** and travel

Figure 15.4 Wave Reflecting from a Fixed Support.

back toward its source. Depending on how the end is held, the reflected wave may be upright, be inverted, or disappear completely. If the end is fixed to a support (see Figure 15.4), the wave exerts a force on the support, which then exerts an equal but opposite force on the rope. This opposite force causes the rope to be displaced in the opposite direction from the original wave. An inverted wave forms and travels back along the rope in the opposite direction from the original wave. If the end is looped around a support so that it is connected but still free to move, an upright wave is reflected. If the end is held in a state somewhere between totally fixed and totally free, the wave disappears completely.

Now, consider what happens if two different ropes, a heavy (more dense) rope and a thin (less dense) rope, are spliced together. See Figure 15.5. One end is tied to a support, and the other end is given a shake so that a wave is produced. When the wave reaches the splice, the wave passes from the thin rope to the heavy rope and is said to be transmitted. However, the transmission is not total because a reflected wave also appears at the splice and travels back down the thin rope. This type of wave behavior is what we see when seismic waves cross a boundary between two distinctly different regions in Earth's interior. This wave behavior was an important clue to Earth's layered internal structure.

Figure 15.5 Transmitted and Reflected Waves at a Splice Between a Heavy and a Light Rope.

If one wave follows another in a rope, the result is a *periodic* wave. See Figure 15.6. In periodic waves a certain waveform, or shape, is repeated at regular intervals. Three quantities are often used to describe periodic waves: wave velocity, wavelength, and frequency. **Wave velocity** is the distance a wave moves each second; **wavelength** is the distance between adjacent crests or troughs; **frequency** is the number of waves that pass a given point in a second. Differences in wavelength, frequency, and the plane in which the waves oscillate are what distinguish different types of seismic waves.

If a periodic wave travels across a boundary between a more dense and a less dense medium at an angle, it is bent, or refracted. See Figure 15.7. This **refraction** occurs because, as the wave crosses the boundary, the front of the wave is moving at a different speed from the rear of the wave. The effect is somewhat like the result when the wheel at one end of an axle turns faster than the wheel at the other end—the path of the axle curves. Here, again, this is what we see when seismic waves cross a boundary in Earth's interior at an angle.

Figure 15.6 Periodic Wave, Showing Waveform, Wavelength, Wave Velocity, and Frequency.

Wave Types

In **transverse waves** the particles of the medium move back and forth in a direction perpendicular to the direction of wave motion. Waves in a stretched rope are transverse waves because any point on the rope moves side to side perpendicularly to the direction in which the wave is moving, as shown in Figure 15.8.

Longitudinal waves occur when the particles of the medium move back and forth in the same direction in which the waves travel. A good analogy is the movement of a compression down a coil spring when the end is pushed in and out. See Figure 15.9.

Seismic Waves

The complex motions that occur during faulting generate several types of periodic seismic waves. Some are transverse waves; others are longitudinal. Seismology took a giant step forward with the invention of the modern seismograph. A **seismograph** is a device

Figure 15.7 Periodic Wave Crossing a Boundary Between a More Dense and a Less Dense Medium and Being Refracted.

Figure 15.8 Transverse Waves, Showing Perpendicular Motion.

Figure 15.9 Longitudinal Waves in a Coil Spring and in Rock. As a longitudinal wave moves through a spring, its coils alternately compress and expand. In a rock, the individual particles are compressed and expanded.

that detects, measures, and records the motions of Earth that are associated with seismic waves.

A seismograph's operation is based upon the law of inertia: Objects that are at rest tend to remain at rest. A simple seismograph consists of a heavy, suspended weight that tends to remain at rest while Earth moves around it. A recording device (e.g., a pen) is attached to the weight. A recording medium, such as paper, is wound around a clockwork-driven drum mounted firmly in bedrock. When a series of seismic wave pulses causes the bedrock to move back and forth, the drum moves with it, but the heavy weight and its attached pen remain motionless for a long time. As the paper moves under the motionless pen, a line is drawn on the paper that records the back and forth motion of the bedrock, as shown in Figure 15.10a. The line recorded on paper by a seismograph is called a **seismogram** (see Figure 15.10b).

Modern seismographs use a laser beam reflected from a mirror on the weight instead of a pen. The laser beam exposes a track along photographic paper that is attached to the drum. When the bedrock vibrates, the track becomes a wavy line. In most earthquake recording stations, at least three seismographs are used, one to measure vibrations in each of three dimensions: x, y, and z (typically north-south, east-west, and up-down).

Figure 15.10 Simple Seismograph and Seismogram. P, S, and L represent P-, S-, and Love waves, respectively. Source: Earth Science on File. © Facts on File, Inc., 1988.

A careful study of seismograms reveals four different types of seismic waves. Two types, *P*-waves and *S*-waves, transport energy through the body of Earth and are therefore called **body waves**. Two other types, Love waves and Rayleigh waves, transport energy along the surface of Earth and are therefore called **surface waves**. Although surface waves play a major role in the damage caused by earthquakes, it is the body waves that reveal the structure of Earth's interior and make it possible to locate an earthquake's focus and epicenter. For this reason we will focus our attention on *P*-waves and *S*-waves.

P-waves
P-waves are longitudinal waves. They are alternating pulses of compression and expansion of rock in the same direction in which the wave travels. Refer back to Figure 15.9. ***P*-waves** are the first (or **P**rimary) waves to reach a seismograph after an earthquake occurs because they travel fastest, at speeds of about 8 kilometers per second in the upper crust. *P*-waves can travel through solids, liquids, and gases because all three can be compressed and expanded. *P*-waves travel faster in these materials because these materials have stronger internal bonds; and, as occurs in a taut rope, the tendency to pull back into their original volume is greater. Since denser materials tend to be more rigid, *P*-wave speed increases in denser rock.

S-waves
S-waves are transverse waves. They rock back and forth, deforming their shape in a direction perpendicular to that of wave travel. Refer back to Figure 15.8. ***S*-waves** (or **S**econdary waves) are the second type to arrive at a seismograph. *S*-waves arrive after *P*-waves because *S*-waves travel at a slower speed—about 4 kilometers per second in the upper crust. Like *P*-waves, *S*-waves travel faster in denser, more rigid rocks. *S*-waves can be transmitted only through solids. These waves cannot travel through liquids and gases because, when they are deformed, they do not return to their original shape.

Analyzing Seismograms

P-wave and *S*-wave recordings from a single event can be used to find the distance to the epicenter of an earthquake. With recordings from three or more seismic recording stations, the location of the epicenter can be determined. The basis for these determinations is the difference in the speeds of *P*-waves and *S*-waves.

When an earthquake occurs, all four types of seismic waves start moving outward from the focus at the same time. However, since they travel at different speeds, they do not all arrive at a seismograph at the same time. The *P*-waves, which travel the fastest, arrive first, followed by the *S*-waves some time later; the surface waves arrive last. The farther a seismograph is from the epicenter, the greater the difference between the arrival times of the *P*-waves and the *S*-waves.

For example, suppose *P*-waves travel 8 kilometers per second and *S*-waves travel 4 kilometers per second. The *P*-waves will arrive at a seismograph 80 kilometers away in 10 seconds. The *S*-waves will reach the same seismograph in 20 seconds. The arrival times at this first station will be 20 – 10 or 10 seconds *apart*. The same *P*-waves will reach a seismograph 200 kilometers away in 25 seconds, and the *S*-waves will arrive in 50 seconds. The arrival times at the farther station will be 50 – 25 or 25 seconds *apart*.

Determining Distance to the Epicenter of an Earthquake

For every distance that seismic waves travel, there is a corresponding difference in arrival times. This relationship between the difference in arrival time between *P*-waves and *S*-waves can be seen clearly in Figure 15.11. The travel time axis is marked off in minutes, and each minute is subdivided into three 20-second segments. The epicenter distance is marked off in thousands of kilometers (10^3 = 1,000) with each 1,000 kilometers subdivided into five 200-kilometer segments. The lines marked *S* and *P* show the times that *S*-waves and *P*-waves, respectively, take to travel a given distance. The vertical difference between the lines is the difference in arrival times at a given distance. It shows, for example, that *P*-waves travel 1,000 kilometers in 2 minutes while *S*-waves take about 4 minutes. The difference in arrival times is 2 minutes.

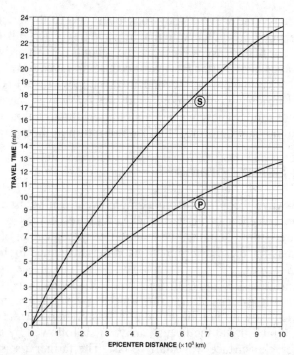

Figure 15.11 Earthquake *P*-wave and *S*-wave Travel Time. Source: The State Education Department, Earth Science Reference Tables, 2011 ed. (Albany, New York; The University of the State of New York).

Using Figure 15.11 and the relationship between difference in arrival times and distance, one can determine the distance between a seismograph and the epicenter of an earthquake as follows:

1. Determine the arrival times of the *P*-waves and the *S*-waves.
2. Calculate the *difference* in arrival times by subtracting the *P*-wave arrival time from the *S*-wave arrival time.
3. Using the travel time axis of the graph, mark off a distance along the edge of a piece of scrap paper equivalent to the *difference* in arrival times.
4. Keeping the edge of the paper vertical, find the point at which the marked-off difference in arrival times corresponds to the vertical distance between the *P*-wave line and the *S*-wave line.
5. Read the epicenter distance corresponding to that location on the graph.

Example:

The seismogram in Figure 15.12 is a record of an earthquake that was recorded by a seismograph. The seismogram does not tell how far away the earthquake occurred, but it does show when the *P*-waves and *S*-waves arrived. How far away from the seismograph was the epicenter of the earthquake?

Figure 15.12 Seismogram.

To determine the epicenter, proceed as follows:

1. Note that the *P*-waves arrived at 08:16:00 and the *S*-waves arrived at 08:21:00.
2. Subtract: 08:21:00 − 08:16:00 = 00:05:00 or 5 min.
3. Using the travel time axis in Figure 15.11, mark off 5 min on the edge of a piece of paper.
4. Keeping the paper vertical, find the place where the marked-off portion of the paper just fits between the *S*-wave and *P*-wave lines.
5. Read the epicenter distance on the axis. It is 3,400 km.

The epicenter distance is 3,400 kilometers. These five steps are illustrated in Figure 15.13.

Determining the Location of the Epicenter

Knowing how far a seismograph station is from the epicenter does not pinpoint the location of the epicenter. In the example above, it could be 3,400 kilometers away in any direction. All of the points that are 3,400 kilometers from the seismograph in station *A* form a circle with a radius of 3,400 kilometers around that station. The epicenter is located somewhere on this circle. See Figure 15.14a.

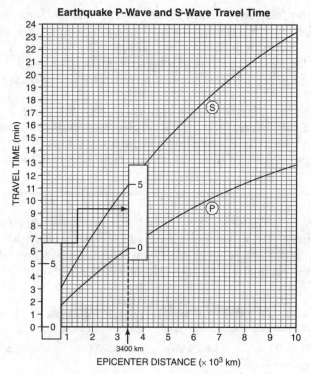

Figure 15.13 Five Steps in Determining Epicenter Distance.

However, if the same earthquake is 2,000 kilometers from a second station, station *B*, a circle can be drawn around that station too. The possible locations are now narrowed down to two points, as shown in Figure 15.14b. A third recording of the same earthquake, at station *C*, will eliminate one of these two points and pinpoint the location of the epicenter, as shown in Figure 15.14c.

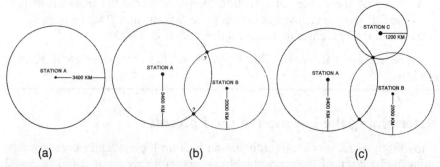

Figure 15.14 Information from Three Seismograph Stations Is Needed to Pinpoint the Location of the Earthquake Epicenter.

Example:

The seismograms shown in Figure 15.15 were recorded at three seismic recording stations for the same earthquake as in the preceding example. How can the epicenter of this earthquake be located?

The epicenter distance from station *A* is known to be 3,400 km. The difference in *P*-wave and *S*-wave arrival times at station *B* is 7 min, yielding an epicenter distance of 5,400 km. The difference in *P*-wave and *S*-wave arrival times at station *C* is 9 min, for an epicenter distance of 7,600 km.

Plotted on a map using the map scale, stations *A*, *B*, and *C* pinpoint the epicenter. See Figure 15.15.

Figure 15.15 Locating the Epicenter. When the three seismograms above are analyzed, the distance from seismograph to the earthquake epicenter can be determined. Plotted on a map, only one place is the correct distance from all three stations.

Measuring Earthquakes

As seismic waves travel farther from the focus, they lose energy and their effects lessen. At a distance of 100 kilometers from the focus, the energy of an earthquake is only about 1/10,000 of what it was at 1 kilometer from the focus. The Richter and the Mercalli scales are used to measure the strengths of earthquakes.

The **Richter Magnitude Scale** (see Figure 15.16) is based upon the energy released by an earthquake, and energy is determined by direct measurements of the motions of the crust using seismic instruments. The greater the energy released by an earthquake, the larger the amplitude (height) of the earthquake waves. The Richter scale consists of numbers ranging from 0 to 8.6. The scale is logarithmic, meaning that the energy of a shock increases by powers of 10 in relation to the Richter magnitude numbers. In other words, each increase of

Larger Magnitude Earthquake

Smaller Magnitude Earthquake

Figure 15.16 The Richter Magnitude Scale. This is based on the motion of bedrock as recorded by seismographs.

1 unit on the Richter scale corresponds to a tenfold increase in the amplitude (height) of the earthquake wave. Also, every tenfold increase in wave height corresponds to a hundredfold increase in energy! Thus, a magnitude-2 earthquake wave is 10 times higher than a magnitude-1 earthquake wave and releases 100 times as much energy.

The actual destructiveness of an earthquake, though, is influenced also by factors other than the amount of energy released. For example, destructiveness depends also upon nearness to the epicenter and the nature of the subsurface materials. Therefore, a magnitude-5 earthquake may cause vastly different amounts of damage in two different places. Since humans are concerned about damage to their land and property, a scale of earthquake damage intensity was developed. The modified **Mercalli Intensity Scale** (see Table 15.1) is based upon observations of earthquake damage and is concerned chiefly with the impact of an earthquake on structures made by humans and on human activities. Engineers and city planners often use this scale in reaching decisions about structural and land-use issues in earthquake-prone areas.

Effects of Earthquakes

The effects of earthquakes depend mainly upon the strength of the seismic waves and the nature of the material through which they pass. Earthquake effects include ground shaking and failure, surface faulting, changes in wells and geysers, and tsunamis.

Ground shaking affects solid rock the least, and loose, water-soaked ground the most. The same shock that causes solid bedrock to sway slightly may cause intense jolting in nearby wet clay or landfill. Tapping a bowl of Jello is a good analogy. The solid bowl vibrates but moves hardly at all, while the Jello jiggles and quivers noticeably. In ground failure, strong vibrations cause loose or water-soaked ground to break apart and settle, forming cracks and fissures. On hillsides these fissures can trigger landslides, slumping, and mudflows.

Surface faulting occurs when movement along a fault causes the surface to be lifted, lowered, or shifted sideways. The result is displacement of surface features and structures.

TABLE 15.1 MODIFIED MERCALLI INTENSITY SCALE

Intensity Value	Description of Effects
I	Not felt.
II	Felt by persons at rest; felt on upper floors because of sway.
III	Felt indoors. Hanging objects swing. Feels like a passing truck.
IV	Hanging objects swing. Feels like heavy truck passing, or a jolt is felt. Standing cars rock. Windows and dishes rattle. Glasses clink. In the upper range of IV, wooden frames and walls creak.
V	Felt outdoors and direction can be estimated. Sleepers are awakened. Liquids are disturbed, some spill. Small, unstable objects upset. Doors swing open and close. Shutters and pictures move.
VI	Felt by all. Persons walk unsteadily, and many are frightened and run indoors. Windows, dishes, and glassware are broken. Furniture is moved or overturned. Pictures fall from walls. Weak plaster cracks. Trees and bushes visibly shaken or heard to rustle.
VII	Difficult to stand. Noticed by drivers of cars. Hanging objects quiver. Furniture is broken. Weak chimneys break at roof line. Fall of plaster, loose bricks, masonry. Waves on ponds and water muddied. Small slides and cave-ins of sand and gravel banks.
VIII	Steering of moving cars is affected. Partial collapse of masonry structures. Chimneys and smokestacks twist and fall. Frame houses move on foundation if not bolted down. Branches broken from trees. Wet ground and steep slopes crack.
IX	General panic. Weak masonry destroyed, stronger masonry cracks, is seriously damaged. Frame structures not bolted shift off foundations. Frames cracked. Reservoirs seriously damaged. Underground pipes broken. Conspicuous cracks in ground.
X	Masonry and frame structures destroyed along with their foundations. Some bridges destroyed. Serious damage to reservoirs, dikes, dams, and embankments. Large landslides. Water thrown out of lakes, canals, rivers, etc. Rails bent slightly.
XI	Rails bent greatly. Underground pipelines completely destroyed and out of service.
XII	Damage nearly total. Large rock masses displaced. Objects thrown into the air. Lines of sight and level are distorted.

Modified from: H. O. Wood and F. Neumann, "Modified Mercalli Intensity Scale of 1931," *Bulletin of the Seismological Society of America*, Vol. 21, No. 4, pp. 277–288.

As seismic waves travel through water-saturated rock or soil, water is compressed and forced through pore spaces. Like a hand alternately squeezing and releasing a soaked sponge in a bucket, seismic waves force water in and out of rock and soil surrounding wells, causing their water levels to fluctuate. Seismic waves have a similar effect on rock fissures, causing geysers to behave erratically.

Tsunamis are immense sea waves caused by earthquakes beneath the ocean floor or by undersea landslides. Tsunamis may be only a few meters high, but they have very long wavelengths and travel much more rapidly than ordinary ocean waves. Tsunamis have been clocked moving faster than 500 kilometers per hour with wavelengths as great as 200 kilometers. When tsunamis reach shallow water, they are slowed by friction with the ocean bottom. In bays and narrow channels, however, their high speed and long wavelength may be funneled into huge breaking waves more than 20 meters high. The force exerted by such a huge, fast-moving mass of water can do extensive damage.

Earthquakes are geologic hazards to humans, but anticipating the hazards earthquakes pose and preparing to meet them can minimize loss of lives and property. Figure 15.17 shows areas of various degrees of seismic risk in the United States.

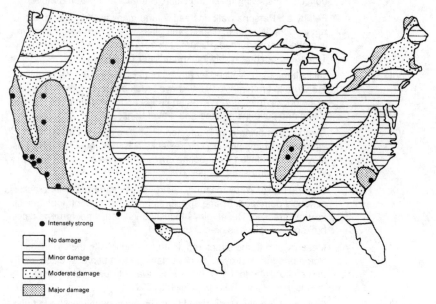

Figure 15.17 Seismic Risk Map of the United States.

Earthquake Zones

If the locations of earthquakes are plotted on a map of Earth, an interesting pattern emerges, as shown in Figure 15.18. Rather than being randomly spread over Earth, earthquakes occur in distinct, narrow zones. These zones include the mid-ocean ridges, the rim of the Pacific Ocean (called the **Ring of Fire** for its many active volcanoes), and the Mediteranean Belt. As you will discover in Chapter 18, the locations of earthquake epicenters reveal the locations of boundaries between the plates of Earth's crust.

Figure 15.18 Worldwide Earthquake Distribution.

THE INFERRED STRUCTURE OF EARTH'S INTERIOR

Whenever an earthquake occurs, its seismic waves are recorded by seismographs at hundreds of seismic recording stations throughout the world. Analysis of these seismograms reveals much about the structure of Earth's interior.

As was discussed earlier, seismic waves are reflected from the surfaces of materials and refracted as they move across boundaries between substances of different densities. If Earth had a homogeneous composition and density, seismic waves would move through it in simple, straight-line paths with reflections only when the surface was reached. They would also travel directly from the earthquake's focus to all points on Earth's surface. Furthermore, if the location of an earthquake's epicenter is known, the arrival times of P-waves and S-waves at all stations should be easy to predict.

This is not the case. P-waves are absent from seismographs at certain distances from the epicenter and are more highly concentrated elsewhere. Predictions of arrival times based upon straight-line travel differ markedly from actual measurements. Moreover, the paths of seismic waves curve. This gradual bending is due to the gradual increase in density with depth, caused by increasing pressure. There are many more reflected waves than would be expected from a homogeneous Earth. There are also places on Earth's surface where P-waves or S-waves, or both, do not arrive directly from the focus of the earthquake. These facts led seismologists to conclude that Earth's interior is not homogeneous.

The Crust

In 1910, Andrija Mohorovicic discovered that two sets of P-waves and S-waves arrived at locations close to an earthquake's epicenter, but only one set arrived farther away. Mohorovicic explained this phenomenon by hypothesizing a boundary between rocks of different densities that would cause the waves to refract. As shown in Figure 15.19, stations close to the epicenter of an earthquake can receive both direct and refracted seismic waves. Farther away, however, waves cannot reach the station without crossing the boundary, so only refracted waves arrive.

This boundary, called the **Moho** in honor of Mohorovicic, marks the boundary between the outermost layer of Earth, the **crust**, and the **mantle** beneath it. The density of the crust varies from about 2.7 grams per cubic centimeter near the surface to about 3.1 grams per cubic centimeter near the Moho. The thickness of the crust varies, but is usually greatest beneath mountains and least beneath oceans.

Figure 15.19 The Moho Boundary.

The Mantle and Core

Between the epicenter and an angle of 103° from the epicenter, P-waves and S-waves arrive directly from the focus, as would be expected. Beyond 103°, however, the S-waves *disappear*! This region is known as the **S-wave shadow zone**. Seismic stations in the S-wave shadow zone receive no S-waves from an earthquake.

Between 103° and 143° from the epicenter, the P-waves *also* disappear. But beyond 143°, the P-waves reappear. Therefore, the region between 103° and 143° from the epicenter is known as the **P-wave shadow zone**. Seismic stations located between 103° and 143° of an earthquake's epicenter fall within both the S-wave shadow zone and the P-wave shadow zone and record no seismic waves from the earthquake. Seismic stations beyond 143° in either direction from the epicenter record only P-waves. That this odd pattern of shadow zones can be explained by Earth's internal structure is shown in Figure 15.20.

Figure 15.20 The *P*-Wave and *S*-Wave Shadow Zones.

A boundary at a depth of about 2,900 kilometers, between rocks of markedly different densities, produces a strong refraction of *P*-waves that cross it. Up to 103°, waves arrive directly from the epicenter. The strong refraction as soon as the boundary is encountered immediately angles the waves away to 143°, so no waves reach the region in between. Since *P*-waves continue to be received past 143°, they are able to pass through the material beneath the boundary. This boundary marks the bottom of the mantle; the layer beneath it is called the **core**.

The complete absence of *S*-waves beyond 103° indicates that they are unable to travel through the material beneath the mantle. Since transverse waves cannot travel through fluids, the disappearance of *S*-waves makes sense if the core is a fluid.

Careful analysis of refraction of *P*-waves that pass through the core and the presence of reflections from inside the core indicate that another boundary inside the core divides it into an **inner core** and an **outer core**. Only the outer core is thought to be a fluid.

Summary of Inferences about Earth's Interior

Figure 15.21 summarizes what has been inferred about Earth's interior based upon studies of seismic waves and other data. The diagram at the top shows a cross section of Earth with the interior layers labeled. Along the right edge, the density of each layer is shown. Notice that the mantle is divided into two regions, the asthenosphere, or plastic mantle, and the stiffer mantle. Notice, too, the sharp increase in density between the mantle and the core, which explains the strong refraction of *P*-waves. Directly beneath the cross section is a graph showing pressure in *millions of atmospheres* at different depths. The dotted lines indicate boundaries between crust, mantle, and outer and inner cores. On the bottom is a similar graph showing temperatures and melting points of rock at different depths. Notice that in the outer core the actual temperature is higher than the melting point of the rocks.

Figure 15.21 Inferred Properties of Earth's Interior. Source: The State Education Department, *Earth Science Reference Tables*, 2011 ed. (Albany, New York; The University of the State of New York).

Meteorites are thought to be fragments of an Earth-like planet. Some are similar in composition to Earth rocks, but others are composed of iron-nickel alloy. An iron-nickel alloy would have the density and rigidity of the core as deduced from seismic wave studies. Therefore, Earth's core is believed to consist of an iron-nickel alloy. An iron-nickel core in which convection currents transfer heat outward (see Chapter 9) would also help to explain Earth's magnetic field.

MULTIPLE-CHOICE QUESTIONS

In each case, write the number of the word or expression that best answers the question or completes the statement.

1. The most frequent cause of major earthquakes is
   (1) faulting
   (2) folding
   (3) landslides
   (4) submarine currents

2. The immediate result of a sudden slippage of rocks within Earth's crust is
   (1) isostasy
   (2) erosion
   (3) an earthquake
   (4) the formation of convection currents

Base your answers to questions 3 and 4 on the diagram below, which shows models of two types of earthquake waves.

Model A

Model B

3. Model *A* best represents the motion of earthquake waves called
   (1) *P*-waves (compressional waves) that travel faster than *S*-waves (shear waves) shown in model *B*
   (2) *P*-waves (compressional waves) that travel slower than *S*-waves (shear waves) shown in model *B*
   (3) *S*-waves (shear waves) that travel faster than *P*-waves (compressional waves) shown in model *B*
   (4) *S*-waves (shear waves) that travel slower than *P*-waves (compressional waves) shown in model *B*

4. The difference in seismic station arrival times of the two waves represented by the models helps scientists determine the
   (1) amount of damage caused by an earthquake
   (2) intensity of an earthquake
   (3) distance to the epicenter of an earthquake
   (4) time of occurrence of the next earthquake

Base your answer to question 5 on the seismogram below. The seismogram was recorded at a seismic station and shows the arrival times of the first P-wave and S-wave from an earthquake.

5. Which part of this seismogram is used to find the distance to the epicenter of the earthquake?
   (1) P-wave arrival time, only
   (2) S-wave arrival time, only
   (3) difference in the arrival time of the P-wave and S-wave
   (4) difference in the height of the P-wave and S-wave

6. How long would it take for the first S-wave to arrive at a seismic station 4,000 kilometers away from the epicenter of an earthquake?
   (1) 5 min 40 sec   (3) 12 min 40 sec
   (2) 7 min 0 sec   (4) 13 min 20 sec

7. Which statement best describes the relationship between the travel rates and travel times of earthquake P-waves and S-waves from the focus of an earthquake to a seismograph station?
   (1) P-waves travel at a slower rate and take less time.
   (2) P-waves travel at a faster rate and take less time.
   (3) S-waves travel at a slower rate and take less time.
   (4) S-waves travel at a faster rate and take less time.

8. How long after receiving the first P-wave from an earthquake centered 4,000 kilometers away does a seismic station receive its first S-wave from the same earthquake?
   (1) 1 minute
   (2) 5 minutes 35 seconds
   (3) 7 minutes
   (4) 12 minutes 40 seconds

9. What is the approximate *P*-wave travel time from an earthquake if the *P*-wave arrives at the seismic station 8 minutes before the *S*-wave?
   (1) 4 minutes 20 seconds
   (2) 6 minutes 30 seconds
   (3) 10 minutes 0 seconds
   (4) 11 minutes 20 seconds

Base your answers to questions 10 through 12 on the data table below, which gives information collected at seismic stations *W*, *X*, *Y*, and *Z* for the same earthquake. Some of the data have been omitted.

Data Table

Seismic Station	P-Wave Arrival Time (h:min:s)	S-Wave Arrival Time (h:min:s)	Difference in Arrival Times (h:min:s)	Distance to Epicenter (km)
W	10:50:00	no S-waves arrived		
X	10:42:00	10:46:40		
Y	10:39:20		00:02:40	
Z	10:45:40			6200

10. Which seismic station was farthest from the earthquake epicenter?
    (1) *W*   (3) *Y*
    (2) *X*   (4) *Z*

11. What is the most probable reason for the absence of *S*-waves at station *W*?
    (1) *S*-waves were not generated at the epicenter.
    (2) *S*-waves cannot travel through liquids.
    (3) Station *W* was located on solid bedrock.
    (4) Station *W* was located on an island.

12. At what time did the *S*-wave arrive at station *Y*?
    (1) 10:36:40   (3) 10:42:00
    (2) 10:39:20   (4) 10:45:20

Base your answers to questions 13 and 14 on the earthquake seismogram below.

13. When did the first *P*-waves arrive as this seismic station?
    (1) 3 minutes after an earthquake occurred 2,600 km away
    (2) 5 minutes after an earthquake occurred 2,600 km away
    (3) 9 minutes after an earthquake occurred 3,500 km away
    (4) 11 minutes after an earthquake occurred 3,500 km away

14. How many additional seismic stations must report seismogram information in order to locate this earthquake?
    (1) one
    (2) two
    (3) three
    (4) four

Base your answers to questions 15 through 18 on the diagram and map below. The diagram shows three seismograms of the same earthquake recorded at three different seismic stations, *X*, *Y*, and *Z*. The distances from each seismic station to the earthquake epicenter have been drawn on the map.

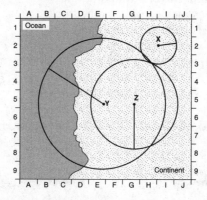

A coordinate system has been placed on the map to describe locations. The map scale has not been included.

15. Approximately how far away from station *Y* is the epicenter?
    (1) 1,300 km
    (2) 2,600 km
    (3) 3,900 km
    (4) 5,200 km

16. The *S*-waves from this earthquake that travel toward Earth's center will
    (1) be deflected by Earth's magnetic field
    (2) be totally reflected off the crust-mantle interface
    (3) be absorbed by the liquid outer core
    (4) reach the other side of Earth faster than those that travel around Earth in the crust

17. Seismic station *Z* is 1,700 kilometers from the epicenter. Approximately how long did the *P*-wave take to travel to station *Z*?
    (1) 1 min 50 sec
    (2) 2 min 50 sec
    (3) 3 min 30 sec
    (4) 6 min 30 sec

18. On the map, which location is closest to the epicenter of the earthquake?
    (1) *E*–5
    (2) *G*–1
    (3) *H*–3
    (4) *H*–8

19. Two different cities experience an earthquake with a magnitude of 5.5 on the Richter Magnitude Scale. However, on the Modified Mercalli Intensity Scale, the earthquake was rated V in one city and VII in the other city. The best explanation for this difference is that
    (1) one city is nearer the equator and the other is near a pole
    (2) the earthquake occurred at 8 P.M. in one city and at 4 A.M. in the other
    (3) one city is built on bedrock while the other is built on loose sediments
    (4) one city is in a drier climate zone than the other

20. An earthquake's magnitude can be determined by
    (1) analyzing the seismic waves recorded by a seismograph
    (2) calculating the depth of the earthquake faulting
    (3) calculating the time the earthquake occurred
    (4) comparing the speed of *P*-waves and *S*-waves

21. Scientists have inferred the structure of Earth's interior mainly by analyzing
    (1) the Moon's interior
    (2) the Moon's composition
    (3) Earth's surface features
    (4) Earth's seismic data

Earthquakes and Earth's Interior

22. The theory that the outer core of Earth is composed of liquid material is best supported by
    (1) seismic studies that indicate that shear waves do not pass through the outer core
    (2) seismic studies that show that compressional waves can pass through the outer core
    (3) density studies that show that the outer core is slightly more dense than the inner core
    (4) gravity studies that indicate that gravitational strength is greatest within the core

Base your answers to questions 23 through 25 on the diagram below and on your knowledge of Earth science. The diagram represents a cut-away view of Earth's interior and the paths of some of the seismic waves produced by an earthquake that originated below Earth's surface. Points A, B, and C represent seismic stations on Earth's surface. Point D represents a location at the boundary between the core and the mantle.

23. Seismic station A is 5,000 kilometers from the epicenter. What is the difference between the arrival time of the first P-wave and the arrival time of the first S-wave recorded at this station?
    (1) 2 minutes 20 seconds
    (2) 6 minutes 40 seconds
    (3) 8 minutes 20 seconds
    (4) 15 minutes 00 seconds

24. Which process prevented *P*-waves from arriving at seismic station *B*?
    (1) refraction
    (2) reflection
    (3) convection
    (4) conduction

25. Only *P*-waves were recorded at seismic station *C* because *P*-waves travel
    (1) only through Earth's interior, and *S*-waves travel only on Earth's surface
    (2) fast enough to penetrate the core, and *S*-waves travel too slowly
    (3) through iron and nickel, while *S*-waves cannot
    (4) through liquids, while *S*-waves cannot

26. Which two Earth layers are separated by the Moho boundary?
    (1) rigid mantle and plastic mantle
    (2) outer core and stiffer mantle
    (3) stiffer mantle and asthenosphere
    (4) crust and rigid mantle

27. What are the inferred pressure and temperature at the boundary of Earth's stiffer mantle and outer core?
    (1) 1.5 million atmospheres pressure and an interior temperature of 4,950°C
    (2) 1.5 million atmospheres pressure and an interior temperature of 6,200°C
    (3) 3.1 million atmospheres pressure and an interior temperature of 4,950°C
    (4) 3.1 million atmospheres pressure and an interior temperature of 6,200°C

28. A huge undersea earthquake off the Alaskan coastline could produce a
    (1) tsunami
    (2) cyclone
    (3) hurricane
    (4) thunderstorm

29. Theories about the composition of Earth's core are supported by meteorites that are composed primarily of
    (1) oxygen and silicon
    (2) aluminum and iron
    (3) aluminum and oxygen
    (4) iron and nickel

Base your answers to questions 30 and 31 on the map below, which shows the risk of damage from seismic activity in the United States.

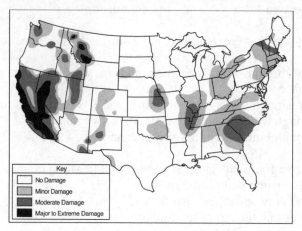

30. In the United States, most of the major damage expected from a future earthquake is predicted to occur near a
(1) divergent plate boundary, only
(2) convergent plate boundary, only
(3) mid-ocean ridge and a divergent plate boundary
(4) transform plate boundary and a hot spot

31. Which New York State location has the greatest risk of earthquake damage?
(1) Binghamton   (3) Plattsburgh
(2) Buffalo   (4) Elmira

32. The data table below shows the origin depths of all large-magnitude earthquakes over a 20-year period.

Data Table

Depth Below Surface (km)	Number of Earthquakes
0–33	27,788
34–100	17,585
101–300	7,329
301–700	3,167

According to these data, most of these earthquakes occurred within Earth's
(1) lithosphere   (3) stiffer mantle
(2) asthenosphere   (4) outer core

33. Which coastal area is most likely to experience a severe earthquake?
 (1) east coast of North America
 (2) east coast of Australia
 (3) west coast of Africa
 (4) west coast of South America

CONSTRUCTED RESPONSE QUESTIONS

Base your answers to questions 34 through 36 on the diagram below, which shows a seismograph that recorded seismic waves from an earthquake located 4,000 kilometers from this seismic station.

34. State *one* possible cause of the earthquake that resulted in the movement of the bedrock detected by this seismograph. [1]

35. Which type of seismic wave was recorded first on the rotating drum? [1]

36. How long does the first *S*-wave take to travel from the earthquake epicenter to this seismograph? [1]

Base your answers to questions 37 and 38 on the diagram below, which shows two seismogram tracings, at stations *A* and *B*, for the same earthquake. The arrival times of the *P*-waves and *S*-waves are indicated on each tracing.

37. Explain how the seismic tracings recorded at station *A* and station *B* indicate that station *A* is farther from the earthquake epicenter than station *B*. [1]

38. Seismic station *A* is located 5,400 kilometers from the epicenter of the earthquake. How much time would it take for the first *S*-wave produced by this earthquake to reach seismic station *A*? [1]

Base your answers to questions 39 through 42 on the map below and on your knowledge of Earth science. The map shows the location of the epicenter, **X**, of an earthquake that occurred on April 20, 2002, about 29 kilometers southwest of Plattsburgh, New York.

39. State the latitude and longitude of this earthquake epicenter. Express your answers to the *nearest tenth of a degree* and include the compass directions. [1]

40. What is the *minimum* number of seismographic stations needed to locate the epicenter of an earthquake? [1]

41. Explain why this earthquake was most likely felt with greater intensity by people in Peru, New York, than by people in Lake Placid, New York. [1]

42. A seismic station located 1,800 kilometers from the epicenter recorded the *P*-wave and *S*-wave arrival times for this earthquake. What was the difference in the arrival time of the first *P*-wave and the first *S*-wave? [1]

EXTENDED CONSTRUCTED RESPONSE QUESTIONS

Base your answers to questions 43 through 45 on the data table and graph below, and on your knowledge of Earth science. The data table shows the velocity of seismic *S*-waves at various depths below Earth's surface. The graph shows the velocity of seismic *P*-waves at various depths below Earth's surface. Letter *A* is a point on the graph.

Data Table

Depth Below Surface (km)	0	100	200	700	800	1800	2900
S-Wave Velocity (km/s)	2.8	4.5	4.2	5.3	6.2	7.0	7.4

43. On the graph, plot the *S*-wave velocity at each depth given on the data table. Connect the plots with a line. [1]

44. What property of Earth's interior causes the *S*-waves to stop at 2,900 km but allows the *P*-waves to continue? [1]

45. State the pressure and temperature of Earth's interior at the depth indicated by point *A* on the graph. [1]

Base your answers to questions 46 through 49 on the information and map below and your knowledge of Earth science.

An earthquake occurred in the southwestern part of the United States. Mercalli scale intensities were plotted for selected locations on the map, as shown below. (As the numerical values of Mercalli ratings increase, the damaging effects of the earthquake waves also increase.)

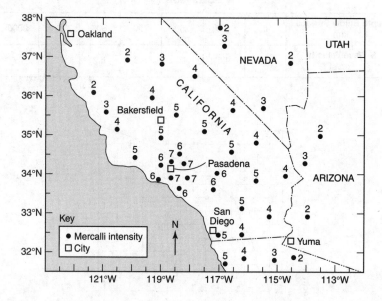

46. Using an interval of 2 Mercalli units and starting with an isoline representing 2 Mercalli units, draw an accurate isoline map of earthquake intensity. [4]

47. State the name of the city that is closest to the earthquake epicenter. [1]

48. State the latitude and longitude of Bakersfield. [2]

49. Using one or more complete sentences, identify the most likely cause of earthquakes that occur in the area shown on the map. [2]

Base your answers to questions 50 through 54 on the modified Mercalli scale of earthquake intensity and the map of Japan shown below, and on your knowledge of Earth science. The modified Mercalli scale classifies earthquake intensity based on observations made during an earthquake. The map indicates the modified Mercalli scale intensity values recorded at several locations in Japan during the March 11, 2011 earthquake, which triggered destructive tsunamis in the Pacific Ocean.

Modified Mercalli Scale of Earthquake Intensity

Intensity Value	Description of Effects
I	Not felt except by a very few under especially favorable conditions.
II	Felt only by a few persons at rest, especially on upper floors of buildings.
III	Felt quite noticeably by persons indoors, especially on upper floors of buildings. Many people do not recognize it as an earthquake. Parked cars may rock slightly. Vibrations similar to the passing of a truck.
IV	Felt indoors by many, outdoors by few during the day. At night, some awakened. Dishes, windows, doors disturbed; walls make cracking sound. Sensation like heavy truck striking building. Parked cars rocked noticeably.
V	Felt by nearly everyone; many awakened. Some dishes, windows broken. Unstable objects overturned. Pendulum clocks may stop.
VI	Felt by all, many frightened. Some heavy furniture moved; a few instances of fallen plaster. Damage slight.
VII	Damage minimal in buildings of good design and construction; slight to moderate in well-built ordinary structures; considerable damage in poorly built or badly designed structures; some chimneys broken.
VIII	Damage slight in specially designed structures; considerable damage with partial collapse in ordinary substantial buildings. Damage great in poorly built structures. Fall of chimneys, factory stacks, columns, monuments, walls. Heavy furniture overturned.
IX	Damage considerable in specially designed structures; well-designed frame structures tilted. Damage great in substantial buildings, with partial collapse. Buildings shifted off foundations.
X	Most masonry and frame structures and foundations are destroyed. Train rails bent.
XI	Few, if any, structures remain standing. Bridges destroyed. Train rails bent greatly.
XII	Damage total. Objects thrown into the air.

March 11, 2011 Earthquake in Japan

[Map of Japan showing Mercalli intensity values V, VI, and VII, with the earthquake epicenter marked east of Ishinomaki in the Pacific Ocean. Cities labeled include Yamagata, Ishinomaki, Fukushima, Nagano, Mito, and Tokyo. Scale: 0–100 miles.]

50. On the map, a line has been drawn to separate regions with Mercalli values of V from regions with Mercalli values of VI. Draw *another* line to separate regions with Mercalli values of VI from regions with Mercalli values of VII. [1]

51. The table below lists some observations that might be made during an earthquake according to the modified Mercalli scale. On the table below, place a check mark (✔) in the box if that observation most likely was recorded at Yamagata during the March 11, 2011 earthquake. More than one box may be checked. [1]

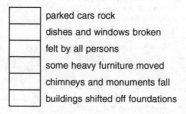

	parked cars rock
	dishes and windows broken
	felt by all persons
	some heavy furniture moved
	chimneys and monuments fall
	buildings shifted off foundations

52. The epicenter of this earthquake was located at 38° N 142° E. Identify the type of tectonic plate boundary that is located nearest to the epicenter of this earthquake. [1]

53. Describe *one* way the *P*-waves and *S*-waves recorded on seismograms at Ishinomaki and Nagano were used to indicate that Ishinomaki was closer to the earthquake epicenter than was Nagano. [1]

54. A 25-foot-high tsunami hit the Japanese city of Ishinomaki. Describe a precaution the city could take now to protect citizens from tsunamis in future years. [1]

CHAPTER 16

VOLCANOES AND EARTH'S INTERNAL HEAT

KEY IDEAS Earth's internal heat engine is powered by heat from the decay of radioactive materials and residual heat from Earth's formation. Measurements of heat flow at and near Earth's surface indicate a steady increase in temperature with depth, reaching temperatures of 6,000°C or more at the core. Heat flowing outward from this fiery interior is largely responsible for a variety of phenomena ranging from Earth's magnetic field to crustal motion and volcanic activity.

KEY OBJECTIVES
Upon completion of this chapter, you will be able to:

- Explain that Earth has internal sources of energy that create heat.
- Explain how the transfer of heat energy within Earth's interior results in the formation of regions of different densities, and how these density differences lead to motion.
- Describe the origin of magma, and identify regions of frequent volcanic activity.
- Compare and contrast intrusive and extrusive volcanic activity.

EARTH'S INTERNAL HEAT ENGINE

Beneath its cool outer crust, Earth is glowing hot. Evidence for a hot interior comes from exploration of the crust, both above and below the surface. At the surface, we find numerous examples of heat escaping from Earth's interior: molten rock pours from volcanoes, and boiling hot water and steam rise from hot springs and jet out of geysers.

Further evidence comes from mines and deep wells. As we drill deeper into the crust, the temperature of the rock gets higher and higher. The deepest oil wells go down about 8 kilometers, and the temperature of the oil from these wells is 150°C. Measurements of rock temperature in mines and boreholes show an average temperature rise of 1°C per 30 meters; this increase in temperature with depth is known as the **geothermal gradient**. The rate of temperature increase, however, falls off rapidly with depth. See Figure 16.1. Although the rate of temperature increase in the mantle is lower, considering that Earth's center is more than 6,000 *kilometers* down, it is not unlikely that temperatures there exceed 6,000°C.

Figure 16.1 The Geothermal Gradient. The solid black line on the graph in the lower part of the diagram is the geothermal gradient, the actual change in temperature with increasing depth. The dotted line indicates the changing melting temperature of rock due to changes in composition and pressure. Source: The State Education Department, *Earth Science Reference Tables*, 2011 ed. (Albany, New York; The University of the State of New York).

Heat flows continuously from Earth's fiery depths toward its cool surface. The flow of heat from Earth's interior is roughly 209 joules (50 calories) per square centimeter per year. This is extremely small compared to the amount of heat that Earth's surface receives from the Sun each year—an average 749,000 joules (179,000 calories) per square centimeter. Therefore, Earth's heat flow from its interior contributes little to its heat balance or to the powering of convection in the atmosphere or oceans. However, this heat does power convective motions within Earth's mantle and core.

Sources of Internal Heat

Original Heat

Some of the heat inside Earth was trapped when it was formed. As discussed in Chapter 9, the heat produced by accretion caused Earth to reach internal temperatures high enough to melt rock. Over time, the outer surface cooled, forming a solid crust. Solid rock is a poor conductor of heat; therefore, once the crust formed, the heat beneath it was trapped. Measurements of the rate at which Earth is losing heat indicate, however, that if this internal store were the only source of heat, Earth would already have totally solidified.

Radioactive Decay

The bodies that accreted to form Earth probably had a composition similar to that of stony meteorites known as chondrites. Among the elements found in chondrites are uranium, thorium, and potassium, each of which has radioactive isotopes. Since these elements have big ions, they don't fit into crystalline structures easily. When Earth's outer layers melted after accretion, these elements tended to stay in the molten magma and floated to the surface. As Earth cooled, they became concentrated in the rocks of the crust, particularly in granite.

The radioactive decay of these elements produces heat energy. It is estimated that the total mass of granite in the crust emits about 20 *trillion* watts of power per year. Volcanic rock derived from the mantle indicates that the concentration of radioactive elements in the mantle is lower. Nevertheless, the mantle is so much larger and more massive than the crust that it supplies about 10 trillion watts of power per year from radioactive decay alone. Thirty trillion watts per year is about 33 calories (140 joules), or more than half of the total heat flow from Earth's interior of 50 calories (209 joules) per square centimeter per year. Thus, more than half of the heat flowing out of Earth is due to radioactive decay within the crust and mantle.

Tidal Friction

Another possible source of ongoing heat production is tidal friction. Gravitational attraction between Earth and the Moon and between Earth and the Sun exert unequal forces on Earth that tend to deform it. The ever-changing configuration of the Earth-Moon-Sun system produces a rhythmic pattern of deformations—the tides. Not only Earth's oceans are affected; the atmosphere and solid earth show detectable tides as well. Measurements made in the 1970s showed that at middle latitudes a surface point on the crust is 30 centimeters farther from the center of Earth at high "earth tide" than at low "earth tide." In other words, tidal forces cause solid rock to bend upward, forming a tidal bulge. The tidal bending and unbending of rock layers in the crust releases heat, just as bending a wire coat hanger back and forth causes it to become hot to the touch.

Internal Heat Transfer

Heat tends to move from areas of high concentration to areas of low concentration. Simply stated, heat moves from materials at a higher temperature to materials at a lower temperature. Heat is transferred in three ways: conduction, radiation, and convection. See Figure 16.2.

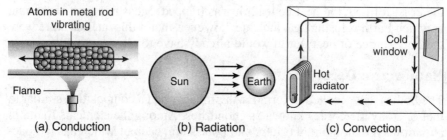

Figure 16.2 Mechanisms of Heat Transfer.

Conduction (Figure 16.2a) is the transfer of heat energy from one molecule to another through collision. Conduction involves actual contact between two materials. Molecules with a lot of heat energy are fast moving. When they collide with cooler, slower moving molecules, an energy transfer occurs. The faster moving molecule slows down, and the slower moving molecule speeds up. Think of a bowling ball hitting the pins at the end of the alley. The ball slows down when it hits the pins, but the pins speed up and go flying. Rock is a poor conductor of heat because its molecules are bound in crystalline structures and cannot move about freely. As a result, heat moves very slowly through rock by conduction.

Radiation (Figure 16.2b) is the transfer of energy by electromagnetic waves. As atoms and molecules vibrate, they send out waves of energy that travel through space. Within rock, however, surrounding atoms and molecules block and trap that energy. Thus, loss of Earth's internal heat by radiation occurs almost entirely at the surface.

Convection (Figure 16.2c) is the actual motion of a volume of hot fluid from one place to another. Heated fluids are less dense than cooler fluids. Thus, when a portion of a fluid is heated, it becomes less dense than the surrounding cooler fluid and rises upward. As the heated fluid rises, it carries upward the heat energy within it. The upward motion of the hot fluid displaces cold fluid in its path, thereby setting up convection currents.

Seismic evidence shows that the outer core of Earth is a fluid. Therefore, convection currents in the outer core can carry heat outward. Enough heating occurs in Earth's mantle to soften the rock and allow convection to remove the heat there as well. Thus, convection is occurring in both the core and the mantle. Plumes of hot material driven by heat from the core interrupt simple convection currents in the mantle. As a result, the nature of convection in the mantle is complicated. As these complex currents of hot material rise upward against the crust, they spread out and move sideways. As you will learn in Chapter 17, it is inferred that the force of these currents dragging along the base of the crust drives the motion of crustal plates.

VOLCANOES

Origin of Magma

Molten rock inside Earth is called **magma**. Melting occurs only in places where heat is concentrated or pressure is reduced. Measurements made on lavas show that some magmas are fluid at about 1,000°C. As shown in Figure 16.1, temperatures high enough to melt rock and form magma probably occur in the upper mantle at depths of 70–200 kilometers. This is the root zone of volcanoes.

Why is Earth's mantle not entirely molten? The answer is that pressure influences melting temperature. Although temperatures within the mantle are high enough to melt rock, the great pressures there hold crystals together and thus prevent the rock from turning into a liquid. For example, the mineral albite melts at 1,104°C at Earth's surface; but at a depth of 100 kilometers, where the pressure is 35,000 times greater, it melts at 1,440°C. Thus, the depth at which a rock melts depends on its composition and on the temperature and pressure at that depth.

The composition of continental crust and of oceanic crust is not the same. Continental crust is granitic, consisting of minerals with rather low melting points. Granitic magma can begin to form at depths of 30–60 kilometers. Ocean crust is basaltic, consisting of minerals with higher melting temperatures. Thus, basaltic magmas begin to form between depths of 100 and 350 kilometers. In either type of rock, some minerals melt at lower temperatures than others, so the magma that forms is a slush of hot, viscous liquid surrounding hot but still solid crystals. Geologists call this magma a "zone of partial melt." Until the liquid portion exceeds half of the melt, this magma behaves more like a solid than a liquid.

Once a body of magma has formed, it will start to rise because it is less dense than the surrounding rock. The rising magma works its way upward through small cracks and fissures. As magma rises into fissures, it widens them and can cause new ones to develop, resulting in earthquakes. Rising also decreases the pressure on the magma, allowing more minerals to remain molten at lower temperatures. Near Earth's surface, magmas can melt surrounding rock, forming underground pools of magma known as **magma chambers**. From these magma chambers, the molten rock may reach the surface through cracks and fissures to erupt, or may cool and solidify underground.

Basaltic magma is more fluid than granitic magma. Therefore, it rises more quickly; and, because of dropping pressure, its melting temperature falls faster than the magma is cooling. Thus, a lot of basaltic magma makes it all the way to Earth's surface as a fluid and erupts as lava. Granitic magma, on the other hand, is much more viscous, so its rate of ascent is slowed. It often cools faster than dropping pressure lowers its melting temperature. Therefore, most granitic magmas solidify underground rather than reaching the surface and forming lavas.

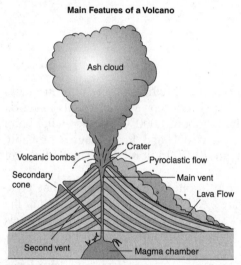

Figure 16.3 A Cross Section of a Volcano. Magma rises from chambers in the upper mantle until it exits the crust through an opening called the *vent*. Thereafter, it is called *lava*, and accumulates in a mound called a *volcano*.

The Formation of Volcanoes

A **volcano** is both the opening in the crust (the vent) through which magma erupts and the mountain built by the erupted material. See Figure 16.3. Volcanoes form where cracks in Earth's crust lead to a magma chamber. Since liquid magma is less dense than the surrounding solid rocks and is under pressure, it rises toward the surface through these cracks. As magma rises, dissolved gases in it expand and are released, giving an upward boost to the magma. The closer to the surface, the less confining pressure there is to overcome and the faster the magma and gases move. Depending upon its viscosity, magma may pour out quietly onto the surface as a flood of molten rock, or spurt out explosively, showering the surrounding terrain with solid rock and globs of molten rock. Once magma emerges from the surface, it is called **lava**.

Volcanic Structures

Erupted materials spread out in all directions around the vent, or central opening, of a volcano. With each new eruption, more material piles up around the vent and a cone-shaped mound forms. The shape of the cone depends upon the viscosity of the magma.

Fluid lava tends to flow quietly through the vent, spreading out in wide, thin sheets. The result is a flat, wide cone called a **shield volcano** (for its resemblance to a shield when viewed from overhead).

Thick, pasty lava containing a lot of dissolved gas does not flow easily through the vent, which tends to become clogged or completely blocked. As a result, gases cannot escape and pressure builds up until a violent explosion takes place. During an explosive eruption, rock and soil surrounding the vent are thrown upward, along with chunks and droplets of lava blown out of the vent. The lava cools as it hurtles through the air and is generally solid by the time it reaches the ground. Collectively, all of the fragments of solidified lava ejected during a volcanic eruption are called **tephra**. This "rain" of solid particles produces a steep, narrow cone called a **cinder cone volcano** (for the particles' resemblance to cinders).

Alternating explosive and quiet eruptions give rise to large, symmetrical **composite cones** made of alternating layers of solidified lava and volcanic rock particles such as ash and cinders.

In some cases, magma emerges through long, open cracks called **fissures**. Lava pouring out of fissures spreads out in wide, thin sheets. Instead of forming a cone as it would around a central vent, it forms a sheet, called a **lava plateau**, that blankets the surrounding land.

The four volcanic structures discussed above are illustrated in Figure 16.4.

Figure 16.4 Typical Volcanic Structures. (a) Broad, gently sloping shield volcanoes form from successive lava flows. (b) Steep-sided cinder cones form from the buildup of ash and cinders. (c) Composite cones of intermediate slope form when periods of lava flows are interspersed with eruptions of ash and cinder. (d) Very fluid lava that erupts along a fissure spreads out to form a lava plateau.

Figure 16.5 Some Types of Plutons.

Intrusive Activities and Structures

Some magma moves around within Earth but never reaches the surface. Underground flows of magma, or **intrusions**, move into and through the countless underground cracks in the rocks of Earth's crust. There, intrusions may solidify into rock structures called **plutons**. See Figure 16.5. Plutons are named for their shapes and include dikes, sills, laccoliths, stocks, and batholiths. **Dikes** are flat, slablike structures that cut across layers of rock like a wall. **Sills** have a similar shape, but lie parallel to layers of rock (as a windowsill is parallel to the ground). **Laccoliths** are large structures with flat bottoms and arched tops. The word laccolith means "lake-rock," and these structures resemble inverted lakes. Laccoliths form when magma is forced between rock layers faster than it can spread out, causing it to push the overlying layers of rock upward to form a dome mountain. Large, irregularly shaped plutons are called *stocks* if they cover less than 75 square kilometers, and *batholiths* if they cover a larger area.

Zones of Volcanic Activity

If the locations of active volcanoes are plotted on a map of Earth, the interesting pattern shown in Figure 16.6 emerges.

Figure 16.6 Earth's Active Volcanoes. Source: Adapted from image at *https://volcano.si.edu/*, website of the Global Volcanism Program, National Museum of Natural History, E-121, Smithsonian Institution, Washington, D.C.

Active volcanoes are not distributed evenly on Earth. Rather, they occur in narrow, elongated zones in some areas and are completely absent from others. For example, active volcanoes surround the Pacific Ocean Basin, a zone known as the **Ring of Fire**, but none of the edges of continents bordering the Atlantic Ocean are volcanic. A narrow zone of volcanoes can also be found near the centers of most oceans, producing the vast chain of undersea mountains known as the mid-ocean ridge. A third zone runs along the northern edge of the Mediterranean Sea; known as the **Mediterranean Belt**, it includes historic volcanoes such as Mt. Etna and Mt. Vesuvius in Italy.

You may have noticed by now a remarkable similarity between the pattern of volcanic activity shown here, and the pattern of earthquake activity described in Chapter 15. As you will discover in Chapter 17, earthquake and volcano zones define the edges of the huge, platelike slabs of rock that make up Earth's crust.

MULTIPLE-CHOICE QUESTIONS

In each case, write the number of the word or expression that best answers the question or completes the statement.

1. The source of energy for the high temperatures found deep within Earth is
   (1) tidal friction
   (2) incoming solar radiation
   (3) decay of radioactive materials
   (4) meteorite bombardment of the Earth

2. What happens to the density and temperature of rock within Earth's interior as depth increases?
   (1) density decreases and temperature decreases
   (2) density decreases and temperature increases
   (3) density increases and temperature increases
   (4) density increases and temperature decreases

3. Which observation provides the strongest evidence for the inference that convection cells exist within Earth's mantle?
   (1) Sea level has varied in the past.
   (2) Marine fossils are found at elevations high above sea level.
   (3) Displaced rock strata are usually accompanied by earthquakes and volcanoes.
   (4) Heat-flow readings vary at different locations in Earth's crust.

4. During which process does heat transfer occur because of density differences?
   (1) conduction   (3) radiation
   (2) convection   (4) reflection

5. The primary cause of convection currents in Earth's mantle is believed to be the
   (1) differences in densities of Earth materials
   (2) subsidence of the crust
   (3) occurrence of earthquakes
   (4) rotation of Earth

6. Which diagram best represents the transfer of heat by convection in a liquid?

7. Which process transfers energy primarily by electromagnetic waves?
   (1) radiation   (3) conduction
   (2) evaporation   (4) convection

8. As depth beneath the surface increases, the temperature at which a particular mineral will melt
   (1) increases
   (2) decreases
   (3) remains the same
   (4) changes from Fahrenheit to Celsius degrees

9. According to the *Reference Tables for Physical Setting/Earth Science*, the melting point of materials in the outer core is
   (1) lower than the interior temperature
   (2) higher than the interior temperature
   (3) lower than that of materials in the crust
   (4) higher than that of materials in the inner core

10. The observed difference in density between continental crust and oceanic crust is most likely due to differences in their
    (1) composition   (3) porosity
    (2) thickness   (4) rate of cooling

11. Magma chambers tend to form in areas beneath Earth's surface where
    (1) heat builds up and pressure is reduced
    (2) pressure builds up and heat is reduced
    (3) pores in Earth are filled with groundwater
    (4) there are extensive deposits of coal, oil, or natural gas

12. Thick, pasty lava with a high gas content tends to erupt
    (1) quietly
    (2) explosively
    (3) from fissures in a broad sheet
    (4) from undersea vents

13. Which conditions can normally be found in Earth's asthenosphere, producing a partial melting of ultramafic rock?
    (1) temperature = 1,000°C; pressure = 10 million atmospheres
    (2) temperature = 2,000°C; pressure = 0.1 million atmospheres
    (3) temperature = 3,500°C; pressure = 0.5 million atmospheres
    (4) temperature = 6,000°C; pressure = 4 million atmospheres

14. Active volcanoes are most abundant along the
    (1) edges of tectonic plates
    (2) eastern coastline of continents
    (3) 23.5° N and 23.5° S parallels of latitude
    (4) equatorial ocean floor

15. The diagrams below represent four rock samples. Which rock was formed by rapid cooling in a volcanic lava flow? [The diagrams are not to scale.]

Bands of alternating light and dark minerals (1) Easily split layers of 0.0001-cm-diameter particles cemented together (2) Glassy black rock that breaks with a shell-shape fracture (3) Interlocking 0.5-cm-diameter crystals of various colors (4)

16. A lava flow that has cooled and solidified into rock is observed to have a surface pockmarked by numerous spherical cavities. These cavities are most likely the result of
    (1) rainwater dissolving out water-soluble mineral crystals
    (2) the corrosive effects of sustained periods of acid rain
    (3) rounded pebbles pried out of the surface by frost action
    (4) gas bubbling out of the lava just as it hardened

Volcanoes and Earth's Internal Heat

17. An igneous intrusion loses heat to its surroundings primarily by
    (1) conduction
    (2) convection
    (3) radiation
    (4) absorption

18. Explosive volcanoes are usually characterized by
    (1) steeply sloped lava cones
    (2) steeply sloped cinder cones
    (3) gently sloped lava cones
    (4) gently sloped cinder cones

19. A flow of magma cools and hardens into rock before it reaches the surface. Which of the following structures most likely formed from the magma?
    (1) a cinder cone
    (2) a basalt plateau
    (3) a geyser
    (4) a dike

20. Which statement concerning a sill is true?
    (1) It may be vertical.
    (2) It must be vertical.
    (3) It may cut across sedimentary strata.
    (4) It must be horizontal.

21. Which statement applies equally to batholiths, laccoliths, dikes, and sills?
    (1) They create folded mountains.
    (2) They are composed of intrusive rocks.
    (3) They are flat sheets of metamorphic material.
    (4) They are composed largely of obsidian.

22. A study of rocks in an extinct volcano indicates a composition largely of cinders and other igneous fragments. This extinct volcano was probably formed as a result of
    (1) lava flows
    (2) igneous intrusions
    (3) explosive eruptions
    (4) magma solidifications

Base your answers to questions 23 through 25 on the passage and cross section below and on your knowledge of Earth science. The cross section represents one theory of the movement of rock materials in Earth's dynamic interior. Some mantle plumes that are slowly rising from the boundary between Earth's outer core and stiffer mantle are indicated.

Hot Spots and Mantle Plumes

Research of mantle hot spots indicates that mantle plumes form in a variety of sizes and shapes. These mantle plumes range in diameter from several hundred kilometers to 1,000 kilometers. Some plumes rise as blobs rather than in a continuous streak; however, most plumes are long, slender columns of hot rock slowly rising in Earth's stiffer mantle. One theory is that most plumes form at the boundary between the outer core and the stiffer mantle. They may reach Earth's surface in the center of plates or at plate boundaries, producing volcanoes or large domes.

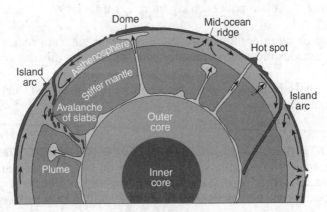

(Not drawn to scale)

23. Compared to the surrounding material, mantle plumes rise toward Earth's surface from the core-mantle boundary because they are
 (1) cooler and less dense
 (2) cooler and more dense
 (3) hotter and less dense
 (4) hotter and more dense

24. At which depth below Earth's surface is the boundary between Earth's outer core and stiffer mantle located?
 (1) 700 km (3) 2,900 km
 (2) 2,000 km (4) 5,100 km

Volcanoes and Earth's Internal Heat

25. The basaltic rock that forms volcanic mountains where mantle plumes reach Earth's surface is usually composed of
    (1) fine-grained, dark-colored felsic minerals
    (2) fine-grained, dark-colored mafic minerals
    (3) coarse-grained, light-colored felsic minerals
    (4) coarse-grained, light-colored mafic minerals

CONSTRUCTED RESPONSE QUESTIONS

26. State three types of evidence that support the statement "There is a tremendous amount of heat within Earth." [3]

27. In one or more complete sentences, explain the difference between magma and lava. [1]

28. Identify three types of volcanic structures. In one or more complete sentences, explain how each structure is formed. [3]

29. In one or more complete sentences, explain why the rock that forms from igneous intrusions typically consists of large mineral crystals. [1]

30. State two differences between a quiet eruption and an explosive eruption. [2]

EXTENDED CONSTRUCTED RESPONSE QUESTIONS

Base your answer to questions 31 through 35 on the diagram below, which represents a cross section of a portion of Earth's crust. Letters *A* through *M* identify rock layers, structures, and boundaries within the cross section.

31. State the rock layer that best represents a dike. [1]

32. State the rock layer that best represents a sill. [1]

33. Place a circle around the rock structure that best represents a volcano. [1]

34. In one or more complete sentences, explain how the igneous intrusion labeled *B* most likely affected the rock adjacent to it in rock layer *H*. [1]

35. State *two* likely sources of the heat originally contained in the igneous rock that formed the intrusions in the cross section. [2]

Base your answers to questions 36 through 41 on the reading passage and maps below and on your knowledge of Earth science. The enlarged map shows the location of volcanoes in Colombia, South America.

Fire and Ice—and Sluggish Magma

On the night of November 13, 1985, Nevado del Ruiz, a 16,200-foot (4,938 meter) snowcapped volcano in northwestern Colombia, erupted. Snow melted, sending a wall of mud and water raging through towns as far as 50 kilometers away, and killing 25,000 people.

Long before disaster struck, Nevado del Ruiz was marked as a trouble spot. Like Mexico City, where an earthquake killed at least 7,000 people in October 1985, Nevado del Ruiz is located along the Ring of Fire. This ring of islands and the coastal lands along the edge of the Pacific Ocean are prone to volcanic eruptions and crustal movements.

The ring gets its turbulent characteristics from the motion of the tectonic plates under it. The perimeter of the Pacific, unlike that of the Atlantic, is located above active tectonic plates. Nevado del Ruiz happens to be located near the junction of four plate boundaries. In this area an enormous amount of heat is created, which melts the rock 100 to 200 kilometers below Earth's surface and creates magma.

Nevado del Ruiz hadn't had a major eruption for 400 years before this tragedy. The reason: sluggish magma. Unlike the runny, mafic magma that makes up the lava flows of oceanic volcanoes such as those in Hawaii, the magma at this type of subduction plate boundary tends to be sticky and slow moving, forming the rock andesite when it cools. This andesitic magma tends to plug up the opening of the volcano. It sits in a magma chamber underground with pressure continually building up. Suddenly, tiny cracks develop in Earth's crust, causing the pressure to drop. This causes the steam and other gases dissolved in the magma to violently expand, blowing the magma plug free. Huge amounts of ash and debris are sent flying, creating what is called an explosive eruption.

Oddly enough, the actual eruption of Nevado del Ruiz didn't cause most of the destruction. It was caused not by lava but by the towering walls of sliding mud created when large chunks of hot ash and pumice mixed with melted snow.

36. What are the names of the *four* tectonic plates located near the Nevado del Ruiz volcano? [1]

37. What caused most of the destruction associated with the eruption of Nevado del Ruiz? [1]

38. What caused the magma to expand, blowing the magma plug free? [1]

39. Vesicular texture is very common in igneous rocks formed during andesitic eruptions. Explain how this texture is formed. [1]

40. Why are eruptions of Nevado del Ruiz generally more explosive than most Hawaiian volcanic eruptions? [1]

41. Describe one emergency preparation that may reduce the loss of life from a future eruption of the Nevado del Ruiz volcano. [1]

CHAPTER 17
PLATE TECTONICS

KEY IDEAS There is much evidence that Earth is a dynamic geologic system whose crust is in nearly constant motion. The theory of plate tectonics is an excellent example of how our ideas about Earth change as new discoveries are made. Building upon earlier theories of continental drift, isostasy, and ocean floor spreading, plate tectonics was able to explain observations that could not be accounted for by these earlier theories.

According to the plate tectonics theory, Earth's crust, or lithosphere, consists of separate plates that rest on the more fluid asthenosphere and move slowly in relationship to one another, creating convergent, divergent, and transform plate boundaries. The outward transfer of Earth's internal heat drives convective circulation in the mantle that moves the lithospheric plates comprising Earth's surface. Plate boundaries are the sites of most earthquakes, volcanoes, and young mountain ranges.

Many of Earth's surface features, such as mid-ocean ridges/rifts, trenches/subduction zones/island arcs, mountain ranges (folded, faulted, and volcanic), hot spots, and the magnetic and age patterns in surface bedrock, are a consequence of forces associated with plate motion and interaction.

KEY OBJECTIVES
Upon completion of this chapter, you will be able to:

- Describe evidence of crustal movements.
- Explain how the theories of continental drift, isostasy, and ocean floor spreading were combined in a single unifying theory—the theory of plate tectonics.
- Explain how differences in density resulting from heat flow within Earth's interior cause movement of the lithospheric plates.
- Relate specific forms of crustal activity, such as earthquakes, volcanoes, and the deformation and metamorphism of rocks during the formation of mountains, to the various types of plate boundaries.
- Describe how many of the processes of the rock cycle are consequences of plate dynamics, including the production of magma, regional meta-

morphism, and the creation of major depositional basins through downwarping of Earth's crust.
- Explain how plate motions have resulted in global changes in geography, climate, and the patterns of organic evolution.

THE DYNAMIC CRUST

Earth's crust, or **lithosphere**, is almost constantly in motion. This motion may be abrupt, as in an earthquake, or so gradual that it is imperceptible to the senses. Movements of the lithosphere exert forces on rock, causing them to be deformed. Evidence that the lithosphere is moving ranges from direct observation of motion to inferences based upon displacement of structures and deformation of rock.

Evidence of Crustal Movements

Earthquakes are unmistakable evidence of crustal movement. During an earthquake, movement of the crust occurs along faults. Every one of the nearly 1 million earthquakes that occur annually involves movement of Earth's crust.

Volcanic eruptions also involve movements of the crust. As magma moves beneath and then out of a volcano, the crust around the volcano rises and sinks measurably, often causing the brittle crust to crack and form fissures. These breaks are direct evidence of movement.

Displaced structures are also evidence of crustal movement. When the crust moves, structures built on its surface move along with it.

Bench marks are permanent metal plaques set in the ground giving the exact locations and precisely determined elevations of points. See Figure 17.1a. Bench marks are used as references in geologic surveys and tidal observations. Measurements of bench mark elevations for more than 100 years reveal that in large areas of the United States the ground is slowly moving upward or downward. See Figure 17.1b. Crustal motion can also be detected and measured using satellite laser technology.

Tilted or folded rock layers are also evidence of crustal movement. Sedimentary strata, lava flows, and tephra are originally deposited in horizontal layers. Therefore, where such materials are observed to be steeply tilted, or bent and folded, it can be inferred that they have been moved from their original positions.

Sedimentary rock layers at high elevations are further evidence of crustal movement. Most layers of sediments are deposited at or below sea level, and may be buried deeper before changing into sedimentary rock. In many places, however, layers of sedimentary rock are found high atop mountains or plateaus. Some are located as much as 2 kilometers above current sea level, far higher than any known body of water, past or present. From this it can be inferred that these rock layers have been uplifted.

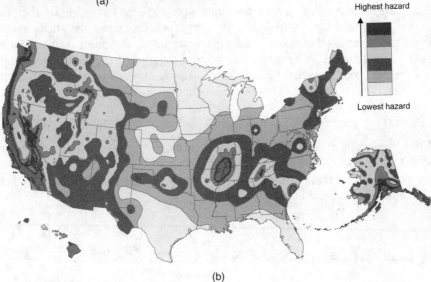

Figure 17.1 (a) A United States Geological Survey Bench Mark. A bench mark is a small plaque, usually mounted in rock or on a concrete base, marked with the precise location and elevation of a point. Source: United States Geological Survey.
(b) U.S. Seismic Hazards Map.

Fossils provide striking evidence of crustal movement. Fossils of marine life, such as corals and clams, found at mountaintops in the Himalayas are evidence of uplift. Fossils of terrestrial or shallow-water organisms beneath the deep ocean bottom indicate sinking, subsidence of the crust.

Thick layers of shallow-water sediments, as much as 15 kilometers thick in some places, are interesting evidence of crustal movement. The average depth of the ocean is 3.8 kilometers with several isolated points having depths of about 10 kilometers. How can 15 kilometers of sediment accumulate in a body of shallow water without filling it completely? One probable inference is that the crust beneath the sediment is sinking while the sediment is accumulating, so sediment layers build up while the water depth remains fairly constant.

Causes and Effects of Crustal Movements

Causes

Unbalanced forces acting on Earth's crust cause crustal movement. Some of the forces acting on the crust are gravity (including gravitational attraction by the Sun and Moon) and forces produced by Earth's rotation, by the expansion and contraction of rock material due to heating and cooling, and by density currents in Earth's mantle.

Stress, Tension, Compression, and Shear

The many forces acting on rock are called **stress**. Stress can act upon rock in several different ways. A rock is under *uniform* stress when the stress in all directions is equal. **Tension** stresses act in opposite directions, pulling rock apart or stretching it. **Compression** stresses act toward each other, pushing or squeezing rock together. **Shear** stresses may act toward or away from each other, but they do so along different lines of action, causing rock to twist or tear.

Effects

When stress acts upon a rock, the rock **strains**, or changes in size, shape, or both. Uniform stress causes rock to change in size, but not in shape. Tension, compression, and shear stresses cause a change in shape and may also cause a change in size. See Figure 17.2.

Figure 17.2 Changes in Rock Shape and/or Size over Time Due to Tension, Compression, Shear, and Uniform Stress.

Deformation and Fracture

When rock is stressed, it goes through a series of changes. The first step is elastic deformation, in which the rock strains, but the change is not permanent. If the stress is removed, the rock will return to its original shape and size. The next step is ductile deformation, which begins when stress reaches a point called the **elastic limit**. At the elastic limit, the stress exceeds the strength of the rock's internal bonding and permanent changes occur. If stress is then removed, the deformation is permanent and the rock no longer returns to its original shape or size. Ductile behavior is similar to what happens when you squeeze or stretch modeling clay without breaking it. The final step is fracture, in which the stress actually causes the rock to break, or fracture. In general, high temperatures and pressures favor ductile behavior and make fracture less likely to occur. See Figure 17.3.

Figure 17.3 Elastic Deformation, Ductile Deformation, and Fracture.

Folds

Ductile deformation of layered rock forms bends or warps called **folds**. Folding is due to compression stresses. An upward-arched fold is an **anticline**; a downward, valleylike fold is a **syncline**. The sides of a fold are called its **limbs**. Folds range from simple **monoclines**, in which only one limb is bent, to **overturned** folds, in which both limbs are tilted in the same direction, to **recumbent** folds, which are bent back on themselves almost horizontally. See Figure 17.4.

Figure 17.4 Various Types of Folds.

Joints and Faults

Joints and faults are fractures that form when rock is stressed until it breaks. **Joints** are fractures along which there has not been any movement of the rock. **Faults** are fractures along which the rock has moved and one side of the fracture is displaced relative to the other side. Faulting is always associated with earthquakes. A fault is classified by the angle of the fracture and the direction of relative movement along it. Five types of faults are illustrated in Figure 17.5.

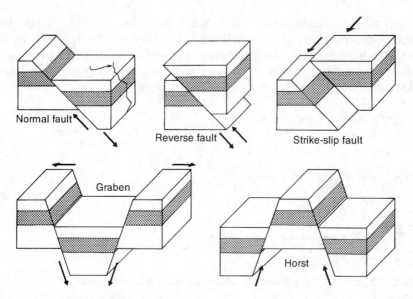

Figure 17.5 Various Types of Faults.

PLATE TECTONICS

The development of the plate tectonics theory is an example of how a theory gains acceptance based upon the evidence that supports it and its ability to explain puzzling observations. Plate tectonics makes sense of such a large range of phenomena that it has become a unifying principle in geology. Two earlier ideas—continental drift and seafloor spreading—were based upon evidence that was later incorporated into a single unifying theory—plate tectonics.

The Theory of Continental Drift: Supporting Evidence

Matching Shorelines The shorelines on both sides of the Atlantic match strikingly, especially the shapes of the Atlantic coasts of Africa and South America. The likeness is even clearer if the edges of the continental shelves are used rather than present shorelines. This match gave rise to the idea that the continents were once joined together but then drifted apart. At first the idea was rejected because it seemed ridiculous to think

Figure 17.6 Pangaea.

that anything as large as a continent could move around. In 1912, however, Alfred Wegener reintroduced the idea of **continental drift**—the concept that continents drift slowly across Earth's surface, sometimes colliding and sometimes breaking into pieces. Wegener and his colleagues presented striking evidence that the world's landmasses had once been joined together, but over time had broken apart into the continents of today and slowly drifted to their present positions. The original landmass was dubbed **Pangaea**, meaning "all land." See Figure 17.6.

Continental and Ocean Crust Statistical studies of elevations and depths indicate that there are two distinct levels for the world's surface—the continents and the ocean floors. These levels alternate and exist side by side with almost no transition between the two, suggesting a series of side-by-side blocks. Some blocks are continents; some are ocean floors. The compositions and thicknesses of the ocean floors and the continents differ. The ocean floors are thin and made of dense basalt, while the continents are thick and consist of less dense granite. See Figure 17.7.

Wegener used a still earlier idea—isostasy—to explain the two levels of the crust. **Isostasy** stated that the crust was floating on hot, fluid rock in the mantle and offered two possible explanations for continents being higher than ocean floors: continents are higher because less dense materials float higher than more dense materials in the same fluid (Pratt's hypothesis), and continents are higher because thick objects float higher than thin objects in the same fluid (Airy's hypothesis). See Figure 17.8. The existence of blocks would explain how continents could "drift"—the blocks simply moved sideways.

Figure 17.7 Compositions and Thicknesses of Continents and Ocean Floors. Source: *The Earth Sciences*, Arthur N. Strahler, Harper & Row, 1971.

Figure 17.8 Isostasy. In both hypotheses, the rocks of the crust float on a denser (3.4 g/cm^3) mantle. In Pratt's hypothesis, mountains are the result of less dense rock (2.5 g/cm^3) floating higher than dense rock (3.2 g/cm^3). In Airy's hypothesis, all the rocks of the crust have the same density (2.7 g/cm^3). Mountains occur where the crust is thicker, and ocean basins occur where it is thinner. Source: Earth Science on File. © Facts on File, Inc., 1988.

Geological Structures Comparison of geological structures on the two sides of the Atlantic Ocean shows a remarkable correspondence. For example, the Sierras near Buenos Aires contain a succession of beds very like those of the Cape Mountains in South Africa. The large gneiss plateau of Africa is strikingly similar to that of Brazil, and matching pockets of igneous rocks and sedimentary rocks can be found in both. The sequence of rock types—older granite, younger granite, alkaline rocks, Jurassic age volcanic rocks, and kimberlite—also matches. The diamond fields of both South Africa and Brazil occur in the kimberlite beds. The Falkland Islands contain rock layers almost indistinguishable from those of the African Cape yet markedly different from layers in nearby Patagonia. Figure 17.9 shows some of the matches of rock types in South America and Africa. Similar matches of North America with Europe, Greenland with North America and Norway, and Madagascar with Africa, as well as between other locations, can also be found.

Fossils and Contemporary Organisms Fossils of identical land animals (*Mesosauroidae*) are found nowhere else but in southern Africa and South America. Fossils of identical trees (*Glossopteris*) are found in Australia, India, and South America. Sixty-four percent of all fossil reptiles from the Carbonaceous Period are identical in Europe and North America.

Contemporary organisms also show a corresponding pattern of distribution. For example, *Lumbricus* earthworms are found from Japan to Spain in Eurasia, but only in the eastern United States. Freshwater perch are found in Europe and Asia, but only in the eastern United States. Similar patterns exist for pearl mussels, mud minnows, and garden snails. Common heather is found only in Europe and Newfoundland. European and eastern North American eels both spawn in the Sargasso Sea off the North American coast. Again, similar patterns can be found between other now-separated continents.

Figure 17.9 Evidence of Continental Drift. Alfred Wegener's correlation of rock type and surface features along the margins of Africa and South America. Source: *The Origins of Continents and Oceans*, Alfred Wegener.

Past Climates There is also unmistakable evidence that 300 million years ago an ice sheet covered South America, southern Africa, India, and southern Australia. This ice sheet was huge, similar to the one covering

Antarctica today. Glacial striations even indicate the direction of ice flow on all four continents. This is difficult to explain without continental drift. If the continents were in their present positions, the ice sheet would have covered all the southern oceans and in places crossed the Equator. In order for this to happen, Earth would have had to be very cold, yet there is no evidence of glaciation at the same time in the Northern Hemisphere. If the continents drifted, though, then 300 million years ago they all could have been joined and some could have been located adjacent to the South Pole, as shown in Figure 18.6. Similarly, places that are today located near the poles contain deposits of coal, which could have formed only in equatorial rain forests. This, too, can be explained by drifting continents.

All of the evidence presented above supports the idea that continents have moved over time. However, it does not suggest a mechanism that would move the continents.

The Theory of Ocean Floor Spreading: Supporting Evidence

In the 1950s the ocean floors were explored extensively by oceanographic research vessels. Much information was gathered and led Professor Harry Hess of Princeton University to propose the idea that the ocean floor is spreading sideways away from the mid-ocean ridges. The evidence supporting this idea suggests a mechanism that could have moved the continents apart.

Mid-Ocean Ridges New instruments enabled scientists to survey the ocean floor with unprecedented accuracy. The ocean basins, once thought to be flat plains, were shown to contain a chain of undersea mountains running down the center of almost every ocean in the world. The mountain chain was split by a deep rift, and the surface around it was riddled with faults. This rugged terrain suggested that movement is occurring on the ocean floors.

Young Rock Surveys of the ocean floors showed that the rocks beneath the oceans are generally younger than the rocks of the continents. Most ocean rocks are only a fraction of the age of some continental rocks. In addition, the age of the rocks on the ocean bottoms increases with distance away from the mid-ocean ridges, suggesting that rocks form at the ridges and then move sideways away from them.

Paleomagnetism Basalt, the major rock of the ocean floors, is rich in iron compounds. Magnetite, pyrrhotite, and hematite are strongly affected by magnetism. When lava is extruded in the mid-ocean rifts, crystals of minerals affected by magnetism align with Earth's magnetic field while the lava cools. When solidified, the igneous rock contains a record of Earth's magnetic field locked in its crystals. Studies of ancient rocks show that

Earth's magnetic field has reversed many times in the past. The reversals seem to take place at irregular intervals ranging from 20,000 to several million years. Research ships carrying sensitive instruments that can measure small changes in magnetism have discovered that magnetism in rocks on the ocean floors reveals a striking pattern of stripes of normal and reversed fields. See Figure 17.10. The fact that the pattern is identical on either side of the mid-ocean ridges also suggests that rocks form at the ridges and then move sideways away from them.

Figure 17.10 Paleomagnetism Patterns on the Ocean Floor.

The proposed mechanism for ocean floor spreading is as follows. Lava is extruded in the rift. The lava hardens to form new ocean floor with Earth's magnetic field "frozen" in its crystals. New lava erupts, splitting the just-formed rock and pushing it aside. The new lava hardens. As this process is repeated, new ocean floor is constantly being formed, and the floor on either side of the mid-ocean ridge is pushed sideways.

Trenches If new ocean floor is constantly being created along the ridges, Earth should be getting larger, but it is not. Therefore, ocean floor is being destroyed at the same rate at which it forms. But *where* is it being destroyed? The answer lies in the trenches. **Trenches** are deep crevices in the ocean floor where it bends downward sharply. Beneath the trenches, earthquakes occur frequently and the positions of their foci show a pattern of increasing depth with distance from one side of the trench. The plunging ocean floor and deepening pattern of earthquake foci beneath the trenches are consistent with a situation where sinking convection currents in the mantle are pulling the ocean floor downward into the mantle, where it melts and is destroyed in a process called **subduction**. See Figure 17.11. With this final piece of evidence the picture was complete, and scientists had not only an explanation for the features and formation of the ocean floor, but also a mechanism for continental drift. Shortly thereafter, continental drift and ocean floor spreading were unified into the theory of plate tectonics.

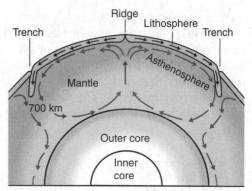

Figure 17.11 Ocean floor is created along the ridges, but is destroyed as it plunges into the mantle (forming trenches) and melts.

The Plate Tectonics Theory

By the 1960s, scientists had enough evidence to believe that continents and seafloors were indeed moving. The evidence showed not only motion, but also motion in many directions. This motion did not agree with the existing model of Earth, which considered the continents and seafloors part of a single, unbroken shell of rocky crust. Clearly, a solid shell cannot move in many different directions at once!

The Lithosphere and the Asthenosphere

As stated earlier, analysis of earthquake waves led to the discovery that Earth's interior consists of layers that have different properties. Based upon density differences, Earth can be divided into the inner and outer core, the mantle, and the crust. If, instead of density, the *rigidity* of rock is considered, another pattern emerges. The crust and upper mantle to a depth of 100 kilometers is strong and rigid, forming a layer called the **lithosphere**. The region of the upper mantle between 100 and 350 kilometers in depth is weak and behaves like a viscous fluid; this layer is called the **asthenosphere**. See Figure 17.12.

Scientists used the lithosphere/asthenosphere pattern to devise a new model of Earth's structure that fit the evidence. They combined aspects of continental drift, isostasy, and ocean floor spreading into a single unifying theory—**plate tectonics**. (*Tectonics* is the branch of geology dealing with the forces affecting the structure of Earth's crust.) In this new model, the lithosphere is seen, not as an unbroken solid, but fragmented and consisting of several huge pieces or **plates**—hence the name *plate* tectonics. The plates are envisioned as "floating" on a layer of fluid rock set in motion by huge convection currents in the mantle. Carried along like bobbing corks on these fiery currents of fluid rock, the plates collide with one another, jostle past one another, or drift apart. Continents and seafloors are in motion because they are part of these plates and move along with them, and it is the interaction between the edges of moving plates that explains all of Earth's features, as well as earthquakes, volcanic activity, and crustal movements.

Figure 17.12   Lithosphere and Asthenosphere.

In simple terms, then, plate tectonics embodies two key ideas: the rigid lithosphere consists of great slabs called plates; and these plates move sideways around Earth, sliding on the fluidlike asthenosphere. As the plates move laterally around Earth, their edges interact in one of three ways: they spread apart, they collide, or they slide past each other. Let us now consider these interactions and see how they account for a wide range of observations.

Interactions at Plate Boundaries

There are three types of plate boundaries. See Figure 17.13.

Divergent boundaries are places where adjacent plates are moving apart. Divergent boundaries create tension stresses, causing rock to fracture. Fractures result in earthquakes and open rifts through which magma can rise to the surface and solidify to form new lithosphere. The mid-ocean rifts are divergent boundaries, and the mid-ocean ridges are volcanic mountains formed by magma emerging from the rifts.

Convergent boundaries are places where adjacent plates are moving toward each other and colliding. When plate edges collide, compression and shear stresses cause rocks to fold, fracture, and move along faults. Plates consisting of ocean crust and plates of continental crust interact differently along divergent boundaries. Since ocean crust is denser than continental crust, ocean crust tends to plunge under continental crust when plates collide, forming a *subduction zone*. In subduction zones, the plunging plate is forced into the mantle and melts. If two plates carrying continental crust collide, they both crumple and fold. The Andes and Himalayan mountains formed along convergent boundaries.

Transform boundaries are places where plates move past each other in strike-slip motions. Shear stresses cause rock to fracture, forming numerous faults. The plates on either side of transform fault margins smash and rub against each other, as two ships in a near-collision grind their sides together.

Figure 17.13 Convergent, Divergent, and Transform Boundaries.

The rock along these faults is intensely shattered, and the sliding motion of the plates causes many earthquakes. If the plates separate a little, some magma may "leak" through the boundary, causing small-scale volcanism. The San Andreas fault is part of a huge transform boundary along which the Pacific plate is moving past the North American plate.

Hot Spots Some volcanoes occur far from plate edges. For example, the Hawaiian volcanoes are located 4,000 kilometers from the nearest plate edge. How does plate tectonics account for volcanic activity far from plate boundaries? To explain the unusual position of these volcanoes and relate them to plate tectonics, Canadian geophysicist J. Tuzo Wilson proposed the concept of hot spots.

According to this concept, there are long-lasting zones of rising hot magma, or **hot spots**, deep within Earth at several places beneath moving plates. Large batches of magma rise from these hot spots and work their way upward through the moving plate. The magma rises because it is more buoyant than the surrounding rock, wedges apart cracks in the plate, and erupts, forming a volcano. As the plate moves along, older volcanoes are carried away in the direction of plate motion and new ones form over the hot spot. See Figure 17.14a.

Wilson's idea explained something that had puzzled scientists for years—why the Hawaiian Islands have active volcanoes only at the southeastern edge of the chain. Furthermore, knowing the rate of plate motion, one should be able to predict the ages of the older volcanic islands. Radioactive dating of the rocks found on various Hawaiian Islands confirmed this hypothesis. See Figure 17.14b.

The geothermal activity—hot springs, geysers, and volcanic rocks—at Yellowstone National Park is the result of another hot spot. The Galapagos, the Azores, and the Society Islands are also examples of volcanic islands formed by hot spots. Thus, while plate tectonics explains the location of belts of volcanic activity, hot-spot volcanoes reveal the direction and rate of plate motion.

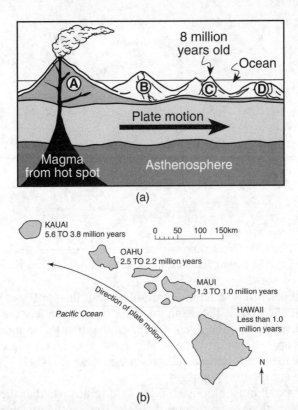

Figure 17.14 (a) Cross Section Showing a Series of Volcanoes Formed by a Plate Moving over a Hot Spot. The extinct volcanoes are eroded to nearly flat surfaces at sea level and are then submerged, forming flat-topped seamounts called *guyots*.
(b) The Progressive Increase in Age of Hawaiian Islands with Distance Away from Currently Active Hawaiian Volcanoes. This supports the hot-spot hypothesis and indicates plate movement toward the northwest.

Key Ideas of the Plate Tectonics Theory

The plate tectonics theory, then, consists of the following ideas.

1. Earth's crust consists of a series of rigid slabs of rock called lithospheric plates. Each plate is about 100 kilometers thick. Earth's surface consists of about six major plates and several smaller pieces. See Figure 17.15.

Plate Tectonics

Figure 17.15 Plate Tectonics Diagram. Source: The State Education Department, *Earth Science Reference Tables*, 2011 ed. (Albany, New York; The University of the State of New York).

2. The plates rest on a relatively fluid layer of the upper mantle and lower crust called the asthenosphere. This fluid layer lets the plates move independently of the deeper rocks in the mantle. Boundaries between plates are regions of volcanic and earthquake activity because they are where the plates collide and scrape against each other. The interiors of plates are relatively quiet.

3. The type of crust atop each plate determines whether the plate carries an ocean floor, a continent, or both. The continents and ocean floors are merely passengers on the moving plates. When the plates change position, the continents and ocean floors move along with them. Past positions of the plates can be inferred from evidence such as fossils, paleomagnetism, and past climates.

4. There are three types of plate boundaries: convergent, divergent, and transform. Plates collide at convergent boundaries, separate at divergent boundaries, and slide past each other at transform boundaries. When an ocean plate collides with a continental plate, the denser ocean plate is pushed under the continental plate, or subducted. Most mountains form at divergent and convergent boundaries.

5. In subduction zones, ocean floor plates are consumed as they are pushed into the hot mantle and melt. As the plate plunges into the mantle, it remains rigid and produces deep-focus earthquakes. Friction between the subducted plate and the adjacent plate melts rock along

the interface and produces volcanic activity directly above and parallel to the trench. For this reason most trenches are bordered by volcanic island arcs or volcanic mountain chains.
6. The creation of ocean floor in the mid-ocean rifts and its destruction in the trenches are in equilibrium. Therefore, Earth remains the same size.
7. Forces exist within Earth that are powerful enough to move the lithospheric plates. Two hypotheses have been proposed to explain plate motion: the plates are pushed by convection currents in the mantle, and they are pulled downward by gravity in subduction zones. Observations of heat flow support the concept of convection in the mantle. However, the scientific community is not yet unified in identifying the specific forces that propel the plates.
8. Convection in the mantle may produce "hot spots" over plumes of rising heat. As plates move over a hot spot, the magma works its way up through the plate and a volcano forms. Volcanic chains in the center of a plate, such as the Hawaiian Islands, are thought to have formed in this way.

Figure 17.16 is a simplified cross section of Earth showing the structures described above.

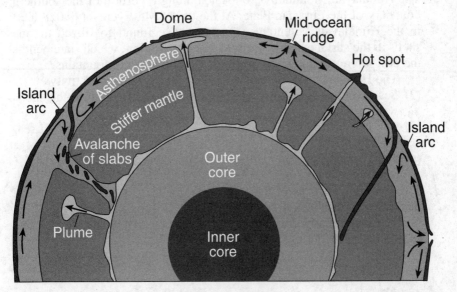

(Not drawn to scale)

Figure 17.16 Cross Section of Earth Showing the Structures Found in Plate Tectonics.

Plate Tectonics

MULTIPLE-CHOICE QUESTIONS

In each case write the number of the word or expression that best answers the question or completes the statement.

1. The diagrams below show cross sections of exposed bedrock. Which cross section shows the *least* evidence of crustal movement?

　　(1)　　　　　　(2)　　　　　　(3)　　　　　　(4)

2. The best evidence of crustal movement is provided by
   (1) dinosaur tracks found in surface bedrock
   (2) marine fossils found on a mountaintop
   (3) weathered bedrock found at the bottom of a cliff
   (4) ripple marks found in sandy sediment

3. The Himalayan Mountains are located along a portion of the southern boundary of the Eurasian Plate. At the top of Mt. Everest (29,028 feet) in the Himalayan Mountains, climbers have found fossilized marine shells in the surface bedrock. From this observation, which statement is the best inference about the origin of the Himalayan Mountains?
   (1) The Himalayan Mountains were formed by volcanic activity.
   (2) Sea level has been lowered more than 29,000 feet since the shells were fossilized.
   (3) The bedrock containing the fossil shells is part of an uplifted seafloor.
   (4) The Himalayan Mountains formed at a divergent plate boundary.

4. Two geologic surveys of the same area, made 50 years apart, showed that the area had been uplifted 5 centimeters during the interval. If the rate of uplift remains constant, how many years will be required for this area to be uplifted a total of 70 centimeters?
   (1) 250　　　　　　(3) 350
   (2) 500　　　　　　(4) 700

5. The photograph below shows an escarpment (cliff) located in the western United States. The directions for north and south are indicated by arrows. A fault in the sedimentary rocks is shown on the front of the escarpment.

The photograph shows that the fault most likely formed
(1) after the rock layers were deposited when the north side moved downward
(2) after the rock layers were deposited when the north side moved upward
(3) before the rock layers were deposited when the south side moved downward
(4) before the rock layers were deposited when the south side moved upward

6. Which block diagram represents the plate motion that causes the earthquakes that occur along the San Andreas Fault in California?

Plate Tectonics

7. Folded sedimentary rock layers are usually caused by
   (1) deposition of sediments in folded layers
   (2) differences in sediment density during deposition
   (3) a rise in sea level after deposition
   (4) crustal movement occurring after deposition

8. Which statement best supports the theory that all the continents were once a single landmass?
   (1) Rocks of the ocean ridges are older than those of the adjacent seafloor.
   (2) Rock and fossil correlation can be made where the continents appear to fit together.
   (3) Marine fossils can be found at high elevations above sea level on all continents.
   (4) Great thicknesses of shallow-water sediments are found at interior locations on some continents.

9. The large coal fields found in Pennsylvania provide evidence that the climate of the northeastern United States was much warmer during the Carboniferous Period. This change in climate over time is best explained by the
   (1) movements of tectonic plates
   (2) effects of seasons
   (3) changes in the environment caused by humans
   (4) evolution of life

10. Compared to the oceanic crust, the continental crust is usually
    (1) thicker, with a less dense granitic composition
    (2) thicker, with a more dense basaltic composition
    (3) thinner, with a less dense granitic composition
    (4) thinner, with a more dense basaltic composition

11. Which statement best supports the theory of continental drift?
    (1) Basaltic rock is found to be progressively younger at increasing distances from a mid-ocean ridge.
    (2) Marine fossils are often found in deep-well drill cores.
    (3) The present continents appear to fit together as pieces of a larger landmass.
    (4) Areas of shallow-water seas tend to accumulate sediment that gradually sinks.

12. According to the Inferred Positions of Earth Landmasses diagram in the *Earth Science Reference Tables*, on what other landmass would you most likely find fossil remains of the late Paleozoic reptile called Mesosaurus, shown below?

(1) North America
(2) Africa
(3) Antarctica
(4) Eurasia

13. Which observation about the Mid-Atlantic Ridge region provides the best evidence that the seafloor has been spreading for millions of years?
(1) The bedrock of the ridge and nearby seafloor is igneous rock.
(2) The ridge is the location of irregular volcanic eruptions.
(3) Several faults cut across the ridge and nearby seafloor.
(4) Seafloor bedrock is younger near the ridge and older farther away.

14. Igneous materials found along oceanic ridges contain magnetic iron particles that show reversal of magnetic orientation. This is evidence that
(1) volcanic activity has occurred constantly throughout history
(2) Earth's magnetic poles have exchanged positions
(3) igneous materials are always formed beneath oceans
(4) Earth's crust does not move

Plate Tectonics

15. Alternating parallel bands of normal and reversed magnetic polarity are found in the basaltic bedrock on either side of the
    (1) Mid-Atlantic Ridge
    (2) Yellowstone Hot Spot
    (3) San Andreas Fault
    (4) Peru-Chile Trench

16. Earth's internal heat is the primary source of energy that
    (1) warms the lower troposphere
    (2) melts glacial ice at lower altitudes
    (3) moves the lithospheric plates
    (4) pollutes deep groundwater with radioactivity

17. In which set are the Earth processes thought to be most closely related to each other because they normally occur in the same zones?
    (1) mountain building, earthquakes, and volcanic activity
    (2) mountain building, shallow-water fossil formation, and rock weathering
    (3) volcanic activity, rock weathering, and deposition of sediments
    (4) earthquakes, shallow-water fossil formation, and shifting magnetic poles

18. According to the *Earth Science Reference Tables*, during which geologic period were the continents all part of one landmass, with North America and South America joined to Africa?
    (1) Tertiary   (3) Triassic
    (2) Cretaceous   (4) Ordovician

Base your answers to questions 19 and 20 on the cross section below and on your knowledge of Earth science. The cross section represents the distance and age of ocean-floor bedrock found on both sides of the Mid-Atlantic Ridge.

19. According to the cross section, every 1 million years, the ocean floor bedrock moves approximately
    (1) 20 km toward the Mid-Atlantic Ridge
    (2) 20 km away from the Mid-Atlantic Ridge
    (3) 40 km toward the Mid-Atlantic Ridge
    (4) 40 km away from the Mid-Atlantic Ridge

20. Which map best represents the pattern of magnetic polarity in the minerals of ocean-floor bedrock on each side of the Mid-Atlantic Ridge?

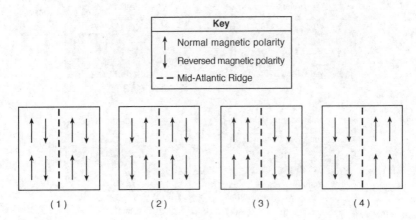

21. Which cross section best represents the convection currents in the mantle beneath the Peru-Chile Trench?

Base your answers to questions 22 through 24 on the map of the Mid-Atlantic Ridge shown below. Points A through D are locations on the ocean floor. Line XY connects locations in North America and Africa.

22. In which cross section do the arrows best show the convection occurring within the asthenosphere beneath line XY?

23. Samples of ocean-floor bedrock were collected at points $A, B, C,$ and D. Which sequence shows the correct order of the age of the bedrock from oldest to youngest?
    (1) $D \to C \to B \to A$
    (2) $A \to D \to B \to C$
    (3) $C \to B \to D \to A$
    (4) $A \to B \to D \to C$

24. The boundary between which two tectonic plates is most similar geologically to the plate boundary at the Mid-Atlantic Ridge?
    (1) Eurasian and Indian-Australian
    (2) Cocos and Caribbean
    (3) Pacific and Nazca
    (4) Nazca and South American

Base your answers to questions 25 through 27 on the passage below.

Crustal Activity at Mid-Ocean Ridges

Mid-ocean ridges are found at one type of tectonic plate boundary. These ridges consist of extensive underwater mountain ranges split by rift valleys. The rift valleys mark places where two crustal plates are pulling apart, widening the ocean basins, and allowing magma from the asthenosphere to move upward. In some cases, mid-ocean ridges have migrated toward nearby mantle hot spots. This explains why mid-ocean ridges and mantle hot spots are found together at several locations.

25. Which type of tectonic plate boundary is located at mid-ocean ridges?
    (1) convergent
    (2) transform
    (3) divergent
    (4) complex

26. Which mantle hot spot is located closest to a mid-ocean ridge?
 (1) Canary Islands   (3) Hawaii
 (2) Easter Island   (4) Tasman

27. The map below shows a part of Earth's surface. Points *A* through *D* are locations on the ocean floor.

At which location is the temperature of the ocean floor bedrock most likely highest?
 (1) *A*   (3) *C*
 (2) *B*   (4) *D*

Plate Tectonics

Base your answers to questions 28 and 29 on the map below, which shows Earth's Southern Hemisphere and the inferred tectonic movement of the continent of Australia over geologic time. The arrows between the dots show the relative movement of the center of the continent of Australia. The parallels of latitude from 0° to 90° south are labeled.

28. The geographic position of Australia on Earth's surface has been changing mainly because
    (1) the gravitational force of the Moon has been pulling on Earth's landmasses
    (2) heat energy has been creating convection currents in Earth's interior
    (3) Earth's rotation has spun Australia into different locations
    (4) the tilt of Earth's axis has changed several times

29. During which geologic time interval did Australia most likely have a warm, tropical climate because of its location?
    (1) Cambrian   (3) Late Permian
    (2) Carboniferous   (4) Eocene

30. Which map best indicates the probable locations of continents 100 million years from now if tectonic plate movement continues at its present rate and direction?

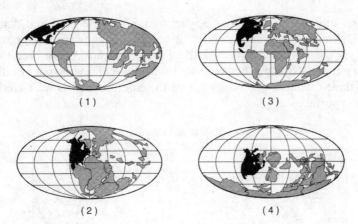

Base your answers to questions 31 through 34 on the cross section below, which shows the boundary between two lithospheric plates. Point *X* is a location in the continental lithosphere. The depth below Earth's surface is labeled in kilometers.

(Not drawn to scale)

31. Between which two lithospheric plates could this boundary be located?
  (1) South American Plate and African Plate
  (2) Scotia Plate and Antarctic Plate
  (3) Nazca Plate and South American Plate
  (4) African Plate and Arabian Plate

32. Compared with the continental crust, the oceanic crust is
  (1) less dense and thinner   (3) more dense and thinner
  (2) less dense and thicker   (4) more dense and thicker

33. The temperature of the asthenosphere at the depth where melting first occurs is inferred to be approximately
  (1) 100°C   (3) 4,200°C
  (2) 1,300°C   (4) 5,000°C

407

Plate Tectonics

34. Point X is located in which Earth layer?
    (1) rigid mantle
    (2) stiffer mantle
    (3) asthenosphere
    (4) outer core

Base your answers to questions 35 through 37 on the map and data table below. The map shows the locations of volcanic islands and seamounts that erupted on the seafloor of the Pacific Plate as it moved northwest over a stationary mantle hotspot beneath the lithosphere. The hotspot is currently under Kilauea. Island size is not drawn to scale. Locations X, Y, and Z are on Earth's surface.

Data Table
Age of Volcanic Features

Volcanic Feature	Distance from Kilauea (km)	Age (millions of years)
Kauai	545	5.6
Nihoa	800	6.9
Necker	1,070	10.4
Midway	2,450	16.2
Suiko seamount	4,950	41.0

35. Approximately how far has location X moved from its original location over the hotspot?
    (1) 3,600 km
    (2) 2,500 km
    (3) 1,800 km
    (4) 20 km

36. According to the data table, what is the approximate speed at which the island of Kauai has been moving away from the mantle hotspot, in kilometers per million years?
    (1) 1
    (2) 10
    (3) 100
    (4) 1,000

37. Which lithospheric plate boundary features are located at *Y* and *Z*?
    (1) trenches created by the subduction of the Pacific Plate
    (2) rift valleys created by seafloor spreading of the Pacific Plate
    (3) secondary plates created by volcanic activity within the Pacific Plate
    (4) mid-ocean ridges created by faulting below the Pacific Plate

38. Crustal formation, which may cause the widening of an ocean, is most likely occurring at the boundary between the
    (1) African Plate and the Eurasian Plate
    (2) Pacific Plate and the Philippine Plate
    (3) Indian-Australian Plate and the Antarctic Plate
    (4) South American Plate and the North American Plate

39. Which two features are commonly found at divergent plate boundaries?
    (1) mid-ocean ridges and rift valleys
    (2) wide valleys and deltas
    (3) ocean trenches and subduction zones
    (4) hot spots and island arcs

40. Oceanic crust is sliding beneath the Aleutian Islands in the North Pacific Ocean, forming the Aleutian Trench at a
    (1) convergent plate boundary between the Pacific Plate and the North American Plate
    (2) convergent plate boundary between the Pacific Plate and the Juan de Fuca Plate
    (3) divergent plate boundary between the Pacific Plate and the North American Plate
    (4) divergent plate boundary between the Pacific Plate and the Juan de Fuca Plate

41. Rifting of tectonic plates in eastern North America during the Jurassic Period was responsible for the
    (1) formation of the Catskill delta
    (2) first uplift of the Adirondack Mountains
    (3) Alleghenian orogeny
    (4) opening of the Atlantic Ocean

42. Which geologic event is inferred to have occurred most recently?
    (1) collision between North America and Africa
    (2) metamorphism of the bedrock of the Hudson Highlands
    (3) formation of the Queenston delta
    (4) initial opening of the Atlantic Ocean

Plate Tectonics

CONSTRUCTED RESPONSE QUESTIONS

Base your answers to questions 43 through 46 on the world map below and on your knowledge of Earth science. The map shows major earthquakes and volcanic activity occurring from 1996 through 2000. Letter *A* represents a volcano on a crustal plate boundary.

Key
● Earthquakes △ Volcanoes

43. Place an **X** on the map to show the location of the Nazca Plate. [1]

44. Explain why most major earthquakes are found in specific zones instead of being randomly scattered across Earth's surface. [1]

45. Identify the source of the magma for the volcanic activity in Hawaii. [1]

46. Identify the type of plate movement responsible for the presence of the volcano at location *A*. [1]

Base your answers to questions 47 and 48 on the block diagram below and on your knowledge of Earth science. The diagram represents the pattern of normal and reversed magnetic polarity of the seafloor bedrock on the east side of a mid-ocean ridge center. The magnetic polarity of the bedrock on the west side of the ridge has been omitted. Arrows represent the direction of seafloor movement on either side of the ridge.

47. Complete the diagram above by shading the pattern of normal polarity on the west side of the ridge center. Assume the rate of plate movement was constant on both sides of the ridge center. Your answer must show the correct width and placement of *each* normal polarity section. [1]

48. Describe the general relationship between the distance from the ridge center and the age of the seafloor bedrock. [1]

49. Identify a process occurring in the plastic mantle that is inferred to cause tectonic plate motion. [1]

Plate Tectonics

Base your answers to questions 50 through 53 on the map below and on your knowledge of Earth science. The map shows the generalized ages of surface bedrock of Iceland, an island located on the Mid-Atlantic Ridge rift. The location of the Mid-Atlantic Ridge rift is indicated. Points *A* and *B* represent locations on the surface bedrock, which is igneous in origin. The ages of the surface bedrock, in million years (my), are indicated in the key.

50. On the map above, identify the *two* tectonic plates, one on each side of the Mid-Atlantic Ridge rift at Iceland, by writing their names on the lines provided on the map. [1]

51. On the map above, draw *one* arrow through point *A* and *one* arrow through point *B* to indicate the relative direction that each plate is moving to produce the Mid-Atlantic Ridge rift. [1]

52. Identify *one* dark-colored, mafic igneous rock with a vesicular texture that is *likely* to be found on the surface of Iceland. [1]

53. Identify *one* feature in the mantle beneath Iceland that causes larger amounts of magma formation in Iceland than at most other locations along the rest of the Mid-Atlantic Ridge rift. [1]

EXTENDED CONSTRUCTED RESPONSE QUESTIONS

Base your answers to questions 54 through 56 on the map below, which is an enlargement of a portion of the *Tectonic Plates* map from the *Reference Tables for Physical Setting/Earth Science*. Points *A* and *B* are locations on different boundaries of the Arabian Plate.

54. Identify the type of tectonic plate boundary located at point *A*. [1]

55. On the map shown, a valley is located south of point *B* and a mountain range north of point *B*. State the tectonic process that is creating these two land features. [1]

56. The block diagram next to the map represents Earth's surface and interior along the East African Rift. Draw *two* arrows, one through point *X* and one through point *Y*, to indicate the relative motion of each of these sections of the continental crust. [1]

Base your answers to questions 57 through 60 on the map and passage below. The map shows the outlines and ages of several calderas created as a result of volcanic activity over the last 16 million years as the North American Plate moved over the Yellowstone Hot Spot. *A* and *B* represent locations within the calderas.

Plate Tectonics

The Yellowstone Hot Spot

The Yellowstone Hot Spot has interacted with the North American Plate, causing widespread outpourings of basalt that buried about 200,000 square miles under layers of lava flows that are a half mile or more thick. Some of the basaltic magma produced by the hot spot accumulates near the base of the plate, where it melts the crust above. The melted crust, in turn, rises closer to the surface to form large reservoirs of potentially explosive rhyolite magma. Catastrophic eruptions have partly emptied some of these reservoirs, causing their roofs to collapse. The resulting craters, some of which are more than 30 miles across, are known as volcanic calderas.

57. Describe the texture and color of the basalt produced by the Yellowstone Hot Spot. [1]

58. Identify *two* minerals found in the igneous rock that is produced from the explosive rhyolite magma. [1]

59. Based on the age pattern of the calderas shown on the map, in which compass direction has the North American Plate moved during the last 16 million years? [1]

60. Calculate, in miles per million years, the rate at which the North American Plate has moved over the Yellowstone Hot Spot between point *A* and point *B*. [1]

Base your answers to questions 61 through 63 on the map below, which shows the inferred position of Earth's landmasses at a particular time in Earth's history. The Taconic Mountains are shown near a subduction zone where they formed after the coast of Laurentia collided with a volcanic island arc, closing the western part of the Iapetus Ocean.

61. On the map, place an **X** to show the approximate location of the remaining part of the Iapetus Ocean. [1]

62. On the map, draw an arrow on the Laurentia landmass to show its direction of movement relative to the subduction zone. [1]

63. Identify the geologic time period represented by the map. [1]

Base your answers to questions 64 through 67 on the map and block diagram below. The map shows the location of North Island in New Zealand. The block diagram shows a portion of North Island. The Hikurangi Trench is shown forming at the edge of the Pacific Plate. Point X is at the boundary between the lithosphere and the asthenosphere.

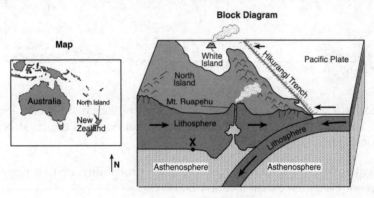

(Not drawn to scale)

64. State the approximate temperature at point X. [1]

65. On what tectonic plate are both North Island and White Island located? [1]

66. Describe the type of tectonic plate motion that formed the Hikurangi Trench. [1]

67. Describe *one* action that people on North Island should take if a tsunami warning is issued. [1]

Plate Tectonics

Base your answers to questions 68 through 71 on the world map shown below and on your knowledge of Earth science. Letters A through H represent locations on Earth's surface.

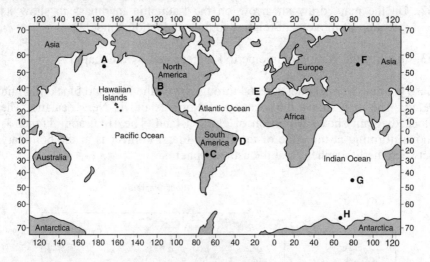

68. Explain why most earthquakes that occur in the crust beneath location B are shallower than most earthquakes that occur in the crust beneath location C. [1]

69. Explain why location A has a greater probability of experiencing a major earthquake than location D. [1]

70. Explain why a volcanic eruption is more likely to occur at location E than at location F. [1]

71. Explain why the geologic age of the oceanic bedrock increases from location G to location H. [1]

Base your answers to questions 72 through 76 on the passage and data tables and map below, and on your knowledge of Earth science. The data tables show trends (patterns) of two lines of Hawaiian island volcanoes, the Loa trend and the Kea trend. For these trends, ages and distances of the Hawaiian island volcanoes are shown. The map shows the locations of volcanoes, labeled with **X**s, that make up each trend line.

Hawaiian Volcano Trends

The Hawaiian volcanic island chain, located on the Pacific Plate, stretches over 600 kilometers. This chain of large volcanoes has grown from the seafloor to heights of over 4,000 meters. Geologists have noted that there appear to be two lines, or "trends," of volcanoes—one that includes Mauna Loa and one that includes Mauna Kea. Loihi and Kilauea are the most recent active volcanoes on the two trends shown on the map.

Loa Trend

Loa Trend Volcanoes	Volcano Age (million years)	Distance from Loihi (km)
Kauai	4.6	575
Waianae	3.7	465
Koolau	2.2	375
West Molokai	1.7	350
Lanai	1.2	300
Kahoolawe	1.1	250
Hualalai	0.3	130
Mauna Loa	0.2	70
Loihi	0	0

Kea Trend

Kea Trend Volcanoes	Volcano Age (million years)	Distance from Kilauea (km)
East Molokai	1.7	256
West Maui	1.5	221
Haleakala	0.9	182
Kohala	0.5	100
Mauna Kea	0.4	54
Kilauea	0.1	0

Volcanoes and Islands of Hawaii

Plate Tectonics

72. The average distance between the volcanoes along the Kea trend is 51.2 kilometers. Place an **X** on the map above to identify the location on the seafloor where the next volcano will most likely form as a part of the Kea trend. [1]

73. Identify the *two* volcanoes, one from each trend, that have the same age. [1]

74. State the general relationship between the age of the volcanoes and the distance from Loihi. [1]

75. Identify the tectonic feature beneath the moving Pacific Plate that caused volcanoes to form in *both* the Loa and Kea trends. [1]

76. Identify the compass direction in which the Pacific Plate has moved during the last 4.6 million years. [1]

Unit Six
WEATHERING, EROSION, AND DEPOSITION

CHAPTER 18
WEATHERING AND SOIL FORMATION

> **KEY IDEAS** In preceding units, we explored the mechanisms by which the materials that comprise Earth's crust were formed. In this unit we will discuss the processes that shape the material exposed at the surface of Earth's crust; a dynamic, changing system of minerals, rocks, weather, water, ice, wind, and living organisms.

KEY OBJECTIVES
Upon completion of this chapter, you will be able to:
- Describe the weathering processes that result in the physical and chemical breakdown of crustal material and describe the products of weathering.
- Explain how weathering and biological activity over long periods of time create soils.

WEATHERING

The surface of Earth's crust, or lithosphere, is constantly exposed to the changing weather of the atmosphere. This environment is very different from the one in which most rocks and minerals were formed. As these materials adjust to their new environment, they change. They break down into smaller pieces, and chemical reactions change them into new substances. Since the processes that produce these changes result from exposure to the weather, they are called weathering. **Weathering** is the breakdown of rocks into smaller particles by natural processes. Whenever rocks are exposed to the air, water, and living things at or near Earth's surface, weathering occurs. Weathering processes are divided into two general types: physical and chemical.

Physical Weathering

Some of the changes caused by weathering involve form only. Weathering may break a large, solid mass of rock into loose fragments varying in size and shape but identical in composition to the original rock. Processes that break down rocks without changing their chemical compositions are called **physical weathering**.

Frost Action

Frost action is the result of an unusual property of water. Most materials expand when heated and contract when cooled. This is true of water except that, when water is cooled from 4°C to 0°C, it *expands*. It expands most when, at 0°C, it solidifies into ice; then its volume increases by 9%. The expansion of water as it cools and solidifies can exert huge forces on anything confining it—forces measuring tens of thousands of pounds per square inch!

Water, in the form of rain, melting snow, or condensation, seeps into any cracks or pores in rock. When the temperature drops below the freezing point of the water in these cracks and pores, the water changes into ice. The expanding ice exerts tremendous pressure against the confining rock. Acting like a wedge, it widens and extends the opening. Then, when the ice thaws, the water seeps deeper into the opening. When the water refreezes, the process is repeated. In this way, the alternate freezing and thawing of water, or **frost action**, breaks rocks apart (see Figure 18.1).

Figure 18.1 Frost Action.

Frost action is particularly effective where bedrock is directly exposed to the atmosphere, moisture is present, and the temperature fluctuates frequently above and below the freezing point of water. These conditions often exist during the winter in temperate climates such as New York State's, and can occur also on mountain tops and at high elevations in spring or fall. Daytime temperatures rise above freezing, causing snow and ice to melt, only to drop below freezing again at night, producing frost action.

Frost action on cliffs of bare rock breaks loose fragments that fall to the base of the cliff. When this takes place rapidly, a pile of fragments called a **talus slope** accumulates at the base of the cliff (see Figure 18.2).

Figure 18.2 A Talus Slope Forms at the Base of a Cliff. The angle of repose is the steepest slope at which the fragments remain stable.

The potholes that threaten motorists in many of our northern states are caused by frost action on exposed road surfaces. Farmers in New England have to clear their fields every spring of rocks and boulders pushed up out of the ground by frost action in soil. As you might expect, frost action is rare in states such as Florida and Hawaii. However, through much of the northern United States, it is probably the primary weathering process, and, throughout the world, probably the most significant physical weathering process.

Abrasion

Rocks can also be broken down by **abrasion**, that is, by rubbing against each other. Rock abrasion occurs mainly when fragments are being carried along by agents of erosion. A typical example occurs in streams. As the fragments are carried along by the water, they bounce off and rub against each other. This abrasion breaks smaller pieces off the surface, and the fragments become still smaller and also more rounded (see Figure 18.3). The fragments abrade the bedrock beneath the stream as well. Wind also weathers rock by abrasion. Anyone who has sat on a sandy beach on a windy day can attest to the abrasive power of wind-driven sand. Abrasion is a major physical weathering process.

Time Increases ———————→

Figure 18.3 Rounding of Particles Due to Abrasion.

Exfoliation

Exfoliation is the scaling off, or peeling, of successive shells from the surface of rocks. Exfoliation generally occurs in coarse-grained rocks that contain the mineral feldspar. Whenever the surface of such rocks becomes wet, moisture penetrates pores and crevices between the mineral grains and reacts with the feldspar. A chemical change occurs, producing a new substance: kaolin. This clay has a greater volume than the feldspar it replaces. The expansion pries loose the surrounding mineral grains, and a thin shell of surface material flakes away. (Note that this is a physical process caused by a chemical change.)

With successive wettings of the rock surface, the process is repeated. Since more surface is exposed at the corners of cracks or joints, these split off at the fastest rate. The result is a rounding in the shape of the rock as seen in Figure 18.4. Feldspar is a common rock-forming mineral; therefore, exfoliation is a significant weathering process.

Figure 18.4 Exfoliation.

Plant and Animal Action

Plants and animals interact with rock in a number of ways that cause the rock to break down into smaller pieces. When a rock develops cracks, small particles of rock and soil are washed into the cracks by rain or blown in by

Weathering and Soil Formation

wind. If a seed finds its way into a crack, it can germinate and begin to grow. As the plant grows, it sends tiny rootlets deeper into the crack in search of water. The growing rootlets thicken and press against the sides of the crack. Like the expansion of ice in frost action, the growing roots widen and extend the crack (see Figure 18.5a). Eventually the rock is broken apart. The roots of tiny plants like mosses and lichens produce a rock-dissolving acid as they grow and decay, thereby further accelerating the breakdown of the rocks.

Although animals (with the exception of humans) do not directly attack rock, they contribute indirectly to its weathering in many ways. More than 100 years ago, Charles Darwin calculated that on 1 acre of land earthworms bring as much as 10 metric tons of particles to the surface each year. Exposed to the atmosphere again, these particles are subjected to further breakdown by weathering processes.

Ants (see Figure 18.5b), termites, woodchucks, moles and other burrowing animals also play a role in weathering. Furthermore, the burrows they create allow air and water to penetrate deeper beneath the surface and to weather the underlying bedrock.

Humans, however, have probably contributed more to the physical weathering of rock than any other single species. Roadcut building, rock quarrying, and strip mining are just a few examples of the human activities that break up rock. In addition, these activities expose vast quantities of fresh rock to other weathering processes.

The contribution of each individual organism to the weathering of Earth's crust may seem small. However, consider the number of living things on Earth and the length of time they have been at work. Taken collectively, the total amount of rock they have affected is tremendous.

Figure 18.5 Plant (a) and Animal (b) Action on Rocks.

Changes in Temperature

Rocks are often exposed to large temperature changes. As rocks heat up during the day, they expand; as they cool off at night, they contract. You might expect this constant change to cause rocks to crack and break up, but experiments have indicated that this is not generally the case. Only extreme temperature changes, such as those resulting from forest and brush fires, cause rocks to crack or flake off at the surface.

Pressure Unloading

Some rocks form deep beneath Earth's surface. There they are under great pressure—millions of pounds per square inch. As the rocks form, stresses build up inside them that cannot be released because of the pressure. When forces within Earth bring these rocks to the surface, the pressure is reduced. The resultant expansion and release of stress cause the rocks to develop large cracks, or joints, at weak points in their structure—a phenomenon known as **pressure unloading**. Pressure unloading can also occur when glaciers melt away and the pressure exerted by the weight of the ice on underlying rock is released.

Chemical Weathering

Chemical weathering breaks down rocks by changing their chemical compositions. Most rocks form in an environment that is very different from that at the surface of Earth. Many of the substances in the atmosphere, for example, were not present in the environment where the rocks were formed. When the minerals in a rock are exposed to these substances, they may react with them to form new compounds with properties different from those of the original minerals. Such changes almost always weaken the structure of the rock, so that it either falls apart or is more easily broken down by physical weathering. Oxygen, water, and carbon dioxide are chiefly responsible for the chemical weathering of rocks.

Oxidation

The atmosphere is about 21 percent oxygen. Oxygen combines with many substances in a reaction called **oxidation**, which is an important chemical weathering process. Oxygen reacts most readily with minerals containing iron, such as magnetite, pyrite, amphibole, and biotite. Oxidation of iron produces compounds of iron and oxygen, iron oxides, of which hematite (Fe_2O_3) and magnetite (Fe_3O_4) are common examples. Since the iron in hematite and magnetite can be extracted economically, these compounds are also considered iron ores.

If water is present during oxidation, another reaction can occur; a compound of iron, oxygen, and water called *goethite* can form. Goethite has a yellowish brown color. If goethite is dehydrated, hematite forms. Many reddish or yellow-brown soils and weathered rocks get their color from the hematite or goethite they contain.

Oxidation of iron oxide in the presence of water to form goethite:
$$4FeO + 2H_2O + O_2 \rightarrow 4(FeO\text{-}OH)$$
iron oxide   water   oxygen   goethite

Dehydration of goethite to form hematite:
$$2(FeO\text{-}OH) \rightarrow Fe_2O_3 + H_2O$$
goethite   hematite   water

How does oxidation break down or weather rocks? When oxygen combines with iron, the chemical bonds between the iron and other substances in the rock are broken, thereby weakening its structure. Compare the strength of a rusty nail with that of a new one. The effect on rock is similar. In addition to iron, other elements in rock, such as aluminum and silicon, can combine with oxygen. The effects are much the same: an oxide is formed, the structure is weakened, and the rock disintegrates.

Hydration, Hydrolysis, and Solution

Most of Earth's surface is covered with water. Trillions of gallons fall on Earth as rain every day. With such abundance it is not surprising that water is an important agent of chemical weathering.

When water combines with another substance, the reaction is called **hydration**. For example, the hydration of anhydrite forms gypsum.

Hydration of anhydrite to form gypsum:
$$CaSO_4 + 2H_2O \rightarrow CaSO_4 \cdot 2H_2O$$
anhydrite   water   gypsum

Water can also form hydrogen ions (H^+) and hydroxide ions (OH^-). When these ions replace the ions of a mineral, the reaction is called **hydrolysis**. Feldspar, amphibole, and biotite are common minerals that can undergo hydrolysis, which causes them to swell and crumble. The product is a powder of clay minerals that are insoluble in water. For example, the hydrolysis of feldspar forms kaolinite, a clay mineral.

Hydrolysis of feldspar to form kaolinite:
$$4\ KAlSi_3O_8 + 4\ H^+ + 2H_2O \rightarrow Al_4Si_4O_{10}(OH)_8 + 8SiO_2$$
K-feldspar   hydrogen ions   water   kaolinite   silica

(Note that the symbol K does not appear on the right. The potassium is released as a positive ion into the soil water.)

Water also weathers rock simply by dissolving it. This process is called **solution**. So many materials dissolve, at least to some extent, in water that water is often called the *universal solvent*. Halite (rock salt) and gypsum are good examples of water-soluble minerals. Water will slowly dissolve these minerals out of a rock, thereby exposing the surrounding minerals to further weathering. In some cases the rock's structure may be so weakened by the empty spaces created that the rock just crumbles. In solution, the dissolved minerals may react with each other to form new substances. If these substances are insoluble in water, they precipitate out.

Carbonation

Carbon dioxide (CO_2) is a colorless, odorless gas that comprises 0.03–0.04 percent of Earth's atmosphere. The chemical combining of carbon dioxide with another substance is called **carbonation**. As a gas, carbon dioxide has almost no effect on rocks. When carbon dioxide comes in contact with water, though, the following carbonation reaction takes place:

Production of carbonic acid by solution of carbon dioxide:
$$CO_2 + H_2O \rightarrow H_2CO_3$$
carbon dioxide   water   carbonic acid

Although carbonic acid is a fairly weak acid, it attacks common rock minerals. Carbonic acid reacts readily with minerals containing the elements sodium, potassium, magnesium, and calcium. The compounds formed when these elements react with carbon dioxide are called *carbonates*.

Carbonic acid is most destructive to the mineral calcite, which is completely dissolved by carbonic acid. Rocks, such as limestones, that are almost entirely composed of calcite are totally dissolved away by carbonic acid in rainwater and groundwater. Spectacular caverns can form as groundwater containing carbonic acid seeps through bedrock composed of calcite and eats huge holes in it (see Figure 18.6).

Figure 18.6 Carbonation Caverns Form as Rainwater Containing Carbonic Acid Seeps into Cracks in Limestone.

Other Chemical Factors

In addition to carbonic acid, other naturally occurring acids attack rocks and minerals. Some of these acids are produced by lightning or the decay of organic material; others, as waste products of certain plants and animals. These acids dissolve in rainwater as it falls through the atmosphere and seeps through the soil. Upon reaching bedrock, they dissolve some rocks and cause others to crumble.

Primitive plants such as lichens can grow on bare rock surfaces. They have limited nutrient needs and can obtain these directly from the rock. Lichens grow when the rock surface is wet and lie dormant when it is dry. Secretions from the lichens eat away at the rock's surface, dissolving out mineral nutrients and loosening mineral particles. These mineral particles accumulate in rock crevices, along with dust from the atmosphere and debris from dead lichens. Seeds and spores may then take hold in these tiny pockets of soil and grow, further increasing the amount of chemical and physical weathering the rock undergoes.

An increasingly important source of acids that weather rock is human activity. Factories, homes, and automobiles release vast quantities of waste gases and other pollutants into the atmosphere. Many, such as the oxides of nitrogen and sulfur, react with water to form very strong, reactive acids.

In some areas surrounding large cities and industrial complexes, the concentration of these acids has reached alarming levels. Rainfall in these areas contains so much acid that it is called **acid rain**. Acid rain accelerates the weathering of rock and the breakdown of structures built by humans, and it damages plant and animal life.

Factors Affecting Weathering

How long does it take for a rock to be broken down by weathering processes? The answer is complex since many factors influence the rate at which a rock will weather. Among the more important factors affecting weathering are climate, particle size, exposure, mineral composition, and time.

Climate

Climate, which is the single most important factor affecting weathering, is the average condition of the atmosphere in a region over a long period of time. It is typically expressed in terms of two factors: temperature and precipitation (see Figure 18.7). Both factors influence the type and the rate of weathering. Warm climates favor chemical weathering; cold climates, physical weathering, principally frost action. In both cases, the more moisture present, the more pronounced the weathering.

Figure 18.7 The Effect of Climate on Weathering Processes. Physical weathering is dominant in areas where rainfall and temperature are both low. High temperature and precipitation favor chemical weathering.

Chemical reactions tend to occur at a faster rate as temperature increases. Many of these reactions require water, which is a reactant in hydration and carbonation and provides a medium in which acid reactions can occur. Hot, moist climates also support increased biological activity, ranging from burrowing to the production of humic acid as plant matter decomposes. Thus, a hot, moist climate is the ideal environment for rapid chemical weathering.

In cold climates, physical weathering by frost action predominates. It is most effective in cold climates that alternately freeze and thaw. Here again, precipitation is a key factor. Without water, ice cannot form and frost action cannot occur. A cold, moist climate will therefore experience strong frost action.

Temperate climates, such as New York's, combine aspects of both extremes. Hot, moist summers followed by moist winters with many freeze-thaw cycles result in an environment in which rocks weather rapidly. A classic example of the destructiveness of this climate is its effect on Cleopatra's Needle, a granite obelisk with hieroglyphics cut into its surface that was moved from Egypt to Central Park in 1880 (see Figure 18.8). In Egypt's hot, dry climate the obelisk stood almost unchanged for 3,000 years. In the moist, changing climate of New York City, however, it began to rapidly deteriorate. Today, its surface cracked, worn, and discolored, the hieroglyphics almost unreadable, it stands as mute testimony to the effect of climate on weathering.

Figure 18.8 Cleopatra's Needle in New York City's Central Park.

Particle Size

The size of rock particles greatly affects the rate at which chemical weathering occurs. Under the same conditions, the smaller the pieces of a particular rock, the faster they will weather. The reason is that a given volume of small particles has more surface area than the same volume of large particles (see Figure 18.9).

(a)

(b)

Figure 18.9 As a Particle Is Broken into Smaller Pieces, Its Total Surface Area Increases. (a) This cube has a surface area of 24 square centimeters. (b) Each of these eight small cubes has a surface area of 6 square centimeters, for a total of 48 square centimeters.

A reaction can take place only when the rock comes into contact with the chemical weathering agent. The more rock surface exposed, the more rock is able to react, and the faster the rate of the reaction.

Size is also a factor in what happens to rock particles after they are produced by weathering. Small particles may be transported to a new location that has a different climate, or to a body of water that contains a variety of reagents.

Exposure

Exposure determines the degree to which a rock comes into contact with weathering agents. Soil, ice, and vegetation can cover a rock and thereby decrease its contact with weathering agents. Rocks thus protected tend to weather more slowly than those completely exposed at the surface. Also, the weathered surface of a rock can itself shield fresh material underneath, thereby slowing the rate of weathering. Another factor that affects exposure is the slope of the land. On steep slopes, loose materials move downhill because of gravity or are carried downhill by erosion, thus continually exposing fresh rock.

Figure 18.10 Results of Exposure. (a) Corners and edges weather fastest, resulting in a rounded particle (b).

Exposure also has its effects on each individual particle. As you can see in Figure 18.10a, the corners of a cube are exposed to weathering on three sides, the edges on two sides, and the faces on one side. The net effect is that the corners of a particle weather away most quickly and the faces most slowly. Over time the result is a spherical particle, as shown in Figure 18.10b.

Mineral Composition

The mineral composition of a rock determines its physical and chemical properties and thus its susceptibility to weathering. Mineral composition is of greatest importance in chemical weathering. Rocks composed of minerals that react readily with acids, water, or oxygen will weather more rapidly than those composed of less reactive minerals. For example, limestone, which is mostly calcite, is dissolved by even mildly acidic rainwater, while granite, which is mostly silicates, is almost unaffected.

Mineral composition also affects physical weathering (see Table 18.1). Softer rocks will abrade more readily than harder rocks. Solid, crystalline rocks have fewer openings into which water can penetrate than rocks composed of cemented-together particles.

TABLE 18.1 RELATIVE RESISTANCES OF THE MAJOR ROCK-FORMING MINERALS TO WEATHERING

Mineral	Resistance to Weathering
Olivine	Least resistance
Pyroxene	
Amphibole	
Biotite	↓
Plagioclase feldspars	
Muscovite	
Orthoclase	
Quartz	Most resistant

Time

Weathering is a slow process. The 3.5-billion-year-old gneisses in Greenland are scarcely weathered. Even easily weathered rocks may take hundreds of years to be completely broken down. The longer a rock is exposed to weathering processes, however, the more it deteriorates and disintegrates. Eventually, all rocks exposed at Earth's surface are completely broken down.

THE PRODUCTS OF WEATHERING

Since Earth formed, the rocks exposed at its surface have been attacked by forces that disintegrate them. Whether these forces are physical or chemical, the result is the same—solid rock is broken into fragments. These fragments range from the tiniest particle dissolved in water to the largest boulders. The fragments produced form sediments and soils.

Sediments

The fragments or particles of rock produced by weathering are called **sediments**. Sediments are named according to their sizes (Table 18.2).

TABLE 18.2 SEDIMENT SIZES AND NAMES

Particle Diameter (cm)	Name
<0.0004	Clay
0.0004–0.006	Silt
0.006–0.2	Sand
0.2–6.4	Pebbles
6.4–25.6	Cobbles
>25.6	Boulders

Soils

Soil is the accumulation of loose, weathered material that covers much of the land surface of Earth. Soil varies in depth, composition, age, color, and texture. Although its chief component is weathered rock, a true soil also contains water, air, bacteria, and decayed plant and animal material (humus).

The rock from which a soil forms is called the **parent material**. Soil

Figure 18.11 The Development of a Residual Soil. Weathering opens bedrock to air, water, and dust. Plant and animal life accelerate the breakdown of rock and allow air and water to penetrate ever more deeply. Eventually a deep layer of soil develops atop the bedrock.

that forms directly from the bedrock beneath it is **residual soil** (see Figure 18.11). If the soil forms from material that was transported to the location by erosion, it is **transported soil**. Transported soil may have a different mineral composition from the underlying bedrock.

As a soil forms, the processes of weathering and plant growth develop recognizable layers, or horizons, in the soil. These horizons differ in structure, composition, color, and texture. A soil that has been forming long enough to

Horizon O (organic horizon)
The material in this horizon is commonly called topsoil and supports plant life. This horizon contains fresh to partly decomposed organic matter. Its color varies from dark brown to black.

Horizon A (leached horizon)
The top of this horizon is highly decomposed organic matter mixed with minerals. Its particles, exposed since the soil began to form, are sand sized or smaller. In a process called *leaching*, soluble minerals and tiny clay particles are carried down to lower layers as water seeps downward through this layer. Leaching is one of the processes that cause horizons to form. This horizon ranges from brown to gray in color.

Horizon B (accumulation horizon)
Materials leached out of horizon A are deposited here. This horizon is richer in clay, has less organic material, and has more and larger particles of bedrock than horizon A. As a result it is less fertile. The clay in this horizon gives it a clumpy or blocky consistency compared to loose, sandy horizon A. The leached materials deposited here (clays and iron oxides) give this horizon a reddish brown or tan color.

Horizon C (partly weathered horizon)
This horizon consists of partly weathered bedrock. Sometimes referred to as subsoil, it is the cracked, broken surface of the bedrock. It marks the final transition from topsoil to unaltered bedrock or parent material.

Parent material (R)
Unaltered bedrock

Figure 18.12 Profile of a Mature Soil.

have developed distinct horizons is called a **mature soil**. In **immature soils**, horizons are indistinct or altogether lacking.

A **soil profile** is a cross section of a soil from surface to bedrock. Figure 18.12 shows the profile of a typical mature soil with a description of each horizon.

Parent material greatly influences the type of soil that forms, especially when the soil is just starting to form or is immature. Over a long period of time, however, climate has a greater influence on the type of soil that forms. Whatever the parent material, the profiles of mature soils that formed in similar climates are very much alike.

MULTIPLE-CHOICE QUESTIONS

In each case, write the number of the word or expression that best answers the question or completes the statement.

1. Which is the best example of physical weathering?
   (1) the cracking of rock caused by the freezing and thawing of water
   (2) the transportation of sediment in a stream
   (3) the reaction of limestone with acid rainwater
   (4) the formation of a sandbar along the side of a stream

2. The diagram below shows granite bedrock with cracks. Water has seeped into the cracks and frozen. The arrows represent the directions in which the cracks have widened due to weathering.

   Which statement best describes the physical weathering shown by the diagram?
   (1) Enlargement of the cracks occurs because water expands when it freezes.
   (2) This type of weathering occurs only in bedrock composed of granite.
   (3) The cracks become wider because of chemical reactions between water and the rock.
   (4) This type of weathering is common in regions of primarily warm and humid climates.

3. At high elevations in New York State, which is the most common form of physical weathering?
   (1) abrasion of rocks by wind
   (2) alternate freezing and melting of water
   (3) dissolving minerals into solution
   (4) oxidation by oxygen in the atmosphere

4. Which weathering process tends to form spherical boulders?
   (1) carbonation
   (2) exfoliation
   (3) frost action
   (4) oxidation

5. Which characteristic would most likely remain constant when a limestone cobble is subjected to extensive abrasion?
   (1) shape
   (2) mass
   (3) volume
   (4) composition

6. The diagram below represents a naturally occurring geologic process.

   Which process is best illustrated by the diagram?
   (1) cementation
   (2) erosion
   (3) metamorphism
   (4) weathering

7. Which event is an example of chemical weathering?
   (1) rocks falling off the face of a steep cliff
   (2) feldspar in granite being crushed into clay-sized particles
   (3) water freezing in cracks in a roadside outcrop
   (4) acid rain reacting with limestone bedrock

8. Landscapes will undergo the most chemical weathering if the climate is
   (1) cool and dry
   (2) cool and wet
   (3) warm and dry
   (4) warm and wet

9. Which factor has the greatest influence on the weathering rate of Earth's surface bedrock?
   (1) local air pressure
   (2) angle of insolation
   (3) age of the bedrock
   (4) regional climate

10. Four samples of the same material with identical composition and mass were cut as shown in the diagrams below. When the samples are subjected to the same chemical weathering, which sample will weather at the fastest rate?

11. Why will a rock weather more rapidly if it is broken into smaller particles?
(1) The mineral structure of the rock has been changed.
(2) The smaller particles are less dense.
(3) The total mass of the rock and particles is reduced.
(4) More surface area is exposed.

12. The two block diagrams below represent the formation of caves.

Which types of weathering and erosion are primarily responsible for the formation of caves?
(1) chemical weathering and groundwater flow
(2) chemical weathering and runoff
(3) physical weathering and groundwater flow
(4) physical weathering and runoff

Base your answers to questions 13 through 16 on the graph below, which shows the effect that average yearly precipitation and temperature have on the type of weathering that will occur in a particular region.

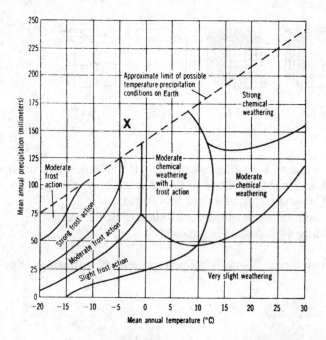

13. Which type of weathering is most common where the average yearly temperature is 5°C and the average yearly precipitation is 45 cm?
    (1) moderate chemical weathering
    (2) very slight weathering
    (3) moderate chemical weathering with frost action
    (4) slight frost action

14. The amount of chemical weathering will increase if
    (1) air temperature decreases and precipitation decreases
    (2) air temperature decreases and precipitation increases
    (3) air temperature increases and precipitation decreases
    (4) air temperature increases and precipitation increases

15. Why is no frost action shown for locations with a mean annual temperature greater than 13°C?
    (1) Very little freezing takes place at these locations.
    (2) Large amounts of evaporation take place at these locations.
    (3) Very little precipitation falls at these locations.
    (4) Very large amounts of precipitation fall at these locations.

16. No particular type of weathering or frost action is given for the temperature and precipitation values at the location represented by the letter X. Why is this the case?
    (1) Only chemical weathering would occur under these conditions.
    (2) Only frost action would occur under these conditions.
    (3) These conditions create both strong frost action and strong chemical weathering.
    (4) These conditions probably do not occur on Earth.

Base your answers to questions 17 through 19 on the flowchart below, which shows a general overview of the processes and substances involved in the weathering of rocks at Earth's surface. Letter X represents an important substance involved in both major types of weathering, labeled A and B on the flowchart. Some weathering processes are defined below the flowchart.

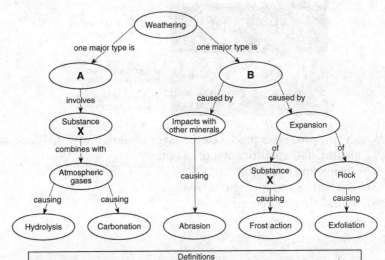

17. Which term best identifies the type of weathering represented by A?
    (1) physical   (3) chemical
    (2) biological   (4) glacial

18. Which substance is represented by X on both sides of the flowchart?
    (1) potassium feldspar   (3) hydrochloric acid
    (2) air   (4) water

Weathering and Soil Formation

19. Which weathering process is most common in a hot, dry environment?
   (1) abrasion
   (2) carbonation
   (3) frost action
   (4) hydrolysis

20. A variety of soil types are found in New York State primarily because areas of the state differ in their
   (1) amounts of insolation
   (2) distances from the ocean
   (3) underlying bedrock and sediments
   (4) amounts and types of human activities

21. The cross section below shows layers of soil.

Dark brown to black soil with a high organic content

Tan to orange soil with a high clay content, some rock fragments

Light gray to black soil, coarse rock fragments

Which two processes produced the layer of dark brown to black soil?
   (1) melting and solidification of magma
   (2) erosion and uplifting
   (3) weathering and biologic activity
   (4) compaction and cementation

CONSTRUCTED RESPONSE QUESTIONS

Base your answers to questions 22 through 24 on the cross section below, which shows limestone bedrock with caves.

22. In the empty box on the left side of the cross section, draw a horizontal line to indicate the level of the water table. [1]

23. The precipitation in this area is becoming more acidic. Explain why acid rain weathers limestone bedrock. [1]

24. Identify one source of pollution caused by human activity that contributes to the precipitation becoming more acidic. [1]

Base your answers to questions 25 through 27 on the diagram below, which shows the soil profile formed in an area of granite bedrock. Four different soil horizons, A, B, C, and D, are shown.

25. Identify the soil horizon containing the highest percentage of organic material. [1]

26. Soil horizon B is rich in clay and soluble minerals.
   (a) State the process by which this soil horizon has become enriched in these materials. [1]
   (b) In one or more complete sentences, compare the fertility of soil in this layer with that of horizon A. [1]

27. State one factor that affects the type of soil that forms in a region. [1]

EXTENDED CONSTRUCTED RESPONSE QUESTIONS

Base your answers to questions 28 through 31 on the given information and the data table below. Samples of three different rock materials, A, B, and C, were placed in three containers of water and shaken vigorously for 20 minutes. At 5-minute intervals, the contents of each container were strained through a sieve. The masses of the materials remaining in the sieve were measured and recorded as shown in the data table below.

MASSES OF MATERIAL REMAINING IN SIEVE

Shaking Time (min)	Rock Material A (gr)	Rock Material B (gr)	Rock Material C (gr)
0	25.0	25.0	25.0
5	24.5	20.0	17.5
10	24.0	18.5	12.5
15	23.5	17.0	7.5
20	23.5	12.5	5.0

28. Using the information in the data table, construct a line graph on the grid provided below, following the directions in parts (a) through (c) below.
 (a) Plot the data for rock sample *A* for the 20 minutes of the investigation. Surround each point with a small circle, and connect the points. [1]
 (b) Plot the data for rock sample *B* for the 20 minutes of the investigation. Surround each point with a small triangle, and connect the points. [1]
 (c) Plot the data for rock sample *C* for the 20 minutes of the investigation. Surround each point with a small square, and connect the points. [1]

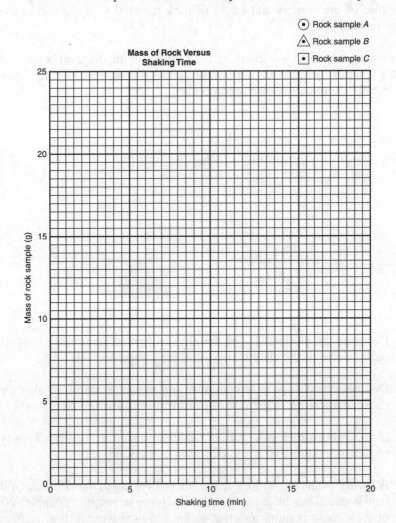

29. Using one or more complete sentences, state the most likely reason for differences in the weathering rates of the three rock materials. [2]

Weathering and Soil Formation

30. Using the directions in parts (a) through (c) below, calculate the average rate of change in the mass of rock material C for the 20 minutes of shaking.
    (a) Write the equation for the rate of change in mass. [1]
    (b) Substitute data into the equation. [1]
    (c) Calculate the rate of change in mass, and label your answer with proper units. [1]

31. Using one or more complete sentences, describe the most likely appearance of the corners and edges of rock material C at the end of the 20 minutes. [2]

Base your answers to questions 32 through 35 on the diagram below, which shows igneous rock that has undergone mainly physical weathering into sand and mainly chemical weathering into clay.

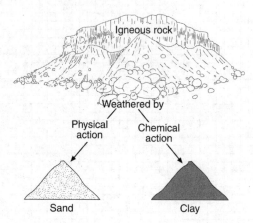

32. Compare the particle size of the physically weathered fragments to the particle size of the chemically weathered fragments. [1]

33. Describe the change in temperature and moisture conditions that would cause an increase in the rate of chemical weathering into clay. [1]

34. If the igneous rock is a layer of vesicular andesite, identify *three* types of mineral grains that could be found in the sand. [1]

35. A family wants to use rock materials as flooring in the entrance of their new house. They have narrowed their choice to granite or marble. Which of these rocks is more resistant to the physical wear of foot traffic and explain why this rock is more resistant. [2]

CHAPTER 19
EROSION

> **KEY IDEAS** Once weathering has broken rock down into smaller particles, the natural agents of erosion—generally driven by gravity—remove, transport, and deposit them. The natural agents of erosion include streams, ocean currents, and wave action (moving water), glaciers (moving ice), wind (moving air) and mass movements. Each of these agents of erosion produces distinctive changes in the material that it transports and creates characteristic surface features and landscapes. In certain erosional situations, destruction of property, personal injury, and loss of life can be reduced by effective emergency preparedness.

KEY OBJECTIVES
Upon completion of this chapter, you will be able to:

- Explain how natural agents of erosion, generally driven by gravity, remove, transport, and deposit weathered rock particles.
- Compare and contrast the distinctive changes each agent of erosion makes in the material it transports, as well as the characteristic surface features and landscapes each produces.
- Recognize the factors that affect erosion and relate them to landforms produced by erosion.
- Cite examples of the impacts of various types of erosion on human activities.

EROSION

What happens to the sediments produced by weathering? In most cases they are moved, often great distances from where they originated. Any process that moves sediments from one place to another on Earth's surface is called **erosion**. As you read these words, erosion is changing Earth in countless ways. Waves crashing against shores are scouring away sand, reshaping the coastline. In deserts, hot winds are moving towering dunes of sand grain by grain. On cold, barren mountaintops fragments of rock broken loose by weathering are tumbling to the ground. Glaciers creeping downhill are tearing huge boulders from the ground and carrying them along. Streams, from tiny trickles to raging torrents, are carving their way into the crust.

Together, weathering and erosion wear away Earth's crust. Weathering breaks down solid rock; erosion carries away the pieces. In this way fresh rock is exposed to weathering, and the cycle repeats itself. Sometimes erosion is rapid and unmistakeable, as in a landslide. Usually, though, the process is gradual, and a long time passes before it is evident that erosion is taking place.

Evidence of Erosion

Any sediment moved from its source is evidence of erosion. At times, this evidence is striking. The owner of a house returning after a mudslide knows that the house was not built with one end hanging over the edge of a cliff. The lack of soil under the house is evidence of erosion. In much the same way, valleys and canyons are evidence of erosion. The Grand Canyon is one of the most spectacular examples of erosion in the United States. Imagine how much sediment had to be carried away to form it!

Quite often, though, the evidence of erosion is subtle. A boulder sitting on bedrock may not immediately seem to indicate erosion. If the boulder is granite and the bedrock is limestone, however, the boulder certainly wasn't derived from the bedrock; it had to be carried there from some other source. Similarly, layers of sediment can often be found overlying bedrock that has an entirely different composition. This, too, is evidence that erosion has occurred. Since the sediment did not come from the bedrock, it must have been carried there from some other source.

AGENTS OF EROSION

Erosion moves sediments. To move anything, a force is needed. The primary driving force of erosion is **gravity**. Gravity can move sediments by acting on them directly. On cliffs and steep slopes, sediments broken loose by weathering move downhill under the direct influence of gravity. Gravity can also move sediments by acting on them indirectly, through **agents of erosion**. For example, water runs downhill under the direct influence of gravity. The running water, in turn, can exert a force on sediments in its path, causing them to move. Thus the running water is an agent of erosion. Some other agents of erosion are waves, currents, winds, and glaciers.

An agent of erosion, together with the driving force (usually gravity) that sets in motion the agent that picks up and transports sediment, comprises an **erosional system**. Erosional systems are like natural conveyor belts, moving sediments from higher to lower elevations.

Mass Wasting

Mass wasting is the downhill movement of sediments under the direct influence of gravity. On most slopes, some kind of downhill movement of sediments is proceeding all the time because, on a slope, gravity acts as if it has two parts, or components. One part, which we will call the normal force (F_N), pulls downward perpendicular to the surface. The other part, which we will call the downhill force (F_D), pulls downhill parallel to the surface. The downhill force gives rise to a frictional force (F_F) that opposes the motion of the object. The frictional force depends on two factors: the normal force holding the two surfaces together and the nature of the two surfaces. The more tightly the two surfaces are pressed together by the normal force, the greater the frictional force between the two. (This explains why an empty box is easier to push across a floor than a similar box filled with something heavy.) As long as the frictional force is greater than the downhill force, the object will remain stationary.

As the slope gets steeper, more of the gravity acts in a downhill direction and less in a direction perpendicular to the surface (i.e., the downhill force increases and the normal force decreases). When the downhill force is greater than the frictional force, the object moves downhill. See Figure 19.1.

Figure 19.1 Forces Acting on an Object Resting on a Slope. In the diagram, F_N = normal force, F_D = downhill force, and F_F = frictional force. As a slope steepens, the downhill force increases while the normal force and frictional force decrease. When the downhill force exceeds the frictional force, the object moves downhill.

The steepest slope angle at which a particular sediment remains stable is called its **angle of repose** (see Figure 19.2). The size, shape, and density of a rock particle affect its angle of repose. Thus, sand, gravel, and clay have different angles of repose. When their angles of repose are exceeded, sediments move downhill and mass wasting occurs. Any perceptible downslope move-

Erosion

ment of rock, soil, or a mixture of the two is commonly referred to as a **landslide**. This is a very general term, however, and may involve many different types of movement of many different types of material. Depending on the slope and the angles of repose of the sediments, mass-wasting processes may be rapid or slow.

Figure 19.2 Angles of Repose in a Pile of Dry Sand.

Rapid Mass Wasting

Rockfalls (see Figure 19.3a) occur when rock fragments broken loose by weathering fall from cliffs or bounce by leaps down steep slopes. Rockfalls are the most rapid of all mass-wasting processes. "Fallen Rock Zone" is a common sign near roadcuts on many highways or in mountainous areas where rockfalls occur frequently. In populated areas or on heavily traveled roads, rockfalls can be very dangerous and have resulted in loss of life. Some localities have spent much money cutting back the rocky slopes of roadcuts or covering them with steel cable nets to prevent rockfalls.

(a) Rockfall   (b) Rockslide   (c) Slump   (d) Mudflow

Figure 19.3 Four Rapid Mass-Wasting Processes.

Rockslides (see Figure 19.3b) occur on less steep slopes when rock masses or debris slide downhill. Most rockslides are triggered by heavy rains or earthquakes. Water acts as a lubricant between particles. The shock of an earthquake knocks particles apart, decreasing the friction between them and the underlying surface. The downhill component of gravity is then greater than the friction holding the particles in place, and the particles move downhill. Rockslides can move enormous amounts of material, often millions of cubic meters. In 1959, twenty-seven people were killed when an earthquake caused an entire mountainside to slide into the Madison River gorge in Montana.

Slump (see Figure 19.3c) is a mass-wasting process in which a huge mass of bedrock or soil slides downward from a cliff in one piece. The mass, or slump block, slides along a curved plane of weakness as shown in the diagram. The slump block rotates and comes to rest with its upper surface tilting toward the cliff. Slump is common where ocean waves or streams undercut cliffs.

A **mudflow** (see Figure 19.3d) is the rapid, downhill flow of a fluid mixture of rock, soil, and water. Mudflows generally occur after heavy rains saturate soil that has no protective covering of vegetation, and usually occur in semiarid regions or on slopes denuded by construction or lumbering. The rain mixes with the soil to form a thin mud, which flows downhill, thickening as more soil and debris are picked up and increasing in speed. Since the mudflow is much heavier and thicker than water, the impact when it hits something in its path is devastating. Entire houses, cars, and trees have been swept away by mudflows. A mudflow comes to rest when it reaches the bottom of the slope or when it has picked up enough material so that it thickens to a point where it can no longer move.

Slow Mass Wasting

In humid regions, slopes are usually covered with vegetation. The vegetation protects the surface from the impact of raindrops, and its root system holds the soil together, inhibiting downhill movement. However, mass wasting occurs even on slopes covered with vegetation, albeit slowly.

On vegetated slopes, rainwater entering the soil loosens the particles and makes them slippery, decreasing their angles of repose. In this situation, an earthflow may occur. In an **earthflow** (see Figure 19.4a), a shallow layer of soil and vegetation, saturated with water, slowly slides downhill. An earthflow may take several hours to ooze its way down a slope. Earthflows are a common cause of road and rail blockages. While they are usually not life threatening since they move at a snail's pace, they can cause considerable property damage.

(a) Earthflow (b) Soil creep

Figure 19.4 Two Slow Mass-Wasting Processes.

The slowest of all mass-wasting processes is soil creep. **Soil creep** (see Figure 19.4b) is the invisibly slow, downhill movement of soil, carrying vegetation with it. The tilting of old poles, fenceposts, or tombstones and bulging or broken retaining walls are all evidence of creep.

Any disturbance of the soil on a slope causes creep. Freezing and thawing, wetting and drying, and trampling or burrowing by animals are just a few of the things that can disturb the soil on a slope. When disturbed, the soil particles shift and settle. As they do so, gravity moves them downhill. Creep is common in regions where the soil alternately freezes and thaws frequently.

Large objects on or in a creeping slope are carried along by the moving soil. Boulders that have crept down a hillside may accumulate at the base of the slope, forming a boulder field. Trees growing on creeping slopes are often misshapen. The tree grows straight up, but the soil that its roots are in slowly creeps downhill.

Mass wasting is usually only the first step in the eroding of Earth. It delivers sediments to the base of slopes. There, agents of erosion, such as streams, can pick up the sediments and carry them farther.

Water in Motion: Runoff and Streams

Runoff

Runoff is precipitation that does not evaporate or sink into the ground. Under the influence of gravity, runoff flows downhill. On even the gentlest slopes, runoff may flow downhill in a thin sheet called **overland flow**. Overland flow exerts a dragging force on the ground. Depending on its speed, it can exert a force great enough to move sediments ranging from fine clay to coarse sand or gravel.

Vegetation greatly decreases erosion by runoff. The leaves and stems of plants absorb the impact of falling raindrops, and the network of plant roots anchors the soil in place. Plant stems also block and slow the downhill movement of water, thereby decreasing the force it exerts on sediments in its path. In arid regions or on slopes denuded of vegetation, however, erosion by runoff can remove vast quantities of loose soil.

Erosion due to runoff is a serious problem on farms. It carries away the rich topsoil essential for healthy plant growth. Contour plowing, terracing, strip cropping, and crop rotation help lessen soil erosion.

On steep, unprotected slopes, erosion by runoff can become intense. Numerous tiny grooves, or rills, form as the water runs downhill, carrying away sediment. If erosion continues, the rills may widen and deepen into a gully. Gullies act as funnels for runoff. They concentrate the force of the flowing water, thus increasing the rate of erosion.

Streams

A **stream** is any body of flowing water that moves downhill under gravity, in a relatively narrow but clearly defined channel, at least part of the year. A stream may be small enough to step across or so wide that the other side is barely visible. Streams are the most important agent of erosion because they affect more of Earth's surface than any other agent. Most of the sediment carried downhill by runoff ends up in streams.

Stream Basics

The long trough-like depression that is normally occupied by the water in a stream is called the **stream channel**. The bottom of the channel is called the **streambed**, and each side is called the **stream bank**. The water carried in a stream channel, or **streamflow**, comes mainly from two sources: runoff and groundwater seepage into the stream, known as **base flow**. A stream's rate of flow over a particular period of time is called stream **discharge** and is usually expressed in cubic meters per second. Stream discharge depends on the volume and velocity of the flow. Where the moving water in a stream comes in contact with the air or the ground, friction causes the water to slow down. As a result of this frictional drag, stream velocity is at a maximum at the center of the channel near the surface and a minimum near the bed and banks. Finally, flow is not always contained within the stream channel. During periods of high stream discharge a stream may overflow its banks, or **flood**.

A stream begins at its **source**, usually in mountains or hills where rainwater or snowmelt collects and drains into low-lying areas between slopes, forming tiny streams. As these tiny streams flow downhill along natural passageways or depressions in the surface to lower and lower levels, they cut into the surface, forming a narrow trench, or gully. Once formed, the gully provides a pathway for all later runoff. Gullies either grow larger when they collect more water and become stream channels themselves or meet streams and add to the water already in the stream. When two streams meet, they merge, and the smaller stream is known as a **tributary**. A stream grows larger as it collects water from more tributaries. The water in most streams ends up in a lake, sea, or ocean. The point at which the stream enters these bodies of water is called the **mouth** of the stream. The change in elevation from a stream's source to its mouth, expressed in degrees or as a distance ratio $\left(\frac{\text{change in elevation}}{\text{distance}}\right)$, is called the stream **gradient**. Figure 19.5 shows a typical stream and its parts.

Figure 19.5 A Stream and Its Parts.

Stream Transport

Water flowing downhill can be very powerful. Have you ever seen whitewater rafters shooting a rapids? The water flowing downhill exerts quite a force! Imagine trying to paddle upstream through the rapids. It would be almost impossible because the force of the flowing water is so great. Now

Erosion

imagine a grain of sand or a pebble in the path of such a stream. Any loose sediments are likely to be carried away.

Material carried by a stream is called **stream load**. A stream's load consists mainly of sediments, which can vary in amount and type, depending on the speed and volume of the stream. The steeper the slope, or gradient, of a stream's channel, the faster the water flows. Water can carry more and larger sediments as its speed increases. The graph in Figure 19.6a shows the sediment water carries at different speeds. Notice that as water velocity increases, the size of the particles it can carry increases. Of course, if water is traveling fast enough to carry pebbles, it is also traveling fast enough to carry any particles smaller than pebbles, too. Thus, the faster the streamflow, the more types of sediment the water can carry and the greater the stream's load. At any speed, the more water there is in a stream, the more load it can carry.

Streams transport different sediments in different ways. See Figure 19.6b. Large sediments, such as pebbles, cobbles, and boulders, are heavy. Even the fastest flowing streams may not be able to pick them up. But sediments can be moved without picking them up! Do you have to pick up a car that has run out of gas in order to move it? Of course not; you can push or drag the car so that it rolls along the road. In much the same way, large particles can be carried downhill by streams. The force of the flowing water in a stream can push or drag large particles downhill by rolling or sliding in a motion called **traction**. If the sediment particles move in a series of bounces, hops, or leaps along the streambed, the motion is called **saltation**.

Figure 19.6 (a) Relationship of Transported Particle Size to Water Velocity.
Source: The State Education Department, *Earth Science Reference Tables*, 2011 ed. (Albany, New York; The University of the State of New York).
(b) Methods of Particle Transport.

Smaller particles, such as silt, sand, and clay, are lighter in weight and therefore require less force to move. Although they are heavier than water, the flowing water can pick up these small sediments; and, whenever they begin to sink, the force of the turbulent water pushes them back up again—they are carried downhill **suspended** in the water.

Some sediments dissolve in water. Water that has combined with carbon dioxide or other gases in the atmosphere to become acidic (see Chemical

Weathering, page 423) is especially effective at dissolving sediments. Sediments that are dissolved are carried downhill **in solution**.

Stream Erosion

Streams not only transport sediments but also change the sediments they carry, wear away the surfaces over which they flow, and create new sediments. Pure water flowing over rock wears it away very little, but water carrying sediment acts like a cutting tool. When the sediments being carried hit rock in the stream channel, chips of rock break off. Rolling and sliding pebbles, cobbles, and boulders collide and knock chips off each other. They crush and grind up smaller particles between them. The process of crushing, grinding, and wearing away rock by the impact of sediments is called **abrasion**. Abrasion wears away a stream's channel and causes the sediments carried by the stream to become rounded.

By abrasion, streams can cut through bedrock over which they flow. The potholes shown in Figure 19.7 were carved out of bedrock by sediments in a swiftly flowing stream. The waterfall formed where a stream flowed over hard bedrock onto soft bedrock. The soft bedrock wore away faster, lowering the streambed, and a waterfall was formed. Prolonged erosion can carve deep canyons and valleys into Earth's surface. The Grand Canyon is an amazing example of what sediment-laden water can do.

Figure 19.7 Waterfall and Pothole. The falling water creates turbulence that swirls sediments against the streambed, forming potholes.

A Stream's Life Cycle

As a stream cuts into the surface and carries away sediment, its channel becomes wider, deeper, and longer. Over time the shape and behavior of the stream change as the surface over which it flows is changed.

Youth. Newly formed streams generally flow down steep slopes. The water flows quickly, often forming rapids. See Figure 19.8a. Over time, the stream wears a long, deep groove, or valley, in the surface. Stream-cut valleys typically have steep sides that form a distinctive **V-shape**. A young stream valley is shown in Figure 19.8b.

Maturity. As a stream erodes the slope over which it flows, the slope becomes less steep. As a result, the water flows more slowly and sediment accumulates at the bottom of the valley, making it flat. As mass wasting and runoff wear away the sides of the valley, the V-shape becomes wider and less

Erosion

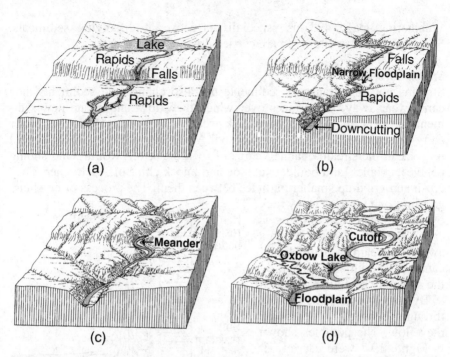

Figure 19.8 The Three Stages in the Life Cycle of a Stream. (a) The relief of the land is steep, and the stream has lakes, rapids, and falls. (b) Continued downcutting of the streambed eliminates falls, and cutting back of the valley walls allows a narrow floodplain to form. (c) Further downcutting and erosion of the valley walls enlarges the floodplain and allows the stream to meander widely between the valley walls. (d) Shifting meanders result in cutoffs and oxbow lakes. Source: *The Earth Sciences*, Arthur N. Strahler, Harper & Row, 1971.

steep. The slower moving water flows around obstacles instead of tumbling over them, and the stream's path forms curving loops called **meanders**. Meanders reflect the way in which a stream minimizes resistance to flow and spreads energy as evenly as possible along its course. Velocity is lowest along the bed and banks of streams because it is there that water encounters the most friction, and therefore the flow is reduced. Along a straight channel segment, water moves the fastest in mid-channel, near the surface. But as water moves around a bend, the zone of high velocity swings to the outside of the channel because a centrifugal force operates in the direction of the outside of the curve. As water rushes past the outer part of the bend, sediment is continuously eroded from the riverbank and is swept downstream. Along the inside of the bend, the slower-moving water deposits sediments. With continuous erosion at the outer banks and deposition at the inner banks, the bending of the meander becomes more pronounced. When runoff increases suddenly, as after a torrential downpour or sudden spring thaw, the stream overflows. Spreading out over the valley floor, the stream slows down and deposits sediments to form a broad, flat **floodplain** adjacent to the stream. Figure 19.8c shows a typical stream at the mature stage.

Old Age. Eventually, the slopes around a stream are worn away almost completely. A flat lowland of gently rolling hills, called a **peneplain**, is formed. Since the slope of the land is very gentle, the water in the stream flows slowly. Its meanders become highly curved and winding and at times may actually loop over one another. During floods, water may gush over the land between meanders, forming a **cutoff**. If the new channel cuts deep enough, part of the meander may become isolated, forming an **oxbow lake**. Figure 19.8d shows a typical stream at the old stage.

Whatever the stage of development, the water in most streams eventually reaches the oceans.

Water in Motion: the Oceans

The ocean never rests. Anyone who has swum in, sailed on, or just sat on the beach beside the ocean knows that it is constantly moving. Waves, currents, and tides all move ocean waters; and, once in motion, ocean water can transport sediments and erode the land.

Ocean Currents

Ocean currents are mainly the result of two interactions with the atmosphere: heating of the ocean and wind. The ocean is able to store more heat than the atmosphere or the land because of water's high specific heat. Differences in the intensity of solar radiation received at the Equator and near the poles results in an uneven distribution of heat in the oceans. The oceans contain more heat near the Equator than near the poles.

Convection Currents

The heat in the oceans near the Equator is transferred toward the poles by convection, or the movement of seawater due to density differences. Dense, cold water formed near the poles sinks and slowly flows toward the Equator. This cold water is oxygen-rich; without it, bottom waters would quickly become oxygen depleted by the oxidation of organic matter that sinks through it. As the cold water moves toward the Equator, it displaces warmer water upward, causing **upwelling**. Upwelling brings nutrients and oxygen to the surface, thereby supporting a rich growth of plants and animals.

Wind-Driven Currents

Uneven heating of Earth's surface and the air above it creates similar circulations in the atmosphere. As warm air at the Equator rises, air from adjacent areas gets sucked in to replace it, creating wind. Winds do not move straight toward the Equator, but curve because of the Coriolis effect. Thus, the winds near the Equator, called the **trade winds**, approach the Equator at about a 45° angle. Over the ocean, with little to block their path, the trade winds are among the steadiest winds on Earth. At middle latitudes, the more variable

prevailing westerlies move in the opposite direction. At high latitudes are the most variable winds, the **polar easterlies**. See Figure 19.9.

Figure 19.9 Planetary Wind and Moisture Belts in the Troposphere. The drawing shows the locations of the belts near the time of an equinox. The locations shift somewhat with the changing latitude of the Sun's vertical ray. In the Northern Hemisphere, the belts shift northward in summer and southward in winter. Source: The State Education Department, *Earth Science Reference Tables*, 2011 ed. (Albany, New York; The University of the State of New York).

Winds exert a force on the ocean surface and produce wind-driven currents. All of the major surface ocean currents are driven by the wind and follow roughly the pattern of surface winds. When pushed by wind, however, the surface water does not move in the same direction as the wind but, because of the Coriolis effect, curves off at about a 45° angle. As a result, the surface currents driven by the trade winds do not approach the Equator at the same 45° angle as the winds that are pushing them, but are themselves bent 45° from the wind direction. Thus, throughout Earth, the trade winds form westward-moving currents, known as **equatorial currents**, that are parallel to the Equator.

In the Atlantic and Pacific oceans, these westward-moving equatorial currents are obstructed by landmasses and are deflected north and south. These deflected currents, which run north and south along the western boundary of the oceans, are among the largest and strongest ocean currents. The Gulf Stream, a western boundary current in the Atlantic Ocean, runs north along the eastern coast of the United States and transports more than 100 times the output of all the rivers in the world.

When the western boundary currents reach the midlatitudes, the prevailing winds, which blow from the west, push them back east across the ocean. The result is a large, circular pattern of motion called a **gyre**. The North and

South Atlantic and Pacific, as well as the Indian Ocean, all have large-scale gyres. In addition, the northern polar regions of the Atlantic and Pacific have smaller gyres. See Figure 19.10.

Figure 19.10 Surface Ocean Currents. Source: The State Education Department, *Earth Science Reference Tables*, 2011 ed. (Albany, New York; The University of the State of New York).

One unfortunate effect of the large-scale gyres is that they carry pollutants from the continental coastlines far out into the oceans. For example, many of the largest cities in the United States are located along the east coast. The sewage, garbage, and industrial waste from these cities have long been dumped into the ocean. The Gulf Stream then carries these pollutants far out into the mid-Atlantic. These practices may have had little impact when populations were small, but today they represent serious threats to the health of ocean life and the future well-being of our oceans.

Waves

Wind-Generated Waves

Wind not only drives surface currents but also creates waves. Wind forms waves on the surface of the ocean by transferring energy from the moving air to the water. The result is a transverse wave that moves through the surface water of the ocean.

The highest part of a wave is called the **crest**; the lowest part, the **trough**. The size of an ocean wave is usually expressed as the **wave height**—the vertical distance between the crest and the trough. Wave crests and troughs can be

Erosion

closer together or farther apart. The distance from crest to crest, or from trough to trough is called the **wavelength**. See Figure 19.11.

The height and the period of wind-generated waves are affected by three factors: the speed of the wind, the length of time it blows, and the *fetch*, or distance of water over which it blows. The higher the wind speed and the longer the wind blows, the higher the wave height. Fetch is important in determining wavelength. The longer the fetch, the longer the wavelength.

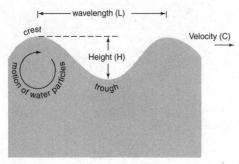

Figure 19.11 A Typical Transverse Wave.

Tsunamis

Tsunamis are waves caused by undersea movements of the crust due to earthquakes, slumping, or volcanic eruptions. The sudden up-and-down movement of the bottom creates a wave that has a long wavelength and travels at high speed. In deep water, tsunamis may have wavelengths up to 700 kilometers and travel at speeds over 350 kilometers per hour, yet may be only several centimeters high. When tsunamis reach a coastline, though, they form huge waves that break against the coast and cause great destruction.

Coastal Processes

Breaking Waves

Breaking waves, or surf, forms as follows. When waves enter shallow water, the bottom of the wave comes into contact with the ground at a depth equal to one-half the wavelength. This contact slows the bottom of the wave and forces the wave upward, while the top continues at its original speed. When the top of the wave outpaces the rest of the wave, it is no longer supported and falls over. See Figure 19.12.

Figure 19.12 Mechanics of a Breaking Wave.

Wave Refraction

Waves that enter shallow water at an angle to the beach are refracted; their direction of travel is changed. This refraction occurs because one part of the wave reaches shallow water and slows down while the rest of the wave is still in deep water and moving faster. Like a rolling log whose one end hits a tree, causing the whole log to swing around, the faster moving end of the wave swings around when the end in shallow water slows down.

On irregularly shaped coastlines, waves converge on headlands and diverge at depressions in the shoreline, as shown in Figure 19.13. The results are erosion of the headland by the force of the waves directed against it and the deposition of sediments in the calmer zone of divergence in inlets. Over time, the irregularly shaped shoreline is straightened out.

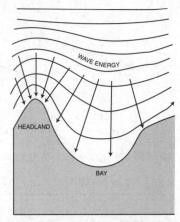

Figure 19.13 Wave Refraction at Headlands and Depressions.

Longshore Currents

When waves break, the water is carried into the surf zone and travels toward the beach. The water that washes up against the beach runs back along the incline under the surf zone. When waves approach the beach at an angle, the water moves upward at an angle and falls straight back. The result is a movement of water parallel to the beach called a **longshore current**. See Figure 19.14.

Figure 19.14 Formation of a Longshore Current. Source: *The Earth Sciences*, Arthur N. Strahler, Harper & Row, 1971.

Erosion

The longshore current and the up-and-back motion of water washing along the beach move sand along the shore. The net result is to transport sediment along the beach and erode it inland. In populated areas, people try to impede this erosion by building structures in the surf zone to block the longshore current and trap beach sediments from being washed away. See Figure 19.15. These attempts are only moderately successful in the long run. Also, the structures often cause increased erosion down-current as the longshore current carries away sediment but brings none in from up-current.

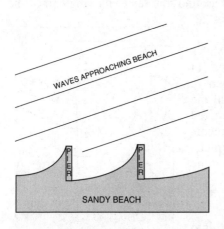

Figure 19.15 Jetties or Groins Built to Trap Sediment.

Shoreline Erosion

Ocean water is constantly in motion. Waves, currents, and tides are examples of the movement of ocean water. Moving ocean water can transport sediments and erode the land.

Waves

Waves can be very powerful. A single wave can send tons of water crashing against a shore. The force of waves can break up rocks on the shore. Loose fragments are stirred up and carried along in the turbulence of breaking waves.

Wave erosion produces a number of shoreline features, some of which are shown in Figure 19.16. When waves erode steep, rocky shores, they remove materials from the base of the slopes. The result is a **wave-cut notch** that undercuts the slope. Overhanging rocks and soil break away, leaving a steep-sided **sea cliff**. When wave erosion penetrates deeply into a cliff, a large hole or **sea cave** may form. As weaker rock is worn away from the cliff, pillars of resistant rock, called **stacks**, may be left behind. As the erosion continues, fragments broken from the cliff are ground up by abrasion in the turbulent waves. This buildup of material forms a flat platform, or **sea terrace**, at the base of the cliff.

Currents

Currents are movements of water. On gently sloping, sandy shores waves mainly shift sediment around. When waves strike perpendicularly to the shore, sediments are moved in and out with the waves. Waves that strike the shore at angles other than 90°, however, are reflected at an angle and interfere with each other, forming a longshore current parallel to the shore. See Figure 19.14. Longshore currents can move sediments along just as streams do. Along some coasts, tides flowing into and out of narrow openings form swift currents. These, too, can move large amounts of sediments.

Figure 19.16 Shoreline Features Formed by Wave Erosion of a Rocky Coast. **P**—platform of abraded sediments, **N**—wave-cut notch at the base of a cliff, **R**—crevice eroded back into the rock, **B**—beach of sediments eroded out of the cliff, **C**—sea cave, **A**—arch, **S**—sea stack. Source: *The Earth Sciences*, Arthur N. Strahler, Harper & Row, 1971.

Wind Erosion

Wind that blows fast enough can pick up and carry loose particles of rock. Wind is also an agent of erosion, though not an important one in most regions. Where rainfall is abundant, plants growing on the surface protect underlying sediments from the full force of the wind. Where the ground is moist, sediment particles tend to stick together and are more difficult for the wind to dislodge and move. (Think of the difference between blowing on a handful of dry, powdery clay and blowing on a handful of wet, sticky clay.) In arid regions, however, there is little plant cover, and the soil, loose sediments, and weathered bedrock are dry and exposed. Under such conditions, wind becomes an important agent of erosion.

Deflation and Abrasion

Wind erodes the land in two ways: deflation and abrasion. In **deflation**, the wind picks up and carries away loose particles much as a stream carries sediments. In **abrasion**, exposed rock is worn away by wind-driven particles. The rock particles worn away by abrasion are then carried away by the wind.

Wind is like a stream of air, and wind carries sediments in much the same way that a stream of water does. However, winds are usually able to carry only small particles, such as sand, silt, clay, and dust, because the winds' medium, air, is not very dense. It exerts less force on a particle than does a denser medium, such as water, moving at the same speed. Imagine the difference between being hit by a balloon filled with air and a balloon filled with water!

Wind carries very small particles, such as clay and dust, in suspension. The finest particles may be carried many meters up into the atmosphere. When volcanoes erupt, dust and ash are often spewed high into the atmosphere, where winds carry them for hundreds or even thousands of miles before they finally settle to the ground. Some particles of dust and ash may reach the jet streams of the stratosphere and remain suspended for years after they are picked up by these fast-moving winds.

Wind carries larger particles closer to the ground. They slide, roll, and bounce, skipping along the surface. As they do so, they collide with exposed bedrock or other particles. In the process of abrasion, these collisions scratch the surface of the exposed rock and cause tiny chips to break off. Like a sandblaster, the wind-borne particles cut and polish the rock surfaces they impact. Over time, rock exposed to wind-driven particles is worn away. The surfaces of both the exposed rock and the particles take on a frosted appearance because of the many tiny scratches inflicted on them.

The amount and the type of erosion caused by wind depend on three factors: the speed of the wind, the size of the particles it carries, and the length of time it blows. The faster the wind blows, the larger the particles it can transport. The larger the particles carried by the wind, the greater the abrasion of exposed surfaces. The longer the wind blows, the more deflation and abrasion that can occur.

Features Produced by Wind Erosion

As deflation removes layer after layer of loose surface materials, a shallow depression, known as a **deflation hollow**, is formed. The semiarid Great Plains of the midwestern United States are dotted with tens of thousands of deflation hollows. In wet years they are covered by grass; and when rainfall is unusually high, they may fill with runoff, forming shallow lakes. In dry years, however, the grass dies off and the wind continues to deflate the bare soil.

When deflation lowers the land surface to the level of the water table, erosion is stopped. The exposed groundwater holds the soil particles together and enables plants to grow. The plants' roots help hold the loose soil in place, and their leaves and stems block the wind, ending any further erosion. A patch of vegetation surrounded by arid wasteland is called an **oasis**.

Deflation usually removes only fine particles. If the soil is composed of a mixture of particle sizes, the larger particles, which are too heavy for the wind to move, are left behind. These larger particles shield the finer particles beneath them from deflation. As exposed fine particles are removed, coarse particles form an increasing percentage of the particles left at the surface. Eventually, a continuous layer of pebbles, gravel, and other coarse particles, called a **desert pavement**, may form, as shown in Figure 19.17.

Abrasion by wind-blown particles produces spectacular features in arid regions. Caves, arches, and bridges may be cut into exposed bedrock. Unusually shaped rock formations may result when rocks of different hardnesses are exposed to abrasion simultaneously.

Figure 19.17 Formation of a Desert Pavement. As deflation removes finer sediments, larger sediments that are left behind eventually form a continuous layer called a desert pavement, which blocks further deflation.

Loose rock particles, too heavy to be transported by the wind, can also be abraded by wind-blown particles. Over time, they develop flat surfaces on the side facing the prevailing winds (see Figure 19.18). These unusual particles are called **ventifacts**, from the Latin word *ventus*, meaning "wind."

Figure 19.18 Ventifact Formation. Abrasion of the exposed surface of a rock fragment by wind-born particles forms a flat-faced ventifact.

Glaciers

A **glacier** is a large mass of ice, formed on land by the compaction and recrystallization of snow, that is moving downhill or outward under the force of gravity. Modern glaciers are found only at high elevations and in polar regions where snowfall exceeds melting over an extended period of time.

Formation of Glaciers

For a glacier to form, the average amount of incoming snowfall must be greater than the average amount lost by melting or evaporation year after year.

When incoming snowfall exceeds snow loss, the net amount of snow remaining on the ground increases each year. Freshly fallen snow is a fluffy mass of delicate ice crystals that consists mostly of pockets of trapped air and therefore has a low density. The situation quickly changes, however, as snow piles up on the ground and sits there for weeks or months. Because of evaporation, the ice crystals become smaller and rounder and form a

Erosion

Figure 19.19 Transformation of Snow Crystals into Glacial Ice. Because of greater surface area, a snowflake's intricate edges melt and evaporate faster than its center. Over time, the snowflake becomes a rounded, compact pellet of ice. Sources: *Earth*, 4th Ed., Frank Press and Raymond Seiver, W.H. Freeman, 1986.

denser, **granular snow**. As more and more snow builds up, alternate melting and refreezing, together with the weight of overlying layers, compact the granular snow into an even denser mass called **firn**. Firn is usually a year or more old and has little or no pore spaces. Eventually, the ice crystals in firn grow together into a solid mass of ice (see Figure 19.19). When snow is converted to ice in this way, some of the dust and gases of the atmosphere are trapped in the ice. Scientists can sample the trapped air and compare it with the current atmosphere to determine what, if any, changes have occurred.

When ice has accumulated to a thickness such that its own weight deforms the ice mass and it moves slowly under pressure, a glacier has formed. Usually this stage is reached when the ice is 100 meters or more thick.

Movement of Ice

Under pressure, ice behaves like a very thick fluid. The ice *flows* downhill over Earth's surface, like thick syrup poured on a tilted plate. If there is just a little syrup, it spreads out into a thin layer and moves very slowly. The rate of flow of syrup can be increased in two ways: by adding more syrup, so that the layer gets thicker and flows more rapidly, and by increasing the tilt of the plate. In much the same way, glaciers flow faster where the glacial ice is thickest and where the slope is steepest.

Variations in the rate of flow of ice cause a glacier to thicken and also to thin out. Where the slope is steep, the ice thins out because it is flowing downslope faster than it is being replenished. The rapid downhill flow exerts a tension on the ice that causes cracks called **crevasses** to form. On extreme slopes the ice may break apart into blocks that cascade downslope in a rock fall. On gentler slopes the ice slows down. The ice from upstream pushes against the slower moving ice, and the compression causes the glacier to bulge and thicken.

These variations in the rate of flow of ice are illustrated in Figure 19.20.

Figure 19.20 Glacial Flow Rates. (a) Ice flows fastest on steep slopes and in places where the ice is thickest. (b) On steep slopes, the ice thins out and may even split, forming crevasses, because it is flowing downhill faster than it is being replenished from farther upstream. On gentle slopes, the ice thickens as ice from upstream piles up behind slower moving ice. Source: *Living Ice: Understanding Glaciers and Glaciation*, Robert P. Sharp, Cambridge University Press, 1988.

Because the ice in a glacier is flowing, the glacier moves in a unique way. Glaciers do *not* move over Earth's surface like a brick sliding down a board. A brick slides down an incline along its base as a single unit; nothing moves around *inside* the brick. In ice, however, both sliding, along the base of the glacier, and flowing, throughout the body of the ice, occur. See Figure 19.21 for a diagrammatic comparison of the two types of motion.

The amount of a glacier's movement that is due to the base sliding along the ground, as compared to the amount caused by internal flow, depends on the temperature of the glacier. Temperature affects the behavior of ice in two key ways:

Figure 19.21 A Brick and a Glacier in Motion. A glacier both slides along its base and flows like a fluid. The relative positions of particles inside the body of the ice change as the glacier flows downhill.

- Ice that is near its melting point is less rigid and is therefore more easily deformed than ice that is well below its melting point. For example, ice at 0°C deforms 10 times faster than ice at –22°C.
- Increasing the pressure on matter causes an increase in temperature. Pressure exerted on ice that is near its melting point may produce enough warming to result in melting.

Warm glaciers, whose ice is near its melting point, slip along their bases more readily than cold glaciers. When warm glaciers encounter an obstacle, the pressure of the ice against the obstacle causes the ice to deform and flow

Erosion

Figure 19.22 Basal Meltwater. High pressure on the upstream side of an obstacle causes ice to melt. The meltwater refreezes on the downstream side of the obstacle, where pressures are lower.

around the obstacle. The glacier may also bypass the obstacle by melting on one side and refreezing on the other. The pressure on the upstream side of the obstacle may cause the ice to melt and then refreeze on the downstream side, where pressure is lower. Meltwater produced by pressure can also fill in cavities and act as a lubricant, allowing the ice to move over the ground more easily. See Figure 19.22.

In cold glaciers moving over bedrock, almost none of the movement is due to slipping of the base along the ground. The bond between the ice and the rock to which it is frozen is greater than the internal strength of the ice. Therefore, under pressure, the ice will deform before it breaks loose from the bedrock.

Types of Glaciers

There are two main types of glaciers: valley glaciers and the much larger continental glaciers. *Valley glaciers* are constrained by topography, and their form and flow are strongly influenced by the shape of the land. *Continental glaciers* submerge the land to the extent that the size and shape of the glacier, rather than the shape of the land, control glacial form and flow.

Valley Glaciers

Valley glaciers (see Figure 19.23) form between mountain slopes at high elevations. They start as snowfields that accumulate in bowl-shaped hollows called **cirques**. As the snowfield deepens and ice forms from firn, the ice flows out of the cirque and down the valley, confined by the surrounding slopes.

As the glacier moves down the valley, it may meet other glaciers from adjacent valleys and the glaciers may merge, forming a larger glacier. Unlike streams, which mix when they merge, glaciers do not mix but rather "weld" together and flow downslope side by side. A large valley glacier may have been fed by dozens of smaller tributary glaciers flowing from high mountain valleys.

Valley glaciers end at a point where there is a balance between ablation and accumulation. This point can be a sensitive indicator of changes in climate. A general warming trend will cause the end of the glacier to melt back up the valley, or retreat. A cooling trend will cause the end of the glacier to move down the valley, or advance.

Figure 19.23 Side View of a Typical Valley Glacier. Snow accumulates in a bowl-shaped cirque and is compacted into firn. The firn recrystallizes into ice and moves downhill; the firn line marks the beginning of the glacial ice. When the glacier reaches warmer temperatures, ablation consumes the ice. The glacier ends in a thin snout edged by a pile of debris called a *moraine*. Source: *The Earth Sciences*, 2nd Ed., Arthur N. Strahler, Harper & Row, 1971.

When a glacier ends on land, streams of meltwater flowing from the glacier often form a river. Large blocks of ice that fall off the face of the glacier may be buried by sediment. A glacier that ends in a body of water forms steep cliffs of ice. Chunks of ice that break off and fall into the water will float since ice is less dense than liquid water. These floating masses of ice, or icebergs, then drift with the currents and gradually melt.

Continental Glaciers
Continental glaciers form at high latitudes where temperatures are always cold. These glaciers are immense masses of ice that spread over the entire land surface, rather than being confined to valleys. Two continental glaciers exist today: one covers Greenland, and the other Antarctica. The Greenland glacier (see Figure 19.24) covers more than 1.7 million square kilometers and is 3.2 kilometers thick at its center. The Antarctic glacier covers more than 12 million square kilometers (about 1.5 times the size of the 48 mainland U.S. states) and is 4 kilometers thick in places. It even extends over the ocean along the margins of the continent of Antarctica, forming several ice shelves.

A continental glacier has a lenslike shape; it is thickest at the center and thinner at the edges. The thick central area is where snow accumulates in a huge firn field and slowly turns to ice. The ice then flows outward in all directions under the pressure of its own weight. The huge weight of a continental glacier actually depresses the surface of the land, and the topography of the land is extensively changed by the movement of the glacier over its surface.

Processes of Glacial Erosion

Glaciers erode the surfaces over which they move by abrasion and plucking. **Abrasion** is the wearing, grinding, scraping, or rubbing away of rock sur-

Erosion

faces by friction. **Plucking** is the process by which rock fragments are loosened, picked up, and carried away by glaciers.

Abrasion

Abrasion is accomplished mainly by fragments of rock frozen in the ice at the bottom of a glacier. These rub and scrape against the underlying bedrock as the base of the glacier slips downslope. Sharp, angular fragments of hard materials such as quartz are very effective abraders. Clay is a good polishing material but doesn't abrade the bedrock very much.

Plucking

Plucking occurs in several ways. The water that forms when pressure on the ice causes melting can seep into cracks and pores in the bedrock. When this water refreezes, it shatters the bedrock and the broken pieces are picked up and carried away by the glacier. Plucking that occurs in this way is probably the most important because it goes on continuously over much of the glacier's bed as it moves along.

Figure 19.24 The Greenland Glacier. Contour lines show the elevations of the ice surface in meters, and a cross-sectional view reveals the lenslike shape of the glacier. Source: *Earth*, 4th Ed., Frank Press and Raymond Seiver, W.H. Freeman, 1986.

Plucking can also occur when glaciers move over bedrock that is broken into large pieces by jointing. Joints are fractures in rock and usually occur in three planes, forming large, rectangular blocks. These blocks freeze to the base of the glacier and are pushed out of the ground and carried along as the ice moves downslope.

A third type of plucking may occur where the glacier is thick and the bedrock is brittle. In such places, the pressure exerted by a glacier may be great enough to crush, shear off, or shatter brittle bedrock, forming fragments that are carried away by the glacier.

Features Resulting from Glacial Erosion

The processes of abrasion and plucking leave behind unmistakable evidence of a glacier's passage over bedrock. Numerous features, ranging from scratched and polished rock surfaces to basins excavated by plucking to streamlined hills of solid rock, are formed by glacial erosion.

Glacial Polish

Fine materials such as clays and silt that are frozen into the ice at the base of a glacier rub against bedrock as the glacier moves. These particles smooth the rock surface as sandpaper smooths wood. The smooth rock surfaces produced by this type of abrasion are said to be **glacially polished**.

Glacial Striations

Glacial striations (see Figure 19.25) are small scratches in the surface of the bedrock. They are created when fragments frozen into the bottom of a glacier scrape against the bedrock as the glacier slides over the bedrock. On smooth surfaces the striations are usually straight and parallel, but on uneven surfaces with protrusions the striations may curve because the ice deformed as it flowed around these obstacles. A striation may also thicken or thin out, depending on what happened to the abrading particle as it scraped against the bedrock. The long axis of a glacial striation indicates the direction of glacial movement.

Figure 19.25 Glacial Striations and the Particles That Made Them. Striations vary in size and shape, depending on the nature of the rock particles that formed them and the way that the particles were dragged over the bedrock.

Glacial Grooves

Glacial grooves are long, deep, U-shaped depressions in bedrock with smooth bottoms and sides and rounded edges. Most grooves are 10–20 centimeters deep, twice as wide, and tens of meters long. Really large grooves, however, can be more than 1 meter deep and hundreds of meters long. Grooves are shaped by abrasion and often form along fractures, soft layers, or other weak zones in the bedrock. Once started, a groove acts as a channel for further ice flow and concentrates the abrasive force of the glacier in this channel.

Rock Drumlins

Drumlins are hills of loose rock and soil molded into a streamlined shape by glaciers moving over them. Rock drumlins are large outcrops of rock that have been abraded into a streamlined shape by glaciers. Rock drumlins are large, often measuring 100 meters in length.

Erosion

U-Shaped Valleys

Glaciers move from high elevations to low elevations along the easiest paths. Those paths are usually stream valleys (see Figure 19.26a) that were eroded before the glacier formed. Stream valleys in mountainous regions tend to be steep-sided and V-shaped and to follow a narrow, twisting path between adjacent slopes (see Figure 19.27a). When a glacier moves down a stream valley, it alters the valley's shape. The ice does not meander back and forth as readily as water, so it often breaks through obstacles to form a straighter path. The ice also cuts back the walls of the valley, making them steeper. The result is a wider, straighter, steeper-sided, **U-shaped valley** (Figure 19.26b).

Figure 19.26 (a) A V-shaped Stream Valley and (b) a U-shaped Glacial Valley.
Source: *Geology of New York: A Simplified Account*, Y. Isachsen et al., eds., New York State Museum Geological Survey, The State Education Department, The University of the State of New York, 1991.

Hanging Valleys

A **hanging valley** (Figure 19.27c) is a valley of a tributary glacier whose floor is above, or hanging over, the floor of the main glacier into which it feeds. The difference in the levels of the valley floors can be hundreds of meters; and if a stream flows down the tributary valley, it may form a spectacular waterfall as it plunges to the floor of the main valley. Hanging valleys may be formed if the main glacier cuts back the valley of a tributary, or if it simply erodes faster and deeper than the tributary.

Cirques, Arêtes, and Horns

A **cirque** is the open, half-bowl-shaped hollow high on a mountainside at the head of a valley glacier. Cirques form when snowfields on mountain slopes melt and refreeze during spring and fall. The repeated freezing and thawing breaks apart the rocks. The rock fragments creep downslope or, when broken into small enough pieces, are carried away by meltwater. In this way, the snowfield creates a hollow in the land on which it rests. As the hollow grows, so does the depth of the snowfield, eventually becoming large and deep enough to form a glacier. Once formed, the glacier plucks rock from the base of the cirque, further deepening it.

Where cirques from adjacent slopes meet, a narrow, knife-edged ridge called an **arête** is formed between them. Where three or four cirques on the flanks of a mountain peak meet, the central peak takes on a pyramidal shape called a *horn*.

These three features—cirque, arête, and horn—are shown in Figure 19.27b.

Some Environmental and Human Impacts of Erosion

Mass Wasting

The sudden failure of slopes usually involves some triggering event. It can be the vibrations of an earthquake, the addition of water by intense or prolonged rainfall, an increase in the steepness of the slope (due to natural processes or human activities) or the loss of stabilizing vegetation.

The movement of rock material down a slope by mass-wasting processes can have a striking impact on the environment and on human activities. For example, a stream that is cutting a valley can create unstable slopes. The resulting landslides into the valley can dam up the stream, creating a natural reservoir that floods surrounding land. In 1983, a landslide in Utah blocked Spanish Fork Canyon, creating a large lake that cut off a transcontinental railway line and a major highway. Debris flows can bury or sweep away large areas of vegetation, leaving massive scars on the landscape.

Figure 19.27 Landscape Before, During, and After Glaciation. (a) Valley before glaciation. (b) Glacial erosion forms cirques in hollows between slopes. Erosion in adjacent cirques forms sharp ridges called arêtes and converts rounded peaks into sharp horns. Eroded material carried along the ice margins forms moraines. (c) Hanging valleys form where tributary glaciers flowed into the main glacier. Source: *Earth*, 4th Ed., Frank Press and Raymond Seiver, W.H. Freeman, 1986.

Unstable slopes are a hazard to human structures built on or below them. Every year, rockfalls along highway cuts cause motorist fatalities. Housing developments built atop slopes are frequently undermined and destroyed by mass wasting. However, with careful planning, building regulations, and zoning laws, and the use of stabilization techniques, the impact of mass wasting on human activities can be reduced.

Hazardous slopes can be stabilized by changing the steepness of the slope, reducing the weight acting on the slope, removing water from the rocks and soil, and planting vegetation. Some commonly used stabilization techniques include (1) the use of retaining devices (steel nets, concrete walls, rock bolts), (2) the removal of water by installing drainage systems or by covering the slope with an impermeable material, (3) grading the slope to reduce its steepness, and (4) building walls to divert flows. All of these preventive measures, however, are costly. The best strategy is to avoid building on unstable slopes in the first place.

Streams

Erosion by moving water affects human lives in many ways. Overland flow may wash away the topsoil your family just bought and spread on the lawn. Splash and sheet erosion of topsoil on farmland decreases fertility and lowers crop yields, resulting in higher prices at the supermarket. Erosion and flooding by streams leave thousands homeless annually. Nevertheless, there are benefits to living near running water, such as fertile soils, access to water, food from marine life, energy, waste absorption and dispersal, and possibly a transportation route.

Flooding is *the* major stream hazard, causing loss of life and structural damage. Flooding is a natural response to an unusually high input of water over a short time, but human activities can influence flooding in many ways. Paving surfaces or constructing buildings changes runoff patterns and infiltration rates. Changes to river channels and dam construction alter a stream's flood patterns.

The primary impacts of flooding are due to actual contact with the flowing water. They include death by drowning, damage to structures, crop loss, and erosion of flooded areas. Floods also impact humans and the environment indirectly by interrupting services, changing river channels, and destroying wildlife habitat.

Regions at risk of flooding and the degree of risk each faces can be determined if accurate maps and records of past floods have been kept. These allow predictions to be made based upon the frequency and extent of past flooding. However, records may not go back long enough to permit precise predictions about rare, severe floods.

The recurrent hazard of flooding has prompted efforts to reduce risk in flood-prone areas by limiting the human population there. Some communities now have floodplain zoning, which prohibits new construction. The Federal Emergency Management Agency (FEMA) requires that, if the owner of a flood-damaged property gets more than 50 percent of its value in compensation, the structure must be demolished and the site rezoned to prohibit future construction. Many lenders place restrictions on mortgages for properties in floodplains, allowing only appropriate activities such as recreation or certain types of farming to be carried out there. In some places there are calls for the buyout of all properties in flood-prone areas and the return of the land to its original wetland status. For urban areas that already exist in floodplains, however, flood hazard will continue to be an issue.

Waves and Currents

Erosion is one of the most important problems in coastal zones, mainly because 85 percent of shorelines is privately owned and rising sea levels are eroding all shorelines. Erosion is most pronounced on beaches and coastal cliffs, where smaller sediments are easily moved by waves and currents. It tends to be less of a problem on rocky coastlines or beaches made of pebbles, which are harder to move.

There are really only three ways to fight erosion of shoreline areas: dredge sediments to fill in where erosion has occurred, dam rivers or build coastal structures such as jetties and breakwaters to decrease the force of eroding waves and currents, and develop coastal dune areas as barriers to inland erosion.

Unfortunately, some of these strategies have proved ineffective. For example, *beach replenishment* involves replacing eroded sand with sand brought in from someplace else. This is an expensive undertaking considering the huge amount of sand needed. Also, most beach replenishment doesn't last very long—40 percent of all replenished beaches are gone in less than 2 years, less than half last for 2–5 years, and only 12 percent last more than 5 years. Beach stabilization is expensive and, as you can see, usually ineffective in the long term. Furthermore, it often changes the nature of the shoreline, and this can have negative effects on the organisms that inhabit it.

Of course, the best solution would be to recognize that coastal erosion is a natural process and allow it to happen without interference. Perhaps the coastal zone is not an environment that is favorable for private ownership.

Wind

Wind erosion presents humans with many problems. In arid regions, poles carrying electric power and telephone lines have been worn through by abrasion, causing them to fall and interrupt service. In times of drought, the precious topsoil of farmland stripped of vegetation must be protected from deflation. Vehicles used in arid regions require special features to keep abrasive particles out of their moving parts and to prevent particles suspended in the air from clogging fuel, air, and cooling lines.

Awareness of wind erosion and its causes enables humans to guard against it. Farmers plant windbreaks of drought-resistant trees. In many arid regions, off-road vehicles, such as dirt bikes and dune buggies, are banned because they destroy the fragile vegetation that protects the land from wind erosion.

Glaciers

Glaciers are responsible for much of North America's topography. Thus, many familiar landforms are the products of glacial erosion. Glacial erosion sculpted the rugged peaks of the Adirondacks and created the beautiful fjord known as the Hudson River Valley. The Great Lakes occupy basins carved out and deepened by glaciers and were once filled with glacial melt-

Erosion

water. Erosion by meltwater from the last glacial period formed most of the Mississippi River's drainage basin.

The modern-day burning of fossil fuels has increased the amount of carbon dioxide in the air. The result has been greenhouse-effect heating that has begun to melt Earth's remaining glaciers and ice sheets; this in turn has led to a rise in sea level and to coastal flooding. The global melting of ice may have other consequences, such as changes in global weather patterns that could have serious effects on agriculture.

MULTIPLE-CHOICE QUESTIONS

In each case, write the number of the word or expression that best answers the question or completes the statement.

1. Which event is the best example of erosion?
    (1) breaking apart of shale as a result of water freezing in a crack
    (2) dissolving of rock particles on a limestone gravestone by acid rain
    (3) rolling of a pebble along the bottom of a stream
    (4) crumbling of bedrock in one area to form soil

2. For which movement of Earth materials is gravity *not* the main force?
    (1) sediments flowing in a river
    (2) boulders carried by a glacier
    (3) snow tumbling in an avalanche
    (4) moisture evaporating from an ocean

3. Unsorted, angular, rough-surfaced cobbles and boulders are found at the base of a cliff. What most likely transported these cobbles and boulders?
    (1) running water   (3) gravity
    (2) wind   (4) ocean currents

4. The diagrams below represent four different examples of one process that transports sediments.

Which process is shown in these diagrams?
    (1) chemical weathering   (3) mass movement
    (2) wind action   (4) rock abrasion

5. On Earth, which agent of erosion is responsible for moving the largest amount of material?
   (1) groundwater
   (2) glaciers
   (3) running water
   (4) wind

6. Compared with an area of Earth's surface with gentle slopes, an area with steeper slopes most likely has
   (1) less infiltration and more runoff
   (2) less infiltration and less runoff
   (3) more infiltration and more runoff
   (4) more infiltration and less runoff

7. During a dry summer, the flow of most large New York State streams generally
   (1) continues because some groundwater seeps into the streams
   (2) increases due to greater surface runoff
   (3) remains unchanged due to transpiration from grasses, shrubs, and trees
   (4) stops completely because no water runs off into the streams

8. Which graph best represents the correct relationship between the discharge of a river and the particle size that can be transported by that river?

9. A river transports material by suspension, rolling, and
   (1) solution
   (2) sublimation
   (3) evaporation
   (4) transpiration

10. Four quartz samples of equal size and shape were placed in a stream. Which of the four quartz samples below has most likely been transported farthest in the stream?

(1)   (2)   (3)   (4)

Erosion

11. Which graph best represents the relationship between the slope of a river and the particle size that can be transported by that river?

12. What is the minimum water velocity necessary to maintain movement of 0.1-centimeter-diameter particles in a stream?
(1) 0.02 cm/s
(2) 0.5 cm/s
(3) 5.0 cm/s
(4) 20.0 cm/s

13. According to the *2011 Edition Reference Tables for Physical Setting/Earth Science*, which stream velocity transports cobbles but not boulders?
(1) 50 cm/s
(2) 100 cm/s
(3) 200 cm/s
(4) 400 cm/s

14. The diagram below shows points A, B, C, and D on a meandering stream.

At which point is erosion greatest?
(1) A
(2) B
(3) C
(4) D

15. The cross section below represents a portion of a meandering stream. Points *X* and *Y* represent two positions on opposite sides of the stream.

Based on the cross section, which map of a meandering stream best shows the positions of points *X* and *Y*?

Base your answers to questions 16 and 17 on the diagrams below. Diagrams *A*, *B*, and *C* represent three different river valleys.

16. Which bar graph best represents the relative gradients of the main rivers shown in diagrams *A*, *B*, and *C*?

Erosion

17. Most sediments found on the floodplain shown in diagram A are likely to be
    (1) angular and weathered from underlying bedrock
    (2) angular and weathered from bedrock upstream
    (3) rounded and weathered from underlying bedrock
    (4) rounded and weathered from bedrock upstream

18. The four streams shown on the topographic maps below have the same volume between X and Y. The distance from X to Y is also the same. All the maps are drawn to the same scale and have the same contour interval. Which map shows the stream with the greatest velocity between points X and Y?

(1)　　　　　(2)　　　　　(3)　　　　　(4)

19. Which interaction between the atmosphere and the hydrosphere causes most surface ocean currents?
    (1) cooling of rising air above the ocean surface
    (2) evaporation of water from the ocean surface
    (3) friction from planetary winds on the ocean surface
    (4) seismic waves on the ocean surface

20. The arrows labeled A through D on the map below show the general paths of abandoned boats that have floated across the Atlantic Ocean.

Which sequence of ocean currents was responsible for the movement of these boats?
(1) South Equatorial → Gulf Stream → Labrador → Benguela
(2) South Equatorial → Australia → West Wind Drift → Peru
(3) North Equatorial → Koroshio → North Pacific → California
(4) North Equatorial → Gulf Stream → North Atlantic → Canaries

21. Which natural agent of erosion is mainly responsible for the formation of the barrier islands along the southern coast of Long Island, New York?
(1) mass movement
(2) running water
(3) prevailing winds
(4) ocean waves

22. The block diagram below shows a part of the eastern coastline of North America. Points *A*, *B*, and *C* are reference points along the coast.

Which list best represents the primary process occurring along the coastline at points *A*, *B*, and *C*?
(1) *A*—folding; *B*—subduction; *C*—crosscutting
(2) *A*—weathering; *B*—erosion; *C*—deposition
(3) *A*—faulting; *B*—conduction; *C*—mass movement
(4) *A*—precipitation; *B*—infiltration; *C*—evaporation

23. Which rock material is most likely to be transported by wind?
(1) large boulders with sets of parallel scratches
(2) jagged cobbles consisting of intergrown crystals
(3) irregularly shaped pebbles that contain fossils
(4) rounded sand grains that have a frosted appearance

Erosion

24. The cross section below shows the movement of wind-driven sand particles that strike a partly exposed basalt cobble located at the surface of a windy desert.

Which cross section best represents the appearance of this cobble after many years of exposure to the wind-driven sand?

25. The photograph below shows a sandstone butte in an arid region.

Which agents of erosion are currently changing the appearance of this butte?
(1) glaciers and mass movement
(2) wave action and running water
(3) wind and mass movement
(4) running water and glaciers

26. Photographs A and B below show two different valleys.

Photograph A

Photograph B

Which list best identifies the agent of erosion that primarily determined the shape of each valley?
(1) photograph A—glacier; photograph B—river
(2) photograph A—river; photograph B—glacier
(3) both photographs—river
(4) both photographs—glacier

27. U-shaped valleys and parallel grooves in bedrock are characteristics of erosion by
(1) mass movement
(2) wave action
(3) running water
(4) glacial ice

28. The photograph below shows scratched and grooved bedrock with boulders on its surface.

Source: www.nr.gov.nl.ca

The scratches and grooves were most likely created when
(1) alternating thawing and freezing of water cracked the bedrock
(2) flooding from a nearby lake covered the bedrock
(3) a glacier dragged rocks over the bedrock
(4) rocks from a landslide slid along the bedrock

Erosion

29. Which landscape surface resulted primarily from erosion by glaciers?

CONSTRUCTED RESPONSE QUESTIONS

Base your answers to questions 30 through 33 on the cross section and block diagram below. The cross section shows an enlarged view of the stream shown in the block diagram. The sediments in the cross section are shown at 60% of actual size. Arrows show the movement of particles in the stream. The block diagram represents a region of Earth's surface and the bedrock beneath the region.

30. After measuring the size of each sediment particle, identify the name of the largest particle shown on the stream bottom in the cross section. [1]

31. What process is responsible for producing the rounded shape of the particles shown on the stream bottom in the cross section? [1]

32. Identify the type of rock shown in the block diagram that appears to be the most easily eroded. [1]

33. How does the shape of a valley eroded by a glacier differ from the shape of the valley shown in the block diagram? [1]

Base your answers to questions 34 through 36 on the block diagrams below, which show three types of streams with equal volumes.

34. Explain how the differences between the type 1 and type 3 stream channels indicate that the average velocities of the streams are different. [1]

35. Explain why the outside of the curve of a meandering channel experiences more erosion than the inside of the curve. [1]

36. Explain how the cobbles and pebbles that were transported by these streams became smooth and rounded in shape. [1]

Base your answers to questions 37 through 39 on the map below and on your knowledge of Earth science. The map shows a retreating valley glacier and the features that have formed because of the advance and retreat of the glacier.

Erosion

37. Describe one piece of evidence likely to be found on the exposed bedrock surfaces that could indicate the direction this glacier moved. [1]

38. Describe one difference between the arrangement of sediment in the moraines and the arrangement of sediment in the outwash plain. [1]

39. Describe the most likely shape of the valley being formed due to erosion by this glacier. [1]

EXTENDED CONSTRUCTED RESPONSE QUESTIONS

Base your answers to questions 40 through 42 on the block diagram below, which represents a landscape drained by a stream system, and on your knowledge of Earth science. The actual sizes and shapes of three rock samples, labeled *A*, *B*, and *C*, and the locations where they were found in the stream are indicated in the diagram. A New York State index fossil is shown in rock sample *A*.

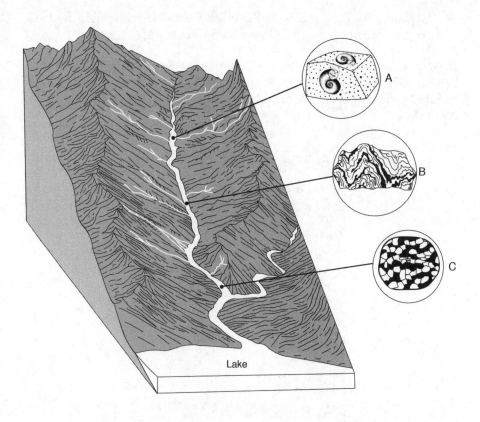

40. Explain how the appearance of rock sample A indicates that the sample has spent very little time being transported by the stream. [1]

41. Rock sample *C* has a diameter of 2 centimeters. Determine the minimum stream velocity needed to transport rock sample *C* to its present location. [1]

42. The stream profile below shows the locations of rock samples *A*, *B*, and *C* in the streambed.

Calculate the stream gradient between the locations of rock sample *A* and rock sample *C*. [1]

43. Complete the table below, by listing *three* agents of erosion and identifying *one* characteristic surface feature formed by *each* agent of erosion. [2]

Agent of Erosion	Surface Feature Formed
(1)	
(2)	
(3)	

Erosion

Base your answers to questions 44 through 47 on the diagram below, which shows several different landscape features. Points *X* and *Y* indicate locations on the streambank.

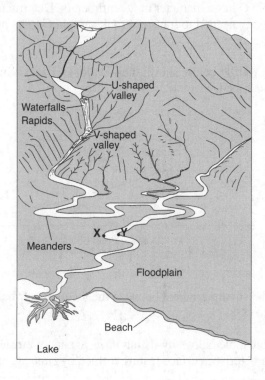

44. Explain why the upper valley in the mountains is U-shaped and the lower valley is V-shaped. [1]

45. Identify which point, *X* or *Y*, has more stream erosion and explain why the amounts of erosion are different. [1]

46. Explain why the stream meanders on the floodplain but *not* in the mountains. [1]

47. The beach consists of particles with diameters from 0.01 cm to 0.1 cm. Identify the sedimentary rock that will form when burial and cementation of these sediments occur. [1]

CHAPTER 20
DEPOSITION

> **KEY IDEAS** Patterns of deposition result from a loss of energy within the transporting system and are influenced by the size, shape, and density of the transported particles. Sediment deposits may be sorted or unsorted.
>
> Sediments of inorganic and organic origin often accumulate in depositional environments. Sedimentary rocks form when sediments are compacted and/or cemented after burial or as the result of chemical precipitation from seawater.
>
> Landscapes are regions in which the physical features of Earth's surface are related by their structure, the processes that formed them, and the way in which they developed. They can be classified by characteristics such as relief, stream patterns, and soil associations. Landscapes with similar characteristics can be grouped into distinctive landscape regions.

KEY OBJECTIVES
Upon completion of this chapter, you will be able to:
- Explain how the characteristics of sediments and the media carrying them affect their deposition.
- Describe the processes by which the various agents of erosion deposit the sediments they are transporting.
- Describe some of the common landforms resulting from erosion and deposition by mass movements, moving water, glaciers, and wind.
- Identify and describe landscape regions.

DEPOSITION

At some point, all agents of erosion begin to deposit sediment. **Deposition** is the process by which transported sediment is dropped in a new place. The same agents that erode sediment also deposit it. Several factors influence when deposition will begin.

Deposition

Figure 20.1 Relationship of Transported Particle Size to Water Velocity. This generalized graph shows the water velocity needed to maintain, but not start, movement. Variations occur because of differences in particle density and shape. Source: The State Education Department, *Earth Science Reference Tables*, 2011 ed. (Albany, New York; The University of the State of New York).

Speed of Medium

The **medium** is the substance that carries sediment. In streams the medium is water, in winds it is air, and in glaciers it is ice. Generally, the slower a medium is moving, the less its carrying power. If the carrying power drops below the level needed to transport a particle, that particle will drop to the ground. Less carrying power results in deposition.

You can see this for yourself. Imagine you are flying a kite and the wind slows down. The slower the wind blows, the less carrying power it has. The kite will begin to drop. If the wind stops blowing, the kite will settle to the ground. Deposition occurs in much the same way. As the medium slows, sediment begins to settle out.

Figure 20.1 shows the sediment that water carries at different speeds. Notice that water carries more sediment sizes when it is moving fast. As water slows down, some sediment is deposited. Large, dense particles will settle first, and light, small particles last. The result is a separation of different particles into layers.

The separation of particles during deposition is called **sorting**. Figure 20.2 shows the sorting that occurs when a stream flows into a larger body of water, such as a lake or an ocean. As the stream flows into the larger body, it slows down. Coarse gravel and pebbles settle first. Farther from the shore, the movement of the water is slower and sand settles out. Still farther from shore, in very slow moving water, the finest particles of silt and clay settle. Sorted layers of particles result.

Figure 20.2 Horizontal Sorting of Particles Carried by a Stream. As a stream decreases in velocity, the heavier particles settle first; the lighter particles are carried farther downstream before settling.

Characteristics of Particles

Particles may vary in size, shape, and density. Each of these characteristics influences the rate at which settling occurs. To learn how particle size influences settling rate, look at Figure 20.3a. Notice that, when all the rock particles are deposited, sorting has occurred. How can that be? The answer is that some particles must settle before others. The particles on the bottom settled fastest, and those on top slowest. Since the larger particles are on the bottom, they settled fastest. The smaller ones, on top, settled slowest.

In general, the larger the particle, the faster it will settle. Very small particles, such as fine clay, take a long time to settle. The muddy appearance of many rivers and lakes is due, in part, to particles that have not yet settled. To see how particle shape affects settling rate, look at Figure 20.3b. Notice that, when all the particles are deposited in water, the bottom layer consists of round particles. These particles must have settled first. Also notice in the figure a top layer of flat particles. These particles must have settled last.

In general, round particles settle faster than flat ones. To understand how particle shape affects settling rate, imagine dropping a crumbled paper and a flat paper to the ground at the same time. The crumbled paper, like a rounded particle, lands faster. Friction between the large surface area of the flat paper and the air slows that paper's fall. A flat shape in rock particles has the same effect in water. Friction between the flat particle and the water will decrease the rate at which the particle settles.

Finally, to see how the density of a particle affects settling, look at Figure 20.3c, Notice, once again, that sorting has taken place. The densest particles form the bottom layer; the least dense particles, the top layer, Thus it can be concluded that the densest particles settled fastest.

To summarize, several factors affect the deposition of sediment. The speed of the medium determines when particles will be deposited. The size, shape, and density of a particle determine the rate at which it will settle. Differences in the settling rates of sediments result in sorting during deposition.

(a) All other factors being the same, large particles settle faster than smaller ones.
(b) All other factors being the same, particles with rounded shapes settle faster than flat particles.
(c) All other factors being the same, particles composed of denser materials settle faster than those composed of less dense materials.

Figure 20.3 Vertical Sorting of Different Particles in a Column. When sediments settle in a quiet medium, they form a series of layers. The fastest settling particles reach the bottom first and form the lowest layer. They are followed by slower settling particles, which form the next layer; the slowest settling particles form the top layer.

Bedding of Sediment

As a result of sorting during deposition, sediment forms layers consisting of different types of particles. Sediment deposited in a quiet body of water will usually form layers similar to those shown in Figure 20.4. Each layer of sediment is called a **bed**, and each bed usually represents a period of deposition. Beds of sediment are commonly found in sedimentary rock.

Figure 20.4 Bedding of Sediment. Rapid deposition results in **graded bedding**, a vertical sorting. Each sequence represents a depositional event.

Hurricane, September
No rainfall, August
Summer storm, July
Periodic gentle rain, May-June
Heavy rains, May
No rainfall, April
Heavy rainfall and spring thaw, March

DEPOSITION BY MASS MOVEMENTS

Deposition by Gravity

During mass movements, sediment is pulled downhill by gravity and is deposited, when it stops moving, at the base of a slope. Sediment deposited by gravity does not settle through a medium, as does sediment deposited by wind or water. Thus, little, if any, sorting takes place. Most deposits resulting from mass movements are unsorted and do not show any distinct layering. In **unsorted deposits**, all types of particles are arranged without any order, as shown in Figure 20.5.

Figure 20.5 **Unsorted Deposit**. Unsorted sediment contains rock fragments of all sizes randomly mixed together.

Deposits by Rapid Mass Movements

Rapid mass movements deposit sediment. As deposition takes place, landforms develop. These landforms are evidence that deposition has occurred.

Rockfalls

During rockfalls and rockslides, fragments slide downhill until they reach the base of a slope, forming a talus. A **talus** is a pile of broken rock that builds up at the base of a cliff or steep slope. Boulders may roll down a talus and onto the flat ground at the base of the slope. As the talus weathers, soil may form on the surface. Then, if the talus is not being constantly covered by new rock fragments, plants may take root and grow. See Figure 20.6a.

Mudflow

During a mudflow, mud moves downhill until it reaches the base of a slope. The mud then spreads out in a thin, wide sheet. See Figure 20.6b. Huge boulders carried by the mudflow may be left to stand isolated on the gently rolling land.

Deposits by Slow Mass Movements

Slow mass movements also deposit sediment. Earthflows, slump, and soil creep all produce sediment deposition from which landforms may develop.

Earthflow

An earthflow deposits a thick pile of sediment. See Figure 20.6c. Earthflow deposits are generally thicker than deposits produced by mudflows. The difference between the two can be demonstrated by spilling a jar of thick honey and a jar of thin tomato sauce next to each other. The thick honey, like an earthflow, produces a small, thick deposit. The thinner sauce, like a mudflow, produces a thin spread. Compare Figures 20.6b and 20.6c.

Slump

Slump forms a deposit that looks like an apron spread on the ground. Notice in Figure 20.6d the steplike appearance of the deposit. The steps are the tops of blocks of material that broke loose from the cliff. Very often, earthflows begin as slump.

Soil Creep

As stated earlier, soil creep is a very slow process. Material deposited by soil creep is often hard to recognize because it is usually covered with plants, and looks like part of the hill on which it formed. Only a person trained to recognize deposition will perceive the series of ripples at the base of the gentle slope as evidence of deposition by soil creep. See Figure 20.6e.

Effects of Mass Movements

Deposition by mass movement along roadsides can cause serious problems. Poorly planned roads may be blocked by deposits. In fact, material deposited by slides, flows, and slump may

Figure 20.6 Deposits Formed by Mass Movements. Source: *The Earth Sciences*, Arthur N. Strahler, Harper & Row, 1971.

make a road impassable. Interstate 40, near Rockwood, Tennessee, is one such road; one stretch has been blocked more than 20 times by major slides. Clearing such roads costs taxpayers millions of dollars each year. In addition, these poorly planned roads pose dangers to motorists.

DEPOSITION BY MOVING WATER

More than 70 percent of Earth's surface is covered by water. Therefore, it is not surprising that most deposition occurs in water. In this section, you will learn how sediment carried by moving water is deposited and how deposition changes the shape of the land.

Deposition by Streams

Rivers and other streams carry large amounts of sediment. Water from the Colorado River is used to irrigate crops; canals carry the water to the fields. In just one of these canals, up to 5,000 metric tons of sediment has to be removed from the water every day.

Streams deposit sediment when they slow down or decrease in volume. Such slowdown or decrease in volume occurs when the slope decreases, runoff and seepage from groundwater decrease, or the stream enters a standing body of water, such as a lake or an ocean. Deposition by streams results in layers that are sorted according to particle size, shape, and density. In addition, these deposited sediments tend to be smoothed and rounded as a result of abrasion while being carried in the stream.

Many landforms develop as a result of stream deposition. These landforms are found in almost all parts of the United States.

Oxbow Lakes

In a preceding chapter you learned how meanders are formed. As meanders become more curved, their ends move closer together. Eventually, the stream erodes its way through the land, separating the meanders (see Figure 20.7). Most of the water then flows through this steeper, straighter shortcut. The slower moving water in the old meander deposits its load. When deposition blocks off the entrance to the old meander and it becomes separated from the stream, an oxbow lake is formed.

Floodplains and Levees

Because of its increased speed and volume, a stream in flood carries much more than the normal amount of sediment. Also, when a stream is in flood, it overflows its banks. Outside the channel, the water is slower moving and shallower. As soon as the flood water leaves the channel, it begins to deposit its sediment load over the surrounding land to form a **floodplain**. The larger

Figure 20.7 Formation of an Oxbow Lake. Water flows faster along the outside curve of a meander than it flows along the inside curve because it has farther to travel in order to make the turn. The faster moving water erodes the bank on the outside curve until just a narrow neck of land (a) separates the loops of the meander. When the meanders erode through the narrow neck, most of the streamflow bypasses the meander. Because less water flows through the meander, water velocity decreases and deposition builds sediment "dams" (c) that separate the meander from the stream, forming an oxbow lake (b). Source: *Earth Science on File.* © Facts on File, Inc., 1988.

particles settle out first, building a **levee**, that is, a low, thick, ridgelike deposit along the banks of both sides of the stream (see Figure 20.8).

Floodplains make good farmland; sediment deposited after each flood renews the soil. For this reason, farmers have been able to raise crops on the floodplains of the Nile River for over 4,000 years!

Figure 20.8 Formation of a Floodplain (a) and Levee (b). During a major flood, the stream's floodplain becomes a lake. Water in the main channel flows rapidly. However, water escaping the channel immediately slows down and deposits fine sand and silt, forming a natural levee. As the water continues to spread out and slow down, clay is deposited on the surrounding lower land. When the flood ends, the levees remain as low ridges along the sides of the channel and the floodplain is covered with wet clay and swampy areas. Source: *Physical Geology*, 2nd Ed., Richard Foster Flint and Brian J. Skinner, John Wiley, 1977.

Deltas and Alluvial Fans

When a stream flows into a quiet body of water such as an inland sea, ocean, or lake, it stops moving. The point at which this occurs is called the **mouth** of the stream. Here most of the stream's sediment is deposited. The result-

Deposition

Figure 20.9 Stages (a)–(f) in the Formation of a Delta. A delta is formed as a stream flows into a standing body of water such as a lake or an ocean. Source: *The Earth Sciences*, by Arthur N. Strahler, Harper & Row, 1971.

ing landform is a **delta**, as seen in Figure 20.9. A delta is a large, flat, fan-shaped pile of sediment at the mouth of a stream. It is composed of sand, silt, and clay and is shaped like a triangle. It gets its name from the Greek letter delta, which is written as Δ.

When streams flowing down steep mountain valleys reach flat, open land, they slow down and deposit much of their coarser load. The resulting landform is an **alluvial fan**, that is, a large mound of coarse sediment deposited by a stream onto open, flat land (see Figure 20.10).

Although an alluvial fan is similar in appearance to a delta, there are some differences. An alluvial fan has steeper sides and is made of coarser sediment than a delta.

Deposition by Oceans

Oceans also deposit sediment. Whenever waves and currents slow down, they deposit their loads. Deposition by waves and currents builds new landforms that change the shape of shorelines.

Beaches

A **beach** is a deposit of sediment along the shoreline of a body of water. Beaches are formed as waves wash up against the land and slow down. Beaches constitute an almost continuous fringe around the continents and most islands. Although most people think of a beach as a strip of fine, white sand along the seashore, beaches may consist of any earth material from fine clay to boulders.

Figure 20.10 Formation of an Alluvial Fan. An alluvial fan forms where a stream flows from a steep gradient onto a gentle gradient. Source: *The Earth Sciences*, Arthur N. Strahler, Harper & Row, 1971.

The nature of beach sediments depends on the source material and the weathering processes at work in the area. Beaches are shaped by waves and are changed daily and monthly by tides. Figure 20.11 shows the general features of a beach. **Berms** are flat areas formed by wave action. The breaking wave loosens sediments upslope and drags them back toward the water as the wave recedes.

Figure 20.11 Beach Features. The usual direction in which breaking waves carry sand is toward the beach. This sand, together with sediments eroded from coastal cliffs, builds a flat area, or a berm. Source: *The Earth Sciences*, Arthur N. Strahler, Harper & Row, 1971.

Figure 20.12 shows the changing shoreline conditions in winter and summer. During stormy winter months, waves tend to be high and to have short periods. The sand picked up by one wave does not have a chance to settle to the bottom before the next wave sets the sand in motion again. As a result, the sand stays in suspension and is carried away from the beach by the backrush of water until it reaches deeper, calmer water and can settle, forming an offshore sandbar. See Figure 20.12a.

In summer months, the weather is calmer and the waves that reach the shore tend to be swells of long periods. The breaking waves pick up sand in shallow water and carry it up onto the beach. Most of the sand is then carried back toward the sea by the backrush of water along the beach. There is time during the backrush, however, for some particles to settle before the next wave again carries sand toward the beach. The overall result is a shifting of sand toward the beach. See Figure 20.12b.

Sandbars and Spits

A **sandbar** is a long, narrow pile of sand deposited in open water. Sandbars may form where receding waves wash beach sediments into deeper, quieter waters and also where the shoreline curves away from

Figure 20.12 Winter (a) and Summer (b) Beach Profiles. High waves of short period generally occur in winter months and drag sediments offshore, forming a sandbar. In summer, waves generally have long periods and move sediment onto the beach. Source: *The Earth Sciences*, Arthur N. Strahler, Harper & Row, 1971.

a longshore current. As the current curves away from the shoreline, it flows into deeper, quieter water and deposits its sediments. Most bars formed in this way are attached at one end to the mainland and are called **spits**.

Deposition

Sandbars that connect an island to the mainland are called **tombolos**. See Figure 20.13.

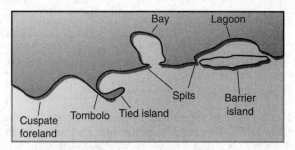

Figure 20.13 Some Coastal Landforms Created by Deposition of Sediments by Ocean Waves and Currents.

Barrier Bars and Lagoons

Eventually, waves deposit enough sand on a bar so that it is above water. Large waves continue to add material, and in time the bar is built up well above sea level. Such large bars are known as **barrier bars**. Barrier bars may completely block off a bay, forming a **tidal marsh**, or may parallel the coast, broken here and there by tidal inlets into elongated **barrier islands**. The calm, protected bodies of water between barrier bars and the mainland are called **lagoons**. See Figure 20.13.

Problems Associated with Deposition by Moving Water

Deposition by moving water can create problems in harbors, which must be deep and have clear openings to the sea to enable large ships to pass through. Occasionally, sandbars and spits block a harbor's entrance. In addition, sediment from streams and rivers deposited in harbors can create areas of shallow water. Ships are then in danger of running aground. To prevent this from happening, sediment must be removed from harbors by dredging.

DEPOSITION BY WIND

Wind is moving air. When a wind slows or stops moving, the particles it is carrying settle to the ground and are deposited. Wind deposits generally contain fine sediments and result in the formation of certain surface features.

Loess

Loess is a thick, unlayered deposit of very fine, buff-colored sediment deposited by wind. Loess deposits may vary in thickness from a few centimeters to several hundred meters. Loess is found throughout the world. There are large deposits in the central United States, Europe, and parts of China. Loess is a source of excellent soil because of its ability to hold large amounts of water.

Dunes

Dunes are mounds of sand deposited by wind. They are often found in barren, desert regions and can also occur on large, sandy beaches. A dune often forms when windblown sand meets an obstacle such as a rock or a bush.

Figure 20.14 shows how a dune develops. On the windward side of the dune, that is, the side facing the wind, sand that strikes an obstacle falls to the ground. In time, the sand forms a mound that slopes gently up toward the tip of the obstacle. Wind then pushes sand grains up this slope to the crest, or top, of the dune.

Figure 20.14 Formation of a Dune. (a) Windblown sand accumulates in low mounds. (b) Steady winds carry sand up the side of the dune, and it rolls over the crest onto the sheltered side. (c) As sand is picked up on the gently sloping side facing the wind and deposited on the steep slip face, the entire dune moves.

On the leeward side of the dune, that is, the side sheltered from the wind, winds are very slow. As a result, the sand grains rolling over the crest fall to the ground because the winds can no longer support them. A deposit of sand forms, creating a steep-sided mound. As sand continues to be deposited, this mound may become unstable. Indeed, sand may slip or slide downhill. For this reason, the steep, leeward side of a dune is called the **slip face**.

As winds blow sand from the windward side of a dune and deposit it on the leeward side, the entire dune moves in the direction in which the wind is blowing. In this way, dunes travel along a desert. Some desert dunes may move from 10 to 25 meters in a year. Moving dunes have been known to engulf forests, farmlands, and buildings.

Deposition

Sand dunes have a variety of heights, shapes, and patterns. Dunes may reach heights of 30–100 meters, depending on wind speed and the amount of sand in the area. Three common shapes and patterns of sand dunes are shown in Figure 20.15. The type of dune that forms in an area depends on the prevailing wind patterns and the amount of sand.

Figure 20.15 Common Shapes and Patterns of Sand Dunes.

In the United States, the dunes of our coastal areas are an important line of defense for inland properties from storm waters. From 1936 to 1940, the National Park Service built nearly 1,000 kilometers of fencing along the coast of North Carolina. The fencing formed an obstacle to the windblown sand, trapping it and thereby forming dunes. These dunes were then planted with grass, shrubs, and over two million trees! Today, the resulting barrier of dunes helps to protect inland regions from serious damage by hurricanes.

DEPOSITION BY GLACIERS

The way in which glaciers transport and deposit sediments is very different from that of any other agent of erosion. To fully understand glacial deposition, let us first consider how glaciers transport their sediment load.

Glacial Transport

Glaciers transport materials suspended in the ice, along the sides and bottom of the ice, on the surface of the ice, and in the area directly alongside the ice. Although it is tempting to picture a glacier as a huge block of ice pushing a pile of debris ahead of it, most glaciers transport very little material by "bulldozing" ahead of the ice. Most of the material carried by a glacier is suspended in the ice.

One of the characteristics that affect the ability of any medium to carry materials in suspension is the viscosity, or resistance to flow, of the medium. The higher the viscosity, the greater is the medium's ability to hold materials in suspension. Ice is millions of times more viscous than liquid water and can carry just about anything in suspension. A boulder that couldn't be budged by the fastest moving stream of water, no less remain suspended in the water, can easily be carried along suspended in the ice of a glacier. See Figure 20.16.

Debris can become suspended in glacial ice in several ways. Many glaciers originate in snowfields high in mountainous regions. Landslides and rockfalls from the steep slopes that surround the glacier deposit debris on top of the snow. When more snow accumulates, the debris is buried and eventually encased in the ice that forms. Also, rock protrusions around which a glacier flows shed fragments that become incorporated into the ice. Where two glaciers merge, the debris along the edges of the glaciers becomes incorporated into the body of the larger glacier that forms. Material also becomes incorporated into the bottom and sides of a glacier as the ice melts and refreezes around loose rock fragments. The 10- to 100-centimeter layer along the base of a glacier is filled with suspended debris.

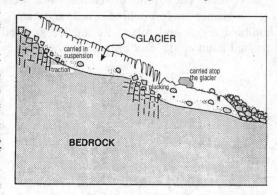

Figure 20.16 Transport of Sediment by a Glacier. A glacier carries sediments on its surface, suspended in the ice, and dragged along its sides and bottom.

Material along the bottom of the glacier is carried along by traction. In **traction**, the particles are not encased in the ice but are dragged along, slipping, sliding, and rolling beneath the moving ice. Material along the sides of the glacier is also moved by traction, and rock fragments may slide or roll off the ice onto ground next to the glacier. Meltwater-soaked debris with the consistency of wet concrete also flows off the sides of the glacier.

Most material carried on the top of a glacier was originally suspended in the ice but has been exposed as melting occurred. Thick debris, however, shields the ice beneath it from melting and results in high spots on the ice surface. Blocks of rock that fall onto a glacier from surrounding slopes can be carried for great distances, conveyor-belt style, atop the glacier.

Types of Glacial Deposits

Part of a glacier's load is carried frozen within the ice or scattered on its surface. The rest is pushed up in piles around the edges of the glacier and carried along as the glacier moves over Earth's surface. When the glacier reaches a warm region, or the climate changes and becomes warm enough, the ice begins to melt. As the glacier melts, the sediments it was carrying are deposited. These deposits may be sorted or unsorted, and each type is formed differently.

Deposition

Unsorted Deposits

At the melting edge of the glacier, sediments within the ice, as well as those carried on top of it, are released. Particles of all shapes and sizes fall to the ground in a confused jumble, forming an unsorted deposit. In addition, as the ice melts back, piles of sediment that were pushed up around the edges of the glacier are left behind. These sediments, too, are unsorted. They were tumbled about and mixed as the ice scraped them from the ground and pushed them along. See Figure 20.17a.

Figure 20.17 Glacial Deposits. (a) Glacial till. As glacial ice melts, the sediments are released and drop to the ground in a confused jumble, forming an unsorted deposit. (b) Sorted varved clays from a glacial metlwater lake deposit. Meltwater streams may sort glacial sediments; or, as in this photograph, the sediments may settle out in sorted layers at the bottom of lakes fed by meltwater.

Sorted Deposits

Some glacial sediment is deposited into streams that originated from a melting glacier. This material is sorted by the running water and then deposited in layers. Fine sediments carried into lakes by meltwater lakes may settle out in sorted layers on the lake bed. See Figure 20.17b.

How Glaciers Deposit Sediments

Some deposition takes place beneath the ice. The base of a warm glacier is continuously melting because of heat radiating from the ground and heat created by friction with the ground. As particles in the base of the glacier are freed by melting, some are rolled or dragged along until they become lodged in the ground. Refreezing of large particles, which jut out more than finer ones, also leaves behind deposits enriched in fine particles.

Materials on top of a glacier are deposited in one of two ways: either meltwater and debris slides carry them off the glacier, or they are set down on the ground as the ice beneath them melts. When meltwater deposits these materials, they are well sorted and somewhat rounded. Materials that slide or fall off the glacier, however, are jumbled masses of large and small angular fragments. By the time a glacier reaches its end, it has already undergone much melting. The remaining material on its surface is dropped off the end of the glacier as the ice beneath it melts, forming a pile called an **end moraine**.

Meltwater carries debris from the glacier beyond the end of the glacier and deposits it over the land. In summer, glacial meltwater streams are in flood and carry huge amounts of sediment, including abundant fine particles. These fine sediments are formed by the crushing pressure and abrasion between particles as they are carried along inside the glacier. If numerous streams spread out over the land and deposit glacial material in a wide sheet, the result is an **outwash plain**. If the meltwater streams are confined in a valley, they may deposit their load on the floor of the valley.

Meltwater often fills closed depressions, forming glacial lakes. If the lake forms along the edge of the glacier, debris can fall directly into the lake. If the water undercuts the ice face, icebergs fall into the lake and drift across its surface. As an iceberg melts, it rains **ice-rafted** debris onto the lake bed. During the winter, the lake freezes over and deposition ceases except for the settling out of the very finest particles of clay. In the spring when the ice thaws, meltwater carries in coarser material, which is deposited atop the clay. The result is a pair of layers known as a **varve**. Each varve represents one seasonal cycle—a year. See Figure 20.17b.

Landforms Resulting from Glacial Deposition

Glacial drift is any material deposited as a result of glacial activity, including material deposited by meltwater and debris slides. **Glacial till** is material deposited directly from ice. The main difference between the two is sorting; drift may be sorted because of the action of meltwater, whereas till is unsorted.

When glacial drift and till are deposited, they form a number of distinctive landscape features, as shown in Figure 20.18.

Moraines

Moraines are deposits of till. A moraine that forms a thin, widespread layer of till is called a **ground moraine**. Ground moraines form gently rolling hills and valleys. Piles of till deposited around the side of a glacier are called **lateral moraines**. Deposits of till at the front of the glacier are **terminal moraines**. Terminal and lateral moraines form long, parallel ridges. The ridges mark the boundaries where a glacier once existed. See Figure 20.19.

Deposition

Drumlins

Drumlins are long, low mounds of till that have a rounded, teardrop shape, as seen in Figure 20.18. By looking at the figure, you can see the direction in which the glacier moved. The rounded part of the drumlin points in the direction from which the glacier advanced. Drumlins are thought to be molded by the ice of a glacier as it slides over previously deposited piles of sediment.

Figure 20.18 Landforms Resulting from Glacial Deposits. Source: Educational Leaflet #28, The New York State Museum, Albany, New York.

Figure 20.19 How Moraines Form. Lateral moraines consist of materials carried along the sides of a glacier. When two glaciers merge, their lateral moraines are welded together between the glaciers, forming a medial moraine. End moraines form where debris melts out of the ice at the end of the glacier. Source: *Earth*, 4th Ed., Frank Press and Raymond Seiver, W.H. Freeman, 1986.

Erratics

Erratics are large, isolated boulders deposited by a glacier. Many erratics are more than 3 meters in diameter and weigh thousands of metric tons. Such boulders are much too large to be carried by wind or water, but glaciers often transport them far from their sources. When the ice melts, the large boulder comes to rest on a surface having a composition different from its own.

Outwash Plains

At the leading edge of a glacier, some melting is almost always taking place. Streams of glacial meltwater carry sediments out beyond the glacier and deposit them in sorted layers. Over time, a broad plain is built up in front of the glacier by deposition of sediments from glacial meltwater. Such a landform is called an **outwash plain** because the sediments it is made of were "washed out" beyond the glacier by meltwater streams.

Kettles and Kettle Lakes

A **kettle** is a pit found in a glacial deposit. There are several steps in the formation of a kettle. First, a chunk of glacial ice breaks loose. Next, sediment from the glacier covers the chunk of ice. After a while, the ice melts, the overlying sediment sinks, and the kettle is formed. See Figure 20.20. A kettle may later fill with glacial meltwater, rainwater, or groundwater, forming a **kettle lake**.

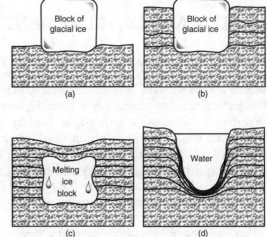

Figure 20.20 Formation of a Kettle. Kettles form when buried blocks of ice melt and the overlying sediments collapse, forming a hole. The melting of deeply buried blocks of ice forms rounder and shallower kettles.

Glacial Lakes

In addition to kettle lakes, glaciers may form lakes in two other ways. A glacier may gouge out a large depression in Earth's surface that may fill with water. Alternatively, a **glacial lake** may form when ice or moraines dam the course of a stream; the flow of water backs up behind the moraine, creating a lake. Long, narrow glacial lakes, such as the ones found in central New York, are known as **finger lakes**.

Ice-dammed lakes can cause disastrous floods when the ice dam melts and releases the lake water. There is much evidence that catastrophic floods caused by ice damming occurred repeatedly during the last ice age, 10,000–20,000 years ago.

Glacial lakes impact society in many ways. They form some of this country's most spectacular scenery. In addition to being beautiful, glacial lakes have both recreational and commercial uses. New York's Finger Lakes are used extensively for fishing, boating, and swimming. The Great Lakes, a system of five huge glacial lakes, are the largest repositories of freshwater on Earth. Native Americans and early explorers used them as key transportation routes. Later, they served as major shipping routes for getting the bountiful harvests of our country's heartland to major population centers back east. Large cities, such as Chicago and Detroit, grew from port cities along this route. The building of the Erie Canal created the link needed for a water route from the Great Lakes to New York City, and catapulted New York City from a small coastal port to the major international center of trade it is today.

LANDSCAPE DEVELOPMENT

Throughout this unit, you have learned about geologic processes that change the surface. Crustal movements cause the surface to be uplifted in some places and to subside in others. Volcanic activity injects new rock beneath the surface in some places while bringing new rock to the surface in others. Weathering and erosion reshape the land by wearing away rock exposed at the surface and transporting sediments downslope. When these sediments are deposited, characteristic features are formed. Together, all of these geologic processes create the **topography** of the land, that is, the landforms that shape the surface of Earth. **Landforms** are physical features of Earth's surface that have characteristic shapes and are produced by natural processes. Landforms include both major forms, such as mountains, plateaus, and plains, and minor forms, such as hills, valleys, drumlins, dunes, and many others.

Characteristics of Landscapes

A **landscape** is region in which the landforms are related by their structures, the processes that formed them, and the ways in which they developed. Landscapes are the product of the interaction of geological processes, climate, and human activities on the land over a long period of time. Landscapes can be classified by characteristics that can be readily observed and measured, such as relief, stream patterns, and soil associations.

Relief

Relief refers to the physical shape or general unevenness of a part of Earth's surface, such as variations in slope or elevation. It is expressed in terms of the vertical difference in elevation between the highest and lowest points in a given region. A region showing great variations in elevation has "high relief," and a region showing little variation in elevation has "low relief." Regions with high relief typically have steep slopes; those with low relief, gentle slopes. Most landscapes can be classified as mountains, plateaus, or plains based upon their reliefs.

Mountains are parts of Earth's crust that project at least 300 meters above the surrounding land. A mountain generally has steep sides, a restricted summit area, and considerable bare-rock surface. Most mountains are formed by crustal movements. Two common types are fold mountains and fault-block mountains. Fold mountains form when layers of rock are folded by compression, wrinkling Earth's crust. Fault-block mountains form when tension forces cause blocks of crust to move along faults. On one side of the fault, rocks rise; on the other side, rocks subside. See Figure 20.21.

(a) (b)

Figure 20.21 Mountains. (a) **Fold mountains.** Source: *The Earth Sciences*, Arthur N. Strahler, Harper & Row, 1971. (b) **Fault-block mountains.** Source: *Macmillan Earth Science*, Eric Danielson and Edward J. Denecke, Jr., Macmillan, 1989.

Plateaus are large areas of flat land at high elevations. A plateau generally has an underlying structure made of horizontal layers of rock that were gently uplifted. Plateaus are often found next to mountain ranges and were probably raised by the same forces that formed the mountains but were not faulted or folded as greatly as the rocks of the mountains. Although made of horizontal rock layers, the surface of a plateau is often not level. Streams and other agents of erosion can cut deep valleys and canyons into the surface of the plateau.

Plains are large areas of flat land at low elevations. Plains are usually formed by the deposition of sediments in horizontal layers at or below sea level. Thus, the underlying structure of a plain consists of horizontal layers of rock. Many plains have undergone uplifting, which has raised their surfaces above sea level, but have not been uplifted as much as plateaus.

Stream Patterns

Stream patterns refer to the patterns formed by the system of streams in an area as they flow down slopes and join with other streams. A stream that flows into and joins another stream is called a **tributary**. The area from which precipitation drains into a stream or system of streams is called a **drainage basin**. See Figure 20.22. Higher areas, called **divides**, separate

Deposition

drainage basins. Streams from neighboring drainage basins may join into larger streams, which, in turn, may join with others to form even larger streams. Together, all of the streams in an area and their tributaries form a **stream drainage system**. Drainage systems often show distinct patterns that reflect the geology of the region. Figure 20.23 shows some of the more common stream drainage patterns that develop on different geological structures.

Figure 20.22 Drainage Basins of New York State.

Stream Pattern	Geologic Structures
(a) Dendritic	Form where underlying rocks have no pronounced differences in erosion rate, so stream flow is equal in all directions.
(b) Radial	Form on rounded surfaces, such as conical volcanoes or dome mountains, where stream flow is away from a central high point.

Stream Pattern	Geologic Structures
(c) Annular	Form around dome mountains with tilted rock layers on their flanks. Differential erosion causes concentric ridges to form. Streams flow in low areas between these ridges.
(d) Rectangular	Form is folded, faulted, or tilted strata with long strips of rock of unequal resistance to erosion. Major streams flow along belts of weak rock, making many right-angle turns.
(e) Trellis	Form in tilted, folded, or faulted strata with long strips of rock of unequal resistance to erosion. Major streams run through long valleys, following belts of weak rock between parallel ridges of stronger rock. Tributary streams enter major streams at right angles.

Figure 20.23 Stream Drainage Patterns. (a) Dendritic (b) Radial (c) Annular (d) Rectangular (e) Trellis.

Soil Associations

Soil associations are groups of two or more soils, occurring together in a characteristic pattern in a given geographical area. Soils are grouped according to characteristics such as composition, organic content, particle size and shape, porosity and permeability, and maturity (i.e., the degree to which the soil has developed horizons). Soil associations form patterns that reflect the soil parent material, underlying bedrock, slope, vegetative cover, and plant litter in a region.

Climate and Landscapes

Climate refers to the characteristic weather of a region over an extended period of time, particularly the region's precipitation and temperature patterns. Temperature and precipitation affect the development of a landscape by influencing its relief, stream patterns, and soil associations.

Climate influences relief by controlling such factors as the type and rate of weathering, the amount and duration of the runoff that removes and transports weathered materials, and the amount, type, and distribution of plant cover.

The development of a landscape's relief depends on the balance between weathering and the removal of weathered materials by erosion. Temperature and precipitation strongly influence the intensity of weathering. Hot, arid regions and cold, arid regions undergo minimal weathering. Chemical weathering dominates in hot, humid regions, while physical weathering (mainly frost action during colder months) dominates in cold, humid regions. Thus, bedrock tends to be broken down at a faster rate in humid climates than in dry regions. Since running water is the predominant agent of erosion on Earth, precipitation is essential to the removal and transport of weathered materials. Thus, precipitation influences the rate at which the relief of the landscape changes. Over the long term, the more abundant the precipitation, the more rapidly the landscape's relief changes.

Climate also influences relief by controlling the amount, type, and distribution of plant cover. In humid climates, abundant rainfall promotes a protective cover of vegetation that protects the soil from rapid erosion by runoff and produces gently rounded slopes. In arid climates, with little protective vegetation, erosion by rapid runoff leads to steep slopes and exposed bedrock.

You may be surprised to learn that, while arid regions may erode more slowly than humid regions in the long term, the lack of plant cover in arid regions allows the exposed soil to be very rapidly eroded during the few periods of heavy rainfall that do occur. Thus, some of the most rapid erosion rates occur in deserts after a storm! However, over the long term, the brief periods of intense erosion in arid regions are outweighed by the prolonged periods of more moderate erosion in humid regions.

Climate influences stream patterns by controlling such factors as the amount and duration of precipitation, differences between the windward and leeward sides of mountains, and the relationship between precipitation and evaporation. For example, in arid climates, smaller streams tend to remain dry for most of the year, while humid climates tend to have more permanent streams. Thus, humid climates often have a denser network of tributary streams than arid climates. The leeward sides of mountains tend to have less annual precipitation than the windward sides. Thus, the network of streams may vary from one side of the mountains to another.

Soil associations are strongly influenced by climate. Arid climates tend to have soils that are thinner and contain less organic matter than those in humid climates. Since physical weathering is dominant, soil particles in arid climates tend to be larger and more angular than those in humid climates. The more abundant rainfall in humid climates results in greater infiltration, which, in turn, speeds up the rate at which soil horizons form. Thus, humid climates develop mature soils much more rapidly than arid climates. Moisture and higher temperatures also favor chemical weathering processes, which also speed the breakdown of weathered materials and the development of soil horizons.

Landscape Regions

Landscapes with similar relief, stream patterns, and soil associations can be grouped into distinctive **landscape regions**. Landscape regions typically have distinctive physical features that set them apart from one another. On a large scale, such as the continental United States, regions with similar landscape regions are grouped into **physiographic provinces** that are classified mainly as mountains, plateaus, and plains. See Figure 20.24.

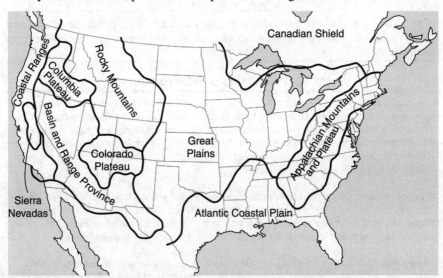

Figure 20.24 Physiographic Provinces of the United States. Source: *UPCO's Review of Earth Science*, Robert B. Sigda, United Publishing Co., 1995.

New York State is divided into a variety of smaller landscape regions based on relief, stream patterns, and the type and structure of the underlying bedrock. See Figure 20.25. The accompanying table gives a brief description of each of the landscape regions in New York State.

Deposition

Figure 20.25 Generalized Landscape Regions of New York State. Source: The State Education Department, *Earth Science Reference Tables*, 2011 ed. (Albany, New York; The University of the State of New York).

LANDSCAPE REGIONS OF NEW YORK STATE

Adirondack Mountains. A circular region that is part of the Grenville Province, a large belt of deeply eroded metamorphic bedrock of Proterozoic age. Erosion has stripped away most of the overlying rock, and it is criss-crossed with faults. The resistant rocks of the Adirondacks eroded more slowly than the rocks of the St. Lawrence and Champlain lowlands around them. The Adirondacks have long, straight valleys that formed along faults, gently curved ridges of resistant rock, and a radial drainage pattern.

St. Lawrence and Champlain Lowlands. The St. Lawrence and Champlain lowlands are part of the Interior Lowlands that extend west through the Great Plains. They consist of layers of sedimentary rock that once covered the Adirondacks but were later eroded away. They now form the low-lying plains occupied by the St. Lawrence River and Lake Champlain.

Taconic Mountains. The Taconic Mountains are a region of intensely folded and faulted metamorphic rocks that were thrust into that area from the east by the collision between a volcanic island arc and the North American continent about 450 million years ago. Ridges and valleys generally run north-south because softer rocks have worn away to produce valleys, while more resistant rocks form the ridges.

Allegheny Plateau. The Allegheny Plateau forms the northern end of the Appalachian Plateau to the southwest. It consists of flat-lying layers of sedimentary rock that were deposited in a warm, shallow sea that covered much of New York State during the Late Silurian and

Devonian. Later the entire area was uplifted to form the plateau. The plateau surface rises to the east until it becomes the Catskills. Streams and their tributaries have cut the surface of the plateau into hilly uplands.

The Catskills. The Catskills are the deeply eroded remains of a huge, apronlike delta formed during the Devonian from sediments that eroded from the ancient Acadian Mountains along the shore of the warm, shallow sea that covered much of New York State. Since then, streams have cut deeply into the sedimentary layers, forming steep-sided valleys that separate rounded "mountains."

Erie-Ontario Lowlands. The Erie-Ontario Lowlands are low, flat areas to the north and west of the Alleghany Plateau, and separated from it by escarpments, or cliffs. Though they are plains, the Erie-Ontario Lowlands are covered with hills of unsorted glacial till and sorted meltwater deposits.

Tug Hill Plateau. East of Lake Ontario, the land rises steadily to form the Tug Hill Plateau, a relatively flat, rocky area separated from the Adirondack Mountains by the Black River Valley. The plateau stands in the path of winter storms that pick up moisture from Lake Ontario. Annual snowfall averages 20 feet, so the plateau is one of the snowiest regions east of the Rocky Mountains.

Hudson Highlands. The Hudson Highlands consist of metamorphic rocks that were originally deposited as sedimentary and volcanic rocks 1.3 billion years ago. They were then metamorphosed into gneiss and marble during a collision of continents 1.1 billion years ago that thrust up an ancient mountain chain—the Grenville Mountains. The Grenville Mountains have since been eroded flat, exposing their roots—the rocks of the Hudson Highlands.

Hudson-Mohawk Lowlands. This region covers most of the Hudson River Valley and the Mohawk River Valley. The bedrock consists of relatively soft sedimentary rocks that are easily eroded. The surrounding highlands are made of more resistant rocks. These lowlands provide the nation's only natural, navigable waterway through the Appalachian Mountains and serve as an important transportation route between the Atlantic and the Great Lakes.

Newark Lowlands. The Newark Lowlands form a gently rolling surface broken by ridges, between the Hudson Highlands and the Manhattan Prong. Formed on layers of igneous and sedimentary rock dating from the Age of Dinosaurs (Triassic-Jurassic), the surface of these lowlands is broken by ridges of igneous rock that is more resistant to erosion. The Palisades Sill intruded into the rocks of the Newark Lowlands 195 million years ago.

Manhattan Prong. The Manhattan Prong consists of metamorphic rock that was folded and faulted by the same collision that formed the Taconic Mountains and was later eroded by glacial ice into a landscape of rolling hills and valleys. Resistant gneiss, schist, and quartzite form the hills, and less resistant marble underlies the valleys.

Atlantic Coastal Plain. The Atlantic Coastal Plain is a flat, low-lying area that slopes toward the Atlantic Ocean. In New York State, it consists of Staten Island and Long Island, both of which are important residential areas. Long Island is composed mainly of glacial sediments. Two terminal moraines form its hilly northern edge while a broad, flat glacial outwash plain stretches from the terminal moraine to the Atlantic Ocean. Long offshore barrier islands with broad beaches make Long Island a popular recreational area.

Deposition

MULTIPLE-CHOICE QUESTIONS

In each case, write the number of the word or expression that best answers the question or completes the statement.

1. Sediment is deposited as a river enters a lake because the
   (1) velocity of the river decreases
   (2) force of gravity decreases
   (3) volume of water increases
   (4) slope of the river increases

2. A river's current carries sediments into the ocean. Which sediment size will most likely be deposited in deeper water farthest from the shore?
   (1) pebble   (3) silt
   (2) sand   (4) clay

3. The largest particles that a stream deposits as it enters a pond are 8 centimeters in diameter. The minimum velocity of the stream is approximately
   (1) 100 cm/sec   (3) 300 cm/sec
   (2) 200 cm/sec   (4) 400 cm/sec

4. A stream is transporting the particles W, X, Y, and Z, shown below.

Density = 3.8 g/mL   Density = 3.8 g/mL   Density = 2.4 g/mL   Density = 2.4 g/mL

   Which particle will most likely settle to the bottom first as the velocity of this stream decreases?
   (1) W (3) Y
   (2) X (4) Z

5. When small particles settle through water faster than large particles, the small particles are probably
   (1) lighter   (3) better sorted
   (2) flatter   (4) more dense

6. Pieces of bedrock material that are broken from a cliff and deposited by a landslide at the base of the cliff are best described as
   (1) rounded and sorted
   (2) rounded and unsorted
   (3) angular and sorted
   (4) angular and unsorted

7. When wind and running water gradually decrease in velocity, the transported sediments are deposited
   (1) all at once, and are unsorted
   (2) all at once, and are sorted by size and density
   (3) over a period of time, and are unsorted
   (4) over a period of time, and are sorted by size and density

8. Which cross section best represents the pattern of sediments deposited on the bottom of a lake as the velocity of the stream entering the lake steadily decreased?

9. The satellite photograph below shows the Mississippi River entering into the Gulf of Mexico. Arrows show the direction of river flow.

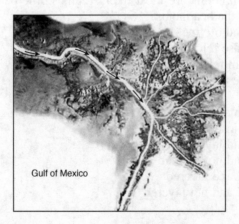

This depositional feature in the Gulf of Mexico is best identified as
   (1) a delta
   (2) a sandbar
   (3) a barrier island
   (4) an outwash plain

Deposition

Base your answers to questions 10 through 13 on the diagram below. The arrows show the direction in which sediment is being transported along the shoreline. A barrier beach has formed, creating a lagoon (a shallow body of water in which sediments are being deposited). The eroded headlands are composed of diorite bedrock. A groin has recently been constructed. Groins are wall-like structures built into the water perpendicular to the shoreline to trap beach sand.

10. The groin structure will change the pattern of deposition along the shoreline, initially causing the beach to become
    (1) wider on the western side of the groin
    (2) wider on the eastern side of the groin
    (3) narrower on both sides of the groin
    (4) wider on both sides of the groin

11. Which two minerals are most likely found in the beach sand that was eroded from the headlands?
    (1) quartz and olivine
    (2) plagioclase feldspar and amphibole
    (3) potassium feldspar and biotite
    (4) pyroxene and calcite

12. The sediments that have been deposited by streams flowing into the lagoon are most likely
    (1) sorted and layered
    (2) sorted and not layered
    (3) unsorted and layered
    (4) unsorted and not layered

13. Which event will most likely occur during a heavy rainfall?
(1) Less sediment will be carried by the streams.
(2) An increase in sea level will cause more sediment to be deposited along the shoreline.
(3) The shoreline will experience a greater range in tides.
(4) The discharge from the streams into the lagoon will increase.

Base your answers to questions 14 through 17 on the contour map below, which shows a hill formed by glacial deposition near Rochester, New York. Letters *A* through *E* are reference points. Elevations are in feet.

14. This glacial deposit is best identified as
(1) a U-shaped valley
(2) a sand dune
(3) a drumlin
(4) an outwash plain

15. Which description best compares the gradients of this hill?
(1) *AE* and *EB* have the same gradient.
(2) *AE* has a steeper gradient than *EB*.
(3) *CE* has a steeper gradient than *ED*.
(4) *CE* and *AE* have the same gradient.

16. Which set of characteristics most likely describes the sediment in this glacial deposit?
(1) sorted and layered
(2) sorted and not layered
(3) unsorted and not layered
(4) unsorted and layered

17. The hill shown on this map is found in which New York State landscape region?
(1) Adirondack Mountains
(2) Catskills
(3) Atlantic Coastal Plain
(4) Erie-Ontario Lowlands

Deposition

Base your answers to questions 18 through 20 on the map of Long Island, New York. *AB*, *CD*, *EF*, and *GH* are reference lines on the map.

18. Which agent of erosion transported the sediments that formed the moraines shown on the map?
    (1) water
    (2) wind
    (3) ice
    (4) mass movement

19. The cross section below represents the sediments beneath the land surface along one of the reference lines shown on the map.

Along which reference line was the cross section taken?
    (1) *AB*
    (2) *CD*
    (3) *EF*
    (4) *GH*

20. A major difference between sediments in the outwash and sediments in the moraines is that the sediments deposited in the outwash are
    (1) larger
    (2) sorted
    (3) more angular
    (4) older

21. What will be the most probable arrangement of rock particles deposited directly by a glacier?
    (1) sorted and layered
    (2) sorted and not layered
    (3) unsorted and layered
    (4) unsorted and not layered

22. Which statement presents the best evidence that a boulder-sized rock is an erratic?
    (1) The boulder has a rounded shape.
    (2) The boulder is larger than surrounding rocks.
    (3) The boulder differs in composition from the underlying bedrock.
    (4) The boulder is located near potholes.

23. The cross sections below show a three-stage sequence in the development of a glacial feature.

Which glacial feature has formed by the end of stage 3?
(1) kettle lake
(2) finger lake
(3) drumlin
(4) parallel scratches

Base your answers to questions 24 through 26 on the diagram below, which shows the edge of a continental glacier that is receding. *R* indicates elongated hills. The ridge of sediments from *X* to *Y* represents a landscape feature.

24. The elongated hills labeled *R* are most useful in determining the
(1) age of the glacier
(2) direction the glacier has moved
(3) thickness of the glacier
(4) rate at which the glacier is melting

25. Which feature will most likely form when the partially buried ice block melts?
(1) drumlin
(2) moraine
(3) kettle lake
(4) finger lake

Deposition

26. The ridge of sediments from X to Y can best be described as
    (1) sorted and deposited by ice
    (2) sorted and deposited by meltwater
    (3) unsorted and deposited by ice
    (4) unsorted and deposited by meltwater

27. The particles in a sand dune deposit are small, are very well sorted, and have surface pits that give them a frosted appearance. This deposit was most likely transported by
    (1) ocean currents   (3) gravity
    (2) glacial ice   (4) wind

28. Which diagram represents a side view of a sand dune most commonly formed as a result of the prevailing wind direction shown?

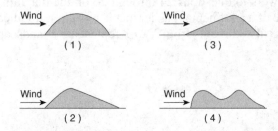

29. New York State landscape regions are identified and classified primarily by their
    (1) surface topography and bedrock structure
    (2) existing vegetation and type of weather
    (3) latitude and longitude
    (4) chemical weathering rate and nearness to large bodies of water

30. The table below describes the characteristics of three landscape regions, A, B, and C, found in the United States.

Landscape	Bedrock	Elevation/Slopes	Streams
A	Faulted and folded gneiss and schist	High elevation Steep slopes	High velocity Rapids
B	Layers of sandstone and shale	Low elevation Gentle slopes	Low velocity Meanders
C	Thick horizontal layers of basalt	Medium elevation Steep to gentle slopes	High to low velocity Rapids and meanders

Which list best identifies landscapes A, B, and C?
(1) A—mountain, B—plain, C—plateau
(2) A—plain, B—plateau, C—mountain
(3) A—plateau, B—mountain, C—plain
(4) A—plain, B—mountain, C—plateau

31. Which sequence shows the order in which landscape regions are crossed as an airplane flies in a straight course from Albany, New York, to Massena, New York?
    (1) plateau→plain→mountain
    (2) plateau→mountain→plain
    (3) plain→mountain→plain
    (4) mountain→plain→plateau

32. The Catskills landscape region is classified as a plateau because it has
    (1) low elevations and mostly faulted or folded bedrock
    (2) low elevations and mostly horizontal bedrock
    (3) high elevations and mostly faulted or folded bedrock
    (4) high elevations and mostly horizontal bedrock

33. Which two New York State landscape regions have surface bedrock that formed about 1,000 million years ago?
    (1) Hudson Highlands and Adirondack Mountains
    (2) Erie-Ontario Lowlands and Atlantic Coastal Plain
    (3) Tug Hill Plateau and Allegheny Plateau
    (4) Newark Lowlands and Manhattan Prong

34. The Adirondacks are classified as mountains because of the high elevation and bedrock that consists mainly of
    (1) deformed and intensely metamorphosed rocks
    (2) glacial deposits of unconsolidated gravels, sands, and clays
    (3) Cambrian and Ordovician quartzites and marbles
    (4) horizontal sedimentary rocks of marine origin

35. The map below shows part of a stream drainage pattern.

Which topographic map best shows the contour lines for this stream drainage pattern?

36. The block diagram below represents the drainage basins of some river systems separated by highland divides, shown with dashed lines. The arrows show the directions of surface-water flow.

The three areas separated by highland divides are called
(1) meanders
(2) floodplains
(3) watersheds
(4) tributaries

Base your answers to questions 37 through 39 on the map below, which shows watershed regions of New York State.

37. On which type of landscape region are both the Susquehanna-Chesapeake and the Delaware watersheds located?
    (1) plain
    (2) plateau
    (3) mountain
    (4) lowland

38. In which watershed is the Genesee River located?
    (1) Ontario-St. Lawrence
    (2) Susquehanna-Chesapeake
    (3) Mohawk-Hudson
    (4) Delaware

39. Most of the surface bedrock of the Ontario–St. Lawrence watershed was formed during which geologic time periods?
    (1) Precambrian and Cambrian
    (2) Ordovician, Silurian, and Devonian
    (3) Mississippian, Pennsylvanian, and Permian
    (4) Triassic, Jurassic, and Cretaceous

Deposition

CONSTRUCTED RESPONSE QUESTIONS

Base your answers to questions 40 through 43 on the block diagram below and on your knowledge of Earth science. The diagram represents a meandering stream flowing into the ocean. Points A and B represent locations along the streambanks. Letter C indicates a triangular-shaped depositional feature where the stream enters the ocean.

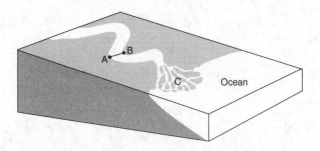

40. The top of the box below represents the stream surface between points A and B. In the box, draw a line from point A to point B to represent a cross-sectional view of the shape of the bottom of the stream channel. [1]

41. Explain how sediments eroded by the water in this stream become smoother and rounder in shape. [1]

42. Identify the triangular-shaped depositional feature indicated by letter C. [1]

43. Identify two factors that determine the rate of stream erosion. [1]

Base your answers to questions 44 through 47 on the map below and on your knowledge of Earth science. The map shows the location of Sandy Creek, west of Rochester, New York. *X* and *Y* represent points on the banks of the stream.

44. On the diagram below, draw a line to represent the shape of the stream bottom from point *X* to point *Y*. [1]

45. Explain why sediments are deposited when Sandy Creek enters Lake Ontario. [1]

46. The symbols representing four sediment particles are shown in the key below. These particles are being transported by Sandy Creek into Lake Ontario. On the cross section below, draw the symbols on the bottom of Lake Ontario to show the relative position where *each* sediment particle is most likely deposited. [1]

47. Record the minimum velocity this stream needs to transport a 2.0-cm-diameter particle. [1]

Deposition

Base your answers to questions 48 through 51 on maps *A*, *B*, and *C* below, which show evidence that much of New York State was once covered by a glacial ice sheet. Map *A* shows the location of the Finger Lakes Region in New York State. The boxed areas on map *A* were enlarged to create maps *B* and *C*. Map *B* shows a portion of a drumlin field near Oswego, New York. Map *C* shows the locations of glacial moraines and outwash plains on Long Island, New York.

48. The arrangement of the drumlins on map *B* indicates that a large ice sheet advanced across New York State in which compass direction? [1]

49. The diagrams below represent three sediment samples labeled *X*, *Y*, and *Z*. These samples were collected from three locations marked with empty boxes (☐) on map *C*.

Sample X — Sorted particle-size range: 0.005–0.09 cm

Sample Y — Unsorted particle-size range: 0.01–62 cm

Sample Z — Sorted particle-size range: 0.1–0.3 cm

(Not drawn to scale)

Write the letter of each sample in the correct box on map *C* to indicate the location from which each sample was most likely collected. [1]

50. The drawing below shows a glacial erratic found on the beach of the north shore of Long Island near the Harbor Hill moraine. This boulder is composed of one-billion-year-old gneiss. Which New York State landscape region has surface bedrock similar in age to this erratic? [1]

Glacial erratic

51. Explain how the effect of global warming on present-day continental glaciers could affect New York City and Long Island. [1]

Deposition

Base your answers to questions 52 through 54 on the map below, which shows the generalized surface bedrock for a portion of New York State that appears in the *2011 Edition Reference Tables for Physical Setting/Earth Science*.

52. Place an **X** on the map to represent a location in the Tug Hill Plateau landscape region. [1]

53. State the longitude of Mt. Marcy, New York, to the nearest degree. The units and compass direction must be included in your answer. [1]

54. Identify the geologic age and name of the surface metamorphic bedrock found at Mt. Marcy. [1]

EXTENDED CONSTRUCTED RESPONSE QUESTIONS

Base your answers to questions 55 through 59 on the map and table below and on your knowledge of Earth science. The map shows the area where the Battenkill River flows into the Hudson River north of Albany, New York. Point A indicates a location within the Battenkill River. The table shows the densities of four common minerals found in Hudson River sediments.

55. Identify the diameter of the largest particle that would be carried at point A when the velocity of the Battenkill River is 50 cm/s. [1]

56. Describe the most likely changes in the size and shape of individual particles of sediment as they are transported downstream by the Battenkill and Hudson Rivers. [1]

57. Describe the arrangement of the sediments being deposited by these rivers. [1]

58. Some of the sediments transported by the Hudson River came from metamorphic rock. Identify *one* foliated metamorphic rock that contains all four minerals listed in the mineral density table. [1]

Deposition

59. Samples of minerals listed in the mineral density table with the same shape and size were removed from the Hudson River and placed in a jar of water. After the jar was shaken, the sediments were allowed to settle. *On the diagram below*, write the mineral name from the table next to the layer in the diagram where *each* mineral is most likely found. [1]

A: _____

B: _____

C: _____

D: _____

Base your answers to questions 60 through 64 on the landscape diagram below and on your knowledge of Earth science. The diagram represents a long river system from its origin (source) in the mountains to its end (mouth) at the ocean.

60. Describe one characteristic of the coastal plain that caused the river to develop meanders. [1]

61. Identify one change that would cause an increase in the rate of stream erosion in the river valley in the mountains. [1]

62. Explain why the sediments deposited in the delta are arranged in layers. [1]

63. List two processes that would change the accumulated sediments in the delta into sedimentary rock. [1]

64. State one reason for the restriction of the construction of buildings near a meandering river on a coastal plain. [1]

Base your answers to questions 65 through 69 on the passage and the cross section below. The passage describes the geologic history of the Pine Bush region near Albany, New York. The cross section shows the bedrock and overlying sediment along a southwest to northeast diagonal line through a portion of this area. Location A shows an ancient buried stream channel, and location B shows a large sand dune.

The Pine Bush Region

The Pine Bush region, just northwest of Albany, New York, is a 40-square-mile area of sand dunes and wetlands covered by pitch pine trees and scrub oak bushes. During the Ordovician Period, this area was covered by a large sea. Layers of mud and sand deposited in this sea were compressed into shale and sandstone bedrock.

During most of the Cenozoic Era, running water eroded stream channels into the bedrock. One of these buried channels is shown at location A in the cross section. Over the last one million years of the Cenozoic Era, this area was affected by glaciation. During the last major advance of glacial ice, soil and bedrock were eroded and later deposited as till (a mixture of boulders, pebbles, sand, and clay).

About 20,000 years ago, the last glacier in New York State began to melt. The meltwater deposited pebbles and sand, forming the stratified drift. During the 5000 years this glacier took to melt, the entire Pine Bush area became submerged under a large 350-foot-deep glacial lake called Lake Albany. Delta deposits of cobbles, pebbles, and sand formed along the lake shorelines, and beds of silt and clay were deposited farther into the lake.

Lake Albany drained about 12,000 years ago, exposing the lake bottom. Wind erosion created the sand dunes that cover much of the Pine Bush area today.

Deposition

65. According to the passage, how old is the bedrock shown in the cross section? [1]

66. What evidence shown at location *A* suggests that the channel in the bedrock was eroded by running water? [1]

67. List, from oldest to youngest, the four types of sediment shown above the bedrock in the cross section. [1]

68. Explain why the till layer is composed of unsorted sediment. [1]

69. How does the shape of the sand dune at location *B* provide evidence that the prevailing winds that formed this dune were blowing from the southwest? [1]

Unit Seven

THE ATMOSPHERE, WEATHER, AND CLIMATE

CHAPTER 21
THE ATMOSPHERE

KEY IDEAS The atmosphere can be thought of as a huge heat engine that converts heat energy into mechanical energy. Earth's atmospheric heat engine is powered primarily by solar energy and is influenced by gravity.

When the various electromagnetic waves that emanate from the Sun strike the atmosphere, only a small percentage are directly absorbed, especially by gases such as ozone, carbon dioxide, and water vapor. Clouds and Earth's surface reflect some energy back into space, and Earth's surface absorbs some. The energy absorbed is then transferred between Earth's surface and the atmosphere by conduction, convection, radiation, and evaporation.

The transfer of heat energy from solar radiation, and of water vapor, into and out of the atmosphere causes regions of different densities to form. The rise or fall of regions of different densities due to the action of gravitational force produces atmospheric circulation, which is affected by Earth's rotation. Atmospheric circulation distributes solar energy over the whole Earth. The interaction of these processes results in the complex atmospheric occurrence known as weather.

KEY OBJECTIVES
Upon completion of this unit, you will be able to:

- Explain why the Sun emits different types of electromagnetic radiation.
- Compare and contrast the different forms of electromagnetic radiation shown in the *Reference Tables for Physical Setting/Earth Science*.
- Describe the mechanisms by which solar energy is absorbed by the atmosphere and the factors that determine the amount of solar energy an area receives.
- Explain how density differences in the atmosphere cause atmospheric circulation.
- Explain how Earth's rotation influences atmospheric circulation (a phenomenon known as the Coriolis effect), and interpret a diagram of the planetary wind and pressure belts.
- Explain how energy can be stored or released during phase changes.

The Atmosphere

SOLAR RADIATION: THE ATMOSPHERE'S PRIMARY ENERGY SOURCE

The Sun is the atmosphere's major source of energy. Energy from the Sun reaches Earth in the form of electromagnetic waves.

Energy from Electromagnetic Waves

A magnet can make a compass needle move from a distance because the magnet is surrounded by an invisible magnetic field. If you move the magnet, the magnetic field will move and the moving magnetic field, in turn, will cause the compass needle to move. In much the same way, a statically charged balloon held near your head will attract your hair from a distance because the balloon is surrounded by an invisible electric field. If you move the balloon, the invisible electric field moves with it and you can feel the effect of the moving field on your hair. Such fields extend outward infinitely in all directions from their sources, but they weaken with distance from these sources (see Figure 21.1).

All matter is composed of atoms. Every atom consists of electrically charged particles such as protons and electrons, which are surrounded by an electric field. Whenever a charged particle moves, a magnetic force is produced and the particle is then also surrounded by a magnetic field. These two fields—an electric field and a magnetic field—exist simultaneously around all electrically charged particles that are moving. Together, they are called an **electromagnetic field**, and they extend outward infinitely in all directions around the particles.

Figure 21.1 An Invisible Magnetic Field Surrounding a Magnet. The lines represent magnetic force; closely spaced, the lines mean stronger magnetic force. Note that the strength of the magnetic field decreases with distance.

When a particle moves back and forth, its electromagnetic field moves with it (see Figure 21.2a). Like a ripple in a pond, this movement spreads out through the field in the form of a transform wave. The way that the wave moves through the electromagnetic field depends on the way in which the particle moves. The faster the particle oscillates, the shorter the wavelength, or distance between successive "ripples" in the electromagnetic field. Every frequency of oscillation produces a wave of a different wavelength (see Figure 21.2b).

A moving wave contains energy. It can exert forces on matter with which it interacts. Since an electromagnetic field does not require a medium to exist, it can extend through space. Disturbances in electromagnetic fields

(a) As a magnet moves, its magnetic field moves too. Like a ripple in a pond, the movement of the field spreads outward in all directions at a rate of 3×10^8 meters per second—the speed of light. When the moving field reaches *another* field (such as the one around a compass needle), it exerts a force of attraction or repulsion that can cause motion of the field and even of the object that produced the field. A similar disturbance would travel through the electric field around an electrically charged particle in motion.

(b) Electromagnetic waves produced by particles vibrating at different rates. The higher the temperature, the faster the particle vibrates, and the shorter the wavelength produced.

Figure 21.2 Electromagnetic Waves Are Produced by Vibrating Particles.

around particles can travel through space as they move outward, or radiate, through the field. In this way, energy is transferred from the Sun to Earth without the existence of a physical medium between the two.

The Electromagnetic Spectrum

Electromagnetic waves are classified by their lengths and range from short waves, such as X-rays, to long waves, such as radio waves. The **electromagnetic spectrum** (see Figure 21.3) is a continuum in which electromagnetic waves are arranged in order, from longest to shortest wavelength. Infrared (heat) waves and visible-light waves fall roughly in the middle range of wavelengths. The wavelength, or distance between "ripples" of visible light waves, falls between 10^{-6} and 10^{-7} meter.

The Atmosphere

Figure 21.3 Electromagnetic Spectrum. This shows the categories into which electromagnetic waves are classified according to their lengths. Source: *Earth Science Reference Tables*—2011 Edition.

Many different waves emanate from the Sun (see Figure 21.4) because it contains particles moving at many different speeds. However, most waves coming from the Sun are in the visible-light and infrared ranges. To move at the speed needed to emit waves of this length, particles at the Sun's surface must have temperatures between 5,000K and 7,000K. Earth's magnetic field and the Van Allen belts of charged particles in the upper atmosphere deflect many of the waves with short lengths. This circumstance is fortunate, since most short-wavelength radiation is harmful to living things.

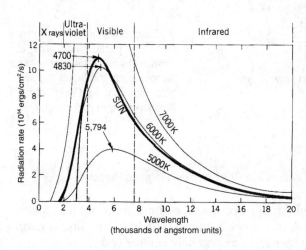

Figure 21.4 Makeup of Solar Radiation. The Sun emits electromagnetic radiation ranging in wavelength from infrared to X-rays. The majority of the waves are in the visible-light range with a peak at 4,700 angstrom units (1 angstrom unit = 1×10^{-9} meter—one billionth of a meter). This is very close to the theoretical output of a body at a temperature at 6,000K, which is why the Sun's surface temperature is thought to be between 5,000 and 7,000K.

The Global Radiation Budget

When Earth intercepts electromagnetic waves from the Sun, the moving waves exert a force on the fields surrounding particles of matter in the atmosphere, hydrosphere, and lithosphere, causing the particles to move. This increased motion shows up as an increase in the level of Earth's heat energy. At the same time, Earth's matter radiates energy in the form of electromagnetic waves back out into space, causing a decrease in the level of Earth's heat energy. Thus, both incoming solar radiation and Earth's outgoing, or terrestrial, radiation pass through the atmosphere. Table 21.1 shows the global radiation budget.

TABLE 21.1 GLOBAL RADIATION BUDGET*

Incoming Solar Radiation	Percent Gain	Percent Loss
Reflection from clouds to space		21
Diffuse reflection (scattering) to space		5
Direct reflection from Earth's surface		6
Net energy loss back to space		32
Absorbed by clouds	3	
Absorbed by molecules, dust, water vapor, and CO_2	15	
Absorbed by Earth's surface	50	
Net incoming energy gained by Earth-atmosphere system	**68%**	

Outgoing Terrestrial Radiation	Percent Gain	Percent Loss
Total infrared radiation emitted by Earth's surface	98	
Absorbed by atmosphere	90	
Lost to space		8
Total infrared radiation emitted by the atmosphere	137	
Absorbed by Earth's surface	77	
Lost to space		60
Net outgoing energy lost from entire Earth-atmosphere system		**68%**
Net energy leaving *Earth's surface* (98% emitted − 77% reabsorbed)		21
Net energy leaving the *atmosphere* (137% emitted − 90% reabsorbed)		47

*Source: *The Earth Sciences*, Arthur N. Strahler, Harper & Row, 1971.

Radiative Balance

Earth's average global temperature depends upon the balance between the energy gained by absorbing sunlight and the energy lost by radiating heat. As you can see in Table 21.1, the Earth-atmosphere system's net energy gain from incoming solar radiation is the same as its net energy loss by terrestrial radiation—68 percent. Over long periods of time, then, the average level of Earth's heat energy remains fairly constant, and the average temperature of Earth's surface hardly changes. Therefore, we say that Earth is in **radiative balance**; it is reradiating as much energy as it absorbs.

This fact, however, *does not mean that all points on Earth's surface are in radiative balance at all times*. Most places on Earth go through temperature changes on both a daily and a yearly basis. These temperature changes indicate that there are times when a place is gaining more energy than it is losing and times when it is losing more energy than it is gaining.

There is also much evidence that over long periods of time Earth experiences warming and cooling trends. These may be triggered by changes in Earth's tilt due to precession as it spins on its axis and to increases or decreases in certain gases in the atmosphere.

Let us now consider some of the factors that influence the way in which Earth absorbs incoming solar radiation.

The Atmosphere

FACTORS AFFECTING INSOLATION

The Sun provides nearly all of the energy received at Earth's surface. Solar radiation reaches the upper atmosphere at a fairly constant rate of about 200 kilocalories per minute per square meter. About one-third of this radiation is reflected back into space, mostly by clouds. As the remaining radiation passes through the atmosphere, some is absorbed by gas molecules. Some is refracted as it crosses boundaries between layers of differing densities. Some is scattered by particles of dust or aerosols in the air. The radiation that reaches Earth's surface, called **insolation**, short for *in*coming *sol*ar radi*ation*, is either reflected or absorbed.

A number of factors control the amount of solar energy that an area absorbs or reflects, including the angle at which insolation strikes the surface, the length of time each day that insolation is received (the duration of insolation), and the nature of the surface.

Angle of Insolation

Latitude

Since Earth is a sphere, insolation does not strike all points on Earth's surface at the same angle. Near the Equator insolation strikes the surface almost vertically, but near the poles it strikes at a more glancing angle. Therefore, the insolation reaching the tropics is more concentrated than that reaching polar regions (see Figure 21.5).

Figure 21.5 Insolation near the Equator and the North Pole. Note that the two beams of sunlight approaching Earth are identical, but the angle at which they strike Earth's surface causes insolation received near the Equator to be more concentrated than that received near the poles. The result is greater heating at the Equator.

Daily and Annual Cycles

The angle at which insolation strikes Earth's surface at any location also varies in both daily and annual cycles. The daily cycle starts at dawn with insolation striking the surface at a very low angle; it then becomes increasingly direct throughout the morning, reaches its greatest directness at noon, and becomes increasingly indirect again throughout the afternoon until sunset.

The annual cycles vary with latitude and Earth's position in its orbit around the Sun. In the United States, insolation is most direct on June 21, the summer solstice, and least direct on December 21, the winter solstice.

Figure 21.6a shows the daily cycle of intensity of solar radiation; Figure 21.6b, the yearly cycle of intensity of solar radiation at noon.

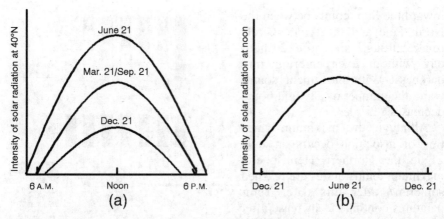

Figure 21.6 Daily and Yearly Cycles of Solar Radiation Intensity. (a) Daily cycle of intensity of solar radiation from 6 A.M. to 6 P.M. at 40° N on the equinox and solstice days. Note that for each day there is a cyclic change from low intensity in early morning to higher intensity at noon and then back to low intensity in late afternoon. (b) Yearly cycle of intensity of solar radiation at noon for an entire year. Note the cyclic change from low intensity at noon on the winter solstice to high intensity at noon on the summer solstice and back to low intensity at the next winter solstice.

Duration of Insolation

The length of time that the surface receives insolation each day, or the **duration of insolation**, depends on the season and the latitude of the location.

Season

The length of daylight varies in an annual cycle with the season (see Figure 21.7). In the United States, the greatest number of hours of insolation is received on June 21, the summer solstice. Each day thereafter, the number of daylight hours decreases, reaching a minimum on December 21, the winter solstice. Then the number of daylight hours increases daily until it peaks again the following June 21. At the fall and spring equinoxes, the number of daylight hours equals the number of hours of darkness. New York State varies from roughly 15 hours of daylight at the summer solstice to only 9 hours at the winter solstice.

Latitude

The length of daylight on any given day also varies with location north or south of the Equator (see Figure 21.8). Since Earth's axis of rotation is tilted, the circles that form the parallels of latitude are also tilted. At different latitudes, different fractions of the parallels are in daylight and darkness. Thus, observers at different latitudes spend different fractions of each 24-hour rotation in daylight and in darkness. When the Northern Hemisphere is tilted

The Atmosphere

toward the Sun, points between the North Pole and the Arctic Circle rotate through an entire 24-hour day without ever entering into darkness. With movement southward, the number of daylight hours decreases.

Although the maximum duration of insolation occurs on June 21 in the Northern Hemisphere, maximum temperatures are reached sometime after this date. Temperatures continue to increase after June 21 because the number of daylight hours still exceeds the number of nighttime hours for many weeks after this date. As long as more energy is received each day than is lost overnight, the temperature will continue to increase.

Nature of Earth's Surface

Earth's surface consists of a wide variety of substances, including ice, water, soil of many types, and vegetation. All interact differently with the insolation that strikes them.

Figure 21.7 Duration of Day and Night at Different Locations on Key Dates Throughout the Year.

Color and Texture

Just as a ball is more likely to bounce off a concrete wall than off a pillow, insolation striking different substances will have a greater or lesser tendency to be absorbed or reflected. Which process will predominate depends on the molecular structure of the substance. The color of a substance is a fairly good indicator of whether the substance absorbs more insolation than it reflects or vice-versa.

The lighter the color of a substance, the more light is being reflected by it; the darker the color, the more light is being absorbed. When insolation strikes substances such as sand and snow, their light color indicates that much of the insolation is being reflected. On the other hand, when insolation strikes rich black soil or deep green leaves, their dark color indicates that much of the insolation is being absorbed. In general, dark-colored substances absorb more insolation than light-colored ones, and the more insolation a substance absorbs the more it is heated.

The word **texture** refers to the smoothness or roughness of a surface. On a smooth surface, insolation is more likely to be reflected away from the surface than on a rough surface. The irregularities and indentations on a rough surface cause some of the insolation to be reflected in such a direction that it hits the surface a second, or even a third, time before leaving (see Figure 21.9). Each time the insolation strikes the surface, a little more of its energy is absorbed. Therefore, if all other characteristics of two surfaces are the same, a rough surface will absorb more insolation than a smooth one.

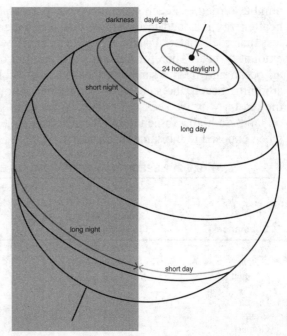

Figure 21.8 Variations in the Lengths of Day and Night Because of Earth's Tilted Axis of Rotation.

Specific Heat

When different substances absorb equal amounts of heat energy, they do not all change temperature by the same number of degrees. **Specific heat** is the number of joules of heat needed to cause a 1°C temperature change in 1 g of a substance. The difference can be thought of as the result of a kind of molecular inertia. Heavy molecules and tightly held molecules require more energy to get them moving than light and loosely held molecules. The temperature of a substance is strictly a result of how fast the molecules are moving. If you put the same amount of energy into two substances, one with tightly held molecules and one with loosely held molecules, the loosely held molecules would move faster and that substance would register a higher temperature.

Figure 21.9 Insolation Striking (a) a Smooth Surface and (b) a Rough Surface.

Table 21.2 shows the specific heat of different Earth materials and also the number of degrees each substance will increase in temperature if 1 joule of heat energy is added to it. Notice that water increases in temperature much less than rock material such as granite or basalt with the addition of the same amount of heat. Therefore, if the same amount of insolation is absorbed by an ocean and by the sand-sized particles of broken rock that make up the adjoining beach, the sand will become much hotter than the water. (This is the reason why we seek out bodies of water to swim in when temperatures on land are high.) In general, land increases in temperature more than water when exposed to the same insolation.

TABLE 21.2 SPECIFIC HEAT OF COMMON EARTH MATERIALS

Substance	Specific Heat (Joules/g • °C)	Temperature Increase if 1 Joule of Heat Is Added to 1 Gram of the Substance (°C)
Water		
Solid	2.11	0.47
Liquid	4.18	0.24
Gas	2.0	0.5
Dry air	1.01	1.0
Basalt	0.84	1.2
Granite	0.79	1.3
Marble	0.88	1.1
Iron	0.45	2.2
Copper	0.38	2.6
Lead	0.13	7.7

Specific heat of common substances

Latent Heat of Water

When water is melting or evaporating, it absorbs energy with no resulting change in temperature. The energy is used to break internal bonds rather than to increase the speed at which water molecules are moving. Since this heat causes no change in temperature, it is called **latent heat** (*latent* means "hidden"). As a result, ice that is melting or water that is evaporating from ocean surfaces absorbs insolation without increasing in temperature. Therefore, ice-covered land and oceans remain cooler than adjacent land areas.

Latent heat is not an issue on land because the substances of which land is composed do not melt or evaporate in the normal range of temperatures at Earth's surface. However, since water is so abundant and exists in all three phases on Earth, latent heat has an important effect on the temperature changes that occur when insolation strikes water.

HOW ENERGY ENTERS THE ATMOSPHERE

As you have learned, the atmosphere does not absorb much solar energy directly. How, then, is the atmosphere heated? Most insolation passes right through the atmosphere to Earth's surface, where it is absorbed and changed into forms of energy that the atmosphere can absorb. Earth's surface, then, heats the atmosphere. Let us now consider how energy moves from Earth's surface into the atmosphere and, once there, how it moves within the atmosphere.

Conduction

When Earth's surface absorbs solar energy, the absorbed energy causes the molecules at the surface to move more rapidly. We perceive this absorbed energy as heat and measure it in terms of the temperature of the substance as shown by a thermometer. Since Earth's surface is in continuous contact with the base of the atmosphere, its surface molecules strike molecules of gases in air. During these collisions, some heat energy from Earth's surface molecules is transferred to the gas molecules of the atmosphere. This kind of energy transfer, in which heat energy moves from molecule to molecule by collisions, is known as **conduction**. Conduction is important because it is the chief method by which energy moves from Earth's surface to the atmosphere.

Convection

Within the atmosphere, heat spreads very slowly by conduction. Therefore, the air touching the ground becomes warmer and warmer. As this air warms, it expands and becomes less dense than the surrounding air. Finally, a bubble of warm air rises, and cooler air settles to the ground in its place, a type of movement known as **convection**. Since the rising bubble of warm air carries the heat it contains upward with it, convection transfers heat energy *within* the atmosphere. Convection is important because it is the chief method by which energy conducted into the atmosphere from Earth's surface is then carried throughout the atmosphere.

Radiation

Energy from the Sun reaches Earth by **radiation**, the release and transfer of energy in the form of electromagnetic waves. As described earlier, all moving atoms and molecules emit electromagnetic radiation, and the wavelength emitted depends upon how fast these particles are moving. However, since both Earth and its atmosphere are much cooler than the Sun, the energy they radiate has much longer wavelengths. Most of the energy radiated by Earth's surface is infrared radiation.

This fact is important because, although the atmosphere is transparent to most of the Sun's short-wavelength radiation, it is not as transparent to infrared radiation. Carbon dioxide, water vapor, methane, and ozone in the atmosphere absorb or reflect most of Earth's infrared radiation; the rest goes

The Atmosphere

through the atmosphere and out into space. Thus, short wavelengths can readily enter the atmosphere, but long wavelengths cannot readily escape, a phenomenon known as the **greenhouse effect** (see Figure 21.10).

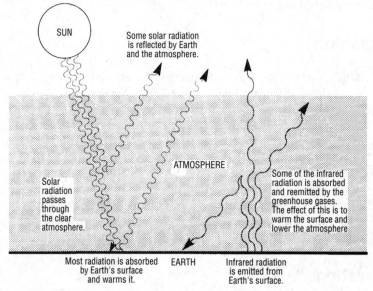

Figure 21.10 The Greenhouse Effect. This sequence of events is called the *greenhouse effect* because the glass in a greenhouse allows light to enter but blocks heat from escaping. Source: *Climate Change*, The Intergovernmental Panel on Climate Change, World Meteorological Organization/UN Environmental Programme.

This term is appropriate because, just as glass surrounds a greenhouse, the atmosphere surrounds Earth. Glass, like the atmosphere, allows visible light rays to pass through it but blocks infrared rays. Similarly, a greenhouse lets sunlight in but does not allow heat to escape. The greenhouse effect keeps the greenhouse—and Earth—warmer than they would otherwise be.

Latent Heat

Heat also enters the atmosphere as energy stored in molecules of water that have evaporated. Try wetting your finger and waving it in the air. Notice that it feels cooler? The reason is that water is evaporating from it. Each molecule of liquid water has to absorb a certain amount of heat from your finger before it can evaporate. Then, when it evaporates and becomes water vapor, each water vapor molecule carries with it into the air the heat energy it absorbed from your finger. The energy stored in water vapor molecules in the air moves along with the air as it circulates in the atmosphere.

Usually, when matter gains heat energy, its temperature increases; and when matter loses heat energy, its temperature decreases. However, when matter changes phase, heat is absorbed or given off in the process of forming or breaking bonds between molecules, rather than in causing molecules to move faster or slower. Thus, as mentioned earlier, the heat energy gained or lost during a phase change does not cause a change in temperature and is therefore latent heat.

The energy *gained* when ice melts into liquid water amounts to about 334 joules per gram of water. The same amount of energy, 334 joules per gram of water, is *released* when liquid water freezes into ice. The energy *gained* when liquid water evaporates into water vapor amounts to about 2,260 joules per gram of water. Here, too, the same amount of energy, 2,260 joules per gram of water, is *released* when the water vapor condenses back into liquid water. The changes of phase that water undergoes and the heat given off or absorbed in each process are summarized in Figure 21.11a.

Water covers more than 70 percent of Earth's surface, so most of the lower atmosphere is in contact with water. Each day solar energy absorbed by this water causes evaporation, sending vast amounts of water vapor into the atmosphere. Every gram of water vapor in the atmosphere contains the 2,260 joules of latent heat energy that was gained during the change of water from a liquid to a gas.

Remember, though, that latent heat stored during evaporation does not show up as a temperature change. The water vapor starts out no warmer than the water from which it formed. How, then, is the atmosphere warmed by latent heat? The heat stored in water molecules when they evaporated is released when the water changes back to water or ice. At the moment the water vapor condenses, it releases its latent heat into the air. As a result, condensation is a warming process that causes the temperature of air to rise as droplets of water in clouds grow in size. This increase in temperature causes the air to expand, become less dense, and rise. Thus, latent heat is an important energy source for violent storms. The release of latent heat during large-scale condensation produces localized heating, which causes updrafts. At the same time, precipitation associated with the heavy condensation causes downdrafts. The result is the violent circulations of air that occur in storms such as hurricanes and tornadoes.

Circulation of the Atmosphere

The transfer of heat energy from solar radiation, and water vapor, into and out of the atmosphere causes regions of different densities to form. The equatorial regions of the Earth are predominantly water and receive more solar energy per year than the polar regions. Air over the Equator becomes warm and moist through contact with the Earth's surface. Air that is warm and moist is less dense than cooler, drier air. Thus, the air at the Equator is surrounded by air of higher density that pushes inward toward the Equator and displaces the warm, moist equatorial air upward. As this air is forced upward it expands outward and cools, resulting in condensation and precipitation. The end result is cooler, dryer air that is denser than the air beneath it and begins to sink back toward the surface. Together, the rising warm, moist air

The Atmosphere

(a) Water gains or loses energy when it undergoes a change in state of matter. Since this gain or loss of heat energy does not result in a change in temperature, it is hidden or *latent heat*.

(b) The heating curve for a sample of water heated from a starting temperature of −100°C to a final temperature of +200°C. The same amount of heat is added to the sample every minute. From A to B the ice increases in temperature until it reaches its melting point. From B to C there is a change in state from solid ice to liquid water with no increase in temperature (the curve is flat, like a plateau). From C to D liquid water increases in temperature until it reaches its boiling point. From D to E there is a change in state from liquid water to water vapor, again with no increase in temperature. From E to F water vapor is increasing in temperature. Note that the flat area from B to C is shorter than the flat area from D to E since less energy is required to change water from a solid to a liquid than to change water from a liquid to a gas.

Figure 21.11 Changes in State of Water (a) and Heating Curve for Water (b).

and sinking cool, dry air form a circular pattern of motion called a **convection cell**. See Figure 21.12. A similar convection cell forms over the poles, although it begins with sinking air, which spreads southward when it reaches the Earth's surface, warms, and rises as shown in Figure 21.13. In between the polar and equatorial convection cells lies a third cell, which is set in motion by westerly winds in the middle latitudes. Thus, three convection cells girdle both Earth's Northern and Southern Hemispheres. This *three-cell theory* of global atmospheric circulation explains how the atmosphere distributes solar energy over the whole Earth. See Figure 21.14. The interaction of these processes results in the complex atmospheric occurrence known as weather.

Figure 21.12 Convection Cells Near the Equator. Air near the Equator is heated by more intense insolation, becomes less dense, and floats upward in the surrounding denser air. As the air rises, it expands and cools, causing it to sink back toward the surface. The result is a circular movement of air, or convection cell.

The Coriolis Effect and Global Atmospheric Circulation

The rise or fall of regions of different densities as a result of the action of gravitational force produces atmospheric circulation, which is affected by Earth's rotation. One of the consequences of Earth's rotation is a tendency of all matter that is in motion on Earth's surface to be deflected to the right from its point of origin in the Northern Hemisphere and to the left in the Southern Hemisphere, called the **Coriolis effect**. It

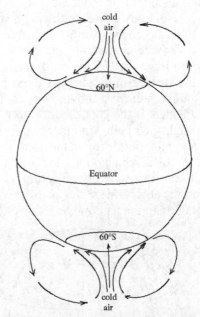

Figure 21.13 Convection Cells Near the Poles. Air near the poles cools because of less intense insolation, becomes more dense, and sinks downward. The sinking air spreads outward at the surface, is warmed as it moves away from the poles, and rises. The result is a circular movement of air, or convection cell.

The Atmosphere

Figure 21.14 The Three-Cell Theory of Convection in the Atmosphere.

is named after the French mathematician Gustave Gaspard Coriolis, who first analyzed it in the nineteenth century. The Coriolis effect is not an actual force but rather is the *apparent* effect of a number of different forces acting on any particles in motion.

The main elements of the Coriolis effect are the curvature of Earth's surface, the rotation of Earth, and the tendency of objects in motion to remain in motion in a straight line (Newton's first law of motion). The east-west path between any two points on Earth's surface is not a straight line, but a curve—a segment of a parallel (line of latitude). As a result, any object following a straight path will appear, to an observer on Earth, to curve from its path. We say "appear" because actually it is the observer who is traveling a curved path while the object travels in a straight line. In Figure 21.15 the solid line represents the actual path of an observer on the surface of a spherical Earth as it rotates. The dotted line represents a straight path for a moving object. The Coriolis effect is most pronounced where there is the greatest difference between the curvature of a parallel and a straight-line path—in other words, near the poles.

A similar deflection in the north-south direction is due to the difference in the speeds at which points on Earth's surface are moving because of rotation. As Earth rotates, every object on its surface rotates with it. In one 24-hour rotation, objects near the poles travel less distance than objects near the Equator and thus are moving slower (see Figure 21.16).

Figure 21.15 Deflection Due to Curvature of Earth's Surface. An object moving in a straight line travels along the dashed line. An observer on Earth's surface moves due east-west as Earth rotates, following a path that curves. To an observer on Earth's surface, the straight-line path seems to veer off, or be deflected, to the right in the Northern Hemisphere. In the Southern Hemisphere, the deflection is to the left.

Figure 21.16 Deflection Due to Different Velocities at Different Latitudes. As Earth rotates, objects on its surface move at different speeds, depending on their latitudes. Objects at the Equator move faster because they cover more distance in one 24-hour rotation than objects near the poles.

Now, let's suppose a rocket at the Arctic Circle (60° N) is aimed due south at New York City (40° N). Before the rocket is even launched, it is moving west to east at the speed of Earth's surface at the Arctic Circle—233 meters per second. The target, New York City, is also moving west to east at the speed of Earth's surface at New York City—356 meters per second. Note that the target is moving west to east faster than the rocket! When the rocket is fired, it moves due south and west to east; but since the target is moving west to east faster than the rocket, the rocket lags behind the target and actually hits the ground behind the target. To an observer, the rocket appears to follow a path that curves to the *right* of the target. Similarly, a rocket fired from New York City toward the Arctic Circle would land ahead of the target because New York City is moving west to east faster than the target. Nevertheless, the deflection would still be to the *right*. As seen in Figure 21.17, because of the Coriolis effect no matter what direction the motion, deflection is always to the right in the Northern Hemisphere.

If we repeat this experimentation in the Southern Hemisphere, we find that the direction of deflection is reversed—the rocket appears to curve to the *left* of the target. Thus, deflection is to the right of the wind's point of origin in the Northern Hemisphere and to the left of its point of origin in the Southern

Figure 21.17 The Coriolis Effect. Objects are deflected to the right from their points of origin in the Northern Hemisphere and to the left from their points of origin in the Southern Hemisphere.

The Atmosphere

Hemisphere, as shown in Figure 21.17. (A good way to recall the direction of deflection is to remember that someone who is *left*-handed is called a *southpaw*.)

For objects following paths that fall somewhere between due north-south and due east-west, the total deflection is a combination of deflection due to curvature and deflection due to differences in rotational speed. Of course, frictional forces within the atmosphere and with Earth's surface modify the Coriolis effect, but it still persists. Over short distances, however, the Coriolis effect is imperceptible, so don't worry about veering off to the right when you walk home from school or drive to the mall in the family car. The Coriolis effect becomes a factor only over great distances, such as those covered by planetary winds.

MULTIPLE-CHOICE QUESTIONS

In each case, write the number of the word or expression that best answers the question or completes the statement.

1. Energy is transferred from the Sun to Earth mainly by
    (1) molecular collisions
    (2) density currents
    (3) electromagnetic waves
    (4) red shifts

2. By which process does most of the Sun's energy travel through space?
    (1) absorption
    (2) conduction
    (3) convection
    (4) radiation

3. In which list are the forms of electromagnetic energy arranged in order from longest to shortest wavelengths?
    (1) gamma rays, X-rays, ultraviolet rays, visible light
    (2) radio waves, infrared rays, visible light, ultraviolet rays
    (3) X-rays, infrared rays, blue light, gamma rays
    (4) infrared rays, radio waves, blue light, red light

4. Which type of electromagnetic radiation has the shortest wavelength?
    (1) ultraviolet
    (2) gamma rays
    (3) radio waves
    (4) visible light

5. Which color of visible light has the shortest wavelength?
    (1) violet
    (2) green
    (3) yellow
    (4) red

6. When is an object in radiative balance?
   (1) when the radiation emitted by the object is equal to that absorbed by the object
   (2) when the radiation emitted by the surroundings is equal to that absorbed by the object
   (3) when the radiation emitted by the object is equal to that absorbed by the surroundings
   (4) when the wavelength of the radiation emitted by the object is equal to that absorbed by the object

7. In which region of the electromagnetic spectrum is most of Earth's outgoing terrestrial radiation?
   (1) infrared   (3) ultraviolet
   (2) visible   (4) X-rays

8. When Earth cools, most of the energy transferred from Earth's surface to space is transferred by the process of
   (1) conduction   (3) refraction
   (2) reflection   (4) radiation

9. What is the usual cause of the drop in temperature that occurs between sunset and sunrise at most New York State locations?
   (1) strong winds   (3) cloud formation
   (2) ground radiation (4) heavy precipitation

10. In New York State, the risk of sunburn is greatest between 11 A.M. and 3 P.M. on summer days because
    (1) the air temperature is hot
    (2) the angle of insolation is high
    (3) Earth's surface reflects most of the sunlight
    (4) the Sun is closest to Earth

11. The average temperature at Earth's equator is higher than the average temperature at Earth's South Pole because the South Pole
    (1) receives less intense insolation
    (2) receives more infrared radiation
    (3) has less land area
    (4) has more cloud cover

The Atmosphere

Base your answers to questions 12 through 16 on your knowledge of Earth science and the graph below, which shows measurements of the insolation above Earth's atmosphere and at Earth's surface on a clear day. [Note that the graph does not show the entire solar spectrum at the longer wavelengths.]

12. Which of the following types of solar radiation has the longest wavelength?
    (1) X-rays
    (2) ultraviolet rays
    (3) visible light rays
    (4) infrared rays

13. The greatest intensity of energy reaching the outer atmosphere of Earth from the Sun has a wavelength of approximately
    (1) 4.5×10^0 angstroms
    (2) 4.5×10^3 angstroms
    (3) 3.0×10^3 angstroms
    (4) 9.1×10^2 angstroms

14. In which portion of the solar spectrum does ozone absorb the greatest amount of energy?
    (1) X-rays
    (2) ultraviolet rays
    (3) visible light rays
    (4) infrared rays

15. What quantity is most likely represented by the area between the "dot-dash" curve (- · -) and the "dash" curve (- - -)?
    (1) the amount of radiation given off by the Sun
    (2) the amount of radiation absorbed in outer space
    (3) the amount of insolation reflected by the atmosphere back into space
    (4) the amount of insolation absorbed by Earth's surface

16. According to the graph, in which portion of the solar spectrum is the greatest total amount of energy absorbed by ozone and other materials in Earth's atmosphere?
    (1) ultraviolet rays and infrared
    (2) X-rays and visible light
    (3) visible light and infrared
    (4) X-rays and infrared

17. The diagram below represents Earth and the Sun's incoming rays. Letters A, B, C, and D represent locations on Earth's surface.

Which two locations are receiving the same intensity of insolation?
(1) A and B (3) C and D
(2) B and C (4) D and B

18. If the tilt of Earth's axis were increased from 23.5° to 30°, summers in New York State would become
    (1) cooler, and winters would become cooler
    (2) cooler, and winters would become warmer
    (3) warmer, and winters would become cooler
    (4) warmer, and winters would become warmer

The Atmosphere

Base your answers to questions 19 through 21 on the graph below, which shows the amount of insolation during one year at four different latitudes on Earth's surface.

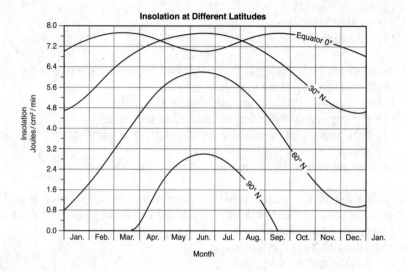

19. This graph shows that insolation varies with
    (1) latitude and time of day
    (2) latitude and time of year
    (3) longitude and time of day
    (4) longitude and time of year

20. Why is less insolation received at the equator in June than in March or September?
    (1) The daylight period is longest at the equator in June.
    (2) Winds blow insolation away from the equator in June.
    (3) The Sun's vertical rays are north of the equator in June.
    (4) Thick clouds block the Sun's vertical rays at the equator in June.

21. Why is insolation 0 joule/cm^2/min from October through February at 90° N?
    (1) Snowfields reflect sunlight during that time.
    (2) Dust in the atmosphere blocks sunlight during that time.
    (3) The Sun is continually below the horizon during that time.
    (4) Intense cold prevents insolation from being absorbed during that time.

22. Compared to a light-colored rock with a smooth surface, a dark-colored rock with a rough surface will
   (1) both absorb and reflect less insolation
   (2) both absorb and reflect more insolation
   (3) absorb less insolation and reflect more insolation
   (4) absorb more insolation and reflect less insolation

23. Equal volumes of the four samples shown below were placed outside and heated by energy from the Sun's rays for 30 minutes.

   The surface temperature of which sample increased at the slowest rate?
   (1) water
   (2) copper pennies
   (3) basaltic sand
   (4) iron fragments

24. Which change would cause a *decrease* in the amount of insolation absorbed at Earth's surface?
   (1) a decrease in cloud cover
   (2) a decrease in atmospheric transparency
   (3) an increase in the duration of daylight
   (4) an increase in nitrogen gas

25. The diagram below shows a student heating a pot of water over a fire. The arrows represent the transfer of heat. Letter *A* represents heat transfer through the metal pot, *B* represents heat transfer by currents in the water, and *C* represents heat that is felt in the air surrounding the pot.

Which table correctly identifies the types of heat transfer at *A*, *B*, and *C*?

Letter	Type of Heat Transfer
A	conduction
B	radiation
C	convection

(1)

Letter	Type of Heat Transfer
A	radiation
B	conduction
C	convection

(3)

Letter	Type of Heat Transfer
A	conduction
B	convection
C	radiation

(2)

Letter	Type of Heat Transfer
A	radiation
B	convection
C	conduction

(4)

26. Which diagram best represents how greenhouse gases in our atmosphere trap heat energy?

(Not drawn to scale)

27. Which component of Earth's atmosphere is classified as a greenhouse gas?
    (1) oxygen
    (2) carbon dioxide
    (3) helium
    (4) hydrogen

28. Deforestation increases the greenhouse effect on Earth because deforestation causes the atmosphere to contain
    (1) more carbon dioxide, which absorbs infrared radiation
    (2) less carbon dioxide, which absorbs shortwave radiation
    (3) more oxygen, which absorbs infrared radiation
    (4) less oxygen, which absorbs shortwave radiation

Base your answers to questions 29 through 31 on the map below, which shows Earth's planetary wind belts.

29. The curving of these planetary winds is the result of
    (1) Earth's rotation on its axis
    (2) the unequal heating of Earth's atmosphere
    (3) the unequal heating of Earth's surface
    (4) Earth's gravitational pull on the Moon

30. Which wind belt has the greatest effect on the climate of New York State?
    (1) prevailing northwesterlies
    (2) prevailing southwesterlies
    (3) northeast trades
    (4) southeast trades

31. Which climatic conditions exist where the trade winds converge?
    (1) cool and wet
    (2) cool and dry
    (3) warm and wet
    (4) warm and dry

The Atmosphere

32. Most of the air in the lower troposphere at the equatorial low-pressure belt is
    (1) warm, moist, and rising
    (2) warm, dry, and rising
    (3) cool, moist, and sinking
    (4) cool, dry, and sinking

33. The Coriolis effect would be influenced most by a change in Earth's
    (1) rate of rotation
    (2) period of revolution
    (3) angle of tilt
    (4) average surface temperature

34. During some winters in the Finger Lakes region of New York State, the lake water remains unfrozen even though the land around the lakes is frozen and covered with snow. The primary cause of this difference is that water
    (1) gains heat during evaporation
    (2) is at a lower elevation
    (3) has a higher specific heat
    (4) reflects more radiation

35. During which phase change will the greatest amount of energy be absorbed by 1 gram of water?
    (1) melting
    (2) freezing
    (3) evaporation
    (4) condensation

36. Which process releases 334 joules (J) of energy for each gram of water?
    (1) melting
    (2) freezing
    (3) vaporization
    (4) condensation

37. Why is the condensation of water vapor considered to be a process that heats the air?
    (1) Liquid water has a lower specific heat than water vapor.
    (2) Energy is released by water vapor as it condenses.
    (3) Water vapor must absorb energy in order to condense.
    (4) Air can hold more water in the liquid phase than in the vapor phase.

CONSTRUCTED RESPONSE QUESTIONS

Base your answers to questions 38 through 41 on the data table below. The data table shows the latitude of several cities in the Northern Hemisphere and the duration of daylight on a particular day.

Data Table

City	Latitude (°N)	Duration of Daylight (hr)
Panama City, Panama	9	11.6
Mexico City, Mexico	19	11.0
Tampa, Florida	28	10.4
Memphis, Tennessee	35	9.8
Winnipeg, Canada	50	8.1
Churchill, Canada	59	6.3
Fairbanks, Alaska	65	3.7

38. On the grid below, plot with an **X** the duration of daylight for each city shown in the data table. Connect your **X**s with a smooth, curved line. [1]

The Atmosphere

39. Based on the data table, state the relationship between latitude and the duration of daylight. [1]

40. Use your graph to determine the latitude at which the Sun sets 7 hours after it rises. [1]

41. The data were recorded for the first day of a certain season in the Northern Hemisphere. State the name of this season. [1]

Base your answers to questions 42 through 45 on the diagram below, which shows Earth as seen from above the North Pole. The curved arrows show the direction of Earth's motion. The shaded portion represents the nighttime side of Earth. Some of the latitude and longitude lines have been labeled. Points *A* and *B* represent locations on Earth's surface.

42. On the diagram, draw a curved arrow, starting at point *B*, showing the general direction that planetary surface winds flow between 30° N and 60° N latitude. [1]

43. If it is 4:00 P.M. at point *B*, what is the time at point *A*? [1]

44. Identify *one* possible date that is represented by the diagram. [1]

45. Explain why the angle of insolation at solar noon is greater at point *B* than at point *A*. [1]

Base your answers to questions 46 and 47 on the passage below.

>   Average temperatures on Earth are primarily the result of the total amount of insolation absorbed by Earth's surface and atmosphere compared to the amount of long-wave energy radiated back into space. Scientists believe that the addition of greenhouse gases into Earth's atmosphere gradually increases global temperatures.

46. Identify *one* major greenhouse gas that contributes to global warming. [1]

47. Explain how increasing the amount of greenhouse gases in Earth's atmosphere increases global temperatures. [1]

Base your answers to questions 48 and 49 on the diagram below and on your knowledge of Earth science. The diagram represents a beaker of water being heated. The curved lines around letters *A* and *B* represent convection cells that have developed in the water.

48. On the diagram *above*, draw *six* arrowheads, one on *each* of the curved lines of the convection cells, to indicate the direction of water movement around letters *A* and *B*. [1]

49. State the amount of heat energy gained by each gram of water that evaporates from the surface of the boiling water in the beaker. [1]

The Atmosphere

Base your answers to questions 50 through 53 on the graph below, which shows the temperatures recorded when a sample of water was heated from −100°C to +200°C. The water received the same amount of heat every minute.

50. In one or more complete sentences, explain why the water did not change temperature between points *B* and *C* or between points *D* and *E* even though the water received the same amount of heat energy each minute. [1]

51. State the lettered points between which the water gained the greatest amount of heat energy. [1]

52. State the lettered points between which the water was changing from a solid to a liquid. [1]

53. The water received the same amount of heat every minute. In one or more complete sentences, explain why the water temperature remained unchanged between points *B* and *C* for less than 2 minutes but remained unchanged between points *D* and *E* for more than 10 minutes. [1]

EXTENDED CONSTRUCTED RESPONSE QUESTIONS

Base your answers to questions 54 through 57 on the graph below and on your knowledge of Earth science. The graph shows changes in hours of daylight during the year at the latitudes of 0°, 30° N, 50° N, and 60° N.

54. Estimate the number of daylight hours that occur on January 1 at 40° N latitude. [1]

55. Identify the latitude shown on the graph that has the earliest sunrise on June 21. Include the units and compass direction in your answer. [1]

56. Explain why all four latitudes have the same number of hours of daylight on March 20 and September 22. [1]

57. The graph *below* shows a curve for the changing length of daylight over the course of one year that occurs for an observer at 50° N latitude. On this same graph, draw a line to show the changing length of daylight over the course of one year that occurs for an observer at 50° S latitude. [1]

Base your answers to questions 58 through 60 on the data table below and on your knowledge of Earth science. The table shows air temperatures recorded under identical conditions at 2-hour intervals on a sunny day. Data were recorded 1 meter above ground level both inside and outside of a glass greenhouse.

Data Table

Time	Inside Air Temperature (°C)	Outside Air Temperature (°C)
8 A.M.	15	15
10 A.M.	18	16
12 noon	21	17
2 P.M.	24	18
4 P.M.	24	17

58. Describe the color and texture of the surfaces inside the greenhouse that would most likely absorb the greatest amount of visible light. [1]

59. Calculate the rate of change in the outside air temperature from 8 A.M. to 2 P.M. in Celsius degrees per hour. [1]

60. Most atmospheric scientists infer that global warming is occurring due to an increase in greenhouse gases. State the names of *two* greenhouse gases. [1]

CHAPTER 22
WEATHER

KEY IDEAS Weather, the present condition of the atmosphere at any given location, is described by the physical characteristic of the atmosphere. Weather patterns become evident when atmospheric variables are observed, measured, and recorded. Atmospheric variables include air temperature, air pressure, moisture (relative humidity and dew point), precipitation (such as rain, snow, hail, and sleet), wind speed and direction, and cloud cover. Atmospheric variables are measured using instruments such as thermometers, barometers, psychrometers, precipitation gauges, anemometers, and wind vanes.

Atmospheric variables can be represented in a variety of formats including radar and satellite images, synoptic weather maps, atmospheric cross-sections, and computer models. Analysis of atmospheric variables reveals that they are interrelated. Atmospheric moisture, temperature and pressure distributions, winds and jet streams, air masses, frontal boundaries, and the movement of cyclonic systems and associated tornadoes, thunderstorms, and hurricanes occur in observable patterns. Loss of property, personal injury, and loss of life can be reduced by effective emergency preparedness.

KEY OBJECTIVES
Upon completion of this unit, you will be able to:

- Define weather, and identify the atmospheric characteristics used to describe it.
- Describe the relationships between important atmospheric characteristics.
- Explain how clouds and precipitation form.
- Construct weather maps, given weather information at numerous locations, and identify the positions of air masses and fronts.
- Predict weather and the movement of weather systems based on current weather maps.
- Describe the weather conditions associated with the various types of frontal boundaries.

- Explain how weather information can be used to make forecasts.
- Identify the conditions that typically lead to hazardous weather such as hurricanes and tornadoes.

HOW CAN WEATHER BE DESCRIBED?

Weather is the present condition of the atmosphere at any location. Weather changes are mainly the result of unequal heating, by solar radiation, of Earth's landmasses, oceans, and atmosphere. As heat energy is transferred at the boundaries between the atmosphere, oceans, and landmasses, layers of different temperatures and densities form in the atmosphere. When gravity acts on these layers of different densities, they rise and fall, producing circulation of air in the atmosphere. This circulation causes the air at any one location to constantly be moved away and replaced by air from a different location—with a resulting change in weather.

The constantly changing weather is described in terms of the characteristics of the atmosphere that change, or atmospheric variables. The **atmospheric variables** used to describe weather are measured with weather instruments and include the temperature, pressure, humidity, movements, and transparency of the air. Atmospheric variables are interrelated, so that a change in one is likely to cause a change in another. This interrelatedness makes weather prediction a complex task. However, measurement of atmospheric variables and the creation of field maps based on these data reveal large-scale patterns that can be used to make predictions.

Air Temperature

The air temperature at any given location is related to the heat energy present in the atmosphere at that location. Solar radiation is the main source of heat energy in the atmosphere; therefore anything that influences solar radiation will ultimately influence air temperature. Although the radiation emitted by the Sun is fairly constant, the amount that reaches Earth's surface and is converted to heat energy in the atmosphere is influenced by many factors. These include the angle at which solar radiation strikes Earth's surface; the number of hours of solar radiation per day; the reflection, refraction, or absorption of solar radiation by the atmosphere (e.g., cloud cover); and the nature of the surface materials absorbing solar radiation. In general, the more direct the solar radiation and the longer it occurs, the higher the temperature.

The angle at which solar radiation strikes Earth's surface varies during the course of a day. Because of Earth's rotation, radiation is received during the day but not at night. Therefore, air temperature often varies in a daily cycle. Similar changes occur as Earth moves through its orbit around the Sun, resulting in seasonal temperature variations.

Temperature is measured with thermometers. A **thermometer** works on a simple principle—matter expands when heated and contracts when cooled, and the amount of expansion or contraction depends on the amount of heat gained or lost. A typical thermometer measures the height of a column of alcohol or mercury in a glass tube. Some modern devices, however, use the change in electrical conductivity of certain metals when heated or cooled to measure temperature. Several scales, named for their creators, have been developed over the years to measure temperature: Fahrenheit, Celsius, and Kelvin. Meteorologists in the United States still use the Fahrenheit scale but are changing over to the Celsius.

Meteorologists use several types of specialized thermometers when measuring air temperature. Simple thermometers are mounted in shaded, vented boxes so that they will measure the temperature of the air and not be heated additionally by absorption of sunlight. Maximum-minimum thermometers have pinched tubes and markers that are pushed or dragged back as the fluid in them rises and falls. This feature enables them to record the highest and lowest temperatures during a time period by the positions of the markers. Continuous temperature readings are made with a thermograph, a thermometer in which a coil of heat-sensitive metal moves a pen against a rotating drum.

Air Pressure

Air is a mixture of gases. A gas consists of many tiny individual molecules that are far apart and moving rapidly. As they whiz to and fro, they are kept from escaping into space by the walls of a container or, in the case of the atmosphere, by Earth's gravity.

Gases exert pressure on surfaces with which they come into contact. **Pressure** is a force that acts on a surface area. The pressure that a gas exerts on a surface is the result of gas particles colliding with the surface. Since gas particles move randomly in all directions, they exert pressure equally in all directions. Figure 22.1 shows gas molecules speeding around in a container and striking surfaces held at different angles.

Air pressure is not an atmospheric variable that you can sense, in the way that you feel temperature or see cloud cover. By itself, knowledge of the air pressure in one location is of little use in weather forecasting. However, *field maps of air pressure in many locations* reveal distinct regions within the atmosphere, and *changes* in air pressure signal movements of these large masses of air and their associated weather conditions. Rising pressure

Figure 22.1 Gas Molecules in a Container. Regardless of the angle at which a surface is held, air molecules collide with it, exerting the same pressure in all directions.

usually signals the approach or continuation of fair weather; falling pressure, the approach of a storm.

Air pressure is measured with a **barometer**. Two commonly used barometers are the mercury barometer and the aneroid barometer.

Figure 22.2 A Mercury Barometer (a) and an Aneroid Barometer (b).

In a mercury barometer (see Figure 22.2a), air pressure forces liquid mercury up a tube. The mercury rises in the tube because the tube contains a vacuum and no pressure is exerted on the mercury inside the tube. The higher the air pressure outside the tube, the higher the mercury is pushed up in the tube. Since pressure is measured by the height to which the mercury is pushed, pressure is often described as "inches or millimeters of mercury." The word *aneroid* means no liquid, and an aneroid barometer (see Figure 22.2b) is a can with no liquid inside it. The can contains a spring scale. As air pressure on the sides of the can increases or decreases, the spring compresses and expands and the scale records the magnitude of the force.

Pressure is measured in units of force per unit area. A variety of units are used to describe atmospheric pressure. The **atmosphere** (atm) is the average pressure exerted by Earth's atmosphere at sea level and is equal to 1,013.2 millibars or 14.7 pounds per square inch. The bar is equal to 1×10^5 newtons per square meter, and the *millibar* is 1/1,000 of a bar or 100 newtons per square meter. (1 newton = 0.225 lb.) In meteorology, atmospheric pressure is measured in millibars. **Normal atmospheric pressure** is 1,013.2 millibars.

Air pressure is often misunderstood because much of our personal experience with pressure involves solids pressing against us, or our bodies pressing against something because of gravity acting on us. Since the molecules of solids are bonded together in a single unit, they act as a single unit. Under the influence of gravity the entire object is pulled downward and exerts a downward pressure. In gases, on the other hand, each tiny molecule acts independently. Gravity pulls the molecule down; but since it is hurtling through space, the effect is like that of gravity acting on a baseball speeding toward home. The ball sinks in a downward trajectory but still exerts a sizable horizontal force, as any professional catcher will attest. Think of gases as millions of tiny baseballs flying in all directions at the same time. Anything that affects the strength and the number of collisions that gas molecules will make with a surface also affects the pressure they exert.

Air Humidity

Humidity is the amount of moisture in the air. Humidity is not liquid water that is suspended in the air. Most of the moisture in the air is in the form of water vapor, a colorless, odorless gas that enters the atmosphere when liquid water evaporates or ice sublimes (changes directly from a solid to a vapor). Humidity is an important weather factor because the water vapor in the atmosphere is the source of the water that condenses to form clouds, fog, and precipitation.

Water molecules are always changing back and forth between phases (the three states of matter: vapor, liquid, and solid). Most water molecules enter the atmosphere from liquid water by evaporation and from the surface of ice and snow by sublimation (molecules leave a solid and enter directly into the vapor phase). Water leaves the atmosphere by condensation or deposition (molecules of water vapor form solid crystals without first becoming a liquid). This constant flow of water molecules between phases results in constantly changing levels of humidity in the atmosphere. By far, the greatest amount of activity occurs at the boundary between the atmosphere and the hydrosphere. Molecules of liquid water in the hydrosphere that gain energy from sunlight and their surroundings may become energetic enough to escape the liquid phase and enter the atmosphere as water vapor. When water vapor molecules in the atmosphere come into contact with a liquid water surface, they may be trapped there, so there is a constant two-way traffic of molecules to and from the liquid. The flow of molecules leaving the surface of the liquid and entering the atmosphere as water vapor (**evaporation**) is strongly influenced by temperature. At higher temperatures, molecules have more energy, so a greater number are likely to escape and enter the vapor phase. The flow of molecules arriving at the surface to exit the atmosphere as a liquid (**condensation**) depends on how densely crowded with molecules the water vapor above the liquid becomes. The more densely crowded the molecules, the more likely they are to come into contact with the surface and be trapped.

If more water molecules are leaving the surface than are arriving, there is net evaporation. If more water molecules are arriving at the surface than leaving, there is net condensation. If there is net evaporation, the amount of water vapor in the atmosphere increases. However, as you just learned, if the amount of water vapor in the atmosphere increases, molecules are more likely to come into contact with the surface and be trapped, so condensation also increases. Eventually, a point is reached where condensation and evaporation are in equilibrium, and the amount of water vapor above the surface remains unchanged. See Figure 22.3. The higher the temperature, the greater the amount of water vapor present in the atmosphere before equilibrium is reached. At 0°C, equilibrium is reached when air contains 5 grams of water vapor per cubic meter; at 20°C the amount is 17 grams per cubic meter, and at 100°C it is 598 grams per cubic meter! If for some reason (such as a sudden drop in temperature) the amount of water vapor exceeds the equilibrium amount, condensation will occur at a faster rate than evaporation until equilibrium is reestablished.

Net evaporation   Equilibrium   Net condensation

Figure 22.3 Net Evaporation and Net Condensation in Equilibrium. Equilibrium occurs when there are equal flows of molecules leaving and arriving at the surface of the liquid. Net evaporation occurs when the water-vapor content of the atmosphere is less than the equilibrium amount. Net condensation occurs when the water-vapor content of the atmosphere is greater than the equilibrium amount.

The amount of water vapor in the atmosphere is often expressed as absolute humidity or relative humidity. **Absolute humidity** is the number of grams of the water vapor molecules in 1 cubic meter of air. Since it is difficult to isolate the water vapor in air and to measure its mass, absolute humidity is seldom directly measured.

Relative humidity is the ratio of the water vapor now present in the atmosphere to the water vapor present when evaporation and condensation are in equilibrium at that temperature, times 100 percent.

$$\text{Relative humidity} = \frac{\text{Water vapor in atmosphere}}{\text{Water vapor in atmosphere at equilibrium}} \times 100\%$$

Relative humidity is a way of comparing the flows of water molecules leaving and arriving at a surface, and is useful in predicting how the air will feel to a person and whether condensation or evaporation is more likely to occur in the atmosphere.

A relative humidity of 20 percent means that the atmosphere currently contains only 20 percent of the water vapor it contains at equilibrium. As you have learned, at a temperature of 0°C, condensation and evaporation reach equilibrium when the amount of water vapor in the air reaches 5 grams per cubic meter. A relative humidity of 20 percent means that the air currently contains only 1 gram per cubic meter (20% of 5 g/m^3 = 1 g/m^3.) Therefore, there will be net evaporation until the air gains an additional 4 grams per cubic meter and reaches equilibrium (at which point there is no net evaporation).

A relative humidity of 100 percent means that the atmosphere contains 100 percent of the water vapor it contains at equilibrium. Therefore, there will be no net evaporation *or* net condensation. The higher the relative humidity, the more uncomfortable people feel because less moisture evaporates from their bodies before equilibrium is reached. Since less water evaporates from the skin, a wet and "sticky" feeling results.

It is important to remember, however, that a relative humidity of 100 percent does not mean that the air is 100 percent water vapor. The air still mostly consists of nitrogen and oxygen. The weight of the water vapor represents only a small fraction of the total weight of all the gases in the air. For example, the weight of water vapor in very warm, moist air may represent only 1/25 of the total weight of all the gases in the air. Nevertheless, if that is all of the water vapor the air contains at equilibrium, the relative humidity will be 100 percent. Similarly, the weight of water vapor in frigid arctic air may be as little as 1/10,000 of the total weight of the gases in the air; yet, if that is all of the moisture the air contains at equilibrium, the relative humidity will also be 100 percent.

High relative humidity also means that the water vapor in the atmosphere is close to its equilibrium amount. A decrease in temperature, which, in turn, decreases the equilibrium amount, could trigger condensation, which forms clouds, fog, or precipitation. A commonly used value in this connection is the **dew point**, that is, the temperature at which the water vapor in the air will begin to condense into liquid water. This is the temperature at which the amount of water vapor currently in the air equals the equilibrium amount of water vapor.

Humidity is measured with a hygrometer or a psychrometer. A **hygrometer** consists of strands of human hair attached to a pointer. Human hair lengthens slightly as humidity increases. As the hair lengthens and shortens because of changing humidity, the pointer changes position.

A **psychrometer** consists of two thermometers, one whose bulb is kept dry and another whose bulb is kept wet by covering it with a cloth wick soaked with water. Evaporation from the wet wick causes cooling because the fastest molecules in the liquid water are the ones with enough energy to escape. The slower molecules left behind have a lower average energy; hence the liquid's temperature decreases as measured on the wet-bulb thermometer. See Figure 22.4. This cooling effect is the key to measuring relative humidity. The lower the relative humidity, the more water can evaporate from the wet-bulb thermometer and the more the thermometer is cooled. The dry-bulb thermometer, however, remains unchanged and registers the temperature of the air. Therefore, the difference in temperature between the wet-bulb thermometer and the dry-bulb thermometer is directly related to the relative humidity of the air.

The fastest molecules escape.

The remaining molecules have lower average energies, hence the liquid temperature is lower.

Figure 22.4 After evaporation, the remaining liquid is cooler than before.

The relationship among dry-bulb temperature, wet-bulb temperature, and relative humidity is shown in Table 22.1. To find the relative humidity, first calculate the difference between the wet-bulb and dry-bulb temperatures. Then find on the table the point where the dry-bulb temperature and the difference between the two temperatures intersect. The number at the intersection is the percent relative humidity.

Example:

Find the relative humidity if the dry-bulb temperature is 10°C and the wet-bulb temperature is 5°C.

To find the relative humidity you need the *dry-bulb temperature* and the *difference between the dry-bulb and wet-bulb temperatures*.

The dry-bulb temperature is given as 10°C.

The difference between the dry-bulb and wet-bulb temperatures must be calculated. It is 10°C − 5°C = 5°C.

The number on the table where these two temperatures intersect is 43. The relative humidity is 43%.

TABLE 22.1 RELATIVE HUMIDITY (%)

Dry-Bulb Temperature (°C)	Difference Between Wet-Bulb and Dry-Bulb Temperatures (C°)															
	0	1	2	3	4	5	6	7	8	9	10	11	12	13	14	15
−20	100	28														
−18	100	40														
−16	100	48														
−14	100	55	11													
−12	100	61	23													
−10	100	66	33													
−8	100	71	41	13												
−6	100	73	48	20												
−4	100	77	54	32	11											
−2	100	79	58	37	20	1										
0	100	81	63	45	28	11										
2	100	83	67	51	36	20	6									
4	100	85	70	56	42	27	14									
6	100	86	72	59	46	35	22	10								
8	100	87	74	62	51	39	28	17	6							
10	100	88	76	65	54	43	33	24	13	4						
12	100	88	78	67	57	48	38	28	19	10	2					
14	100	89	79	69	60	50	41	33	25	16	8	1				
16	100	90	80	71	62	54	45	37	29	21	14	7	1			
18	100	91	81	72	64	56	48	40	33	26	19	12	6			
20	100	91	82	74	66	58	51	44	36	30	23	17	11	5		
22	100	92	83	75	68	60	53	46	40	33	27	21	15	10	4	
24	100	92	84	76	69	62	55	49	42	36	30	25	20	14	9	4
26	100	92	85	77	70	64	57	51	45	39	34	28	23	18	13	9
28	100	93	86	78	71	65	59	53	47	42	36	31	26	21	17	12
30	100	93	86	79	72	66	61	55	49	44	39	34	29	25	20	16

Source: The State Education Department, *Earth Science Reference Tables*. 2011 ed. (Albany, New York; The University of the State of New York).

The difference between the wet-bulb temperature and the dry-bulb temperature is also directly related to the dew point of the air. The higher the relative humidity, the closer the air is to being filled to its capacity. Cooling air decreases its capacity to hold moisture. The closer the air is to being filled to capacity, the less it has to be cooled to reach the point of maximum moisture. Table 22.2, which shows the relationship among wet-bulb temperature, dry-bulb temperature, and dew point, can be used to find the dew point. The number on the table at the intersection of the dry-bulb temperature and the difference between the wet-bulb and dry-bulb temperatures is the dew point in degrees Celsius.

Example:

Find the dew point if the dry-bulb temperature is 10°C and the wet-bulb temperature is 5°C.

The dry-bulb temperature is given as 10°C.

The difference between the dry-bulb and wet-bulb temperatures must be calculated. It is 10°C − 5°C = 5°C.

The number on the table where these two temperatures intersect is −2. The dew point is −2°C.

TABLE 22.2 DEW POINT TEMPERATURES (°C)

Dry-Bulb Temperature (°C)	Difference Between Wet-Bulb and Dry-Bulb Temperatures (C°)															
	0	1	2	3	4	5	6	7	8	9	10	11	12	13	14	15
−20	−20	−33														
−18	−18	−28														
−16	−16	−24														
−14	−14	−21	−36													
−12	−12	−18	−28													
−10	−10	−14	−22													
−8	−8	−12	−18	−29												
−6	−6	−10	−14	−22												
−4	−4	−7	−12	−17	−29											
−2	−2	−5	−8	−13	−20											
0	0	−3	−6	−9	−15	−24										
2	2	−1	−3	−6	−11	−17										
4	4	1	−1	−4	−7	−11	−19									
6	6	4	1	−1	−4	−7	−13	−21								
8	8	6	3	1	−2	−5	−9	−14								
10	10	8	6	4	1	−2	−5	−9	−14	−28						
12	12	10	8	6	4	1	−2	−5	−9	−16						
14	14	12	11	9	6	4	1	−2	−5	−10	−17					
16	16	14	13	11	9	7	4	1	−1	−6	−10	−17				
18	18	16	15	13	11	9	7	4	2	−2	−5	−10	−19			
20	20	19	17	15	14	12	10	7	4	2	−2	−5	−10	−19		
22	22	21	19	17	16	14	12	10	8	5	3	−1	−5	−10	−19	
24	24	23	21	20	18	16	14	12	10	8	6	2	−1	−5	−10	−18
26	26	25	23	22	20	18	17	15	13	11	9	6	3	0	−4	−9
28	28	27	25	24	22	21	19	17	16	14	11	9	7	4	1	−3
30	30	29	27	26	24	23	21	19	18	16	14	12	10	8	5	1

Source: The State Education Department, *Earth Science Reference Tables*. 2011 ed. (Albany, New York; The University of the State of New York).

Air Movements

In the atmosphere, air circulates because of density differences. The vertical movements of rising and sinking air are called **air currents**. When sinking air reaches Earth's surface, it spreads out horizontally. When rising air expands, it also spreads out horizontally. Horizontal movements of air parallel to Earth's surface are called **winds**. Wind is described by both its speed and its direction.

Wind speed is measured with an **anemometer**. An anemometer consists of three or four cups mounted on a vertical axis and driven by the wind, causing the axis to spin. The speed of rotation of the axis varies with wind speed. Wind speed is measured in knots, a nautical measure of speed; 1 knot = 1.85 kilometers per hour, or 1.15 miles per hour.

Wind direction is determined with a **wind vane**. A wind vane is a pointer mounted on an axis that is attached to a compass rose. The tail of the pointer is larger in surface area than the tip. Thus, the wind exerts more pressure on the tail, causing it to swing around so that the tip points in the direction from which the wind is blowing. A wind is named according to the direction from which it blows. Just as a person who comes from the South is called a Southerner, a wind that blows from the south is called a south wind.

Atmospheric Transparency

All of the gases of which air is composed are transparent. However, there are many substances that become suspended in the atmosphere and decrease its transparency. These substances include dust, volcanic ash, smoke, salt particles, aerosols (droplets of liquids), ice particles, and water droplets. By far, water droplets have the greatest effect on atmospheric transparency. Clouds and fog consist of tiny, but visible, droplets of water. Atmospheric transparency is expressed in three ways: visibility, cloud ceiling, and cloud cover.

Visibility is the horizontal distance through which the eye can distinguish objects. It is usually expressed in miles.

Cloud ceiling is the base height of cloud layers. It is measured with a ceilometer, which uses pulses of light and a photoelectric telescope.

Cloud cover is the fraction of the sky that is obscured by clouds.

WHAT ARE THE RELATIONSHIPS AMONG THE SEVERAL ATMOSPHERIC VARIABLES?

Atmospheric variables are interrelated, so that a change in one is likely to cause a change in another. Since the atmosphere is gaseous, this interrelatedness is best understood in light of the kinetic theory of gases. According to the kinetic theory, gases consist of many tiny, individual molecules that do not interact with each other except when they collide. The molecules are far apart and are in constant random motion, and the temperature of the gas is proportional to the speed at which they are moving.

Air Pressure and Air Temperature

As explained earlier, air pressure is the result of the forces exerted by gas molecules colliding with a surface or each other. Air temperature depends on the speed of the gas molecules. Air pressure and temperature are related because both involve the motion of molecules. The higher the temperature, the faster the gas molecules are moving. The faster the gas molecules are moving, the more force they exert in a collision. Thus, you would expect that an increase in temperature would result in an increase in pressure, but in the atmosphere this is not the case. The reason is that the atmosphere is not contained by rigid walls but is an open system free to expand.

When the temperature of the atmosphere increases, its molecules move faster and collide more vigorously. As a result, the molecules bounce farther apart and the atmosphere expands. Even though the molecules are moving faster, and each individual impact exerts greater force, the total number of collisions decreases because the molecules are spread farther apart. The decrease in collisions far exceeds any increase in force due to faster moving

Weather

Cold air

Warm air

Figure 22.5 Relationship Between Air Temperature and Air Pressure. Even though the molecules are moving slower and collide with less force, the cold air on the left exerts more pressure because its closely spaced molecules collide much more frequently.

molecules, so the net effect is a *decrease* in air pressure when air temperature increases. See Figure 22.5.

| Air temperature ↑, Air pressure ↓ | Air temperature ↓, Air pressure ↑ |

Now let us consider what happens to temperature if pressure is changed. See Figure 22.6. The main cause of pressure changes in the atmosphere is the rising or sinking of air in convection currents. When a parcel of air rises, the surrounding air exerts less confining pressure on the parcel and it expands. Conversely, when a parcel of air sinks, the surrounding air exerts more pressure on the parcel and it is compressed.

Any gas that is allowed to expand because the surrounding pressure is reduced will become cooler, and any gas compressed into a smaller volume will become warmer. Such temperature changes, which occur without heat being added from outside sources or heat being lost to the surroundings, are known as **adiabatic** changes.

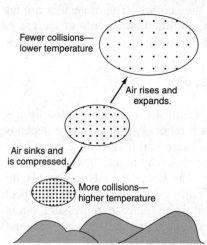

Figure 22.6 Relationship Between Air Pressure and Air Temperature. As air rises, pressure decreases, air molecules spread out and collide less frequently, and temperature decreases. As air sinks, pressure increases, air molecules crowd closer together and collide more frequently, and temperature increases.

In adiabatic temperature changes, the change in the temperature of the air reflects a change in the degree of crowding of its molecules. If a gas is allowed to expand into a larger volume, its molecules spread farther apart, and collide with one another less frequently. If a thermometer is placed in such air, fewer molecules will collide with it. Fewer collisions means less energy transferred to the thermometer, causing less expansion of the fluid in the thermometer, and the thermometer registers a lower temperature. Thus, a decrease in pressure causes a decrease in temperature.

Compression forces the air molecules to crowd more closely together and causes more frequent collisions. If a thermometer is placed in the compressed air, more molecules will collide with it, and it will register a higher temperature. Thus an increase in pressure causes an increase in temperature.

| Air pressure ↑, Air temperature ↑ | Air pressure ↓, Air temperature ↓ |

Air Pressure and Humidity

The greater the mass of the molecules, the more force they exert during a collision. The mass of a molecule in air depends on the element(s) of which it is composed. Nitrogen and oxygen molecules have more mass than water molecules (molecular mass of $N_2 = 28$, of $O_2 = 32$, of $H_2O = 18$). As water vapor builds up in the atmosphere, lighter water molecules displace heavier nitrogen or oxygen molecules. Since the lighter water molecules exert less force during a collision than the heavier nitrogen and oxygen molecules, air pressure decreases as humidity increases. (Don't be confused by thinking of moist air as containing *liquid* water, which is denser than any gas in air; moist air contains water *vapor*, which is less dense than most gases in air.) Figure 22.7 shows the displacement of heavier gas molecules by lighter water vapor molecules as humidity increases.

| Humidity ↑, Air pressure ↓ | Humidity ↓, Air pressure ↑ |

Total Weight of All Molecules = 352
Dry Air

Total Weight of All Molecules = 308
Moist Air

Oxygen M.W. = 32
Nitrogen M.W. = 28
Water M.W. = 18

Figure 22.7 Relationship Between Molecular Mass and Density. A volume of dry air has more mass than an equal volume of moist air because water molecules have less mass than the nitrogen and oxygen molecules they displace. Therefore, moist air is less dense than dry air.

Air Temperature and Humidity

The higher the temperature, the greater the amount of water vapor present in the atmosphere before equilibrium is reached. Thus, warm air contains more water vapor at equilibrium than cool air. Why is this so? Imagine a sealed jar of dry air half filled with water and kept at a certain temperature. All along the water surface, water molecules evaporate and enter the air as water vapor. At the same time, some of the molecules of water vapor in the air lose energy, drop back to the water surface, and return to the liquid. At first, more water molecules leave by evaporation than return to the liquid, and the amount of water vapor in the jar increases. However, over time, the number of water vapor molecules returning to the liquid balances the number of water molecules leaving the liquid. Then, the amount of water vapor in the jar no longer increases. At any given temperature, a fixed amount of water vapor will be present in the air when equilibrium is reached. See Figure 22.8a.

(a) Equilibrium at low temperature

(b) Equilibrium at high temperature

Figure 22.8 At lower temperatures (a) less water vapor is present in the atmosphere when equilibrium is reached than is present at higher temperatures (b).

Now let's consider what happens if we heat the sealed jar. In order for molecules of water in the liquid to break free and become water vapor, they require energy. At a higher temperature, more of the molecules in the liquid water will have enough energy to break free and become a gas. As the water vapor becomes more crowded with molecules, it becomes more likely that some will be trapped at the liquid surface and condense. Thus, the amount of water vapor in the warmed jar increases until a new balance between water molecules leaving and water molecules returning to the liquid is reached. The amount of water vapor in the jar at the higher temperature is greater than it was at the lower temperature. The same volume of space contains more water vapor at equilibrium! See Figure 22.8b. Roughly speaking, for every 10°C increase in temperature, the amount of water vapor in a given volume of air at equilibrium is doubled.

Since the amount of water vapor in the air at equilibrium changes with temperature, the relative humidity is also influenced by temperature. If the temperature increases, causing the equilibrium amount to increase, but the moisture in the air remains the same, the relative humidity will decrease. Conversely, if the temperature decreases, causing the equilibrium amount to decrease, but the moisture in the air remains the same, the relative humidity will increase.

Figure 22.9 Cooling Decreases the Equilibrium Amount of Water Vapor, Leading to Condensation. As air is cooled, the amount of water vapor the air contains approaches the equilibrium amount. At the dew point, evaporation and condensation are in equilibrium. If the air is cooled below the dew point, net condensation results in formation of clouds and precipitation.

As air temperature decreases, it eventually reaches the **dew point**, the temperature at which the moisture the air contains equals the equilibrium amount. At that point, the relative humidity is 100 percent. If the temperature decreases further, moisture in the air will exceed the equilibrium amount, and condensation will occur at a faster rate than evaporation until equilibrium is reestablished. The condensed moisture forms droplets of water or directly forms ice crystals, which, in turn, may form clouds or precipitation. Therefore, as air temperature approaches the dew point, precipitation becomes more likely. See Figure 22.9.

Air Pressure and Winds

Winds blow from regions of high air pressure to regions of low air pressure. This is easy to understand if you think of two people pushing against each other. The person pushing harder will advance against the person pushing with less force. In much the same way, air exerting high pres-

Figure 22.10 Air Pressure and Winds. Winds Blow from High to Low. High-pressure air exerts more force than low-pressure air and pushes the low-pressure air ahead of it, resulting in a horizontal movement of air—a wind.

sure will advance against air exerting low pressure. The net movement of the air is from high pressure toward low pressure. See Figure 22.10.

Figure 22.11 Land and Sea Breezes. (a) During the day, land surfaces heat up faster than water. The higher temperatures over the land result in lower air pressures, while the lower temperatures over the water result in higher air pressures. A wind develops that blows from the high-pressure area over the water toward the low-pressure area over the land. (b) At night, land surfaces cool off faster than water and the situation reverses, causing a wind to blow from the land toward the water.

Land and Sea Breezes

Land and sea breezes illustrate this air movement from high to low pressure regions very clearly. Differences in air pressure between adjacent regions occur wherever land and water meet. During the day, when exposed to sunlight, land surfaces increase in temperature more than water surfaces because water has a higher specific heat than soil. As a result, air over land surfaces will have a higher daytime temperature than adjacent air over water surfaces. Since warm air exerts less pressure than cool air, the pressure over the land will be lower and the pressure over the water will be higher. The result is a movement of air from the water toward the land. Since a wind is named for the direction or place from which it is blowing, this wind is called a **sea breeze**. See Figure 22.11a.

At night, though, land cools off faster than water. The air over the land becomes cooler than the air over the water. Thus, the air pressure is greater over the land than over the water, and air moves from the land toward the water. This wind is called a **land breeze** because it comes from the land. See Figure 22.11b.

Global Wind Belts

Differences in pressure between adjacent regions of air also occur on a global scale. In Chapter 21, you learned that unequal heating of the atmosphere at different latitudes causes three large atmospheric convection cells to form in the Northern and Southern Hemispheres. Rising air in these convection cells produces low-pressure belts that circle Earth at the Equator (0°), the Arctic Circle (60° N), and the Antarctic Circle (60° S). Sinking air produces high-pressure belts at the Poles (90° N, 90° S) and at the Tropics of Cancer (30° N) and Capricorn (30° S). Thus, a series of alternating high- and low-pressure belts circle Earth, giving rise to a series of **global wind belts** as air moves from regions of high pressure toward regions of low pressure. See Figure 22.12. The global winds that blow from the high pressure belts

Figure 22.12 Planetary Wind and Moisture Belts in the Troposphere. The drawing shows the locations of the belts near the time of an equinox. The locations shift somewhat with the changing latitude of the Sun's vertical ray. In the Northern Hemisphere, the belts shift northward in summer and southward in winter. Source: The State Education Department, *Earth Science Reference Tables*, 2011 ed. (Albany, New York; The University of the State of New York).

at 30° North and South latitude toward the low-pressure belt at the Equator are known as the **trade winds**. The Coriolis effect causes these large-scale winds to curve to the right in the Northern Hemisphere (so they blow from the northeast) and to the left in the Southern Hemisphere (so they blow from the southeast).

Global winds also blow from the high-pressure belts at 30° North and South latitudes toward the low-pressure belts at 60° North and South latitudes and are known as the **prevailing winds**. The Coriolis effect causes these large-scale winds to blow from the southwest in the Northern Hemisphere and from the southeast in the Southern Hemisphere.

The global winds that blow from the high-pressure regions over the poles toward the low pressure regions at 60° N and 60° S are known as the **polar easterlies**. The Coriolis effect causes these winds to blow from the northeast in the Northern Hemisphere and from the southeast in the Southern Hemisphere.

HOW DO CLOUDS AND PRECIPITATION FORM?

Clouds and precipitation form when air is cooled below its dew point and water vapor condenses into tiny water droplets or ice crystals. The conditions under which this cooling takes place determine what type of cloud or precipitation will form.

Condensation is the change of phase from gas to liquid. To condense into a liquid, a gas must have a surface to condense on. In the atmosphere, this surface is provided by particles suspended in the air, such as salt or ice crystals, dust, or smoke, called **condensation nuclei**. Water vapor in the air condenses when the air is cooled to the dew point and there are condensation nuclei on which condensation can form. **Deposition** is the change of phase from a gas directly to a solid. Water vapor may form ice crystals if moisture is released from the air when its temperature is below the freezing point of water.

As you can see from Figure 22.13, temperature drops steadily with altitude in the troposphere. In the upper portion of the troposphere, temperatures below –40°C are the norm year-round. For this reason, even in the summer, cirrus clouds are composed of tiny ice crystals, and thunderstorms whose cloud tops extend into the upper troposphere may produce hail. Clouds are important components of the atmosphere for three reasons. They are the producers of precipitation, such as rain, snow, sleet, and hail. They are also major reflectors of solar radiation, and they serve as key indicators of overall weather conditions.

As shown in Figure 22.14, clouds are categorized by the altitudes at which they exist and their degrees of vertical development.

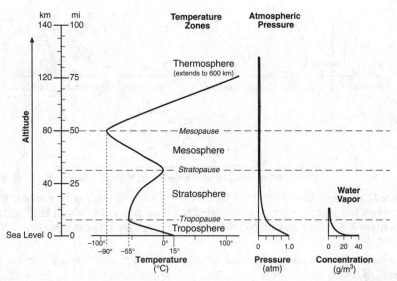

Figure 22.13 Physical Properties of the Atmosphere. Source: The State Education Department, *Earth Science Reference Tables*, 2011 ed. (Albany, New York; The University of the State of New York).

Figure 22.14 Cloud-Form Classification. Clouds are classified into families according to altitude and degree of vertical development. Source: Arthur N. Strahler, *The Earth Sciences*, Harper & Row, 1971.

Air that is warmed by contact with the Sun-heated surface of Earth expands, becomes less dense, and floats upward in the denser surrounding air. As the air rises, it continues to expand, causing a decrease in pressure and temperature. Warm air will also expand and cool if it is forced upward as wind pushes it over a natural obstacle, such as a mountain, or by cooler, denser air pushing its way underneath it. These three conditions that cause air to rise are diagrammed in Figure 22.15.

Figure 22.15 Three Conditions That Cause Air To Rise. Air rises when it is heated and becomes less dense, when it is forced upward by winds over rising land surfaces, and when it is forced upward by colder, denser air pushing under it.

The cooling of air as it rises results in condensation when the air temperature drops to the dew point and condensation nuclei are present. This condensation generates a **cloud** made of a great many tiny but visible water droplets or ice crystals. The **cloud base** indicates the elevation at which the rising air reached its dew point. Since this depends on the amount of moisture in the air and the starting temperature, clouds may form at many different elevations.

Fog is a cloud whose base is at ground level. It forms when moist air at ground level is cooled below its dew point. When moist air comes into contact with a surface such as grass, or with a car or other object that has cooled below the dew point because heat has radiated out of it overnight, droplets of water called **dew** condense on the surface. If the temperature is below freezing, tiny ice crystals called **frost** will form instead of dew.

The water droplets or ice crystals in clouds are very tiny (1/50 millimeter) and often remain suspended in the air. Thus, all clouds do not form precipitation. **Precipitation** occurs only when cloud droplets or ice crystals join together and become heavy enough to fall. Table 22.3 summarizes the different types of precipitation and their origins.

As precipitation falls through the atmosphere, condensation nuclei and other suspended material that adhere to it are brought down to Earth's surface. In this way dust and many pollutants are removed from the atmosphere by precipitation. However, some pollutants react with the water in precipitation to form harmful solutions such as acid rain, which has had many negative effects on the environment.

TABLE 22.3 TYPES OF PRECIPITATION AND ORIGINS

Name	Description	Origin
Rain	Droplets of water up to 4 mm in diameter	Cloud droplets coalesce when they collide.
Drizzle	Very fine droplets falling slowly and close together	Cloud droplets coalesce when they collide.
Sleet	Clear pellets of ice	Raindrops freeze as they fall through layers of air at below-freezing temperatures.
Glaze	Rain that forms a layer of ice on surfaces it touches	Supercooled raindrops freeze as soon as they come in contact with below-freezing surfaces.
Snow	Hexagonal crystals of ice, or needlelike crystals at very low temperatures	Water vapor sublimes, forming ice crystals on condensation nuclei at temperatures below freezing.
Hail	Balls of ice ranging in size from small pellets to as large as a softball, with an internal structure of concentric layers of ice and snow	Again and again hailstones are hurled up by updrafts in thunderstorms and then fall through layers of air that alternate above and below freezing. Each cycle adds a layer to the hailstone. The more violent the updrafts, the larger and heavier the hailstone can become before falling.

WHAT INFORMATION IS SHOWN BY WEATHER MAPS?

Synoptic Weather Maps

One of the most widely used methods of forecasting weather is known as **synoptic forecasting**. Synoptic forecasting is based on examination of a synopsis, or summary of the total weather picture at a particular time.

A **synoptic weather map** is made by measuring atmospheric variables at thousands of weather stations around the world four times a day. These data are then used to create field maps, which reveal large-scale weather patterns. By looking at a sequence of synoptic weather maps, meteorologists can perceive the developments and movements of weather systems and make predictions.

Synoptic weather maps use a symbol called a **station model** to show a summary of the weather conditions at a particular weather station. Figure 22.16 shows a typical station model and explains what each of its elements represents.

Weather

Figure 22.16 A Station Model. The symbol uses shorthand notation to summarize atmospheric variable for a particular location on a weather map. Source: The State Education Department, *Earth Science Reference Tables*, 2011 ed. (Albany, New York; The University of the State of New York).

Field Maps

Station models plotted on a map summarize a wide range of weather data and can be used to create many different field maps. A **field** is defined as a region of space that has a measurable quantity at every point. Field maps can be used to represent any quantity that varies in a region of space. One way to represent field quantities on a two-dimensional field map is to use **isolines**, which connect points of equal field value. For example, a temperature field map contains lines connecting points of equal temperature, or **isotherms**. A pressure field map contains lines connecting points of equal pressure, or **isobars**. Field maps clearly show the weather patterns in the atmosphere.

Figure 22.17 shows a synoptic weather map, followed by the same map with isotherms and isobars drawn in (Figures 22.17b and 22.17c).

As you can see in Figure 22.17a, there are distinct regions in the atmosphere that have similar conditions. For example, a region of high pressure, cool temperature, and clear skies is centered near Salt Lake City, Utah. You can also see a region of low pressure, slightly warmer temperature, and rain centered near Cincinnati, Ohio.

Within a field, a field value changes as you move from place to place. The rate at which the field value changes is called its **gradient**. Rapid changes in field values, or steep gradients, show up as closely spaced isolines on a field map. In Figure 22.17c, you can see that the isobars are widely spaced within each of the two air masses, but are closely spaced between them. This means that there is little change in pressure within each of the regions, but there is a boundary between them where the pressure changes rapidly. You can see the same pattern repeated for temperature: there are regions of the atmosphere with relatively uniform characteristics, which have boundaries separating them from adjacent regions.

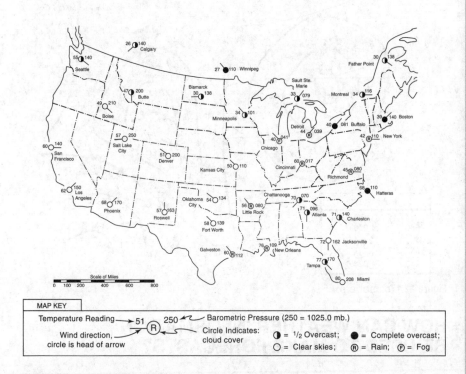

Figure 22.17a Synoptic Weather Map of the United States.

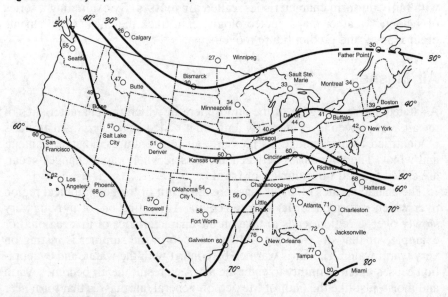

Figure 22.17b Isotherms.

Weather

Figure 22.17c Isobars.

HOW CAN WEATHER INFORMATION BE USED TO MAKE FORECASTS?

Most weather forecasts are based on the movements of large regions of air with fairly uniform characteristics, called **air masses**. By examining a series of synoptic weather maps, meteorologists can track the current movements of air masses and predict future movements.

Air Masses

As stated above, a region of the atmosphere in which the characteristics of the air at any given level are fairly uniform is called an *air mass*. An air mass is identified by its average air pressure, air temperature, moisture content, and winds. The boundaries between air masses, where rapid changes occur, are called **frontal boundaries**, or **fronts**.

The characteristics of an air mass are the result of the geographical region over which it formed, or its **source region**. Air resting on, or moving very slowly over, a region tends to take on the characteristics of that region. For example, air that sits over the Gulf of Mexico in the summer is resting on very warm water. The air is warmed by contact with the water, and evaporation causes much humidity to enter the air. As a result the air becomes warm and moist, just like the Gulf of Mexico. In general, air masses that form near the poles are cold, and those that form near the Equator are warm. Air masses that form over water are moist; those that form over land, dry. The longer the air remains stationary over a region, the larger it becomes and the closer it matches the characteristics of the region. On weather maps, air masses are named for their source regions and are designated by a letter or letters as shown in Table 22.4.

TABLE 22.4 AIR-MASS NAMES AND SOURCES

	Arctic A	Polar P	Tropical T
	Formed over extremely cold, ice-covered regions	Formed over regions at high latitudes where temperatures are relatively low	Formed over regions at low latitudes where temperatures are relatively high
maritime m Formed over water, moist		mP—cold, moist Formed over North Atlantic, North Pacific	mT—warm, moist Formed over Gulf of Mexico, middle Atlantic, Caribbean, Pacific south of California
continental c Formed over land, dry	cA—dry, frigid Formed north of Canada	cP—cold, dry Formed over northern and central Canada	cT—warm, dry Formed over southwestern United States in summer

Air masses are moved by planetary winds and jet streams (study Figure 22.12 carefully). In general, arctic and polar air masses are moved south and west over the United States by the northeast wind belt near the North Pole, and tropical air masses are moved to the north and east by the southwest wind belt between 30° and 60° N. As the air masses are carried over Earth's surface, they slowly change and take on the characteristics of the regions over which they are moving. For example, a cP air mass moving south out of Canada will gradually become warmer as it moves farther and farther south.

Air masses that affect U.S. weather and the regions in which they form are shown in Figure 22.18.

Figure 22.18 Air Masses That Influence Weather in the United States and Their Source Regions.

Weather Fronts

At any given point in time, several air masses will be moving across the United States. They generally move from west to east, driven by the prevailing southwesterlies. When different air masses meet, very little mixing of

Weather

Figure 22.19 Perspective View of a Frontal Surface and a Front Line on the Ground. A *frontal surface* forms along the boundary between two different air masses. The line along which the boundary touches the surface is the *front*.

the air takes place and a sharp transition zone forms between them. This zone is termed a **weather front** because it marks the leading edge, or front, of an air mass that is pushing against another air mass. The whole surface along which the air masses meet is called the **frontal surface**, and the line on the ground marking the transition from one air mass to the other is the **front**. See Figure 22.19. Fronts are areas of rapid changes in weather conditions and are often sites of unsettled and rainy weather. Different types of frontal surfaces, such as cold, warm, stationary, and occluded fronts, will form, depending on which type of air mass is advancing against another.

Cold Fronts

A **cold front** forms along the leading edge of a cold air mass that is advancing against a warmer air mass. The cold air mass is denser than the warmer air ahead of it, so it pushes against and under it like a wedge. This *forces* the warmer air upward rapidly and results in turbulence, rapid condensation forming heavy, vertically developed clouds, and heavy precipitation or thunderstorms. Cold fronts are often preceded by a zone of thunderstorms in the summer and of snow flurries in the winter. After the front passes, the temperature drops sharply and the pressure rises rapidly. Figure 22.20 shows a typical cold front.

Figure 22.20 A Typical Cold Front.

Warm Fronts

A **warm front** forms where a warm air mass overrides the trailing edge of a cold air mass ahead of it. The less dense warm air mass rides up and over the denser cold air. As the warm air mass rises above the denser cold air mass, it expands and cools, causing condensation to occur over the wide, gently sloping boundary. The results are thickening, lowering clouds and widespread precipitation. Figure 22.21 shows a typical warm front.

Figure 22.21 A Typical Warm Front.

Stationary Fronts

A **stationary front** forms along the boundary between a warm air mass and a cold air mass when neither moves appreciably in any direction. A stationary front slowly takes on the shape and characteristics of a warm front as the denser cold air slowly slides beneath the less dense, warmer air. Although stationary fronts develop the same widespread rain and cloudiness as do warm fronts, the rain and clouds may persist for a longer period, perhaps many days, until another air mass comes along with enough impetus to get the stalled air masses moving.

Occluded Fronts

Cold air is denser and exerts more pressure than warm air. Therefore, cold air masses and the cold fronts associated with them tend to move faster than warm air masses and warm fronts. If a slow-moving warm front is followed by a fast-moving cold front, the cold front will sometimes overtake the warm front. Then the cold air lifts the warm air entirely aloft, forming an **occluded front**. The lifting of the warm air mass causes large-scale condensation and precipitation, resulting in widespread rain and thunderstorms. Then, depending on whether the cold air mass coming from behind is colder or warmer than the cold air mass now ahead, a cold-front occlusion or a warm-front occlusion may form, as shown in Figure 22.22.

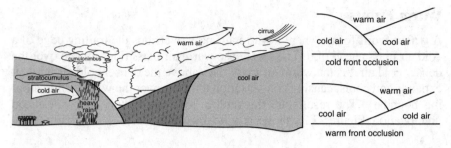

Figure 22.22 An Occluded Front. Two types of occlusions can form, depending on whether the advancing cold air mass is warmer or cooler than the cold air mass ahead of the warm air mass.

Cyclones, Anticyclones, and Wave Cyclones (Frontal Cyclones)

Regions of low pressure surrounded by higher pressure form winds that blow toward the center of the warm air mass but, because of the Coriolis effect, are deflected to the right. The result is a **cyclone**, a counterclockwise flow of air, that moves in a curved path toward the center of the low pressure.

Similarly, regions of high pressure surrounded by lower pressure form winds that blow outward from the center. The winds are deflected by the Coriolis effect into a flow of air, an **anticyclone**, that is clockwise and outward from the center.

A cyclone and an anticyclone are diagrammed in Figure 22.23.

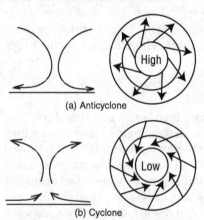

Figure 22.23 An Anticyclone (a) and a Cyclone (b).

Most of the weather systems that move across the United States are *mid-latitude cyclones*, or *just plain cyclones*. **Mid-latitude cyclones** are warm and cold fronts that move in a counterclockwise direction around a low-pressure center. They typically form in the westerly wind belts along the boundary between polar and tropical air, called the **polar front** (see Figure 22.12). In the United States, this boundary falls just about along our northern border with Canada.

A mid-latitude cyclone forms when southwesterly winds push warm, low-pressure air against cold air along the polar front (see Figure 22.24). The result is to lower the pressure along the polar front, and the boundary bends to form a wave of cold air advancing from west to east against the warm air—a cold front. The cold air pushes the warm air ahead of it into the cold air on the other side of it, forming a warm front. The end result is a wave-like structure consisting of a cold front overtaking a warm front. The counterclockwise flow in the low-pressure center causes the cold front to swing around the equatorial side of the low-pressure center as the wave moves from west to east. The mid-latitude cyclone may even break loose from the polar front and be carried, swirling along, by the jet stream. In the United States, mid-latitude cyclones generally move from west to northeast and get carried out over the north Atlantic. The stormy weather associated with a mid-latitude cyclone may affect millions of square miles and last for several days.

Figure 22.24 Formation of a Mid-latitude Cyclone.

HOW CAN WE FORECAST THE WEATHER?

Until relatively recently, weather forecasts were based on local observations made directly by human senses. Through first-hand experience and passed-down weather lore, people learned to recognize certain patterns of weather changes associated with cloud types or shifts in wind direction. It was not until the invention of the barometer and the thermometer in the 1600s, however, that accurate measurements of pressure and temperature changes could be made. Even then, forecasting was not practicable until communication among many points by telegraph made possible the rapid collection and sharing of weather data. The first government weather forecasts were issued in the United States in 1870.

Synoptic Weather Forecasting

The basis of most weather forecasting is the systematic collection of weather data from a network of weather-reporting sites that blanket the globe. At the simplest level, these data are summarized on a synoptic weather map using shorthand symbols called *station models*. At their most sophisticated, computer programs compile, store, analyze, and plot weather data, plot isolines for various weather factors, and can display weather data in a variety of graphic formats.

Synoptic weather maps are made four times a day—at midnight, noon, 6 A.M., and 6 P.M. Greenwich Mean Time. By looking at a sequence of these maps, a meteorologist can detect the development and movement of weather systems. The movements of these systems are projected into the future, and predictions of weather for various locations are made. Computer-drawn maps show wind, temperature, air pressure, and humidity patterns at many different atmospheric levels. Statistical analysis is then used to map projections of weather factors.

Synoptic forecasting is fairly accurate for large-scale, short-term forecasts. However, there are always locally unique conditions that modify the general behavior of weather systems. For example, cities tend to absorb and reradiate more heat energy than rural areas, creating urban "heat islands." Local meteorologists must take these specific conditions into account when making forecasts, for they have a significant effect on such factors as the exact path a storm will take locally, the times when rain will begin and end, and the extremes that high and low temperatures will reach. For this reason, most local meteorologists have a network of weather hobbyists who feed them weather data in addition to those collected by official Weather Service stations.

Statistical Weather Forecasting

Statistical weather forecasting is based on the past behavior of the atmosphere. Mathematical equations are used to predict the probability that the atmosphere will behave in a certain way. Basically the system works something like this: Suppose you have math first period on Mondays, Tuesdays, and Wednesdays, and science first period on Thursdays. A person observing you would soon see a pattern in which three days of first-period math are followed by a first period of science. However, there would be exceptions. There might be a holiday on a Monday, so the sequence would be science after *two* maths. Or Thursday might be a holiday, so the sequence would be six maths before the next science. However, such deviations in the schedule would probably be the exception rather than the rule. Let's assume that in 10 weeks there was one deviation. Statistical forecasting would then say that for the next week there was a 9 in 10, or 90 percent, chance that three maths would be followed by a science.

Now let's apply this reasoning to weather. Suppose we observed that, nine out of the past ten times the wind shifted to blow from the northeast, we had rain within 6 hours. Then, if the wind shifted to the northeast, statistical forecasting would say that there was a 90 percent chance of rain in the next 6 hours.

The process of statistical forecasting is threefold: maintaining detailed historical records, identifying all previous occurrences of a pattern, and then calculating the probabilities. However, there is no way to anticipate unusual behavior, nor is past behavior a guarantee of future behavior.

Numerical Weather Forecasting

Numerical forecasting is based on fluid dynamics. Theoretically, if the initial states of the atmosphere, land surfaces, and oceans are known, we should be able to predict the behavior of the atmosphere based on the laws of physics governing heat and moisture transfer and fluid behavior. However, such complete information is not currently available.

The sheer quantity of data, coupled with the number and complexity of the required calculations, made numerical forecasting impracticable until computers capable of high-speed calculations were developed in the 1940s. Now, computers solve six equations simultaneously—one for each of the three dimensions of motion, and one each for the conservation of moisture, heat, and mass. These equations are solved for thousands of points on a map grid, and changes at each point are projected. These changes are then fed back into the loop for another round of calculations, producing a projection further into the future. As new actual data are entered into the equations throughout the day, the revised calculations serve to correct the predictions. Calculations for several levels in the atmosphere, projecting changes over the next 2 days, are made every 12 hours. A 5-day forecast is calculated once a day.

Long-Range Weather Forecasting

No matter what method is used, however, forecasting reliability decreases as the time frame increases. Short-term forecasts (1–3 days) of weather are usually far more accurate than long-term forecasts (a week or more). The decrease in forecast reliability over time is due to a number of factors: the wide spacing of initial data points, the unreliability in measurements taken over many areas, and a lack of understanding of the behavior of the atmosphere. A small error in an initial measurement can cause errors in computer-calculated forecasts.

For example, let's make a simple mathematical model that assumes that temperature increases or decreases in direct proportion to the ratio of day to night:

$$\text{Temperature today} \times \frac{\text{Length of night}}{\text{Length of day}} = \text{Temperature tomorrow}$$

Let's compare the temperatures predicted by this model, using a correct temperature reading and one in which an error was made:

Starting Temperature	Prediction When Daylight = 13 hours, Night = 11 hours	Prediction When Daylight = 14 hours, Night = 10 hours
21°C (correct)	21 × 13/11 = 24.8°C	24.8 × 14/10 = 34.7°C
22°C (incorrect)	22 × 13/11 = 26°C	26 × 14/10 = 36.4°C

As you can see, an initial error of 1°C has almost *doubled* as it was compounded in the calculations. Errors like these multiply as forecasts are extended farther into the future, and at some point the forecast becomes useless.

Chaos theory shows that small, unpredictable changes can result in large-scale changes in a system over time. For example, a forest fire set by lightning may cause a localized increase in temperature that may set in motion large-scale changes in air-flow patterns. For this reason, chaos theory seems to indicate that long-term forecasts may never be reliable.

WHERE AND HOW ARE THE MOST HAZARDOUS WEATHER SITUATIONS LIKELY TO OCCUR?

Thunderstorms

One of the most common weather hazards is a thunderstorm (see Figure 22.25). A thunderstorm begins as a convection cell in the atmosphere. Warm air rises, expands, cools, and then sinks back to Earth's surface. If the convection in the cell is strong, because of intense heating, the air may rise very rapidly, reaching heights of many kilometers in just a few minutes. A cloud forms in the updraft of the convection cell and grows as more and more warm, moist air is carried upward. The strong updrafts in the rising air support water droplets and ice crystals in the cloud so that they grow in size. When the updrafts can no longer support the moisture, it falls as rain or even hail. The falling rain sets up downdrafts that cause internal friction with updrafts, and the internal friction builds up static electric charges that may discharge as lightning. Thunderstorms are likely to occur wherever and whenever there is strong heating of Earth's surface.

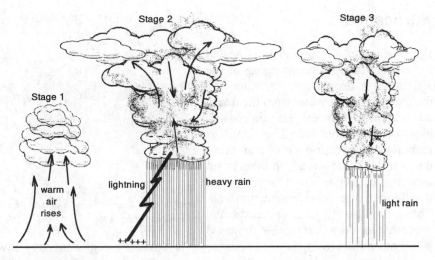

Figure 22.25 Formation of a Thunderstorm. Stage 1: Intense heating causes warm air to rise rapidly, forming tall, vertically developed clouds. Stage 2: Strong updrafts keep precipitation aloft until it is heavy. Falling rain creates downdrafts, and internal friction with updrafts causes buildup of static charge, leading to lightning and thunder. Stage 3: Downdrafts cause air to cool and updrafts subside, so rain becomes lighter.

Tornadoes

Late in the day, when Earth's surface is warmest and convection is strongest, a tornado may form (see Figure 22.26). Tornadoes are small (most are less than 100 meters in diameter), brief (most last only a few minutes) disturbances that usually develop over land from intense thunderstorms. When heating is very intense, warm air rises in strong convection currents. The upward movement of the air causes a sharp decrease in pressure. Air rushes into the low-pressure region from the sides and, deflected by the Coriolis effect, is given a spin. As air moves toward the center of the updraft, it reaches high speeds because of the big difference in pressure. The rapidly moving air decreases the pressure even more, further feeding the updraft. The whole process spirals upward in intensity, and a funnel forms that eventually touches the ground. Wind speeds near the center of a tornado may reach 500 kilometers per hour or more.

Figure 22.26 Formation of a Tornado.

Hurricanes

When air over warm oceans is heated by solar radiation, the warming of the air causes a decrease in pressure and the evaporation of water from the ocean adds humidity to the air, decreasing pressure even further. The result is numerous convection cells and thunderstorms. Widespread thunderstorm activity can merge into a large updraft that is part of a huge convection cell. The rising air further decreases the pressure, as does latent heat released when the moisture in the air condenses. As a result a region of very low pressure is created. Winds begin to blow toward the low-pressure center as air moves in from surrounding areas of higher pressure. The Coriolis effect deflects the winds, and a cyclone forms. Heat, released when fresh moisture brought in by winds condenses, feeds the convection cell with new energy, and it grows larger and stronger. When the winds in the cyclone exceed 119 kilometers per hour, we have a hurricane (see Figure 22.27). Fully formed hurricanes are huge cyclones, often exceeding 500 kilometers in diameter.

Figure 22.27 Cross Section of a Hurricane. Source: Earth Science on File. © Facts on File, Inc., 1988.

Hurricanes tend to form over oceans near the Equator during summer months, when the ocean surface is the warmest. Once the hurricane has formed, heat and moisture from the ocean provide the energy that maintains it. Anything that cuts off the supply of heat or moisture will diminish the strength of the hurricane, so hurricanes weaken as they move over land or cool water. Therefore a hurricane is usually most destructive when it first moves over land. However, even after moving over land, hurricanes can last for many days and cause much damage because of high winds and flooding.

Planetary winds tend to move hurricanes to the west until they reach 30° N; then the prevailing southwesterlies move them from southwest to northeast (see Figure 22.28). However, air masses that a hurricane encounters may deflect its path.

Figure 22.28 Typical Tracks of Tropical Cyclones—Hurricanes—near the United States.

Emergency Preparedness for Hazardous Weather

Hazardous weather, such as tornadoes, severe thunderstorms, and hurricanes, claims many lives and causes much property damage each year in the United States. Hazardous weather happens; there's not much humans can do about that. Measures can be taken, however, to reduce the risk of damage and loss of life. Reducing the risk is called *mitigation*. Some forms of mitigation are expensive and complicated, such as moving to a safer location or making structural changes to a house. Some measures are easier, such as developing an emergency plan, having disaster supplies on hand, protecting windows with shutters, and securing loose outdoor items that could cause damage when hurled by strong winds.

Emergency preparedness is an important responsibility of local government. Communities can help reduce risks by taking such actions as developing disaster plans, establishing evacuation routes and public shelters, enacting zoning regulations that limit development of high-risk areas, and enforcing building codes that require structures to withstand hazardous weather forces.

Watches and Warnings

Advance planning and quick response are the keys to surviving hazardous weather. Toward this end, the National Weather Service issues two levels of alerts about life-threatening weather such as tornadoes, severe thunderstorms, and hurricanes: watches and warnings. *Watche*s are issued for tornadoes and severe thunderstorms when weather conditions are such that these hazards are likely to develop in your area. Hurricane watches are issued when there is a threat of hurricane conditions within 24–36 hours. *Warnings* are issued when a tornado or severe thunderstorm has been sighted or indicated on radar in your area, or when hurricane conditions are expected in 24 hours or less.

When a severe thunderstorm or tornado is coming, there is very little time to make life-or-death decisions, so the National Weather Service issues a warning as soon as they are spotted. On the other hand, hurricanes are easier to track because they are so large and last so long. A 1-mile-wide tornado is huge and may last for minutes or hours; a 100-mile-wide hurricane is small, and hurricanes usually last for a week or more. Although hurricane warnings are easier to issue, the decisions involved are more complicated. An average hurricane warning along several hundred miles of United States coastline costs tens of millions of dollars spent for boarding up homes, closing down schools, businesses, and manufacturing plants, and evacuating people. Then, just a small deviation in a hurricane's path may mean the difference between a disaster and a very rainy day for a city under a hurricane warning. Forecasting is improving, however, and in the next few years meteorologists hope to make 30-hour forecasts as accurate as current 24-hour forecasts.

What to Do During Watches and Warnings

If a watch is issued for your area, you should stay alert to the weather by listening to the radio or television and be prepared to take shelter. This is the time to check emergency supplies and to remind family members of disaster plans and ways to respond during an emergency. Check for hazards outdoors and secure your home.

If a warning is issued, the danger is immediate and very serious, and everyone should go to a safe place. Once there, listen to a battery-operated radio or television for further instructions and for an official "all clear" before resuming normal activities. The Federal Emergency Management Agency has developed severe thunderstorm, tornado, and hurricane fact sheets that describe specific steps to be taken when watches and warnings are issued for these weather hazards. These fact sheets are available online at *www.fema.gov*.

MULTIPLE-CHOICE QUESTIONS

In each case write the number of the word or expression that best answers the question or completes the statement.

1. Which set of instruments is correctly paired with the weather variables that they measure?

2. A psychrometer is used to determine which weather variables?
   (1) wind speed and wind direction
   (2) percentage of cloud cover and cloud height
   (3) air pressure and air temperature
   (4) relative humidity and dewpoint

3. A barometric pressure of 1021.0 millibars is equal to how many inches of mercury?
   (1) 29.88   (3) 30.25
   (2) 30.15   (4) 30.50

4. A temperature of 80° Fahrenheit would be approximately equal to how many degrees on the Celsius scale?
   (1) 34
   (2) 299
   (3) 27
   (4) 178

5. Which sequence of events affecting moist air within Earth's atmosphere causes cloud formation?
   (1) rising → expanding → cooling → condensation
   (2) rising → contracting → warming → evaporation
   (3) sinking → expanding → warming → condensation
   (4) sinking → contracting → cooling → evaporation

6. Under which atmospheric conditions will water most likely evaporate at the fastest rate?
   (1) hot, humid, and calm
   (2) hot, dry, and windy
   (3) cold, humid, and windy
   (4) cold, dry, and calm

7. The diagram below represents the wet-bulb and dry-bulb temperatures on a sling psychrometer.

What was the relative humidity of the air when these temperatures were recorded?
   (1) 5%
   (2) 17%
   (3) 20%
   (4) 63%

8. If the air temperature is 20°C and the relative humidity is 58%, what is the dewpoint?
   (1) 5°C
   (2) 12°C
   (3) 15°C
   (4) 38°C

9. What is the difference between the dry-bulb temperature and the wet-bulb temperature when the relative humidity is 28% and the dry-bulb temperature is 0°C?
   (1) 11°C
   (2) 2°C
   (3) 28°C
   (4) 4°C

Weather

10. Earth's surface winds generally blow from regions of higher
   (1) air temperature toward regions of lower air temperature
   (2) air pressure toward regions of lower air pressure
   (3) latitudes toward regions of lower latitudes
   (4) elevations toward regions of lower elevations

11. Which graph best represents the change in air pressure as air temperature increases at Earth's surface?

12. Air pressure is usually highest when the air is
   (1) cool and humid   (3) warm and humid
   (2) cool and dry   (4) warm and dry

13. The table below shows the air temperature and dewpoint at each of four locations, A, B, C, and D.

Location	A	B	C	D
Air temperature (°F)	80	60	45	35
Dewpoint (°F)	60	43	35	33

Based on these measurements, which location has the greatest chance of precipitation?
   (1) A   (3) C
   (2) B   (4) D

14. Which statement best explains why an increase in the relative humidity of a parcel of air generally increases the chance of precipitation?
   (1) The dewpoint is farther from the condensation point, causing rain.
   (2) The air temperature is closer to the dewpoint, making cloud formation more likely.
   (3) The amount of moisture in the air is greater, making the air heavier.
   (4) The specific heat of the moist air is greater than the drier air, releasing energy.

15. The upward movement of air in the atmosphere generally causes the temperature of that air to
  (1) decrease and become closer to the dewpoint
  (2) decrease and become farther from the dewpoint
  (3) increase and become closer to the dewpoint
  (4) increase and become farther from the dewpoint

16. Which weather condition most directly determines wind speeds at Earth's surface?
  (1) visibility changes
  (2) amount of cloud cover
  (3) air-pressure gradient
  (4) dewpoint differences

17. The table below shows air-pressure readings taken at two cities, in the same region of the United States, at noon on four different days.

Air-Pressure Readings

Day	City A Air Pressure (mb)	City B Air Pressure (mb)
1	1004.0	1004.0
2	1000.1	1002.9
3	1000.2	1011.1
4	1010.4	1012.3

The wind speed in the region between cities *A* and *B* was probably the greatest at noon on day
  (1) 1
  (2) 2
  (3) 3
  (4) 4

Base your answers to questions 18 and 19 on the weather map below, which shows a low-pressure system centered near Poughkeepsie, New York. Isobars shown are measured in millibars.

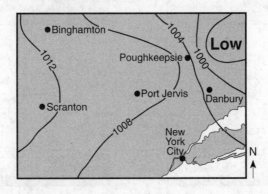

Weather

18. Which city is most likely experiencing winds of the greatest velocity?
    (1) New York City
    (2) Binghamton
    (3) Poughkeepsie
    (4) Scranton

19. Surface winds are most likely blowing from
    (1) Danbury toward New York City
    (2) Poughkeepsie toward Scranton
    (3) Binghamton toward Danbury
    (4) Port Jervis toward Binghamton

20. Which cross section below best shows the locations of high air pressure and low air pressure near a beach on a hot, sunny, summer afternoon?

21. Which cross section best represents how surface winds form by midafternoon near a shoreline on a hot summer day? [Diagrams are not drawn to scale.]

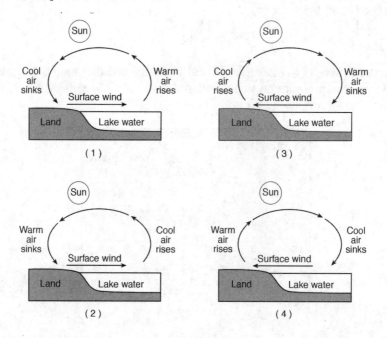

22. Most of the hurricanes that affect the east coast of the United States originally form over the
    (1) warm waters of the Atlantic Ocean in summer
    (2) warm land of the southeastern United States in summer
    (3) cool waters of the Atlantic Ocean in spring
    (4) cool land of the southeastern United States in spring

23. The striped areas on the map below show regions along the Great Lakes that often receive large amounts of snowfall due to lake-effect storms.

These storms generally develop when
    (1) cold air moves to the east over warmer lake water
    (2) cold air moves to the west over warmer land regions
    (3) warm air moves to the east over colder lake water
    (4) warm air moves to the west over colder land regions

24. The seasonal shifts of Earth's planetary wind and moisture belts are due to changes in the
    (1) distance between Earth and the Sun
    (2) amount of energy given off by the Sun
    (3) latitude that receives the Sun's vertical rays
    (4) rate of Earth's rotation on its axis

Weather

Base your answers to questions 25 through 28 on the diagram below, which represents the planetary wind and moisture belts in Earth's Northern Hemisphere.

25. The climate at 90° north latitude is dry because the air at that location is usually
    (1) warm and rising
    (2) warm and sinking
    (3) cool and rising
    (4) cool and sinking

26. Which wind belt has the greatest effect on the climate of New York State?
    (1) prevailing northwesterlies
    (2) prevailing southwesterlies
    (3) northeast trades
    (4) southeast trades

27. Which climatic conditions exist where the trade winds converge?
    (1) cool and wet
    (2) cool and dry
    (3) warm and wet
    (4) warm and dry

28. The tropopause is approximately how far above sea level?
    (1) 12 mi
    (2) 12 km
    (3) 60 mi
    (4) 60 km

29. Condensation of water vapor in the atmosphere is most likely to occur when a condensation surface is available and
    (1) a strong wind is blowing
    (2) the temperature of the air is below 0°C
    (3) the air temperature reaches the dewpoint
    (4) the air pressure is rising

30. Which is a form of precipitation?
    (1) frost   (3) dew
    (2) snow   (4) fog

31. What form of precipitation results from rain that falls through a freezing air mass?
    (1) fog   (3) sleet
    (2) snow   (4) frost

32. Liquid water sometimes turns into ice when it comes in contact with Earth's surface. Which present weather symbol on a station model represents this type of precipitation?

Base your answers to questions 33 through 35 on the station models below, which show various weather conditions recorded at the same time on the same day at four different cities.

33. Which wind speed was recorded at Detroit?
    (1) 15 knots   (3) 35 knots
    (2) 25 knots   (4) 45 knots

34. Which city had the lowest relative humidity?
    (1) Chicago   (3) Buffalo
    (2) Detroit   (4) Utica

35. Which weather symbol best represents the type of precipitation that was most likely occurring in Utica?

36. Air masses are identified on the basis of temperature and
    (1) type of precipitation
    (2) wind velocity
    (3) moisture content
    (4) atmospheric transparency

37. The properties of an air mass are mostly determined by the
    (1) rate of Earth's rotation
    (2) direction of Earth's surface winds
    (3) source region where the air mass formed
    (4) path the air mass follows along a land surface

38. Which area is the most common source region for cold, dry air masses that move over New York State?
    (1) North Atlantic Ocean
    (2) Gulf of Mexico
    (3) central Canada
    (4) central Mexico

39. Which type of air mass is associated with warm, dry atmospheric conditions?
    (1) cP
    (2) cT
    (3) mP
    (4) mT

Base your answers to questions 40 through 44 on the weather maps below and on your knowledge of Earth science. The weather maps show the eastern United States on two consecutive days. Some isobars are labeled in millibars (mb). Letter **X** represents a location on Earth's surface on December 8, 2009.

40. What was the barometric pressure for location **X** on December 8?
 (1) 1016 mb (3) 1008 mb
 (2) 1012 mb (4) 1004 mb

41. Which map best shows the general surface wind pattern around the high-pressure system on December 8?

42. In which direction did the high-pressure center move from December 8, 2009, to December 9, 2009?
 (1) southwest (3) northwest
 (2) southeast (4) northeast

43. Which type of front was located just south of New York City on December 9?
 (1) cold (3) stationary
 (2) warm (4) occluded

44. Which information shown on the weather maps best indicates that wind speeds in New York State were greater on December 9 than on December 8?
 (1) The isobars were closer together on December 9.
 (2) The fronts were closer together on December 9.
 (3) The air pressure over New York State was lower on December 9.
 (4) The air pressure over New York State was higher on December 9.

Weather

Base your answers to questions 45 through 47 on the weather map below and on your knowledge of Earth science. The map of a portion of eastern North America shows a high-pressure center (**H**) and a low-pressure center (**L**), frontal boundaries, and present weather conditions.

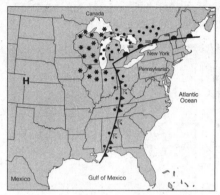

45. Which weather condition is shown along the cold front?
    (1) fog
    (2) snow
    (3) haze
    (4) thunderstorms

46. What was the most likely source region for the air mass over Pennsylvania?
    (1) New York State
    (2) Pacific Ocean
    (3) Gulf of Mexico
    (4) Canada

47. The general surface wind circulation associated with the high-pressure center (**H**) is most likely
    (1) clockwise and outward
    (2) clockwise and inward
    (3) counterclockwise and outward
    (4) counterclockwise and inward

48. Present-day weather predictions are based primarily upon
    (1) land and sea breezes
    (2) cloud heights
    (3) ocean currents
    (4) air mass movements

49. The winds shift from southwest to northwest as heavy rains and hail begin to fall in Albany, New York. These changes are most likely caused by the arrival of
    (1) an mT air mass
    (2) a cT air mass
    (3) a cold front
    (4) a warm front

50. Hurricane season in the North Atlantic Ocean officially begins in June and ends in November. Which ocean surface conditions are responsible for the development of hurricanes?
    (1) warm water temperatures and low evaporation rates
    (2) warm water temperatures and high evaporation rates
    (3) cool water temperatures and low evaporation rates
    (4) cool water temperatures and high evaporation rates

51. Most tornadoes in the Northern Hemisphere are best described as violently rotating columns of air surrounded by
    (1) clockwise surface winds moving toward the columns
    (2) clockwise surface winds moving away from the columns
    (3) counterclockwise surface winds moving toward the columns
    (4) counterclockwise surface winds moving away from the columns

52. Which weather map symbol is used to represent violently rotating winds that have the appearance of a funnel-shaped cloud?

    (1)   (2)   (3)   (4)

CONSTRUCTED RESPONSE QUESTIONS

Base your responses to questions 53 through 56 on the station models below and on your knowledge of Earth science. The changing weather conditions at a location in New York State during a winter storm are recorded on the station models.

Time and Day	Actual Barometric Pressure (mb)	Cloud Cover (%)	Wind Direction From the
12 noon Thursday			

53. Complete the table above by recording the weather data shown on the station model for 12 noon Thursday. [1]

54. State the relative humidity at this location at 8 P.M. Thursday. [1]

55. From 12 noon Thursday until 8 P.M. Thursday, the total amount of snowfall was 12 inches. Calculate the snowfall rate, in inches per hour. [1]

56. As this storm approached, the National Weather Service issued a winter storm warning. Identify two items that should be included in emergency preparedness supplies for a winter storm. [1]

Base your answers to questions 57 through 59 on the cross section below, which shows two weather fronts moving across New York State. Lines X and Y represent frontal boundaries. The large arrows show the general direction the air masses are moving. The smaller arrows show the general direction warm, moist air is moving over the frontal boundaries.

57. Which type of front is represented by letter X? [1]

58. Explain why the warm, moist air rises over the frontal boundaries. [1]

59. Which type of front forms when front X catches and overtakes front Y? [1]

Base your answers to questions 60 through 63 on the weather map below and on your knowledge of Earth science. The weather map shows atmospheric pressures, recorded in millibars (mb), at locations around a low-pressure center (L) in the eastern United States. Isobars indicate air pressures in the western portion of the mapped area. Point A represents a location on Earth's surface.

60. On the weather map *above*, draw the 1012 millibar and the 1008 millibar isobars. Extend the isobars to the east coast of the United States. [1]

61. Identify the weather instrument that was used to measure the air pressures recorded on the map. [1]

62. Identify the compass direction toward which the center of the low-pressure system will move if it follows a typical storm track. [1]

63. Convert the air pressure at location A from millibars to inches of mercury. [1]

Base your answers to questions 64 through 66 on the weather map below. The weather map shows a low-pressure system in New York State during July. The **L** represents the center of the low-pressure system. Two fronts extend from the center of the low. Line *XY* on the map is a reference line.

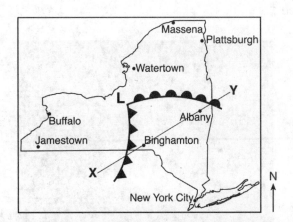

64. The cross section shows a side view of the area along line *XY* on the map. On lines 1 and 2 in the cross section, place the appropriate two-letter air-mass symbols to identify the most likely type of air mass at *each* of these locations. [1]

Weather

65. The forecast for one city located on the map is given below:

"In the next hour, skies will become cloud covered. Heavy rains are expected with possible lightning and thunder. Temperatures will become much cooler."

State the name of the city for which this forecast was given. [1]

66. Identify *one* action that people should take to protect themselves from lightning. [1]

EXTENDED CONSTRUCTED RESPONSE QUESTIONS

Base your answers to questions 67 through 71 on the weather satellite photograph of a portion of the United States and Mexico provided below. The photograph shows the clouds of a major hurricane approaching the eastern coastline of Texas and Mexico. The calm center of the hurricane, the eye, is labeled.

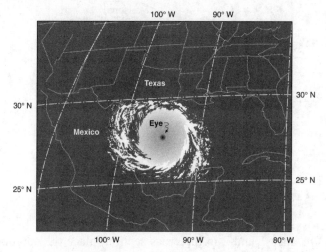

67. This hurricane has a pattern of surface winds typical of all low-pressure systems in the Northern Hemisphere. On the satellite photograph provided, draw *three* arrows on the clouds to show the direction of the surface wind movement outside the eye of the hurricane. [1]

68. Cloud droplets form around small particles in the atmosphere. Describe how the hurricane clouds formed from water vapor. Include the terms "dewpoint" and either "condensation" or "condense" in your answer. [1]

69. State the latitude and longitude of the hurricane's eye. The compass directions must be included in the answer. [1]

70. At the location shown in the photograph, the hurricane had maximum winds recorded at 110 miles per hour. Within a 24-hour period, the hurricane moved 150 miles inland and had maximum winds of only 65 miles per hour. State why the wind velocity of a hurricane usually decreases when the hurricane moves over a land surface. [1]

71. (a) State *two* dangerous conditions, other than hurricane winds, that could cause human fatalities as the hurricane strikes the coast. [2]
    (b) Describe *one* emergency preparation humans could take to avoid a problem caused by one of these dangerous conditions. [1]

72. The map provided below shows six source regions for different air masses that affect the weather of North America. The directions of movement of the air masses are shown. Using the standard two-letter air-mass symbols from the *Earth Science Reference Tables*, label the air masses by writing the correct symbol in each circle on the map. [2]

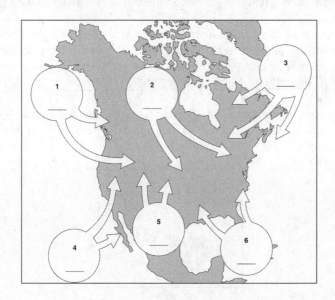

Weather

Base your answers to questions 73 and 74 on the cross section provided below, which represents a house at an ocean shoreline at night. Smoke from the chimney is blowing out to sea.

73. Label the *two* lines provided on the cross section to show where air pressure is relatively "high" and where it is relatively "low." [1]

74. Assume that the wind blowing out to sea on this night is caused by local air-temperature conditions. Label the *two* lines provided on the cross section to show where Earth's surface air temperature is relatively "warm" and where it is relatively "cool." [1]

Base your answers to questions 75 and 76 on the diagram below, which represents water molecules attached to salt and dust particles within a cloud in the atmosphere.

(Not drawn to scale)

75. Explain why salt and dust particles are important in cloud formation. [1]

76. State *one* natural process that causes large amounts of dust to enter Earth's atmosphere. [1]

Base your answers to questions 77 through 79 on the weather station model below and on your knowledge of Earth science. The model shows atmospheric conditions at Oswego, New York.

77. In the spaces below, fill in the correct information for each weather variable listed for this station model. [1]
    Air temperature: _____ °F
    Dewpoint: _____ °F
    Wind speed: _____ knots
    Cloud cover: _____ %

78. Explain how the data on the station model indicate a high relative humidity. [1]

79. Convert the coded air pressure shown on the station model into the actual millibars of air pressure. [1]

Weather

Base your answers to questions 80 through 83 on the map below and on your knowledge of Earth science. The map shows the path of a tornado that moved through a portion of Nebraska on May 22, 2004 between 7:30 P.M. and 9:10 P.M. The path of the tornado along the ground is indicated by the shaded region. The width of the shading indicates the width of destruction on the ground. Numbers on the tornado's path indicate the Fujita intensity at those locations. The Fujita Intensity Scale (F-Scale), in the left corner of the map, provides information about wind speed and damage at various F-Scale intensities.

80. On the map *above*, place an **X** at a location where the tornado damage was greatest. [1]

81. State a possible wind speed of the tornado, in kilometers per hour (km/h), when it was moving through the town of Bennet. [1]

82. Identify the weather instrument usually used to measure wind speed. [1]

83. Describe *one* safety precaution that should be taken if a tornado has been sighted approaching your home. [1]

CHAPTER 23
CLIMATE

> **KEY IDEAS** *Climate* is the term used to describe the general weather conditions in a given area over a long period of time. Global climate is determined by the interaction of solar energy with Earth's surface and atmosphere. Dynamic processes such as cloud formation and Earth's rotation, as well as the positions of mountain ranges and oceans, influence this energy transfer. Climate is important in determining the habitability of a location and the most reasonable land uses within the region. One of the most important effects of climate is the availability of water at the surface and within the ground.

KEY OBJECTIVES
Upon completion of this unit, you will be able to:

- Differentiate between weather and climate.
- Identify the various factors that control climate.
- Describe the role that the cycling of water and energy in and out of the atmosphere plays in determining climatic patterns.
- Describe several patterns of global climatic change.

WHAT IS CLIMATE?

Definition of Climate

Climate is the characteristic weather of a region, particularly **temperature** and **precipitation**, averaged over an extended period of time. We recognize that weather phenomena, when viewed against the long-term, stable backdrop of global climate, are temporary, local changes.

The climate of a region cannot be determined by looking at the weather over a single year. In 1988, the Midwest suffered a nearly rainless summer. Fields of grain in the nation's heartland turned brown and shriveled, water levels dropped in the Mississippi River, and wildfires blazed through millions of acres of forested land. Yet, in the summer of 1993, many of these same places were drenched by torrential rains, farmers watched floodwaters wipe out their fields, and the Mississippi River flooded communities all along its course. Many people wondered, are we seeing Earth's global climate change? One or two extreme summers can't answer that question. Our picture of climate develops slowly as we watch scores of seasons pass. Some winters are warmer than others, some summers dryer, some falls colder. We can gain a sense of the shifting patterns of climate only by comparing measurements taken over many years and even decades.

At any given point in time, a wide variety of climates can exist on Earth simultaneously. Earth's climates range from steamy rain forests to icy glaciers, hot deserts, and cool forests of conifers. However, as shown by the summers of 1988 and 1993 in the Midwest described above, any given region may also be subject to wide seasonal changes in temperature and rainfall. One of the challenges in studying climate is deciding just how to measure it.

The Main Elements of Climate: Temperature and Precipitation

Climate is the result of the interplay of a number of factors. One of the most important elements in determining the climate of a region is energy. As you know, Earth's main source of energy is sunlight, which warms the land, which, in turns, heats the atmosphere. Globally, the amount of heating that occurs depends on the amount of sunlight reaching Earth's surface. A key way of tracking energy flow in a region is by monitoring temperatures. Therefore, scientists who study climate keep track of air temperatures over land and sea, air temperatures at various altitudes, and ocean water temperatures around the globe.

Another major climate element is water. Water is important to all living things, and controls to a large extent the types of plant and animal life that can inhabit a region. It also plays an important role in weathering and is the main agent of erosion and deposition on Earth. Thus, the distribution of atmospheric moisture and the amount of precipitation are important factors in characterizing climates. Some other measurements used to monitor climates and climate changes are the amount of snow and ice cover on land, the extent of sea ice, and the concentrations of various gases in the atmosphere.

CLIMATE FACTORS

You have learned that the main elements of a region's climate are its temperature and precipitation patterns. However, various factors influence the cycling of water and energy in Earth's atmosphere and thereby produce different climates. Temperature and precipitation patterns are controlled by such factors as latitude, the distribution of land and water, high- and low-pressure belts and prevailing winds, monsoons, ocean currents, vegetative cover, elevation, mountain ranges, clouds, and cyclonic storm activity.

Latitude

Latitude is the primary factor of temperature control in climate because it determines both the angle and duration of insolation. Much more insolation is received in low latitudes near the Equator than in high latitudes near the poles.

Only the areas that lie between 23½° N (the Tropic of Cancer) and 23½° S (the Tropic of Capricorn) ever receive the vertical rays of the Sun. The average angle of insolation decreases toward the poles, causing the insolation to be spread over a larger area and therefore to be less intense. As the angle of insolation decreases, the solar energy passes through more of the atmosphere, so more is reflected or absorbed and insolation is further decreased.

While the average length of daylight, or duration of insolation, is 12 hours everywhere on Earth at the time of the equinoxes, at other times it varies. It is always 12 hours at the Equator; but as one moves north or south of the Equator, the duration of insolation varies in a cyclic manner. It increases toward the pole tilted toward the Sun, ranging from 12 hours at the Equator to 24 hours at the pole. It decreases toward the pole tilted away from the Sun, ranging from 12 hours at the Equator to 0 hours at the pole. The result is that the higher the latitude, the greater the cyclic change in duration of insolation over the course of a year. The poles range from 24 hours of insolation in the summer to no insolation in the winter. At the latitudes of New York State, insolation ranges from about 15 hours in the summer to about 9 hours in the winter.

Differences in the angle and duration of insolation with latitude result in three main temperature zones: the always-hot **torrid zone** near the Equator, the always-cold **frigid zone** near the poles, and the seasonally changing, intermediate **temperate zone** in between.

Distribution of Land and Water

The irregular distribution of land and water surfaces on Earth is another major factor controlling climate. As discussed earlier, land surfaces increase in temperature more than water surfaces when insolation strikes them, and they also cool off more rapidly than water surfaces. As a result, air temperatures are warmer in the summer and cooler in the winter over landmasses than they are over oceans at the same latitude. Large bodies of water tend to moderate the temperatures of nearby landmasses by warming them in winter and cooling them in summer. For this reason cities in the interior United States have a greater annual temperature range than coastal cities (see Figure 23.1).

Large areas permanently covered by ice also affect climate. The light color, high specific heat capacity, and latent heat of fusion of ice combine to keep it from melting completely in certain areas. The air over ice surfaces is chilled by contact with the ice, and this cold air, in turn, chills the land and water surrounding these surfaces.

Climate

	J	F	M	A	M	J	J	A	S	O	N	D
San Francisco, CA	49	51	53	56	58	61	63	63	64	61	55	50
St. Louis, MO	32	35	43	55	64	74	78	77	70	58	44	35

Figure 23.1 Average Monthly Temperatures at St. Louis, Missouri, an Inland City, and San Francisco, California, a Coastal City, Both at Approximately the Same Latitude.

Pressure and Wind Belts

The huge convection cells that form in the atmosphere because of unequal heating give rise to global pressure and wind belts. A belt of low pressure caused by rising warm air lies over the Equator. Low pressure favors evaporation, so the air there tends to be both warm and moist. The rising air spreads outward from the Equator, cools, loses moisture because of condensation, and then sinks back to Earth. This descending air creates a belt of high pressure on either side of the equatorial low between 25° and 30° north or south latitude. As the air sinks toward the ground, it is compressed and warms slightly. This warm, dry air is able to absorb moisture, and land surfaces in these regions tend to be dry. Most of Earth's deserts are located in these subtropical, high-pressure belts.

The adjacent high- and low-pressure belts give rise to winds that carry warm, moist air outward from the Equator. These are known as the tropical easterlies or **trade winds**.

The poles are regions of high pressure because of their frigid, dry air. As the frigid air moves outward from the poles, it warms up and begins to rise, creating the subpolar low-pressure belts at 60° north and south latitude. The 30° high-pressure belts and the 60° low-pressure belts give rise to the moist but cooler prevailing westerlies.

In general, the high-pressure belts create regions of dry air, and the low-pressure belts create regions of moist air. The planetary wind belts carry moist or dry air over landmasses, greatly influencing their climates. These winds also transfer heat energy from the Equator toward the poles, modifying temperature patterns.

Figure 23.2 shows the planetary wind and pressure belts and their climatic effects.

Monsoons

Some landmasses are so large that they create their own seasonal convection cells. A **monsoon** is a seasonally reversing system of surface winds caused by temperature differences between land and ocean. A monsoon is a large-scale version of the type of circulation seen in land and sea breezes. The largest monsoon circulation occurs in southern Asia.

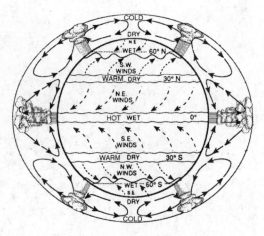

Figure 23.2 The Climatic Effects of Convection Cells in the Atmosphere.

In land and sea breezes, we see *daily* changes in wind direction along a coastline. During the day, land heats up more quickly than water. Air above the land warms up and rises, forming a convection cell that draws cooler air from over the sea inland—a sea breeze. During the night, the land cools off more quickly than the water. The air above the warmer water rises, forming a convection cell that draws cooler air from over the land out to sea—a land breeze. (Refer to Figure 22.11: Land and Sea Breezes.)

In a monsoon, we see *seasonal* changes in wind direction between a large landmass and surrounding oceans. During the summer, the large Asiatic landmass warms up and a low-pressure center develops in the southern part of the continent. This low-pressure center results in winds that blow inland from the surrounding oceans. These moist summer monsoon winds bring clouds and heavy rains. During the winter, intense cooling produces a mass of very cold, very dry air over Siberia. The result is a high-pressure center in the northern part of the landmass that causes the winds to reverse direction and blow from the continental interior outward to the coast. Although these winter monsoon winds may warm up as they descend toward the coast, they pick up little moisture over the land and they bring clear, dry weather. See Figure 23.3.

A monsoon wind system also affects northern Australia. Since the seasons in Australia are opposite to those in Asia, however, winds blow in from the sea in January and outward from the continental interior in July. There is also some indication of a seasonal monsoon in the central part of the United States. During the summer, moist air generally flows from the Gulf of Mexico northward into the central plains. In the winter, cool, dry winds from Canada reverse this flow most of the time.

Climate

(a) Summer Monsoon Winds

(b) Winter Monsoon Winds

Figure 23.3 A Monsoon. A monsoon acts like a large-scale sea breeze in the summer (a) and a land breeze in the winter (b). Source: *The Earth Sciences*, Arthur N. Strahler, Harper & Row, 1971.

Ocean Currents

Ocean currents control climates by transferring heat from the Equator toward the poles, thereby cooling the equatorial regions, warming the polar regions, and influencing the temperature of adjacent landmasses these currents pass. The major surface ocean currents (see Figure 23.4) follow the prevailing winds that blow out of the subtropical high-pressure belts, giving rise to a clockwise flow in the Northern Hemisphere and a counterclockwise flow in the Southern Hemisphere.

Figure 23.4 The Major Surface Ocean Currents of the World. Source: The State Education Department, *Earth Science Reference Tables,* 2011 ed. (Albany, New York: The University of the State of New York).

Wherever the ocean currents flow toward the poles, they carry warm water from the Equator. Wherever they flow toward the Equator, they carry cold water from the polar regions. Ocean currents such as the Gulf Stream and the North Atlantic Current carry warm water northward and eastward. The warm air over these ocean currents moderates the climate of the eastern coast of the United States, the Azores, and western Europe. The cold air over the icy waters of the Peru Current modifies the climate all along the western coast of South America, causing it to be much cooler and dryer than places with the same latitudes on the eastern coast of South America, where the Brazil Current and South Equatorial Current bring warm, moist air with their waters.

Vegetative Cover

Climate influences a region's natural vegetation; conversely, natural vegetation influences climate by affecting rainfall and temperature. Vegetation affects the water cycle by influencing the processes of evapotranspiration and surface runoff, which, in turn, influence rainfall. **Evapotranspiration** is a process that combines evaporation and transpiration, the two mechanisms by which liquid water is returned to the atmosphere as water vapor. **Transpiration** is the process by which plants release water vapor into the atmosphere through their leaves. It has been estimated that approximately half the rainfall in the Amazon Basin is derived from local evapotranspiration. Without the vegetation, the region would have a much drier climate.

Vegetation also influences temperature. Vegetation reflects less insolation back into space than bare surfaces, which tend to warm the climate. This effect is small, however, compared to the cooling that results when vegetation absorbs atmospheric carbon dioxide and thereby decreases the greenhouse effect. Thus, the net effect of deforesting land is to warm the climate.

Elevation

There is a gradual decrease in average temperature with elevation. This is due to the decrease in air pressure with elevation, which causes air to expand and cool. The fact that temperature decreases approximately 1°C for every 100-meter rise in elevation explains why high mountains may have tropical vegetation at their bases but permanent ice and snow at their peaks.

Mountain Ranges

Relief, or the differences in the heights of landforms in an area, is another factor that controls climate. For example, mountain ranges serve as barriers to outbreaks of cold air. In this way, the Alps protect the Mediterranean coast and the Himalayas protect India's lowlands.

In a process called the **orographic effect**, the windward side of a mountain facing winds carrying moisture-laden air usually receives much more precipitation than the leeward side. Mountain ranges force the air that is blown over them by winds to rise, and cool by expansion. This decreases the air's capacity to hold moisture and causes condensation, which forms clouds and precipitation. By the time the air reaches the top of the mountain, it has lost much of its moisture. When the air descends on the leeward side of the mountain, it is warmed by compression, its capacity to hold moisture is decreased, and precipitation is less likely. This orographic effect explains why cities along the Oregon coast west of the Cascades (e.g., Tillamook) have much precipitation while cities in the state's interior (e.g., Bend) have a dry climate (see Figure 23.5).

Clouds and Cyclonic Storm Tracks

Clouds control climate by reducing the amount of insolation gained or lost by a region. During daylight, clouds block sunlight and keep the temperature from rising as high as it would if the skies were clear. At night, the water droplets and water vapor in clouds absorb long-wavelength heat being radiated into space and thus keep the region from getting as cold as it would without them.

Figure 23.5 The **Orographic Effect**. On the windward side of the mountains, moist air is forced to rise and then cools by expansion, resulting in clouds and precipitation. On the leeward side, the cool air, which has lost much of its moisture, descends and warms by compression, resulting in warm, dry air that creates desertlike conditions.

The warm and cold fronts of mid-latitude cyclones that form along the polar front produce the changing weather of middle latitudes. These fronts produce precipitation and carry it over a wide area, causing changes in temperature as both warm and cold fronts pass through.

TYPES OF CLIMATES: THREE PATTERNS OF CLASSIFICATION

Climates are classified by moisture, temperature, and vegetation patterns. There are several systems in use today. The accompanying table summarizes terms often used to describe different patterns of moisture, temperature, and vegetation.

TABLE 23.1 DESCRIPTIVE TERMS FOR CLIMATIC PATTERNS

Moisture	Temperature	Vegetation
Arid	Polar	Desert
Semiarid	Subpolar	Grassland, steppe, taiga
Subhumid	Subtropical	Deciduous forest
		Coniferous forest
Humid	Tropical	Rain forest

Figure 23.6 is a map of the major climates of the world.

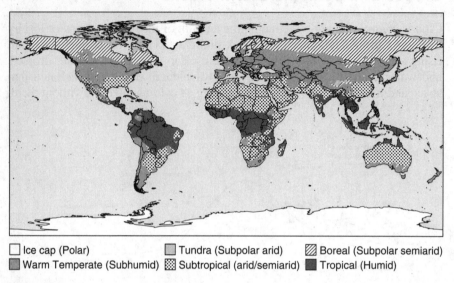

- ☐ Ice cap (Polar)
- ■ Tundra (Subpolar arid)
- ▨ Boreal (Subpolar semiarid)
- ■ Warm Temperate (Subhumid)
- ▦ Subtropical (arid/semiarid)
- ■ Tropical (Humid)

Figure 23.6 Major Climates of the World.

EFFECTS OF CLIMATE

Climate has its greatest effect on vegetation. Individual plant species can generally survive only in a certain range of sunlight, temperature, precipitation, humidity, soil type, and wind. There is such a strong relationship between climate types and vegetation that climate regions are often named for their dominant vegetations.

Climate also affects the type of soil that will develop in an area. Warm, wet climates break down rock into soil faster than cool, dry climates. Soils in wet climates tend to be less fertile, though, because heavy precipitation and infiltration dissolve out nutrients needed by plants. Climate also affects soil by determining the types of plants that will become part of the soil when they eventually decay.

Another effect of climate is the shape into which landforms are eroded. In humid climates, where abundant rainfall makes running water the primary agent of erosion, landforms tend to have rounded contours. In arid climates, where wind is dominant and water is likely to erode Earth's surface in violent bursts, landforms are likely to be more jagged and angular. See Figure 23.7.

Figure 23.7 The Effects of Climate on Landforms. (a) In humid climates, abundant rainfall promotes a vegetative cover that protects the soil from rapid erosion by runoff and produces gently rounded slopes. (b) In arid climates, with little protective vegetation, erosion by rapid runoff and wind leads to steep slopes and exposed bedrock.

GLOBAL CLIMATIC TRENDS

Earth is an active planet. Plate motions move landmasses to different latitudes and change topography. Global temperatures change because of a variety of factors, including changes in the output of energy by the Sun, variation in the tilt of Earth's axis of rotation, and the blocking of incoming solar radiation by some pressure in the atmosphere. As a result, climatic zones shift over time. Some of the climatic trends noted in recent years include desertification, global warming, and the cyclic climate changes associated with El Niño.

Desertification

Desertification is the rapid development of deserts caused by the impact of human activities. Desertification does not occur because of forces originating within the desert. Rather, this patchy changing of dry, habitable land to unhabitable desert is the result of unwise land use by humans, accelerated by natural factors such as drought. Vegetation in dry lands is naturally scarce. Nevertheless, it is a valuable resource that protects the soil from erosion and provides food for people and for livestock. It may also be used for shelter and fuel.

When the sparse natural vegetation of dry land is severely disturbed by overgrazing, clearing for crops, or drought, the land deteriorates. Erosion removes fertile topsoil, leading to a drop in soil fertility. With fewer plant roots to break it up, the soil crusts over in the baking Sun and becomes less permeable to water. When precipitation does fall, less infiltrates and more runs off, leading to increased erosion and a decrease in soil moisture and groundwater. The result is a permanent conversion of marginal dry lands to desert. Natural cycles of drought do play a role in desertification; but without human influence, the arid land is less severely damaged and its natural systems generally recover when the drought ends. Some projections suggest that within the next 20 years human activities will have caused one-third of the world's once-arable land to become useless for growing food crops.

Global Climate Change

Modern burning of fossil fuels has been increasing the amount of carbon dioxide in the air. In the past century, carbon dioxide levels in the atmosphere have increased by an estimated 25 percent, and these levels continue to climb. It is feared that, if the heat trapped by carbon dioxide increases with its concentration, rising levels of this gas will cause increased greenhouse-effect heating of the atmosphere. Global temperatures have shown an upward trend since 1850; it is difficult to tell, however, whether this increased warming is due to human impact or is part of the natural fluctuations in global temperatures.

One of the best known changes in global temperatures in modern times was the "little Ice Age," which lasted from 1450 to 1850. Paintings from this time show alpine glaciers in the Alps advancing dramatically. The high mountains of Ethiopia were covered in snow. There were times when a person could walk across New York Harbor from Manhattan to Staten Island. Then, about 1850, steady warming began.

Only in the mid-1990s did climatologists believe that they had enough evidence to state that greenhouse-effect heating has occurred and that the warming trend has moved beyond the probable range of normal changes in global temperatures. Global warming has probably accelerated desertification. It also has the potential to severely affect agriculture worldwide, increase sea level as glacial ice caps melt, and change global weather patterns. Recent projections suggest that soil-moisture levels in U.S. farmlands could drop by 40 percent if carbon dioxide levels double—an alarming prospect for farmers in areas that get barely enough rainfall now.

The El Niño-Southern Oscillation

The term **El Niño** originally referred to the changes in surface currents along the coasts of Peru and Chile. Trade winds blowing north along these coasts produce upwelling, making these waters rich fisheries. Every year, usually in December, the trade winds subside, upwelling slacks off, and the water becomes warmer. Local fisherman have known for centuries about this change in currents, which marks the end of the peak fishing season. Since the change comes around Christmas, they call it El Niño—"The Child." Every few years, however, the change is greater than usual. The surface water gets much warmer, the upwelling of nutrients ceases, and fish that depend on these food sources disappear.

The warming of the surface waters during El Niño causes a large region of low pressure to form over the southeastern Pacific Ocean. This, in turn, causes changes in the jet stream winds, which profoundly affect global weather patterns.

People in India depend on the summer monsoon to water their crops. When the monsoon fails, famine is widespread. In the early 1900s, analysis of weather records in an attempt to predict when India's summer monsoon would fail led to an important discovery: the air pressure over the Indian Ocean fluctuates. When it is extremely low, the normal monsoon flow carrying moist air from the ocean over the land is disrupted. Also, when the air pressure is unusually high over the Pacific Ocean, it tends to be unusually low over the Indian Ocean, and vice-versa. Over time, the air pressure seesaws back and forth over these two regions, a phenomenon called the **Southern Oscillation**. The changes in air pressure bring dramatic changes in wind and rainfall, and failure of the summer monsoon in India. Exactly what triggers the process is not yet clear; but when the air pressure starts to seesaw, failure of the monsoon can be expected.

By the 1950s it was clear that El Niño and the Southern Oscillation were not isolated events, but part of a complex interaction between ocean and atmosphere that spans half the globe. See Figure 23.8. The El Niño-Southern Oscillation, or **ENSO**, is not really an abnormal event, but simply one extreme of a regular climate cycle. Nevertheless, the weakening or reversal of wind and ocean current patterns drastically changes normal regional weather systems.

The ENSO has had profound effects on weather, wildlife, and society. El Niño years bring drought, forest fires, and heat waves to some regions and storms and floods to others. One of the major El Niño events in the last century occurred in 1982–1983; milder events followed in 1986–1987 and

Climate

1991–1995. By analyzing oceanographic and atmospheric data, scientists detected signs of an El Niño early in 1997 and predicted that another event was on its way. The 1997–1998 event turned out to be even stronger than the 1982–1983 El Niño. In Peru, Chile, and Ecuador, it brought torrential rains, flooding, and failure of the fishing industry. There were also floods in China, East Africa, and southern California. At the other extreme, Indonesia, Australia, southern Africa, and Central America suffered droughts. The weather changes were so severe that Florida, which rarely experiences tornadoes, was hit by a series of killer tornadoes. Estimated losses from the 1997–1998 El Niño totaled more than $20 billion, but could have been much higher. However, scientists' predictions in both of these major events gave the world advance warning and allowed officials in many countries to build up food reserves and prepare for impending disasters.

Figure 23.8 The El Niño-Southern Oscillation (ENSO). (a) Normal Pacific pattern: Equatorial winds gather warm water pool toward the west. Cold water upwells along South American coast. (b) El Niño conditions: Warm water pool approaches the South American coast. The absence of cold upwelling increases warming. (c) La Niña conditions: The Eastern Pacific is cooler than usual, and unusually cool water extends farther westward than is usual.

MULTIPLE-CHOICE QUESTIONS

In each case, write the number of the word or expression that best answers the question or completes the statement.

1. The type of climate in a particular location can be determined by comparing long-term monthly averages for
   (1) precipitation and temperature
   (2) wind speed and wind direction
   (3) infiltration and runoff
   (4) stream discharge and stream velocity

2. Which factors have the *least* effect on the climate of a region?
   (1) mountain barriers and nearness to large bodies of water
   (2) longitude and population density
   (3) latitude and elevation
   (4) wind belts and storm tracks

3. Which single factor generally has the *greatest* effect on the climate of an area on Earth's surface?
   (1) the distance from the equator
   (2) the extent of vegetative cover
   (3) the month of the year
   (4) the degrees of longitude

4. Which two 23.5°-latitude locations are influenced by cool surface ocean currents?
   (1) the east coast of North America and the west coast of Australia
   (2) the east coast of Asia and the east coast of North America
   (3) the west coast of Africa and the east coast of South America
   (4) the west coast of North America and the west coast of South America

5. Which climate condition generally results from both an increase in distance from the equator and an increase in elevation above sea level?
   (1) cooler temperatures
   (2) warmer prevailing winds
   (3) increased precipitation
   (4) increased air pressure

Climate

6. Large oceans moderate the climatic temperatures of surrounding coastal land areas because the temperature of ocean water changes
   (1) rapidly due to water's low specific heat
   (2) rapidly due to water's high specific heat
   (3) slowly due to water's low specific heat
   (4) slowly due to water's high specific heat

Base your answers to questions 7 and 8 on the map below, which represents an imaginary continent. Locations A and B are on opposite sides of a mountain range on a planet similar to Earth. Location C is on the planet's equator.

7. Compared with the climate at location A, the climate at location B would most likely be
   (1) warmer and more humid
   (2) warmer and less humid
   (3) cooler and more humid
   (4) cooler and less humid

8. Location C most likely experiences
   (1) low air pressure and low precipitation
   (2) low air pressure and high precipitation
   (3) high air pressure and low precipitation
   (4) high air pressure and high precipitation

9. Near which two latitudes are most of Earth's dry climate regions found?
   (1) 0° and 60° N
   (2) 0° and 30° S
   (3) 30° N and 60° N
   (4) 30° N and 30° S

10. The map below shows four coastal locations labeled A, B, C, and D.

The climate of which location is warmed by a nearby major ocean current?
(1) A
(2) B
(3) C
(4) D

11. Most scientists infer that increasing levels of carbon dioxide in Earth's atmosphere are contributing to
(1) decreased thickness of the troposphere
(2) depletion of ozone
(3) increased absorption of ultraviolet radiation
(4) increased global temperatures

12. The diagram below shows the flow of air over a mountain, from location A to B to C.

Which graph best shows how the air temperature and probability of precipitation change during this air movement?

13. Arrows on the maps below show differences in the direction of winds in the region of India and the Indian Ocean during January and July. Isobar values are recorded in millibars.

Heavy monsoon rains usually occur in India during
(1) January, when winds blow from the land
(2) January, when winds blow toward high pressure
(3) July, when winds blow from the ocean
(4) July, when winds blow toward high pressure

14. The map below shows a portion of the western United States and Canada. Two cities in Canada, Vancouver and Winnipeg, are labeled on the map.

Which graph best represents the average monthly air temperatures for Vancouver and Winnipeg?

Climate

Base your answers to questions 15 through 18 on the climate graphs below, which show average monthly precipitation and temperatures at four cities, $A, B, C,$ and D.

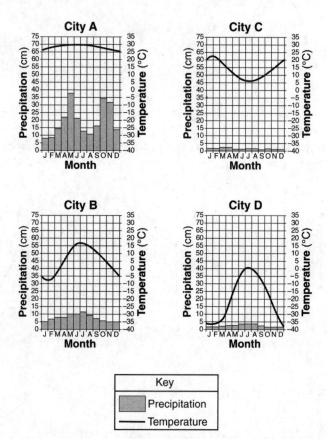

15. City A has very little variation in temperature during the year because city A is located
    (1) on the dry side of a mountain
    (2) on the wet side of a mountain
    (3) near the center of a large landmass
    (4) near the equator

16. During which season does city B usually experience the month with the highest average precipitation?
    (1) spring (3) fall
    (2) summer (4) winter

17. It can be concluded that city *C* is located in the Southern Hemisphere because city *C* has
    (1) small amounts of precipitation throughout the year
    (2) large amounts of precipitation throughout the year
    (3) its warmest temperatures in January and February
    (4) its warmest temperatures in July and August

18. Very little water will infiltrate the soil around city *D* because the region usually has
    (1) a frozen surface
    (2) nearly flat surfaces
    (3) a small amount of runoff
    (4) permeable soil

Base your answers to questions 19 through 22 on the passage and cross section below and on your knowledge of Earth science. The cross section represents a generalized region of the Pacific Ocean along the equator during normal (non-El Niño) conditions. The relative temperatures of the ocean water and the prevailing wind direction are indicated.

El Niño

Under normal Pacific Ocean conditions, strong winds blow from east to west along the equator. Surface ocean water piles up on the western part of the Pacific due to these winds. This allows deeper, colder ocean water on the eastern rim of the Pacific to be pulled up (upwelling) to replace the warmer surface water that was pushed westward.

During an El Niño event, these westward-blowing winds get weaker. As a result, warmer water does not get pushed westward as much, and colder water in the east is not pulled toward the surface. This creates warmer surface ocean water temperatures in the east, allowing the thunderstorms that normally occur at the equator in the western Pacific to move eastward. A strong El Niño is often associated with wet winters along the northwestern coast of South America and in the southeastern United States, and drier weather patterns in Southeast Asia (Indonesia) and Australia. The northeastern United States usually has warmer and drier winters in an El Niño year.

Climate

Normal Pacific Ocean Conditions (non-El Niño years)

(Not drawn to scale)

19. Which statement best describes the planetary wind belts that produce the winds represented in the cross section above?
   (1) Southwest and northwest winds diverge at the equator and blow toward the west.
   (2) Southwest and northwest winds diverge at the equator and blow toward the east.
   (3) Northeast and southeast winds converge at the equator and blow toward the west.
   (4) Northeast and southeast winds converge at the equator and blow toward the east.

20. Compared to non-El Niño years, which climatic conditions exist near the equator on the western and eastern sides of the Pacific Ocean during an El Niño event?
   (1) The western Pacific is drier and the eastern Pacific is wetter.
   (2) The western Pacific is wetter and the eastern Pacific is drier.
   (3) The western and the eastern Pacific are both wetter.
   (4) The western and the eastern Pacific are both drier.

21. Which cross section best represents the changed wind conditions and Pacific Ocean temperatures during an El Niño event? [Diagrams are not drawn to scale.]

22. During an El Niño year, winter climatic conditions in New York State will most likely be
    (1) colder and wetter
    (2) colder and drier
    (3) warmer and wetter
    (4) warmer and drier

CONSTRUCTED RESPONSE QUESTIONS

23. Describe the effect that global warming most likely will have on *both* present-day glaciers and sea level. [1]

Base your answers to questions 24 through 27 on the graph and map below and on your knowledge of Earth science. The average monthly temperatures for Eureka, California, and Omaha, Nebraska, are plotted on the graph. The map indicates the locations of these two cities.

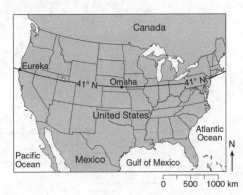

24. Calculate the rate of change in the average monthly temperature for Omaha during the two-month period between October and December, as shown on the graph. [1]

25. Explain why Omaha, which is farther inland, has a greater variation in temperatures throughout the year than Eureka, which is closer to the ocean. [1]

26. Identify the month with the greatest difference in the average temperature between these two cities. [1]

27. Identify the surface ocean current that affects the climate of Eureka. [1]

Climate

EXTENDED CONSTRUCTED RESPONSE QUESTIONS

Base your answers to questions 28 and 29 on the map of Australia below and on your knowledge of Earth science. Points A through D on the map represent locations on the continent.

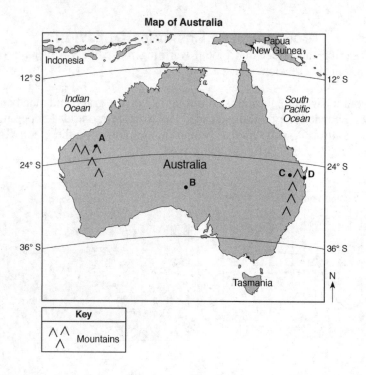

28. Explain why location A has a cooler average yearly air temperature than location B. [1]

29. The cross section below represents a mountain between locations C and D and the direction of prevailing winds.

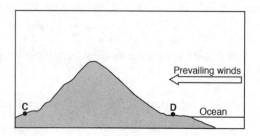

Explain why location D has a wetter climate than location C. [1]

Base your answers to questions 30 through 32 on the data table below and on your knowledge of Earth science. The table shows the area, in million square kilometers, of the Arctic Ocean covered by ice from June through November. The average area covered by ice from 1979 to 2000 from June to November is compared to the area covered by ice in 2005 for the same time period.

Data Table

Month	Average Area Covered by Ice 1979–2000 (million km^2)	Area Covered by Ice 2005 (million km^2)
June	12.2	11.3
July	10.1	8.9
August	7.7	6.3
September	7.0	5.6
October	9.3	8.5
November	11.3	10.5

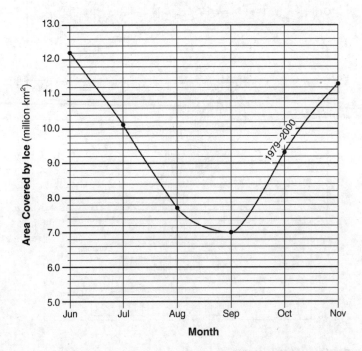

30. Use the information in the data table to construct a line graph. On the grid *above*, plot the data for the area covered by ice in 2005 for *each* month shown on the data table and connect the plots with a line. The average area covered by ice for 1979–2000 has been plotted and labeled on the grid. [1]

Climate

31. Scientists have noted that since 2002, the area of the Arctic Ocean covered by ice during these warmer months has shown an overall decrease from the long-term average (1979–2000). State *one* way in which this ice coverage since 2002 and the ice coverage shown in the 2005 data above provide evidence of global warming, when compared to this long-term average. [1]

32. Identify *one* greenhouse gas that is believed to cause global warming. [1]

Base your answers to questions 33 through 35 on the generalized climatic moisture map of North America below and on your knowledge of Earth science. Areas are classified as generally dry or generally wet, and then ranked by relative moisture conditions. Glacial and mountain climate areas are also shown on the map. Points *A*, *B*, *C*, *D*, and *E* indicate locations on Earth's surface.

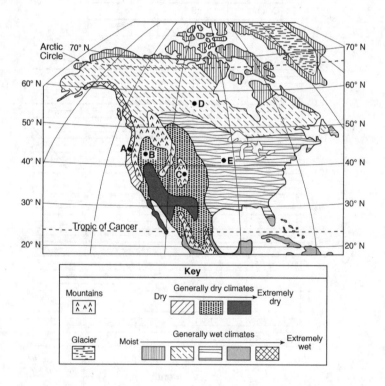

33. Explain why the climate at location *A* is more moist than the climate at location *B*. [1]

34. State the climate factor that causes a cold climate at location *C*. [1]

35. Explain why location *D* has a cooler climate than location *E*. [1]

Appendix A
ANSWERS TO REVIEW QUESTIONS

CHAPTER 1

Multiple-Choice Questions

1.	3	5.	2	9.	4	13.	2	17.	4
2.	2	6.	1	10.	1	14.	2		
3.	4	7.	4	11.	1	15.	3		
4.	4	8.	1	12.	3	16.	4		

Constructed Response Questions

18. Acceptable responses include but are not limited to: rotation; spinning/turning on its axis.
19. 4 h
20. An arrow must be within or touching the zone shown that points away from the North Pole and is generally aligned with Earth's axis.

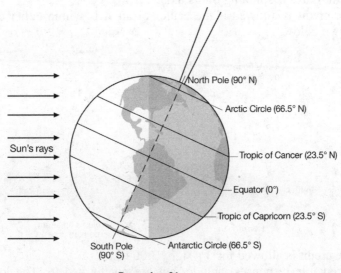

21. 3:00 P.M.

Appendix A: Answers to Review Questions

22.

23. North
24. Hercules: down and to the left (west) *and* Perseus: up and to the right (east)

Extended Constructed Response Questions

25. Acceptable responses include but are not limited to: *Polaris* is not overhead; at the North Pole; the altitude of *Polaris* is 90°; all compass directions are shown, the Sun's path is tilted.
26. Acceptable responses include but are not limited to: the rotation of Earth; Earth is spinning on its axis.
27. One credit is allowed if the center of an **X** is within either clear box shown below.

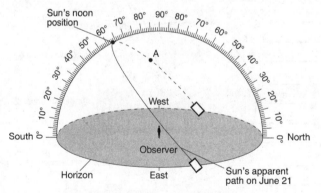

28. One credit is allowed for 3 P.M. *or* 3:00 P.M.
29. Acceptable responses include, but are not limited to: The longer the Sun's path, the longer the duration of daylight; The shorter the Sun's path, the shorter the daylight will be; direct relationship

CHAPTER 2

Multiple-Choice Questions

1.	1	5.	4	9.	1	13.	1
2.	1	6.	3	10.	2	14.	1
3.	4	7.	3	11.	2	15.	1
4.	1	8.	2	12.	1	16.	1

Constructed Response Questions

17. (a) **3** (b) **4** (c) **7** (d) **2** (e) **5**
18. Jupiter
19. Earth's rotation causes day and night.
20. The line graph shows a direct relationship.

21. The geocentric model has Earth in the center. In a geocentric model, Earth does not rotate. Instead, planets revolve around Earth instead of the Sun.
22. The distance from Earth to the Sun varies. There are two foci instead of one center. Earth's orbit is an oval shape; Earth's eccentricity of orbit is 0.017.
23. The force of gravity decreases and then increases.
24. Earth's orbit would become more eccentric, i.e., more elliptical.
25. 0.333
26. The given ellipse has a higher eccentricity than the orbit of Mars.

Extended Constructed Response Questions

27.

28. Earth has an elliptical orbit; therefore, the distance between the Sun and Earth varies in a cyclic manner.
29. The comet orbits the Sun; the comet doesn't orbit Earth.
30. The comet's average distance from the Sun is greater. Therefore, the comet has a larger orbit. During most of its orbit, the comet is moving slower than Earth.

31.   32.

CHAPTER 3

Multiple-Choice Questions

1.	4	6.	2	11.	3	16.	2	21.	3
2.	2	7.	1	12.	1	17.	2		
3.	2	8.	2	13.	4	18.	3		
4.	1	9.	4	14.	3	19.	1		
5.	3	10.	4	15.	1	20.	4		

Constructed Response Questions

22.
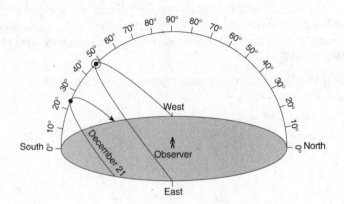

23. Any date from June 19 to June 23
24. Acceptable responses include: gravitational attraction; gravity; pull of the Moon/Sun.
25. Any value from 6 h to 6.25 h
26.

27. As latitude increases, the duration of daylight decreases.
28. 56° ± 1° N
29. winter
30. The **X** may be located anywhere along the line representing the Tropic of Capricorn (23.5° S).
31. Acceptable responses include: parallelism of Earth's axis; the North Pole always points toward *Polaris*; revolution of Earth; location of the Sun's vertical ray; duration/intensity of insolation; angle of insolation.
32. 66.5° N *or* 66½° N *or* 66° 30′ N *or* at the Arctic Circle

Extended Constructed Response Questions

33. The axis line should be drawn through Earth at location *A* within the stippled areas shown below and with the North Pole correctly labeled.
34. The arrow at location *D* should show a general west to east rotation as shown below.

    Example of a 2-credit response for questions 33 and 34:

    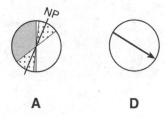

            A   **D**

35. Any value from 88 d to 94 d
36. Earth's distance to the Sun is increasing.
37. The nighttime side of Earth at location *A* faces a different region of space than at location *C*.
38. Position 6 should be circled.
39. Shading should show a gibbous Moon, shaded generally on the right side of the diagram. The shaded area must be less than half of the circle as shown below.

40. The Moon's period of rotation equals its period of revolution.
41. The Moon is closer to Earth, so the Moon's gravity has a greater effect on the ocean tides.
42. Any time from 6:33 P.M. to 6:45 P.M. *or* any military time from 18:33 to 18:45
43. 6 months *or* six months
44. Acceptable responses include: New York State is receiving higher angles of insolation; the Northern Hemisphere is experiencing a longer duration of insolation; the North Pole axis or Northern Hemisphere is tilted toward the Sun; the Sun appears higher in the sky.

45.

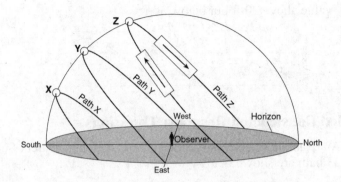

46. Path *X*: Dec. 20 *or* Dec. 21 *or* Dec. 22, winter solstice, first day of winter; Path *Y*: March 20 *or* March 21 *or* March 22, Sept. 21 *or* Sept. 22 *or* Sept. 23, autumnal equinox or fall equinox, vernal equinox *or* spring equinox, an equinox, first day of spring *or* first day of fall; Path *Z*: June 20 *or* June 21 *or* June 22, summer solstice, first day of summer
47. 15°/h

CHAPTER 4

Multiple-Choice Questions

1.	1	6.	2	11.	2	16.	3	21.	3
2.	1	7.	4	12.	1	17.	1	22.	4
3.	4	8.	2	13.	4	18.	3	23.	2
4.	2	9.	2	14.	2	19.	2		
5.	2	10.	1	15.	1	20.	3		

Constructed Response Questions

24. 15°
25. 90°
26. 12 hours
27.

28. Any value between 18 and 23 ft/mi
29. Any depth greater than 600 ft and less than 700 ft
30. The isolines are more closely spaced.

Appendix A: Answers to Review Questions

31. Any value above 20 ft but below 30 ft
32.

Extended Constructed Response Questions

33. 34. An example of correctly drawn arrows and the apparent September 23 path are shown below.

35. 23°30′ N (±1°) or 23.5° N (±1°) or 23½° N (±1°)
36. 15 or 15° E or 15° W
37. One credit is allowed if all three contour lines are drawn correctly as shown below.
38. One credit is allowed for a line starting at location E, ending at Spruce Creek, and within the white region shown below.

39. Acceptable responses include but are not limited to: has a gentler gradient; it is flatter; section II is lower in elevation; section IV is steeper.
40. Any value greater than 1,400 ft but less than 1,500 ft
41. Any value from 323 ft/mi to 345 ft/mi
42.

CHAPTER 5

Multiple-Choice Questions

1.	1	7.	4	13.	4	19.	4	25.	3
2.	4	8.	1	14.	2	20.	3	26.	1
3.	3	9.	1	15.	4	21.	4	27.	4
4.	3	10.	3	16.	1	22.	2		
5.	2	11.	1	17.	1	23.	2		
6.	1	12.	1	18.	3	24.	4		

Constructed Response Questions

28. Three answers are acceptable. Earth is revolving around the Sun. Tilt of Earth's axis. Parallelism of Earth's axis.
29. The center of the **X** must be between the brackets indicated on Earth's orbit, as shown in the diagram below.

30. Acceptable responses include but are not limited to these examples. The North Pole is tilted toward the Sun in the summer. In summer, the Sun is higher in the sky due to the tilt of Earth's axis. New York State receives higher angles of insolation in summer when Earth is farthest from the

Sun. New York State receives lower angles of insolation in winter when Earth is closest to the Sun. There is a greater duration of insolation.

Extended Constructed Response Questions
31.

32. Two answers are acceptable. As the latitude of the observer increases, the azimuth decreases. As the latitude increases, the sunrise is farther north of east.
33. March 19 *or* March 20 *or* March 21 *or* March 22
34. Any value from 23.4° S to 23.5° S. The acceptable unit and compass direction must be included.
35. 24 h
36. Acceptable responses include but are not limited to: Earth's axis is tilted 23.5 degrees from a line perpendicular to the plane of Earth's orbit; axis is tilted; Earth's axis is always parallel to itself at any other place in Earth's orbit; parallelism of Earth's axis; Earth's axis is always aligned with the North Star (*Polaris*) as Earth orbits the Sun.

CHAPTER 6

Multiple-Choice Questions

1.	1	6.	3	11.	4	16.	3	21.	1
2.	3	7.	1	12.	4	17.	1		
3.	2	8.	2	13.	3	18.	2		
4.	1	9.	2	14.	2	19.	1		
5.	4	10.	1	15.	4	20.	4		

Constructed Response Questions

22. As temperature increases, luminosity increases.
23. Red giants or giants
24.

Star	Color	Classification
Sun	yellow	main sequence
Procyon B	white	white dwarf

25.

universe	galaxy	star

Largest ————————————————→ Smallest

26.

Stars	Temperature		Luminosity	
	Hotter	Cooler	Brighter	Dimmer
Procyon B	X			X
Barnard's Star		X		X
Rigel	X		X	

27. *Betelgeuse* is larger. *Betelgeuse* is more massive than *Aldebaran*.
28. Stars with larger masses reach the main sequence faster.
29. The luminosity will decrease.
30. Gravity or gravitational

Extended Constructed Response Questions

31. *Barnard's Star* has a lower surface temperature than the Sun. *Barnard's Star* is less luminous than the Sun.
32. (oldest) universe—*Barnard's Star*—Sun (youngest)
33. *Barnard's Star* is a smaller, less massive star than the Sun. The Sun has more mass.
34. Gravity *or* gravitational attraction
35. Diameter: increases, becomes larger; Luminosity: increases, higher rate of energy emission, the star appears brighter

36. The center of the **X** should be within or touching the box shown below.

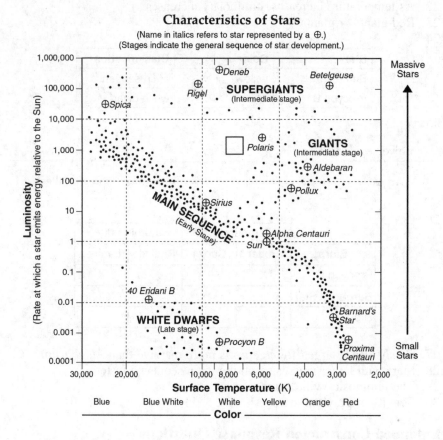

37. Credit is allowed for two acceptable responses. Acceptable responses include but are not limited to: *Spica*; *Sirius*; *Alpha Centauri*; *Barnard's Star*; *Proxima Centauri*.
38. Acceptable responses include but are not limited to: more massive/larger/giant sized/supergiant; *Spica* emits energy at a greater rate than *Barnard's Star*, hotter/greater surface temperature; *Spica* is a blue-colored star.
39. *Both* the relative surface temperature change and the relative luminosity change must be acceptable. Acceptable responses include but are not limited to: Relative surface temperature: *Aldebaran's* surface temperature will increase; it will get hotter. Relative luminosity: Its luminosity will be reduced; luminosity will decrease.

CHAPTER 7

Multiple-Choice Questions

1.	2	6.	1	11.	3	16.	1	21.	3
2.	4	7.	4	12.	2	17.	3	22.	4
3.	2	8.	1	13.	2	18.	2	23.	4
4.	4	9.	1	14.	4	19.	1	24.	3
5.	2	10.	4	15.	4	20.	2	25.	1

Constructed Response Questions

26. The center of the **X** is drawn in or touches the box shown below.

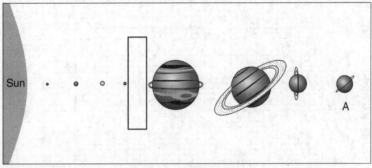

(Not drawn to scale)

27. 16 h
28. Any value equivalent to 4,600 million years ago
29. Any value from 115 to 115.003305 times larger
30. Acceptable responses include but are not limited to: fusion; nuclear fusion; conversion of hydrogen to helium/H to He.

31. The center of the **X** for Neptune should be plotted so that it is within or touches the grid square that is circled as shown below.

32. Any value indicating a diameter from 1.6 cm to 1.9 cm
33. *Both* answers must be correct. Acceptable responses include but are not limited to: Jovian periods of revolution: longer, greater, more time; Jovian periods of rotation: shorter, less time.
34. Mars

Extended Constructed Response Questions

35. The tops of *all eight* bars must end within the acceptable range rectangles indicated below.

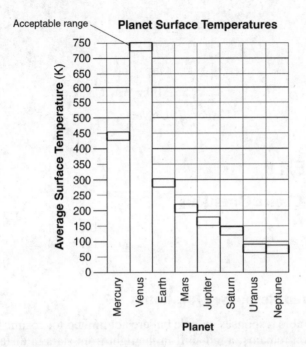

36. Acceptable responses include but are not limited to: carbon dioxide traps heat in the atmosphere; carbon dioxide absorbs infrared radiation and reradiates it back to Venus; carbon dioxide is a greenhouse gas.
37. The line should show a negative slope.

38. Acceptable responses include but are not limited to: Earth's distance to the Sun changes in a cyclic pattern; gravity is greater when Earth is closer to the Sun; Earth moves slower when it is farther from the Sun; Earth has an elliptical/slightly eccentric orbit.
39. Any value from 25,000 light-years to 35,000 light-years

40. Acceptable responses include but are not limited to: a spiral galaxy; a dense center of stars with spiral arms; pinwheel shaped.
41. *All five* astronomical features must listed in the correct order as shown below.

CHAPTER 8

Multiple-Choice Questions

1. **2**   4. **4**   7. **4**   10. **4**
2. **2**   5. **1**   8. **2**   11. **4**
3. **1**   6. **1**   9. **1**   12. **2**

Constructed Response Questions

13. Acceptable responses include but are not limited to: cosmic background radiation remains; a red shift in the light from stars in distant galaxies; the apparent expansion of the universe; and more-distant stars are moving away from Earth at a greater rate than nearby stars.
14. 1300 (±200) million years
15. Background radiation; all galaxies receding
16. (a) Universe contracts, leading to another big bang
    (b) Universe continues to expand and cool indefinitely

Extended Constructed Response Questions

17.

18. Acceptable responses include but are not limited to: as the distance increases, the velocity increases; the farther from Earth, the faster it moves; the farther a galaxy is from Earth, the greater the velocity; and direct relationship.
19. Any value from 3,900 to 4,300 million light-years
20. Acceptable responses include but are not limited to these examples. A shift of light from distant galaxies toward the red end of the spectrum shows that galaxies are moving away from Earth. The red shift shows that the universe is expanding.
21. Acceptable responses include but are not limited to these examples. Earth and our solar system are younger than the Milky Way galaxy. The estimated age of Earth and our solar system is 4.6 billion years, and these distant galaxies are 12 billion years old. Our solar system is about 5 billion years old, much younger than these 12-billion-year-old galaxies.

CHAPTER 9

Multiple-Choice Questions

1.	2	5.	3	9.	3	13.	1
2.	1	6.	2	10.	4	14.	4
3.	2	7.	1	11.	3	15.	1
4.	3	8.	1	12.	3	16.	4

Constructed Response Questions

17. Differentiation occurred while Earth's interior was a molten fluid. In a fluid, two of the forces acting on an object are weight and buoyancy. Weight is a downward force because gravity is attracting the object's mass toward Earth's center. The denser the object, the greater its weight. Buoyancy is an upward force and is equal to the weight of the fluid displaced by the object. If an object is denser than the surrounding fluid, the downward force of weight will be greater than the upward force of buoyancy, and the object will sink downward. If the object is less dense than the surrounding fluid, the downward force of weight will be less than the upward force of buoyancy, and the object will float upward. Thus, denser materials (e.g., iron) sank toward the center of the molten Earth, while less dense materials (e.g., water) floated toward the surface.
18. It is highly unlikely that Earth would have captured such a large object and even less likely that it would go into a circular orbit.
    Or
    Moon rocks are more similar to Earth's mantle than to meteorites like those that would have accreted to form an object the size of the Moon.
19. See Figure 9.2.

Appendix A: Answers to Review Questions

Extended Constructed Response Questions

20. (a) 1. Crust   2. Mantle   3. Core
    (b) 1. Core—3,470 km   2. Mantle—2,860–2,890 km
        3. Crust—10–40 km
    (c) When Earth was mostly molten, denser materials sank toward the center while less dense materials floated toward the surface—a process called differentiation.
21. 1 credit for *each* of two correct responses. Acceptable responses include but are not limited to these examples:
    Friction from Earth's atmosphere causes many meteors to burn up.
    Many meteors fall into the ocean, so craters aren't visible.
    Old craters are eroded away.
    Vegetative growth may hide evidence of a crater.
    Plate tectonics has destroyed some craters.
    Deposition has buried some craters.

CHAPTER 10

Multiple-Choice Questions

1.	**4**	5.	**1**	9.	**3**	13.	**3**	17.	**2**
2.	**4**	6.	**4**	10.	**1**	14.	**2**	18.	**2**
3.	**1**	7.	**4**	11.	**3**	15.	**1**		
4.	**3**	8.	**1**	12.	**2**	16.	**4**		

Constructed Response Questions

19. Acceptable responses include but are not limited to cooling and solidification are processes that form igneous rocks; as early Earth cooled and solidified, igneous rocks were formed; the once-molten Earth formed igneous rocks as it solidified.
20. Hematite
21. Stratosphere
22.

Gases in Earth's Troposphere (%)

Appendix A: Answers to Review Questions

Extended Constructed Response Questions

23. Stratosphere
24. Acceptable responses include but are not limited to the following. The ozone layer absorbs some of the harmful ultraviolet radiation from the Sun. The layer decreases the amount of ultraviolet radiation reaching Earth. The ozone protects humans from skin cancer and eye damage.
25. Acceptable responses include but are not limited to June 20, 21, or 22; the first day of summer; and the summer solstice.
26.

27. Acceptable responses include but are not limited to these examples. Inverse relationship, indirect relationship, or as elevation increases, air pressure decreases.

CHAPTER 11

Multiple-Choice Questions

1.	1	7.	3	13.	1	19.	1	25.	2
2.	4	8.	3	14.	2	20.	1	26.	2
3.	3	9.	3	15.	1	21.	4		
4.	4	10.	1	16.	4	22.	1		
5.	1	11.	1	17.	3	23.	4		
6.	3	12.	3	18.	4	24.	4		

Constructed Response Questions

27. Liquid. The molecules are more closely packed in the liquid phase than they are in either the solid or the gas phase.

Appendix A: Answers to Review Questions

28. The molecules would be spaced even farther apart.
29. As the water changes from water to ice, the volume of the water sample increases.
30. Condensation
31. Acceptable responses include but are not limited to: Soil permeability: high, the soil is unsaturated, soil allows water to seep through easily or rapidly, the surface of the soil is not frozen, a very permeable soil, loosely packed large particles; Land surface slope: a gentle slope, a slope that is not steep, a level slope, flat/a flat plain.
32. Acceptable responses include but are not limited to: the heavy rainfall will infiltrate the ground, causing the water table to rise closer to the surface; infiltration will occur; the ground becomes more saturated; the saturated zone will increase; the water table will rise; erosion of the land surface.
33. The center of the **X** should touch the boundary between the saturation zone and the aeration zone.
34. Acceptable responses include but are not limited to: expansion; condensation; cooling.
35. Acceptable responses include but are not limited to: ice has filled the pore spaces; the ice has produced impermeable soil; the ground is frozen.

Extended Constructed Response Questions

36. Acceptable responses include but are not limited to: when precipitation increases, the water table will rise (or get closer to the surface); the level of the water table above the bedrock will increase with greater precipitation; less precipitation will cause a lower water table; there is a direct relationship between the amount of precipitation and the height of the water table above the impermeable bedrock.
37. Acceptable responses include but are not limited to: an increase in temperature; the stream's surface area increased; increase in wind.
38.

Appendix A: Answers to Review Questions

39. Any value from 14 s to 16 s
40. Acceptable responses include but are not limited to the following. Larger particles have larger pore spaces between them. Larger particles have less total surface area than smaller particles and, therefore, less friction with the moving water.

CHAPTER 12

Multiple-Choice Questions

1.	1	10.	4	19.	1	28.	1	37.	4
2.	4	11.	1	20.	3	29.	4	38.	3
3.	1	12.	1	21.	2	30.	3	39.	4
4.	4	13.	3	22.	3	31.	2	40.	4
5.	1	14.	4	23.	1	32.	3	41.	2
6.	3	15.	1	24.	3	33.	3		
7.	4	16.	2	25.	3	34.	3		
8.	3	17.	4	26.	2	35.	3		
9.	3	18.	3	27.	3	36.	3		

Constructed Response Questions

42. Letters __W__, __X__, __Y__, __V__, __Z__

    Oldest ————————————→ Youngest

43. Acceptable responses include but are not limited to: the grain size of rock layer C is smaller; smaller sediment is deposited in deeper water; shale is made of clay-sized particles/clay; rock layer A contains larger sediments.
44. Acceptable responses include but are not limited to: rock layers A through D are tilted/slanted; the sedimentary rock layers are not horizontal; sedimentary rocks that formed under water have been uplifted above sea level; an intrusion/extrusion has occurred; an unconformity/erosional surface has been buried by igneous rock.
45. One credit is allowed if *both* the circle fossil symbol *and* the evidence are correct. Acceptable evidence includes but is not limited to: the fossil was found only in the Devonian layer/one layer in each outcrop; the fossil was geographically widespread; the fossil indicates a short existence in geologic time/limited time interval.
46. One credit is allowed if *both* outcrop 2 is stated *and* the evidence is correct. Acceptable evidence includes but is not limited to: the rock layers of the same age as those shown in outcrop 2 are all found in New York State; Permian Period rock is not present in New York State but is shown in outcrops 1 and 3.

Appendix A: Answers to Review Questions

47. Acceptable responses include but are not limited to: carbon-14 has a short half-life; these rock layers are too old to contain measurable carbon-14; carbon-14 is used to date recent remains; no organic material remains in the rock.
48. Acceptable responses include but are not limited to: the bedrock in the outcrops formed during the Paleozoic Era and *Coelophysis* lived during the Mesozoic Era; the youngest rock layer is from the Permian and *Coelophysis* did not exist yet; *Coelophysis* lived at a much later time; no Triassic bedrock is shown; layers containing *Coelophysis* have been removed by erosion.
49.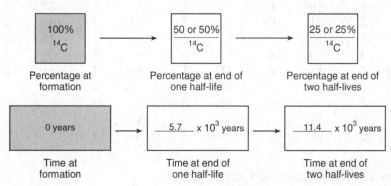

50. Acceptable responses include but are not limited to: ^{14}N; nitrogen-14; N-14; nitrogen/N; $^{14}C \rightarrow {}^{14}N$.
51. Acceptable responses include but are not limited to: carbon-14 has a short half-life; after 1,000,000 years, there would not be enough C-14; ^{14}C decays quickly; the organic remains are too old to be dated with C-14; too little of the original radioactive sample would remain.
52. Acceptable responses include but are not limited to: uranium-238; ^{238}U; U-238, uranium/U.
53. Acceptable responses include but are not limited to: uplift/emergence; erosion; submergence/subsidence; weathering; deposition; burial.
54. *Tetragraptus*
55. Acceptable responses include but are not limited to: intrusion of the Palisades sill; the breakup of Pangaea.

Appendix A: Answers to Review Questions

56. One credit is allowed for one arrow pointing downward on the left side of line *XY* and one arrow pointing upward on the right side of line *XY*.

57. F
58. Any value greater than 420 million years ago but less than 454 million years ago.

Extended Constructed Response Questions

59. One credit is allowed if *two* **X**s are located on *two* of the three boundaries shown below. Credit is *not* allowed if both **X**s are along the same unconformity.

60. Acceptable responses include but are not limited to Devonian Period or Mississippian Period, and Carboniferous Period or Devonian Period.
61. Acceptable responses include but are not limited to *C* is on top of *D* and *C* metamorphosed *D*.
62. *G*
63. Acceptable responses include but are not limited to Cretaceous Period, Paleogene Period, Neogene Period, and Quaternary Period.

Appendix A: Answers to Review Questions

64. Acceptable responses include but are not limited to the fossils are too old for carbon-14 dating and carbon-14 has a very short half-life.
65. One credit is allowed if *all four* bars are drawn in the correct columns and the ends of the bars are within or touch the rectangular areas shown at the end of each bar.

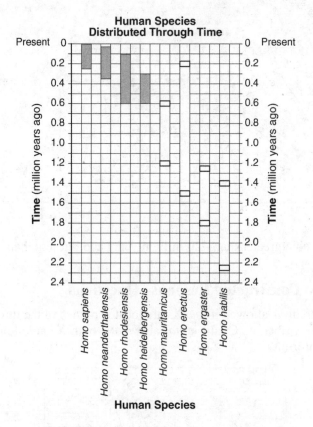

66. *Homo habilis* or *habilis* or *H. habilis*.
67. One credit is allowed for *two* correct species from the following list: *Homo rhodesiensis; Homo heidelbergensis; Homo erectus.*
68. Pleistocene epoch
69. 5,700 years *or* 5.7×10^3 years
70. Acceptable responses include but are not limited to: ^{238}U has a longer half-life; ^{238}U can be used to date older geologic events; ^{14}C is used to date organic remains while ^{238}U is not.
71. Middle Cambrian epoch
72. Acceptable responses include but are not limited to: They were rapidly buried by sediment deposition; oxygen was lacking; it was a deep-water environment.
73. 4, 5, 7, *or* 9

CHAPTER 13

Multiple-Choice Questions

1. **3**
2. **4**
3. **3**
4. **1**
5. **4**
6. **2**
7. **2**
8. **4**
9. **2**
10. **3**
11. **3**
12. **3**
13. **4**
14. **3**
15. **2**
16. **3**

Constructed Response Questions

17.

18. 50 grams
19. Many minerals have the same color; few minerals have exactly the same cleavage.
20. Minerals form by natural processes over long periods of time. Once used, they may not form again during your lifetime or even during a thousand lifetimes.
21.

22. One credit is allowed for quartz and an acceptable reason. Acceptable reasons include but are not limited to the following. Quartz is the hardest mineral shown. Quartz has a hardness of 7. Quartz has the same hardness as garnet, which is used as an abrasive.
23. Luster
24. The color of the dust or powdered form of the mineral; the color of the mark left when a mineral is rubbed on an unglazed porcelain tile
25. Acceptable responses include but are not limited to quartz, garnet, diamond, and pyrite.

Extended Constructed Response Questions

26. One credit is allowed for *two* acceptable responses. Acceptable responses include but are not limited to: pencil lead; lubricants; graphite is a component in composite materials in cars, aircraft, and sports equipment.
27. Acceptable responses include but are not limited to: they have different internal arrangements of atoms; diamonds form at greater depths than graphite; they form under different conditions.
28. One credit is allowed if the table has been correctly completed with one acceptable response in each unshaded box.

Property	Diamond	Graphite
		— silver — gray — black
		— metallic
	— hard — 10	— soft — 1 — 2 — 1–2

29. Acceptable responses include but are not limited to: jewelry; abrasives; gemstone.
30. Garnet or quartz
31. One credit is allowed for *one* of the following minerals: talc; biotite mica/biotite; pyroxene; potassium feldspar/orthoclase; olivine.
32. Chromium or Cr
33. Acceptable responses include but are not limited to hardness, luster, and crystal shape.
34. Marble
35. Eocene Epoch
36. Acceptable responses include but are not limited to convergent plate boundary; subduction zone; and collision boundary.

Appendix A: Answers to Review Questions

CHAPTER 14

Multiple-Choice Questions

1.	1	9.	4	17.	1	25.	2	33.	1
2.	3	10.	3	18.	3	26.	3	34.	4
3.	3	11.	1	19.	4	27.	1	35.	3
4.	1	12.	4	20.	4	28.	4	36.	2
5.	4	13.	4	21.	3	29.	3	37.	1
6.	4	14.	2	22.	1	30.	1		
7.	2	15.	1	23.	2	31.	2		
8.	2	16.	3	24.	4	32.	2		

Constructed Response Questions

38. Acceptable responses include but are not limited to: *A* is cooling slower than *B*; *B* is cooling faster than *A*; intrusive rock forms from molten rock that cools slowly; and extrusive rock forms from molten rock that cools rapidly.
39. 1 mm to 10 mm or 1 to 10 mm. Correct units must be included in the answer.
40. Acceptable responses include but are not limited to: obsidian; basaltic glass; pumice; and vesicular basalt glass.
41. Acceptable responses include but are limited to: it shows banding; the rock is foliated; the minerals are segregated into layers; and distortion.
42. Any *two* of the following three responses must be included: pyroxene (augite), mica (biotite), and amphibole (hornblende).
43. Garnet and one acceptable use. Acceptable responses include but are not limited to jewelry and abrasives.
44. Acceptable responses include but are not limited to: the plagioclase feldspar in the basaltic rock is more calcium rich; the plagioclase feldspar in the granitic rock contains more sodium; less sodium in basaltic plagioclase feldspar; the basaltic rock is more calcium rich.
45. Acceptable responses include but are not limited to: the minerals crystallize at different temperatures; olivine is the first to crystallize and quartz is the last; quartz crystallizes at a lower temperature than olivine; and olivine forms at a higher temperature.
46. One credit is allowed if *both* responses are correct. Acceptable responses include but are not limited to: Similarity: both form at lower temperatures; the rocks have similar mineral compositions; the minerals have similar densities; similar color. Difference: andesite is extrusive and diorite is intrusive; andesite has a finer texture; crystal size/grain size; cooling rates; environment of formation.
47. Acceptable responses include but are not limited to: intrusive; plutonic; underground.

48. One credit is allowed if *both* the color and density are correct. Acceptable responses include but are not limited to: Color—lighter; whiter; pinker. Density—lower; less dense.
49. Acceptable responses include but are not limited to: foliated; mineral alignment; flattened crystals.
50. less than 0.0004 cm or any number given that is less than 0.0004 cm
51. slate
52. Acceptable responses include but are not limited to nonvesicular, coarse, and large crystal.
53. One credit is allowed for marble *or* hornfels.
54. Acceptable responses include but are not limited to: color/light color/dark color; density/low density/high density; mineral composition; rich in Al, Si, *or* rich in Fe, Mg; presence/absence of quartz/potassium feldspar/pyroxene/olivine. **Note:** Credit is not allowed for "composition" alone because it is stated in the question.

Extended Constructed Response Questions

55. One credit is allowed if *all five* lines are drawn from granite to the minerals quartz, potassium feldspar, plagioclase feldspar, amphibole, and biotite mica.
56. Acceptable responses include but are not limited to: Group *A*: lighter colored; more felsic; lower density; lacks magnesium/Mg/iron/Fe; rich in silicon/Si/aluminum/Al.
57. One credit is allowed for quartz *or* pyroxene.
58. Acceptable responses include but are not limited to: the particles are layered; the sedimentary rock may have fossils; there are no intergrown crystals; the sedimentary rock may have rounded or angular fragments; the grains are cemented together; the rock contains different sediments; sedimentary rock contains fragments.
59. Acceptable responses include but are not limited to: large crystals form from slow cooling deep underground; the crystals in syenite formed in an intrusion or an intrusive environment; the texture is coarse; syenite formed by solidification of magma; large interlocking crystals; syenite formed inside of Earth. **Note:** Credit is *not* allowed for "texture," "crystal," or "interlocking crystals" alone because these terms also describe volcanic igneous rock.
60. Acceptable responses include but are not limited to: anorthosite is made of plagioclase feldspar, which is white to gray in color; anorthosite is made of light-colored minerals; plagioclase feldspar is white to gray; because of anorthosite's mineral composition. **Note:** Credit is not allowed for "anorthosite is felsic rich" because plagioclase feldspar is contained in both felsic-rich and mafic-rich igneous rocks.
61. One credit is allowed if *both* the density and composition of gabbro are correct. Acceptable responses include but are not limited to: Density of gabbro: higher; greater. Composition of gabbro: mafic; rich in Fe and Mg; presence of pyroxene and/or olivine; absence of quartz and/

or potassium feldspar. **Note:** Credit is *not* allowed if you list all of the minerals in gabbro because the question asks how the composition of gabbro is different from that of granite.
62. One credit is allowed for limestone *or* marble.
63. One credit is allowed for *two* of the three correct responses: silicon *or* Si; oxygen *or* O; aluminum *or* Al.
64. One credit is allowed for diorite.
65. One credit is allowed for *two* correct responses. Acceptable responses include but are not limited to: melting; cooling; solidification/crystallization/hardening; and heating.

CHAPTER 15

Multiple-Choice Questions

1.	1	8.	2	15.	2	22.	1	29.	4
2.	3	9.	3	16.	3	23.	2	30.	4
3.	1	10.	1	17.	3	24.	1	31.	3
4.	3	11.	2	18.	3	25.	4	32.	1
5.	3	12.	3	19.	3	26.	4	33.	4
6.	3	13.	2	20.	1	27.	1		
7.	2	14.	2	21.	4	28.	1		

Constructed Response Questions

34. Acceptable responses include but are not limited to: tectonic plate movement; movement along a fault; and volcanic eruption.
35. Acceptable responses include but are not limited to: *P*-wave; primary wave; and compressional wave.
36. Any response from 12 min 30 sec to 12 min 50 sec
37. Acceptable responses include but are not limited to the following. The arrival time of the *P*-wave at station *A* is later than the arrival time of the *P*-wave at station *B*. The arrival time difference between the *P*-wave and *S*-wave is greater at station *A*. The amplitudes of the *P*-wave and *S*-wave tracings are greater on the seismogram at station *B*.
38. 15 minutes 50 seconds (±10 seconds)
39. 44.5° N (latitude) and 73.7° W (longitude). The correct compass directions and degrees must be included for both latitude and longitude to receive credit.
40. 3 or three
41. Peru is closer to the epicenter.
42. 3 min 0 sec (±20 sec)

Appendix A: Answers to Review Questions

Extended Constructed Response Questions

43. One credit is allowed if the centers of *all seven* plots are within or touch the circles shown below and are correctly connected with a line that passes within or touches the circles.

44. Acceptable responses include but are not limited to: the outer core is a liquid; the interior temperature of Earth is above the melting point; the outer core absorbs *S*-waves; *S*-waves cannot travel through a liquid; and they cannot travel through a fluid.

45. One credit is allowed if both responses are correct. Pressure: any value from 0.7 million atmospheres to 0.9 million atmospheres; Interior temperature: any value from 4,100°C to 4,300°C.

46.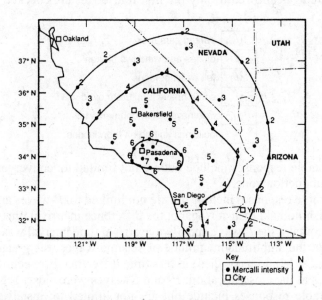
47. Pasadena
48. 35°30′ N, 119° W
49. Movement along the San Andreas fault or movement along the boundaries of the North American and Pacific Plates
50. One credit is allowed for any acceptable line separating all values of VI from VII. **Note:** Credit is allowed even if the line passes through water. Credit is *not* allowed if the line touches or passes through any Mercalli value.

Appendix A: Answers to Review Questions

51. One credit is allowed if only the first four boxes are checked as shown below.

√	parked cars rock
√	dishes and windows broken
√	felt by all persons
√	some heavy furniture moved
	chimneys and monuments fall
	buildings shifted off foundations

52. Acceptable responses include but are not limited to: convergent boundary; subduction zone.
53. Acceptable responses include but are not limited to: P-waves arrived earlier at Ishinomaki than at Nagano; the difference in arrival times was less at Ishinomaki; the P-wave and S-wave arrival time interval was greater at Nagano; the amplitude/magnitude of seismic waves was greater/bigger/stronger at Ishinomaki; there is less time difference between the P- and S-waves at the closer location; P- and S-waves were closer together.
54. Acceptable responses include but are not limited to: install a tsunami monitoring and warning system; build a seawall/barricade/barrier; build tall structures on stronger foundations; designate or plan evacuation routes; prepare emergency kits/supplies; relocate buildings to higher ground. **Note:** Credit is *not* allowed for an action indicating an imminent tsunami (e.g., evacuate to higher ground).

CHAPTER 16

Multiple-Choice Questions

1.	3	6.	1	11.	1	16.	4	21.	2
2.	3	7.	1	12.	2	17.	1	22.	3
3.	4	8.	1	13.	2	18.	2	23.	3
4.	2	9.	1	14.	1	19.	4	24.	3
5.	1	10.	1	15.	3	20.	4	25.	2

Constructed Response Questions

26. Volcanic eruptions, geothermal activity (geysers, hot springs, etc.), temperature readings taken in deep mines and bore holes
27. Magma is molten rock beneath Earth's surface. Lava is molten rock at Earth's surface.
28. Cinder cone—solid particles ejected during a volcanic eruption build up around the vent in a steep, narrow cone.
Composite cone—large, symmetrical cones consisting of alternate layers of solidified lava and volcanic rock particles are formed by alternating explosive and quiet eruptions.

Appendix A: Answers to Review Questions

Shield cone—a broad, flat cone is formed when lava flows quietly from the vent, spreading out in wide, flat sheets.

Lava plateau—lava erupts from a fissure and spreads out in wide, thin sheets that blanket the surrounding land.

29. The longer molten rock takes to cool, the more time mineral crystals have to grow in size. Igneous intrusions are insulated by surrounding rock and cool slowly, allowing large mineral crystals to form.
30. The magma associated with explosive eruptions has a composition similar to that of granite and is thick and pasty. The magma associated with quiet eruptions has a composition similar to that of basalt and is thinner and more fluid.

    The material ejected during explosive eruptions typically consists of particles of molten rock that solidify as they "rain" down on the surrounding land. The material ejected during quiet eruptions is fluid rock that spreads out over the surrounding land in sheets and then solidifies.

Extended Constructed Response Questions

31. Layer *I*
32. Layer *B*
33.

34. The high temperature of the molten rock that intruded to form layer *B* most likely caused the rock in layer *H* to undergo contact metamorphism.
35. Radioactive decay; heat transferred outward from Earth's core
36. (1) South American Plate; (2) Cocos Plate; (3) Caribbean Plate; (4) Nazca Plate
37. Acceptable responses include but are not limited to: mass movement of mud down the mountain; a mud avalanche; it melted snow, causing mudslides; and hot ash and pumice melted snow, creating landslides.
38. Acceptable responses include but are not limited to: a drop in pressure on the magma; steam and gases that were dissolved in the magma violently expanded; cracks in Earth's crust lowered pressure on the magma; and magma pressure cracked the overlying rocks, releasing the gases.

Appendix A: Answers to Review Questions

39. Acceptable responses include but are not limited to these examples. Escaping gas bubbles are trapped in the rapidly cooling magma. Gas/air pockets form in the rock as it cools.
40. Acceptable responses include but are not limited to these examples. Hawaiian magma is mafic, and the magma of Nevado del Ruiz volcano is andesitic. Hawaiian magma is runny, and the magma of Nevado del Ruiz is thick and slow moving. Hawaii is located at a hot spot in the center of the Pacific Plate. Nevado del Ruiz is near a subduction plate boundary.
41. Acceptable responses include but are not limited to these examples. Geologists should monitor conditions and provide early warning. People should leave their houses when early warning of an eruption is given. Avoid building homes in valleys. People should be discouraged from building near the volcano. Evacuation routes should be publicized. Predicted mudslide routes should be identified.

CHAPTER 17

Multiple-Choice Questions

1.	2	10.	1	19.	2	28.	2	37.	1
2.	2	11.	3	20.	2	29.	1	38.	3
3.	3	12.	2	21.	3	30.	1	39.	1
4.	4	13.	4	22.	2	31.	3	40.	1
5.	1	14.	2	23.	2	32.	3	41.	4
6.	4	15.	1	24.	3	33.	2	42.	4
7.	4	16.	3	25.	3	34.	1		
8.	2	17.	1	26.	2	35.	3		
9.	1	18.	3	27.	3	36.	3		

Constructed Response Questions

43. The **X** should be drawn somewhere on the Nazca Plate shaded below.

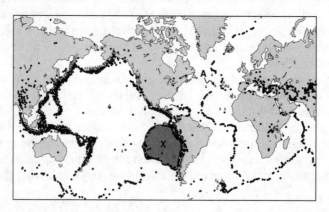

44. Acceptable responses include but are not limited to the following. Most major earthquakes occur at tectonic plate boundaries. Most earthquakes occur at the location of major fault zones. Crustal movement at plate boundaries causes frequent earthquake activity.
45. Acceptable responses include but are not limited to: a hot spot; a magma plume; and the mantle.
46. Acceptable responses include but are not limited to: divergent, diverging lithospheric plates; seafloor spreading; and rifting.
47. One credit is allowed if the width and placement of the shading have been correctly indicated on either the surface and/or the side view as shown below:

48. Acceptable responses include but are not limited to: as distance from the ridge increases, the age of the bedrock increases; the farther away from the ridge, the older the age of the bedrock; the youngest bedrock is near the ridge center; direct relationship; and bedrock nearer the continents is older than bedrock nearer the ridge.
49. Acceptable responses include but are not limited to: convection; convective circulation; slab pull; subduction; convection currents; and magma rising.
50. One credit is allowed for North American Plate to the west of the ridge and Eurasian Plate to the east of the ridge as shown below.
51. One credit is allowed if the *two* arrows are drawn pointing away from the Mid-Atlantic Ridge rift to indicate a divergent plate boundary as shown below.

Appendix A: Answers to Review Questions

Example of 2-credit response for questions 50 and 51.

52. Vesicular basalt *or* scoria. **Note:** No credit is given for "basalt" alone because some basalts do not have a vesicular texture.
53. Acceptable responses include but are not limited to: Iceland is over a mantle hot spot; a mantle plume rises in the region; Iceland hot spot. **Note:** No credit is given for "diverging plate," "convection currents," or "rising magma" because they occur mostly along the rest of the Mid-Atlantic Ridge rift.

Extended Constructed Response Questions

54. Acceptable responses include but are not limited to: transform boundary and transform fault.
55. Acceptable responses include but are not limited to: subduction of Arabian Plate and convergence.

Appendix A: Answers to Review Questions

56.

57. One credit allowed if both texture and color are correct. Acceptable responses include but are not limited to the following. Texture—fine grained; nonvesicular or vesicular; glassy; noncrystalline; grain size less than 1 mm. Color—dark colored; black; green.
58. One credit allowed if both minerals are correct. Acceptable responses include but are not limited to: plagioclase feldspar; potassium feldspar (orthoclase); quartz; amphibole (hornblende); and biotite (mica).
59. Acceptable responses include but are not limited to: WSW; SW; and southwest.
60. Any value from 23.5 to 26.5 miles per million years
61. One credit allowed if the center of the **X** is within the stippled area shown in the example below.
62. One credit allowed for an arrow that shows Laurentia moving to the southeast, south, or east as shown in the example below.

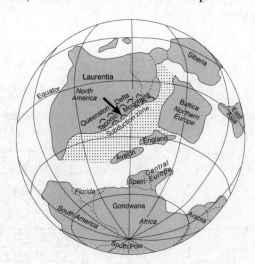

63. Ordovician Period
64. Any value from 500°C to 1,200°C
65. Indian-Australian Plate

Appendix A: Answers to Review Questions

66. Acceptable responses include but are not limited to: subduction; convergence; and plate collision.
67. Acceptable responses include but are not limited to move to: higher ground; evacuate; and move inland.
68. Acceptable responses include but are not limited to the following. Location B is at a transform fault, and location C is at a subduction boundary. Location B has horizontal plate movement, but location C has vertical plate movement. There is a transform plate boundary at B. There is a subducting plate at C.
69. Acceptable responses include but are not limited to the following. A is located at a plate boundary, and D is not located at a plate boundary. Crustal plates are colliding at A, and no plate collision is occurring at D. Location A is on the Pacific Ring of Fire.
70. Acceptable responses include but are not limited to the following. E is located above a mantle hot spot. E is the Canary Islands Hot Spot. F is near the center of a tectonic plate.
71. Acceptable responses include but are not limited to the following. Location G is at the ridge and is presently forming while H was at the ridge in the past. Location H is moving away from the new crust forming in the region at G. The youngest ocean floor bedrock is at the mid-ocean ridge. The plates are diverging at the Southeast Indian Ridge.
72. One credit is allowed if the center of the **X** is drawn within or touches the box shown below.

Volcanoes and Islands of Hawaii

73. One credit is given for *both* West Molokai and East Molokai.

74. Acceptable responses include but are not limited to: as distance increases, age increases; direct relationship; the oldest volcanoes are farthest from Loihi; the younger the volcano, the closer it is to Loihi; they both increase.
75. Acceptable responses include but are not limited to: Hawaii Hot Spot; mantle plume; hot spot; rising magma. **Note:** "Convection" is not accepted because this is a process, not a tectonic feature.
76. Acceptable responses include but are not limited to: to the northwest; NW; from the SE toward the NW; NNW; west northwest.

CHAPTER 18

Multiple-Choice Questions

1.	1	6.	4	11.	4	16.	4	21.	3
2.	1	7.	4	12.	1	17.	3		
3.	2	8.	4	13.	4	18.	4		
4.	2	9.	4	14.	4	19.	1		
5.	4	10.	2	15.	1	20.	3		

Constructed Response Questions

22.

23. Acceptable responses include but are not limited to the following. The acid rain dissolves the limestone. The calcite in limestone chemically reacts with the acid.
24. Acceptable responses include but are not limited to: burning fossil fuels; exhaust emission from automobiles; and smoke from factories.
25. Soil horizon A
26. (a) Leaching
    (b) The soil in horizon B is less fertile than the soil in horizon A because it contains less organic matter and its high clay content inhibits the penetration of water and air.
27. Climate; type of bedrock

Extended Constructed Response Questions

28. (a–c)

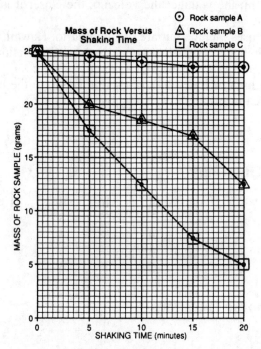

29. Sample A weathered more slowly than samples B and C because sample A is harder and more resistant to erosion.

30. (a) Rate of change in mass = $\dfrac{\text{change in mass}}{\text{change in time}}$

    (b) Rate of change in mass = $\dfrac{25.0 \text{ g} - 5.0 \text{ g}}{20 \text{ min}}$

    (c) Rate of change in mass = 1.0 g/min

31. During the process of abrasion, any part of a rock surface that juts out, such as an edge or corner, is more likely to be struck by another rock particle and be broken off. Thus, after 20 minutes of shaking, the corners and edges of rock material C will be smoother and more rounded than they were before shaking.

32. Acceptable responses include but are not limited to these examples. The physically weathered sediments are larger in particle size than the chemically weathered particles. The sand fragments are larger than the clay fragments. The sand fragments range from 0.006 cm to 0.2 cm in diameter, and the clay fragments are less than 0.0004 cm in diameter.

33. Moisture and temperature should both increase.

34. Answers must have *three* of the following mineral grains: plagioclase feldspar, biotite, amphibole, quartz, or pyroxene.
35. The answer must state granite and describe the greater hardness of the minerals found in granite. Acceptable responses include but are not limited to this example. Granite is composed mainly of quartz and feldspar, which are resistant to abrasion because of their hardness (7 and 6, respectively), while marble is made of calcite, which is softer (hardness of 3).

CHAPTER 19

Multiple-Choice Questions

1.	3	7.	1	13.	3	19.	3	25.	3
2.	4	8.	2	14.	3	20.	4	26.	1
3.	3	9.	1	15.	4	21.	4	27.	4
4.	3	10.	4	16.	2	22.	2	28.	3
5.	3	11.	2	17.	4	23.	4	29.	3
6.	1	12.	3	18.	2	24.	2		

Constructed Response Questions

30. Pebble
31. Acceptable responses include but are not limited to: abrasion; weathering; erosion; and particles were worn down as they were scraped along the bedrock.
32. Shale
33. Acceptable responses include but are not limited to the following. A glacier forms a U-shaped valley. Glaciers form U-shaped valleys, and streams form V-shaped valleys.
34. Acceptable responses include but are not limited to the following. The type 3 stream meanders more. The type 3 stream occupies a wider floodplain. The type 1 stream has a straighter course.
35. Acceptable responses include but are not limited to the following. Stream velocity is greater on the outside of the meandering channel. Stream flow is slower on the inside of the meandering channel. Water is moving faster on the outside of a meandering curve.
36. Acceptable responses include but are not limited to the following. These tumbling cobbles and pebbles were abraded against other transported rocks and the stream channel. Abrasion occurred as the rocks bounced and rolled along the bottom of the streambed. Sharp corners and edges were knocked off, scraped, and/or worn down. They ground against other sediment and rocks. **Note:** Credit is *not* allowed for statements that describe the water alone as the primary cause of rounding.

37. Acceptable responses include but are not limited to: scratches/striations on the bedrock surface; grooves in bedrock; a boulder transported from a more northerly outcrop on the bedrock; an erratic; drumlin.
38. One credit is allowed if *both* responses are correct. Acceptable responses include but are not limited to: Moraines: unsorted sediments/mixed particles, unlayered; Outwash plain: sorted deposits, layered sediments.
39. Acceptable responses include but are not limited to: the valley would have a U-shaped appearance; flat bottom and steep sides; rounded shape.

Extended Constructed Response Questions

40. Acceptable responses include but are not limited to: the sample is angular in appearance; it is not rounded; the edges are not worn off. **Note:** The responses "not very weathered" or "no abrasion" are *not* acceptable because they do not describe an appearance or observable characteristic of the rock sample.
41. Any value from 70 cm/s to 110 cm/s
42. Any value from 48 m/km to 52 m/km
43. Two credits are earned for correctly listing three agents of erosion and identifying a characteristic surface feature formed by each of the three agents of erosion. Only one credit is earned for correctly listing only two agents of erosion and identifying a characteristic surface feature formed by each of the two agents of erosion. Only one credit is earned for correctly listing three agents of erosion even if the surface feature is incorrectly identified. Acceptable responses include but are not limited to:

Agent of Erosion	Surface Feature Formed
Waves	beach, sandbars, barrier islands
Wind	loss of topsoil, dunes
Glacier	U-shaped valley, moraines, drumlins
Running water (streams)	V-shaped valley, deltas, meanders
Mass movement	landslides, slumps

44. Acceptable responses include but are not limited to: U-shaped: it was eroded by glaciers; a glacier formed the valley; formed by glacial ice; V-shaped: running water cut the V-shaped valley; a stream formed the valley.
45. One credit is allowed for X and a correct explanation. Acceptable explanations include but are not limited to: point X is on the outside of a meander curve; stream velocity is greater at point X; more deposition occurs at Y.
46. Acceptable responses include but are not limited to: the stream began to flow over a nearly flat landscape; stream velocity decreased; gradient

decreases from the mountains to the floodplain; the stream flows more slowly on the floodplain; the floodplain is composed of loose sediment.
47. Sandstone

CHAPTER 20

Multiple-Choice Questions

1.	**1**	10.	**2**	19.	**3**	28.	**3**	37.	**2**
2.	**4**	11.	**2**	20.	**2**	29.	**1**	38.	**1**
3.	**2**	12.	**1**	21.	**4**	30.	**1**	39.	**2**
4.	**1**	13.	**4**	22.	**3**	31.	**3**		
5.	**4**	14.	**3**	23.	**1**	32.	**4**		
6.	**4**	15.	**2**	24.	**2**	33.	**1**		
7.	**4**	16.	**3**	25.	**3**	34.	**1**		
8.	**1**	17.	**4**	26.	**3**	35.	**1**		
9.	**1**	18.	**3**	27.	**4**	36.	**3**		

Constructed Response Questions

40. One credit is allowed if the line is drawn from point *A* to point *B* and shows that the stream channel is deeper near side *A*.

41. Acceptable responses include but are not limited to: the particles were weathered by abrading with other particles; rolling and bouncing along the streambed breaks off corners and polishes rocks; the sediments scrape against the streambed; they rub against one another; abrasion; weathering due to collision of particles. **Note:** Credit is *not* allowed for water or erosion, acting alone, to smooth rocks because this is restating the question, and water alone, without sediments, does not abrade rock.
42. Delta *or* any specific type of delta
43. Acceptable responses include but are not limited to: stream velocity/speed; gradient/slope of the stream; location within a meander/stream channel; volume/amount of stream discharge; shape of stream channel (straight vs. meandering); water depth; material found in the stream or along the streambed (vegetation, trees, sediments); type of bedrock; particle size/shape/density.

Appendix A: Answers to Review Questions

44. One credit for a streambed that is deeper near the X.

45. Acceptable responses include but are not limited to: water velocity decreases, causing some sediment to be dropped; the stream slows down as it enters the lake.
46. One credit is allowed if the relative positions of the symbols or particle names are in the order shown.

47. Any value from 80 cm/s to 100 cm/s
48. Acceptable responses include but are not limited to: southeastward, from NW to SE; south-southeastward; and from N to S.
49.

50. Hudson Highlands or Adirondack Mountains
51. Acceptable responses include but are not limited to the following. Global warming will cause glaciers to melt, which will raise the sea level. New York City and Long Island could be flooded when the sea level rises.

52. One credit is allowed if the center of the **X** is located within the shaded area shown. Credit is also allowed if a symbol other than **X** is used.

53. 74° W Both the correct unit and compass direction must be included in the answer.
54. One credit is allowed if both the name of the geologic age and the name of the bedrock are correct. Acceptable responses include but are not limited to:
Geologic age—Proterozoic, Middle Proterozoic, Precambrian, about 1,000 million years.
Name of bedrock—anorthosite, anorthositic.

Extended Constructed Response Questions

55. Any value from 0.8 cm to 1.1 cm
56. Credit is allowed if both responses are acceptable. Acceptable responses include but are not limited to: Changes in size—smaller; decreases: Changes in shape—they become more rounded; rounder; less angular.
Note: Credit is not allowed for "smoother" alone because this denotes a texture, not a shape.
57. Acceptable responses include but are not limited to: sediments deposited by a river are arranged in layers/stratified layers; river deposits are sorted; sorted by size, shape, and density.
58. Acceptable responses include but are not limited to: gneiss; schist; phyllite.
59. *A*: feldspar, *B*: quartz, *C*: amphibole, *D*: garnet
60. Acceptable responses include but are not limited to: the slope is less steep; gentle slope; the land is nearly flat.
61. Acceptable responses include but are not limited to: increase in stream discharge/velocity; precipitation; snowmelt; higher/steeper slope; uplift

of the mountains or plateau; cutting down into softer bedrock layers; deforestation.
62. Acceptable responses include but are not limited to: the stream velocity varied over time, producing layered deposits; sediments deposited in water are sorted by size and density; large particles settle faster; the smallest particles were dropped last; sediments were deposited by a stream; sediments are sorted.
63. One credit is allowed for two acceptable responses. Acceptable responses include but are not limited to: compaction/compression/pressure; cementation; burial; deposition/deposition of more sediment on top; dewatering. **Note:** Credit is not allowed for "weathering and erosion" because weathering and erosion do not change the already accumulated sediments into sedimentary rock.
64. Acceptable responses include but are not limited to: buildings could be damaged or destroyed by flooding; the river's course changes due to erosion and deposition; it's on the floodplain; the ground might be unstable; the ground can become saturated/a swamp; the meanders change positions over time.
65. Any response from 490 to 443 million years
66. Acceptable responses include but are not limited to: the channel at A has a V shape, and running water produces V-shaped channels.
67. 1. Till, 2. stratified drift, 3. clay and silt, 4. windblown sand
68. Acceptable responses include but are not limited to: glacial deposits are unsorted and till is a direct ice deposit.
69. Acceptable responses include but are not limited to the following. The gentle slope of the dune is on the southwest side. The windward side has a less steep slope. The steeper side is leeward.

CHAPTER 21

Multiple-Choice Questions

1.	3	9.	2	17.	4	25.	2	33.	1
2.	4	10.	2	18.	3	26.	1	34.	3
3.	2	11.	1	19.	2	27.	2	35.	3
4.	2	12.	4	20.	3	28.	1	36.	2
5.	1	13.	2	21.	3	29.	1	37.	2
6.	1	14.	2	22.	4	30.	2		
7.	1	15.	3	23.	1	31.	3		
8.	4	16.	1	24.	2	32.	1		

Appendix A: Answers to Review Questions

Constructed Response Questions

38. One credit is allowed if the center of all **X**s are plotted within the circles shown and are correctly connected with a smooth, curved line that passes through the circles.

39. Acceptable responses include but are not limited to the following. As latitude increases, the duration of daylight decreases. Higher latitudes have shorter daylight periods. Lower latitudes have longer daylight periods. It is an inverse relationship.
40. One credit is allowed for a correct answer ±1° based on the student-drawn graph. For example, on the graph shown, the answer should be 56° ±1° N.
41. Winter
42. One credit is allowed for an arrow beginning at point *B* and curving to the northeast or curving to the right. Credit is allowed even if the arrow is not curved or if it does not start at point *B* but is drawn from southwest to northeast.

43. 12 noon or noon or 12 P.M.
44. Acceptable responses include but are not limited to: March 20, 21, or 22; September 21, 22, 23, or 24; and equinox.
45. Acceptable responses include but are not limited to the following. *B* is located at a lower latitude. *B* is located closer to the equator. *A* is farther

north. *A* is located at a greater distance from the latitude receiving direct Sun rays on this day.
46. Acceptable responses include but are not limited to: methane; water vapor; carbon dioxide (CO_2); and ozone.
47. Acceptable responses include but are not limited to the following. Greenhouse gases absorb the longer-wave radiation coming from Earth's surface. Greenhouse gases trap the heat given off by Earth. Greenhouse gases absorb infrared energy.
48. Credit is allowed if *all six* arrowheads are drawn anywhere on each line to indicate the correct direction of water movement around letters *A* and *B* as shown below:

49. 2,260 J/g
50. The energy absorbed during a phase change, as between points *B* and *C* or between points *D* and *E*, is used to break molecular bonds rather than to make molecules move faster. Since molecular speed doesn't increase, temperature doesn't increase.
51. *D* and *E*
52. *B* and *C*
53. More energy (2,260 J/g) is required to evaporate water between points *D* and *E* than to melt water (334 J/g) between points *B* and *C*.

Extended Constructed Response Questions

54. Any value from 8.0 h to 9.5 h
55. 60° N
56. Acceptable responses include but are not limited to: Earth's North Pole is not tilted toward or away from the Sun on those dates; Earth's axis is perpendicular to sunlight on those two dates; the Sun's direct rays at noon are over the equator; the Sun rises directly east and sets directly west on those dates; these dates are equinoxes; and these dates are the

first day of spring and the first day of fall. **Note:** "Earth is not tilted" alone is *not* acceptable because Earth is always tilted on its axis with respect to its orbital plane.

57. [1] Any line that extends from the beginning of January to the end of December, and is completely within the clear band shown below.

58. One credit is allowed if *both* responses are acceptable. Acceptable responses include but are not limited to: Color—black, dark; Texture—rough, bumpy, uneven, jagged, coarse.
59. 0.5 C°/h. **Note:** Credit not allowed for negative values or for any fraction other than ½.
60. One credit is allowed if *both* responses are acceptable. Acceptable responses include but are not limited to: carbon dioxide/CO_2; methane/CH_4; water vapor/H_2O; chlorofluorocarbons/CFCs; nitrous oxide/N_2O; ozone/O_3.

CHAPTER 22
Multiple-Choice Questions

1.	2	12.	2	23.	1	34.	1	45.	4
2.	4	13.	4	24.	3	35.	4	46.	3
3.	2	14.	2	25.	4	36.	3	47.	1
4.	3	15.	1	26.	2	37.	3	48.	4
5.	1	16.	3	27.	3	38.	3	49.	3
6.	2	17.	3	28.	2	39.	2	50.	2
7.	4	18.	3	29.	3	40.	2	51.	3
8.	2	19.	3	30.	2	41.	4	52.	3
9.	4	20.	3	31.	3	42.	4		
10.	2	21.	4	32.	1	43.	2		
11.	2	22.	1	33.	2	44.	1		

Constructed Response Questions

53. One credit is allowed if all three weather variables for 12 noon Thursday are correctly recorded.

Time and Day	Actual Barometric Pressure (mb)	Cloud Cover (%)	Wind Direction From the
12 noon Thursday	1001.2	100	SW or SSW

54. 100%
55. 1.5 in./h or 1½ in./h. **Note:** Credit is not allowed for 12/8 in. *or* 3/2 in./h because these do not show a complete calculation.
56. One credit is awarded for two acceptable items. Acceptable responses include but are not limited to: first aid kit; blankets; batteries; radio; flashlight; bottled water; food; generator; and necessary medications.
57. cold front
58. Acceptable responses include but are not limited to the following. The warm, moist air is less dense. The warm, moist air is lighter. Warm air is overriding the more dense cold air.
59. occluded front
60. One credit is allowed if both isobars are correctly drawn to the east coast of the United States or to the edge of the map.

61. Barometer or barograph
62. Acceptable responses include but are not limited to: NE; northeast; east; ENE.
63. Any value from 30.00 to 30.01 in. of Hg
64. One credit is allowed if both air masses are correct: (1) cP (2) mT.
65. Binghamton
66. Acceptable responses include but are not limited to: move indoors, do not use electrical equipment or telephones; and do not stand under tall objects.

Extended Constructed Response Questions

67.

68. Acceptable responses include but are not limited to these examples. Rising air cools to the dewpoint, and water vapor condenses. Condensation occurs when the dewpoint is reached.
69. 27°30′ N or 27.5° N (±1°) and 95° W (±1°)
70. Acceptable responses include but are not limited to these examples. Over land there is less energy from evaporating water. Winds decrease in strength due to friction with the land.
71. (a) Response should include two dangerous conditions. Acceptable responses include but are not limited to these examples: flooding and tornadoes; storm surge and collapsing structures; hail and lightning; and downed electrical wires and flying debris.
(b) The response must be an emergency preparation that can be taken before the approaching hurricane hits the area. Acceptable responses include but are not limited to these examples. Evacuate to a higher elevation. Take shelter. Board up windows. Build a seawall.

72.

73. One credit is allowed for showing "high" air pressure over the land and "low" air pressure over the ocean, as shown below.
74. One credit is allowed for showing "warm" air temperature over the ocean and "cool" air temperature over the land, as shown below.

75. Acceptable responses include but are not limited to: the following. Water droplets form on the surfaces provided by the salt and dust particles. Salt and dust particles are condensation nuclei, allowing the water vapor to change into liquid drops and to form clouds.
76. Acceptable responses include but are not limited to: dust particles can be blown into the atmosphere by winds; a volcanic eruption; and a forest fire.
77. One credit is allowed if all four weather variables are correct. Air temperature: 31°F; Dewpoint: 29°F; Wind speed: 10 knots; Cloud cover: 100%.
78. One credit is allowed for an acceptable response. Acceptable responses include but are not limited to: the dewpoint and air temperature are nearly the same; snow is falling in Oswego; there is 100% cloud cover; air pressure is low.
79. 999.5 mb

80. One credit is allowed if the center of the **X** is within or touches any of the clear areas along the path of the tornado shown below.

81. Any value from 254 km/h to 332 km/h
82. Anemometer or wind speed meter
83. Acceptable responses include but are not limited to: go into a basement or underground storm shelter; go to an interior room; stay away from windows; get under something sturdy. **Note:** Credit is not allowed for a response that indicates a safety precaution to prepare for a future tornado.

CHAPTER 23

Multiple-Choice Questions

1.	1	6.	4	11.	4	16.	2	21.	2
2.	2	7.	2	12.	2	17.	3	22.	4
3.	1	8.	2	13.	3	18.	1		
4.	4	9.	4	14.	4	19.	3		
5.	1	10.	3	15.	4	20.	1		

Constructed Response Questions

23. One credit allowed if *both* responses are correct. Acceptable responses include but are not limited to: Glaciers—will melt, will retreat; decrease in size, become smaller, shrink, the rate of melting will increase; Sea level—will rise, increase, higher, coastal flooding.
24. 8C°/month *or* –8C°/mo

Appendix A: Answers to Review Questions

25. One credit is allowed for an acceptable response. Acceptable responses include but are not limited to: Omaha is surrounded by land, which has a low specific heat; the Pacific Ocean moderates the temperature/climate of Eureka; large bodies of water change temperature more slowly than land does; water has a higher specific heat than land; the relatively drier air around Omaha has a lower specific heat than the moist air around Eureka. **Note:** Credit is not allowed for "Eureka is closer to water so temperatures remain constant" because this just restates the question without explaining the role that water plays in causing constant temperatures.
26. January *or* Jan.
27. California Current

Extended Constructed Response Questions

28. One credit is allowed for an acceptable response. Acceptable responses include but are not limited to: the higher elevation at A has a cooler temperature; location A is at a higher elevation; location A is in the mountains; location B is not as high in elevation.
29. One credit is allowed for an acceptable response. Acceptable responses include, but are not limited to: location D has air that is rising, expanding, and cooling to the dewpoint; location D is on the windward side of the mountain; location D is closer to the ocean; location C is on the leeward side of the mountain.
30. One credit is allowed if the centers of all six plots are within the circles shown and are correctly connected with a line that passes within each circle.

Appendix A: Answers to Review Questions

31. One credit is allowed. Acceptable responses include but are not limited to: the area covered by ice in 2005 was less than the average area covered by ice from 1979 to 2000; the area covered by ice was less, showing evidence of global warming; more ice melted in 2005 than the average that melted from 1979 to 2000; the ice caps were melting, causing less surface ice in 2005; there was less ice in 2005.
32. One credit is allowed. Acceptable responses include but are not limited to: carbon dioxide *or* CO_2; methane *or* CH_4; water vapor *or* H_2O gas; nitrous oxide *or* N_2O; ozone *or* O_3; chlorofluorocarbons *or* CFCs.
33. One credit is allowed. Acceptable responses include but are not limited to: location A is on the windward side of mountains; location A receives prevailing winds off the ocean; location A is closer to the ocean; location B is on the leeward side of a mountain range; adiabatic warming occurs in descending air at location B after losing most of its moisture on the windward side of a mountain/orographic effect; the prevailing southwest winds bring moist air to location A.
34. One credit is allowed. Acceptable responses include but are not limited to: elevation; high altitude; mountains.
35. One credit is allowed. Acceptable responses include but are not limited to: higher latitude; farther north of the equator; lower angle of insolation; location E is closer to the equator. **Note:** Credit is not allowed for "latitude" alone because it is not specific enough.

Appendix B

REFERENCE TABLES FOR PHYSICAL SETTING/ EARTH SCIENCE

The University of the State of New York • THE STATE EDUCATION DEPARTMENT • Albany, New York 12234 • www.nysed.gov

Reference Tables for Physical Setting/EARTH SCIENCE

Radioactive Decay Data

RADIOACTIVE ISOTOPE	DISINTEGRATION	HALF-LIFE (years)
Carbon-14	$^{14}C \rightarrow {}^{14}N$	5.7×10^3
Potassium-40	$^{40}K \rightarrow {}^{40}Ar$, $^{40}K \rightarrow {}^{40}Ca$	1.3×10^9
Uranium-238	$^{238}U \rightarrow {}^{206}Pb$	4.5×10^9
Rubidium-87	$^{87}Rb \rightarrow {}^{87}Sr$	4.9×10^{10}

Specific Heats of Common Materials

MATERIAL	SPECIFIC HEAT (Joules/gram • °C)
Liquid water	4.18
Solid water (ice)	2.11
Water vapor	2.00
Dry air	1.01
Basalt	0.84
Granite	0.79
Iron	0.45
Copper	0.38
Lead	0.13

Equations

$$\text{Eccentricity} = \frac{\text{distance between foci}}{\text{length of major axis}}$$

$$\text{Gradient} = \frac{\text{change in field value}}{\text{distance}}$$

$$\text{Rate of change} = \frac{\text{change in value}}{\text{time}}$$

$$\text{Density} = \frac{\text{mass}}{\text{volume}}$$

Properties of Water

Heat energy gained during melting	334 J/g
Heat energy released during freezing	334 J/g
Heat energy gained during vaporization	2260 J/g
Heat energy released during condensation	2260 J/g
Density at 3.98°C	1.0 g/mL

Average Chemical Composition of Earth's Crust, Hydrosphere, and Troposphere

ELEMENT (symbol)	CRUST Percent by mass	CRUST Percent by volume	HYDROSPHERE Percent by volume	TROPOSPHERE Percent by volume
Oxygen (O)	46.10	94.04	33.0	21.0
Silicon (Si)	28.20	0.88		
Aluminum (Al)	8.23	0.48		
Iron (Fe)	5.63	0.49		
Calcium (Ca)	4.15	1.18		
Sodium (Na)	2.36	1.11		
Magnesium (Mg)	2.33	0.33		
Potassium (K)	2.09	1.42		
Nitrogen (N)				78.0
Hydrogen (H)			66.0	
Other	0.91	0.07	1.0	1.0

2011 EDITION°
This edition of the Earth Science Reference Tables should be used in the classroom beginning in the 2011–12 school year. The first examination for which these tables will be used is the January 2012 Regents Examination in Physical Setting/Earth Science.

Eurypterus remipes
New York State Fossil

°Note that the 2011 edition of the *Earth Science Reference Tables* has the same content as the 2010 edition but it does not have the ruler.

Physical Setting/Earth Science Reference Tables

Physical Setting/Earth Science Reference Tables

Appendix B: Earth Science—Physical Setting Reference Tables

Appendix B: Earth Science—Physical Setting Reference Tables

Physical Setting/Earth Science Reference Tables

Appendix B: Earth Science—Physical Setting Reference Tables

Physical Setting/Earth Science Reference Tables

Appendix B: Earth Science—Physical Setting Reference Tables

Physical Setting/Earth Science Reference Tables

Rock Cycle in Earth's Crust

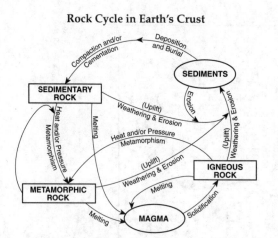

Relationship of Transported Particle Size to Water Velocity

This generalized graph shows the water velocity needed to maintain, but not start, movement. Variations occur due to differences in particle density and shape.

Scheme for Igneous Rock Identification

Physical Setting/Earth Science Reference Tables

Scheme for Sedimentary Rock Identification

INORGANIC LAND-DERIVED SEDIMENTARY ROCKS					
TEXTURE	GRAIN SIZE	COMPOSITION	COMMENTS	ROCK NAME	MAP SYMBOL
Clastic (fragmental)	Pebbles, cobbles, and/or boulders embedded in sand, silt, and/or clay	Mostly quartz, feldspar, and clay minerals; may contain fragments of other rocks and minerals	Rounded fragments	Conglomerate	
			Angular fragments	Breccia	
	Sand (0.006 to 0.2 cm)		Fine to coarse	Sandstone	
	Silt (0.0004 to 0.006 cm)		Very fine grain	Siltstone	
	Clay (less than 0.0004 cm)		Compact; may split easily	Shale	

CHEMICALLY AND/OR ORGANICALLY FORMED SEDIMENTARY ROCKS					
TEXTURE	GRAIN SIZE	COMPOSITION	COMMENTS	ROCK NAME	MAP SYMBOL
Crystalline	Fine to coarse crystals	Halite	Crystals from chemical precipitates and evaporites	Rock salt	
		Gypsum		Rock gypsum	
		Dolomite		Dolostone	
Crystalline or bioclastic	Microscopic to very coarse	Calcite	Precipitates of biologic origin or cemented shell fragments	Limestone	
Bioclastic		Carbon	Compacted plant remains	Bituminous coal	

Scheme for Metamorphic Rock Identification

TEXTURE	GRAIN SIZE	COMPOSITION	TYPE OF METAMORPHISM	COMMENTS	ROCK NAME	MAP SYMBOL
FOLIATED — MINERAL ALIGNMENT	Fine	MICA, QUARTZ, FELDSPAR, AMPHIBOLE, GARNET, PYROXENE	Regional (Heat and pressure increases)	Low-grade metamorphism of shale	Slate	
	Fine to medium			Foliation surfaces shiny from microscopic mica crystals	Phyllite	
				Platy mica crystals visible from metamorphism of clay or feldspars	Schist	
FOLIATED — BANDING	Medium to coarse			High-grade metamorphism; mineral types segregated into bands	Gneiss	
NONFOLIATED	Fine	Carbon	Regional	Metamorphism of bituminous coal	Anthracite coal	
	Fine	Various minerals	Contact (heat)	Various rocks changed by heat from nearby magma/lava	Hornfels	
	Fine to coarse	Quartz	Regional or contact	Metamorphism of quartz sandstone	Quartzite	
		Calcite and/or dolomite		Metamorphism of limestone or dolostone	Marble	
	Coarse	Various minerals		Pebbles may be distorted or stretched	Metaconglomerate	

Appendix B: Earth Science—Physical Setting Reference Tables

Physical Setting/Earth Science Reference Tables

GEOLOGIC HISTORY

Appendix B: Earth Science—Physical Setting Reference Tables

Physical Setting/Earth Science Reference Tables

OF NEW YORK STATE

Time Distribution of Fossils (including important fossils of New York)

The center of each lettered circle indicates the approximate time of existence of a specific index fossil (e.g. Fossil Ⓐ lived at the end of the Early Cambrian).

Fossil groups shown: NAUTILOIDS, DINOSAURS, MAMMALS, BIRDS, AMMONOIDS, CRINOIDS, VASCULAR PLANTS, CORALS, GASTROPODS, BRACHIOPODS, TRILOBITES, EURYPTERIDS, GRAPTOLITES, PLACODERM FISH

Index fossils labeled: A, B, C, D, E, F, G, H, I, J, K, L, M, N, O, P, Q, R, S, T, U, V, W, X, Y, Z

Important Geologic Events in New York

- Advance and retreat of last continental ice
- Sands and clays underlying Long Island and Staten Island deposited on margin of Atlantic Ocean
- Dome-like uplift of Adirondack region begins
- Initial opening of Atlantic Ocean North America and Africa separate
- Intrusion of Palisades sill
- Pangaea begins to break up
- **Alleghenian orogeny** caused by collision of North America and Africa along transform margin, forming Pangaea
- Catskill delta forms
- Erosion of Acadian Mountains
- **Acadian orogeny** caused by collision of North America and Avalon and closing of remaining part of Iapetus Ocean
- Salt and gypsum deposited in evaporite basins
- Erosion of Taconic Mountains; Queenston delta forms
- **Taconian orogeny** caused by closing of western part of Iapetus Ocean and collision between North America and volcanic island arc
- Widespread deposition over most of New York along edge of Iapetus Ocean
- Rifting and initial opening of Iapetus Ocean
- Erosion of Grenville Mountains
- **Grenville orogeny:** metamorphism of bedrock now exposed in the Adirondacks and Hudson Highlands

Inferred Positions of Earth's Landmasses

- 59 million years ago
- 119 million years ago
- 232 million years ago
- 359 million years ago
- 458 million years ago

Index Fossils

- O — Mastodont
- O — Beluga Whale
- P — Cooksonia
- Q — Aneurophyton
- Q — Naples Tree
- R — Bothriolepis
- S — Condor
- T — Lichenaria
- U — Cystiphyllum
- U — Pleurodictyum
- V — Maclurites
- W — Platyceras
- X — Eospirifer
- Y — Mucrospirifer

ESC/BW/TN (2009)

703

Appendix B: Earth Science—Physical Setting Reference Tables

Inferred Properties of Earth's Interior

Physical Setting/Earth Science Reference Tables

Appendix B: Earth Science—Physical Setting Reference Tables

Dewpoint (°C)

Dry-Bulb Temperature (°C)	Difference Between Wet-Bulb and Dry-Bulb Temperatures (C°)															
	0	1	2	3	4	5	6	7	8	9	10	11	12	13	14	15
-20	-20	-33														
-18	-18	-28														
-16	-16	-24														
-14	-14	-21	-36													
-12	-12	-18	-28													
-10	-10	-14	-22													
-8	-8	-12	-18	-29												
-6	-6	-10	-14	-22												
-4	-4	-7	-12	-17	-29											
-2	-2	-5	-8	-13	-20											
0	0	-3	-6	-9	-15	-24										
2	2	-1	-3	-6	-11	-17										
4	4	1	-1	-4	-7	-11	-19									
6	6	4	1	-1	-4	-7	-13	-21								
8	8	6	3	1	-2	-5	-9	-14								
10	10	8	6	4	1	-2	-5	-9	-14	-28						
12	12	10	8	6	4	1	-2	-5	-9	-16						
14	14	12	11	9	6	4	1	-2	-5	-10	-17					
16	16	14	13	11	9	7	4	1	-1	-6	-10	-17				
18	18	16	15	13	11	9	7	4	2	-2	-5	-10	-19			
20	20	19	17	15	14	12	10	7	4	2	-2	-5	-10	-19		
22	22	21	19	17	16	14	12	10	8	5	3	-1	-5	-10	-19	
24	24	23	21	20	18	16	14	12	10	8	6	2	-1	-5	-10	-18
26	26	25	23	22	20	18	17	15	13	11	9	6	3	0	-4	-9
28	28	27	25	24	22	21	19	17	16	14	11	9	7	4	1	-3
30	30	29	27	26	24	23	21	19	18	16	14	12	10	8	5	1

Relative Humidity (%)

Dry-Bulb Temperature (°C)	Difference Between Wet-Bulb and Dry-Bulb Temperatures (C°)															
	0	1	2	3	4	5	6	7	8	9	10	11	12	13	14	15
-20	100	28														
-18	100	40														
-16	100	48														
-14	100	55	11													
-12	100	61	23													
-10	100	66	33													
-8	100	71	41	13												
-6	100	73	48	20												
-4	100	77	54	32	11											
-2	100	79	58	37	20	1										
0	100	81	63	45	28	11										
2	100	83	67	51	36	20	6									
4	100	85	70	56	42	27	14									
6	100	86	72	59	46	35	22	10								
8	100	87	74	62	51	39	28	17	6							
10	100	88	76	65	54	43	33	24	13	4						
12	100	88	78	67	57	48	38	28	19	10	2					
14	100	89	79	69	60	50	41	33	25	16	8	1				
16	100	90	80	71	62	54	45	37	29	21	14	7	1			
18	100	91	81	72	64	56	48	40	33	26	19	12	6			
20	100	91	82	74	66	58	51	44	36	30	23	17	11	5		
22	100	92	83	75	68	60	53	46	40	33	27	21	15	10	4	
24	100	92	84	76	69	62	55	49	42	36	30	25	20	14	9	4
26	100	92	85	77	70	64	57	51	45	39	34	28	23	18	13	9
28	100	93	86	78	71	65	59	53	47	42	36	31	26	21	17	12
30	100	93	86	79	72	66	61	55	49	44	39	34	29	25	20	16

Appendix B: Earth Science—Physical Setting Reference Tables

Physical Setting/Earth Science Reference Tables

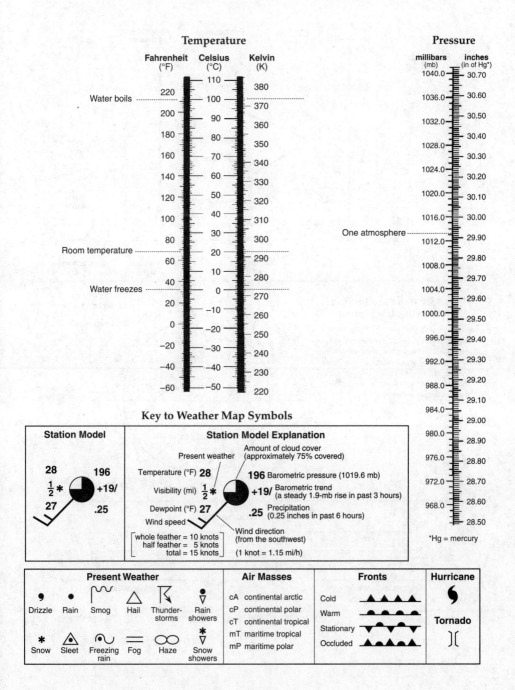

Appendix B: Earth Science—Physical Setting Reference Tables

Physical Setting/Earth Science Reference Tables

Selected Properties of Earth's Atmosphere

Planetary Wind and Moisture Belts in the Troposphere

The drawing on the right shows the locations of the belts near the time of an equinox. The locations shift somewhat with the changing latitude of the Sun's vertical ray. In the Northern Hemisphere, the belts shift northward in the summer and southward in the winter.

(Not drawn to scale)

Electromagnetic Spectrum

(Not drawn to scale)

Physical Setting/Earth Science Reference Tables

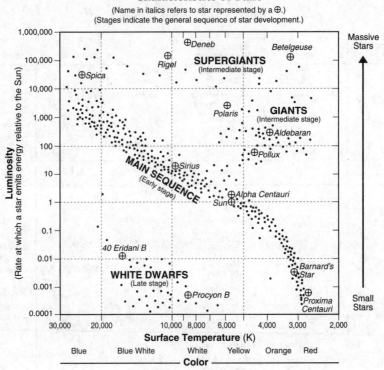

Solar System Data

Celestial Object	Mean Distance from Sun (million km)	Period of Revolution (d=days) (y=years)	Period of Rotation at Equator	Eccentricity of Orbit	Equatorial Diameter (km)	Mass (Earth = 1)	Density (g/cm³)
SUN	—	—	27 d	—	1,392,000	333,000.00	1.4
MERCURY	57.9	88 d	59 d	0.206	4,879	0.06	5.4
VENUS	108.2	224.7 d	243 d	0.007	12,104	0.82	5.2
EARTH	149.6	365.26 d	23 h 56 min 4 s	0.017	12,756	1.00	5.5
MARS	227.9	687 d	24 h 37 min 23 s	0.093	6,794	0.11	3.9
JUPITER	778.4	11.9 y	9 h 50 min 30 s	0.048	142,984	317.83	1.3
SATURN	1,426.7	29.5 y	10 h 14 min	0.054	120,536	95.16	0.7
URANUS	2,871.0	84.0 y	17 h 14 min	0.047	51,118	14.54	1.3
NEPTUNE	4,498.3	164.8 y	16 h	0.009	49,528	17.15	1.8
EARTH'S MOON	149.6 (0.386 from Earth)	27.3 d	27.3 d	0.055	3,476	0.01	3.3

Appendix B: Earth Science—Physical Setting Reference Tables

Physical Setting/Earth Science Reference Tables

Properties of Common Minerals

LUSTER	HARD-NESS	CLEAVAGE	FRACTURE	COMMON COLORS	DISTINGUISHING CHARACTERISTICS	USE(S)	COMPOSITION*	MINERAL NAME
Metallic luster	1–2		✔	silver to gray	black streak, greasy feel	pencil lead, lubricants	C	Graphite
Metallic luster	2.5	✔		metallic silver	gray-black streak, cubic cleavage, density = 7.6 g/cm³	ore of lead, batteries	PbS	Galena
Metallic luster	5.5–6.5		✔	black to silver	black streak, magnetic	ore of iron, steel	Fe_3O_4	Magnetite
Metallic luster	6.5		✔	brassy yellow	green-black streak, (fool's gold)	ore of sulfur	FeS_2	Pyrite
Either	5.5–6.5 or 1		✔	metallic silver or earthy red	red-brown streak	ore of iron, jewelry	Fe_2O_3	Hematite
Nonmetallic luster	1	✔		white to green	greasy feel	ceramics, paper	$Mg_3Si_4O_{10}(OH)_2$	Talc
Nonmetallic luster	2		✔	yellow to amber	white-yellow streak	sulfuric acid	S	Sulfur
Nonmetallic luster	2	✔		white to pink or gray	easily scratched by fingernail	plaster of paris, drywall	$CaSO_4 \cdot 2H_2O$	Selenite gypsum
Nonmetallic luster	2–2.5	✔		colorless to yellow	flexible in thin sheets	paint, roofing	$KAl_3Si_3O_{10}(OH)_2$	Muscovite mica
Nonmetallic luster	2.5	✔		colorless to white	cubic cleavage, salty taste	food additive, melts ice	NaCl	Halite
Nonmetallic luster	2.5–3	✔		black to dark brown	flexible in thin sheets	construction materials	$K(Mg,Fe)_3$ $AlSi_3O_{10}(OH)_2$	Biotite mica
Nonmetallic luster	3	✔		colorless or variable	bubbles with acid, rhombohedral cleavage	cement, lime	$CaCO_3$	Calcite
Nonmetallic luster	3.5	✔		colorless or variable	bubbles with acid when powdered	building stones	$CaMg(CO_3)_2$	Dolomite
Nonmetallic luster	4	✔		colorless or variable	cleaves in 4 directions	hydrofluoric acid	CaF_2	Fluorite
Nonmetallic luster	5–6	✔		black to dark green	cleaves in 2 directions at 90°	mineral collections, jewelry	$(Ca,Na)(Mg,Fe,Al)$ $(Si,Al)_2O_6$	Pyroxene (commonly augite)
Nonmetallic luster	5.5	✔		black to dark green	cleaves at 56° and 124°	mineral collections, jewelry	$CaNa(Mg,Fe)_4(Al,Fe,Ti)_3$ $Si_6O_{22}(O,OH)_2$	Amphibole (commonly hornblende)
Nonmetallic luster	6	✔		white to pink	cleaves in 2 directions at 90°	ceramics, glass	$KAlSi_3O_8$	Potassium feldspar (commonly orthoclase)
Nonmetallic luster	6	✔		white to gray	cleaves in 2 directions, striations visible	ceramics, glass	$(Na,Ca)AlSi_3O_8$	Plagioclase feldspar
Nonmetallic luster	6.5		✔	green to gray or brown	commonly light green and granular	furnace bricks, jewelry	$(Fe,Mg)_2SiO_4$	Olivine
Nonmetallic luster	7		✔	colorless or variable	glassy luster, may form hexagonal crystals	glass, jewelry, electronics	SiO_2	Quartz
Nonmetallic luster	6.5–7.5		✔	dark red to green	often seen as red glassy grains in NYS metamorphic rocks	jewelry (NYS gem), abrasives	$Fe_3Al_2Si_3O_{12}$	Garnet

*Chemical symbols:
Al = aluminum   Cl = chlorine   H = hydrogen   Na = sodium   S = sulfur
C = carbon   F = fluorine   K = potassium   O = oxygen   Si = silicon
Ca = calcium   Fe = iron   Mg = magnesium   Pb = lead   Ti = titanium

✔ = dominant form of breakage

Glossary of Earth Science Terms

ablation The melting and evaporation of glacial ice.

ablation debris Material on the surface of a glacier that was suspended in the ice but is now exposed due to ablation.

abrasion The breaking down of rocks by rubbing together.

absolute humidity The number of grams of water vapor in a cubic meter of air.

abyssal plain Wide, flat areas adjacent to the mid-ocean ridges.

accretion Process by which particles of matter are drawn together by gravitational attraction and accumulate to form larger particles. Causes gradual growth of a celestial object as particles of matter attracted to the object by its gravity accumulate and become part of the project.

acid rain Rain that is acidic due to substances that have become dissolved in it.

adiabatic A change that does not involve the addition or removal of heat.

adiabatic lapse rate The rate at which the temperature of air changes due to a decrease in air pressure rather than the addition or removal of heat.

aerobic atmosphere An atmosphere with sufficient oxygen to support aerobic respiration by living things.

agent of erosion A medium set in motion by gravity that can exert a force on sediments causing them to be moved.

air current Air that is moving in a vertical direction.

air mass A large region of air with uniform characteristics at any given level.

air pressure The force exerted on a unit of area by the air.

alluvial fan A flat, fan-shaped mound of sediment deposited where a stream slows down because its gradient suddenly becomes less steep (i.e., where a mountain valley empties onto a plain).

altitude 1. The angle between the horizon and an object seen in the sky with the observer at the vertex. 2. Distance above the earth's surface.

anemometer An instrument that measures wind speed.

angle of repose The greatest angle at which a pile of sediments is stable.

angstrom (Å) A unit of length. There are 10^8 angstroms in a centimeter.

angular unconformity A place where sedimentary layers at one angle are in contact with sedimentary layers at a different angle.

annular eclipse A solar eclipse in which the moon's shadow does not actually reach the surface of Earth; to an observer on Earth, the sun's disk is obscured by the Moon except for a narrow ring around the perimeter.

anticline An upward, arched fold.

anticyclone A flow of air that is clockwise and outward from a high pressure center.

apparent brightness The brightness of a celestial object as seen by human eyes on Earth.

apparent daily motion The apparent motion of all celestial objects along a circular path at a constant rate of 15° per hour, or one complete circle every day.

aquiciude A rock or sediment layer that is impermeable to water.

aquifer A rock or sediment layer in which the pore spaces are saturated with water.

arrival time The precise time at which a seismic wave reaches a seismograph and is recorded.

artesian well A well in which water rises to the surface unassisted.

asteroid Solid body having no atmosphere that orbits the sun.

asthenosphere A region of the upper mantle between 100 and 350 kilometers in depth that behaves like a fluid.

astronomical unit The average distance from Earth to the Sun; 149,598,000 kilometers.

atmosphere The thin shell of gases bound to Earth by gravity; it consists of gases, water, ice, dust, and other particles.

atmospheric variables The characteristics of the atmosphere that change.

aurora Glowing, often shimmering light forms seen near the poles that are caused by radiation from high altitude air molecules excited by collisions with particles from the Sun.

axis of rotation The central line around which all parts of an object move during rotation.

banding A structure in a metamorphic rock of nearly parallel bands of different textures or minerals; i.e., the nearly parallel bands of light and dark minerals seen in gneiss.

barrier bar A sandbar running parallel to a shoreline that shelters the water behind it from waves.

batholith A large, irregularly shaped pluton covering an area greater than 75 square kilometers.

beach A strip of loose materials along the shore of a body of water formed and washed by waves.

bed A layer of sediment with a particular physical or structural character.

bedrock A general term for the solid rock underlying the soil or loose material covering Earth's surface.

bench mark A permanent metal marker set into the ground that gives the exact location and precisely measured elevation of that point.

big bang theory The idea that the universe started out with all of its matter in a small volume and then expanded outward in all directions.

body waves Seismic waves that travel through the body of the earth; P-waves and S-waves are body waves.

capillary fringe A narrow region just above the water table in which narrow pore spaces are filled with water that is drawn up by capillary action.

capillarity The action by which a fluid, such as water, is drawn upward in narrow spaces as a result of surface tension.

carbonaceous chondrite A stony meteorite containing hydrated claylike minerals and a great variety of organic compounds.

carbonation A chemical reaction in which carbon dioxide combines with a substance.

celestial meridian A meridian on the celestial sphere that passes through the zenith of a given place and terminates at the celestial poles.

celestial objects Objects that can be seen in the sky, but that are beyond Earth's atmosphere.

celestial sphere An imaginary sphere of infinite radius around Earth's center upon whose "inner surface" all celestial objects appear to be projected. The apparent dome of the visible sky forms half of the celestial sphere.

cementation The binding together of sediment particles by a cementing material.

chaos theory The idea that small, unpredictable changes can cause large-scale changes in a system over time.

chemical sediments Particles (usually crystals) that crystallized out of solutions at or near Earth's surface.

chemical weathering Processes that break down rock by changing its chemical composition.

cinder cone volcano A narrow, steep-sided volcano formed from the accumulation of tephra around the vent.

clastic sediments Rock or mineral fragments formed by the breakdown of rock due to weathering.

cleavage The tendency of a mineral to break along flat surfaces parallel to atomic planes in its crystalline structure.

climate The weather in a region averaged over a long period of time.

closed universe A model of the universe of fixed volume with curved boundaries. If closed, the universe will eventually contract and give rise to another big bang.

cloud base The elevation of the base of a cloud; the elevation at which condensation begins.

cloud ceiling The elevation of the base of a layer of cloud.

cloud cover The fraction of the sky obscured by clouds.

coastal zone The part of a continent that is adjacent to the ocean; it includes features such as the coast, shoreline, beaches, marshes, estuaries, lagoons, and deltas.

comet Bodies composed of meteoroids embedded in ice that revolve around the Sun in highly elliptical orbits.

comet coma The cloud of gas surrounding a comet's nucleus when ice in the nucleus sublimes as the comet approaches the Sun.

comet nucleus The solid center of a comet thought to consist of meteoroids embedded in ice.

comet tail The elongation of the cloud of gas surrounding a comet caused by collisions with particles emitted by the Sun.

compaction The reduction in volume or thickness of a sediment layer due to overlying weight or pressure.

composite cone volcano A large symmetrical volcano consisting of alternating layers of lava flows and tephra.

compound A substance that consists of two or more elements joined together in a definite proportion and have properties different from the elements that compose it.

compression Forces acting toward each other along the same line of action causing rock to be squeezed.

conchoidal fracture A tendency to break along a curved, concave surface that looks somewhat like a mussel shell.

condensation The process by which a gas changes into a liquid.

condensation nuclei Particles suspended in the air that provide a surface upon which condensation can occur.

conic section A family of figures obtained by cutting a circular cone with a plane.

conservation The act of decreasing the consumption of a natural resource, usually by more efficient use.

constellation A group of stars in an area of the sky that form imaginary patterns, supposedly in the form of an animal, person, or thing.

contact metamorphism Changes in rock that result from the extreme heat produced by contact with magma or lava.

continental drift The theory that the continents drift slowly across Earth's surface, sometimes colliding and breaking into pieces.

continental glacier Immense masses of ice that form at high latitudes where temperatures are always cold and that spread out over the entire land surface rather than being confined to valleys.

continental rise A gently sloping feature between the continental slope and the abyssal plain.

continental shelf The shallow part of the ocean floor adjacent to the coastal zone.

continental slope The portion of the ocean floor that slopes steeply from the continental shelf toward the abyssal plain.

contour interval The difference in elevation between two adjacent contour lines.

convection cell A circular pattern of motion in which warm, moist air rises, expands outward, cools, and sinks back to the surface.

convection current A fluid current set in motion by differences in density within the fluid.

convergent boundary Places where edges of adjacent plates are colliding.

coordinates Two numbers that describe the position of any point on a map in terms of the intersection of two lines.

Glossary

coordinate system A system for assigning two numbers to every point on a surface.

core The innermost zone of Earth's interior.

Coriolis effect A deflection of motion from a straight-line path due to the rotation of Earth.

correlation The process of determining that rock layers in different areas are the same age.

cosmic background radiation A remnant of radiation left over from the original big bang that fills the universe.

cosmology The study of the universe.

crevasse A deep, nearly vertical fissure in glacial ice.

cross-bedding Layers of sediment lying at an angle to the plane in which sediment layers are normally deposited.

crystal A solid whose internal structural pattern is expressed as plane faces that can be seen with the unaided eye.

crystalline Having a definite internal structural pattern.

crystallization The process of forming crystals.

cutoff A new channel formed when a stream cuts through a narrow strip of land between adjacent meanders.

cyclone A flow of air in which motion is counterclockwise and toward a low pressure center.

decarbonation A chemical reaction in which carbon dioxide is given off.

decay product The new element formed when a radioisotope decays.

deferent In Ptolemy's geocentric model of the universe, a circle with Earth near its center. The center of a planet's epicycle moves around the circumference of a deferent.

deflation The pick up and removal of fine sediments from Earth's surface by wind.

deflation hollow A shallow depression formed by the removal of sediments by wind.

degassing The process by which water is released from the molecules of minerals when heated to high temperatures.

dehydration A chemical reaction in which water is given off.

delta A flat, fan-shaped mound of sediment deposited where a stream enters a body of standing water and slows down.

density The amount of matter in a given amount of space; usually expressed as the mass per unit of volume of a substance.

desertification The rapid development of deserts caused by the impact of human activities.

desert pavement A continuous layer of sediments too heavy to be moved by wind that remains after finer sediments have been removed by wind, and shields underlying materials from further deflation.

dew Droplets of water that condense on ground surfaces due to radiative cooling of the ground overnight.

dew point The temperature at which the water vapor in air will begin to condense.

differentiation The process by which planets develop concentric layers or zones due to density differences.

dike A pluton that cuts across existing rock layers.

disconformity An irregular erosional surface between parallel layers of rock.

divergent boundary Places where edges of adjacent plates are spreading apart.

Doppler effect The shift in wavelength of a spectrum line from its normal position due to relative motion between the source and the observer.

drainage basin The area from which precipitation drains into a stream or system of streams.

drift A general term for all sediments transported and deposited by a glacier.

drizzle Very fine droplets of water falling slowly and close together.

drumlin Low mounds of till having an elongated teardrop shape formed when a glacier rides over a previously deposited pile of sediment.

dry bulb temperature The temperature recorded by a thermometer whose bulb is kept dry.

ductile deformation A permanent change in rock that occurs when stress exceeds the strength of a rock's internal bonds.

dune A mound of sand deposited by wind.

earthflow The slow, downhill movement of a water-saturated layer of soil and vegetation.

earthquake A sudden trembling of the ground.

eccentricity The flatness of an ellipse, expressed as the ratio of the distance between the foci of the ellipse to the length of its major axis.

ecliptic The apparent annual path of the Sun around the sky; the intersection of the plane of Earth's orbit and the celestial sphere.

elastic deformation A temporary change in which rock strains due to stress but returns to its original condition after the stress is removed.

electromagnetic spectrum A continuum in which electromagnetic waves are arranged in order of wavelength.

electromagnetic wave An oscillation in the electromagnetic field surrounding a particle.

element A substance that cannot be broken down into simpler substances by ordinary chemical means.

ellipse A closed figure obtained by cutting a circular cone with a plane; its appearance is that of a flattened circle.

El Niño A current of warm water flowing southward along the coast of Ecuador, which about every seven to ten years, extends to the coast of Peru, where it interrupts the upwelling of nutrient-rich water, causing plankton and fish to die.

end moraine A pile of till that forms at the melting edge of a glacier.

eon The largest unit of geologic time.

epicenter The point on Earth's surface directly above the focus of an earthquake.

epicycle In Ptolemy's geocentric model of the universe, the circular orbit of a planet. The center of an epicycle travels around the circumference of another circle called the deferent.

epoch The unit of geologic time into which periods are divided.

equator A circle on which all points are midway between Earth's poles.

equatorial diameter Diameter of Earth measured from a point on the Equator through the center of Earth to the Equator (12,756 kilometers).

equinox Either of the two times during a year when the Sun crosses the celestial Equator and when the length of day and night are approximately equal.

era The unit of geologic time into which eons are divided.

erosion Any process that transports sediments.

erratic An isolated boulder deposited by a glacier.

escape velocity The smallest velocity a body must attain in order to escape from the gravitational attraction of another body.

escarpment A steep slope or long cliff separating two relatively level areas of land at different elevations, often at the edge of a plateau.

evaporite Sediments, rocks, or minerals formed when water containing dissolved minerals evaporates.

evapotranspiration The combined processes of evaporation and transpiration, the two mechanisms by which water is returned to the atmosphere.

evolution The idea that current life-forms have developed from earlier, different life-forms.

exfoliation The scaling off, or peeling, of successive layers from the surface of rocks.

extrusive igneous rock A rock formed by the rapid cooling of lava at the surface of Earth.

faulting A sudden movement of rock along planes of weakness in Earth's crust called faults.

felsic Rocks rich in the minerals feldspar and silica (quartz).

field A region of space in which a measurable quantity is present at every point.

field map Represents any quantity that varies in a region of space.

floodplain A broad, flat strip of land adjacent to a stream consisting of sediments deposited when the stream is in flood.

focus The point where rock first breaks or moves in an earthquake.

focus of an ellipse One of two points along the major axis of an ellipse with the characteristic that the total distance from one focus to any point on the ellipse and back to the other focus is always the same.

fog A cloud whose base is at ground level.

fold A bend or warp in layered rock.

fossil Any remains, trace, or imprint of a living thing that has been preserved in Earth's crust.

fossil fuel A fuel derived from the remains of once-living things; e.g., coal, petroleum, and natural gas.

Foucault pendulum A freely swinging pendulum whose path appears to change direction relative to Earth's surface in a predictable manner due to Earth's rotation.

fracture The tendency not to break along any particular direction.

fragmental sedimentary rock A rock composed of rock fragments cemented or compacted together in a solid mass.

frame of reference A specific point in time or space that serves as the basis of comparison for describing change; in a coordinate system, the coordinate axes, in terms of which position is specified.

frequency The number of waves that pass a given point in a unit of time.

front The boundary between two adjacent air masses.

frost Ice crystals that form on ground surfaces by sublimation due to radiative cooling of the ground overnight.

frost action The breaking down of rock by the force exerted by water expanding as it freezes into ice.

galaxy A system consisting of hundreds of billions of stars.

gas giant Planets that have thick atmospheres surrounding a small, rocky core.

geocentric model A model of the universe in which Earth is at the center and all other objects revolve around Earth.

geologic column The correlation of rocks worldwide into a single sequence showing their relative ages.

geologic time scale The division of Earth's past into a sequence of time units.

geothermal gradient The rate at which temperatures within Earth rise with depth.

glacial advance The forward and downslope movement of a glacier that results when accumulation exceeds ablation.

glacial basin Depressions formed when fast-moving glaciers remove large quantities of rock material plucked from the bedrock surface.

glacial polish A bedrock surface smoothed by abrasion as a glacier flows over it.

glacial retreat A decrease in the length of a glacier that results from ablation exceeding accumulation.

glacial striations Long, narrow grooves inscribed in bedrock by rock fragments embedded in a glacier as the glacier moves over the bedrock.

glacial till Unsorted sediment deposited by a glacier.

glacier A solid mass of ice formed by the compaction of accumulated snow that flows downslope and outward under the influence of gravity.

glaze Rain that forms a layer of ice as soon as it comes into contact with below-freezing surfaces.

global wind belt A region of prevailing winds caused by a difference in air pressure between adjacent regions on a global scale.

graded bedding A layer of sediments in which the particle size changes gradually from large on the bottom to small on the top.

gradient The rate at which a field changes; usually measured in terms of change in field value per unit of distance.

gravitational differentiation The process by which the material from which the solar system formed became separated into layers according to density, because atoms of lighter elements do not exert as strong a gravitational attraction as atoms of heavier elements.

gravity The force of attraction that exists between all matter.

great circle A circle that bisects the Earth.

greenhouse effect The process by which solar radiation that is re-radiated at a longer wavelength is absorbed by carbon dioxide, water vapor, and other gases in the atmosphere rather than escaping into space.

ground moraine A thin, widespread layer of till deposited beneath a glacier.

groundmass A mass of fine mineral crystals surrounding the large, isolated phenocrysts in a porphyry.

gully A long, narrow trough eroded in soil by running water.

hail Balls of ice with an internal structure of concentric layers of ice and snow formed when ice pellets fall through turbulent air and are alternately carried through layers of air that are above and below freezing.

half-life The time required for half of a mass of radioactive atoms to decay.

hanging valley A valley of a tributary glacier whose floor is above, or hanging over, the floor of the main glacier that it feeds.

hardness A mineral's ability to resist being scratched.

heliocentric model A model of the universe in which the Sun is at the center of our solar system and the planets revolve around the Sun.

Hertzsprung-Russell Diagram A plot of the luminosity versus spectral type for a group of stars; also called the H-R diagram and is titled Characteristics of Stars chart on the reference tables.

horizon The visible line of demarcation between land or sea and sky.

horizontal sorting The gradual change in the size of sediments deposited horizontally along the bed of a stream as the water slows down.

humidity The amount of water vapor in the air.

humus The dark, highly decomposed organic matter in soil.

hurricane Large, cyclonic storms in which the winds exceed 119 kilometers per hour.

hydration A chemical reaction in which water combines with a substance.

hydrogen bonding A weak bond between adjacent water molecules due to the attraction between the slightly positive end of one water molecule and the slightly negative end of another water molecule.

hydrolysis A reaction in which water separates into hydrogen ions and hydroxide ions that replace ions in a mineral.

hydrosphere All of the water that rests on the lithosphere; it includes all oceans, lakes, streams, groundwater, and ice.

Glossary

ice age A period of time during which conditions exist that cause ice sheets more than 1 million square kilometers in area and thousands of meters thick to develop on nonpolar continents.

iceberg A mass of floating ice that detached from a glacier into a body of water.

igneous intrusion A mass of magma that penetrates an opening in an existing rock and solidifies.

igneous rock Rock formed by the cooling and crystallization of molten minerals (magma or lava).

immature soil A soil in which boundaries between forming horizons are indistinct.

impact crater A circular depression formed on a surface by the impact of a projectile, such as a meteorite or a comet.

in solution Sediments that are transported as a consequence of being dissolved in the water flowing in a stream.

index fossil A fossil of an organism that had distinctive body features and was abundant over a wide geographical area, but only existed for a short period of time.

infiltrate The process by which water seeps into the ground and through interconnected pore spaces.

inorganic Non-living and not derived from a living thing.

insolation Solar radiation that reaches the surface of Earth; short for incoming solar radiation.

interglacial Time periods during an ice age when glaciers retreat or even disappear.

international date line The 180° meridian; it marks the transition from one date to another.

intrusive igneous rock A rock formed by the slow cooling of magma within the earth.

ion An atom or molecule in which the electrical charges are unbalanced resulting in a net positive or negative electrical charge.

isobar A line connecting points of equal air pressure.

isoline Lines connecting points of equal field value.

isostasy The idea that Earth's crust floats on hot fluid rock in the mantle.

isotherm A line connecting points of equal temperature on a field map.

isotope Varieties of the same element that differ slightly in mass.

joint A fracture in a rock along which movement has not occurred.

kettle A depression formed when a block of ice buried in till melts.

key bed Well-defined, easily identifiable rock layers that have distinctive characteristics or fossil content that allow(s) them to be used in correlation.

Kuiper Belt A thick belt of debris left over from the formation of the outer planets, extending from Neptune's orbit (29 AU) to the outer edge of Pluto's orbit (50 AU). Considered the probable source of short-period comets, over 60 objects with diameters over 100 kilometers have been detected orbiting the Sun within this region. Pluto and its moon Charon are considered Kuiper Belt objects.

laccolith A pluton formed when magma intrudes between rock layers and pushes upward and solidifies creating a rock structure that has a flat bottom and an arched top.

lagoon The sheltered body of water between a barrier bar and the mainland.

land breeze A local wind blowing from land toward an adjacent body of water because nighttime cooling increases the air pressure over the land.

landform Physical features of Earth's surface that have a characteristic shape and are produced by natural processes.

landscape A region in which the landforms are related by their structure, the processes that formed them, and the way in which they developed.

landscape region A grouping of landscapes with similar relief, stream patterns and soil associations.

latent heat of fusion The amount of heat energy given off or absorbed in causing a change between the liquid and solid states that does not result in a change in temperature.

latent heat of vaporization The amount of heat energy given off or absorbed in causing a change between the liquid and gaseous states that does not result in a change in temperature.

lateral moraine A mound of till deposited along the sides of a glacier.

latitude The angular distance north or south of the equator.

Laurentide ice sheet The glacier that advanced several times across the northern United States during the Pleistocene epoch.

lava Magma (molten rock) that has emerged onto Earth's surface.

lava plateau A flat sheet of igneous rock formed when a very fluid lava erupts from a long fissure instead of from a central vent.

leeward The side of a feature facing away from the wind.

lens A transparent object of regular form that changes the direction of travel of light waves that pass through it.

levee A low, thick, ridgelike deposit along the banks of a stream consisting of coarse sediments deposited when a stream overflows its banks.

light year The distance light travels in one year; roughly 9.5 trillion kilometers, or 6 trillion miles.

lithification The process of converting sediments into coherent, solid rock.

lithosphere Earth's solid outer layer of soil and solid brittle rock extending from the surface to a depth of about 100 kilometers; it includes the crust and the upper mantle.

loess A fine, buff-colored sediment deposited by wind.

longitude The angular distance east or west of the prime meridian.

longitudinal waves Waves in which the medium oscillates parallel to the direction of wave motion.

longshore current An ocean current flowing parallel to a shore caused by the approach of waves at an angle to the coast.

luminosity The total amount of energy radiated by a star in 1 second.

lunar eclipse An event in which the Moon passes into Earth's shadow during the full moon phase.

luster The way light reflects from the surface of a mineral.

mafic Rocks rich in minerals containing magnesium and Fe (iron).

magma Molten rock beneath Earth's surface.

magma chamber A pocket of molten magma in the crust or upper mantle.

magnetosphere The region surrounding Earth that is influenced by its magnetic field

main sequence A diagonal line across the H-R diagram on which the majority of stars lie; the main phase of the life of a star.

mantle The zone of Earth's interior between the crust and the core.

map A model of Earth's surface.

map scale The ratio between distance on a map and actual distance on the ground; usually expressed as a ratio.

mass The amount of matter in an object.

mass wasting The downhill movement of sediments under the direct influence of gravity.

mature soil A soil in which distinct horizons have formed.

meander A curving loop in a stream channel.

meridian Semicircles on which all points have the same longitude.

mesopause The boundary between the mesosphere and the thermosphere.

Glossary

mesosphere The layer of the atmosphere ranging from 30 to 50 kilometers; temperatures in this layer decrease with altitude.

metamorphic rock Rocks that form as a result of physical and chemical changes in existing rocks.

metamorphism The processes that change the physical and chemical properties of existing rocks.

meteor The streak of light seen in the sky as a meteoroid is vaporized by frictional heating as it falls through the atmosphere.

meteorite The portion of a meteoroid that survives to strike the ground.

meteoroid A chunk of matter moving through space.

mid-latitude cyclone Warm and cold fronts that move in a counterclockwise direction around a low pressure center.

mid-ocean ridge A chain of undersea mountains running through the center of an ocean.

Milky Way galaxy The galaxy in which our solar system is located.

mineral A naturally occurring, inorganic, crystalline solid with a composition that can be expressed by a chemical symbol or a formula.

mineral composition The minerals present in a rock.

mixture Consisting of one or more substances that retain their characteristic properties and can be separated by relatively simple means.

modified Mercalli scale A system that measures the strength of an earthquake based upon perception of motion by human observers and damage to structures built by humans.

Moho The boundary between the dense rock of the mantle and the less dense rock of the crust. It was named for the seismologist Andrija Mohorovicic, who recognized its existence after analyzing seismic wave behavior inside Earth.

Mohorovicic discontinuity The boundary between Earth's crust and mantle. It marks the interface between rocks within Earth of two different average densities.

monocline A fold in which only one side of the fold is inclined.

moraine A deposit of unsorted sediments deposited directly from glacial ice.

mountain Any part of Earth's crust that projects at least 300 m (1,000 ft) above the surrounding land, has a limited summit area (as opposed to a plateau), steep sides, and considerable bare-rock surface.

mudflow The rapid downhill flow of a mixture of rock fragments, soil, and water.

natural selection The process by which organisms that are better suited to their environment survive and reproduce while poorly suited organisms do not.

neap tide A tidal cycle in which there is very little difference between the water level at high and low tides; caused by the canceling out effect of the Moon's gravity acting at right angles to the Sun's.

nebula A cloud of interstellar matter.

neutrino A fundamental particle produced in certain nuclear reactions that has no charge and no mass, but which does have momentum.

neutron star An incredibly dense (10^{16}–10^{18} kg/m^3) star whose interior is composed mostly of neutrons

nonconformity A place where sedimentary rock lies directly on another type of rock, such as igneous or metamorphic rock.

nuclear fusion A nuclear reaction in which the nuclei of atoms combine; during fusion, some of the mass of the combining nuclei is converted into energy.

numerical forecasting Weather forecasting based upon mathematical models of the physical behavior of the atmosphere.

oasis A place where deflation has lowered the surface in a desert to the water table allowing plants to take root and grow.

oblate spheroid A slightly flattened sphere; Earth's shape.

ocean-floor spreading The idea that the ocean floors were spreading sideways away from the mid-ocean ridges.

Oort Cloud A vast shell of icy objects (i.e., comets) surrounding the solar system between 50,000 and 100,000 AU; thought to be the source of long-period comets.

open universe A model of the universe with infinite volume and no boundaries. If open, the universe will continue expanding forever.

orbit The path along which a satellite revolves around a primary.

orbital velocity The smallest velocity a body must attain in order to enter a stable orbit around another body.

ordinary well A hole dug in the ground so that it penetrates the water table in which water accumulates and can be pumped to the surface.

organic sediments Particles produced by the life activities of plants or animals.

origin time The time at which an earthquake occurred.

original horizontality, law of The observation that sediments deposited in water form flat level layers.

orographic effect The net increase in temperature of a parcel of air that is forced over a mountain range due to adiabatic cooling on the windward side followed by adiabatic heating on the leeward side.

outgassing The release of gases and water vapor from molten rocks, leading to the formation of Earth's atmosphere and oceans.

outwash plain A flat plain next to the leading edge of a glacier formed by the deposition of sediments from meltwater streams.

oxbow lake A crescent-shaped body of water formed when a cutoff isolates the water in a meander from the main channel of the stream.

oxidation A chemical reaction in which a substance combines with oxygen.

oxidizing atmosphere An atmosphere that contains enough oxygen to cause elements and compounds to oxidize, but not enough to support aerobic life.

ozone A gas whose molecules consist of three atoms of oxygen joined together.

***P*-waves** Longitudinal seismic waves produced by an earthquake that travel the fastest and are therefore the first to be recorded by a seismograph.

Pacific ring of fire A narrow zone of active volcanoes surrounding the Pacific Ocean.

paleomagnetism A record of the orientation of Earth's magnetic field preserved in the crystalline structure of igneous rock containing magnetic minerals.

parallel Circles on which all points have the same latitude north or south of the Equator.

parting The tendency to break along flat surfaces that follow structural weaknesses.

path of totality The path of the Moon's umbra over the surface of Earth during a solar eclipse.

pedalfer Soils rich in aluminum and iron compounds.

pedocal Soils rich in calcium compounds.

penumbra The portion of a shadow in which only part of the light has been blocked, so the light is dimmed but not totally absent.

period The unit of geologic time into which eras are divided.

period of revolution The length of time it takes for a satellite to complete one revolution around its primary.

permeability The rate at which water can infiltrate a material.

phases of matter The form in which matter may exist; solid, liquid, and gas.

phases of the Moon The changing shape of the illuminated portion of the Moon that can be seen by an observer on Earth.

phenocryst Isolated large crystals in a rock.

photosynthesis The process by which cells convert water and carbon dioxide into carbohydrates and oxygen using light as an energy source.

physical weathering Processes that break down rock without changing their chemical composition.

physiographic province A large-scale region in which all parts are similar in geologic structure and climate, and thus has had a unified history of landform development.

phytoplankton Plankton that carry out photosynthesis.

plain A flat area at low elevation.

planet A body that is at least partly solid, orbits the Sun, and emits mainly reflected sunlight.

planetesimal One of the small bodies from which planets formed.

plant action The breaking down of rock by the force exerted by the growth of plant roots and stems or by chemicals produced as a by-product of plant growth.

plateau A large, flat region elevated more than 150–300 m (500–1,000 ft) above the surrounding land or above sea level.

plate tectonics The theory that Earth's crust consists of large, rigid slabs of rock, or plates, resting on a fluidlike layer of denser rock and that these plates slide around causing their edges to collide, separate, and slide past each other.

Pleistocene The time period, which began 1.6 million years ago, during which climates grew colder worldwide.

plucking The tearing loose of rock frozen to the base of a glacier as the glacier moves along.

pluton A rock structure formed by the solidification of an igneous intrusion.

polar diameter Diameter of Earth measured from a pole through the center of Earth to the opposite pole (12,714 kilometers).

polar front A boundary between polar and tropical air that forms in the prevailing westerly global wind belts.

Polaris A star whose position is almost directly over the Earth's north geographic pole.

pollutant The particular materials or forms of energy that are harmful to humans.

pollution The concentration of any material or energy form that is harmful to humans.

porosity The percentage of empty space in a material.

porphyry A rock consisting of large crystals called phenocrysts embedded in a mass of fine mineral grains.

potential evapotranspiration The maximum amount of water that could be returned to the atmosphere by evapotranspiration under a given set of conditions.

precession The very slow change in the orientation of a rotating body's axis of rotation.

precipitate A solid that crystallizes out of a solution and settles to the bottom of the solution.

precipitation Condensed moisture that falls to the ground.

pressure unloading The release of stress on rock as the stress of the weight of overlying rock is removed as the latter is worn away, or the removal of the stress as the weight of the ice in a glacier is removed as the glacier melts away.

primary The body a satellite orbits.

prime meridian The zero meridian, or starting point, for measuring the angular distance east or west of any other meridian.

profile What a cross section of the land between two points would look like viewed from the side.

protostar A collapsing cloud of dust and gas destined to become a star.

radiation curve A plot of intensity versus wavelength of the radiation emitted by a source, e.g., a star. The resulting graph illustrates the rate at which a source is emitting energy at different wavelengths.

radiative balance The condition in which Earth's average level of heat energy remains constant over a long period of time.

radioactive decay The process by which unstable isotopes break apart spontaneously giving off energy, subatomic particles, or both.

radioisotope Isotopes that are unstable and undergo radioactive decay.

recharge Water that seeps into the ground and replaces soil moisture that was lost during a period of drought.

red giant A large red star that has consumed most of its hydrogen and developed a helium core; heat from contraction of the helium core triggered fusion in the outer shell of hydrogen, causing the star to expand to hundreds of times its initial size, and its outer surface to cool.

reducing atmosphere An atmosphere with so little oxygen that elements and compounds do not oxidize.

refraction The bending of light as it passes from one medium to another.

regional metamorphism Changes in rock over an extensive area that occur due to the pressure and high temperatures associated with either deep burial or movements of Earth's crust.

relative age The age of an object or event relative to another object or event.

relative humidity The ratio of water vapor in the air to the maximum amount of water vapor the air can hold at that temperature.

relief The differences in the height of landforms in an area.

renewable resource A material that can be renewed, or restored to its original form after being used, by means of a natural process; i.e., pure water that is polluted by human use is purified by the natural processes of the water cycle.

replacement reaction A chemical reaction in which one element replaces another element in a molecule.

residual soil A soil formed by the weathering of bedrock in place.

respiration The oxidation of food within living cells by which the chemical energy of food molecules is released in a series of metabolic steps involving the consumption of oxygen and the liberation of carbon dioxide and water.

retrograde motion An apparent motion of a planet in a direction opposite to its normal motion.

revolution The motion of one body around another body.

Richter scale A system that measures the strength of an earthquake by the amount of motion of Earth's crust that occurs during the earthquake.

rill Tiny grooves eroded in soil by a trickling stream of water flowing downhill.

rock The naturally formed, solid material that makes up Earth's crust.

rock cycle A sequence of events that shows how rocks are formed, changed, destroyed, and reformed from the same parent material.

rock-forming minerals The one hundred or so common minerals that make up more than 95 percent of the rock in Earth's crust.

rockfall Mass wasting process in which rock fragments broken loose by weathering fall from cliffs or bounce by leaps down steep slopes.

rockslide Mass wasting process in which rock fragments slide downhill.

rotation Motion in which every part of an object is moving in a circular path around a central line called the axis of rotation.

runoff Runoff is precipitation that does not evaporate or sink into the ground, but runs downhill along Earth's surface.

***S*-waves** Transverse seismic waves produced by an earthquake that travel slower than *P*-waves and are therefore the second to be recorded by a seismograph.

salinity The total amount of dissolved matter in seawater expressed in parts per thousand.

saltation A type of motion in which sediments transported in a stream move along the streambed in a series of bounces, hops, or leaps.

sandbar A long, narrow pile of sand deposited in open water.

satellite A solid body that orbits another body; i.e., the Moon is a satellite of the Earth.

sea breeze A local wind blowing from a body of water toward land that develops because daytime heating lowers the air pressure over the land.

sea cliff A steep cliff formed by the undercutting of steep slopes by waves and subsequent collapse of the overhanging material.

sea stack An isolated pillar of resistant rock that was detached from a headland by wave erosion.

sea terrace A flat platform of sediments deposited by wave action on the ocean floor adjacent to a steep coastline.

sea waves Wind-generated sea waves that are directly affected by the wind.

sediment Solid particles that have been transported and then deposited by agents of erosion.

sedimentary rock Rock formed by the lithification of sediments.

seismic wave A vibration of Earth's crust produced by faulting.

seismogram A line recorded on paper by a seismograph that represents the motion during an earthquake.

seismograph A device that detects the motion of Earth's crust during an earthquake.

seismology The study of earthquakes and their effects.

shadow zone A region in which no seismic waves from an earthquake are received by seismograph stations because the waves have been refracted or blocked as they traveled through Earth.

shear Forces acting along different lines of action causing rock to twist or tear.

shield volcano A wide, gently sloping volcano formed from successive layers of solidified lava flows.

sidereal month The time it takes the Moon to complete one 360° orbit around Earth, measured relative to a distant star; 27.32166 mean solar days, or 27 days, 7 hours, forty-three minutes and 11 seconds.

silicates Minerals composed of compounds of silicon and oxygen together with other elements; the most abundant mineral group on Earth.

silicon tetrahedron An ion of silicon joined with four ions of oxygen in the shape of a tiny tetrahedron (pyramid-like shape).

sill A pluton that forms between, and parallel to, existing rock layers.

sleet Clear pellets of ice that form when raindrops freeze as they fall through layers of air at below-freezing temperatures.

slump Mass wasting process in which a mass of bedrock or soil slides downward along a curved plane of weakness.

snow Hexagonal crystals of ice or needlelike ice crystals formed by the sublimation of water vapor in the atmosphere.

soil The accumulation of loose, weathered material that covers much of the land surface of Earth.

soil association A group of two or more soils occurring together in a characteristic pattern in a given geographical area.

soil creep The invisibly slow, downhill movement of soil.

soil horizon Recognizable layers in soil with characteristic properties.

soil moisture storage Water that is retained in the soil after water seeps into the ground.

soil moisture zone A moist layer of soil in which water clinging to soil particles supplies plant roots with water.

soil profile A cross section of soil from surface to bedrock.

solar day The time it takes Earth to rotate from one solar noon until the next solar noon.

solar eclipse An event in which the Moon passes directly between Earth and the Sun, casting a shadow on the Earth and partly or completely blocking an observer's view of the Sun.

solar noon The moment when the Sun is at its highest point in the sky for the day.

solar wind The flow of gas from the Sun out through the solar system; near Earth the solar wind travels at speeds of 600–1,000 km/s.

solstice The two times of the year when the Sun's altitude at noon reaches a maximum or a minimum. The summer solstice in the Northern Hemisphere occurs about June 21, when the Sun is in the zenith at the Tropic of Cancer; the winter solstice occurs about December 21, when the Sun is over the Tropic of Capricorn. The summer solstice is the longest day of the year and the winter solstice is the shortest.

source region The geographical region in which an air mass originated.

specific gravity The density of a mineral compared to the density of water.

specific heat capacity The number of degrees the temperature of one gram of a material will increase if one calorie of heat is added to it.

spectrum The band of colors formed when a beam of white light is passed through a prism; each wavelength of light in the beam is refracted at a slightly different angle, causing the light to be arrayed in order of its constituent wavelengths.

spit A sandbar attached at one end to the mainland.

spring A place where groundwater seeps out of the ground.

spring tide A tidal cycle in which there are very high and very low tides; caused by the additive effect of the Sun and Moon's gravity acting together along the same line of action.

station model A shorthand symbolic representation of the weather conditions at a particular location.

statistical forecasting Weather forecasting based upon the statistical analysis of historical weather data.

stock An irregularly shaped pluton covering an area less than 75 square kilometers.

strain The change in size or shape of a rock due to stress.

stratopause The boundary between the stratosphere and the mesosphere.

stratosphere The layer of the atmosphere ranging from about 6 to 30 kilometers above Earth's surface; it contains ozone that absorbs ultraviolet light; temperatures in this layer rise with altitude.

streak The color of a powdered mineral; created by rubbing a mineral against an unglazed piece of porcelain.

stream A body of water that moves under the influence of gravity to progressively lower levels in a narrow, but clearly defined channel.

stream banks The sides of the channel containing the water of a stream.

streambed The bottom of the channel containing the water of a stream.

stream channel Clearly defined path along which the water in a stream flows.

stream gradient The angle between the stream bed and the horizontal; the slope of the stream.

stream load The sediments transported by the water flowing in a stream.

stream mouth The site at which a stream flows into a standing body of water.

stream pattern The pattern formed by the system of streams in an area as they flow down slopes and join with other streams, e.g., dendritic, trellis, radial, rectangular, annular.

stream source The site at which the water in a stream originates.

stress Forces acting on a rock.

subduction The sliding of a denser ocean plate beneath a less dense continental plate resulting in the melting of the ocean plate as it plunges into the hot mantle.

sublimation The process by which a solid changes directly into a gas without passing through a liquid stage.

supercluster A cluster of galaxy clusters.

supergiant star An exceptionally luminous star that is 10 to 1,000 times the Sun's diameter.

supernova A rare stellar explosion in which the entire outer envelope of a star is blown away in a violent outburst, leaving behind a dense core.

superposition The idea that the bottom layer of a sedimentary series is the oldest, unless it was overturned or had older rock thrust over it, because the bottom layer was deposited first.

suspended load The sediments carried in suspension in the water flowing in a stream.

syncline A downward, valleylike fold.

synodic month The time between successive new moon phases; 29.530588 mean solar days, or 29 days, 12 hours, 44 minutes and three seconds.

synoptic forecasting Weather forecasting based upon analysis of a sequence of synoptic weather maps and charts.

synoptic weather map A map that summarizes weather data from many locations using station models.

system A group of things or processes that interact to perform some function.

talus A mound of rock fragments that accumulates at the base of a rock cliff that is weathering.

technology The application of knowledge to do things.

telescope An instrument for collecting electromagnetic radiation and producing magnified images of distant objects.

temperature A specific degree of hotness or coldness as indicated on or referred to a standard scale.

tension Forces acting away from each other along the same line of action causing rock to stretch.

tephra A collective name for particles formed when globs of lava ejected during an eruption cool and solidify as they fall to the ground.

terminal moraine A mound of till deposited along the leading edge of a glacier.

terrigenous Sediments derived from the land or a continent.

texture The size, shape, and arrangement of mineral crystals or grains in a rock.

thermocline A layer of seawater in which temperature drops sharply with depth.

thermosphere The uppermost layer of the atmosphere ranging from about 50 to 75 kilometers in altitude where it grades into space; this layer contains the ionosphere, which consists of charged particles or ions that deflect harmful radiation and reflect radio waves.

thunderstorm A vigorous, turbulent convection cell in the atmosphere resulting in a transient, violent storm of thunder and lightning, often accompanied by rain and sometimes hail.

time zone A region in which the same time is used throughout, instead of the local time at each place within it.

Glossary

topographic map A field map in which the field value is elevation above sea level; it shows the three-dimensional shape of Earth's surface in two dimensions.

tornado A rotating column of air whirling at speeds of up to 800 kilometers per hour, usually accompanied by a funnel-shaped downward extension of a cumulonimbus cloud.

transform boundary Places where edges of plates are sliding laterally past each other.

transpiration The process by which plants release water into the atmosphere through their leaves.

transported soil A soil that formed from material that was deposited in a region after being transported from another place.

transverse waves Waves in which the medium oscillates perpendicular to the direction of wave motion.

travel time The time it takes for a seismic wave to travel from the epicenter to a seismograph.

trench A deep crevice in the ocean floor produced by downward warping of the ocean floor where plates collide.

tropopause The boundary between the troposphere and the stratosphere.

troposphere The layer of the atmosphere ranging from Earth's surface to about 6 kilometers in altitude; it contains most of the air in the atmosphere; temperatures in this layer decrease with altitude.

tsunami Waves caused by undersea movements of Earth's crust such as slumping, earthquakes, or volcanic eruptions.

umbra The portion of a shadow in which all light has been blocked.

unconformity A break or gap in the sequence of a series of rock layers.

uniformitarianism The idea that Earth's features have formed gradually by natural processes that are still occurring, not by instantaneous creation or by catastrophic events.

usage Moisture lost from the soil during a period of drought.

valley glacier A glacier that forms between mountain slopes at high elevations.

velocity The speed and direction in which something is moving.

vent The central opening of a volcano out of which lava erupts.

ventifact A rock particle that has developed a flat side facing the prevailing winds due to abrasion of its exposed surface.

vertical sorting The separation and deposition of sediments in successive layers caused by differences in the settling rate of the sediment particles; particles that settle fastest form the bottom-most layer, whereas those that settle most slowly form the top layer.

vesicular A rock texture characterized by cavities formed by gas bubbles escaping from a lava as it cools and solidifies.

visibility The horizontal distance through which the eye can distinguish objects.

volcanic glass A rock formed by cooling that was so rapid that no crystalline structure was able to form; such rocks have a glassy appearance.

volcanism The processes by which magma rises into the crust and is extruded onto Earth's surface.

volcano The opening in Earth's crust through which lava emerges and the moundlike structure formed by the accumulation of lava.

walking the outcrop Physically following layers of rock from one place to another in order to correlate them.

water cycle The process by which Earth's water cycles in and out of the atmosphere.

watershed The area drained by a stream and its tributaries.

water table The top of the subsurface zone in which the pore spaces in a soil are saturated with water.

wavelength The distance between two successive wave crests or troughs.

weather The present condition of the atmosphere at any location.

weathering The breakdown of rocks into smaller particles by natural processes.

weight The gravitational force exerted on an object by Earth.

Wentworth Scale A scheme for classifying and naming sediment particles according to their size.

wet-bulb temperature The temperature recorded by a thermometer whose bulb is kept wet; it is usually lower than dry-bulb temperature due to evaporative cooling.

white dwarf A star, below the main sequence in the H-R diagram, which is small and dense; a dying star that has collapsed to about Earth's size and is slowly cooling off.

wind Air that is moving horizontally.

windward The side of a feature facing the wind.

zenith A point directly overhead to an observer.

zone of aeration The subsurface zone in which the pore spaces in a soil are filled mainly with air.

zone of saturation The subsurface zone in which the pore spaces in a soil are saturated with water.

Examination June 2019
Physical Setting/Earth Science

PART A
Answer all questions in this part.

Directions (1–35): For *each* statement or question, choose the word or expression that, of those given, best completes the statement or answers the question. Some questions may require the use of the *2011 Edition Reference Tables for Physical Setting/Earth Science*. Record your answers in the space provided.

1 The map below shows four major time zones of the United States. The locations of Boston and San Diego are shown.

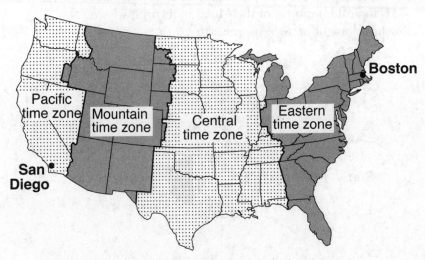

What is the time in Boston when it is 11 A.M. in San Diego?

(1) 8 A.M.
(2) 2 P.M.
(3) 3 P.M.
(4) noon

1 _____

2 The diagram below represents the spectral lines from the light emitted from a mixture of two gaseous elements in a laboratory on Earth.

If the same two elements were detected in a distant star that was moving away from Earth, how would the spectral lines appear?

(1) The entire set of spectral lines would shift toward the red end.
(2) The entire set of spectral lines would shift toward the blue end.
(3) The spectral lines of the shorter wavelengths would move closer together.
(4) The spectral lines of the longer wavelengths would move closer together.

2 ____

3 The diagram below shows Earth in orbit around the Sun, and the Moon in orbit around Earth. M_1 and M_2 indicate positions of the Moon in its orbit where eclipses might be seen from Earth.

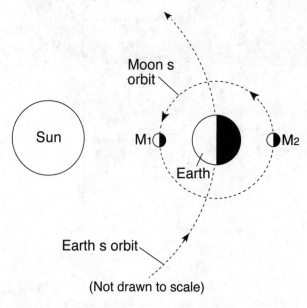

(Not drawn to scale)

Which table correctly matches each type of eclipse with the orbital position of the Moon and the cause of each eclipse?

Type of Eclipse	Moon's Position	Cause of Eclipse
Solar	M_1	Earth's shadow falls on Moon
Lunar	M_2	Moon's shadow falls on Earth

(1)

Type of Eclipse	Moon's Position	Cause of Eclipse
Solar	M_1	Moon's shadow falls on Earth
Lunar	M_2	Earth's shadow falls on Moon

(2)

Type of Eclipse	Moon's Position	Cause of Eclipse
Solar	M_2	Earth's shadow falls on Moon
Lunar	M_1	Moon's shadow falls on Earth

(3)

Type of Eclipse	Moon's Position	Cause of Eclipse
Solar	M_2	Moon's shadow falls on Earth
Lunar	M_1	Earth's shadow falls on Moon

(4)

3 _____

4 The timeline below represents the entire geologic history of Earth. The lettered dots on the timeline represent events in Earth's history.

Which lettered dot best indicates the geologic time when humans first appeared on Earth?

(1) A
(2) B
(3) C
(4) D

4 ____

5 Compared to our solar system, the universe is

(1) younger, smaller, and contains fewer stars
(2) younger, larger, and contains more stars
(3) older, smaller, and contains fewer stars
(4) older, larger, and contains more stars

5 ____

6 Which motion allows an observer on Earth to view different constellations throughout the year?

(1) Earth orbiting the Sun
(2) constellations orbiting Earth
(3) Earth orbiting the constellations
(4) constellations orbiting the Sun

6 ____

7 What is the approximate location of the Tasman Hot Spot in the Pacific Ocean?

(1) 36° N 160° W
(2) 36° S 160° E
(3) 160° N 36° W
(4) 160° S 36° E

7 ____

8 Most ozone is found in a region of Earth's atmosphere between 10 and 20 miles above Earth's surface. This temperature zone of the atmosphere is known as the

(1) thermosphere
(2) mesosphere
(3) stratosphere
(4) troposphere

8 ____

9 The Coriolis effect, which causes the curving of planetary winds, is a direct result of the

(1) distance between Earth and the Sun
(2) inclination of Earth's axis
(3) orbiting of Earth around the Sun
(4) spinning of Earth on its axis

9 ____

10 What is the dewpoint if the dry-bulb temperature is 18°C and the relative humidity is 64%?

(1) 14°C
(2) 11°C
(3) 9°C
(4) 4°C

10 ____

11 When one gram of liquid water at its boiling point is changed into water vapor

(1) 334 J/g is gained from the surrounding environment
(2) 334 J/g is released into the surrounding environment
(3) 2260 J/g is gained from the surrounding environment
(4) 2260 J/g is released into the surrounding environment

11 ____

12 The cross section below represents some processes of the water cycle. Arrows represent the infiltration of water. The dashed line labeled X represents the uppermost level of Earth material that is saturated by groundwater.

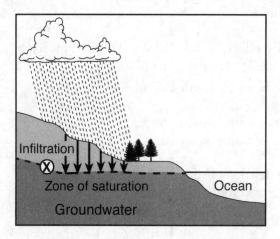

What is indicated by the dashed line labeled X?

(1) watershed
(2) water table
(3) impermeable bedrock
(4) impermeable soil

12 _____

13 Which index fossil in sedimentary surface bedrock most likely indicates that a marine environment once existed in the region where the sediments were deposited?

(1) Mastodont   (3) *Eospirifer*
(2) Condor   (4) *Coelophysis*

13 _____

14 The three diagrams below represent three frontal boundaries with the surface locations of the fronts labeled *A*, *B*, and *C*. Arrows indicate direction of air movement.

Which table correctly matches each letter with the type of frontal boundary it represents?

Letter	Type of Frontal Boundary
A	Cold Front
B	Warm Front
C	Occluded Front

(1)

Letter	Type of Frontal Boundary
A	Warm Front
B	Cold Front
C	Occluded Front

(2)

Letter	Type of Frontal Boundary
A	Cold Front
B	Warm Front
C	Stationary Front

(3)

Letter	Type of Frontal Boundary
A	Warm Front
B	Cold Front
C	Stationary Front

(4)

14 _____

15 The cross section below indicates the geologic ages of the bedrock beneath the state of Michigan. These rocks formed from sediments deposited in an ancient depositional basin. This region is called the Michigan Basin. Glacial deposits cover most of the surface.

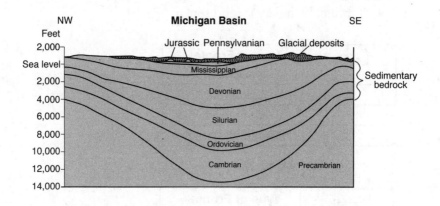

Which process most likely caused the formation of the Michigan Basin?

(1) uplift
(2) faulting
(3) metamorphism
(4) downwarping

15 _____

16 The cross section below represents prevailing winds, shown by arrows, moving over a coastal mountain range. Letters A through D represent locations on Earth's surface.

Which location would most likely have cloud cover and precipitation?

(1) A (3) C
(2) B (4) D   16 ____

17 Which Earth surface would most likely absorb the greatest amount of insolation on a sunny day if all of these surfaces have equal areas?

(1) blacktop parking lot
(2) white sand beach
(3) surface of a calm lake
(4) snow covered mountain slope   17 ____

18 Most of the infrared radiation given off by Earth's surface is absorbed in Earth's atmosphere by greenhouse gases such as water vapor, carbon dioxide, and

(1) hydrogen   (3) oxygen
(2) nitrogen   (4) methane   18 ____

19 The X on the map below indicates the region where the state of Washington is located on the present-day North American continent.

During which geologic period was the region of Washington State closest to the equator?

(1) Cretaceous   (3) Mississippian
(2) Triassic   (4) Ordovician   19____

20 A climate event that occurs when surface water in the eastern equatorial area of the Pacific Ocean becomes warmer than normal and may cause a warm, dry winter in New York State is

(1) an air mass formation
(2) the Doppler effect
(3) an El Niño
(4) a monsoon   20____

21 In the New York State rock record, there is *no* bedrock from the Permian Period. Which two other entire geologic periods have no sediment or bedrock in the New York State rock record?

(1) Cretaceous and Quaternary Periods
(2) Paleogene and Neogene Periods
(3) Triassic and Jurassic Periods
(4) Mississippian and Pennsylvanian Periods   21____

22 The cross sections below represent two widely separated bedrock outcrops, 1 and 2. Letters A, B, C, and D identify some rock layers. Line XY represents a fault. The rock layers have *not* been overturned.

Which lettered rock layer is the youngest?

(1) A   (3) C
(2) B   (4) D   22 _____

23 The block diagram below represents the surface features in a landscape region.

Which diagram best represents the general stream drainage pattern of this entire region?

(1)

(3)

(2)

(4)

23 ____

24 Compared to a landscape that develops in a cool, dry climate, a landscape that develops in a warm, rainy climate will most likely weather and erode

(1) slower, so the landforms are more angular
(2) slower, so the landforms are more rounded
(3) faster, so the landforms are more angular
(4) faster, so the landforms are more rounded   24 ____

25 Earth's crustal bedrock at the Mid-Atlantic Ridge is composed mostly of

(1) basalt, with a density of 2.7 g/cm^3
(2) basalt, with a density of 3.0 g/cm^3
(3) granite, with a density of 2.7 g/cm^3
(4) granite, with a density of 3.0 g/cm^3 25 ____

26 A seismic recording station located 4000 kilometers from the epicenter of an earthquake received the first P-wave at 7:10:00 P.M. (h:min:s). Other information that could be determined from this recording was that the earthquake occurred at approximately

(1) 7:03:00 P.M., and the S-wave arrived at this station at 7:12:40 P.M.
(2) 7:03:00 P.M., and the S-wave arrived at this station at 7:15:40 P.M.
(3) 7:17:00 P.M., and the S-wave arrived at this station at 7:12:40 P.M.
(4) 7:17:00 P.M., and the S-wave arrived at this station at 7:15:40 P.M.

27 Which graph best represents the percentage by mass of elements of Earth's crust?

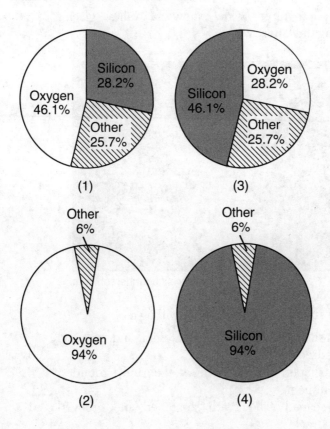

28 The aerial photograph below shows an elongated hill near the Finger Lakes formed by glacial deposition in New York State.

This elongated hill is best identified as a

(1) kettle
(2) moraine
(3) sand dune
(4) drumlin

28 _____

29 The photograph below shows a valley glacier altering the landscape.

www.alicehenderson.com

The result of this glacial action will be a

(1) U-shaped valley with scratched and grooved bedrock
(2) U-shaped valley with well-sorted, rounded sediment
(3) V-shaped valley with scratched and grooved bedrock
(4) V-shaped valley with well-sorted, rounded sediment

29 _____

30 Which element is always found in bioclastic sedimentary rocks?

(1) iron
(2) sodium
(3) carbon
(4) sulfur

30 ____

31 What are two major factors that control the development of soil in a given location?

(1) vegetative cover and slope
(2) tectonic activity and elevation
(3) erosion and transpiration
(4) bedrock composition and climate

31 ____

32 The cross section below represents rock units within Earth's crust. The geologic ages of two of the layers are shown.

Which sequence of geologic events occurred after the formation of the Cambrian limestone and before the formation of the Silurian limestone?

(1) uplift → weathering → erosion → subsidence
(2) uplift → subsidence → erosion → weathering
(3) subsidence → weathering → erosion → uplift
(4) subsidence → erosion → uplift → weathering

32 ____

33 Which two rocks contain the mineral quartz?

(1) gabbro and schist
(2) dunite and sandstone
(3) granite and gneiss
(4) pumice and scoria

33 ____

34 Which chemical formula represents the composition of a mineral that usually exhibits fracture, *not* cleavage?

(1) $Mg_3Si_4O_{10}(OH)_2$ (3) $CaCO_3$
(2) $NaCl$ (4) $(Fe,Mg)_2SiO_4$ 34____

35 Which mineral is commonly used as the "lead" in pencils?

(1) pyrite (3) galena
(2) graphite (4) fluorite 35____

PART B–1
Answer all questions in this part.

Directions (36–50): For *each* statement or question, choose the word or expression that, of those given, best completes the statement or answers the question. Some questions may require the use of the *2011 Edition Reference Tables for Physical Setting/Earth Science*. Record your answers in the space provided.

Base your answers to questions 36 and 37 on the graph below and on your knowledge of Earth science. The graph shows changing ocean water levels, over a 3-day period, at a shoreline location at Kings Point, New York on Long Island.

36 Based on the graph, the first low tide on December 26 occurred at approximately

(1) 6 A.M. (3) 6 P.M.
(2) 11 A.M. (4) 11 P.M.   36 ____

37 These Long Island tides show a pattern that is

(1) cyclic and predictable
(2) cyclic and unpredictable
(3) noncyclic and predictable
(4) noncyclic and unpredictable   37 ____

Base your answers to questions 38 through 41 on the diagram below and on your knowledge of Earth science. The diagram represents Earth orbiting the Sun. Four positions of Earth in its orbit are labeled A, B, C, and D. Letter N represents the North Pole. Distances are indicated for aphelion (Earth's farthest position from the Sun around July 4) and perihelion (Earth's closest position to the Sun around January 3). Arrows indicate directions of movement.

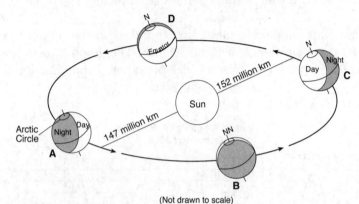

(Not drawn to scale)

38 During which season in the Northern Hemisphere is Earth at aphelion?

(1) winter
(2) spring
(3) summer
(4) fall

38 _____

39 Between which pair of lettered positions is the Sun's vertical ray moving from the equator southward to the Tropic of Capricorn?

(1) A and B
(2) B and C
(3) C and D
(4) D and A

39 _____

40 Approximately how many times does Earth rotate as it moves in its orbit from position A back to position A?

(1) 1 time
(2) 15 times
(3) 24 times
(4) 365 times

40 _____

41 What is the tilt of Earth's rotational axis relative to a line perpendicular to the plane of Earth's orbit?

(1) 15°
(2) 23.5°
(3) 66.5°
(4) 90°

41 _____

Base your answers to questions 42 through 44 on the map below and on your knowledge of Earth science. The map shows a composite of Doppler radar images. Darker shadings indicate the precipitation pattern of a large storm system over the eastern United States.

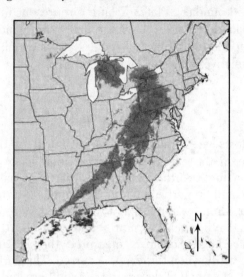

42 The surface wind circulation pattern around the center of this storm system is

(1) inward and clockwise
(2) inward and counterclockwise
(3) outward and clockwise
(4) outward and counterclockwise   42 _____

43 The best evidence on a weather map to indicate high-speed winds near the center of this storm system would most likely be

(1) cloud cover of 100%
(2) type of precipitation
(3) temperature and dewpoint values
(4) isobars drawn close together   43 _____

44 As this storm system follows a normal storm track, it will most likely move toward the

(1) southeast   (3) northeast
(2) southwest   (4) northwest   44 _____

EXAMINATION June 2019

Base your answers to questions 45 through 48 on the passage and map below and on your knowledge of Earth science. The map shows the location of the epicenter (✱) of a major earthquake that occurred about 1700 years ago. Point A represents a location on a tectonic plate boundary. Plates X and Y represent major tectonic plates. The island of Crete; the Anatolian Plate, which is a minor tectonic plate; and the Hellenic Trench have been labeled. Arrows indicate the relative directions of plate motion.

Crete Earthquake

Scientists have located the geological fault, off the coast of Crete in the Mediterranean Sea, that likely shifted, causing a huge earthquake in the year 365 that devastated life and property on Crete. The southwestern coastal region of Crete was uplifted, as evidenced by remains of corals and other sea life now found on land 10 meters above sea level. Scientists measured the age of these corals to verify when this event occurred. This earthquake caused a tsunami that devastated the southern and eastern coasts of the Mediterranean Sea. It is estimated that earthquakes along the fault, associated with the Hellenic Trench, may occur about every 800 years.

45 Which type of plate boundary is represented at point *A*?

(1) divergent
(2) convergent
(3) transform
(4) complex

45 ____

46 What are the names of the major tectonic plates *X* and *Y*?

(1) *X* = Eurasian Plate; *Y* = African Plate
(2) *X* = Eurasian Plate; *Y* = Arabian Plate
(3) *X* = Indian-Australian Plate; *Y* = African Plate
(4) *X* = Indian-Australian Plate; *Y* = Arabian Plate

46 ____

47 Which two New York State index fossils are most closely related to the corals that were radioactively dated in this study?

(1) *Eucalyptocrinus* and *Ctenocrinus*
(2) *Elliptocephala* and *Phacops*
(3) *Maclurites* and *Platyceras*
(4) *Lichenaria* and *Pleurodictyum*

47 ____

48 Which activity could best prepare residents along the Mediterranean coast to reduce the loss of human life during a future tsunami?

(1) Board up windows.
(2) Remove heavy objects from the walls in homes.
(3) Plan evacuation routes to higher ground.
(4) Build reinforced basements.

48 ____

Base your answers to questions 49 and 50 on the photograph below and on your knowledge of Earth science. The photograph shows a sandstone erosional feature that formed near the Grand Canyon, in southwestern United States.

www.nationalgeographic.com

49 Which erosional agent is most likely sandblasting this rock formation?

(1) wind
(2) waves
(3) running water
(4) moving ice

49 _____

50 What is the range of grain sizes that are most commonly found in rock making up this feature?

(1) 0.0004 cm–0.006 cm
(2) 0.006 cm–0.2 cm
(3) 0.2 cm–6.4 cm
(4) 6.4 cm–25.6 cm

50 _____

PART B–2
Answer all questions in this part.

Directions (51–65): Record your answers in the spaces provided. Some questions may require the use of the *2011 Edition Reference Tables for Physical Setting/Earth Science*.

Base your answers to questions 51 through 53 on the data table below, on the graph below, and on your knowledge of Earth science. The data table shows the projected percentages of radioactive isotope X remaining and its disintegration product Z forming over 6.5 billion years. The graph shows the disintegration of radioactive isotope X.

Radioactive Isotope X (%)	Disintegration Product Z (%)	Time (billion years)
100	0	0
50	50	1.3
25	75	2.6
12.5	87.5	3.9
6.25	93.75	5.2
3.125	96.875	6.5

51 On the grid *below*, construct a line graph by plotting the percentages of disintegration product Z forming over 6.5 billion years. Connect *all six plots* with a line. The percentages of radioactive isotope X have already been plotted. [1]

52 Identify radioactive isotope X. [1]

_____

53 Calculate the amount, in grams, of an original 300-gram sample of radioactive isotope X remaining after 3.9 billion years. [1]

_____ g

Base your answers to questions 54 through 57 on the weather map below and on your knowledge of Earth science. The map shows the location of a low-pressure system over New York State during summer. Isobar values are recorded in millibars. The darker shading indicates areas of precipitation. Some New York State locations are indicated.

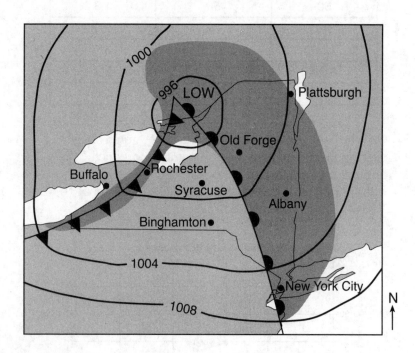

54 Describe the change in air pressure that will most likely occur at Rochester by the time that the cold front has reached Syracuse. Then describe what will most likely happen to the amount of cloud cover in Rochester with this change in air pressure and location of the cold front. [1]

Change in air pressure: _____

Amount of cloud cover: _____

55 The station model below represents the weather conditions at Buffalo, New York, at the time that this map was prepared.

Buffalo, New York

In the table *below*, record the weather data represented on this station model. [1]

Air Temperature (°F)	Barometric Pressure (mb)	Wind Direction from the	Wind Speed (knots)

56 State the relative humidity at Albany when the air temperature is equal to the dewpoint. [1]

_____ %

57 Identify the name of the weather instrument used to measure the wind speed at Plattsburgh. [1]

_____

EXAMINATION June 2019

Base your answers to questions 58 through 61 on the diagram and passage below and on your knowledge of Earth science. The diagram represents the orbits of Earth, Comet Tempel-Tuttle, and planet *X*, another planet in our solar system. Arrows on each orbit represent the direction of movement.

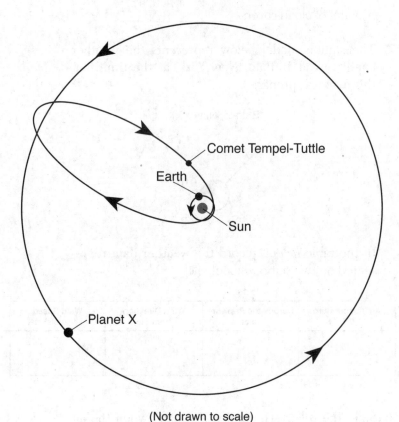

Orbit of Comet Tempel-Tuttle

(Not drawn to scale)

Comet Tempel-Tuttle

Comet Tempel-Tuttle orbits our Sun and is responsible for the Leonid meteor shower event observed from Earth. This meteor shower occurs every year in November and is visible in the night sky as Earth passes through the debris left in space by this comet. The debris from the comet produces meteors that are smaller than a grain of sand, which enter Earth's atmosphere and burn up in the mesosphere temperature zone. Comet Tempel-Tuttle's orbital distance from the Sun ranges from about 145 million kilometers at its closest approach to 2900 million kilometers at its farthest distance. Its two most recent closest approaches to the Sun occurred in 1965 and one revolution later in 1998.

58 Identify the name of the object located at one of the foci of the elliptical orbit of Comet Tempel-Tuttle. [1]

_____

59 Identify the solar system planet represented by planet X, which orbits near Comet Tempel-Tuttle's farthest distance from the Sun. [1]

_____

60 Determine the year in which Comet Tempel-Tuttle will next make its closest approach to the Sun. [1]

_____

61 Identify the force that causes debris from the comet to fall through Earth's atmosphere. [1]

_____

_____

Base your answers to questions 62 through 65 on the diagram below and on your knowledge of Earth science. The diagram represents the apparent path of the Sun across the sky as seen by an observer on Earth's surface on June 21. Points A, B, C, and D represent positions of the Sun at different times of the day. The angle of *Polaris* above the horizon as seen in the nighttime sky is indicated.

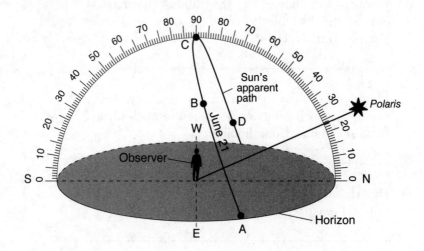

62 Describe the changes in the length of the observer's shadow as the Sun appears to move from position A to position D. [1]

63 Describe *one* piece of evidence from the diagram that supports the inference that the observer is located at the Tropic of Cancer. [1]

64 State the number of daylight hours at this location on September 23. [1]

_____ h

65 The Sun's apparent movement through the sky is caused by Earth's rotation. Identify the device that was used to first demonstrate that Earth rotates. [1]

_____

EXAMINATION June 2019

PART C
Answer all questions in this part.

Directions (66–85): Record your answers in the spaces provided. Some questions may require the use of the *2011 Edition Reference Tables for Physical Setting/Earth Science*.

Base your answers to questions 66 through 68 on the models below and on your knowledge of Earth science. The models represent cutaway views of four planets in our solar system, showing their inferred interior structures. Each planet is shown in relation to the size of Earth.

Seeds, Michael and Backman, Dana. 2011. *The Solar System.*

66 Determine how many times larger Jupiter's equatorial diameter is, compared to Earth's equatorial diameter. [1]

_____ **times larger**

67 Explain why Jupiter appears brighter in the night sky than Mercury, despite Jupiter's greater distance from Earth. [1]

_____

_____

68 Identify *two* terrestrial planets shown in the models. Explain why they are considered terrestrial planets. [1]

Terrestrial planets:_____ and _____

Explanation: _____

_____

EXAMINATION June 2019

Base your answers to questions 69 through 72 on the diagram below and on your knowledge of Earth science. The diagram represents several streams converging and eventually flowing into a lake. Points X and Y indicate locations on either side of a meander in the stream. Points A and B indicate locations in the streams where the stream discharge was measured in cubic meters per second. The circled region labeled C represents a depositional feature.

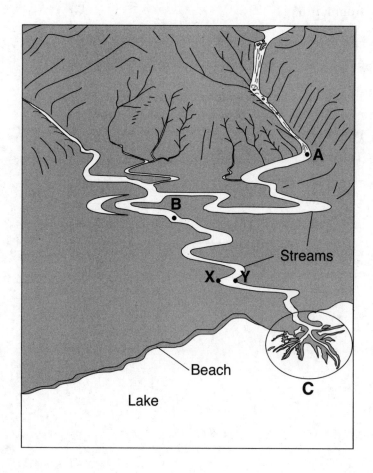

69 Identify the name of the depositional feature labeled *C*. [1]

_____

70 In the box *below*, draw a cross-sectional view of the general shape of the stream bottom from point *X* to point *Y*. [1]

```
X          Stream Surface         Y
●─────────────────────────────────●
│                                 │
│                                 │
│                                 │
│                                 │
│                                 │
└─────────────────────────────────┘
```

71 Describe how the size and shape of sediments will most likely change as they are transported downstream. [1]

Size: _____

Shape: _____

72 The stream velocity at location *A* is 100 centimeters per second, and the stream velocity at location *B* is 10 centimeters per second. Identify *one* possible particle diameter that would most likely be deposited between points *A* and *B*. [1]

_____ **cm**

Base your answers to questions 73 through 76 on the map and graphs below and on your knowledge of Earth science. The map shows the locations of two cities, Hastings, Nebraska, and Riverhead, New York. The graphs show average monthly air temperatures for Hastings and Riverhead.

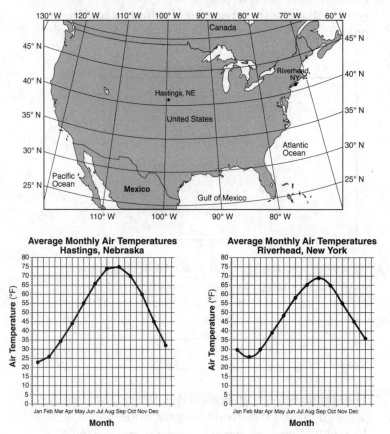

73 Explain why Hastings has a greater annual temperature range than Riverhead. [1]

_____

_____

74 Explain why the angle of insolation at both locations is approximately the same at solar noon on any given day. [1]

_____

_____

EXAMINATION June 2019

75 Identify the planetary wind belt that primarily influences the climates of *both* Hastings and Riverhead. [1]

_____

76 Name the ocean current that most likely has the greatest effect on the climate of Riverhead. [1]

_____ **Current**

_____

Base your answers to questions 77 through 80 on the block diagram below and on your knowledge of Earth science. The block diagram represents a region of sedimentary rock that has been intruded by magma, which has since solidified. Points X and Y identify locations at the boundary between the igneous intrusion and surrounding sedimentary rock layers. Letters A and B represent specific rock units. Letter C represents rock formed from the lava flow of the nearby volcano. The rock layers have *not* been overturned.

Adapted from www.brocku.ca/earthsciences

77 Identify *two* processes that formed the sedimentary rock layers represented in the diagram. [1]

1: _____ 2: _____

78 Describe *one* piece of evidence shown in the diagram that indicates that rock unit *A* is younger than rock unit *B*. [1]

_____

_____

79 Explain why the igneous rock that formed at location *C* is composed of crystals less than 1 millimeter in size. [1]

_____

_____

80 State the names of *two different* metamorphic rocks that are most likely found in the zone of contact metamorphism at locations *X* and *Y*. [1]

Location *X*: _____

Location *Y*: _____

EXAMINATION June 2019

Base your answers to questions 81 through 85 on the passage and map below, the field map below, and on your knowledge of Earth science. The map shows the location of Crater Lake in Oregon in the western United States. The field map shows lake depths and some isolines in Crater Lake recorded in meters. Line *AB* and line *CD* are reference lines. Letter *X* represents a location on the lake bottom.

Crater Lake

Crater Lake is the deepest lake in the United States. The lake formed in the crater at the top of volcanic Mount Mazama after it exploded in a violent eruption approximately 7700 years ago. The rim of the crater is approximately 2300 meters (7500 feet) above sea level and is mostly composed of the rock andesite. The average yearly air temperature at the lake is 38°F, and snowfall often occurs from October through June. Hydrothermal activity (heating of the water) is ongoing under the lake, indicating that this region is still volcanically active.

EXAMINATION June 2019

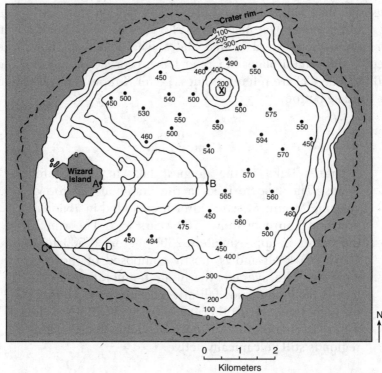

Field Map of Crater Lake

81 On the field map *above*, draw the 500-meter depth isoline. [1]

82 On the grid *below*, construct a profile along line AB by plotting the lake depth of each isoline that crosses line AB. Connect *all* plots with a line to complete the profile. [1]

Lake Depth

83 Determine the gradient, in meters per kilometer, between points C and D. [1]

_____ **m/km**

84 Determine *one* possible depth, in meters, of Crater Lake at location X. [1]

_____ **meters**

85 Line YZ on the diagram below represents the mineral composition of an andesitic rock taken from the bottom of Crater Lake.

Identify the percent by volume of each of the three minerals in this andesitic rock. [1]

Plagioclase feldspar: _____ %

Biotite: _____ %

Amphibole: _____ %

Answers June 2019

Physical Setting/Earth Science

Answer Key

PART A

1. 2	8. 3	15. 4	22. 1	29. 1
2. 1	9. 4	16. 2	23. 3	30. 3
3. 2	10. 2	17. 1	24. 4	31. 4
4. 1	11. 3	18. 4	25. 2	32. 1
5. 4	12. 2	19. 4	26. 2	33. 3
6. 1	13. 3	20. 3	27. 1	34. 4
7. 2	14. 1	21. 2	28. 4	35. 2

PART B–1

36. 1	39. 4	42. 2	45. 3	48. 3
37. 1	40. 4	43. 4	46. 1	49. 1
38. 3	41. 2	44. 3	47. 4	50. 2

In **PARTS B–2 and C** you are required to show how you arrived at your answers. For sample methods of solutions, see Barron's *Regents Exams and Answers* book for Earth Science—Physical Setting.

Index

A

Abrasion, 421, 449, 457–458, 463–464
Absolute age, 239
Absolute humidity, 564
Acid rain, 426
Actual age, 239
Adiabatic changes, 570
Age, 239–241
  determining, 247–253
Air composition, 203
Air currents, 568
Air humidity, 563–568
Air masses, 582–583
Air movements, 568
Air pressure
  description of, 561–563
  humidity and, 571
  temperature and, 569–571
  winds and, 573–576
Air temperature
  description of, 560–561
  humidity and, 572–573
Alfonsine Tables, 23
Alluvial fans, 489–490
Ancient environments, 255–256
Andesite, 306
Anemometer, 568
Angle of insolation, 48, 532–533
Angle of repose, 443
Angular unconformities, 241–242
Animal action, 421–422
Annual cycles, 532–533
Annual traverse of the constellations, 45
Annular eclipse, 53
Annular stream pattern, 503
Anticline, 385
Anticyclones, 586–587
Apparent brightness, 126
Apparent daily motion, 5
Apparent motion, 5–6
Aquicludes, 226–227
Aquifers, 226
Archean eon, 244–246
Aristotle, 19–21
Artes, 466–467
Artesians, 226–227
Asterism, minerals, 282
Asteroids, 155
Asthenosphere, 392–393, 396
Astronomy
  early, 1–18
  celestial coordinate system, 8–9
    meridian, 8
    motion, 4–6
    objects, 3–4
    sphere, 6–11
    constellations, 5
    Dubhe, 8
    horizon, 7
    Jupiter, 10
    Mars, 10
    Merak, 8
    Mercury, 10
    Moon, 10
    planets, 10–11
    Polaris, 8
    Saturn, 10
    sky, 1–6
    stars, 8, 10
    Sun, 9–10
    Venus, 10
    zenith, 8
    zodiac, 9–10
  modern, 19–40
    Alfonsine Tables, 23
    Aristotle, 19–21
    astronomical unit, 28–29
    Brahe, Tycho, 25
    Copernicus, 23–25
    Galilei, Galileo, 29
    gravity, 29–32
    heliocentric model, 19–40
    Kepler, Johannes. *See* Kepler, Johannes
    Moon, 29
    Newton, Isaac, 29–32
    orbital motion, 30–32
    planetary motion
      Kepler's law, 26–29
      velocity, 27–28
    Ptolemy, 19, 21–23
    retrograde motion, 21–22, 24–25
    revolution, period of, 28–29
    spheres, 20–21
    universal gravitation, 29
  tools for studying, 122–125
    Mount Palomar Observatory, 123
    observation instruments, 124–125
    spectroscopes. *See* Spectroscopes
Atmosphere, 527–558
  aerobic, 202
  circulation, 539–541
  conduction, 537
  convection, 537
  Coriolis effect, 541–544
  definition of, 562
  development, 201–203
  entering, 536–544
  formation, 200–201
  insolation. *See* Insolation
  latent heat, 538–539
  layers, 204–205
  lost, 199
  origin, 199–203
  oxidizing, 201–202
  radiation, 537–538
  reducing, 201
  solar radiation. *See* Solar radiation
  structure, 203–205
  transparency, 569
  variables, 560
Atmospheric pressure, 562
Atoms, 249
Aurora Borealis, 190
Axis, 74
Axis of rotation, 42

B

Banding, 315
Barite, 287
Barometer, 562
Barrier bars, 492
Barrier islands, 492
Basalt, 303, 306
Basaltic glass, 305
Batholiths, 372
Beaches, 490–491
Bed, 486
Bench marks, 382–383
Berms, 490
Big Bang Theory, 178–179
Biogenic sediments, 310
Biological evolution, 254
Body waves, 339
Borates, 287
Boundaries, 393–394

Index

Brahe, Tycho, 25
Breaking waves, 454
Breezes, 574–575
Brightness, stellar, 126

C

Calcite, 287
Callisto, 153
Capillarity, 217
Capillary action, 225–226
Capillary fringe, 225
Carbon-14, 250–253
Carbonate ion, 287
Carbonates, 287
Carbonation, 425
Cartesian system, 74
Catastrophic theory, 145
Celestial coordinate system, 8–9
Celestial meridian, 8
Celestial motion, 4–5
Celestial objects, 3–4
Celestial sphere, 6–11
Celestite, 287
Celsius, 561
Cementation, 307
Cenozoic era, 245–246
Ceres, 150
Chaos theory, 590
Charon, 148
Chatoyancy, minerals, 282
Chemical formulas, 277
Chemical sediments, 310–311
Chemical symbol, 277
Chemical tests, 282
Chemical weathering, 423–426
Chromates, 287
Cinder cone volcano, 371
Cinnabar, 287
Circles, great, 76–78
Cirques, 462, 466–467
Clastic sediments, 310
Cleavage, minerals, 280
Climate, 426–427, 613–638
  continental drift theory and, 389–390
  definition, 613
  effects, 622–623
  factors, 614–621
  landscapes and, 504–505
  types of, 621–622
Climatic trends, 623–626
Cloud(s), 576–579, 621
Cloud base, 578
Cloud ceiling, 569
Cloud cover, 569
Clusters, 174–175
Coastal processes, 454–456
Cold fronts, 584

Color
  description of, 534–535
  of minerals, 279
  of stars, 127–128
Coma, 157
Cometary impacts, 200–201
Comets, 156–157
Compaction, 307
Composite cone, 371
Compounds, 277–278
Compression stress, 384
Condensation, 222, 563–564, 576
Condensation nuclei, 576
Conduction, 368, 537
Constellations, 5
Contact metamorphism, 314
Continental drift theory, 386–390
Continental glaciers, 463
Contour interval, 81
Contour lines, 81–82
Convection, 368–369, 397, 537
Convection cell, 541
Convection currents, 451
Convergent boundaries, 393–394, 396
Coordinate system, 8–9, 74–75
Copernicus, 23–25
Core, 348–349
Coriolis effect, 43–44, 541–544, 576
Correlation, rock layers, 241–242
Corundum, 287
Cosmic background radiation, 179
Cosmology, 173–186
  Big Bang Theory, 178–179
  clusters, 174–175
  cosmic background radiation, 179
  Doppler effect, 176–177
  light-year, 175
  Olber's Paradox, 174
  superclusters, 175
  universe. See Universe
Crevasses, 460
Cross-bedded sediments, 308–309
Crust, 188–189
  definition of, 348
  dynamic, 382–386
  lithosphere, 382
Crustal movements
  causes, 384
  effects, 384–385
  evidence, 382–383
Crystalline form, 278

Crystalline spheres, 20–21
Crystalline structure, 219–220
Crystalline texture, 314–315
Crystals, 302
Currents, 456
  convection, 451
  equatorial, 452
  longshore, 455–456
  ocean, 451–453
  wind-driven, 451–453
Cutoff, 451
Cyclones, 586–587
Cyclonic storm tracks, 621

D

Daily cycles, 532–533
Day, 97–103
  solar, 100
Deferents, 22
Deflation, 457–458
Deflation hollow, 458
Degassing, 213
Deltas, 489–490
Dendritic stream pattern, 502
Density, 315
Deposition, 483–526
  definition of, 576
  glaciers, 494–500
  gravity, 486–488
  moving water, 488–492
  particles, characteristics, 485
  sediment, bedding, 486
  sorting, 484
  speed of medium, 484
  wind, 492–494
Deposits, 495–496
Desert pavement, 458–459
Desertification, 624
Dessication, 307
Dew, 578
Dew point, 565, 568, 573
Diabase, 303
Dikes, 372
Diorite, 306
Disconformities, 241–242
Displaced structures, 382
Divergent boundaries, 393–394, 396
Divides, 501
Domolmite, 287
Doppler effect, 176–177
Double refraction, minerals, 282
Drainage basin, 501–502
Drizzle, 579
Drumlins, 498
Ductile deformation, 384–385
Dunes, 493–494
Dunite, 306

Index

Duration of insolation, 48
Dwarf planets, 148–150

E

Earth, 149–151
　aerobic, 202
　air of, 203
　　cometary impacts, 200–201
　　development, 201–203
　　formation, 200–201
　　layers, 204–205
　　lost, 199
　　origin, 199–203
　　outgassing, 200
　　oxidizing, 201–202
　　reducing, 201
　　structure, 203–205
　atmosphere of, 199–210
　birth of, 187–188
　coordinate system, 69–96
　core of, 188–189, 348–349
　crust of. *See* Crust
　crustal movements. *See* Crustal movements
　differentiation of, 188
　hydrosphere, 211–236
　interior structure, 347–351
　layers of, 188–189
　magnet, 189–191
　mantle, 188–189, 348–349
　motions
　　effects, 46–55
　　heliocentric. *See* Heliocentric motions
　origins of, 187–198
　shape of, 70–73
　　evidence, 71–73
　　photographs from space, 72
　　Polaris, altitude, 72–73
　　roundness, 70–71
　　smoothness, 71
　size, determining, 73–74
　surface, 534–536
　mapping, 74–85
Earthflow, 445, 487
Earthquakes, 333–364, 382
　causes, 334
　definition, 333–334
　effects, 344–346
　elastic rebound theory, 334
　epicenter. *See* Epicenter
　measuring of, 343–344
　　Mercalli scales for, 344–345
　　Richter magnitude scale for, 343–344
　P-waves, 339–340
　seismic waves, 334, 337–339
　seismogram, 338–343
　S-waves, 339–340
　waves, 335–339
　　frequency, 336
　　length, 336
　　longitudinal waves, 337
　　motion, 334
　　periodic waves, 336
　　refraction, 336
　　transverse waves, 337
　　types, 337
　　velocity, 336
　　zones, 346–347
Eccentricity, 26–27
Eclipse, 51–54
　annular, 53
　evidence of Earth's shape, 71
　lunar, 53–54
　solar, 51–53
Ecliptic, 9–10
El Niño, 623, 625–626
Elastic deformation, 384–385
Elastic limit, 384
Elastic rebound theory, 334
Electromagnetic field, 528
Electromagnetic spectrum, 529–530
Electromagnetic waves, 528–529
Elements, 277
Elevation, 620
Elevation field maps, 81
Ellipse, 26–27
End moraine, 497
Energy, in atmosphere, 536–544
　circulation, 539–541
　conduction, 537
　convection, 537
　Coriolis effect, 541–544
　latent heat, 538–539
　radiation, 537–538
Environments, ancient, 255–256
Eons, 244–246
Epicenter, 334
　distance to, 340–341
　location of, 341–343
Epicycles, 22–23
Epochs, 245–247
Equator, 75
Equatorial currents, 452
Equinoxes, 106–107
Eras, 245–246
Eris, 150
Erosion, 441–482
　agents of. *See* Erosion agents
　evidence of, 442
Erosion agents, 442–470
　coastal processes, 454–456
　convection currents, 451
　equatorial currents, 452
　glaciers, 459–467, 469–470
　gravity, 442
　　angle of repose, 443
　　earthflow, 445
　　mass wasting, 443–446, 467–468
　　rapid, 444–445
　　slow, 445–446
　　mudflow, 445
　　rockfalls, 444
　　slump, 444–445
　　soil creep, 445–446
　longshore currents, 455–456
　oceans, 451–467
　　currents, 451–453
　raindrops, 446
　runoff, 446
　shoreline erosion, 456–457
　streams, 446–451, 468
　　erosion, 449
　　life cycle, 449–451
　　transport, 447–449
　tsunamis, 454
　water in motion, 446–457
　waves, 453–456, 469
　　crest, 453
　　height, 453
　　length, 454
　　refraction, 455
　　trough, 453
　　wind-driven currents, 451–453
　　wind-generated, 453–454
　wind, 457–459, 469
Erratics, 499
Europa, 153
Evaporation, 218, 221, 563–564
Evaporative cooling, 219
Evaporites, 311
Evapotranspiration, 620
Evolution, 254
Exfoliation, 421
Exposure, 428
Extrusive rocks, 302–303
Eyepiece, 122

F

Fahrenheit, 561
Fall equinox, 49
Faulting, 334
Faults, 240, 385–386
Feldspar, 304
Felsic, 304
Field, 79, 580
Field maps, 79–85, 580–582
　contour lines, 81–82
　conventions, 85
　direction, 82

Index

elevation field maps, 81
  gradient, 80–81
  isolines, 80
  profiles, 84–85
  scale, 82, 84
  symbols, 82–83
  topographic, 81–85
Finger lakes, 499
Firn, 460
Fissures, 371
Flame color, minerals, 282
Flooding, 468
Floodplains, 450, 488–489
Focus, 26–27
Fog, 578
Folded rock layers, 382
Folds, 385
Foliation, 315
Fossils, 383
  continental drift theory, 388
  fuels, conservation, 288–289
  index, 243
Foucault Pendulum, 42–43
Fracture
  minerals, 280–281
  rock, 384–385
Frame of reference, 74
Freezing, 219
Frequency, 336
Frigid zone, 615
Frontal cyclones, 586–587
Fronts, 582, 583–586, 584
Frost, 578
Frost action, 420–421
Fusion, 130–131

G

Gabbro, 302–303, 306
Galaxies, 124–125, 144
Galena, 287
Galilei, Galileo, 29
Ganymede, 153
Geocentric model, 1–18, 7, 21
Geocentric universe, 20–23
Geologic column, 244
Geologic time scale, 244–247
Geological record, 238–256
Geological structures, 388
Geothermal gradient, 365–366
Glacial deposits, 495–496
Glacial drift, 497
Glacial erosion, 463–467
Glacial grooves, 465
Glacial lakes, 499–500
Glacial polish, 465
Glacial striations, 465
Glacial till, 497
Glacial transport, 494–495

Glaciers, 459–467, 469–470
  deposition, 494–500
  landforms, 497–500
  types, 462–463
Global climatic trends, 623–626
Global radiation budget, 530–531
Global warming, 624–625
Global wind belts, 575–576
Goethite, 423
Graded bedding, 486
Gradient, 80–81, 580
Granite, 302, 306
Gravitational differentiation, 155
Gravity, 29–32, 132–133
  agent of erosion, 442
Greenhouse effect, 538
Greenwich Mean Time, 101
Groundmass, 303
Groundwater, 222, 223–227
  aquicludes, 226–227
  artesians, 226–227
  capillary action, 225
  capillary fringe, 225
  permeability, 224–225
  porosity, 223–224
  soil moisture zone, 225
  water table, 225–226
  wells, 226
  zone of saturation, 225
Gypsum, 287
Gyres, 43, 452

H

Hachure marks, 81–82
Hail, 579
Half-life, 250–253
Halides, 288
Halite, 311
Halley's Comet, 148
Hanging valleys, 466
Hardness, minerals, 279–280
Heat, internal
  conduction, 368
  convection, 368–369
  original heat, 367
  radiation, 368
  radioactive decay, 367
  sources of, 367–368
  tidal friction, 367–368
  transfer of, 368–369
Heat capacity, 218
Heft, 282
Heliocentric model, 19–40
Heliocentric motions, 41–68
  constellations, annual traverse, 45
  Coriolis effect, 43–44

Earth-Moon-Sun System, 50–55
eclipses, 51–54
Foucault Pendulum, 42–43
Moon, phases, 50–51
parallelism, Earth's axis of rotation, 44
Polaris, 45
precession, 45
revolution, 44–45
rotation, 42–44
  parallelism, 44
Sun's path, 46–50
tides, 54–55
Vega, 45
Hematite, 287
Herschel, William, 154
Hertzsprung-Russell Diagram, 128–130
Horizon, 7
Horns, 466–467
Hot spots, 394–395, 397
Hours, 98–100
Human perception, 122
Humidity, 563–568
  air, 563–568
  relative, 564–566
Hurricanes, 592
Hutton, James, 238
Huygens, Christian, 153
Hydration, 424–425
Hydrogen bonds, 217
Hydrolysis, 424–425
Hydrosphere
  groundwater, 222, 223–227
  origin, 211–236
    groundwater, 222
    outgassing, 213
    runoff, 222
    seawater. *See* Seawater
  water, 212–214
    liquid, 217–218
    solid, 219–220
    states, 217–220
    structure, 216–217
    vapor, 218–219
Hydroxides, 287
Hygrometer, 565

I

Ice movement, 460–462
Igneous rock, 240–241, 300
  characteristics, 301
  formation, 300–301
  identification, 304–306
  mineral composition, 304
  porphyrys, 303–304
  texture, 302–304
Immature soil, 431

772

Index

Impact crater, 158
Impact events, 158–159
Index fossils, 243
Inorganic matter, 276
Insolation, 47–48, 221,
    532–536
  angle of, 532–533
  annual cycles, 532–533
  color, 534–535
  daily cycles, 532–533
  duration, 533–534
  Earth's surface, 534–536
  latent heat of water, 536
  latitude, 532–534
  season, 533
  specific heat, 535–536
  texture, 534–535
Internal heat transfer, 368–369
International Date Line,
    102–103
Interplanetary bodies, 155–159
Intrusions, 372
Intrusive rocks, 302
Io, 153
Iron-nickel core, 351
Isobars, 580, 582
Isolines, 80, 580
Isostasy, 387
Isotherms, 580–581
Isotopes, 249–253

J
Joints, 385
Jovian planets, 145,
    149, 152–154
Jupiter, 10, 147, 149–150,
    152–153

K
Kelvin, 561
Kepler, Johannes, 25–29
  law of equal areas, 28
  law of planetary motion,
    26–29
Kettle lake, 499
Kettles, 499
Key beds, 244
Kuiper belt, 147–148

L
Laccoliths, 372
Lagoons, 492
Land breezes, 574–575
Land distribution, 615
Landforms, 500
Landscape development,
    500–507
  annular stream pattern, 503

climate and landscapes,
    504–505
dendritic stream pattern, 502
drainage basin, 501–502
landforms, 500
landscape regions, 505–507
  New York State, 506–507
mountains, 501
plains, 501
plateaus, 501
radial stream pattern, 502
rectangular stream pattern, 503
relief, 500–501
soil associations, 503–504
stream drainage system, 502
stream patterns, 501–503
topography, 500
trellis stream pattern, 50
tributary, 501
Landslide, 444
Latent heat, 538–539
  of fusion, 220
  of vaporization, 219
  of water, 536
Latitude, 532–534, 615
  determining, 78
  Sun's path changes, 49–50
Latitude-longitude coordinate
    system, 75–78
Lava, 300, 370–371
Lava plateau, 371
Laws
  equal areas, 28
  original horizontality, 239
  planetary motion, 26–29
Layers, distortion, 315
Levees, 488–489
Lichens, 425–426
Life
  evolution, 254
  geological record, 238–256
  natural selection, 254–255
Light-year, 175
Limbs, 385
Lithification, 307–308
Lithosphere, 382, 392–393, 395
Load, 448
Local time, 100–103
Loess, 493
Longitude, 75–79
Longitudinal waves, 337
Long-range weather
    forecasting, 589–590
Longshore currents, 455–456
Lord Kelvin, 130
Luminescence, minerals, 282
Luminosity, 126–127, 128–129
Lunar eclipse, 53–54
Luster, minerals, 279

M
Mafic, 304
Magma, 300, 369–370
Magnetic dynamo, 189
Magnetism, 282
Magnetite, 287
Magnetosphere, 189–191
Main sequence, 129
Mantle, 348–349
Map
  definition of, 74
  direction, 82
  profiles, 84–85
  symbols, 82–83
Map scale, 82, 84
Mapping, 69–96
  axis, 74
  Cartesian system, 74
  coordinate systems, 74–75
  equator, 75
  field maps. *See* Field maps
  frame of reference, 74
  great circles, 76–78
  latitude, 78
  latitude-longitude coordinate
    system, 75–78
  longitude, 75–79
    determining, 79
  meridians, of, 77
  meridians, 75–78
  origin, 74
  parallels, 75–76
  polar coordinate system,
    74–75
  prime meridian, 78
Marcasite, 287
Mariner spacecraft, 151–152
Mars, 10, 149–152
Mass, 133–134
Mass movements, 486–488
Mass wasting, 443–446,
    467–468
  rapid, 444–445
  slow, 445–446
Mature soil, 431
Meanders, 450
Mediterranean Belt, 373
Merak, 8
Mercalli scales, 344–345
Mercury, 10, 149–151
Meridians, 75–78
  celestial, 8
  of longitude, 77
  prime, 78
Mesozoic era, 245–246
Metamorphic rock, 300,
    312–317
  characteristics, 314–316
  classification, 316

Index

formation, 312–313
  chemical activity, 313
  heat, 312
  pressure, 313
  identification, 316–317
  texture, 314–315
Metamorphism, 313–314
Meteor, 155
Meteorite, 155, 351
Meteoroids, 155–156
Mid-latitude cyclones, 586
Mid-ocean ridges, 390
Milky Way, 124–125, 144
Minerals, 275–298. *See also specific mineral*
  building blocks of rocks, 283–284
  changes, 315–316
  characteristics of, 276–278
  chemical
    formulas, 277
    symbol, 277
    tests, 282
  cleavage, 280
  color, 279
  compounds, 277–278
  conservation, 288
  crystalline form, 278
  definition, 276
  double refraction, 282
  elements, 277
  flame color, minerals, 282
  fracture, 280–281
  groups of, 284–288
  hardness, 279–280
  identifying, 278–283
  inorganic matter, 276
  luminescence, 282
  luster, 279
  magnetism, 282
  minor groups of, 287–288
  oxygen, 287
  parting, 280
  piezoelectricity, 282
  play of colors, 282
  polyatomic ions, 287
  pyroelectricity, 282
  resource distribution, 289–290
  special properties, 282
  specific gravity, 281–282
  streak, 279
Moho, 348
Mohorovicic, Andrija, 348
Monoclines, 385
Monsoons, 617–618
Month, 103–105
  sidereal, 103–105
  synodic, 103–105

Moon, 10, 29, 151–152
  eclipse, 51–54
  origins, 191–193
    Apollo missions, 191–192
    giant impact, 192–193
  phases, 50–51
Moraines, 497–498
Motion
  apparent, 5–6
  celestial, 4–5
  Newton's laws, 29–32
  orbital, 30–32
  planetary, Kepler's law, 26–29
  real, 5–6
  retrograde, 21–22, 24–25
  sky, 2
Mount Palomar Observatory, 123
Mountains, 501, 620
Mt. Etna, 373
Mt. Vesuvius, 373
Mudflow, 445, 486–487

N

Natural selection, 254–255
Neap tides, 55
Nebula, 125, 132–133
Nebular theory, 146
Neptune, 147, 149–150, 154
Newton, Isaac, 29–32, 123
Nitrates, 287
Nonconformities, 241–242
Nuclear fusion, 130–131
Numerical weather forecasting, 589

O

Oasis, 458
Objective, 122
Oblate spheroid, 70
Observations, instruments, 124–125
Obsidian, 303, 305
Occluded fronts, 585–586
Oceans, 451–467
  currents, 451–453, 619
  deposition, 490–492
  floor spreading theory, 390–392
Olber's Paradox, 174
Oort cloud, 148
Orbital motion, 30–32
Orbits, 25–27
Organic sediments, 310
Original heat, 367
Original horizontality, 239
Orion, 149
Orographic effect, 620–621

Outcrop, walking, 242
Outgassing, 200, 213
Outwash plains, 497, 499
Overland flow, 446
Overturned folds, 385
Oxbow lakes, 451, 488–489
Oxidation, 423–424
Oxides, 287
Oxygen, 287

P

Paleomagnetism, 390–391
Paleozoic era, 245–246
Pangaea, 386–387
Parallelism, Earth's axis of rotation, 44
Parallels, 75–76
Parent material, 430
Parent rock, 314
Particle, 427–428, 485
Parting, minerals, 280
Path of totality, 52
Pegmatite, 302, 306
Peneplain, 451
Penumbra, 52
Perception, human, 122
Peridotite, 306
Periodic waves, 336
Periods, 245–246
  of revolution, 28–29
Permeability, 224–225
Phanerozoic eon, 244–246
Phenocrysts, 303
Phosphates, 287
Physical weathering, 419–423
  abrasion, 421
  animal action, 421–422
  exfoliation, 421
  frost action, 420–421
  plant action, 421–422, 425–426
  pressure unloading, 423
  temperature, changes, 423
Piezoelectricity, minerals, 282
Plains, 501
Planet(s), 10–11, 149–150
  dwarf, 148–150
  Earth, 149–151
  jovian, 147, 149–150, 152–154
  Jupiter, 147, 149–150, 152–154
  Mars, 149–152
  Mercury, 149–151
  Neptune, 147, 149–150, 154
  Pluto, 148, 150
  Saturn, 147, 149–150, 153–154
  terrestrial, 150–152
  Uranus, 147, 149–150, 154

Index

Venus, 149–151
  word origin of, 20
Planetary motion
  Kepler's law, 26–29
  velocity, 27–28
Planetesimals, 146
Plant action, 421–422, 425–426
Plate boundaries, 393–394
Plate tectonics, 381–418
  asthenosphere, 392–393, 396
  boundaries, 393–394
  hot spots, 394–395
  key ideas, 395–397
  lithosphere, 392–393, 395
  rigidity, 392
Plateaus, 501
Play of colors, minerals, 282
Plucking, 464
Pluto, 148, 150
Plutonic rock, 302
Plutons, 372
Polar coordinate system, 74–75
Polar easterlies, 452, 576
Polar front, 586
Polaris, 8, 45
  altitude, 72–73
Pole, 74
Polyatomic ions, 287
Porosity, 223–224
Precambrian eon, 245–246
Precession, 45
Precipitation, 222, 578–579,
  613–614
Predictions, 23
Pressure
  air. *See* Air pressure
  belts, 616–617
  unloading, 423
Prevailing westerlies, 452
Prevailing winds, 576
Prime meridian, 78
Principle of superposition, 239
Prism, 123
Proterozoic eon, 244–246
Proton-proton reaction, 130–131
Protostar, 132–133
Psychrometer, 565
Ptolemy, 19, 21–23
Pumice, 303, 305–306
P-wave shadow zone, 348–349
P-waves, 339–340, 347–349
Pyrite, 287
Pyroelectricity, minerals, 282

Q

Quartz, 304

R

Radial stream pattern, 502
Radiation, 368, 537–538

Radiation curve, 127–128
Radiative balance, 531
Radioactive decay, 250, 367
Radioisotopes, 250–253
Rain, 579
Raindrops, 446
Rapid mass movements,
  486–487
Real motion, 5–6
Rectangular stream pattern, 503
Recumbent folds, 385
Red dwarfs, 133
Reflected waves, 335–336
Reflecting telescope, 122
Refracting telescope, 122
Refraction, 336
Regional metamorphism, 314
Relative age, 239–241
Relative humidity, 564, 567
Relief, 500–501, 620
Residual soil, 430
Resource distribution, 289–290
Retrograde motion, 21–22,
  24–25
Revolution, 44–45
  period of, 28–29
Rhyolite, 303, 306
Richter magnitude scale,
  343–344
Ring of Fire, 346, 373
Rock(s), 299–332
  age, 247–253
  banding, 315
  crystalline texture, 314–315
  cycle, 317–318
  definition, 300
  density, 315
  extrusive rocks, 302–303
  folded layers, 382
  foliation, 315
  igneous. *See* Igneous rock
  intrusive rocks, 302
  lava, 300, 302
  layers, 241–242, 315
  magma, 300, 302
  metamorphic. *See*
    Metamorphic rock
  metamorphism, 313–314
  mineral
    building blocks, 283–284
    composition, 301, 315–316,
      428–429
    parent rock, 314
    physical characteristics,
      matching, 242–243
    rigidity, 392
    sedimentary. *See* Sedimentary
      rock
    tilted layers, 382
    types, 300–317

volcanic glasses, 303
Wentworth scale, 310
young, 390
Rock drumlins, 465
Rockfalls, 444, 486
Rockslide, 444
Rotation, 42–44
  axis of, 42
  parallelism, 44
  Coriolis effect, 43–44
  evidence of, 42–44
  Foucault Pendulum, 42–43
Roy G. Biv, 123
Runoff, 222, 446

S

Salinity, 214
Saltation, 448
Sandbar, 491–492
Satellites, 44, 149
Saturn, 10, 147, 149–150,
  153–154
Scale, map, 82, 84
Sea breezes, 574–575
Sea cave, 456–457
Sea cliff, 456–457
Sea terrace, 456
Season, 533
  Sun's path, 48–49, 105–106
Seawater, 214–216
  composition, 214–216
  dissolved gases, 216
  inorganic matter, 215
  organic matter, 216
  origin, 214
  particulates, 216
Sediment
  bedding, 486
  cross-bedded, 308–309
  definition of, 307, 429
  shallow water, 383
Sedimentary rock, 300,
  307–312
  cementation, 307
  characteristics, 308–309
  chemical sediments, 310–311
  compaction, 307
  desiccation, 307
  identification, 311–312
  layers, high elevation, 382
  lithification, 307–308
  organic sediments, 310
  particles, types, 309–311
Seismic waves, 334, 337–339
Seismogram, 338–343
Seismograph, 337–338
Shear stress, 384
Shield volcano, 371
Shoreline erosion, 456–457
Short-lived giants, 134

775

Index

Sidereal month, 103–105
Silicates, 284–286
Silicon tetrahedron, 284
Sills, 372
Sky, 1–6
Sleet, 579
Slip face, 493
Slow mass movements, 487–488
Slump, 444–445, 487
Snow, 579
Soil, 430–431
  associations, 503–504
  creep, 445–446, 487
  formation of, 419–440
  immature, 431
  mature, 431
  parent material, 430
  residual, 430
  sediments, 429
  transported, 430
Soil moisture zone, 225
Soil profile, 431
Solar day, 100
  mean, 103–104
Solar eclipse, 51–53
Solar radiation, 528–531
  electromagnetic spectrum, 529–530
  electromagnetic waves, 528–529
  global radiation budget, 530–531
  radiative balance, 531
Solar system, 143–172
  asteroids, 155
  catastrophic theory of, 145
  comets, 156–157
  evolution of, 145–149
  exploration, 159
  galaxy, 144
  gravitational differentiation, 155
  impact events, 158–159
  interplanetary bodies, 155–159
  Kuiper belt, 147–148
  meteor, 155
  meteorite, 155
  meteoroids, 155–156
  Milky Way, 144
  Moon, 151–152
  nebular theory of, 146
  origin, 145–149
  planetesimals, 146
  planets. *See* Planet(s)
  structure, 149–159
Solstices, 49, 105–106

Solution, 424–425
Sorted deposits, 496
Sorting, 484
Southern Oscillation, 625–626
Space, photographs from, 72
Special properties, of minerals, 282
Specific gravity, of minerals, 281–282
Specific heat, 535–536
Spectroscopes, 123
  prism, 123
Spectrum, 123
Speed of medium, 484
Sphalerite, 287
Spheres, 20–21
Spits, 491
Spring equinox, 49
Spring tides, 55
Stacks, 456–457
Stars, 10, 124, 126–134
  brightness of, 126
  characteristics of, 126–130
  color, 127–128
  evolution of, 130–134
  gravity, 132–133
  Hertzsprung-Russell Diagram, 128–130
  luminosity of, 126–127, 128–129
  main sequence, 129
  mass of, 133–134
  nebula, 132–133
  nuclear fusion, 130–131
  origin of, 130–134
  pointer, 8
  protostar, 132–133
  radiation curve, 127–128
  red dwarfs, 133
  short-lived giants, 134
  Sun-class, 134
  temperature, 127–128
Station model, 579–580
Station models, 588
Stationary fronts, 585
Statistical weather forecasting, 588–589
Stellar brightness, 126
Stibnite, 287
Stocks, 372
Strain, 384
Streak, minerals, 279
Streams, 446–451, 468
  deposition, 488–490
  drainage system, 502
  erosion, 449
  life cycle, 449–451
  patterns, 501–503
  transport, 447–449

Stress, 384
Structures, displaced, 382
Subduction, 392
Subduction zone, 393, 396
Sulfate ion, 287
Sulfates, 287
Sulfides, 287
Summer solstice, 49
Sun
  description of, 9–10, 149
  eclipses, 51–53
  path of, 9–10
    changing, 47–50, 107–108
    seasons, 105–106
Sun-class stars, 134
Sunrise, 99–100
Sunset, 98–100
Superclusters, 175
Superposition, 239
Surf, 454
Surface tension, 217
Surface waves, 339
S-wave shadow zone, 348–349
S-waves, 339–340, 347–349
Symbols, map, 82–83
Syncline, 385
Synodic months, 103–105
Synoptic forecasting, 579
Synoptic weather, 579, 588

T

Talus, 486
Telescopes, 122–123
Temperate zone, 615
Temperature, 613–614, 616
  air, 560–561
  changes, 423
  star, 127–128
Tension stress, 384
Tephra, 371
Terrestrial planets, 149–152
Tetrahedron, 284
Texture, 534–535
Thermometer, 560
Thunderstorms, 590–591
Tidal friction, 367–368
Tidal marsh, 492
Tides, 54–55
  neap, 55
  spring, 55
Tilted rock layers, 382
Time, 97–120, 429
  day, 97–103
  eons, 244–246
  epochs, 245–247
  equinoxes, 106–107
  eras, 245–246
  geologic scale of, 244–247
  Greenwich Mean Time, 101

Index

hours, 98–100
International Date Line, 102–103
  local, 100–103
  month, 103–105
  periods, 245–246
  seasons, Sun's path, 105–106
  sidereal month, 103–105
  solar day, 100
    mean, 103–104
  solar noon, 100
  solstices, 49, 105–106
  sunrise, 99–100
  Sun's path, 105–106, 107–108
  sunset, 98–100
  synodic months, 103–105
  year, 105–109
  zones, 101
Titan, 154
Tombolos, 492
Topographic maps, 81–85, 85
Topography, 500
Tornadoes, 591
Torrid zone, 615
Totality, path of, 52
Traction, 448, 495
Trade winds, 451, 576, 616
Transform boundaries, 394, 396
Transpiration, 223, 620
Transported soil, 430
Transverse waves, 337
Trellis stream pattern, 503
Trenches, 391–392
Tributary, 447, 501
Tropopause, 204
Tsunamis, 345–346, 454

U
Umbra, 52
Unconformities, 241–242
Uniform stress, 384
Uniformitarianism, 238
Universal gravitation, 29
Universe
  closed, 180
  expanding, 176–177
  future, 179
  open, 179–180
  size and structure of, 125–126
  structure, 174–175
  theories of origin, 173–186
Unsorted deposits, 486, 496
Upwelling, 451
Uranus, 147, 149–150, 154
U-shaped valleys, 466–467

V
Valley glaciers, 462–463

Valleys, 466–467
Van Allen belts, 190
Varve, 497
Vega, 45
Vegetative cover, 620
Velocity, planetary motion, 27–28
Ventifacts, 459
Venus, 10, 149–151
Vesicles, 303
Vesicular rocks, 306
Visibility, 569
Volcanic eruptions, 382
Volcanic glasses, 303
Volcanic rock, 302–303
Volcanoes, 365–380
  batholiths, 372
  cinder cone, 371
  composite cone, 371
  dikes, 372
  formation, 370
  geothermal gradient, 365–366
  internal heat of the Earth, 365–369
  intrusions, 372
  intrusive activities, 372
  laccoliths, 372
  lava of, 370
  magma of, 369–370
  Mt. Etna, 373
  Mt. Vesuvius, 373
  plateau, 371
  plutons, 372
  shield volcano, 371
  sills, 372
  structures, 370–372
  tephra, 371

W
Warm fronts, 585
Wasting, mass, 443–446, 467–468
Water
  capillarity of, 217
  condensation of, 222
  crystalline structure of, 219–220
  distribution, 615
  evaporation of, 218, 221
  evaporative cooling, 219
  freezing of, 219
  heat capacity of, 218
  hydrogen bonds, 217
  latent heat of
    fusion, 220
    vaporization, 219
  liquid, 217–218
  motion, erosion, 446–457
  origin of, 212–214

precipitation, 222
solid, 219–220
states of, 217–220
structure of, 216–217
surface tension of, 217
transpiration, 223
universal solvent, 220–221
Water cycle, 221–223
Water table, 225–226
Water vapor, 218–219
Wavelength, 336, 454
Wave(s), 453–456, 469
  behavior, 335–337
  breaking, 454
  crest, 453
  cut notch, 456–457
  cyclones, 586–587
  height, 454
  motion, 334
  refraction, 455
  trough, 453
  types, 337
  velocity, 336
  wind-generated, 453–454
Weather
  adiabatic changes, 570
  air humidity, 563–568
  air masses, 582–583
  air movements, 568
  air pressure. *See* Air pressure
  air temperature, 560–561
    and humidity, 572–573
  anticyclones, 586–587
  atmospheric
    transparency, 569
    variables, 560
  barometer, 562
  breezes, 574–575
  Celsius, 561
  chaos theory, 590
  clouds, 569, 576–579
  cold fronts, 584
  condensation, 563–564, 576
  Coriolis effect, 576
  cyclones, 586–587
  dew, 578
  dew point, 565, 568
  drizzle, 579
  evaporation, 563–564
  Fahrenheit, 561
  field maps, 580–582
  fog, 578
  forecasts, 582–590
  frontal cyclones, 586–587
  fronts, 582, 583–586, 584
  frost, 578
  global wind belts, 575–576
  gradient, 580
  hail, 579

Index

hazardous, emergency preparedness, 593–594
humidity, 563–568
hurricanes, 592
isobars, 580, 582
isolines, 580
isotherms, 580–581
kelvin, 561
land breezes, 574–575
long-range weather forecasting, 589–590
maps, 579–582
numerical weather forecasting, 589
occluded fronts, 585–586
precipitation, 578–579
prevailing winds, 576
rain, 579
relative humidity, 564, 567
sea breezes, 574–575
sleet, 579
snow, 579
stationary fronts, 585
statistical weather forecasting, 588–589
synoptic forecasting, 588
synoptic maps, 579
thermometer, 560
thunderstorms, 590–591
tornadoes, 591
trade winds, 576
visibility, 569
warm fronts, 585
warnings, 593–594
watches, 593–594
wave cyclones, 586–587
weather, 559–612
Weathering, 419–440
chemical. *See* Chemical weathering
factors affecting, 426–429
physical. *See* Physical weathering
products of, 429–431
soils, 430–431
Wegener, Alfred, 242, 386–387
Westerlies, 452
Wind belts, 616–617
Wind deposition, 492–494
Wind erosion, 457–459, 469
Wind vane, 568
Wind-driven currents, 451–453
Winter solstice, 49

Y

Year, 105–109
Young rock, 390

Z

Zenith, 8
Zodiac, 9–10
Zone of aeration, 225
Zone of saturation, 225
Zone of soil moisture, 225